Lecture Notes in Computer Science 4683

Commenced Publication in 1973
Founding and Former Series Editors:
Gerhard Goos, Juris Hartmanis, and Jan van Leeuwen

Lishan Kang   Yong Liu   Sanyou Zeng (Eds.)

Advances in Computation and Intelligence

Second International Symposium, ISICA 2007
Wuhan, China, September 21-23, 2007
Proceedings

 Springer

Volume Editors

Lishan Kang
China University of Geosciences
School of Computer Science
Wuhan, Hubei 430074, China
E-mail: kang_whu@yahoo.com

Yong Liu
The University of Aizu
Tsuruga, Ikki-machi, Aizu-Wakamatsu, Fukushima 965-8580, Japan
E-mail: yliu@u-aizu.ac.jp

Sanyou Zeng
China University of Geosciences
School of Computer Science
Wuhan, Hubei 430074, China
E-mail: sanyou-zeng@263.net

Library of Congress Control Number: 2007933208

CR Subject Classification (1998): C.1.3, I.2, I.2.6, I.5.1, H.2.8, J.3

LNCS Sublibrary: SL 1 – Theoretical Computer Science and General Issues

ISSN   0302-9743
ISBN-10   3-540-74580-7 Springer Berlin Heidelberg New York
ISBN-13   978-3-540-74580-8 Springer Berlin Heidelberg New York

Springer is a part of Springer Science+Business Media

springer.com

© Springer-Verlag Berlin Heidelberg 2007
Printed in Germany

Typesetting: Camera-ready by author, data conversion by Scientific Publishing Services, Chennai, India
Printed on acid-free paper   SPIN: 12114597   06/3180   5 4 3 2 1 0

Preface

We are proud to introduce the proceedings of the 2nd International Symposium on Intelligence Computation and Applications (ISICA 2007) held in China University of Geosciences (Wuhan), China, September 21–23, 2007. ISICA 2007 successfully attracted nearly 1000 submissions. After rigorous reviews, 71 high-quality papers were included in the proceedings of ISICA 2007.

The 1st International Symposium on Intelligence Computation and Applications (ISICA 2005) held in Wuhan, April 4–6, 2005 was organized by the School of Computer Science, China University of Geosciences. It was a great success with over 100 participants, including a number of invited speakers. The proceedings of ISICA 2005 have a number of special features including uniqueness, newness, successfulness, and broadness. The proceedings of ISICA 2005 have also been accepted in the Index to Scientific and Technical Proceedings.

Following the success of ISICA 2005, ISICA 2007 focused on research on computational intelligence in analyzing and processing massive real-time data. ISICA 2007 featured the most up-to-date research on multiobjective evolutionary optimization, evolutionary algorithms and operators, evolutionary optimization, evolutionary learning, neural networks, ant colony and artificial immune systems, particle swarm optimization, pattern recognition, data mining, intelligent systems, and evolutionary design. ISICA 2007 also provided a venue to foster technical exchanges, renew everlasting friendships, and establish new connections.

On behalf of the Organizing Committee, we would like to thank warmly the sponsors, China University of Geosciences and Chinese Society of Astronautics, who helped in one way or another to achieve our goals for the conference. We express our appreciation to Springer, for publishing the proceedings of ISICA 2007. We would also like to thank the authors for submitting their research work, as well as the Program Committee members and reviewers for their enthusiasm, time, and expertise.

The invaluable help of active members from the Organizing Committee, including Qiuming Zhang, Siqing Xue, Ziyi Chen, Yan Guo, Xuesong Yan, Xiang Li, Guang Chen, Rui Wang, Hui Wang, Hui Shi, Tao Hu, Zhenhua Cai, and Gang Liu, in setting up and maintaining the online submission systems, assigning the papers to the reviewers, and preparing the camera-ready version of the proceedings was highly appreciated and we would like to thank them personally for their efforts to make ISICA 2007 a success.

We wish to express our gratitude to Alfred Hofmann, the Executive Editor, Computer Science Editorial, Springer-Verlag, for his great support of the conference. We also wish to acknowledge the dedication and commitment of the LNCS editorial staff.

September 2007   Lishan Kang
                                                                   Yong Liu
                                                                 Sanyou Zeng

Organization

ISICA 2007 was organized by the School of Computer Science and Research Center for Space Science and Technology, China University of Geosciences, and sponsored by China University of Geosciences and Chinese Society of Astronautics.

General Chair

Yanxin Wang   China University of Geosciences, China

Program Chair

Lishan Kang   China University of Geosciences, China

Advisory Board

Guoliang Chen	University of Science and Technology of China, China
Pingyuan Cui	Harbin Institute of Technology, China
Kalyanmoy Deb	Indian Institute of Technology Kanpur, India
David B. Fogel	Natural Selection, Inc., USA
Erik Goodman	Michigan State University, USA
Xinqui He	Peking University, China
Licheng Jiao	Xidian University, China
Zbigniew Michalewicz	University of Adelaide, Australia
Yongqiang Qiao	Astronautics Science and Technology Group Time Electron Company, China
Marc Schoenauer	University Paris Sud, France
Hans-Paul Schwefel	University of Dortmund, Germany
Zhongzhi Shi	Institute of Computing Technology, Chinese Academy of Sciences, China
Adrian Stoica	Jet Propulsion Laboratory, USA
Mei Tan	Astronautics Science and Technology Group Five Academe, China
Tieniu Tan	Institute of Automation, Chinese Academy of Sciences, China
Jiaqu Tao	Astronautics Science and Technology Group Nine Academe, China
Edward Tsang	University of Essex, UK
Jiaying Wang	China University of Geosciences, China
Xin Yao	University of Birmingham, UK

Zongben Xu   Xi'an Jaotong University, China
Jianchao Zeng   Taiyuan University of Science and
                                    Technology, China
Ba Zhang   Tsinghua University, China

General Co-chairs

Yong Liu   The University of Aizu, Japan
Sanyou Zeng   China University of Geosciences, China

Program Co-chairs

Bob McKay   Seoul National University, South Korea

Program Committee

Hussein A. Abbass   University of New South Wales, Australia
Tughrul Arslan   The University of Edinburgh, UK
Wolfgang Banzhaf   Memorial University of Newfoundland, Canada
Zhihua Cai   China University of Geosciences, China
Guoliang Chen   University of Science and Technology of China,
                                    China Academician, The Chinese Academy
                                    of Sciences, China
Ying-Ping Chen   National Chiao Tung University, Taiwan, China
Carlos A. Coello Coello   LANIA, Mexico
Guangming Dai   China University of Geosciences, China
Kalyanmoy Deb   Indian Institute of Technology, India
Lixin Ding   Wuhan University, China
Candida Ferreira   Gepsoft
Garry Greenwood   Portland State University, Portland, USA
Jun He   University of Birmingham, UK
Xingui He   Peking University, China Academician,
                                    the Chinese Academy of Engineering, China
Zhenya He   Eastsouth University, China Academician,
                                    the Chinese Academy of Sciences, China
Tetsuya Higuchi   National Institute of Advanced Industrial
                                    Science and Technology, Japan
Houkuan Huang   Beijing Jiaotong University, China
Zhangcan Huang   Wuhan University of Technology, China
Hisao Ishibuch   Osaka Prefecture University, Japan
Licheng Jiao   Xidian University, China
John R. Koza   Stanford University, USA
Lawrence W. Lan   National Chiao Tung University, Taiwan, China
Yuanxiang Li   Wuhan University, China
Guangxi Liang   Chinese University of Hong Kong, China

Jiajun Lin	East China University of Science and Technology, China
Bob Mckay	Seoul National University, South Korea
Zbigniew Michalewicz	University of Adelaide, Australia
Erkki Oja	University of Technology Helsinki, Finland
Ping-Feng Pai	National Chi Nan University, Taiwan, China
Peter Ross	Napier University, UK
Wei-Chiang Samuelson Hong	Oriental Institute of Technology, Taiwan, China
Marc Schoenauer	University of Paris Sud, France
Zhongzhi Shi	Institute of Computing Technology, China
Hsu-Shih Shih	Tamkang University,Taiwan, China
Dianxun Shuai	East China University of Science Technology, China
Huai-Kuang Tsai	Institute of Information Science, Academia Sinica, Taiwan, China
Edward Tsang	University of Essex, UK
Jiaying Wang	China University of Geosciences, China
Shaowei Wang	Nanjing University, China
Zhijian Wu	Wuhan University, China
Tao Xie	National University of Defense Technology, China
Zongben Xu	Xi'an Jiaotong University, China
Shengxiang Yang	University of Leicester, UK
Xin Yao	University of Birmingham, UK
Jianchao Zeng	Taiyuan University of Technology, China
Sanyou Zeng	China University of Geosciences, China
Ba Zhang	Tsinghua University, China Academician, The Chinese Academy of Sciences, China
Huajie Zhang	University of New Brunswick, Canada
Jun Zhang	Sun Yat-Sen University, China
Qingfu Zhang	University of Essex, UK
Jinhua Zheng	Xiangtan University, China
Zhi-Hua Zhou	Nanjing University, China
Xiufen Zou	Wuhan University, China

Local Chair

Yadong Liu	China University of Geosciences, China

Local Co-chairs

Zhihua Cai	China University of Geosciences, China
Guangming Dai	China University of Geosciences, China
Hui Li	China University of Geosciences, China
Sifa Zhang	China University of Geosciences, China

Local Committee

Ziyi Chen	China University of Geosciences, China
Yan Guo	China University of Geosciences, China
Shuanghai Hu	China University of Geosciences, China
Xiang Li	China University of Geosciences, China
Zhenhua Li	China University of Geosciences, China
Siqing Xue	China University of Geosciences, China
Xuesong Yan	China University of Geosciences, China
Li Zhang	China University of Geosciences, China
Qiuming Zhang	China University of Geosciences, China

Table of Contents

Multiobjective Evolutionary Optimization

Evolutionary Algorithms and Operators

Evolutionary Optimization

Evolutionary Learning

Neural Networks

Ant Colony, Particle Swarm Optimization and Artificial Immune Systems

Pattern Recognition

Data Mining

Intelligent Systems

Evolutionary Design

A New Evolutionary Decision Theory for Many-Objective Optimization Problems

Zhuo Kang[1], Lishan Kang[2,3,*], Xiufen Zou[4], Minzhong Liu[5], Changhe Li[2],
Ming Yang[2], Yan Li[1], Yuping Chen[3], and Sanyou Zeng[2]

[1] Computation Center, Wuhan University, Wuhan 430072, China1
kang_whu@yahoo.com
[2] School of Computer Science, China University of Geosciences(Wuhan) Wuhan,
430074, China
[3] State Key Laboratory of Software Engineering, Wuhan University, Wuhan 430072, China
[4] School of Mathematics and Statistics, Wuhan University, Wuhan 430072, China
[5] School of Computer Science, Wuhan University, Wuhan 430072, China

Abstract. In this paper the authors point out that the Pareto Optimality is unfair, unreasonable and imperfect for Many-objective Optimization Problems (MOPs) underlying the hypothesis that all objectives have equal importance. The key contribution of this paper is the discovery of the new definition of optimality called ε-optimality for MOP that is based on a new conception, so called ε-dominance, which not only considers the difference of the number of superior and inferior objectives between two feasible solutions, but also considers the values of improved objective functions underlying the hypothesis that all objectives in the problem have equal importance. Two new evolutionary algorithms are given, where ε- dominance is used as a selection strategy with the winning score as an elite strategy for search -optimal solutions. Two benchmark problems are designed for testing the new concepts of many-objective optimization problems. Numerical experiments show that the new definition of optimality is more perfect than that of the Pareto Optimality which is widely used in the evolutionary computation community for solving many-objective optimization problems.

Keywords: Many-objective optimization; Pareto optimality; ε-optimality; ε-dominance.

1 Introduction

Most optimization problems in nature have many objectives (normally conflicting with each other) that need to be optimized at the same time. These problems are called Multi-Objective Problems (MOPs), which are studied in economies, sciences and engineering. People realize the importance of solving MOPs and the development of

---

* Corresponding author.

L. Kang, Y. Liu, and S. Zeng (Eds.): ISICA 2007, LNCS 4683, pp. 1–11, 2007.

theory and methodology to deal with this kind of problems become an important area of computational intelligence. Because of the conflicting nature of their objectives, MOP does not normally have a single solution and, in fact, they even require the definition of a new notion of "optimality." The most commonly adopted notion of optimality in MOPs is that originally proposed by Edgeworth [1] and later generalized by Pareto [2]. Such a notion is called Edgeworth-Pareto Optimality or, more commonly, Pareto Optimality[3].

There are two ways for solving MOPs to find their Pareto optimal sets. One is using the weighted objectives summed in one objective. Another is using the population search strategies. The former is called "One Many" and the later is called "Many Once"[4]. Why evolutionary multi-objective optimization algorithms are increasingly inappropriate as the number of objectives increases? Can we find other ways for solving many-objective optimization problems? Recently, Maneeratana, Boonlong, and Chaiyaratana[5] proposed a method called Compressed-Objective Genetic Algorithm (COGA) for solving an optimization problem with a large number of objectives by transforming the original objective vector into a two-objective vector during survival selection is presented. The transformed objectives, referred to as preference objectives, consist of a winning score and a vicinity index. The winning score, a maximization criterion, describes the difference of the number of superior and inferior objectives between two solutions. The vicinity index, a minimization criterion, describes the level of solution clustering around a search location which is used to encourage the results to spread throughout the Pareto front. The new conception in that paper is the definition of Winning Score, a preference objective which concerns numbers of superior and inferior objectives between a pair of two non-dominated solutions. But the goal of the authors is still to get a solution set which is coverage and closeness to the true Pareto front.

In 2003, Zhou, Kang, Chen and Huang[6] proposed a new definition (Evolving Solutions) for MOP to answer the essential question: what is a multi-objective optimal solution and advance an asynchronous evolutionary model, Multiple Island Net Training Model (MINT model), to solve MOPs, especially to solve many-objective optimization problems. The new theory is based on their understanding of the natural evolution and the analysis of the difference between natural evolution and MOP, thus it is not only different from the tradition methods of weighted objective sum, but also different from Pareto Optimization.

Benedelti, Farina, and Gobbi [7] mentioned three reasons of Pareto definition of optimality which can be unsatisfactory to a large number of objective optimization.

P1, The number of improved objective functions values is not taken into account.
P2, The (normalized) relevance of the improvements is not taken into account.
P3, No preference among objectives is considered.

They give two new definitions to meet above issues. One is called k Optimality when taking into account number of improved objectives. Another is called Fuzzy Optimality when taking into account size of improvements and including parameters to be provided by decision maker with the underlying hypothesis that all objectives have equal importance.

We must point out that the definition of Pareto Optimality is fair, reasonable and perfectly correct for two-objective optimization problems, but is unfair, unreasonable and imperfect for more than two objectives which are in strict conflict of each other, especially for a large number of objectives, if all objectives have equal importance.

Kang [8] gives an example:

Minimize $F(x) \underline{\Delta} (f_1(x), f_2(x), \cdots, f_m(x))$

where m is a large number,

and $f_i(x) = (x - i)^2$, $i = 1, 2, ..., m$, $1 \leq x \leq m$.

It is clear that its Pareto Optimal Set is [1, m] which is the exact decision space of the MOP. It is not clear at all what a decision maker can do with such a large result of a Pareto Optimal Set in practice!

In this paper, a new definition of optimality for MOPs is given. It not only considers the number of improved objectives, but also considers the convergence of improved values of objective functions. The remainder of this paper is organized as follows. The new definitions of optimality are presented in section 2. The new evolutionary algorithms are given in Section 3. Section 4 presents experimental design, benchmarks and results for the new definitions of optimality. Section 5 is conclusions and new research directions for future MOPs research with the new concepts proposed by this paper.

2 What Is the Optimality for MOP?

A MOP can be described as follows.

Let $D \subseteq R^n$ be decision variable space and $O \subseteq R^m$ be objective space.
Let F: $D \rightarrow O$, $g : D \rightarrow R^p$ and $h : D \rightarrow R^q$ be vector functions.
A nonlinear constrained multi-objective optimization problem (MOP) is defined as

$$\min_{x \in S} f(x) \underline{\Delta} (f_1(x), f_2(x), \cdots, f_m(x)) \tag{1}$$

where

$$S = \{X \in D \mid g(X) \leq 0 \wedge h(X) = 0\}$$
$$D = \{L \leq X \leq U\},$$

where $L, U \in R^n$ are given.

We first normalize the MOP (1) as follows.

Denote

$$\min_{x \in S} f_i(x) = f_i^{min} \text{ and } \max_{x \in S} f_i(x) = f_i^{max} \quad \forall i \in M = \{1, 2, \cdots, m\}$$

Let $F_i(x) = (f_i(x) - f_i^{min}) / (f_i^{max} - f_i^{min}) \quad \forall i \in M$ be transformations.

The MOP (1) is transformed into

$$\min_{x \in S} F(x) \underline{\Delta} (F_1(x), F_2(x), \cdots, F_m(x)) \tag{2}$$

where   F: $D \to O = [0,1]^m$,   the objective space.

For each pair of variables X_1, $X_2 \in S$, three integer functions are defined as follows:

$$B_t(X_1, X_2) = \left|\{i \in M \;\; F_i(X_1) < F_i(X_2)\}\right|$$
$$W_s(X_1, X_2) = \left|\{i \in M \;\; F_i(X_1) > F_i(X_2)\}\right|$$
$$E_q(X_1, X_2) = \left|\{i \in M \;\; F_i(X_1) = F_i(X_2)\}\right|$$

where B_t, W_s and E_q mean Better Than, Worse Than and Equivalent To, respectively.

It is clear that $B_t + W_s + E_q = m$ (the number of objectives).

The ideal objective vector of the MOP (2) in objective space is the origin which corresponds to a non-existent solution in general.

The distance from any point $F(X)$ in objective space O to the ideal objective vector (the origin) is defined as

$$\|F(X)\| = \sqrt{\sum_{i=1}^{m} (F_i(X))^2}$$

For any pair $X_1, X_2 \in S$, we give following definitions:

Definition 1 (ε-dominance): If ($B_t - W_s = \varepsilon > 0$) $\wedge$ ($\|F(X_1)\| < \|F(X_2)\|$) then X_1 is said

to ε-dominate X_2 (denoted as $X_1 \prec_e X_2$).

If $W_s(X_1, X_2) = 0$, then ε-dominance is equivalent to Pareto dominance.

Definition 2 (ε-Optimality): $X^* \in S$ is a ε-optimal solution if there is no $X \in S$ such that X ε-dominates X^*. (Note: the concept of **ε-Optimality** is different to the ε optimal solutions in [13].)

Definition 3 (ε-Optimal Set and Front): The set of all **ε**-optimal solutions is called **ε**-optimal set denoted as S_ε and its range of mapping $F(S_\varepsilon)$ is called **ε**-optimal front denoted as F_ε.

In definition 1, **ε** > 0 means that the number of superior objectives is great than the number of inferior objectives between two feasible solutions: X_1 and X_2. Because the importance of all objectives are equal, so the definition of ε-dominance is fair and reasonable, and the distance condition: $\| F(X_1) \| < \| F(X_2) \|$ not only substantiates the meaning of dominance, but also can be used to control the values of inferior objectives. We must point out that the new definitions of optimality for MOPs are imperfect, too, because the transitive property is not existent so the pair $([0,1]^m, \preceq_\varepsilon)$ can not form a partially ordered set. For overcoming this difficulty, we use the winning score[5] to rank the individuals of population and archives.

Definition 4 (Winning Score): Let the size of a set A be L, the winning score WS_i of the ith element X_i in A is given by

$$WS_i = \sum_{j=1}^{L} [B_i(X_i, X_j) - W_s(X_i, X_j)], \quad i = 1, 2, \cdots, L$$

For X_i, $i = 1, 2, \ldots, L$, we rank them by their winning score from the biggest one to the smallest as $0, 1, 2, \ldots, k$, $k \leq L-1$. The element X with biggest WS has RANK=0. If there are many elements in A having the same winning score, then they have the same rank.

Definition 5 (minimal element): Element $x \in A$ is called a minimal element of A, if its rank is 0.

All minimal elements of A form a minimal element set M(A,RANK=0).

3 New Evolutionary Algorithms for Solving MOPs

New algorithms for solving many-objective optimization problems are given as follows.

```
Algorithm I
begin
        Generate a population P(0) uniformly from D ⊂ Rⁿ;
           A(0):= M(P(0),RANK=0);
           t: = 0;
           repeat
              P(t+1): = search-optimizer P(t);
              A(t+1): = M(A(t)∪P(t+1),RANK=0);
              t: = t+1;
           until stopping criterion fulfilled
  end
```

where search-optimizer can be any efficient optimizer which is used for producing offspring. We use the Fast Particle Swarm Optimization Algorithm[12] as search-optimizer which is based on Particle Swarm Optimization with Cauchy mutation and an ε- optimality based selection strategy.

```
Fast Particle Swarm Optimizer

    Begin

       For   i = 1 to N do

           Vi:= Vi +η1 rand( ) (Xi best-Xi) +η2 rand( ) (Xg
best-Xi);
```

$$X'_i := X_i + V_i;$$
$$V'_i := V'_i \exp(\tau_1 \delta_1 + \tau_2 \delta_2);$$
$$X_i := X_i + V'_i \delta_3;$$
$$\text{if } (better(X'_i, X_i) \wedge (X_i \prec_\varepsilon X')) \text{ then } X_i := X'_i;$$

```
       endfor

  end
```

where δ_1, δ_2 and δ_3 are Cauchy random numbers, and η_1, η_2, τ_1 and τ_2 are parameters,

$$better(X,Y) = \begin{cases} true & \|h(X)\| + \overline{g}(X) < \|h(Y)\| + \overline{g}(Y) \\ false & \|h(X)\| + \overline{g}(X) > \|h(Y)\| + \overline{g}(Y) \\ true & (\|h(X)\| + \overline{g}(X) = \|h(Y)\| + \overline{g}(Y)) \wedge (\|F(X)\| < \|F(Y)\|) \\ false & (\|h(X)\| + \overline{g}(X) = \|h(Y)\| + \overline{g}(Y)) \wedge (\|F(X)\| \geq \|F(Y)\|) \end{cases}$$

where

Algorithm I is almost the same as the Base Algorithm VV in [8].

```
  Algorithm II
  begin

         Generate a population P(0) uniformly from D ⊂ Rⁿ;

           A(0):= M(P(0),RANK=0);

           t: = 0;

           repeat

               P(t+1): = search-optimizer P(t);

               A(t+1): = M(A(t) ∪ P(t+1),RANK=0);

               A(t+1):=Sieve(A(t+1);

               t: = t+1;
```

```
        until stopping criterion fulfilled
end
```

where the Fast Particle Swarm Optimization Algorithm[12]is used as search-optimizer, and a Sieve procedure is used for control the diversity.

```
Sieve algorithm

begin

for i = 1 to | A(t+1)| do

    if (d(i)<δ) then delete element Xi from A(t+1);

endfor

end
```

where $d(i)$ is the distance between two nearest points in $A(t+1)$. For more detail, please refer to [11].

Using Algorithm II, a finite well-distributed subset of $M([0,1]^m, RANK = 0)$ can be obtained, because the number of nodes on an δ-net in $[0, 1]^m$ is finite.

4 Numerical Experiments

In order to test the correctness of the new definitions and the effectiveness of Algorithm II, we give two benchmarks[8] with more than three objectives and some numerical results to compare with the results obtained by using Pareto-Optimality.

4.1 Benchmark Problems

Benchmark Problem 1
Minimize $F(X) = (F_1(X), F_2(X), ... , F_m(X))$

where $F_i(X) = (X - i)^2$, $i = 1,2,...,m$, $X \in [1, m]$.

The theoretical Pareto optimal set S_p is the whole feasible domain S= [1, m], so it gives us too much information for decision making. However, if we consider ε-optimal set S_ε, then we can easily get the following theoretical results:

(1) If m is even, then its ε-optimal set $S_\varepsilon = [m/2, (m+1)/2]$.

(2) If m is odd,   then its ε- optimal set $S_\varepsilon = \{ (m+1)/2\}$, a unique optimal solution.

We will illustrate these results with numerical experiments.

Benchmark Problem 2
Minimize $F(X) = (f_1(x), f_2(x), \cdots, f_9(x))$, $-2 \leq x \leq 2$, $-2 \leq y \leq 2$,

where

$$f_1(x) = -(1/(1+(((x^2-y^2)((x^2-y^2)-12x^2y^2)-1)^2 + 4x^2y^2(3(x^2-y^2)^2-4\ x^2y^2)^2)^{1/2}$$
$$+t\sin(x)), \text{ where parameter } t \in [-1,1],$$

$$f_2(x) = \sqrt{(x+1.01)^2 + y^2}$$

$$f_3(x) = \sqrt{x^2 + (y+1.01)^2}$$

$$f_4(x) = \sqrt{x^2 + (y-1.01)^2}$$

$$f_5(x) = \sqrt{(x+0.4825)^2 + (y+0.8283)^2}$$

$$f_6(x) = \sqrt{(x+0.4825)^2 + (y-0.8283)^2}$$

$$f_7(x) = \sqrt{(x-0.4825)^2 + (y-0.8283)^2}$$

$$f_8(x) = \sqrt{(x-0.4825)^2 + (y+0.8283)^2}$$

The first objective $f_1(x)$ with parameter $t=0$ is a multi-modal function, which has eight valleys respectively at points : (0, -1) , (0,1), (-1, 0) , (1,0), (0.47245126882998, -0.81830960171391), (0.47245126882998, 0.81830960171391), (-0.47245126882998, -0.81830960171391), (-0.47245126882998, 0.81830960171391)

4.2 Numerical Results

Benchmark Problem 1. When $m=4$ and m=5, the numerical results are depicted on Fig.1 and Fig.2 respectively. From the figures, we can clearly see that Pareto-optimal set S_p is uniformly distributed in the whole feasible domain S= [1, m] ; ε-optimal set S_ε is consistent with the above theoretical results .

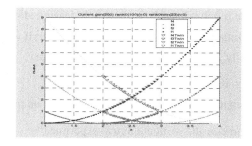

Fig. 1. Comparison of Pareto-front and ε-front for Benchmark Problem 1 with four objectives. (black-dot denote the Pareto-front , red- triangle denote the ε-front.)

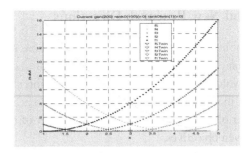

Fig. 2. Comparison of Pareto-front and ε-front for Benchmark Problem 1 with five objectives

Benchmark Problem 2

From the numerical results(Fig.1, Fig.2, Fig.3(b) and Fig.4(b)), it is found that the number of ε-optimal solutions depends on the number of objectives been odd or even. But compared with the number of Pareto optimal solutions, no matter the number of objectives is odd or even, the number of ε-optimal solutions is quite smaller than that of Pareto optimal solutions.

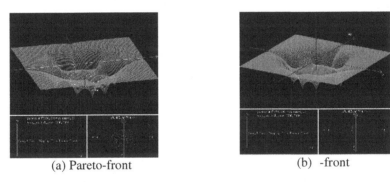

(a) Pareto-front   (b) -front

Fig. 3. Comparison of Pareto-front (a) and ε-front (b) with 8 conflict objectives

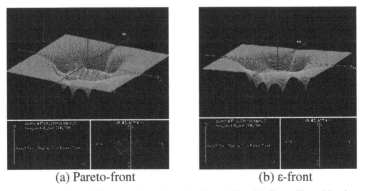

(a) Pareto-front   (b) ε-front

Fig. 4. Comparison of Pareto-front (a) and ε-front (b) with 7 conflict objectives

5 Conclusions and New Research Directions

The key contribution in this paper is the new conception of optimality, called ε-optimality, for many-objective optimization. It can be used instead of Pareto-optimality in evolutionary multi-objective optimization algorithms. In the new concept of optimality, the number of objectives is taken into account. It is found that for ε = 1 if the number of objectives is odd then the number of ε-optimal solutions is fewer than that with even objectives.

As we mentioned before that the definition of -optimality is imperfect. There exist many difficulties waiting to overcome. For example, how to prove the convergence of the new algorithms? It is believed that the new concept will have a profound and lasting influence in the field of optimization as the applications of the -optimality based evolutionary algorithms being used widely.

Acknowledgements. This work was supported by the National Natural Science Foundation of China (No.60473081) and the Natural Science Foundation of Hubei Province(No. 2005ABA234). Thanks especially give to the anonymous reviewers for their valuable comments.

References

1. Edgeworth, F.Y.: Mathematical Psychics. P. Keagan, London, UK (1881)
2. Pareto, V.: Cours D'Economie Politique. F. Rouge, Lausanne, Switzerland (1896)
3. Coello Coello, C.A.: Guest Editorial: Special issue on evolutionary multi objective optimization. IEEE Transactions on Evolutionary Computation 7(2), 97–99 (2003)
4. Hughes, E.J.: Evolutionary many-objective optimization: Many once or one many? In: Proceedings of 2005 Congress on Evolutionary Computation, pp. 460–465. IEEE-Press, NJ, New York (2005)
5. Maneeratana, K., Boonlong, K., Chaiyaratana, N.: Compressed-objective genetic algorithm. In: Runarsson, T.P., Beyer, H.-G., Burke, E., Merelo-Guervós, J.J., Whitley, L.D., Yao, X. (eds.) Parallel Problem Solving from Nature - PPSN IX. LNCS, vol. 4193, pp. 473–482. Springer, Heidelberg (2006)
6. Zhou, A., Kang, L., Chen, Y., Huang, Y.: A new definition and calculation model for evolutionary multi-objective optimization. Wuhan University of Natural Sciences 8(1B), 189–194 (2003)
7. Benedetti, A., Farina, M., Gobbi, M.: Evolutionary multiobective industrial design: The case of a racing car tire-suspension system. IEEE Transactions on evolutionary Computation 10(3), 230-244 (2006)
8. Kang, Z.: Evolution Programming and Its Application on Modeling and Forecasting, Ph. D. dissertation, Wuhan University, Wuhan, China, 36–38 (2006)
9. Rudolph, G., Agapie, A.: Convergence properties of some multi-objection evolutionary algorithms. In: CEC 2000. Proceedings of the Congress on Evolutionary Computation, pp. 1010–1016. IEEE Press, NJ, New York (2000)

10. Guo, T., Kang, L.: A new evolutionary algorithm for function optimization. Wuhan University Journal of Natural Sciences 4(4), 409–414 (1999)

11. Zou, X.F., Liu, M.Z., Kang, L.S., He, J.: A high performance multi-objective evolutionary algorithm based on the principle of thermodynamics. In: Yao, X., Burke, E.K., Lozano, J.A., Smith, J., Merelo-Guervós, J.J., Bullinaria, J.A., Rowe, J.E., Tiňo, P., Kabán, A., Schwefel, H.-P. (eds.) Parallel Problem Solving from Nature - PPSN VIII. LNCS, vol. 3242, pp. 922–931. Springer, Heidelberg (2004)

12. Li, C.H., et al.: A fast particle swarm optimization algorithm with Cauchy mutation (to be published)

13. Laumanns, M., Thiele, L., Deb, K., Zitzler, E.: Combining convergence and diversity in evolutionary multiobjectvie optimization Evolutionary Computation, vol. 10(3), pp. 263–282 (2002)

A Multi-Objective Genetic Algorithm Based on Density

Jinhua Zheng[1], Guixia Xiao[1], Wu Song[1], Xuyong Li[1], and Charles X. Ling[2]

[1] College of Information Engineering, Xiangtan University, 411105, Hunan, P.R. China
[2] Department of Computer Science, University of Western Ontario, London, Canada
jhzheng@xtu.edu.cn, xgx800805@126.com,
evol_song@yahoo.com.cn, cling@csd.uwo.ca

Abstract. This paper presents a new kind of MOEA, namely DMOGA (Density based Multi-Objective Genetic Algorithm). After discussing the influence function and the density function, we employ density of a solution point as its fitness in order to make the DMOGA perform well on diversity. And then, we extend our discussions to fitness assignment and computation, pruning procedure when the non-dominated set is bigger than the size of evolutionary population, and selection from the environmental selection population. To make DMOGA more efficient, we propose to construct the non-dominated set with the Dealer's Principle. We compare our DMOGA with two popular MOEAs, and the experimental results are satisfactory.

Keywords: Density, Dealer's Principle, Evolutionary Computation, Genetic Algorithm.

1 Introduction

Many researchers have been focusing on Multi-Objective Evolutionary Algorithms (MOEA) in recent years, mainly because MOEAs have the abilities to solve complicated optimization tasks such as discontinuous, multi-modal and noisy function evaluations and to find a widespread of Pareto optimal solutions in a single run. There are several kinds of MOEAs, including VEGA (Vector Evaluated Genetic Algorithm) [10], Fonseca and Fleming's MOGA [7], NPGA (A Niched Pareto Genetic Algorithm for Multi objective Optimization) [8,9], PESA (The Pareto envelope based selection algorithm for multi objective optimization) [1,2], SPEA (the Strength Pareto Evolutionary Algorithm for Multiobjective Optimization) [12,13], and NSGA (Non-Dominated Sorting Genetic Algorithm for Multi Objective Optimization) [4,5]. The popular MOEAs without parameters are PESA-II [2], SPEA2 [13], and NSGA-II [4]. There are different features among the three popular MOEAs. For example, PESA-II has good records in efficiency, and SPEA2 performs well on diversity of solutions.

In this paper, we present a new kind of MOEA namely DMOGA (Density based Multi Objective Genetic Algorithm). The main loop of DMOGA is introduced in section 3, and some discussions about pruning procedure, fitness assignment, and selection are given in this section too. We describe the influence function, the density function and the fitness computation in section 4. To make DMOGA more efficient, we discuss to construct the non-dominated set with the Dealer's Principle in section 5.

L. Kang, Y. Liu, and S. Zeng (Eds.): ISICA 2007, LNCS 4683, pp. 12–25, 2007.

We compare our DMOGA with three popular MOEAs, and the experimental results are shown in section 6.

2 Related Work

As mentioned above, there are several kinds of MOEAs. We only introduce the most recent three kinds of MOEAs: PESA-II, SPEA2, and NSGA-II in this section.

In PESA [1, 2], there are two populations: internal population (IP) and external population (EP). The crowding strategy in PESA works by forming an implicit hyper-grid that divides objective space into hyper-boxes. Each individual in the population is associated with a particular hyper-box. Here, a squeeze factor of an individual is the number of crowing points in a hyper-box. A candidate solution may enter into the archive EP if it is non-dominated within IP, and not dominated by any member of EP. Once a candidate solution has entered into the archive EP, one individual must therefore be removed from EP. The choice is made by the first finding of the maximal squeeze factor in the population, and removing an arbitrary individual that has the squeeze factor. The selection is on the basis of squeeze factor that is used for selective fitness. When binary tournament selection is adopted, two individuals are taken at random from EP, and the one with lower squeeze factor is chosen.

In SPEA2 [13], each individual i in the archive and population is assigned a strength value $S(i)$, representing the number of solutions it dominates. The raw fitness $R(i)$ is determined by the strength of its dominators in both archive and population. Very simple density information is incorporated to discriminate between individuals with identical raw fitness values. The inverse of the distance to the k-th nearest neighbor is simply used as density estimate, denoted as $D(i)$. Thus, the fitness $F(i)$ of an individual i is the sum of $R(i)$ and $D(i)$. The individual with lower fitness is selected firstly in the evolutionary process. In the truncation procedure, the individual which has minimum distance to another individual is removed from the archive, and this operation has been done iteratively until the size of the archive is satisfactory.

The main contribution in NSGA-II (Non-dominated Genetic Algorithm) is to propose a method to construct the non-dominated set based on multi-layer classifiers [4]. The NSGA-II calculates n_i and s_i of each individual iteratively, where n_i records the sum of individuals that dominate individual i, and s_i is a set of individuals which are dominated by i. Then the algorithm classifies the population by formula $P_k = \{ i \mid n_i - k + 1 = 0 \}$, where P_k is a non-dominated set when the rank k=1, and a dominated set when rank k>1. The time complexity to construct a non-dominated set is $O(n^2)$, here n is the size of population. The NSGA-II keeps its diversity by a crowding procedure in which the average distance between two neighboring individuals and each of the objectives are calculated. The average distance is a density estimation of solutions surrounding a particular point in population. Thus, the individual with lower rank is preferred. If both points belong to the same front, then the one with lower average distance is preferred.

In the above three MOEAs, not only the squeeze factor in PESA-II but the distance between two individuals in SPEA2 and NSGA-II are the density estimation for crowding strategy factually. The problem is that the density estimations in the above three MOEAs are more simple. There are often a number of individuals surrounding a particular point and influencing its density. However, only a few of individuals are considered in the above three density estimation methods. The density is really important to maintain the diversity of solutions in MOEAs. In this paper, we suggest a method of density estimation for crowding strategy to maintain the diversity of solutions. In addition, to make MOEAs more efficient, we introduce a better algorithm than existing methods to construct non-dominated set.

3 Density Based Multi-Objective Genetic Algorithm

In general, a multi-objective optimization is defined as follows:

Find the vector $X^* = [X_1^*, X_2^*, ..., X_n^*]^T$, to optimize the function:

$$Min \quad f(X) = [f_1(X), f_2(X), ..., f_m(X)]^T \tag{1}$$

And at the same time it should satisfy p inequality constraints (2), and q equality constraints (3).

$$g_i(X) \geq 0 \quad i = 1, 2, ..., p \tag{2}$$

$$h_i(X) = 0 \quad i = 1, 2, ..., q \tag{3}$$

Here $X = [X_1, X_2, ..., X_n]^T$ is the vector of decision variables. We say that a point $X^* \in F$ is Pareto optimal if $\forall X \in F$ either (4), or there is at least one $i \in I$ satisfy (5). Where, $I = \{1, 2, ..., m\}$.

$$\bigcap_{i \in I}(f_i(X) = f_i(X^*)) \tag{4}$$

$$f_i(X) > f_i(X^*) \tag{5}$$

$$F = \{X \in R^n \mid g_i(X) \geq 0, i = 1, 2, ..., k;$$
$$h_j(X) = 0, j = 1, 2, ..., l\}$$

The concept of Pareto optimum was formulated by Vilfredo Pareto in 1896. In general, there is more than one Pareto optimal solution. The basic function of MOEAs is to construct a non-dominated set, and to make the non-dominated set approximate to the optimal solutions continually.

To discuss our algorithm conveniently, some concepts are defined as follows.

Definition 1: Assuming the size of set P is n, every individual in P has r attributes, and $f_k()$ is an evaluation function ($k = 1, 2, ..., r$). The relationship between individuals in P is defined as follows:

(1) Dominated relationship $\forall x$, $y \in P$, if $f_k(x) \leq f_k(y)$, (k=1,2, ..., r), and $\exists l \in \{1,2,\cdots,r\}$ make $f_l(x) < f_l(y)$, then the relationship is that x dominates y, or $x \succ y$. Here "$\succ$" represents dominated relationship, and we say x is a non-dominated individual, and y is a dominated individual.

(2) Irrelevance: $\forall x$, $y \in P$, if x and y do not dominate each other, or there is no relationship between x and y, then we say that x and y are irrelevant.

Definition 2: $\exists x \in P$, if $\neg \exists y \in P, y \succ x$, then x is a non-dominated individual of P. The set including all non-dominated individuals of P is called the non-dominated set of P.

It is known that the Pareto optimal front is approximated by the non-dominated set going close to it continually. Therefore, one of our discussion problems is how to construct a non-dominated set with respect to a specific evolution population, and the detail investigation is depicted in section 5. When the size of the non-dominated set exceeds the specified value, we have to employ a method, called pruning procedure, to reduce its size and maintain the diversity of solutions at the same time. On the other hand, we also have to use a method to select or generate some individuals (dominated) to fill the mating selection population when the number of the non-dominated individuals in the mating selection population is less than the specified value. The fitness assignment is also an important problem in our discussion.

The main loop of DMOGA (Density-based Multi-Objective Genetic Algorithm) is briefly described in Algorithm 1. Here, $|P_t| = N$ denotes the size of the mating selection population P_t, and the size of the environmental selection population R_t is 2N.

Algorithm 1: Density-based Multi-Objective Genetic Algorithm

Step 1: Generate an initial population R_0, and create an empty P_0. Set t=0.
Step 2: Construct the non-dominated set of R_t with Dealer' Principle-based algorithm, and copy all non-dominated individuals to P_{t+1} and R_{t+1}.
Step 3: If the size of Pt+1 exceeds N then assign fitness to each individual of Pt+1, and reduce the size of Pt+1 by means of a pruning procedure;
If the size of P_{t+1} is less than N, then select (n-$|P_{t+1}|$) individuals from (R_t- P_{t+1}) or generate (n- $|P_{t+1}|$) new individuals randomly to fill up the mating selection population P_{t+1}. And then assign fitness to each individual of P_{t+1}.
Step 4: If the termination criterion is satisfied, stop and return the P_{t+1} as a result.
Step 5: Perform binary tournament selection on P_{t+1}.
Step 6: Apply the recombination and mutation operation on P_{t+1}, and copy P_{t+1} to R_{t+1}.
Step 7: Increment generation counter, or let t=t+1; and then go to step 2.

Discussions about Pruning Procedure: In our research, all non-dominated individuals are copied to P_{t+1} when they are found from the environmental selection population R_t. The set P_{t+1} must be reduced when its size exceeds a specified value N in step 3. This process is called pruning procedure, and the pruning procedure is very important to make the MOEAs perform well on diversity of solutions. Since all individuals are not dominated each other, the density is calculated for each individual and defined as one's fitness with which different individual can be distinguished. The individual with the biggest density is selected and removed from the mating selection population P_{t+1}. The operation of selecting and removing is repeated until the size of P_{t+1} is equal to N. If all the non-dominated individuals have the same density, one can be selected by random from the mating selection population P_{t+1}.

The density must be adjusted (re-calculate) by moving the influence function value away from the others when an individual is removed from the mating population in the pruning procedure.

Discussions about Fitness Assignment: To make our algorithm more efficient, the procedure of fitness computation and assignment is evoked only when it is needed. In step 3 of the Algorithm 1, when the size of mating population P_{t+1} exceeds N, the fitness assignment procedure is needed to be evoked before the pruning procedure. In step3, if we employ a method to select individuals from $(R_t\text{-}P_{t+1})$ to fill up the mating population, the individuals in $(R_t\text{-}P_{t+1})$ need to be assigned fitness; And after filling the mating population P_{t+1}, each individual of P_{t+1} must be assigned fitness too. In step 6, all individuals of mating population P_{t+1} must be assigned fitness before performing binary tournament selection. In fact, the fitness assignment operation has been completed in step 3 or step5 before performing binary tournament selection in step 6.

Discussions about Selection: In step 3 of Algorithm 1, it is very simple and easy to generate $(N\text{-}|P_{t+1}|)$ new individuals by random to fill up the mating population P_{t+1}. But if a method is employed to select some individuals from the environmental selection population $(R_t\text{-}P_{t+1})$ in step 3, there are two tasks in our research. One task is to assign fitness to each individual of the environmental selection population $(R_t\text{-}P_{t+1})$. Another task is to select $(N\text{-}|P_{t+1}|)$ individuals from the environmental selection population $(R_t\text{-}P_{t+1})$. The individual with the smallest density is selected, and this selection operation is repeated until all the $(N\text{-}|P_{t+1}|)$ individuals are selected.

4 Fitness Computation

Before discussing the fitness assignment, let's investigate influence function and density function. Assume that the m-dimensional feasible domain in the objective space of a multi-objective optimization problem is denoted by $F^m \subseteq R^m$. Similarly to our human society in a group, every two solution points must be influenced by each other. The difficulty is how to quantify the influence degree of one solution point to another. Therefore, we define an influence function to depict this problem. The influence function of the i-th solution point to solution point y is defined as follows:

$$\psi(l_{i \to y}) : R \to R \qquad (6)$$

Such that ψ is a decreasing function of the distance $l(i, y)$ from i-th solution point to solution point y. As shown in Fig. 1. , similarly to $l(i, y)$, $l(j, y)$ is a scalar that represents the Euclidean distance from j-th solution point to solution point y. There, y denotes an individual in objective space, and i and j have the same mean as y. Different function forms can be employed as an influence function, such as parabolic, square wave or Gaussian. A Gaussian influence function has the general form of:

$$\psi(r) = \left(1/(\sigma\sqrt{2\pi})\right)e^{-r^2/2\sigma^2} \tag{7}$$

There σ is the standard deviation of the distribution. A very small σ can cause the influence function to decay very rapidly. In this case, the density function is insensitive to the distribution of the solution points. It is possible that even two neighboring points do not significantly influence each other. In contrast, a very large σ indicates a flat influence function, and the density function is still not very sensitive to the distribution of solution points. Therefore, it is very important to set an appropriate value of σ .In our experiment we decide the σ =1.0/20.0.

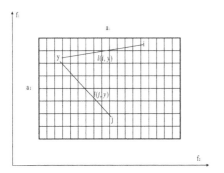

Fig. 1. Objective space with grids

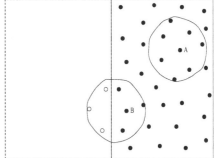

Fig. 2. Compensation to boundary effect

Now let's consider how to calculate the distance between two individuals. As a two-dimensional objective space shown in Fig. 1., the important thing is how to determine the values of a_1 and a_2. The grid size of each box becomes an indifferent region in which any solution points are considered to be the same after the quantities a_1 and a_2 are decided. In general, the appropriate grid size is problem-dependent and differs from one situation to another. As a rule, we can determine the grid size by the available computational power and desired accuracy. The density of an arbitrary point $y \in F^2$ can be calculated as follows:

$$D(y) = \sum_{i=1}^{N}\psi(l(i,y)) = \sum_{i_1=1}^{a_1}\sum_{i_2=1}^{a_2}\psi(l(<i_1,i_2>,y)) \tag{8}$$

Without losing generality, assume that there are m objectives. The feasible region can be depicted by a hyper-grid of size $a_1 \times a_2 \times \cdots \times a_m$ boxes. We suggest that the number of all boxes is the minimum value that is bigger than or equal to N, or $a_1 \times a_2 \times \cdots \times a_m \geq N$.

The density at each point of objective space is defined as the aggregation of the influence function values from all solution points. Assume there are N solution points ($N>0$), the density of an arbitrary point $y \in F^m$ can be captured as follows:

$$D(y)=\sum_{i=1}^{N}\psi(l(i,y))=\sum_{i_1=1}^{a_1}\sum_{i_2=1}^{a_2}\cdots\sum_{i_i=1}^{a_m}\psi(l(<i_1,i_2,\cdots;i_m>,y) \tag{9}$$

The density calculation for one solution point has a computational complexity of $O(N)$. We define $D(y)$ as the fitness of individual y. When there is an evolutionary population involving N individuals, the calculation for fitness assignment has a computational complexity of $O(N^2)$. The computational complexity may be very expensive, so the density estimation is very slow or even computational infeasible.

To make the density estimation computationally available, we set $b_k = \lfloor\sqrt{a_k}\rfloor$, $(k=1, 2, \ldots, m)$. Assume that the corresponding hyper-box of individual y is $<u_1, u_2, \cdots, u_m>$. Set $c_k = \min\{u_k + b_k, a_k\}$, and $d_k = \max\{u_k - b_k, 1\}$. Then the density estimation of an arbitrary point $y \in F^m$ can be obtained as follows:

$$D(y)=\sum_{i=1}^{N}\psi(l(i,y))=\sum_{i_1=d_1}^{c_1}\sum_{i_2=d_2}^{c_2}\cdots\sum_{i_i=d_m}^{c_m}\psi(l(<i_1,i_2,\cdots,i_m>,y) \tag{10}$$

We needn't consider the boundary effect in the above density estimation. In fact, the boundary effect is significant in the proximity of boundaries, and gradually vanishes as our view moves from the boundary to the middle of the feasible region. From a closer look of Fig. 2., we cam find that there is different density estimation between point A and point B. This is because that the density function at the point A in Fig. 2. is the aggregation of influence functions of all solution points in its vicinity, locating along different directions around it. On the contrary, point B of Fig. 2., which is located very close to the boundary, is mainly influenced by the solution points in the feasible region of its right half space while there are a few of solution points in the infeasible region of its left half space. In Fig. 2., the dashed rectangle shows a modification to the construction of the density estimation in order to compensate for the above discussed boundary effect. The points in the feasible region are mirrored by the boundary to create a set of virtual points in the infeasible region. Thus, the density estimation at point B is defined the aggregation of the influence functions of all real and virtual solution points. Note that even this simplified approach is difficult to implement, and the computational complexity is very expensive. The boundary effect is usually small, and can be negligible for the efficiency purpose of algorithms [6].

We define the density function as the fitness for a particular solution point. On this basis, we can perform the selection operation, and further perform the operations of recombination and mutation.

5   Construct Non-dominated Sets with Dealer's Principle

In the MOEAs based on Pareto optimum, it is very important to construct a non-dominated set efficiently because the construction speed affects the convergent speed of an algorithm directly. This section will introduce a fast approach, namely Dealer's Principle, to construct non-dominated sets.

The process to construct a non-dominated set with dealer's Principle can be described as follows. Firstly, a dealer, or a reference individual, is selected from the candidate solutions in each round of comparison (the dealer is usually the first individual of the current evolutionary population). Then the dealer is compared with the other members of the current candidate solutions one by one. The individuals that the dealer dominates must be removed after each round of comparison. If the dealer is not dominated by any other individuals, it is joined into the non-dominated set; otherwise the dealer must be removed. This process is repeated until there is no candidate solution. This approach has no backtracking when constructing non-dominated sets. In other words, a new non-dominated individual needs not to be compared with these non-dominated individuals that already exist. Therefore, we call the method the Dealer's Principle.

Assume P is an evolutionary population, and Q is a set of candidate solutions. Starting with an empty Nds of a non-dominated set, let Q=P. One individual x is picked up at random from Q (in fact, x is removed from Q in this case) and compared with other individuals in Q one by one, and then the individuals dominated by x are deleted. If x is not dominated by any other members of Q, then it is a non-dominated individual, and it is joined into Nds. We continue this process until Q is empty.

The Dealer's Principle of constructing a non-dominated set of P is briefly described in Algorithm 2.

Algorithm 2: Construct a non-dominated set with Dealer's Principle

```
Function Nds(Pop: population)
{Q=Pop;
 while (Q≠ ∅) do
  {x∈ Q; let Q=Q-{ x };
x-is-undominated=.T.;
   for ( all  y∈ Q )
{if ( x≻ y)
then Q=Q-{y }
else if (y≻ x)
    then x-is-undominated=.F.}
if (x-is-undominated) then Nds=Nds∪{ x };
   }
}
```

It is proved that the Nds is a non-dominated set of P, and is the biggest non-dominated set of P. Details about the proof are shown in reference [11].

The computational complexity is less than $O(n^2)$ in Algorithm 2, and the analysis is as follows. Suppose that n is the size of set P, and there are m non-dominated individuals in P. As a general rule, there have been k rounds of comparison altogether, $m \le k \le n$. There are k individuals are removed naturally for k rounds of comparison, and one for each comparison. Note that there are only m non-dominated individuals, and the other (k-m) members are dominated individuals. There are (n-k) dominated individuals removed from the set of candidate solutions Q for k rounds of comparison. Suppose that the probability of removing (n-k) dominated individuals is equal: (n-k)/k, the computational complexity of k rounds of comparison is:

(n-1)+ (n-2-(n-k)/k) +(n-3-2(n-k)/k) +...+(n-k-(k-1)(n-k)/k)

=[(n-1)+ (n-2) +...+(n-k)]-[(n-k)/k+2(n-k)/k+...+k(n-k)/k]+k (n-k)/k

=[(n-1)+ (n-k)]k/2-[(n-k)/k+k(n-k)/k]k/2+ (n-k)

=k(n-1)/2+ (n-k)/2= $O(kn)$ $<n^2$

From the above analysis, the time complexity of the algorithm to construct a non-dominated set with the Dealer's Principle is better than that in SPEA2 and NSGA-II. Especially in the early stage of algorithm execution, m is always much less than n (or m<<n), so the algorithm is more efficient.

6 Experiment

In this section, we compare the DMOGA with some state-of-art MOEA (NSGA-II and SPEA2) on three-objective test proplems DLZ1 and DLZ2 [3]. As we all known, the final solutions for each MOEAs' single run aims at three goals: i.convengence to the true Preto optimal front. ii .distribution over the the true Preto optimal front.iii.lower the computational cost.

Here, we use the Generational distance (GD) to evaluate the convengence and in diversity metric the Spacing (SP) is used. GD is indicates the mean distance between the solutions that MOEA obtain and is defined as:

$$GD = \sqrt{\sum_{i=1}^{n} d_i} / n \qquad (11)$$

Where d_i is the shortest distance between the non-dominated set that obtained and the Ture Pareto front.And the SP is used to estimate the spread of the solutions and is defined as follow:

$$SP = \sqrt{1/(n-1) * \sum_{i=1}^{n} (\bar{d} - d_i)^2} \qquad (12)$$

$$d_i = \min_j (|f_1^i(\vec{x}) - f_1^j(\vec{x})| + |f_2^i(\vec{x}) - f_2^j(\vec{x})|)$$

Where i , j=1,..,n. $\bar{d}$ is the mean of all d_i .

Next we introduce some parameter settings of the algorithm. In those algrithms binary code is used with 20 bits by paremeters, and we use a crossover probability of 0.8, a mutation probility of 1/L(L is the string length).The population size is 200. Each run is executed 500 gernerations. For the test proplems with each alogrithm we do 30 experiments, and for each experiment we set 200 to the number of non-dominated solutions that the alogrithm obtaied in max-generation.

DLZ1 test problem

The difficulty in this proplem is to converge to the hyper-palne, the search space contains $(11^k - 1)$ local Pareto optimal fronts [3].The true Pareto front lie on the linear hyper-plane: $\sum_{m=1}^{M} f_m = 0.5$. (Where M is the number of objectives, and in this experment, M=3).

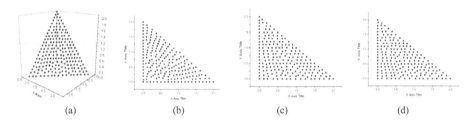

| (a) | (b) | (c) | (d) |

Fig. 3. DLZ21 optimization with DMOGA single run (a) DLZ1 optimization front (b) f1*f2 plane projection (c) f2*f3 plane projection (d) f1*f3 plane projection

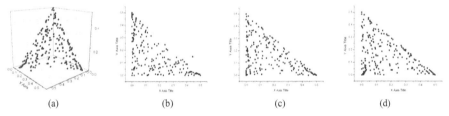

| (a) | (b) | (c) | (d) |

Fig. 4. DLZ1 optimization with NSGA-II single run (a) DLZ1 optimization front (b) f1*f2 plane projection (c) f2*f3 plane projection (d) f1*f3 plane projection

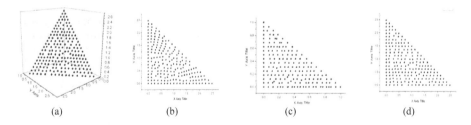

| (a) | (b) | (c) | (d) |

Fig. 5. DLZ1 optimization with SPEA2 single run (a) DLZ1 optimization front (b) f1*f2 plane projection (c) f2*f3 plane projection (d) f1*f3 plane projection

Tables 1-3 show the performance metrics of all approaches. And Fig.3-5 show the DLZ1 shape which algrithms obtained. For DLZ1 optimization DMOGA achieves the best distribution of solutions, and in this metric NSGA-II is the worst, SPEA2 is similar to DMOGA. In convengence metric NSGA-II is the best, and DMOGA follows. And about efficiency, the cpu time of DMOGA is faster than SPEA2, however slower than NSGA-II.

Table 1. Comparision result of GD

	EMOGA	NSGA-II	SPEA2
Best	0.03177	0.02039	0.08003
Worst	0.46512	0.15331	0.66416
Average	0.16325	**0.09449**	0.2849
Median	0.16076	**0.10411**	0.24846
Std.Dev	0.12012	0.04457	0.22684

Table 4. Comparision result of GD

	EMOEA	NSGA-II	SPEA2
Best	3E-6	5.3296E-6	1.5065E-4
Worst	3.22E-4	3.4142E-5	0.00136
Average	7.6266E-5	**2.1877E-5**	4.8026E-4
Median	5E-5	**2.1955E-5**	3.9638E-4
Std.Dev	9.0103E-5	9.8141E-6	4.4776E-4

Table 2. Comparision result of SP

	EMOGA	NSGA-II	SPEA2
Best	0.02644	0.03120	0.01475
Worst	0.16375	0.22916	0.14213
Average	**0.04362**	0.10568	0.06378
Median	**0.02644**	0.102339	0.05466
Std.Dev	0.04101	0.05326	0.05854

Table 5. Comparision result of SP

	EMOEA	NSGA-II	SPEA2
Best	0.01429	0.03171	0.01249
Worst	0.01898	0.04088	0.01901
Average	0.01622	0.03761	**0.01582**
Median	0.01613	0.03767	**0.01573**
Std.Dev	7.6918E-4	0.00261	0.00207

Table 3. Comparision result of CPU time

	EMOGA	NSGA-II	SPEA2
Best	39.6517	15.421	313.375
Worst	76.03974	16.109	372.328
Average	55.05135	**15.76945**	331.9788
Median	37.01888	**15.781**	324.3825
Std.Dev	14.13329	0.19447	22.15925

Table 6. Comparision result of CPU time

	EMOEA	NSGA-II	SPEA2
Best	43.02065	22.171	437.281
Worst	44.76925	22.828	503.078
Average	43.8284	**22.5075**	453.7698
Median	43.71061	**22.462**	443.607
Std.Dev	0.55231	0.20171	25.2992

DLZ2 test problem

We consider the three-objective test problem of DLZ2, the Pareto optimal front satisfying $f_1^2 + f_2^2 + f_3^2 = 1$ in DLZ2 [3]. Seen from the Fig. 6-8, all the alogrithms can converge to the true Pareto front, and the distribution of DMOGA in DLZ2 optimal front is better than NSGA-II, and this is similar to SPEA2. From the table 4-6, the convergence metric of the DMOGA is also better than SPEA2, but worse than NSGA-II. The cpu time of DMOGA is faster than SPEA2 and that of NSGA-II is the fastest of all.

For these test proplems we can draw a conclusion that DMOGA is outperform than SPEA2 in all performance metrics, and it is also better than NSGA-II in diversity metric. That the diversity metric of DMOGA is best proved that the density is really important to maintain the diversity of solutions.

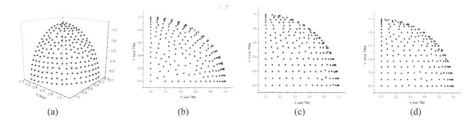

Fig. 6. DLZ2 optimization with DMOGA single run (a) DLZ2 optimization front (b) f1*f2 plane projection (c) f2*f3 plane projection (d) f1*f3 plane projection

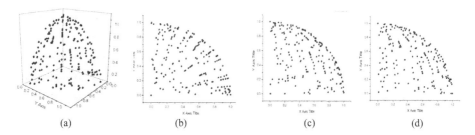

Fig. 7. DLZ2 optimization with NSGA-II single run (a) DLZ2 optimization front (b) f1*f2 plane projection (c) f2*f3 plane projection (d) f1*f3 plane projection

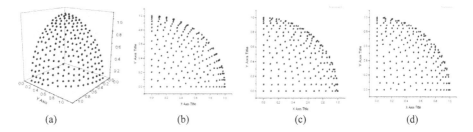

Fig. 8. DLZ2 optimization with SPEA2 single run (a) DLZ2 optimization front (b) f1*f2 plane projection (c) f2*f3 plane projection (d) f1*f3 plane projection

7   Conclusion

After discussing some concepts about MOEAs based on Pareto Optimum, this paper proposes a new kind of MOEA, namely DMOGA (Density based Multi-Objective Genetic Algorithm). And then by surrounding with DMOGA, we extend our discussions to pruning procedure, fitness assignment, and selection. In order to make DMOGA perform well on wide spread of solutions, we have discussed the influence function, the density function and the fitness computation. To make DMOGA more efficient, we have discussed to construct the non-dominated set with the Dealer's Principle.

In this paper, three of the MOEA (DMOGA, SPEA2 and NSGA-II) are tested on two test problems: they are DLTZ1 and DLTZ2 with three objectives. And we can see from the result of comparing experiments, DMOGA performs well on convergence, diversity and efficiency, especially the diversity metric is outperform than other algorithms. This proved that the density is really important, and could achieve more distribution solutions.

Acknowledgments. Authors acknowledge the supports from the Natural Science Foundation of Hunan Province (05JJ30125), and the Keystone Science Research Project of the Education Office of Hunan Province (06A074).

References

1. Corne, D.W., Knowles, J.D., Oates, M.J.: The Pareto envelope based selection algorithm for multi-objective optimization. In: Proceedings of the Parallel Problem Solving from Nature VI Conference, pp. 839–848 (2000)
2. Corne, D.W., Jerram, N.R., Knowles, J.D., Oates, M.J.: PESA-II: Region-based Selection in Evolutionary Multiobjective Optimization. In: GECCO-2001. Proceedings of the Genetic and Evolutionary Computation Conference, pp. 283–290. Morgan Kaufmann Publishers, San Francisco (2001)
3. Deb, K., Thiele, L., Laumanns, M., Zitzler, E.: Scalable Test Problems for Evolutionary Multi-Objective Optimization. KanGAL Report Number 2001001 (2001)
4. Deb, K., Pratap, A., Agrawal, S., Meyrivan, T.: A Fast and Elitist Multi-objective Genetic Algorithm: NSGA-II. IEEE Transactions on Evolutionary Computation 6(2), 182–197 (2002)
5. Deb, K., Mohan, M., Mishra, S.: A Fast Multi-objective Evolutionary Algorithm for Finding Well-Spread Pareto-Optimal Solutions. KanGAL Report No. 2003002 (February 2003)
6. Farhang-Mehr, A., Azarm, S.: Diversity Assessment of Pareto Optimal Solution Sets: An Entropy Approach. In: CEC'2002. Congress on Evolutionary Computation, vol. 1, pp. 723–728. IEEE Service Center, Piscataway, New Jersey (May 2002)
7. Fonseca, C.M., Fleming, P.J.: Genetic Algorithms for Multiobjective Optimization: Formulation, discussion and Generalization. In: Forrest, S. (ed.) Proceedings of the Fifth International Conference on Genetic Algorithms, pp. 416–423. University of Illinois at Urbana-Champaign, Morgan Kauffman Publishers (1993)
8. Horn, J., Nafpliotis, N.: Multiobjective Optimization using the Niched Pareto Genetic Algorithm, Technical Report IlliGAl Report 93005. University of Illinois at Urbana-Champaign, Urbana, Illinois, USA (1993)
9. Horn, J., Nafpliotis, N., Goldberg, D.E.: A Niched Pareto genetic Algorithm for Multiobjective Optimization. In: Proceeding of the first IEEE Conference on Evolutionary Computation, pp. 82–87. IEEE Computer Society Press, Los Alamitos (1994)
10. Schaffer, J.D.: Multi objective optimization with vector evaluated genetic algorithms. In: Grefenstette, J. (ed.) Proceedings of an International Conference on Genetic Algorithms and their Applications, pp. 93–100 (1985)

11. Zheng, J., Shi, Z., Ling, C.X., Xie, Y.: A New Method to Construct the Non-Dominated Set in Multi-Objective Genetic Algorithms. IIP 2004, Beijing, 2004.10 (to appear)
12. Zitzler, E., Thiele, L.: Multi-objective evolutionary algorithms: A comparative case study and the strength pareto approach. IEEE Transactions on Evolutionary Computation 3(4), 257–271 (1999)
13. Zitzler, E., Laumanns, M., Thiele, L.: SPEA2: Improving the Strength Pareto Evolutionary Algorithm for Multi-objective Optimization. EUROGEN 2001 - Evolutionary Methods for Design, Optimization and Control with Applications to Industrial Problems (September 2001)

Interplanetary Trajectory Optimization with Swing-Bys Using Evolutionary Multi-objective Optimization

Kalyanmoy Deb, Nikhil Padhye, and Ganesh Neema

Department of Mechanical Engineering
Indian Institute of Technology Kanpur, PIN 208016, India
{deb,npadhye}@iitk.ac.in

Abstract. Interplanetary trajectory optimization studies mostly considered a single objective of minimizing travel time between two planets or launch velocity of spacecraft at the departure planet or maximizing delivered payload at the destination planet. Despite a few studies, in this paper, we have considered a simultaneous minimization study of both launch velocity and time of travel between two specified planets with and without the use of gravitational advantage (swing-by) of some intermediate planets. Using careful consideration of a Lambert's approach with the Newton-Raphson based root finding procedure of developing a trajectory dictated by a set of variables, a number of derived parameters, such as time of flight between arrival and destination planet, date of arrival, and launch velocity, are computed. A commonly-used evolutionary multi-objective optimization algorithm (NSGA-II) is then employed to find a set of trade-off solutions. The accuracy of the developed software (we called GOSpel) is demonstrated by matching the trajectories with known missions.

1 Introduction

The interplanetary mission design is a challenging task. These missions are complicated due to dynamics of our solar system. As spacecraft travels through our solar system it may encounter many celestial bodies, and may get influenced by their gravitational fields (swing-by of planets). These gravitational fields may be used in a constructive way to help in reducing energy requirement of a flight. Sending satellites to interplanetary trajectory is risky and expensive. There can be various trajectories which a spacecraft may follow. But there has to be an optimal trajectory which when followed, gives high performance boost either in energy requirement or in time required for the mission.

Genetic Algorithms (GAs) have been used for over 20 years in various applications of optimization. GAs are successfully applied in many complex real-world optimization problems where the function to be optimized is highly non-linear and discrete. In the recent past, they have been adapted very successfully in solving multi-objective optimization problems involving more than one conflicting objectives. Non-dominated sorting algorithm-II (NSGA-II) [7] is an example of such GA based multi-objective optimizers. In this paper, we develop NSGA-II-based trajectory optimization problems in a software, we called GOSpel. The time of window in which optimal solutions are sought, type of the transfer, choice of mission and information about the swing-by planets is accepted from the user. The software displays a number of trade-off optimized solutions and allows user to investigate the trajectory through a graphical user interface.

L. Kang, Y. Liu, and S. Zeng (Eds.): ISICA 2007, LNCS 4683, pp. 26–35, 2007.
© Springer-Verlag Berlin Heidelberg 2007

2   Previous Studies on Interplanetary Trajectory Optimization

The celestial mechanics of the 'swing-by' is known to astronomers for at least 150 years. A lot of research work have been done in the area of Interplanetary Trajectory Optimization (IPTO), for both direct transfer and with swing-by of other planets. In 1925, Walter Hohmann [9] designed the transfer trajectory for two non intersecting orbits, which later has been used for direct transfer from one planet to another planet. Cornelisse [6] studied the various methods for computation of interplanetary trajectory and showed that interplanetary trajectories were most efficiently accomplished by the method of patched conics. Miele and Wang [10] presented fundamental properties of optimal orbital direct transfer using the cases of transfer from Earth orbit to Mars orbit and from Earth orbit to Venus orbit. They focused on compromised trajectories between flight time and propellant mass. Biesbroek and Ancarola [4] studied genetic algorithm setting for trajectory optimization. They used swing-by calculator [3] for their work for finding trajectory of rockets and interplanetary trajectories.

There are some softwares available for optimizing interplanetary trajectories. One of them is "Orbital Mechanics with Numerit PRO" [2]. The major disadvantage of the software is that the user has to decide certain parameters like, to use swing-by or not, which planet to be used as swing-by and at what altitude etc. The software also uses grid search technique (exhaustive search) to optimize the trajectory. Another disadvantage is that, at most only one planet can be used for swing-by. Another software "Swing-By Calculator" provided by Jaquar Space Engineering [3] adds extra feature for performing multiple swing-bys. It is dependent on user choices like which planet to take swing-by and at what dates. In this paper, we try to overcome these deficiencies and develop a Genetically Optimized Spacecraft Launcher (GOSpel).

3   Proposed Approach with Swing-By

In flyby or swing-by assisted missions, a spacecraft first have to go to one or more flyby planets and then to the destination planet. This way there may exist a conflict between the energy requirement and the time to complete for the mission. Here, we formulate the above-mentioned problem into a multi-objective optimization problem and solve using the NSGA-II procedure [7]. The objective function considered here are (i) minimization of launch velocity and (ii) minimization of the total time of travel. The launch velocity is directly associated with the fuel (energy) requirement, which in turn effects the cost and technical aspects. By minimizing time of flight, one can achieve missions to complete quickly.

The trajectory optimization problem involves a number of decision variables and constraints, a description of which requires us to first understand the transfer of space-craft from one planet to another. The standard orbital mechanics [5] uses parameters like state vectors and velocity maneuvers for fixing a trajectory. For computing a transfer from one planet to another, a patched conic model [5] is usually employed. In this model, the motion takes place along a plane. In practice, the transfer can involve swing-by planets or can be a direct one.

3.1 Direct Transfer

Let us now explain how to compute the motion between a pair of planets. Say, the spacecraft moves from first planet to the second planet. This involves knowing the location of both planets at the start and at the end when the spacecraft reaches the second planet. Moreover, assume that we fix a transfer time t for reaching second planet from the first one and investigate if such a transfer is possible from the location information of both planets. The Lambert's approach [8] helps us determine the velocity vectors required at the first (v_1) and second (v_2) planet in order to materialize such a transfer time. Lambert's approach involves an iterative procedure of adjusting the velocities so that the desired transfer time t is achieved. Thus, for a direct transfer, the departure date and transfer time are the two variables of the optimization problem.

3.2 Swing-By Transfer

It is clear that for S swing-by planets, there are $(S + 1)$ pairs of transfer needed. For a transfer time t_i for the i-th transfer, Lambert's approach can be used to find velocity vectors of the spacecraft v_{i-1}^+ (+ mean outgoing from the planet) and v_i^- (− means incoming to the planet) near $(i-1)$-th and i-th planets, respectively. Thus, for a transfer time t_i, the Lambert's approach computes a pair of velocity vectors of the spacecraft near the participating planets. Now, for the i-th swing-by planet, we have an incoming velocity v_i^- computed from the i-th transfer (with i-th planet as the second planet) and outgoing velocity v_i^+ computed from the $(i + 1)$-th transfer (with i-th planet as the first planet). The difference between these two velocity vectors introduce a change in plane from one transfer to the other. This requires that the spacecraft generates some thrust so that the required change of plane is achieved. This energy will be taken into account by adding it to the launch energy objective. Thus, the departure date, S swing-by planets, and $(S + 1)$ transfer times are the variables for an optimization problem.

For the swing-by case, the feasibility of an overall transfer from launch till final arrival planet needs to be checked. It is mentioned above that by Lambert's approach the incoming and outgoing velocity vectors of each swing-by planet can be computed for a given set transfer time values of each pair of successive planets. In order to have the overall transfer feasible, the difference between the incoming and outgoing speed ($|v_i^-|$ and $|v_i^+|$) of every swing-by planet must be as small as possible. In practice, we construct one equality constraint for each swing-by planet:

$$|v_i^+| - |v_i^-| = 0, \quad i = 1, 2, \ldots, S. \tag{1}$$

Since the above two velocities are computed from two consecutive transfer times (t_{i-1} and t_i), these transfer times can be adjusted so that the above equality constraint is met. Thus, for S swing-by planets, we shall have S such equality constraints involving $(S + 1)$ speed values. This gives us some flexibility of choosing the speed values for a feasible overall transfer. However, there is a difficulty with this approach.

Recall that in the Lambert's approach, the velocity and location of spacecraft at two participating planets are the outcome and the departure date and transfer time are input parameters. Thus, by adjusting $(S + 1)$ transfer time values and using the Lambert's approach S times (one for each transfer), we can try to come up with a set of $(S + 1)$

velocity vectors which would satisfy all S equality constraints mentioned above. To convert the above problem into a root-finding problem, we introduce another equality constraint involving the altitude (h_i) of the spacecraft at the first swing-by planet. For the first swing-by planet, we introduce the following equality constraint:

$$h_1 - h_1^d = 0, \qquad (2)$$

where h_1^d is the desired altitude of the first swing-by planet. For all other swing-by planets, ideally we should ensure that the altitude is positive, but in this study we consider negative altitude solutions as well by assuming them to be 'powered swing-bys'. We add an equivalent energy component in the launch velocity to take into account of not colliding with the swing-by planet.

3.3 Handling Using NSGA-II

First, we discuss the representation scheme for the decision variables within the NSGA-II framework [7]. We fix a maximum of three swing-by planets, thereby leaving us with four options: (i) direct flight (no swing-by), (ii) one planet swing-by, (iii) two planet swing-by and (iv) three planet swing-by. We use a two-bit substring for representing these four options with 00, 01, 10 and 11, respectively. Thereafter, we have three substrings of three bits each. Each three-bit substring represents a swing-by planet (one of the first eight planets of the solar system coded as 000: Mercury, 001: Venus, 010: Earth, etc. Depending on the first two-bit substring dictating the number of swing-by planets we pick the corresponding planets from the string. These $2 + 3 \times 3$ or 11-bit strings give us information about which and how many planets are used in the trajectory determination.

The next set of 4 variables are coded as real-valued variables and represent transfer times between departure planet to first swing-by planet, first to second swing-by planet, second to third swing-by planet and third swing-by to arrival planet. Here again, depending on the number of swing-by planets (S) dictated by the first two-bit substring, we consider only the first ($S + 1$) transfer times. A typical NSGA-II solution may look like the following:

10 000 100 101 16/6/2005 124 205 580 425

The solution signifies that there are two swing-by planets and they are the first planet (Mercury) and fifth planet (Jupiter) between departure and arrival planets (need not be represented in NSGA-II, as they are fixed for all feasible solutions). Thus, we ignore the third swing-by planet mentioned in the solution. The next decision variable is the departure date (16 June 2005). This date is actually represented using the Julian day (which is an integer value). The next four real-valued values are transfer times and we only pick the first three values as the transfer time between the departure and the first swing-by planet and so on. Once again, the transfer time of 425 is a useless parameter for this solution, since only two swing-bys are considered dictated by the first two-bit substring 10. In the case of direct transfers, only two variables (the departure date and the first real-parameter value indicating the transfer time) are used in the evaluation procedure. Thus for a single-planet swing-by six, for a two-planet swing-by eight and for three-planet swing-by all ten variables are to be considered.

Staring with the departure date and transfer time values in a NSGA-II string, incoming and outgoing speed values are calculated using Lambert's approach for each transfer. The transfer time values and the altitude of the first swing-by planet are adjusted by using the Newton-Raphson method till the equality constraints are satisfied using a small ϵ value. Thereafter, the original transfer time values are replaced with the ones computed using the Newton-Raphson method. Objective values are then computed for the solution. Variable bounds on transfer times are checked and any violation is assigned as the 'constraint violation' of the solution and the solution is declared infeasible. We combine the evaluation scheme with NSGA-II and develop a user-friendly software for practice. The GA Optimized Spacecraft Launcher (GOSpel) software is capable of handling the following features:

- An option of direct or one to three swing-bys individually,
- An option of simultaneous consideration of direct or to a maximum of three planet swing-bys.
- A true optimization varying departure date over a large launch window as a real-valued parameter (and not with a finite step of one full day, as used in many commercial softwares)
- An option of flyby out of arrival planet or orbital around the arrival planet

4 Proof-of-Principle Results

To demonstrate the correctness of our implementation of the trajectory optimization procedure and evaluation of solutions, we first apply our developed code to a number of known missions, taken from the web and literature. The comparison of our obtained solutions with those computed by other means is then performed.

4.1 Earth-Venus-Mercury Mission

First, we consider a Earth-to-Mercury mission with a possible swing-by from Venus. This problem was studied by using an exhaustive search procedure for the minimization of launch velocity. For the years 2001 and 2002, the launch possibility and corresponding launch velocity needed for the mission to Mercury via Venus was calculated with a step size of one day for departure. The best solution found is shown in Table 1.

To validate our procedure, we use GOSpel during this two-year departure window to find Pareto-optimal solution for the minimization of launch velocity and time of flight. We use the option for using one or no swing-by and the option of an orbital motion to the destination planet. A population of size 200 and a maximum generation of 200 are fixed. The GOSpel software uses the SBX operator with $p_c = 0.9$ and $\eta_c = 10$ and the polynomial mutation operator with $p_m = 1/n$ and $\eta_m = 20$. Figure 1 shows the corresponding frontier. It is interesting to note that there are two disconnected fronts: (i) trajectories with swing-by and (ii) trajectories with direct transfer. For minimum launch velocity trajectories, it is recommended to use the swing-by from Venus and for minimum time trajectories it is better to go straight to Mercury from Earth. Table 1 shows a

Table 1. Earth-Venus-Mercury trajectories using GOSpel and exhaustive search

		Exhaustive Search	GOSpel
Earth Departure Date	(mm/dd/yy)	08/05/02	08/05/02
Venus Swing-By Date	(mm/dd/yy)	12/05/02	12/05/02
Mercury Arrival Date	(mm/dd/yy)	02/13/03	02/07/03
Altitude at Venus	(km)	-938.9	-978.84
Total Time	(Days)	192	186
Launch Velocity	(km/s)	2.79	2.78

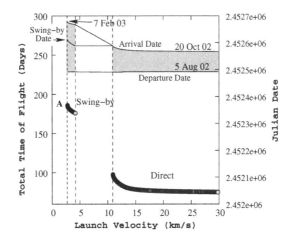

Fig. 1. Trade-off solutions for the Earth-Venus-Mercury mission using GOSpel

closest solution to the exhaustively searched solution for the minimum launch velocity objective. The GOSpel solution is closer to the exhaustive search solution. In fact, since no finite step is used in GOSpel, a better launch-velocity solution than the exhaustive search method (with a step size of one day) is found. The best launch velocity solution demands a slightly smaller value than the exhaustive search solution. The matching of our results with the exhaustive search solution by an independent study provides confidence to our developed software.

Before we leave this proof-of-principle study, we also plot the departure, swing-by, and arrival dates of all obtained solutions by GOSpel. Figure 1 marks the Julian dates of these trajectories (values marked on the right axis). The following features of trajectories are gathered:

1. All Pareto-optimal missions must start at the same date: 5 October 2002, irrespective of whether the mission involves a swing-by or not.
2. With an increase in launch velocity requirement, the arrival becomes quicker. It seems that if departed from Earth on 5 October 2002, there exist a number of plausible missions trading-off launch velocity and travel time.

The use of an evolutionary multi-objective optimization algorithm to find a set of Pareto-optimal solutions allows one to look for such important information about the solutions. Before executing this study, it would be difficult to predict that on a two-year long span of plausible departure dates, there exist one particular day (5 October 2002) which opens up enormous and optimal opportunities of several launching.

4.2   Earth-Mars-Venus-Mercury Mission

Next, we apply GOSpel and compare its results with another study performed by a commercial software, Swing-by Calculator (SC) [3] on a mission involving two planet swing-bys: Mars and Venus. The destination planet is Mercury and the departure window is kept within 1 Jan 2005 for a year. Table 2 shows the obtained SC result obtained for minimum time of flight. The dates of arrival at Mars and Venus and the corresponding arrival and de-

Table 2. Earth-Mars-Venus-Mercury trajectories. Dates are in mm/dd/yy.

	SC	GOSpel
Earth Departure Date	08/14/05	08/14/05
Mars swing-by Date	10/26/05	10/26/05
Mars: v_∞ incoming (km/s)	15.9737	15.9465
Mars: v_∞ outgoing (km/s)	15.7722	15.9465
Venus swing-by Date	02/01/06	01/31/06
Venus: v_∞ incoming (km/s)	8.7439	8.9958
Venus: v_∞ outgoing (km/s)	8.9459	8.9958
Mercury Arrival	03/31/06	03/31/06
Flight Time (Days)	229	228.326
Launch Velocity (km/s)	11.176	11.145

parture velocities are also shown in the table. The GOSpel solutions are shown in Figure 2 for both objectives. In this case, we allow only two-planet swing-by trajectories to be considered. Thus, direct or one-planet swing-by option is not considered. All solutions found involve two swing-bys, but providing a trade-off between time of flight and energy requirement. The solution on the Pareto-optimal front closest to the SC solution is tabulated in Table 2. The comparison of both solutions again indicates the accuracy of GOSpel procedure.

Interestingly, the SC solution does not seem to be the minimum-time solution. The figure shows that there exists a solution with a smaller time of flight. Figure 3 shows the trajectory of the two-planet swing-by solution of the solution shown in the table. The travel from earth to Mars (outward) and then back to Venus (inward) and finally to Mercury is interesting.

4.3   Direct Transfer

Next, we consider a direct transfer scenario from Earth to Mars. For this purpose, we use solutions from two softwares Numerit and SC. The departure dates within 1 June 2013 to 1 June 2014 are considered. This scenario is an example test problem reported in users' manual of the NUMERIT software. Both Numerit and SC softwares use a grid search strategy in which solutions at a step of one day is considered one at a time and the solution having the minimum launch velocity is found. The table shows that SC

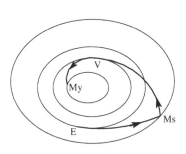

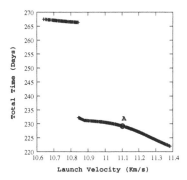

Fig. 2. Trade-off solutions for the Earth-Mars-Venus-Mercury mission using GOSpel

Fig. 3. Trajectory for the specific Earth-Mars-Venus-Mercury mission shown in the table

Table 3. Earth-Mars direct transfer trajectories

	NUMERIT	SC	GOSpel	
			Min. Vel.	Min. Time
Dep. date	01/10/14	12/05/13	12/06/13	03/20/14
Arr. date	08/19/14	09/26/14	09/27/14	05/10/14
Time (days)	221.29	295	295.132	50
Vel. (km/s)	7.65	6.2436	6.244	41.7006

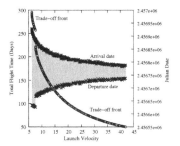

Fig. 4. Trade-off solutions for the Earth-Mars direct mission using GOSpel

solution is better than the Numerit solution. Next, we apply GOSpel with a population size of 100 and run till 300 generations. The two extreme solutions are shown in table. It is interesting to note that the minimum launch velocity solution obtained by GOSpel is almost similar to that obtained by SC. Both Numerit and SC do not allow to find a solution corresponding to minimum time of flight. The minimum time solution obtained by GOSpel takes only 50 days to reach Mars from Earth, by requiring about seven times more launch velocity. Figure 4.3 shows the Pareto-optimal frontier obtained by GOSpel. The arrival and departure times for each of these solutions are also shown in the same figure. In this case, the departure date gradually increases towards the upper limit for a quicker time of flight. Missions starting in January of 2014 results in smaller energy requirements but at the expense of flight time and missions in February/March of 2014 requires smaller time of flight but at the expense of larger energy requirement. Such is a trade-off often occurs in interplanetary missions and the studies of this paper amply demonstrates the ability of GOSpel in finding a number of them. It then becomes a matter of higher-level decision-making task to choose one solution for implementation, which we do not discuss in this paper.

4.4   Cassini Mission

Next, we a mission having three swing-by planets. We found that the Cassini-Huygens mission has four swingbys [1]. But since our software is limited to a maximum of three swing-by planets, we has the first three of four swing-by planets in this study. The mission type is set to be a flyby type at the destination planet.

Table 4. Earth-Venus-Venus-Earth-Jupiter transfer (part of Cassini mission) using SC and GOSpel

	Web [1]	GOSpel	SC
Departure Date	10/15/97	10/15/97	10/29/97
Venus Date	04/26/98	05/19/98	05/11/98
Earth-to-Venus Time (Days)	193	216.318	194
Venus Date	06/24/99	06/22/99	06/26/99
Venus-to-Venus Time (Days)	424	399	411
Earth Date	08/18/99	08/17/99	08/18/99
Venus-to-Earth Time (Days)	55	55.940	53
Jupiter Arrival Date	12/30/00	12/22/00	02/02/01
Earth-to-Jupiter Time (Days)	500	493.46	534
Total Flight Time (Days)	1172	1164.72	1192
Launch Vel. (km/s)	NA	6.805	4.46

Thus, the complete mission for the case is departure from Earth, swing-by from Venus, another swing-by from Venus, third swing-by from Earth and the final arrival to Jupiter. A typical trajectory (taken from [1]), GOSpel and SC solutions are compared in Table 4. From the table, it can be observed that solution found by GOSpel and SC are non-dominated GOSpel solution but time taken are more close to actual mission dates.

5   Earth-Mercury Mission

Here, we consider three different optimization studies: (i) direct (ii) swing-by from Venus and (iii) optional direct or Venus swing-by. The departure time window considered here is 21/9/2002 till 22/9/2002 (just a day). 200 population members are used for 200 generations. Figure 5 shows the Pareto-optimal front obtained by a direct transfer. The reason for a break in the continuity in the Pareto-optimal front is due to the availability of two different opportunity windows for an optimal mission. Next, we apply GOSpel on the same depar-

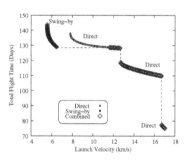

Fig. 5. Comparison of three transfer cases

ture date window and obtain solutions with forced swing-bys. Figure 5 also shows the Pareto-optimal solutions for this case. Due to the swing-by option, the required launch velocity is much smaller than that obtained with the direct transfer.

Finally, we consider both options (direct and swing-by) in GOSpel and obtain a combined Pareto-optimal front. Interestingly, this front is found to be identical to a combined non-dominated front of the two fronts obtained earlier. The Pareto front is divided into two discrete parts. One part belong to flyby cases where swing-by planet is Venus. The other part shows direct transfer. From Figure 5, it can be said that there exists a clear conflict, whether to take a swing-by or not. If swing-by is taken then, smaller launch velocity is required, whereas with direct transfer the flight time is less.

The figure shows that a saving of about 3 km/s launch velocity occurs between the two type of transfers with a mission of 130 days.

6 Conclusions

In this paper, we have discussed the development of a multi-objective optimization software (GOSpel) for finding optimal interplanetary trajectories between any two planets for a dual minimization of travel time and launch velocity which is directly related to the fuel consumption. The software is capable of considering a maximum of three swing-by of intermediate planets to assist in reducing the fuel consumption. The use Pareto-optimality concept and genetic algorithms has demonstrated that the proposed approach can be used to find a set of trade-off solutions which match with the existing solutions of known missions. Thereafter, the developed code is applied to a number of complex case studies and interesting solutions have been obtained. This paper has amply shown the usefulness and flexibility of such a code for real-time application of EMO for interplanetary trajectory optimization.

Acknowledgement

Authors appreciate the support by the ISRO-IITK Cell for executing this study. Discussions with Dr. V. Adimurthy and R. V. Ramanan are appreciated.

References

1. Cassini-Huygens mission to Saturn, http://saturn.jpl.nasa.gov/operations/present-position.cfm
2. Numerit Software, http://www.numerit.com
3. Swingby Calculator, http://www.jaqar.com/swingby.html
4. Biesbroek, R.G.J., Ancarola, B.P.: Optimization of launcher performance and interplanetary trajectories for pre-assessment studies. In: 53rd International Astronautical Congress The World Space congress (2002)
5. Brown, C.D.: Spacecraft Mission Design. AIAA (1992)
6. Cornelisse, J.W.: Trajectory analysis for interplanetary missions. ESA Journal 2, 131–143 (1978)
7. Deb, K., Agrawal, S., Pratap, A., Meyarivan, T.: A fast and elitist multi-objective genetic algorithm: NSGA-II. IEEE Transactions on Evolutionary Computation 6(2), 182–197 (2002)
8. Gooding, R.H.: A procedure for the solution of lambert's orbital bundary-value problem. Celestial Mechanics and Dynamical Astronomy 48(2), 145–165 (1990)
9. Hohmann, W.: Erreichbarkeit der himmelskorper. R. Oldenbourg, Munich and Berlin (1925)
10. Miele, A., Wang, T.: Fundamental properties of optimal orbital transfers. In: 54th International Astronautical Congress, pp. 127–131 (2003)

A Hybrid Evolutionary Multi-objective and SQP Based Procedure for Constrained Optimization

Kalyanmoy Deb, Swanand Lele, and Rituparna Datta

Kanpur Genetic Algorithms Laboratory (KanGAL)
Indian Institute of Technology Kanpur
Kanpur, PIN 208 016, India
{deb,swanand,rdatta}@iitk.ac.in

Abstract. In this paper, we propose a hybrid reference-point based evolutionary multi-objective optimization (EMO) algorithm coupled with the classical SQP procedure for solving constrained single-objective optimization problems. The reference point based EMO procedure allows the procedure to focus its search near the constraint boundaries, while the SQP methodology acts as a local search to improve the solutions. The hybrid procedure is shown to solve a number of state-of-the-art constrained test problems with success. In some of the difficult problems, the SQP procedure alone is unable to find the true optimum, while the combined procedure solves them repeatedly. The proposed procedure is now ready to be tested on real-world optimization problems.

Keywords: Reference point based NSGA-II, EMO, Constrained optimization, SBX, SQP, hybrid procedure, multi-objective optimization.

1 Introduction

In practice, optimization problems always have some constraints involving physical, geometric or other limitations restricting the search. To handle such constraints, various methodologies are used. Of these, the penalty parameter based approaches are commonly used. One difficulty in these methods is to set an appropriate value of the penalty parameters. To overcome this difficulty, researchers suggested various methods, of them the use multi-objective optimization in minimizing constraint violation as additional objective(s) was found to be a viable approach. In this paper, we suggest one such procedure in which two objectives are used. In addition to the objective function, the approach constructs the overall normalized constraint violation as the second objective for minimization. In such a bi-objective problem, the constrained minimum lies on the Pareto-optimal front. A recently proposed reference-point approach is employed to find only a preferred set of Pareto-optimal solutions near the constrained minimum. Such a bi-objective optimization makes the search flexible so that constrained minimum can be found more reliably.

In numerical optimization literature, the sequential quadratic programming (SQP) procedure is popularly used as a local optimization procedure. The procedure computes the gradients numerically and solves approximated quadratic programming problems sequentially till a local optimal solution is found. In this paper, we also combine the

L. Kang, Y. Liu, and S. Zeng (Eds.): ISICA 2007, LNCS 4683, pp. 36–45, 2007.
© Springer-Verlag Berlin Heidelberg 2007

evolutionary bi-objective optimization procedure with the SQP procedure and develop a hybrid procedure which inherits the good aspects of both optimization procedures. The working principle of the hybrid procedure is demonstrated by solving a number of constrained and unconstrained optimization problems.

2   Description of Proposed Procedure

The algorithm used here is derived from the reference point based evolutionary multi-objective optimization (EMO) algorithm. Here, we have used NSGA-II as an EMO procedure. The detail description of the proposed algorithm is given in following sections.

2.1   Conversion of Constrained Problem to Multi-objective Optimization Problem

Let us consider the following constrained single objective problem:

$$\text{Minimize } f(\mathbf{x}),$$
$$\text{Subject to } g_j(\mathbf{x}) \geq 0, \quad j = 1, 2, \ldots, J. \tag{1}$$

Here we do not consider equality constraints, but they can also be handled by the proposed procedure as well. We defer such a consideration till a later paper. The variable bounds, if any, can also be considered as inequality constraints. The above single-objective constrained optimization problem is converted into multi-objective unconstrained problem. Original objective forms one objective and summation of all normalized constraint violations as second objective:

$$\text{Minimize } f(\mathbf{x}),$$
$$\text{Minimize } \text{CV}(\mathbf{x}) = \sum_{j=1}^{J} \langle \bar{g}_j(\mathbf{x}) \rangle, \tag{2}$$

where $\langle \bar{g}_j(\mathbf{x}) \rangle = -\bar{g}_j(\mathbf{x})$ if $\bar{g}_j(\mathbf{x}) < 0$ and zero, otherwise. The constraints $g_j(\mathbf{x})$ are normalized to form $\bar{g}_j(\mathbf{x})$ before computing the violation. Consider Figure 1, which makes a sketch of the two-objective interaction mentioned above. All solutions (feasible or infeasible) will lie on the shaded region. All feasible solutions will correspond to a constraint violation (CV) equal to zero and hence lie on the y-axis. Thus, the solution with the minimum function value ($f(\mathbf{x}^*)$) on the y-axis is the constrained minimum solution. The two-objective interaction creates a Pareto-optimal set (without weak Pareto-optimal solutions), of which the solution with CV=0 (constrained minimum) is an extreme solution. All other Pareto-optimal solutions are infeasible. The other extreme solution is the unconstrained minimum.

Although the complete range of solutions from constrained to unconstrained minima is Pareto-optimal for minimizing $f(\mathbf{x})$ and CV, in order to find the constrained minimum of $f(\mathbf{x})$, we should concentrate the search effort in a region closer to the constrained minimum. The guided domination based method [1] or other preferred EMO methodologies can be used for this purpose. Here, we suggest and use a reference point based methodology.

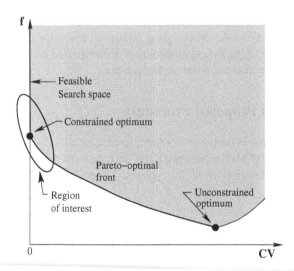

Fig. 1. A sketch of two-objective reformulation of constrained optimization problem

2.2 Modified Reference-Point Based NSGA-II Approach

To emphasize the search in the 'region of interest' shown in Figure 1 on the Pareto-optimal front, we suggest the use of a dynamic reference-point based multi-objective search algorithm. In addition, we use a self-adaptive recombination operator in the NSGA-II procedure [3], which we discuss in the following subsections.

Dynamic Reference-Point Based NSGA-II. In an earlier study [5], the classical reference point based multi-criterion decision-making strategy [7] was combined with the NSGA-II procedure. In this procedure, one or more reference points (or aspiration points) are supplied and instead of finding the complete Pareto-optimal front, the region on the Pareto-optimal front which corresponds to the solution of a scalarizing achievement function was found. In some sense, this is the Pareto-optimal region closer to the supplied reference point. For multiple reference points, a union of such Pareto-optimal regions were found. The main difference between reference point based NSGA-II approach and Wierzbicki's reference point approach [7] is that in the NSGA-II approach we find a set of solutions *close* to the reference points, rather than a single point, thereby providing an idea of the preferred region of interest.

We use this concept for handling constraints using a bi-objective problem formulation. We choose a reference point *dynamically* with generation counter. For every second generation, we identify the the best non-dominated front in the NSGA-II population and choose the feasible solution having the smallest function value as the reference point. In the event of a population with no feasible solution, the solution with the least CV is chosen as the reference point. Since this solution already lies on the current non-dominated front, the reference point based NSGA-II emphasizes the non-dominated solutions on either side of the reference point. In later generations (when we already have feasible solutions in the population), we consider weak domination principle to determine non-dominated solutions. By this method, solutions having CV=0 are included in the non-dominated front with a probability 0.8, however the best feasible solution is always included. This way, the presence of infeasible and feasible solutions in the population provides a better search towards the constrained minimum.

To emphasize maintenance of diversity among non-dominated solutions, we use an ϵ-domination method. For every non-dominated front, we rank the solutions according to increasing normalized objective difference from the chosen reference point having function value of f_{RP}:

$$\bar{f}_i = \frac{f_i - f_{RP}}{f_{max} - f_{min}}.$$

The solution with the smallest objective difference is assigned a rank one. Thereafter, all solutions within an ϵ objective difference is assigned a large rank so as to discourage them to be selected. The solution having the next smallest objective difference is then assigned rank two and all other solutions within ϵ objective difference from the rank-two solution is assigned a large rank. This procedure is continued till all solutions are assigned a rank.

Self-Adaptive Recombination Operator. In the simulated binary crossover (SBX) operator [2], two parent solutions are blended to create two new offspring solutions by using a bimodal probability distribution with a parameter η_c. In most applications, the parameter η_c is kept fixed throughout a simulation run. In a recent study [4], it was shown that an judicious update of this parameter based on the relation merit of offspring solutions with respect to parent solutions allows a better and self-adaptive search in single as well as multi-objective optimization problems. In our procedure, each child is compared with both of its parents. If the child dominates any of the parents then it is said to be winner. If child gets dominated by any of the parent then it is termed as a loser. If both parents and child are feasible, then the child is termed as winner if its function value is better than any of the parents. If both parents and the child are infeasible and non-dominated to each other and if child is nearer to reference point than the parent solutions, then the child is termed as a winner. In all other cases, the child is a loser. If the child is a winner, we exploit the region further by assigning an η_c value which is smaller than that of parents. On the other hand, if the child is a loser, a larger η_c value is assigned. We use identical parameter values ($\alpha = 1.5$) for η_c update, as used in the original study [4]. All initial population members use $\eta_c = 2$ and this value changes adaptively as discussed above.

3   Hybrid NSGA-II and SQP Procedure

GAs, in general, have the global perspective, whereas in solving unimodal problems a more efficient classical approach may be better. Thus, we suggest a combined approach here. After each 10 generations of NSGA-II, a randomly chosen solution from the best 15% population members is modified using the SQP procedure. Gradients are computed numerically within SQP. By this process, the selection solution reaches its nearest local-optimal solution after SQP. Thereafter, the modified solution is put back to NSGA-II and search continues for another 10 generations. The SQP operation can be considered as a special mutation operator applied only after 10 generations to one of the best 15%

population members. The parameters (15% solutions and steps of 10 generations) for update are obtained by some trial-and-error simulation runs on test problems of this study. We are now ready to present simulation results.

4   Simulation Results

First we consider, constrained single-objective optimization test problems taken from the literature [6]. 13 problems used in the study are described in the appendix. These problems were chosen in an open competition arranged in a major international conference (CEC-06 held in Vancouver in July 2006) and results of top five papers were announced. Of the 24 problems in the competition, we select 13 problems which did not have any equality constraint.

The results of 13 problems are tabulated in Tables 1, 2, and 3. In all problems, we have used a population of size 100, $p_c = 0.9$ and polynomial mutation with $p_m = 0.05$ and a distribution index $\eta_m = 10$. Three tables show the error in the best obtained function value with the known optimal value for all the functions for $5(10^3)$, $5(10^4)$ and $5(10^5)$ function evaluations. A total of 25 runs are performed for each test problem and the best, median and worst error values are presented. To really investigate the effect of the proposed bi-objective reference-point based approach, we perform a separate SQP run alone. In this case, the best solution in the initial population is identified and only SQP is run alone. The tables also show the best, median, and worst error values obtained using the sole SQP method. are attempted to solve using SQP. Results of SQP are presented in bottom row of each table. Function evaluations required for SQP are shown in bracket. The best results reported with existing methodologies [6] found inferior solutions after 5,000 evaluations, as shown in the tables. However, after $5(10^5)$ evaluations, at least one of the five existing methods finds the true optimum. The solution with a negative error is an infeasible solution.

Table 1. Error values achieved for problems 1 to 6

FES		Prob1	Prob2	Prob4	Prob6
	Best	1.200E-03	3.645E-01	1.208E-09	1.173E-10
$5(10^3)$	Median	2.612	5.189E-01	2.597E+01	3.128E-01
	Worst	7.356	5.582E-01	5.845E+01	1.077E+02
Existing	Best	2.953	0.329	1.274	1.220E-02
	Best	0.0	1.876E-07	3.638E-12	1.182E-11
$5(10^4)$	Median	0.0	1.634E-01	3.638E-12	1.182E-11
	Worst	3.719	5.315E-01	3.638E-12	2.001E-11
	Best	0.0	2.109E-15	3.638E-12	1.182E-11
$5(10^5)$	Median	0.0	1.054E-02	3.638E-12	1.182E-11
	Worst	2.547	3.459E-02	3.638E-12	2.001E-11
	Best	0.0 (84)	0.304 ($5(10^5)$)	3.638E-12 (128)	2.000E-11 (45)
SQP (FEV)	Median	1.172 ($5(10^5)$)	0.494 ($5(10^5)$)	3.638E-12 (170)	2.000E-11 (122)
	Worst	4.891 ($5(10^5)$)	0.611 ($5(10^5)$)	3.638E-12 (270)	2.000E-11 (122)

Table 2. Error values achieved for problems 7 to 10

FES		Prob7	Prob8	Prob9	Prob10
	Best	5.529E-01	3.238E-07	2.206E-01	3.3121E+02
$5(10^3)$	Median	1.769E+01	3.238E-07	2.683E+00	1.989E+03
	Worst	5.351E+01	4.601E-05	1.615E+01	3.853E+03
Existing	Best	3.885	0.0	0.061	923.630
	Best	9.990E-04	3.238E-07	2.274E-13	1.128E+01
$(5(10^4)$	Median	9.990E-04	3.238E-07	3.411E-13	2.567E+02
	Worst	9.990E-04	3.238E-07	8.187E-10	6.533E+02
	Best	9.990E-04	3.238E-07	2.274E-13	4.063E0.0
$(5(10^5)$	Median	9.990E-04	3.238E-07	2.274E-13	5.397E+1
	Worst	9.990E-04	3.238E-07	5.684E-13	4.351E+2
	Best	9.990E-4 (269)	3.238E-07 (122)	5.684E-13 (1297)	feasible
SQP (FEV)	Median	9.990E-04 (630)	3.238E-07 (148)	5.684E-13 (10,297)	solution
	Worst	9.990E-04 (1,540)	0.070 ($5(10^5)$)	5.684E-13 (59,157)	not found

It is interesting to note that in most problems, the sole application of SQP is found to be better than the hybrid approach. SQP is capable of handling a few local optima and in most cases is found to be successful. However, SQP alone is not able solve problems 1 and 2 adequately. The combined use of bi-objective reference point method and SQP is able to solve all 13 problems without tuning of any parameter. Problems 10 and 16 are known to be a difficult ones to solve to optimality. Our procedure comes close to the true optimum solution for problem 10 and finds the optimum solution for problem 16.

Figures 2 and 3 show the convergence pattern for problems 2 and 4, respectively, using both SQP and the proposed hybrid procedures. In the first case, SQP gets stuck to a sub-optimal solution, whereas the hybrid approach is able to find a better convergence. On the other hand, for problem 4, the reverse happens. To demonstrate the distribution of solutions at the end of simulation, we plot the two-objective interaction for problems 7 and 18 in Figures 4 and 5, respectively. The figures clearly show the Pareto

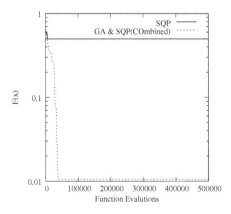

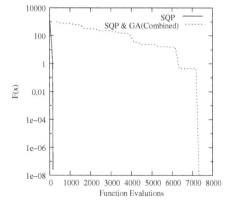

Fig. 2. Convergence pattern for problem 2   **Fig. 3.** Convergence pattern for problem 4

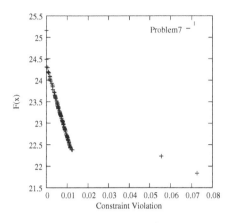

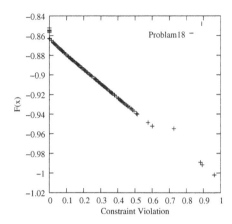

Fig. 4. Obtained trade-off solutions for problem 7

Fig. 5. Obtained trade-off solutions for problem 18

optimal fronts obtained in both these cases. In both cases, the constrained minimum is included in the Pareto-optimal front.

Table 3. Error values achieved for problems 12 to 24

FES		Prob12	Prob16	Prob18	Prob19	Prob24
	Best	0.0	9.082E-03	−8.465E-01	3.666E+00	8.704E-14
$5(10^3)$	Median	0.0	8.191E-02	1.946E-01	1.390E+02	2.639E-08
	Worst	0.0	3.928E-01	7.080E-01	7.029E+02	8.107E-06
Existing	Best	0.0	1.178E-03	1.965E-14	29.147	0.0
	Best	0.0	1.290E-05	4.441E-16	4.974E-14	8.704E-14
$5(10^4)$	Median	0.0	8.469E-05	9.302E-03	7.105E-14	8.704E-14
	Worst	0.0	1.427E-04	3.660E-01	1.776E-13	8.704E-14
	Best	0.0	3.830E-07	4.441E-16	4.974E-14	8.704E-14
$5(10^5)$	Median	0.0	4.119E-06	9.302E-03	7.105E-14	8.704E-14
	Worst	0.0	3.899E-05	3.661E-01	1.776E-13	8.704E-14
	Best	0.0 (600)	Feasible	4.0E-16 (1,356)	5.0E-14 (432)	8.7E-14 (18)
SQP (FEV)	Median	5.6E-3 $(5(10^5))$	solution	0.366 $(5(10^5))$	5.0E-14 (473)	8.7E-14 (48)
	Worst	3.0E-2 $(5(10^5))$	not found	0.860 $(5(10^5))$	5.0E-14 (1,014)	1.090 $(5(10^5))$

5   Solving Unconstrained Single-Objective Optimization Problem

The success of the proposed hybrid procedure on difficult constrained optimization problems brings out opportunities in trying such a procedure for multi-modal unconstrained optimization problem-solving tasks. To investigate, we choose the multi-modal Rastrigin's test problem and use the proposed procedure. There is no constraint in this problem, thus there is no constraint violation. However, the same modified NSGA-II procedure with a reference point can still be used on a single objective function, in which case all solutions lie on the y-axis.

Table 4. Performance on Rastrigin's function. SQP alone is run for a maximum of $5(10^5)$ function evaluations.

Number	With $1(10^{-8})$ accuracy		
of var.	Best	Median	Worst
20	54,519	98,787	125,333
SQP $(f(\mathbf{x}))$	71.630	231.824	606.061
40	6,09,182	7,33,653	9,67,740
50	1,700,279	2,049,322	2,399,957

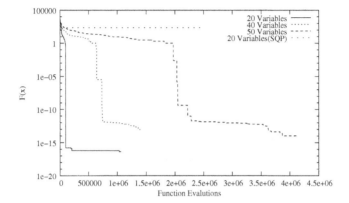

Fig. 6. Obtained trade-off solutions for Rastrigin's function

The proposed algorithm is used to solve the problem with $n = 20$, 40, and 50 variables with population size $N = 5n$. The optimum corresponds to $x_i = 0$ in this problem. The population is initialized out of bounds (variable bounds: $x_i \in [10, 15]$), thereby making sure that the initial population does not bracket the global minimum. Thereafter, the solutions are allowed to be anywhere in the range $x_i \in [-10, 15]$.

The simulation results are shown in Table 4. To have a global search, initially we perform NSGA-II for 7,000 generations and then SQP is employed as before after every 20 generations with best 15 population members. A comparison of sole SQP procedure on the 20-variable version of the problem shows that SQP alone cannot overcome multitude of local optima and eventually converge to the global optima with a function value equal to zero. On the other hand, the use of the hybrid approach is able to overcome all the local optima and converge near the true global optimum. The convergence pattern on these problems are shown in Figure 6. It is clear that SQP gets stuck, while the proposed procedure is able to converge to global basin for all three problems.

6   Discussion of Results

From the results it is clear that for all problems the proposed algorithm converges very fast as well as avoids local optima quite satisfactorily. SQP alone works well for problems with single optima (or with a few optima) as is the case of most of the test problems used in this study. But in case of multiple local optima, like in problems 1, 2 and Rastrigin's function, the SQP alone prematurely converges near the nearest local optimum from the initial point. The proposed hybrid algorithm works well for all these functions including the unimodal or few-modal problems and Rastrigin's function up to 50 variables.

7   Conclusions

In this paper, we have presented a new approach to handle constraints by converting them into an aggregate objective of constraint violation. A reference point approach has been used to emphasize preferred solutions near the constrained minimum on the resulting Pareto-optimal front. Further, this approach has been combined with the SQP procedure to accelerate the convergence. Simulation results on 13 constrained test problems and one unconstrained yet multi-modal test problem have shown superior performance of the proposed procedure.

The judicious use of the local search approach with a more global NSGA-II procedure with preference by reference point strategy combines good aspects of three different optimization concepts into one algorithm. In simpler problems, SQP provides an advantage in bringing solutions close to optimal solution, whereas for multi-modal and complex problems, the use of preferred NSGA-II and SQP keeps enough diversity in the population so as to avoid local optimal regions and eventually converge to the global basin of attraction, despite presence of many local optima.

The procedure involves a few parameters which we have set by trial-and-error simulation runs. We are currently working on self-adaptive procedures so that no additional parameters are needed to be fixed by the user. This paper shows how different optimization concepts can be put together for a better and more efficient algorithm. More such studies must be made to perfect the act of global and local optimization adaptively.

References

1. Branke, J., Kauβler, T., Schmeck, H.: Guidance in evolutionary multi-objective optimization. Advances in Engineering Software 32, 499–507 (2001)
2. Deb, K.: Multi-objective optimization using evolutionary algorithms. Wiley, Chichester, UK (2001)
3. Deb, K., Agrawal, S., Pratap, A., Meyarivan, T.: A fast and elitist multi-objective genetic algorithm: NSGA-II. IEEE Transactions on Evolutionary Computation 6(2), 182–197 (2002)
4. Deb, K., Karthik, S., Okabe, T.: Self-adaptive simulated binary crossover for real-parameter optimization. In: GECCO-2007. Proceedings of the Genetic and Evolutionary Computation Conference, July 7-11, The Association of Computing Machinery (ACM), New York (2007) (in press)
5. Deb, K., Sundar, J., Uday, N., Chaudhuri, S.: Reference point based multi-objective optimization using evolutionary algorithms. International Journal of Computational Intelligence Research (IJCIR) 2(6), 273–286 (2006)
6. Liang, J.J., Runarsson, T.P., Mezura-Montes, E., Clerc, M., Suganthan, P.N., Coello Coello, C.A., Deb, K.: Special session on constrained real-parameter optimization ( ) (2006), http://www.ntu.edu.sg/home/epnsugan/

7. Wierzbicki, A.P.: The use of reference objectives in multiobjective optimization. In: Fandel, G., Gal, T. (eds.) Multiple Criteria Decision Making Theory and Applications, pp. 468–486. Springer, Berlin (1980)

A   Test Problems Used in the Study

The nature of the test problems used in this study is described in Table 5. All problem formulations can be found from [6].

Table 5. Description of test problems

Prob.	No. Of variables	no. of constraints	no. of active constraints	Nature of objective function	nature of constraints	Fraction of feasible region
1	13	9	6	quadratic	linear	0.0111
2	20	2	1	non-linear	non-linear	99.9971
4	5	6	2	polynomial	non-linear	52.1230
6	2	2	2	cubic	non-linear	0.0066
7	10	8	6	quadratic	linear+non-linear	0.0003
8	2	2	0	non-linear	non-linear	0.8560
9	7	4	2	polynomial	non-linear	0.5121
10	8	6	6	linear	linear+non-linear	0.001
12	3	1	0	quadratic	non-linear	4.7713
16	5	38	4	non-linear	non-linear	0.0204
18	9	13	6	quadratic	non-linear	0.000
19	15	5	0	non-linear	non-linear	33.4761
24	2	2	2	linear	non-linear	79.6556

Rastrigin Problem

$$\text{Minimize } 10n + \sum_{i=1}^{n} x_i^2 - 10\cos(2\pi x_i),$$

where $10 \leq x_i \leq 15$ for $i = 1, 2, \ldots, n$. The global optimal solution is $x_i^* = 0$ with $f(\mathbf{x}^*) = 0$.

Study on Application of Multi-Objective Differential Evolution Algorithm in Space Rendezvous

Lei Peng, Guangming Dai, Fangjie Chen, and Fei Liu

School of Computer Science
China University of Geosciences, Wuhan 430074, P.R. China
penglei0114@gmail.com, gmdai@cug.edu.cn

Abstract. As the development of human missions and space station, space rendezvous and docking technology is the key to modern space exploration. There are a lot of multi-objective optimization problems in aerospace field. At present, polymerization technology is often used to change multi-objective to single objective. This method makes the problem easier but gives one solution only which is not suitable for project application. In this paper, we introduce an extension of DE(SMODE) to cope with the spacecraft rendezvous problem. The experiment results indicate that SMODE is successful to locate the real Pareto front for the spacecraft rendezvous problem. Also, the effect of PopSize–population size and Max_gen–maximum number of generations of SMODE is studied.

1   Introduction

As the development of human missions and space station, space rendezvous and docking technology is the key to modern space exploration. It can be used for personnel exchange, payload handling, in-orbit repair, rescue activity, special assemble and construction and a series of tasks between station and spacecraft, spacecraft and spacecraft or arbitrary two spatial flight vehicles.

Optimal rendezvous research is of great significance in practical applications. Because of the complexity of the issues, in recent years, evolutionary algorithm has been used to solve the optimization problem in the aerospace, and achieved better results. Different from the work[2][3][4] before ,in this paper, we directly use a multi-objective differential evolution algorithm to solve the space optimal rendezvous problem instead of simplifying the multi-objectives problems to single-objective. Compared with the single objective method, multi-objective method can find a solution set rather than a single solution. Thus we can choose the optimal solution according to the requirements of specific project.

2   Optimized Model of Spacecraft Rendezvous

In the spacecraft rendezvous problem, the spacecraft flying passive in the orbit is called target spacecraft and the one flying to target spacecraft with maneuver is called tracking spacecraft.

Studying the relative motion of the two spacecraft, we choose coordinate OXYZ which is called orbit coordinates. As shown in Figure 1. We consider the Earth, the target spacecraft, and the tracking spacecraft are in an identical plane.

L. Kang, Y. Liu, and S. Zeng (Eds.): ISICA 2007, LNCS 4683, pp. 46–52, 2007.

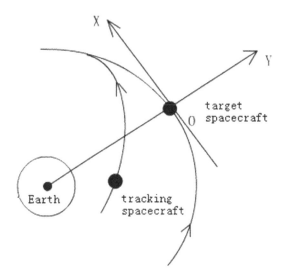

Fig. 1. Orbit Coordinates

To simplify the space rendezvous and docking problem, we consider the orbits of the target spacecraft and the tracking spacecraft are coplanar circular orbits. The rendezvous takes place at the appropriate moment. The equation of the relative motion which is also called Hill equation[1] is as follows.

Suppose the initial time t_0, the rendezvous time t and the position and velocity of the tracking spacecraft at t_0 as well as the terminate condition are all fixed, we will be able to calculate the two impulses in the process according to formulas below.

$$\ddot{x} + 2\omega\dot{y} = f_x$$
$$\ddot{y} - 2\omega\dot{x} - 3\omega^2 y = f_y \qquad (1)$$
$$\ddot{z} + \omega^2 z = f_z$$

$$\underline{v}_1(0) = \omega \begin{pmatrix} a_{11}/b & a_{12}/b & 0 \\ a_{21}/b & a_{22}/b & 0 \\ 0 & 0 & a_{33} \end{pmatrix} \underline{\rho}(0) - \underline{v}_\rho(0) \qquad (2)$$

$$\underline{v}_2(T) = -\left\{ \omega \begin{pmatrix} 4\cos\omega T - 3 & -2\sin\omega T & 0 \\ 2\sin\omega T & \cos\omega T & 0 \\ 0 & 0 & \cos\omega T \end{pmatrix} \begin{pmatrix} a_{11}/b & a_{12}/b & 0 \\ a_{21}/b & a_{22}/b & 0 \\ 0 & 0 & a_{33} \end{pmatrix} + \begin{pmatrix} 0 & -6\omega(1-\cos\omega T) & 0 \\ 0 & 3\omega\sin\omega T & 0 \\ 0 & 0 & -\omega\sin\omega T \end{pmatrix} \right\} (\underline{\rho}(0)) \qquad (3)$$

$$a_{11} = \sin\omega T, a_{12} = 14(1-\cos\omega T) - 6\omega T\sin\omega T$$
$$a_{21} = -2(1-\cos\omega T), a_{22} = 4\sin\omega T - 3\omega T\cos\omega T$$
$$a_{33} = -\cot\omega T, b = 3\omega T\sin\omega T - 8(1-\cos\omega T)$$

ω is angular velocity of the target spacecraft, T is the time interval of the rendezvous. $\underline{v}_1(0) = [\dot{x}_1(0)\ \dot{y}_1(0)\ \dot{z}_1(0)]^T$ is the required velocity impulse to get the tracking spacecraft reach the target planet during T time interval. Upon arrival at the target spacecraft, we have to give another impulse in the opposite direction so that the relative velocity magnitude of the two spacecrafts will be zero. The impulse magnitude is $\underline{v}_2(T) = [\dot{x}_2(T)\ \dot{y}_2(T)\ \dot{z}_2(T)]^T$. $\underline{v}_\rho(0) = [\dot{x}_\rho(0)\ \dot{y}_\rho(0)\ \dot{z}_\rho(0)]^T$ and $\underline{\rho}(0) = [x(0)\ y(0)\ z(0)]^T$ are the initial velocity and the initial position of the tracking spacecraft. The two-impulse energy consumption is:

$$E = |\underline{v}_1(0)| + |\underline{v}_2(T)|$$

The key of this method is the selection of t_0 and T. The paper considered this problem with fixed t_0 but unrestricted T. In this situation, we will be able to obtain $\underline{v}_1(0)$ and $\underline{v}_2(T)$ according to equation (2) and (3).

3   Self-adaptive Multi-Objective Differential Evolution

Differential Evolution (DE) is a branch of evolutionary algorithms developed by Rainer Storn[6] and Kenneth Price for optimization problems over continuous domains. It is significantly faster and robust at numerical optimization and is more likely to find a function's true global optimum. The DE's advantages are its simple structure ,ease of use ,speed and robustness. A few researchers[7][8][9][10][12] [13]have studied the extension of DE to solve multi-objective optimization problem in continuous domain. In our previous studies[5] , Original DE is designed to carry out the optimum solution of spacecraft rendezvous problem. In the spacecraft rendezvous problem, there are two optimized objectives which are the time interval of the rendezvous and two-impulse energy. In order to simplify the problem, we transform the two multi-objectives into single-objective through polymerization method .In this paper, we propose a self-adaptive multi-objective differential evolution(SMODE) to solve the spacecraft rendezvous problem. This idea of SMODE comes from the SPDE[7] and DEMO [9].

When using DE to solve MOPs, there are many difficulties. The most main problems are how to preserve an uniform spread front of nondominated solutions and when to replace the parent individual with the candidate solution. In this paper, we use the concept of dominance. If the trial vector dominates the target, the trial vector replaces the target one.

Firstly, we set the several key parameters .CR–crossover constant , F–scaling factor, PopSize–population size, Max_gen–maximum number of generations. Because F has great effect on the final results , we adopt the method of Gaussian$(0, 1)$[7].And then , individuals are generated randomly within the bounds of decision variables and the fitness functions are evaluated. At a given generation of the evolutionary search, the population is sorted into several ranks based on dominance concept. Secondly, DE operations are carried out over the individuals of the population. The fitness functions of the trial vectors, thus formed, are evaluated. Then the concept of dominance is used. If the trial vector dominates the target one, the trial vector replaces the target one. And if the target one dominates the trial vector, the trial vector is discarded. Otherwise, the trial vector is added in the population. After that, the size of the new population is between

PopSize and 2*PopSize. So we must choose PopSize individuals from the population. These act as the parent vectors for the next generation. The choice consists of sorting the individuals based on its ranking and crowding distance which come from elitist nondominated sorting genetic algorithm(NSGA-¢ò)[11].

Self-Adaptive Multi-Objective Differential Evolution

3.1 Initialization of the Population

$$T_{ij}(0) = rnd(0,1)(T_{ij}^U - T_{ij}^L) + T_{ij}^L$$

$$i = 1, 2, \cdots, PopSize; j = 1, 2, \cdots, n$$

where PopSize is the population size, n is the solution's dimension, and $T_{ij}(0)$ is the j-th variable in the i-th solution vector . In the initial generation G = 0, $T_{ij}(0)$ is initialized within its boundaries (T_{ij}^L, T_{ij}^U).

3.2 Mutation Operation

We use the DE/BEST/1/bin scheme

$$h_{ij}(t+1) = T_{bestj}(t) + F(T_{r1j} - T_{r2j})$$

$F \in$ Gaussian$(0, 1)$ is scaling factor. $T_{bestj}(t)$ is the best performing vector of the current generation. T_{r1} £¬ T_{r2} $(r1 \neq r2)$ are two vectors that randomly chosen from the population.

3.3 Crossover Operation

The perturbed individual, $h_{ij}(t+1)$, and the current $T_{ij}(t)$ population member, are then subject to the crossover operation, which finally generates the population of candidates, or "trial" vectors, $V_{ij}(t+1)$, as follows:

$$V_{ij}(t+1) = \begin{cases} h_{ij}(t+1) \ rand_{ij}[0,1] \leq CR \text{ or } j = rand(i) \\ T_{ij}(t) \ rand_{ij}[0,1] > CR \text{ or } j \neq rand(i) \end{cases}$$

$CR \in [0,1]$ is the crossover factor which is often set by the user.

3.4 Auxiliary Operation

After crossover, if one or more variables in the new solution are outside their boundaries, the following auxiliary operation is applied

$$V_{ij}'(t+1) = \begin{cases} T_{bestj}(t) - rand_{ij}[0,1](T_{bestj}(t) - T_{ij}^L) if(V_{ij}(t+1) < T_{ij}^L) \\ T_{bestj}(t) + rand_{ij}[0,1](T_{ij}^U - T_{bestj}(t)) if(V_{ij}(t+1) > T_{ij}^U) \\ V_{ij}(t+1) \text{otherwise} \end{cases}$$

$T_{bestj}(t)$ is the best performing vector of the current generation. (T_{ij}^L, T_{ij}^U) is the boundary.

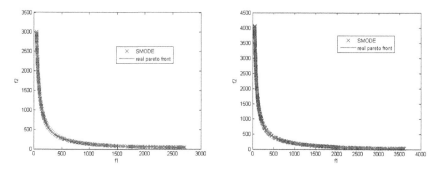

Fig. 2. Nondominated solutions of the final population obtained by SMODE in the initial condition $\underline{v}_\rho(0) = [40, 30, 0]$, $\underline{\rho}(0) = [-70000, -30000, 0]$ and $\underline{v}_\rho(0) = [20, 0, 0]$, $\underline{\rho}(0) = [-100000, -20000, 0]$

3.5   Selection Operation

$$T_i(t+1) = \begin{cases} V_i'(t+1) if(V_i'(t+1) \text{dominates} T_i(t)) \\ T_i(t) if(T_i(t) \text{dominates} V_i'(t+1)) \\ V_i'(t+1) \text{otherwise} \end{cases}$$

$V_i'(t+1)$ is the trial vector that has been adjusted by auxiliary operation . $T_i(t)$ is the target vector. If the trial vector dominates the target, the trial vector replaces the target one. And if the target vector dominates the trial one ,the trial vector is discarded. Otherwise, the trial vector is added in the population.

4   Experimental Results

The self-adaptive multi-objective differential evolution is used to optimize the real spacecraft rendezvous problem mentioned above. This problem consists of two objectives to be minimized as shown in equation (4).

$$\begin{aligned} Minimize\ f_1 &= T \\ Minimize\ f_2 &= |v_1(0)| + |v_2(T)| \\ Where\ 50 &\le T \le 5000 \end{aligned} \tag{4}$$

To the spacecraft rendezvous problem, the orbits of the target spacecraft and the tracking spacecraft are coplanar circular orbits. The height of the target orbit is 400km.When the two spacecrafts finish rendezvous, the relative velocity and position magnitude of the two spacecrafts will be zero.

We discuss the problem under two different initial conditions. One kind is $\underline{v}_\rho(0) = [40, 30, 0]$ and $\underline{\rho}(0) = [-70000, -30000, 0]$, another kind is $\underline{v}_\rho(0) = [20, 0, 0]$ and $\underline{\rho}(0) = [-100000, -20000, 0]$.

In this paper, parameters used for SMODE: PopSize is 300,Max_gen is 200,F uses Gaussian$(0, 1)$,CR is 0.3.

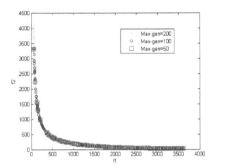 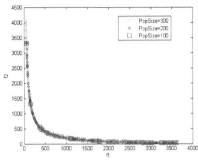

Fig. 3. Pareto Optimal Front for the spacecraft rendezvous problem using different Max_gen and $PopSize$

As simulation results shown in Figure 2 , we find that SMODE can find the real Pareto front of the spacecraft rendezvous problem under the two initial conditions. And Figure 2 show that the non-dominated solutions that SMODE find have a much uniform distribution on the real Pareto front.

5 Discussion

Various experiments have been carried out to test the proposed algorithm by studying the effect of PopSize and Max_gen. The key parameters used are,CR=0.3,F=Gaussian$(0, 1)$, $\underline{v}_\rho(0) = [20, 0, 0]$ and $\underline{\rho}(0) = [-100000, -20000, 0]$.We find that PopSize and Max_gen have great effect on the final Pareto Set. Figure 3 shows that when $PopSize = 200$ is not changed, with Max_gen increasing, the non-dominated solutions have a much better uniform distribution on the real Pareto front and when $Max\_gen = 100$ is not altered, it is clear that with increase in $PopSize$, the non-dominated solutions have a much better uniform distribution on the real Pareto front. After $Popsize = 300$,there is no significant improvement in the final Pareto Set. According to the experiments, an appropriate $PopSize$ seems to be about 300,and Max_gen is 200.

6 Conclusion

There are a lot of multi-objective optimization problems in aerospace field. Polymerization technology is often used to change multi-objective to single objective. This method makes the problem easier but gives one solution only which is not suitable for project application. In this paper, we deal with the extension of DE(SMODE) to cope with the spacecraft rendezvous problem and give the optimal solution set. The results indicate that SMODE is able to locate the real Pareto front for the spacecraft rendezvous problem. And $PopSize$ and Max_gen have great influence on the final results. Compared with the single objective method, multi-objective method can find a solution set rather than a single solution. Thus we can choose the optimal solution according to the requirements of specific project. This is very significant in the aerospace field. The experiment shows that multi-objective optimization algorithm has a good prospect in the aerospace field. And further study will be done in future work.

References

1. Zhao, J., Liu, T.: Spacecraft Dynamics. Press of Harbin Institute of Technology (2003)
2. Wang, H., Tang, G.: Study on Appliaction of Genetic Algorithm in Spacecraft Optimal Rendezvous. Journal of Astronautics Control 1, 16–21 (2003)
3. Wang, S., Zhu, K.-j., Dai, J.-h., Ren, X.: Solving orbital transformation problems based on EA. Journal of Astronautics 23(1), 73–75 (2002)
4. Tang, Y., Chen, S., Xu, M., Wan, Z.: A genetic algorithm (GA) method of orbit interception with finite thrust. Journal of Northwestern Polytechnical University 23(5), 671–674 (2005)
5. Peng, L., Wu, Y., Hu, H.: Solving spacecraft rendezvous problems based on DE. In: The third Chinese astronautics institute deep-space committee academic conference, pp. 81–86 (2006)
6. Storn, R., Price, K.: Differential evolution-A simple and efficient heuristic for global optimization over continuous spaces. Journal of Global Optimization 11, 341–359 (1997)
7. Abbass, H.A.: The self-adaptive pareto differential evolution algorithm. In: Congress on Evolutionary Computation (CEC2002), vol. 1, pp. 831–836. IEEE Service Center, Piscataway, New Jersey (2002)
8. Xue, F., Sanderson, A.C., Graves, R.J.: Pareto-based multi-objective differential evolution. In: CEC2003. Proceedings of the 2003 Congress on Evolutionary Computation, Canberra, Australia, vol. 2, pp. 862–869. IEEE Press, NJ, New York (2003)
9. Robi, T., Filipic, B.: DEMO: Differential Evolution for Multi-objective Optimization. In: Coello Coello, C.A., Hernández Aguirre, A., Zitzler, E. (eds.) EMO 2005. LNCS, vol. 3410, pp. 520–533. Springer, Heidelberg (2005)
10. Babu, B.V., Jehan, M.M.L.: Differential Evolution for Multi-Objective Optimization. In: CEC'2003, Canberra, Australia, vol. 4, pp. 2696–2703 (December 2003)
11. Deb, K., Pratap, A., Agarwal, S., Meyarivan, T.: A fast and elitist multiobjective genetic algorithm: NSGACII. IEEE Transactions on Evolutionary Computation 6, 182–197 (2002)
12. Angira, R., Babu, B.V.: Non-dominated sorting differential evolution (NSDE):an extension of differential evolution for multi-objective optimization. In: IICAI-05. 2nd Indian international conference artificial intellgence, pp. 1428–1443 (2005)
13. Madavan, N.K.: Multiobjective optimization using a Pareto differential evolution approach. In: Proc. of IEEE Congress on Evolutionary Computation, pp. 1145–1150. IEEE Computer Society Press, Los Alamitos (2002)

The Multi-objective ITO Algorithms

Wenyong Dong[1], Dengyi Zhang [1], and Yuanxiang Li[2]

[1] Computer School, Wuhan University
430079,Wuhan, China
{dwy77, dengyiz}@public.wh.hb.cn
[2] State Key lab of Softeate Engineering, Wuhan University
430072,Wuhan, China
yxli@whu.edu.cn

Abstract. The Multi-objective ITO algorithm, which is based on the Ito process and is a new intelligent meta-heuristic algorithm, is proposed in this paper. Two new operators, named excursion (or trend) ratio and volatility ratio, are introduced in this algorithm. Integrated displacement equation of Einsteinian ideal gas huge molecules system to design trend ratio and volatility ratio, the Ito Algorithm can effectively deal with contradictions between the 'exploration' and 'exploitation'. Some famous benchmarks of MOPs are employed to verify the performance of MITO, and comparisons perform to the convergence of MITO, SPEA, and NSGA. The experiments show that the MITO algorithm is simple and effective.

Keywords: Multi-objective Optimization, ITO algorithm, Wiener Process.

1 Introduction

The MOPs are wildly applied in economics, socialites and engineering etc, and it is more difficult to solve the Multi-Objective Problems (MOPs) than the Single Objective Problems (SOPs). Until to now many new MOP algorithms are proposed, and they can divide into two kinds: the traditional methods and the Pareto methods[1,2,3]. In the traditional methods the objective functions of MOPS are combined into one objective function, and then conquered by the SOPs algorithms[3]. There are many shortages in the traditional methods, such as how to choose the weights of different objective functions is in a dilemma. So the Pareto methods are proposed. The Pareto methods try to find the Pareto fronts in the objective landscape[3]. In this paper a new multi-objective ITO algorithm is proposed to find the set of Pareto Solutions. So first we review the main steams of Pareto methods and denote their main shortages.

There are many doctorial dissertations and journal papers discuss the Pareto Methods. In 1984, The Vector Evaluated Genetic Algorithm (VEGA)[1] is firstly proposed by David Scharerto deal with MOPs. Since then many Multi-objective Optimization Evolutionary Algorithms (MOEAs), most of them are based on Pareto techniques, are proposed. The main shortage of VEGA is that the asymptotic convergence can't be ensured. In order to overcome the shortages, the elitism strategy, which is proposed by E. Zitzler et al. in 1999, is employed[2]. In general, in

L. Kang, Y. Liu, and S. Zeng (Eds.): ISICA 2007, LNCS 4683, pp. 53–61, 2007.

the elitism strategy, the current popular MOEAs employ explicit or implicit archive to store the non-dominated solutions. There are many outstanding Elitism MOEAs, such as SPEAs/SPEAs2(Strength Pareto Evolutionary Algorithm)[3], PSOPOs(Particle Swarm Optimization for Pareto Optimization with Enhanced Archiving Techniques), NSGA(Non-dominated Sorting Genetic Algorithm)[1], PAES(Pareto Archived Evolution Strategy)[2], HPMOEA(High Performance MOEA)[3], NPGA(Niched-Pareto Genetic Algorithm)[3] and so on. So the main techniques of current MOEAs based on the Pareto Partial Order could be depicted as follow: (1) How to generate the new solutions? (2) How to storage the Pareto solutions? And (3) How to judge the solutions are non-dominate solutions or not? The first problem is to design a more efficient algorithm to utilize the characters of MOPs in the production of new individuals; the second problem is to design a kind of efficient storage structure; in general the non-dominate solutions set is unknown, so the third problem is that verify the archived solutions is belong to the Pareto Sets or not during the search process.

In this paper, we propose a novel multi-objective optimization algorithm, which combines a new searching algorithm named ITO algorithm and a new storage structure for the potential Pareto solutions. The experimental results show that this algorithm runs much faster than SPEA, NSGA and can obtain approximate Pareto fronts. This paper is organized as following: the basic idea of ITO algorithm is introduced in the second part; How to design the Multi-Objective ITO algorithm is depicted in the third part; the comparisons between ITO algorithm and the other meta-heuristic algorithms are introduced in the forth part; in the fifth part there are the experiments results and conclusion.

2   The Basic Idea of ITO Algorithm

The Brownian motions, which descript the irregular movement of flower powder on the surface of liquid, are one of the classical scientific experiments that were proposed by the English biologist R.Brown in 1827. Einstein gave the mathematical description to the Brownian motions in 1905 and in 1918, respectively. Wiener proposed the Wiener process, which described the motional trajectory of the Brownian motions, and then he proposed the definition of measure and integral in the space of Brown motion. After that, Weiner process had been successfully applied in explanation of the heat motion and gas motion. Also, Brown motion can be used to model some natural phenomena and social phenomena. Two American professors, Robert C. Merton and Myron S. Scholes, have won the Nobel Prize for Economics for their mathematical formula for pricing options. Nowadays, the extended forms of Brown motion have been introduced into many scientific areas such as physics, economics, management and social science.

In 1940's, a Japanese mathematician Ito Kiyosi extended Weiner's research results and proposed Ito process, which introduced the volatility item in the stochastic differential equation for the Brown motion and is descript in Definition 1.

Definition 1: Suppose X= {X(t), t≥0} satisfying Ito integral, for all $0 \leq t_0 < t < T$,

$$dX(t) = b(t, X(t))dt + \delta(t, X(t))dB(t) \qquad (1)$$

then X is called Ito stochastic process.

We can easily extend the scalar Ito process to the form of vector, the detail of m-dimensions Ito process will be found in some literatures about applied stochastic process. The Ito process has been used extensively in the economical and control domains. Actually, it is the most popularly model used in the stock market. In practice, the item of b(t,X(t))dt in the Ito differential equation is often viewed as expected trend or excursion within unit time. The item b(t,X(t)) is called as trend or excursion rate and denotes the general tendency of stochastic process. The item of $\delta(t, X(t))$ denotes the track's fluctuation of variable X and it is called fluctuation(or volatility) rate. Because the Ito process considers the Brown motion as stochastic disturbance, thus it endows Brown motion the most general meaning. Ito differential equation is the dynamics equation with stochastic item, hence its solution describes the probability of X(t) in the future time t. Also, Ito Kiyosi defines the stochastic integral based on the particle trajectory of Brown motion directly and hence builds up his stochastic integral theory. In the applied domain, we can define trend rate and volatility rate with different meanings to adapt different areas.

3   Design of Multi-objective ITO Algorithm

Here, we will propose an intelligent heuristic optimization algorithm based on Ito process (ITO : Ito Optimal algorithm). We classify the ITO algorithm into two subtypes: one is called continual ITO algorithm which is suitable to solve the continual optimization problems and the other one is called discrete ITO algorithm which is suitable to the discrete optimization problems. In this paper, we will focus on the continual ITO algorithm. The key of the algorithm is how to define the trend rate and volatility rate.

The equation of ideal gas huge molecule proposed by Einstein and P. Langevin is as following:

$$x = \frac{KT}{3\pi a\eta}t \tag{2}$$

From equation (2) we will see that the higher is the liquid temperature the far the moving distance. So we can design that the volatility rate decreases along with the temperature. Hence the algorithm deals with conflicts between exploration and exploitation. Similarly, the design method of trend rate also affects the performance of algorithm. In our algorithm, the trend rate denotes the moving tendency of a particle. In general, it is reasonable to define that a particle moves toward to the better solutions.

Generally, the MOPs can be stated by below formula:

$$\max_{X \in R^n} F(X) = (f_1 \cdots f_p) \tag{3}$$

Subject to:.

$$h_i(X) = 0, i = 1, 2, \cdots, l \tag{4}$$

$$g_j(X) \le 0, j = 1, 2, \cdots, m \tag{5}$$

In order to design the Multi-objective ITO algorithm, we firstly design the storage structure for the Pareto Set. Maybe the Histogram method is a choice[3]. The main ideas of Histogram method is divide the solutions space into some subspace according one or more objective functions, in each subspace storage one solution. When new solution is generated, it must be decided which subspace is belonging to and compare with the solution in the subspace. The main ideas are shown in the Fig.1, and the detail about Histogram method will be fond in reference [3].

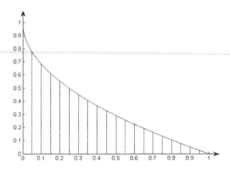

Fig. 1. One Histogram Methods: the space is divided equally into some subspace by one objective function

But there is a cute shortage in the Histogram methods: the solution in each subspace may not be the non-dominated solution. So we design a new storage structure named GPS (Geometric Position Structure), whose main ideas can be fond in reference [3].

The GPS method can be depicted as follows:

1. Use some techniques to estimate Pareto front, let $f_1 \in [l_1, u_1]$ and $f_2 \in [l_2, u_2]$. Here f_1 and f_2 are the objective functions.

2. Choose a constant integer θ_0, here θ_0 is an angle constant which splits the Pareto front equally.

3. Choose a point X which is far away from Pareto front. Calculate the maximized angle and the minimized angle, denoted by $\theta_{max}, \theta_{min}$,respectively.

4. Divide the space into floor($(\theta_{max} - \theta_{min}) / \theta_0$) subspaces, which makeup the archive A to store the potential Pareto Solutions and in each subspace a solution will be storied.

In the following we will design the ITO algorithm. Expediently we take the model with two dimensions into consideration. The results are easily extended to the case with multiple dimensions.

The tendency item b(t,X(t)) in the ITO differential equation (2) is the function of time and location; hence the potential design of tendency can be:

$$b(t, X(t)) = \beta \cdot r \cdot (X_l(t) - X(t))$$ (6)

Where r denotes [0,1] uniform distributing; β denotes the coefficient of tendency item and it is usually set to a value between [0,1],item Xl(t) denotes the current particles in the archive A[pos].

The location variable pos is decided by the following formula:

$$pos = \left\lfloor \frac{f_2(X(t)) - f_2(X^*)}{f_1(X(t)) - f_1(X^*)} \right\rfloor$$ (7)

The volatility rate denotes the movement of particle alone the its history trajectory, where dB(t) denotes the Weiner process with independent increment and it satisfies the normal distribution. And the volatility rate, according to the theory of Einstein and P. Langevin, can be designed a function of current temperature. At the early searching period of the algorithm, the volatility rate of a particle will be large due to the high temperature. While the searching time goes on, the temperature of liquid gradually becomes low and hence the volatility rate of a particle declines.

Thus, we design the volatility rate as:

$$\sum(t, X(t)) = (1 - \exp(\frac{-d(X_l(t) - X(t))}{T_0})) \exp(\frac{-1}{\alpha^t T_0})$$ (8)

The movement of a particle of will be

$$dX(t) = \beta \cdot r \cdot (X_l(t) - X(t))dt + (1 - \exp(\frac{-d(X_l(t) - X(t))}{T_0})) \exp(\frac{-1}{\alpha^t T_0})dB(t)$$ (9)

Here T_0 the initial temperature. We set T_0=1000. The parameter r satisfies [0,1] normal distribution. According to the discussion above, the framework of Multi-objective ITO algorithm can be descript as following:

- Step0: Estimate Pareto front and divide the space into subspace, initialize the population of particles, and construct the Archive A; set t=0 ;
- Step1: If the terminal condition satisfied, stop and exit. Or, go to step 2;
- Step2: Repeat the process below for N_0 times:
- (1) Compute the increment of each particle by the equation (9) and update the location of each particle;
- (2) Update the Archive.
- Step3: Go to step1 ;

4   The ITO Algorithm and Other Meta-heuristic Algorithms

Similar with other meta-heuristic algorithms such as Genetic Algorithm, Immune Algorithm, Ant Algorithm, Swarm Algorithm, etc. ITO algorithm is also based on population and supports the concept of generation. The most important difference between ITO and other meta-heuristic rest on that ITO algorithm deal with the conflict between exploration and exploitation fairly well. The ITO algorithm is based on the Ito process and hence the theory conclusions of Ito process can be used directly to do some theoretic analysis for ITO algorithms. Considering the corresponding origin of natural phenomena, ITO algorithm is inspired by irregular movement of flower powder; Genetic Algorithm is inspired by the evolution process of creature; Ant Algorithm is inspired by the action of ants; Swarm Algorithm is inspired by the flight of birds and Immune Algorithm is inspired by the creature immune system. As for ITO algorithm, there is no need extra storage space to guarantee global convergence. The current best solution will be always kept in the population until another newly produced better solution appears. The last point, ITO algorithm is relatively suitable to solve the stochastic optimization problems because the ITO algorithm would become pure stochastic algorithm if the tendency item is deleted from the momentum equation. In that case, ITO algorithm and Weiner process have the same dynamics.

5 Experiments

In the following experiments some famous MOP functions are choose as benchmarks, the performance of MITO(the Multi-objective ITO algorithm) is firstly studied. And then the ZDT function is choose to compare our algorithm with SPEA and NSGA, of course the generation number and size of population (the same evaluation times) are set to same for all the algorithms.

5.1 The Performance of Multi-objective ITO Algorithm

In the following experiments the parameters of MITO are set as: Size of population is 100, the terminal condition is 2000 generations, the parameter α is 0.2 and the β is 0.8.

Test Case 1: CUBE function

$$Min f_1(X) = \frac{1 - e^{-4x_1}}{1 - e^{-4}}$$

$$Min f_2(X) = (x_2 + 1)\left[1 - \left(\frac{f_1(x_1)}{x_2 + 1}\right)^{0.5}\right]$$

$$Min f_3(X) = (x_3 + 1)\left[1 - \left(\frac{f_1(x_1)}{x_3 + 1}\right)^{0.5}\right]$$

$$s.t.: \qquad 0 \le x_1, x_2, x_3 \le 1$$

(10)

This CUBE problem is consisted of three objective functions, and the Pareto front is a surface. The results obtained by the multi-objective ITO algorithm is shown in Fig.2

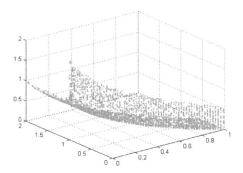

Fig. 2. The Pareto front of Cube

Test Case 2: VNT function

$$Min\, f_1(X) = 0.5(x_1^2 + x_2^2) + \sin(x_1^2 + x_2^2)$$

$$Min\, f_2(X) = \frac{(3x_1 - 2x_2 + 4)^2}{8} + \frac{(x_1 - x_2 + 1)^2}{27} + 15$$

$$Min\, f_3(X) = \frac{1}{(x_1^2 + x_2^2 + 1)} - 1.1\exp(-x_1^2 - x_2^2)$$

$$s.t.: \qquad -3 \le x_1, x_2 \le 3$$

(11)

The Pareto front of VNT problem is a curve in the 3-dimmensionspace and the results obtained by the multi-objective ITO algorithm is shown in Fig.3.

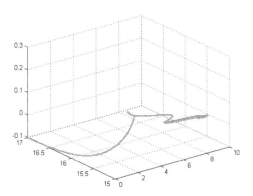

Fig. 3. One the Pareto front of VNT

Form the three experiments we can see that the MITO works well, and the performance is excellent.

5.2 Comparisons

In the following we compare MITO (the Multi-objective ITO algorithm) with SPEA and NSGA.

Test Case 3: ZDT function

$$Min\ f_1(X) = x_1$$

$$Min\ f_2(X) = g(X)\left[1 - \sqrt{\frac{x_1}{g(X)}} - \frac{x_1\sin(10\pi x_1)}{g(X)}\right]$$

$$g(X) = 1 + 10(n-1) + \sum_{i=2}^{n}(x_i^2 - 10\cos(4\pi x_i))$$

$$s.t.:\qquad 0 \le x_i \le 1\qquad n = 30$$

(12)

The Pareto front curve isn't continue and the results of the MITO, SPEA and NSGA are shown in the Fig.4.

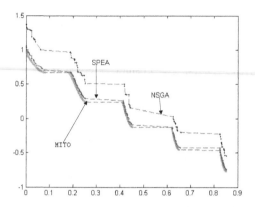

Fig. 4. Comparisons of Pareto front by the three algorithms

From the Fig.4, we can see that this algorithm can obtain very fine curves in a once run, and experiment results show that the Pareto Set of SPEA and NSGA is not the true Pareto set. The Pareto front of MITO is better than the SPEA and NSGA and the sampling points of MITO are well-distributed along the Pareto front. The convergence is also compared. For play fair all the algorithms are carried out on the same machine with Intel PIV 2.4G CPU, and each algorithm runs 30 times, the sizes of populations are set to 100, the numbers of generation are set to 2000. The results are listed in Table 1.

Table 1. The running time of the three algorithms *above* the tables

Algorithm	Average running time of 30 times
MITO	1.534
SPEA	6.339
NSGA	7.910

From Table 1 we can see that MITO runs much faster than SPEA and NSGA, so we can conclude that the performance of MITO is better than SPEA and NSGA, and the convergent rate is also better than SPEA, NSGA.

6 Summary and Future Works

In this paper, we proposed a totally new algorithm to solve MOPS: MITO algorithm. We are inspired by a kind of natural phenomenon: the irregular Brown motion of flower powder which has been well study and has much theory conclusion about the ITO process. To the author's best knowledge, it is the first time to present an algorithm framework based on Ito process theory. Moreover, we applied ITO algorithm into the MOPs field and get fairly good results both in effectiveness and speed. There are only there main parameters in ITO algorithm. Hence, the algorithm's structure will be relatively simple which helps the algorithm applied in various fields easily. At the same time, we also did some comparison research and experiments. The experiments' results prove that our algorithm has a better convergence speed. As we have mentioned above, MITO algorithm is a totally new algorithm. Obviously, there are much space left about the algorithm theory and its applications. That's our near future research plans.

Acknowledgments. The authors gratefully acknowledge the financial support of the Wuhan Chenguang Project of China under Grant No. 20065004116-03, and the Specialized Research Fund for the Doctoral Program of Higher Education of China under Grant No. 20030486049.

References

1. Pan, Z.-j., Kang, L.-S., Chen, Y.-P.: Evolutionary Computation. Tsinghua University Press, Beijing (1998)
2. Xin, Y., Yong, X.: Recent Advances in Evolutionary Computation. Computer Science & Technology, 1–18 (2006)
3. Zheng, B.J.: A Novel Multi-Objective Evolutionary Algorithm. In: International Conference on Computer Sciences, Part IV, vol. 4490, pp. 1029–1036. Springer, Heidelberg (2007)
4. Li, M.-Q., Kou, J.-S., Lin, D., Li, S.-Q.: Basic Theory And Application Of Genetic Algorithm. Science Press, Beijing (2002)
5. Wang, X., Cao, L.: Genetic Algorithm – Theory application and Software realization. Xi'an JiaoTong University Press, Shanxi (2002)
6. Tu, C., Zeng, Y.: A New Genetic Algorithm Based Upon Globally-Optimal Choosing and Its Practices. Engineering Science, 28–29 (2003)
7. Liu, L.-Y., Zheng, G.-M.: General zoology, 3rd. Higher Education Press, Beijing (1997)
8. Guo, T., Kang, L-S., Li, Y.: A New Evolutionary Algorithm for Function Optimization. Wuhan University Journal of Natural Sciences, 771–775 (1999)
9. Li, K.-S., Li, Y.-X., Tang, M.-D., Zheng, B.-J.: A Particle Dynamical Evolutionary Algorithm and Its Application on solving Single-Object Problems [J]. Journal of System Simulation, 595–598 (2005)

An Evolutionary Algorithm
for Dynamic Multi-Objective TSP

Ming Yang[1], Lishan Kang[1,2], and Jing Guan[1]

[1] School of Computer Science, China University of Geosciences, Wuhan,
430074, Hubei, China
[2] State Key Laboratory of Software Engineering, Wuhan University, Wuhan,
430072, Hubei, China
yangming0702@gmail.com, kang_whu@yahoo.com,
g_jing0414@yahoo.com.cn

Abstract. Dynamic multi-objective TSP (DMOTSP), a new research filed of evolutionary computation, is an NP-hard problem which comes from the applications of mobile computing, mobile communications. Currently, only a small number of literatures related to the research of static multi-objective TSP and dynamic single objective TSP. In this paper, an evaluation criterion of the algorithms for DMOTSP called Paretos-Similarity is first proposed, with which can evaluate the Pareto set and algorithms' performance for DMOTSP. A dynamic multi-objective evolutionary algorithm for DMOTSP, DMOTSP-EA, is also proposed, which embraces an effective operator, Inver-Over, for static TSP and dynamic elastic operators for dynamic TSP. It can track the Pareto front of medium-scale dynamic multi-objective TSP in which the number of cities is between 100 and 200. In experiment, taking CHN144+5 with two objectives for example, the algorithm is tested effective and the evaluation criterion, Paretos-Similarity, is available.

Keywords: dynamic multi-objective TSP optimization, evolutionary algorithm, Paretos-Similarity.

1 Introduction

In many application areas of scientific management and economic decision making, there are a lot of multi-objective optimization problems. In a specific TSP, often have to consider some goals simultaneously, such as the shortest route, the minimum time, the lowest risk, minimum cost and many other factors. How to get a fair and reasonable solution is a complicated matter. In real life, a large number of multi-objective optimization problems are time-dependent and the environment of the problems changes with the time. So in addition to these optimization problems with multi-objective nature, but also has a dynamic nature.

Since Psaraftis [1] first introduced dynamic TSP (DTSP) in 1988, some research works have touched on this area [2-10] and a wide variety of algorithms have been proposed for DTSP, such as ant algorithms [5,7], competitive algorithms (on-line algorithms) [8], dynamic Inver-Over evolutionary algorithms [9] and some effective

L. Kang, Y. Liu, and S. Zeng (Eds.): ISICA 2007, LNCS 4683, pp. 62–71, 2007.
© Springer-Verlag Berlin Heidelberg 2007

algorithms composed of several operators [10]. And some effective algorithms solving static TSP can be changed to solve DTSP [11,12]. The evaluation criteria of optimization algorithms for DTSP have been proposed in [9].

In contrast to a lot of continuous multi-objective function optimization problems [14] whose Pareto optimal front is a continuous or discontinuous figure, multi-objective TSP (MOTSP) [15] is discrete. Based on the theory of Pareto optimality [16], there have been a few algorithms for MOTSP [15,17,18]. The distribution of Pareto front depends greatly on the cost matrix between cities of each objective.

Dynamic multi-objective TSP (DMOTSP) is a combination of dynamic TSP and multi-objective TSP, which was first proposed as a mathematical model of aerospace communication between ground base, aero base and space base in 2004. As far as we are aware, there is no report of theoretical results and algorithms in this area. It is harder than classical TSP, DTSP and MOTSP and demands to track the dynamic Pareto optimal front in real time. Researching DMOTSP is of great theoretical significance and research value, because it would promote a development in network's topology and routing optimization problems such as mobile computing, mobile communications and aerospace communication.

2   What Is DMOTSP

The cost matrixes between cities in DMOTSP change with time. It can be described by the following equation:

$$D_k(t) = \{d_{ij}^k(t)\}_{n(t) \times n(t)}, \ k = 1, 2, ..., m \tag{1}$$

where $d_{ij}^k(t)$ is the cost of objective k from $city_i$ to $city_j$ at time t, $n(t)$ is the number of cities at time t, m is the number of objectives and t is the real word time.

Definition1 Dynamic Objective Fitness Function of Route (DOFFR): Given $n(t)$ cities and $D_k(t)$, $k = 1, 2, ..., m$, the fitness function of route $\pi(t)$ for objective k at time t can be described by the following equation:

$$f_k(\pi(t)) = \sum_{j=1}^{n(t)} d_{\pi_j \pi_{j+1}}^k(t), \ k = 1, 2, ..., m, \ \pi_{n(t)+1} = \pi_1 \tag{2}$$

Defintion2 Dynamic Pareto Optimal Solution (DPOS): Given m objectives, $f_k(\pi(t))$, $k = 1, 2, ..., m$, for route $\pi^*(t)$, if not exist any route $\pi(t)$ to make at least one of the following inequalities strictly true, then $\pi^*(t)$ is the dynamic Pareto optimal solution at time t.

$$f_k(\pi(t)) \le f_k(\pi^*(t)), \ k = 1, 2, ..., m \tag{3}$$

where at least one inequality strictly sets up.

$S^*(t)$, a set composed of $\pi^*(t)$, is the Pareto optimal set at time t.

Defintion3 Dynamic Pareto Optimal Front (DPOF): Given m objectives, the Pareto optimal front at time t, $F^*(t)$, is a graph of $(f_1(\pi^*(t)), f_2(\pi^*(t)), ..., f_m(\pi^*(t)))$, $\forall \pi^*(t) \in S^*(t)$ in the objectives' space of m dimensions.

Definition4 Dynamic Multi-Objective TSP (DMOTSP): At time t, given $n(t)$ cities and $D_k(t)$, $k = 1, 2, ..., m$, to get $S^*(t)$ and $F^*(t)$.

Suppose that the number of cities is n for a certain DMOTSP problem and the number of solutions of $S^*(t)$ is N^*. If this DMOTSP is symmetric for all objectives, the number of all the combination routes is $(n-1)!$ and $N^* \in [1, (n-1)!]$. If this DMOTSP is asymmetric for all objectives, the number of all the combination routes is $\dfrac{(n-1)!}{2}$ and $N^* \in [1, \dfrac{(n-1)!}{2}]$. The value of N^* is limited and $F^*(t)$ is discrete.

DMOTSP can be divided into the global dynamic TSP and local dynamic TSP according to the degree of change in $D_k(t)$. If $D_k(t + \Delta t)$ is much different from $D_k(t)$, this dynamic TSP is a global dynamic TSP for objective k. If $D_k(t + \Delta t)$ is not much different from $D_k(t)$, this dynamic TSP is a local dynamic TSP. DMOTSP can be divided into conflict and unanimity TSP according to the degree of conflict between objectives. If there exits one route $\pi(t)$ that when the fitness of $\pi(t)$ for objective a better, better for objective b consequentially, the objectives, a and b, are unanimity and this TSP is a unanimity TSP. If when the fitness of each route for objective a better, worse for objective b consequentially, the objectives, a and b, are conflict and this TSP is a conflict TSP. So DMOTSP can be divided into the following four kinds: (1) the global dynamic and objective conflict TSP (GDOC-TSP); (2) the global dynamic and objective unanimity TSP (GDOU-TSP); (3) the local dynamic and objective conflict TSP (LDOC-TSP); (4) the local dynamic and objective unanimity TSP (LDOU-TSP).

3 DMOTSP Optimization Algorithms

3.1 DMOTSP Optimization

In Def.4, the change of DMOTSP's cost matrixes with time is deemed as a continuous process. Practically, the algorithms for DMOTSP discretize this change process. Thus, a DMOTSP becomes a series of static multi-objective TSP problems: for each route $\pi_1(t_i)$ belonging to the Pareto set $S(t_i)$ got by algorithms, not exit $\pi_2(t_i) \in S(t_i)$ to make:

$$f_k(\pi_2(t_i)) \leq f_k(\pi_1(t_i)), \ k = 1, 2, ..., m, \ i = 1, 2, ..., N \tag{4}$$

where at least one inequality strictly sets up.

The optimization goal of DMOTSP becomes a series of static multi-objective TSP optimization problems whose goals are to make $S(t_i)$ close to $S^*(t_i)$ most possibly and we describe it by the following equation:

$$S(t_i) \rightarrow S^*(t_i), \; i = 1, 2, ..., N \tag{5}$$

with time fragment $[t_{i-1}, t_i]$, where m is the number of objectives, $\{t_i\}_0^N$ are the sampling points of time, $S(t_i)$ is the Pareto set got by a DMOTSP solver in time fragment $[t_{i-1}, t_i]$ and $S^*(t_i)$ is the dynamic Pareto optimal set at $t = t_i$.

3.2 How to Evaluate a DMOTSP Algorithm

The goal of DMOTSP algorithm is to meet Equ.5, so how to evaluate an algorithm is to see how much $S(t_i)$ close to $S^*(t_i)$. Firstly give the following definitions.

Definition5 Clustering of Pareto Set: Given m objectives and a Pareto set which has N solutions, the clustering of Pareto set is the process of sorting these solutions into several classes composed of some similar solutions. The solutions of the same class have similar objective fitness of route and the ones of different classes are different. The notion of clustering can be seen in [19].

Definition6 Barycenter of Class: If a Pareto set is clustered into n classes, the barycenter of class k is:

$$(x_1', x_2', ..., x_m')_k = \frac{1}{n_k} \sum_{j=1}^{n_k} (f_1, f_2, ..., f_m)_j, \; k = 1, 2, ..., n \tag{6}$$

where m is the number of objectives, f_i is dynamic objective fitness of route (see Equ.2) and n_k is the number of solutions of class k, $\sum_{k=1}^{n} n_k = N$.

Definition7 Clustering Barycenter: If a Pareto set is clustered into n classes, the clustering barycenter of Pareto set is:

$$(x_1, x_2, ..., x_m)_c = \frac{1}{n} \sum_{j=1}^{n} (x_1', x_2', ..., x_m')_j \tag{7}$$

where m is the number of objectives and $(x_1, x_2, ..., x_m)_j$ is the barycenter of class j (see Equ.6).

Definition8 Barycenter of Pareto Set: Given m objectives and a Pareto set which has N solutions, the barycenter of Pareto set is:

$$(x_1, x_2, ..., x_m)_p = \frac{1}{N} \sum_{j=1}^{N} (f_1, f_2, ..., f_m)_j \tag{8}$$

where f_i is dynamic objective fitness of route (see Equ.2).

Based on the above definitions, give the following equation:

$$Paretos-Similarity = \left(\frac{1}{1+|N-N^*|}\right)^\alpha \left(\frac{1}{1+d_1}\right)^\beta \left(\prod_{k=1}^{m} \frac{f_k^*}{f_k}\right)^\delta \left(\frac{1}{1+|d_2-d_2^*|}\right)^\lambda \tag{9}$$

In Equ.9, the values ranges of α, β, δ and λ are [0,1], but they can not be 0 simultaneously. *Paretos-Similarity* is monotonic decreasing function of α, β, δ and λ. N and N^* are respectively the numbers of solutions in $S(t_i)$ and $S^*(t_i)$. m is the number of objectives. d_1 is the distance between berycenters of $S(t_i)$ and $S^*(t_i)$ (see Equ.10). f_k and f_k^* are respectively the best fitness of routes for objective k in $S(t_i)$ and $S^*(t_i)$. d_2 is the average value of distances between barycenter of classes and clustering barycenter for $S(t_i)$ and d_2^* is for $S^*(t_i)$ (see Equ.11).

$$d_1 = \sqrt{\sum_{k=1}^{m}(x_k - x_k^*)_p^2} \tag{10}$$

$$d_2 = \frac{1}{n}\sum_{j=1}^{n}\sqrt{\sum_{k=1}^{m}(x'_{k,j} - x_{k,c})^2} \quad d_2^* = \frac{1}{n^*}\sum_{j=1}^{n^*}\sqrt{\sum_{k=1}^{m}(x''_{k,j} - x_{k,c}^*)^2} \tag{11}$$

where m is the number of objectives, n and n^* are respectively the numbers of classes for $S(t_i)$ and $S^*(t_i)$ after clustering, $(x_1, x_2,..., x_m)_p$ and $(x_1^*, x_2^*,..., x_m^*)_p$ are respectively the barycenters of $S(t_i)$ and $S^*(t_i)$ (see Equ.8), $(x'_1, x'_2,..., x'_m)_j$ and $(x''_1, x''_2,..., x''_m)_j$ are respectively the barycenters of class j in $S(t_i)$ and $S^*(t_i)$ (see Equ.6) and $(x_1, x_2,..., x_m)_c$ and $(x_1^*, x_2^*,..., x_m^*)_c$ are respectively the clustering barycenter of $S(t_i)$ and $S^*(t_i)$ (see Equ.7).

Paretos-Similarity can evaluate how much $S(t_i)$ close to $S^*(t_i)$, which is from the number of solutions in Pareto set, the best approximate degree of each objective in Pareto set, the average approximate degree of Pareto set and the uniform and diversity of distribution of solutions in Pareto set. The greater the value of *Paretos-Similarity*, the closer $S(t_i)$ to $S^*(t_i)$. The first three can be from an overall perspective to evaluate how much $S(t_i)$ toward $S^*(t_i)$, but not from the distribution of Pareto front.

In Equ.9, $\dfrac{1}{1+|d_2 - d_2^*|}$ can reflect how the distribution of Pareto front of $S(t_i)$ to $S^*(t_i)$. From Equ.9, it can be seen that the value range of *Paretos-Similarity* is $(0,1]$. Are the weights of the four above and if the value of one is 0, the value of *Paretos-Similarity* has nothing to do with it.

If the value of *Paretos-Similarity* is greater, it can be educed that $S(t_i)$ is closer to $S^*(t_i)$ and the performance of algorithm of DMOTSP is better.

But the Pareto optimal sets $S^*(t_i)$ are not known for most problems. It can replace $S^*(t_i)$ with $\bar{S}^*(t_i)$ which is the best Pareto set got by a solver without time restriction. Then Equ.9 is changed to be Equ.12 and the significances of the parameters are similar to the ones in Equ.9.

$$\overline{Paretos-Similarity} = \left(\frac{1}{1+\mid N-\overline{N^*}\mid}\right)^{\alpha} \left(\frac{1}{1+\overline{d_1}}\right)^{\beta} \left(\prod_{k=1}^{m} \frac{\overline{f_k^*}}{\overline{f_k}}\right)^{\delta} \left(\frac{1}{1+\mid d_2-\overline{d_2^*}\mid}\right)^{\lambda} \quad (12)$$

4 Dynamic Multi-objective Evolutionary Algorithm for DMOTSP (DMOTSP-EA)

DMOTSP-EA composed of Inver-Over [11] and Dynamic Elastic Operators (Insert Operator, Delete Operator and Change Operator) [9] with the greed rules for generating offspring population can get a Pareto approximate optimal set in short time.

4.1 Rules for Generating Offspring Population

Suppose that the fixed size of population is n, the number of solutions of Pareto set in iteration is m and the set of the individuals which have the best fitness for every objective in current population is W. In the evolution process, if the offspring individual R^* dominates the parent individual R $(R^* \succeq R)$, then replace R with R^*. If R dominates R^* $(R \succeq R^*)$, there is no change to population. If R is noncomparable to R^* $(R^* \sim R)$, then R and R^* are all selected into offspring population (the concept of "dominate" and "noncomparable" seen in [16]). The rules are described as the followings:

1. If $m=n$, then take the individuals of Pareto set as the offspring population;
2. If $m>n$, then search the two individuals the distance between which is the minimum and they do not belong to W, suppose that they are R_1 and R_2. For R_1 (R_2), search the minimum distance D_1 (D_2). If $D_1<D_2$, then delete R_1 from Pareto set; else delete R_2. Repeat the above process until $m=n$. Finally take the individuals of Pareto set as the offspring population;
3. If $m<n$, suppose that $n-m=k$, then select randomly k individuals which do not belong to Pareto set from population and take these k individuals and the ones of Pareto set together as the offspring population.

4.2 The Outline of DMOTSP-EA

Suppose that Dlist is a list of storing the information of cities' state change.

1. initialize population randomly and suppose the size of population is M;
2. if Dlist is empty, go to step 4; otherwise go to step 3;
3. for each individual of population
      Dynamic Elastic Operators;
4. Inver-Over Operator;
5. generate offspring population according to the rules in section 4.1;
6. if not termination condition, go to step 2; otherwise end.

5 Experiments and Conclusions

Fig.1 shows the experiment model for DMOTSP. In addition to the cities of CHN144 in [9] but their coordinates are three-dimension, there is a Chinese island and four satellites. One of the satellites is static and stops over China, the other three are mobile. One of the three satellites orbits Earth in equatorial orbit, the other two ones orbit Earth in polar orbit.

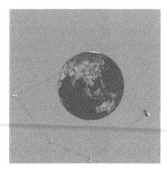

Fig. 1. Experiment model: the cities in CHN144, an island and four satellites

There are two objectives, objective$_1$ and objective$_2$. The costs between cities for objective$_1$ are the actual distances between them. The costs between cities for objective$_2$ are generated by selecting two costs of objective$_1$ randomly at the probability p and exchanging them. In the whole process, the costs between cities and satellites and the ones between satellites are the actual distances between them. It is the dynamic nature of the problem that the distances between cities and satellites and the ones between satellites are changing as satellites orbit Earth.

Sample 120 points in a cycle of satellite orbiting Earth. For each sample point, test DMOTSP-EA with the $\overline{Pareto-Similarity}$ (see Equ.12).

The clustering uses fuzzy mathematical method in [20]. The experiment's environments are that CPU is Intel C4 1.7GHz and RAM is 256MB.

The parameters of algorithm and Equ.12 are set as Table 1, where p_1 is the parameter in Inver-Over operator.

Table 1. Algorithm's Parameters setting

p	p_1	M	α	β	δ	λ
0.3	0.001	30	0.01	0.02	1.0	0.01

Suppose ΔT is the average time interval of state changing. Fig.2 shows the Pareto front of a certain sample point with different ΔT, from which it can be seen that if ΔT is greater the dynamic Pareto front is closer to the static Pareto front.

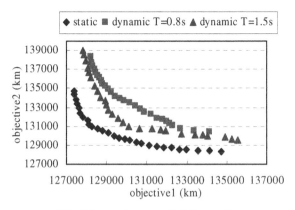

Fig. 2. Pareto fronts with different ΔT

Suppose that $\left(\dfrac{1}{1+|N-\overline{N^*}|}\right)^{\alpha}=A,\ \left(\dfrac{1}{1+\overline{d_1}}\right)^{\beta}=B,\ \left(\displaystyle\prod_{k=1}^{m}\dfrac{\overline{f_k^*}}{\overline{f_k}}\right)^{\delta}=C,\ \left(\dfrac{1}{1+|d_2-\overline{d_2^*}|}\right)^{\lambda}=D$

and $\overline{Paretos-Similarity}=E$ (see Equ.12). Fig.3 shows A, B, C, D, and E for 120 sample points in the condition that $\Delta T=1.5$s. Fig.4 shows the average results running algorithm ten times with different values of ΔT. From Fig.3 and Fig.4, it can be seen that the changing trends of B, C, D and E are same on the whole and if ΔT is greater the value of E is greater too, matching the fact that if ΔT is greater the dynamic Pareto front is closer to the static Pareto front. This can prove that Equ.12 used to evaluate Equ.5 is available.

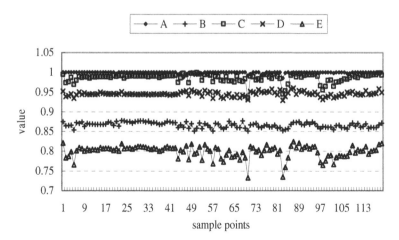

Fig. 3. The change of A, B, C, D and E with $\Delta T=1.5$s

From the above experiments, it can be seen that DMOTSP-EA solver can get an approximate Pareto optimal set with Inver-Over and dynamic elastic operators. The rules for generating offspring population can speed up the convergence of the

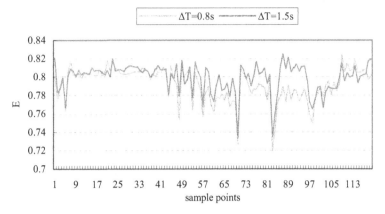

Fig. 4. *Paretos − Similarity* with different ΔT

algorithms, because the offspring is selected from the Pareto set first. The evaluation criterion, Paretos-Similarity, is available to evaluate the performance of algorithms for DMOTSP.

Acknowledgements

This work was supported by the National Natural Science Foundation of China (No.60473081) and the Natural Science Foundation of Hubei Province (No.2005ABA234). Thanks the anonymous reviewers very much for their valuable comments.

References

1. Psaraftis, H.N.: Dynamic vehicle routing problems. In: Golden, B.L., Assad, A.A. (eds.) Vehicle Routing: Methods and Studies, pp. 223–248. Elsevier Science Publishers, Amsterdam (1988)
2. Durbin, R., Willshaw, D.: An Analogue Approach to the Traveling Salesman Problem Using an Elastic Net Method. Nature 326(6114), 689–691 (1987)
3. Stone, J.V.: The Optimal Elastic Net: Finding Solutions for the Travelling Salesman. Artificial Neural Networks 2, 1077–1080 (1992)
4. Power, W., Jaillet, P., Odoni, A.: Stochastic and Dynamic Networks and Routing. In: Ball, M., Maganti, T., Monma, C., Namhauser, G. (eds.) Network Rougting, North-Holland, Amsterdam (1995)
5. Guntsch, M., Branke, J., Middendorf, M., Schmeck, H.: ACO Strategies for Dynamic TSP. In: Dorrigo, M., et al. (eds.) Abstract Proceedings of ANTS'2000, pp. 59–62 (2000)
6. Huang, Z.C., Hu, X.L., Chen, S.D.: Dynamic Traveling Salesman Problem based on Evolutionay Computation. In: Congress on Evolution Computation (CEC'01), pp. 1283–1288. IEEE Computer Society Press, Los Alamitos (2001)

7. Eyckelhof, C.J., Snoek, M.: Ant Systems for a Dynamic TSP-Ants caught in a traffic jam. In: ANTS2002. 3rd International Workshop on Ant Algorithms, pp. 88–99. Springer, Brussels, Belgium (2002)

8. Ausiello, G., Feuestein, E., Leonardi, S., Stougie, L., Talamo, M.: Algorithms for the On-line Traveling Salesman. Algorithmica 29(4), 560–581 (2001)

9. Kang, L., Zhou, A., McKay, B., et al.: Benchmarking Algorithms for Dynamic Travelling Salesman Problems. In: Proceedings of the Congress on Evolutionary Computation, Portland, Oregon (2004)

10. Li, C., Yang, M., Kang, L.: A New Approach to Solving Dynamic Traveling Salesman Problems. In: Wang, T.-D., Li, X., Chen, S.-H., Wang, X., Abbass, H., Iba, H., Chen, G., Yao, X. (eds.) SEAL 2006. LNCS, vol. 4247, pp. 236–243. Springer, Heidelberg (2006)

11. Guo, T., Michalewicz, Z.: Inver-Over operator for the TSP. In: Eiben, A.E., Bäck, T., Schoenauer, M., Schwefel, H.-P. (eds.) Parallel Problem Solving from Nature - PPSN V. LNCS, vol. 1498, pp. 803–812. Springer, Heidelberg (1998)

12. Helsgaun, K.: An Effective Implementation of the Lin-Kernighan Traveling Salesman Heuristic. Eur. J. Oper. Res. 126, 106–130 (2000)

13. Rudolph, G.: On a Multi-objective Evolutionary Algorithm and Its Convergence to the Pareto Set. In: Proceedings of the 1998 IEEE International Conference on Evolutionary Computation, pp. 511–516. IEEE Press, Piscataway (NJ) (1998)

14. Deb, K.: A Fast and Elitist Multi-objective Genetic Algorithm: NSGA-II. IEEE Transactions on Evolutionary Computation 6(2), 182–197 (2002)

15. Yan, Z.Y., Zhang, L.H., Kang, L.S.: A New MOEA for Multi-objective TSP and Its Convergence Property Analysis. In: Fonseca, C.M., Fleming, P.J., Zitzler, E., Deb, K., Thiele, L. (eds.) EMO 2003. LNCS, vol. 2632, pp. 342–354. Springer, Heidelberg (2003)

16. Pareto, V.: Cours D' Economie Politique, vol. 1. F. Rouge, Lausanne (1896)

17. Ji, Z., Chen, A., Subprasom, K.: Finding Multi-Objective Paths in Stochastic Networks: A Simulation-based Genetic Algorithm Approach. In: Congress on Evolution Computation (CEC2004), pp. 174–180. IEEE Press, NJ, New York (2004)

18. Marwaha, S., Srinivasan, D., Tham, C.K., et al.: Evolutionary Fuzzy Multi-Objective Routing For Wireless Mobile Ad Hoc Networks. In: Congress on Evolution Computation (CEC2004), pp. 1964–1971. IEEE Press, NJ, New York (2004)

19. Han, J., Kamber, M.: Data Mining Concepts and Techniques. China Machine Press, pp. 223–225 (2005)

20. Xie, J., Liu, C.: Fuzzy Mathematics and Its Applications, 2nd edn., pp. 81–118. Huzhong University of Science and Technology Publication (Ch) (2002)

The Construction of Dynamic Multi-objective Optimization Test Functions

Min Tang[1], Zhangcan Huang [2], and Guangxi Chen[1]

[1] Guilin University of Electronic Technology, Guilin China
[2] School of Science, Wuhan University of Technology, Wuhan China
dengtangmin@126.com

Abstract. Dynamic Multi-objective Optimization Problems (DMOPs) gradually become a difficult and hot topic in Multi-objective Optimization area. However, there is lack of standard test functions for Dynamic Multi-objective Optimization Algorithms now. Firstly this paper proves the existence of Pareto optimal set of a class of a special non-dynamic two-objective optimization problem theoretically. Based on this result, we present one method of constructing dynamic two-objective and scalable multi-objective optimization problems, and then providing the test suites which are easy to be constructed and have known Pareto Optimal set and Pareto optimal front.

Keywords: test functions, multi-objective optimization, dynamic.

1 Introduction

Multi-objective Optimization Problems (MOPs) originated from many design and programming for real-world complex systems which comprise most popular application fields, including engineering, industrial and scientific applications for examples, aeronautical engineering, electrical engineering, robotics, transport engineering, scheduling, computer science, and so on. Commonly two or more objective functions need to be optimized simultaneously, instead of single objective. As a consequence, there is no unique solution to multi-objective optimization problems, but instead, we aim to find all / a sampling of the good trade-off solutions available. Evolutionary Algorithms (EAs) are kind of heuristic algorithms that use natural selection as their search engine to solve problems and become very popular in solving MOPs.

The Evolutionary Multi-objective Optimization (EMO) literature contains a large number of static test problems covering different types of difficulties which may be encountered by an EMO algorithm when converging toward the Pareto optimal front, but several other important real-world applications require a time-dependent (on-line) multi-objective optimization, in which either the objective function and constraint or the associated problem parameters or both vary with time. Now there is lack of standard dynamic test functions and dynamic evolutionary multiobjective optimization algorithms.

In this paper, we obtain the exact location of Pareto optimal set and Pareto front of a class of a special non-dynamic two-objective test functions theoretically, based on this result, we present a method to construct dynamic two-objective test functions.

L. Kang, Y. Liu, and S. Zeng (Eds.): ISICA 2007, LNCS 4683, pp. 72–79, 2007.
© Springer-Verlag Berlin Heidelberg 2007

The paper is organized as follows: Section 2 introduces key concepts used in the field of multi-objective optimization and dynamic multi-objective optimization. Section 3 gives an overview of the construction of a special non-dynamic two-objective test functions. Section 4 presents a method to construct dynamic two-objective and scalable test functions, and give some examples. The last section offers concluding remarks and future perspectives.

2 Basic Concepts

Definition 1: Multi-objective optimization problems (MOPs) [3]

$$Minimize \ \ [f_1(\bar{x}), f_2(\bar{x}), \cdots, f_k(\bar{x})] \tag{1}$$

Subject to the m inequality constrains:

$$g_i(\bar{x}) \leq 0, \ i = 1, 2, \cdots, m \tag{2}$$

And the p equality constrains:

$$h_i(\bar{x}) = 0, \ i = 1, 2, \cdots, p \tag{3}$$

Where k is the number of objective functions $f_i : R^n \to R$. We call $\bar{x} = [x_1, x_2, \cdots, x_n]^T \in \Omega$ the vector of decision variables. We wish to determine from among the set Ω of all vectors which satisfy (2) and (3) the particular set of values $x_1^*, x_2^*, \cdots, x_n^*$ which yield the optimum values of all the objective functions.

Definition 2 [4]
Let t be the time variable. V and W be n-dimensional and M-dimensional continuous or discrete vector spaces, g and h be two functions defining inequalities and constraint equalities, and f be a function from $V \times t$ to W. A dynamic multi-objective problem with M objectives is defined as

$$\min_{v \in V} f = \{f_1(\bar{v}, t), \cdots, f_M(\bar{v}, t)\}$$

$$s.t. \ \ g(\bar{v}, t) \leq 0, h(\bar{v}, t) = 0$$

In the problem defined above, some variables are available for optimization ($\bar{v}$), and some others (t) is imposed parameters that are independent from the optimization variables.

Definition 3 [4]: We call the Pareto optimal set at time t ($S_p(t)$) and the Pareto optimal front at time t ($F_p(t)$) the set of Pareto optimal solutions at time t in decision variable and objective spaces, respectively.

3 A Special Non-dynamic Two-Objective Optimization Problems

Kalyanmoy Deb [1] presents a method to construct N-variable non-dynamic two-objective problem:

Minimize $f_1(\bar{x}) = f_1(x_1, x_2, \cdots, x_m)$

Minimize $f_2(\bar{x}) = g(x_{m+1}, \cdots, x_N) \times h(f_1(x_1, \cdots, x_m), g(x_{m+1}, \cdots, x_N))$ \hfill (4)

The function f_1 is a function of $m(< N)$ variables ($\bar{x}_1 = (x_1, \cdots, x_m)$), and the function f_2 is a function of all N variables. The function g is a function of $(N - m)$ variables $\bar{x}_{II} = (x_{m+1}, \cdots, x_N)$ which do not appear in the function f_1. The function h is a function of f_1 and g function value directly. We take only positive values in the search space. If the following two properties of h are satisfied, the global Pareto-optimal set will correspond to the global minimum of the function g and to all values of the function f_1:

(1) The function h is a monotonically non-decreasing function in g for a fixed value of f_1, i.e., if $g(\bar{x}_{II}^{(1)}) > g(\bar{x}_{II}^{(2)})$,

$$h(f_1(\bar{x}_1), g(\bar{x}_{II}^{(1)})) \geq h(f_1(\bar{x}_1), g(\bar{x}_{II}^{(2)})), \; f_1(\bar{x}_1) \in (0, +\infty). \tag{5}$$

(2) The function h is a monotonically decreasing function of f_1 for a fixed value of g, i.e., if $f_1(\bar{x}_1^{(1)}) > f_1(\bar{x}_1^{(2)})$,

$$h(f_1(\bar{x}_1^{(1)}), g(\bar{x}_{II})) < h(f_1(\bar{x}_1^{(2)}), g(\bar{x}_{II})), \; g(\bar{x}_{II}) \in (0, +\infty). \tag{6}$$

We can prove the above conclusion theoretically as follows:

THEOREM: if $\bar{x}_{II}^*$ is the solution corresponding to the minimum value of the function g, $g(\bar{x}_{II}^*)$ is the minimum value of the function g, then Pareto optimal set of above question is $(\bar{x}_1^{\Delta}, \bar{x}_{II}^*)$, Pareto optimal front is $(f_1(\bar{x}_1^{\Delta}), f_2(\bar{x}_{II}^*))$, where $(\bar{x}_1^{\Delta}, \bar{x}_{II}^*) \in \Omega$, $(\bar{x}_1, \bar{x}_{II}) \in \Omega$.

PROOF: $\forall (\bar{x}_1, \bar{x}_{II})$, there exists

$$g(\bar{x}_{II}) > g(\bar{x}_{II}^*). \tag{7}$$

There are three possible type relationships between $f_1(\bar{x}_1^{\Delta})$ and $f_1(\bar{x}_1)$, we prove $(\bar{x}_1^{\Delta}, \bar{x}_{II}^*)$ is not dominated by $(\bar{x}_1, \bar{x}_{II})$ through following three steps:

(1) If $f_1(\bar{x}_1^{\Delta}) = f_1(\bar{x}_1)$, according to (5), (7), $h(f_1(\bar{x}_1^{\Delta}), g(\bar{x}_{II})) \geq h(f_1(\bar{x}_1), g(\bar{x}_{II}^*))$, there exists $f_2(\bar{x}) > f_2(\bar{x}^*)$, i.e., solution $(\bar{x}_1^{\Delta}, \bar{x}_{II}^*)$ dominates solution $(\bar{x}_1, \bar{x}_{II})$.

(2) If $f_1(\bar{x}_1^{\Delta}) < f_1(\bar{x}_1)$, apparently solution $(\bar{x}_1, \bar{x}_{II})$ cannot dominate solution $(\bar{x}_1^{\Delta}, \bar{x}_{II}^*)$.

(3) If $f_1(\bar{x}_1^{\Delta}) > f_1(\bar{x}_1)$, according to (5), (7), there exists

$$h(f_1(\bar{x}_1), g(\bar{x}_{II})) \geq h(f_1(\bar{x}_1), g(\bar{x}_{II}^*)). \tag{8}$$

And according to (6), there exists

$$h(f_1(\bar{x}_1), g(\bar{x}_{II}^*)) > h(f_1(\bar{x}_1^\Delta), g(\bar{x}_{II}^*)). \tag{9}$$

According to (8), (9), there exists $f_2(\bar{x}) > f_2(\bar{x}^*)$, i.e., solution $(\bar{x}_1^\Delta, \bar{x}_{II}^*)$ dominates solution $(\bar{x}_1, \bar{x}_{II})$.

To sum up, $\forall(\bar{x}_1^\Delta, \bar{x}_{II}^*) \in \Omega$ which is not dominated by any other solution in decision space.

4   The Construction of Dynamic Two-Objective Optimization Test Problems

4.1   The Types of Dynamic Multi-objective Optimization Test Problems

We are dealing with two distinct yet related spaces: decision variable space and objective space. There are four possible ways that a problem can demonstrate a time-varying change [4]:

Type I : The S_P changes, whereas the F_P does not change.

Type II : Both S_P and F_P change.

Type III : S_P does not change, whereas F_P changes.

Type IV : Both S_P and F_P do not change, although the problem can change.

4.2   The Construction of Dynamic Two-objective Optimization Test Problems

In chapter 3 we study a special non-dynamic two-objective optimization problem. Based on the result, we present a method to construct dynamic problem:

Minimize $f_1(\bar{x}) = f_1(x_1, x_2, \cdots, x_m)$

Minimize $f_2(\bar{x}) = u(t)(g((x_{m+1}, \cdots, x_N)v(t)) \times h(f_1(x_1, \cdots, x_m), g((x_{m+1}, \cdots, x_N)v(t))))$

Where $u(t)$ and $v(t)$ are functions of time t.

By choosing an appropriate $u(t)$ and $v(t)$, we can construct various dynamic test problems which have known Pareto optimal set and Pareto optimal front, describing as follows:

(1) set $u(t) = 1$, $v(t)$ changes with time, and the resulting problem becomes a sample of Type I. $(\bar{x}_1^\Delta, \bar{x}_{II}^{**})$ is Pareto optimal set of such question, where $\bar{x}_{II}^{**} = \bar{x}_{II}^* / v(t)$.

(2) set $v(t) = 1$, $u(t)$ changes with time, and the resulting problem becomes a sample of Type III. $(f_1(\bar{x}_1^\Delta), f_2^*(\bar{x}^*))$ is Pareto optimal front of such question, where $f_2^*(\bar{x}^*) = f_2(\bar{x}^*)u(t)$.

(3) set $u(t)$ and $v(t)$ changes with time, and the resulting problem becomes a sample of Type II. $(\bar{x}_1^\Delta, \bar{x}_{II}^{**})$ is Pareto optimal set, where $\bar{x}_{II}^{**} = \bar{x}_{II}^* / v(t)$, and $(f_1(\bar{x}_1^\Delta), f_2^*(\bar{x}^*))$ is Pareto optimal front of such question, where $f_2^*(\bar{x}^*) = f_2(\bar{x}^*)u(t)$.

We avoid complications by choosing $u(t)$ and $v(t)$ and can be easy to visualize, we can choose linear function $u(t) = v(t) = t$, otherwise, $u(t)$ and $v(t)$ can be adopted more complex functions.

4.3 The Construction of Scalable Dynamic Multi-objective Optimization Test Problems

The following is a generic scalable dynamic problem to that described in equation 10, having M objectives.

$$
\begin{aligned}
Minimize\ f_1(\vec{\theta}, r) &= u(t)((1 + g(r \cdot v(t)))\cos(\theta_1)\cos(\theta_2)\cdots\cos(\theta_{M-2})\cos(\theta_{M-1})) \\
Minimize\ f_2(\vec{\theta}, r) &= u(t)((1 + g(r \cdot v(t)))\cos(\theta_1)\cos(\theta_2)\cdots\cos(\theta_{M-2})\sin(\theta_{M-1})) \\
Minimize\ f_3(\vec{\theta}, r) &= u(t)((1 + g(r \cdot v(t)))\cos(\theta_1)\cos(\theta_2)\cdots\sin(\theta_{M-2})) \\
&\vdots \qquad \vdots \\
Minimize\ f_{M-1}(\vec{\theta}, r) &= u(t)((1 + g(r \cdot v(t)))\cos(\theta_1)\sin(\theta_2)) \\
Minimize\ f_M(\vec{\theta}, r) &= u(t)((1 + g(r \cdot v(t)))\sin(\theta_1)) \\
& 0 \le \theta_i \le \pi/2, i = 1,2,\cdots,(M-1) \\
& g(r \cdot v(t)) \ge 0
\end{aligned}
\tag{10}
$$

The decision variables are mapped to the meta-variable vector $\vec{\theta}$ (of size $(M-1)$) as follows:

$$
\theta_i = \frac{\pi}{2} x_i, \quad for \ \ i = 1,2,\cdots,(M-1).
$$

The above mapping and the condition on θ_i restrict each of the above x_i to lie within $x_i \in [0,1]$. The Pareto-optimal surface occurs for the minimum of the g function, or at $x_M^* = 0$ and the function values must satisfy the following condition:

$$
\sum_{i=1}^{M}(f_i^*)^2 = 1
$$

As mentioned earlier, the difficulty of the above test problem can also be varied by using different functional for f_i and g.

4.4 Examples

Using the method of chapter 3, we can construct a non-dynamic multi-objective optimization problem named test problem 1, describing as follows:

$$
Minimize\ f_1(\vec{x}) = 1 - \exp(-4x_1)\sin^6(6\pi x_1).
$$

$$
Minimize\ f_2(\vec{x}) = g \times (1 - (\frac{f_1}{g})^2)
$$

$$
where\ g = 1 + 9 \times (\frac{\sum_{i=2}^{n} x_i}{n-1})^{0.25}, x_i \in [0,1], i = 1,\cdots,10, n = 10
$$

f_1 is a multimodality function, the test problem 1 is corresponding to a non-uniformity Pareto optimal front (POF).

We construct dynamic test problem 1 by choosing a non-linear function $u(t) = t^2$, and $v(t) = 1$, i.e., it belongs to Type III, describing as follows:

$$Minimize \quad f_1(\vec{x}) = 1 - \exp(-4x_1)\sin^6(6\pi x_1)$$

$$Minimize \quad f_2(\vec{x}) = t^2(g \times (1 - (\frac{f_1}{g})^2))$$

$$where \quad g = 1 + 9 \times (\frac{\sum_{i=2}^{n} x_i}{n-1})^{0.25}, x_i \in [0,1], i = 1, \cdots, 10, n = 10$$

Figure 1 is function f_1 of test problem 1, where horizontal coordinate denotes variable x_1, vertical coordinate denotes function f_1, figure 2 is the Pareto optimal front of test problem 1, figure 3 is the Pareto optimal front of dynamic test problem 1 for 4 time steps, in figure 2 and figure 3, horizontal coordinate denotes function f_1, vertical coordinate denotes function f_2.

Another example: (test problem 2)

$$Minimize \quad f_1(\vec{x}) = x_1.$$

$$Minimize \quad f_2(\vec{x}) = g \times (1 - \sqrt{\frac{f_1}{g}})$$

$$where \quad g(\vec{x}) = 91 + \sum_{i=2}^{10} (x_i^2 - 10\cos(4\pi x_i)), \quad 0 \le x_1 \le 1, \ -5 \le x_2, \cdots, x_{10} \le 5$$

Test problem 2 is corresponding to a convex Pareto optimal front.

We use the same manner to construct dynamic problem named test problem 2 except for choosing a linear function $u(t) = t$, describing as follows:

$$Minimize \quad f_1(\vec{x}) = x_1.$$

$$Minimize \quad f_2(\vec{x}) = t(g \times (1 - \sqrt{\frac{f_1}{g}}))$$

$$where \quad g(\vec{x}) = 91 + \sum_{i=2}^{10} (x_i^2 - 10\cos(4\pi x_i)), \quad 0 \le x_1 \le 1, \ -5 \le x_2, \cdots, x_{10} \le 5$$

Figure 4 is function f_1 of test problem 2, figure 5 is the Pareto optimal front of test problem 2, figure 6 is the Pareto optimal front of dynamic test problem 2 for 5 time steps.

5   Summary and Future Work

In this paper we prove the existence of Pareto optimal set of a class of special non-dynamic two-objective optimization problems theoretically. Based on this result, we present one method of constructing dynamic multi-objective optimization benchmark functions and providing the test function suites which are easy to construct and have known Pareto Optimal set and Pareto front.

After tacking static problems with two or more objective functions, the next logical step is to develop MOEAs that can deal with dynamic test functions. We should expect that more complex dynamic test functions and algorithms appear in the literature in the next few years.

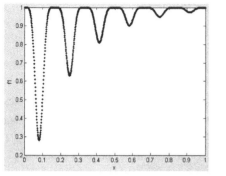

Fig. 1. function f_1 of test problem 1

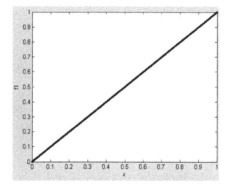

Fig. 2. POF of test problem 1

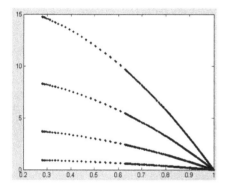

Fig. 3. POF of dynamic test problem 1

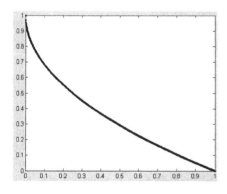

Fig. 4. function f_1 of test problem 2

Fig. 5. POF of test problem 2

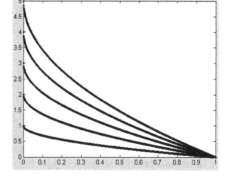

Fig. 6. POF of dynamic test problem 2

Acknowledgments. The work is supported by the Soft Science Foundation of Guilin University of Electronic Technology.

References

1. Deb, K.: Multi-Objective Genetic Algorithms: Problem Difficulties and Construction of Test Problems. Evolutionary Computation 7(3), 205–230 (1999)
2. Coello Coello, C.A.: Short Tutorial on Evolutionary Multiobjective Optimization. In: Zitzler, E., Deb, K., Thiele, L., Coello Coello, C.A., Corne, D.W. (eds.) EMO 2001. LNCS, vol. 1993, pp. 21–40. Springer, Heidelberg (2001)
3. Coello Coello, C.A.: Recent Trends in Evolutionary Multiobjective Optimization. In: Abraham, A., Jain, L., Goldberg, R. (eds.) Evolutionary Multiobjective Optimization: Theoretical Advances And Applications, pp. 7–32. Springer-Verlag, London (2005)
4. Farina, M., Deb, K., Amato, P.: Dynamic Multiobjective Optimization Problems: Test cases, Approximation, and Applications. In: Fonseca, C.M., Fleming, P.J., Zitzler, E., Deb, K., Thiele, L. (eds.) EMO 2003. LNCS, vol. 2632, pp. 311–326. Springer, Heidelberg (2003)
5. Deb, K., Thiele, L., Laumanns, M., Zitzler, E.: Scalable Test Problems for Evolutionary Multi-Objective Optimization. TIK-Report No.112, Computer Engineering and Networks Laboratory (TIK), Swiss Federal Institute of Technology (ETH) Zurich (July 2001)

An Effective Dynamical Multi-objective Evolutionary Algorithm for Solving Optimization Problems with High Dimensional Objective Space[*]

Minzhong Liu[1,**], Xiufen Zou[2], and Lishan Kang[3]

[1]College of Computer Science, Wuhan University, Wuhan 430072, China
minzhongliu@163.com
[2]College of Mathematics and Statistics, Wuhan University, Wuhan 430072, China
[3]Department of Computer Science & Technology, China University of Geosciences,
Wuhan 430074, China

Abstract. An effective dynamical multi-objective evolutionary algorithm (DMOEA) based on the principle of the minimal free energy in thermodynamics was proposed in the paper. It provided a new fitness assignment strategy based on the principle of free energy minimization of thermodynamics for the convergence of solves, introduced a density-estimate technique for evaluating the crowding distance between individuals and a new criterion for selection of new individuals to maintain the diversity of the population. By using multi-crossover operator, it improved the search efficiency and the robustness. The test example results proves the validity of the algorithm in its rapidly convergence and maintaining diversity.

Keywords: evolutionary algorithm, multi-objective optimization, minimal free energy, convergence and diversity.

1 Introduction

The multi-objective optimization problems (MOPs) are generally difficult to handle and have received considerable attention in operations research.After the doctoral study of David Schaffer on vector evaluated genetic algorithm (VEGA) in 1984 [1], Goldberg's suggestion of the use of non-dominated sorting along with a niching mechanism [2] generated an overwhelming interest on multi-objective evolutionary algorithms (MOEAs). A number of Pareto-based techniques and elitist algorithms have been proposed in the last decade [3,4], such as Pareto-based ranking procedure (FFGA) [5], niched Pareto genetic algorithm (NPGA) [6], Pareto-archived evolution strategy (PAES) [7], nondominated sorting genetic algorithm (NSGA) [8], and NSGA II [9], the strength Pareto evolutionary algorithm (SPEA) [10], and SPEA2 [11], thermodynamical genetic algorithm (TDGA) [12]. Although these techniques performed well in different comparative studies, there is still a large room for improvement as recent studies have shown [9,13,14].

---

[*] This work was supported by Chinese National Natural Science Foundation grant No. 60573168 and 50677046.
[**] Corresponding author.

L. Kang, Y. Liu, and S. Zeng (Eds.): ISICA 2007, LNCS 4683, pp. 80–89, 2007.

In 15, we propose a dynamical multi-objective evolutionary algorithm (DMOEA) based on the principle of the minimal free energy in statistical physics, for solving multi-objective optimization problems (MOPs). In this paper, we successfully apply DMOEA to MOPs with high dimensional objective space.

The paper is structured as follows: Section 2 provides descriptions of the proposed algorithm. In Section 3, numerical experiments are conducted, the DMOEA is applied to solve the engineering problems and comparisons with other methods are made. Finally, some conclusions and future work are addressed in Section 4.

2   Description of the DMOEA for Multi-objective Optimization

Without loss of generality, we consider the following multi-objective minimization problem with n decision variables(parameters), M objectives and k constrained conditions:

$$\text{minimize} \quad y = H(\vec{x}) = (f_1(\vec{x}), f_2(\vec{x}), \cdots, f_M(\vec{x}))$$

$$\text{subject to} \quad g_i(\vec{x}) \le 0, \ i = 1, 2, \cdots, k; \tag{1}$$

2.1   The New Fitness Assignment Strategy

The DMOEA is based on partial order ranking. The rank value of individual i is the number of solution n_i that dominates solution i [5]. The rank values can be computed as follows:

$$R(i) = |\Omega_i|, \text{ where } \Omega_i = \{\vec{x}_j \mid \vec{x}_j \prec \vec{x}_i, 1 \le j \le N, j \ne i\}. \tag{2}$$

With the rank value, the algorithm is independent of the true objective value of MOPs.

In statistical physics, the Gibbs distribution models a system in thermo-dynamical equilibrium at a given temperature. Further, it is also known that this distribution minimizes the free energy F defined by:

$$F = <E> - TS . \tag{3}$$

Where $<E>$ is the mean energy of the system, S is the entropy and T is the temperature. The minimization of F means minimization of $<E>$ and maximization of TS. It is called "the principle of the minimal free energy".

Such a statistical framework has been introduced into many fields. Since the minimization of the objective function (convergence towards the Pareto-optimal set) and the maximization of diversity in obtained solutions are two key goals in the multi-objective optimization, the working principle of a MOEA and the principle of finding the minimum free energy state in a thermodynamic system is analogous. In order to plunge a multi-objective optimization problem into such a statistical framework, we combine the rank value $R(i)$ calculated by Pareto-dominance relation with Gibbs entropy $S(i)$ to assign a new fitness $F(i)$ for each individual i in the population, that is

$$F(i) = R(i) - TS(i) . \tag{4}$$

where $R(i)$ is the rank value of individual i .In this way, the R(i)=0 corresponds to a non-dominated individual, while a high Rank(i) means that i is dominated by many individuals.

$$S(i) = -p_T(i) \log p_T(i) . \tag{5}$$

where $p_T(i) = \dfrac{1}{Z} \exp(-\dfrac{R(i)}{T})$ is the analogue of the Gibbs distribution,

$Z = \displaystyle\sum_{i=1}^{N} \exp(-\dfrac{R(i)}{T})$ is called the partition function, and N is the population size.

From the expression (4), we easily observe that it is difficult to distinguish the different individuals when their Rank values are equal. Therefore, we use a density estimation technique (which proposed by Deb et al9) to compute an individual's crowding distance. A solution with a smaller value of this distance measure is, in some sense, more crowded by other solutions. So we use the crowding distance to correct the expression of fitness value (4):

$$fitness(i) = R(i) - TS_i - d(i) . \tag{6}$$

In DMOEA, the fitness values are sorting in increasing order. The individual in population which the fitness value is smallest is called "the best individual", and the individual in population which the fitness value is largest is called "the worst individual".

2.2   The New Selection Criterion

In every generation, we always obtain new individuals by genetic operator, but it is worthwhile discussing that we use what kind of way to accept the new individuals, or eliminate the old individuals, and form new population at next generation. Since the DMOEA is based on the thermodynamical principle, we attempt to employ the Metropolis criterion of simulated annealing algorithm (SA)16 and the crowding distance to guide the select process, that is,

(1) If $R(X_{new}) < R(X_{worse})$, then $X_{worst} := X_{new}$

(2) If $R(X_{new}) = R(X_{worse})$ and $d(X_{new}) > d(X_{worse})$, then $X_{worst} := X_{new}$ \hfill (7)

(3) else if $\exp(\dfrac{R_{worst} - R_{new}}{T}) >$ random $(0,1)$, then $X_{worst} := X_{new}$.

Where R_{worst} and R_{new} are respectively the Rank value of the worst individual and the new individual.

3 The Numerical Experiments

3.1 Test Problems

(1) SPH-3

This problem has three objective functions defined on one hundred decision variables.

$$\min \quad f_1(\vec{x}) = (x_1 - 1)^2 + \sum_{i=2}^{n} x_i^2$$

$$\min \quad f_2(\vec{x}) = x_1^2 + (x_2 - 1)^2 + \sum_{i=3}^{n} x_i^2$$

$$\min \quad f_3(\vec{x}) = x_1^2 + x_2^2 + (x_3 - 1)^2 + \sum_{i=4}^{n} x_i^2 , \qquad \text{n=100,}$$

$$subject \quad to \quad x_i \in [-1000, 1000].$$

(2) Water

This problem has five objective functions defined on three decision variables with seven inequation restrictions.

$$\min \quad f_1(\vec{x}) = 106780.37(x_2 + x_3) + 61704.67$$

$$\min \quad f_2(\vec{x}) = 3000x_1$$

$$\min \quad f_3(\vec{x}) = (305700)2289x_2 / (0.06 \times 2289)^{0.65}$$

$$\min \quad f_4(\vec{x}) = (250)2289 \exp(-39.75x_2 + 9.9x_3 + 2.74)$$

$$\min \quad f_5(\vec{x}) = 25(1.39/(x_1 x_2) + 4940x_3 - 80)$$

$$subject \quad to: \quad g_1(\vec{x}) = 0.00139/(x_1 x_2) + 4.94x_3 - 0.08 - 1 \le 0$$

$$g_2(\vec{x}) = 0.000306/(x_1 x_2) + 1.082x_3 - 0.0986 - 1 < 0$$

$$g_3(\vec{x}) = 12.307/(x_1 x_2) + 49408.24x_3 + 4051.02 - 50000 \le 0$$

$$g_4(\vec{x}) = 2.098/(x_1 x_2) + 8046.33x_3 - 696.71 - 16000 \le 0$$

$$g_5(\vec{x}) = 2.138/(x_1 x_2) + 7883.39x_3 - 705.04 - 10000 \le 0$$

$$g_6(\vec{x}) = 0.417/(x_1 x_2) + 1721.26x_3 - 136.54 - 2000 \le 0$$

$$g_7(\vec{x}) = 0.164/(x_1 x_2) + 631.13x_3 - 54.48 - 550 \le 0$$

$$x_1 \in [0.01, 0.45], x_2, x_3 \in [0.01, 0.10].$$

(3) KANG

This problem has four objective functions defined on two decision variables.

$$\min imize\left\{f_1(\vec{x}), f_2(\vec{x}), ..., f_4(\vec{x})\right\} \tag{2}$$

$$f_1(\vec{x}) = -\left(\frac{1}{1+\sqrt{\left((x^2-y^2)\left((x^2-y^2)-12x^2y^2\right)-1\right)^2 + 4x^2y^2\left(3(x^2-y^2)^2 -\right.}}\right.$$

$$f_2(x) = \sqrt{(x+1.01)^2 + y^2}$$

$$f_3(x) = \sqrt{x^2 + (y+1.01)^2}$$

$$f_4(x) = \sqrt{x^2 + (y-1.01)^2}$$

$$x_1 \in [-2, 2], x_2 \in [-2, 2].$$

(4) DTLZ2a

This problem has four objective functions defined on eight decision variables.

$$\min \quad f_1(\vec{x}) = (1+g)\cos(x_1 \pi/2)\cos(x_2 \pi/2)\cos(x_3 \pi/2),$$

$$\min \quad f_2(\vec{x}) = (1+g)\cos(x_1 \pi/2)\cos(x_2 \pi/2)\sin(x_3 \pi/2),$$

$$\min \quad f_3(\vec{x}) = (1+g)\cos(x_1 \pi/2)\sin(x_2 \pi/2),$$

$$\min \quad f_4(\vec{x}) = (1+g)\sin(x_1 \pi/2), \tag{3}$$

$$g = \sum_{i \in \{4, \cdots, 8\}} (x_i - 0.5)^2,$$

$$x_i \in [0, 1], i \in \{1, \cdots, n\}, n = 8$$

3.2 Results and Discussion

The algorithm has been coded in C++ language and implemented on a Pentium PC 1.6GHz 256MB RAM. The EMOEA has been executed in Pentium III. For each problem, 20 runs with different random seeds have been carried out. The main parameter setting is: population size N=100, the maximum generation: g=200 (20000 function evaluations), Temperature T=10000.

The Pareto-Optimization set of problem SPH-3 is 1/8 spherical surface, our optimized solutions distribute uniformly in it(see figure1).

The problem WATER has five objects with seven inequation restrictions, its optimized solutions as in figure 2. Ray et al. 17 studied the problem WATER, and normalized   the   objective   functions   in   the   following   manner:

$f_1/8(10^4), f_2/1500, f_3/3(10^6), f_4/6(10^6), f_5/8000.$ Since there are five objective functions in the problemWATER, we observe the range of the normalized objective function values of the obtained nondominated solutions. Table 1 shows the comparison with NSGA-II and Ray–Tai–Seow's algorithm. In most objective functions,DMOEA has found a better spread of solutions than NSGA-II and Ray–Tai–Seow's approach.

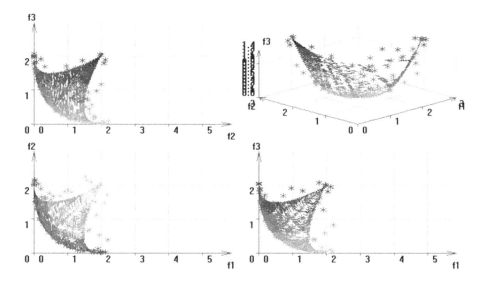

Fig. 1. Optimized solutions obtained using DMOEA for the SPH-3 problem with 4000 generation

Figure3 and figure4 show the obtained non-dominated solutions by DMOEA for the KANG problem. Figure5 shows the obtained solutions for the DTLZ2a problem. The figures shows that DMOEA can obtain widely spread solutions.

Table 1. Minimize and maximize value of non-dominated solutions of Water

Method	f1	F2	f3	f4	f5
DMOEA	0.798	0.02775	0.09512	0.03062	0.0009028
	0.9181	0.90	0.9512	1.036	3.124
NSGA-II	0.798	0.027	0.095	0.031	0.001
	0.920	0.900	0.951	1.110	3.124
Ray-Tai-	0.810	0.046	0.067	0.036	0.211
Seow	0.956	0.834	0.934	1.561	3.116

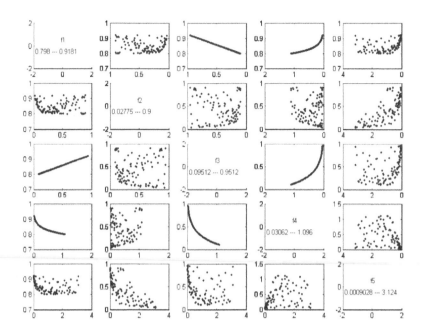

Fig. 2. Optimized solutions obtained using DMOEA for the WATER problem with 500 generation

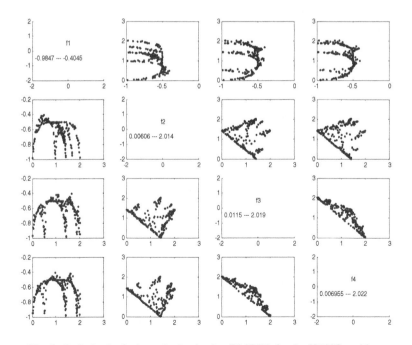

Fig. 3. Optimized solutions obtained using DMOEA for the KANG problem

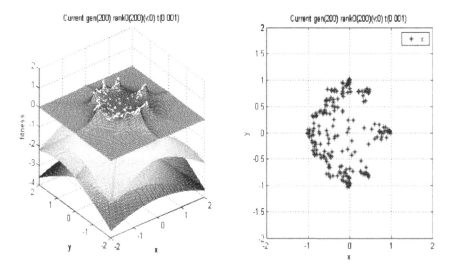

Fig. 4. Optimized solutions obtained using DMOEA for the KANG problem

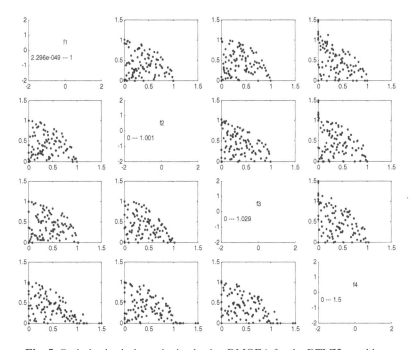

Fig. 5. Optimized solutions obtained using DMOEA for the DTLZ2a problem

4   Conclusions and Future Work

In this paper, we have presented DMOEA, an effective dynamical evolutionary algorithm for multiobjective optimization problems that employs a new fitness assignment strategy and a new accepting criterion based on the principle of the minimal free energy. We have conducted experiments by computer simulation to explore the applicability of DMOEA to the complex problems of multiobjective optimization. Our computational experience shows that the proposed DMOEA can obtain very promising results. In addition, the procedure can be easily extended to solve other real-world application problems with different objectives. Therefore in the future, the applicability of DMOEA to more difficult and complex types of real-world problems, including the combinatorial optimization problems must be studied.

References

1. Schaffer, J.D.: Some Experiments in Machine Learning Using Vector Evaluated Genetic Algorithms, Ph.D.Thesis. Vanderbilt University, Nashville,TN (1984)
2. Goldberg, D.E.: Genetic Algorithms for Search, Optimization,and Machine Learning. Addison-Wesley, Reading, MA (1989)
3. Kalyanmoy, D.: Multi-Objective Optimization using Evolutionary Algorithms. John Wiley & Sons, Chichester,U.K (2001)
4. Coello Coello, C.A., et al.: Evolutionary Algorithms for Solving Multi-Objective Problems. Plenum. Pub. Corp. (2002)
5. Fonseca, C.M., Fleming, P.J.: Genetic algorithms for multi-objective optimization: Formulation, discussion and generalization. In: Forrest, S. (ed.) Proceedings of the Fifth International Conference on Genetic Algorithms, San Mateo, California, pp. 416–423 (1993)
6. Horn, J., Nafploitis, N., Goldberg, D.: A Niched Pareto Genetic Algorithm for Multi-objective Optimization. In: Michalewicz, Z. (ed.) Proceedings of the first IEEE Conference on Evolutionary Computation, pp. 82–87. IEEE Press, Piscataway,NJ (1994)
7. Knowles, J., Corne, D.: The Pareto archived evolution strategy: A new baseline algorithm for multiobjective optimization. In: Proceedings of the 1999 Congress on Evolutionary Computation, pp. 98–105. IEEE Press, Piscataway, NJ (1999)
8. Srinivas, N., Deb, K.: Multiobjective function optimization using nondominated sorting genetic algorithms. Evolutionary Computation 2(3), 221–248 (1995)
9. Deb, K., Pratap, A., Agarwal, S., Meyarivan, T.: A Fast and Elitist Multi-objective Genetic Algorithm:NSGA II. IEEE Transaction on Evolutionary Computation 6(2), 182–197 (2002)
10. Zitzler, E.: Evolutionary algorithms for multiobjective optimization:Methods and Applications, Doctoral dissertation ETH 13398, Swiss Federal Institute of Technology (ETH), Zurich, Switzerland (1999)
11. Zitzler, E., Laumanns, M., Thiele, L.: SPEA2: Improving the Strength Pareto Evolutionary Algorithm. TIK-Report 103, ETH Zentrum, Gloriastrasse 35, CH-8092 Zurich, Switzerland (2001)
12. Kita, H., Yabumoto, Y., Mori, N., Nishikawa, Y.: Multi-objective Optimization by means of Thermodynamical Genetic Algorithm. In: Ebeling, W., Rechenberg, I., Voigt, H.-M., Schwefel, H.-P. (eds.) Parallel Problem Solving from Nature - PPSN IV. LNCS, vol. 1141, pp. 504–512. Springer, Heidelberg (1996)
13. Zitzler, E., Deb, K., Thiele, L.: Comparison of Multi-objective Evolutionary Algorithms: Empirical Results. Evolutionary Computation Journal 8, 125–148 (2000)

14. Zitzler, E., Thiele, L., Laumanns, M., et al.: Performance Assessment of Multiobjective Optimizers: An Analysis and Review. IEEE Transactions on Evolutionary Computation 7, 117–132 (2003)
15. Zou, X.F., et al.: A High Performance Multi-objective Evolutionary Algorithm Based on the Principles of Thermodynamics. In: Yao, X., Burke, E.K., Lozano, J.A., Smith, J., Merelo-Guervós, J.J., Bullinaria, J.A., Rowe, J.E., Tiňo, P., Kabán, A., Schwefel, H.-P. (eds.) Parallel Problem Solving from Nature - PPSN VIII. LNCS, vol. 3242, pp. 922–931. Springer, Heidelberg (2004)
16. Aarts, E.H.H., Korst, J.H.M.: Simulated annealing and Boltzmann machines. John Wiley and Sons, Chichester (1989)
17. Ray, T., Tai, K., Seow, C.: An evolutionary algorithm for multiobjective optimization. Eng. Optim. 33(3), 399–424 (2001)

Operator Adaptation in Evolutionary Programming

Yong Liu

The University of Aizu
Aizu-Wakamatsu, Fukushima 965-8580, Japan
yliu@u-aizu.ac.jp

Abstract. Operator adaptation in evolutionary programming is investigated from both population level and individual level in this paper. The updating rule for operator adaptation is defined based on the fitness distributions at population level compared to the immediate reward or punishment from the feedback of mutations at individual level. Through observing the behaviors of operator adaptation in evolutionary programming, it is discovered that a small-stepping operator could become a dominant operator when other operators have rather larger step sizes. Therefore, it is possible that operator adaptation could lead to slow evolution when operators are adapted freely by themselves.

1   Introduction

Evolutionary programming (EP) is a population-based variant of generate-and-test algorithms [4], in which search operators, such as mutation, are applied to generate new solutions, and a selection scheme is used to decide which solutions to survive to the next generation. Therefore, mutation is a key search operator which generates new solutions from the current ones. How likely an offspring will be generated from a parent or parents is determined by the search bias of a search operator.

Search bias of an evolutionary search operator includes its step size and search directions. The search step size is crucial for the performance of a search operator. In general, different search operators have different search step sizes, and thus are appropriate for different problems as well as different evolutionary search stages for a single problem. Since the global optimum is unknown in real world applications, it is impossible to know a priori what search biases should be used in EP. Operator adaptation is one way to get around this problem by using a variety of different biases and allowing evolution to find out which one(s) are more promising than others [1].

Operator adaptation in evolutionary algorithms could be classified into population level, individual, and component level operator adaptation based on how operators are adapted [5]. At population level, an operator is evaluated from all offspring whom it has generated from different parents, and operators are adapted globally over the whole population. At individual level, operators are

L. Kang, Y. Liu, and S. Zeng (Eds.): ISICA 2007, LNCS 4683, pp. 90–99, 2007.
© Springer-Verlag Berlin Heidelberg 2007

changed specifically for each individual in the population. The difference between the two operator adaptation schemes is that operator updating rules are defined for the whole population in the population-level operator adaptation rather than for each individual in the individual-level operator adaptation. The component-level operator adaptation is done independently in each component of an individual. A typical example of the component-level operator adaptation is self-adaptive mutation used in EP in which each individual is formed by a vector of real components. A parameter is assigned on each component to specify the magnitude of mutation on the component, the value of the parameter is adapted in the evolution process.

Different operator adaptation methods have long been developed for evolutionary algorithms [1]. The results are rather controversial. Some results favor the operator adaptation while others show worse performance led by the operator adaptation. Although more results of operator adaptation have been obtained for genetic algorithms and genetic programming with varieties of operators, few results have been done for EP on function optimization with real values [6,3,2]. One possible reason of having not attracted much attention of operator adaptation in EP might lie in having a fewer operators in EP than other evolutionary algorithms.

The purpose of this paper is to systematically discuss the operator adaptation in EP from both population level and individual, and explain their relations to the operator adaptation at the component level. Fitness distributions are used to measure the performance of operators over the whole population, and to define the updating rule for operator adaptation at population level. The immediate rewards or punishments are given to the operators from the feedback of comparisons between parents and offsprings in operator adaptation at individual level.

The rest of this paper is organized as follows. Section 2 describes the major steps of EP used in this paper. Section 3 gives ideas how operators are adapted at the two levels in EP. Section 4 presents the experimental results and discussions. Finally, Section 5 concludes with some remarks and future research directions.

2   Function Optimization by Classical Evolutionary Programming

A global minimization problem can be formalized as a pair (S, f), where $S \subseteq R^n$ is a bounded set on R^n and $f : S \mapsto R$ is an n-dimensional real-valued function. The problem is to find a point $\mathbf{x}_{min} \in S$ such that $f(\mathbf{x}_{min})$ is a global minimum on S.

According to the description by Bäck and Schwefel [7], the classical EP (CEP) for function optimization is implemented as follows:

1. Generate the initial population of μ individuals, and set $k = 1$. Each individual is taken as a pair of real-valued vectors, $(\mathbf{x}_i, \eta_i)$, $\forall i \in \{1, \cdots, \mu\}$, where $\mathbf{x}_i$'s are objective variables and η_i's are standard deviations for Gaussian mutations (also known as strategy parameters in self-adaptive evolutionary algorithms).

2. Evaluate the fitness score for each individual $(\mathbf{x}_i, \eta_i)$, $\forall i \in \{1, \cdots, \mu\}$, of the population based on the objective function, $f(\mathbf{x}_i)$.
3. Each parent $(\mathbf{x}_i, \eta_i)$, $i = 1, \cdots, \mu$, creates a single offspring $(\mathbf{x}_i', \eta_i')$ by: for $j = 1, \cdots, n$,

$$\eta_i'(j) = \eta_i(j) \exp(\tau' N(0,1) + \tau N_j(0,1)) \tag{1}$$
$$x_i'(j) = x_i(j) + \eta_i(j) N_j(0,1), \tag{2}$$

where $x_i(j)$, $x_i'(j)$, $\eta_i(j)$ and $\eta_i'(j)$ denote the j-th component of the vectors $\mathbf{x}_i$, $\mathbf{x}_i'$, η_i and η_i', respectively. $N(0,1)$ denotes a normally distributed one-dimensional random number with mean 0 and standard deviation 1. $N_j(0,1)$ indicates that the random number is generated anew for each value of j. The factors τ and τ' are commonly set to $\left(\sqrt{2\sqrt{n}}\right)^{-1}$ and $\left(\sqrt{2n}\right)^{-1}$ [7].

4. Calculate the fitness of each offspring $(\mathbf{x}_i', \eta_i')$, $\forall i \in \{1, \cdots, \mu\}$.
5. Conduct pairwise comparison over the union of parents $(\mathbf{x}_i, \eta_i)$ and offspring $(\mathbf{x}_i', \eta_i')$, $\forall i \in \{1, \cdots, \mu\}$. For each individual, q opponents are chosen uniformly at random from all the parents and offspring. For each comparison, if the individual's fitness is no smaller than the opponent's, it receives a "win."
6. Select the μ individuals out of $(\mathbf{x}_i, \eta_i)$ and $(\mathbf{x}_i', \eta_i')$, $\forall i \in \{1, \cdots, \mu\}$, that have the most wins to be parents of the next generation.
7. Stop if the halting criterion is satisfied; otherwise, $k = k+1$ and go to Step 3.

3   Two Levels of Operator Adaptation

3.1   Five Mutations

In CEP described in Section 2, there is only one Gaussian mutation operator defined in Eq. 2. Besides Gaussian mutation, four other mutations are introduced into EP with operator adaptation:

1. Cauchy mutation.
$$x_i'(j) = x_i(j) + \eta_i(j)\delta_j \tag{3}$$
where δ_j is a Cauchy random variable with the scale parameter $t = 1$, and is generated anew for each value of j.
2. Two Lévy mutations:

$$x_i'(j) = x_i(j) + \eta_i(j) L_j(\alpha), \tag{4}$$

where $L_j(\alpha)$ is a random number generated anew for for each value of j from the Lévy distribution with the parameter α. Two Lévy mutations with α at 1.3 and 1.7, respectively, are used in this paper.
3. Single component Gaussian mutation. In this mutation, only one component $x_i(k)$ in x_i, and $\eta_i(k)$ in η_i are randomly selected to be changed by Gaussian mutation while all other components in x_i and η_i remain unchanged:

$$\eta_i'(k) = \eta_i(k) \exp(\tau' N(0,1) + \tau N_k(0,1)), \tag{5}$$
$$x_i'(k) = x_i(k) + \eta_i(k) N_k(0,1), \tag{6}$$

η_i in all mutations except for single component Gaussian mutation are changed in the same way defined in Eq. 1. For simplicity, five operators, i.e. Gaussian mutation, Cauchy mutation, Lévy mutation with $\alpha = 1.3$, Lévy mutation with $\alpha = 1.7$, and single component Gaussian mutation, are denoted as Gaussian, Cauchy, LevyS, LevyL, and OneG, respectively.

3.2   Fitness Distributions

In order to develop an updating rule for operator adaptation in EP, some measurements of performance made by different mutations should be defined. One way to define such measurements is from fitness distributions [9].

Let Ω denote the set of the five mutations defined in Section 3.1, and ω a mutation operator in Ω. The expected improvement EI_ω can be defined as:

$$EI_\omega = E[f(\omega(x)) - f(x)] \tag{7}$$

where $E[.]$ denotes the expectation, f is the objective function to be minimized, and x is a random individual. EI_ω gives the average change in fitness between the parent and offspring generated by ω. By calculating the negative values of EI_ω only when $f(\omega(x))$ is smaller than $f(x)$, the benefit B_ω can be defined as

$$B_\omega = E[min\{0, f(\omega(x)) - f(x)\}] \tag{8}$$

Note that EI_ω only measures the improvement from the parents to the offspring. It does not necessarily show evolvability of a mutation operator that is the ability of producing better offspring from the parents than the best one from the current population. By measuring the improvement made by ω compared to the best individual x_{best}, the absolute benefit AB_ω of ω can therefore be defined:

$$AB_\omega = E[min\{0, f(\omega(x)) - f(x_{best})\}] \tag{9}$$

3.3   Operator Adaptation at Population Level

The idea of operator adaptation at population level is to update the rates of operator based on the performance of operators over the whole population during evolution. At each generation g, the performance of an operator ω could be evaluated by its absolute benefit from Eq.9:

$$AB_\omega(g) = \frac{1}{|S_\omega(g)|} \sum_{x \in S_\omega(g)} min\{0, f(\omega(x)) - f(x_{best}(g))\} \tag{10}$$

where $S_\omega(g)$ denotes the subset of population at generation g that are chosen to be mutated by operator ω, and $x_{best}(g)$ is the best individual in the population at generation g.

At the beginning of EP, all five operators are assumed to be applied at the same rate $R_\omega = \frac{1}{|\Omega|}$. At each generation g, the rate $R_\omega(g+1)$ of operator ω at the next generation $(g+1)$ could be re-defined from the rate $R_\omega(m)$ of operator ω at the previous m generations:

$$R'_\omega(g) = \alpha \times max \left\{ R_{min}, \frac{\sum_{k=1}^{m} AB_\omega(g-k)}{\sum_{\gamma \in \Omega} \sum_{k=1}^{m} AB_\gamma(g-k)} \right\} + (1-\alpha) \times R_\omega(g) \qquad (11)$$

$$R_\omega(g+1) = \frac{R'_\omega(g)}{\sum_{\gamma \in \Omega} R'_\gamma(g)} \qquad (12)$$

where a momentum term α with a small positive value less than 1 is introduced in order to prevent from fluctuations of rates. A lower bound of rate $R_{min} = 0.1/|\Omega|$ is also used to avoid the complete elimination of an operator [8].

In operator adaptation at population level, the updating rule for the rates of operators are designed over the whole population so that all individuals in the population share one common table of rates that contain the rates R_γ of all operators.

3.4 Operator Adaptation at Individual Level

In operator adaptation at individual level, each individual has a table of rate R_γ for all mutation operators. At each generation, each individual selects an operator ω randomly based on the table of rates to generate an offspring. If the operator ω generates a better offspring than its parent, the rate of ω is increased while the rates of other operators γ are decreased at the next generation:

$$R_\omega(g+1) = \alpha \times (1.0 - (m-1) \times R_{min}) + (1-\alpha) \times R_\omega(g) \qquad (13)$$

$$R_\gamma(g+1) = \alpha \times R_{min} + (1-\alpha) \times R_\gamma(g) \qquad (14)$$

If the the operator ω generates a worse offspring than its parent, the rate of ω is decreased while the rates of other operators γ are increased at the next generation:

$$R_\omega(g+1) = \alpha \times R_{min} + (1-\alpha) \times R_\omega(g) \qquad (15)$$

$$R_\gamma(g+1) = \alpha \times (1 - R_{min})/(m-1) + (1-\alpha) \times R_\gamma(g) \qquad (16)$$

4 Experimental Studies

In this section, five different EP algorithms, including EP with operator adaptation at population level (called EPOAP in short), EP with operator adaptation at individual level (named as EPOAI), EP with non-adaptive operator adaptation but simple mixed operators at the fixed rate (named as EPNOA), CEP, and Fast EP (FEP) [6], on a suite of benchmark functions. The same $\alpha = 0.2$ was used in both EPOAP and EPOAI. In EPNOA, each of five mutations is applied at a fixed rate $1/5$ at each generation.

Twenty-three benchmark functions used in [4,7,6] were tested in our experimental studies. Among the twenty-three benchmark functions, there are unimodal functions f_1 to f_7, multimodal functions f_8 to f_{13} with many local minima, and low-dimensional multimodal functions f_{14} to f_{23} with a few local minima.

In all experiments, the same self-adaptive method (i.e., Eq.(1) or Eq.(5)), the same population size $\mu = 100$, the same tournament size $q = 10$ for selection, the same initial $\eta = 3.0$, and the same initial population were used for all five EP algorithms. The initial population was generated uniformly at random in the specified ranges in each test function.

4.1   Performance Comparison

The average results of 50 independent runs of five EP algorithms at generation 2000 are summarized in Table 1. Unexpectedly, for all high-dimensional functions f_1 to f_{13} except f_5, neither EPOAP nor EPOAI could achieve better results than the best among other three EP algorithms with single operator or with the mixed operators at the fixed rates. Among five EP algorithms, FEP had better performace than the rest of them on the most of high-dimensional functions. All five EP algorithms had shown the similar performance on the rest of low-dimensional functions.

Because of no different performance on the low-dimensional functions, the comparisons should be focused on the different performance among the five EP algorithms on the high-dimensional functions f_1 to f_{13}. Between EPOAP and EPOAI, EPOAP performed consistently better than EPOAI. Between EP with operator adaptation (i.e. EPOAP and EPOAI), and EP with mixed operators at the fixed rate (i.e. EPNOA), EPOAP had found better results than EPNOA while EPOAI showed worse results than EPNOA, on most of the high-dimensional functions. Between EP with multiple operators (i.e. EPOAP, EPOAI, and EPNOA) and EP with single operator (i.e. CEP and FEP), EP with single operator had evolved better results on nearly all high-dimensional functions except f_1 and f_5. Particularly, FEP found better results oftener than CEP.

In order to find out why EP with operator adaptation had performed rather poorly, it is worth examining how different operators had been accurately adapted at both the population level and the individual level in the evolution process.

4.2   Operator Adaptation in EPOAP

Fig. 1 shows the adaptive rates of five operators in the EPOAP's evolution process for f_1 and f_{10}.

For f_1, at earlier generations, Cauchy, LevyS, and LevyL with larger step sizes were applied at higher rates than Gaussian and OneG with smaller step sizes. After about 900 generations, the rates of all three large step size mutations started to decay while the rates of Gaussian and OneG started to rise. No sooner had the rates of Gaussian reached their highest value than they fell sharply, and finally became as low as those of Cauchy, LevyS, and LevyL. Meanwhile, OneG started to dominate the mutations among five mutations and most of offspring were generated by OneG.

Five mutations were adapted differently in the evolution process for f_{10}. At the beginning, Gausian mutation obtained rather higher rates while other 4

Table 1. Comparison among EPOAP, EPOAI, EPNOA, CEP, and FEP. The results are the mean best function values found at generation 2000 over 50 runs.

	EPOAP	EPOAI	EPNOA	CEP	FEP
f_1	1.9×10^{-4}	94.73	2.5×10^{-6}	2.9×10^{-5}	2.8×10^{-4}
f_2	0.268	1.362	1.736	0.022	0.067
f_3	573.79	1833.33	870.88	244.01	129.2
f_4	5.02	6.5	6.08	10.89	2.37
f_5	62.32	613.5	66.48	64.76	68.68
f_6	14.84	12.94	0.02	77.46	0
f_7	0.151	0.275	0.363	0.039	0.013
f_8	-10581.0	-10342.3	-10539.2	-8048.8	-11079.0
f_9	2.86	16.8	19.47	85.79	1.76
f_{10}	1.869	2.982	1.131	7.758	0.013
f_{11}	0.051	0.25	0.046	0.117	0.021
f_{12}	0.083	1.574	0.481	0.591	3.2×10^{-6}
f_{13}	0.916	4.19	2.24	0.70	4.8×10^{-5}
f_{14}	1.34	1.38	1.32	1.82	1.24
f_{15}	1.0×10^{-3}	9.8×10^{-4}	1.0×10^{-3}	9.9×10^{-4}	1.1×10^{-3}
f_{16}	-1.03	-1.03	-1.03	-1.03	-1.03
f_{17}	0.398	0.398	0.398	0.398	0.398
f_{18}	3.0	3.0	3.0	3.0	3.0
f_{19}	-3.86	-3.86	-3.86	-3.86	-3.86
f_{20}	-3.24	-3.22	-3.22	-3.25	-3.25
f_{21}	-5.06	-5.06	-5.06	-5.06	-5.06
f_{22}	-5.19	-5.09	-5.09	-5.09	-5.19
f_{23}	-5.34	-5.34	-5.34	-5.45	-5.13

operators remained at very low rates. Once the rate of Gausian mutation climbed to its highest rate, it started to decline quickly. At the same time, OneG was applied more and more. In the whole evolution process, the rates of Cauchy, LevyS, and LevyL were kept as low as about 0.1.

One major reason causing such difference in operator adaptation on minimizing f_1 and f_{10} lies in the different landscapes of f_1 and f_{10}. For f_1 in which there is only one optimum on its landscape, operators with larger step sizes could make bigger improvement from the current population when the current solutions were not so good at the beginning so that adaptation based on the absolute fitness benefit in EPOAP preferred to choose large step size operators such as Cauchy, LevyS, and LevyL. When the current solutions became better after moving closer to the optimum, the chances for large step-size operator to find better solutions became slimmer. By contrast, smaller step size operators such as Gaussian and OneG were still able to find the finer solutions at the higher chances. By mutating one component only rather than changing the whole vector as the four other mutations, OneG had much higher chances to find better offspring than other 4 mutations when the current solutions were already relatively good. It was the reason that OneG could dominate the operator adaptation at later evolution process in f_1.

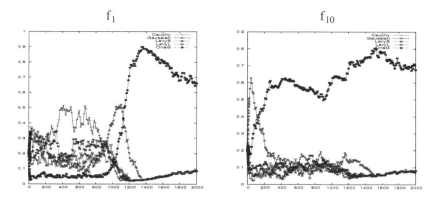

Fig. 1. Operator adaptation in EPOAP on minimizing f_1 and f_{10}. The vertical axis is the probabilities of five operators and the horizontal axis is the number of generations. The left figure show the results on f_1, and the right figure gives the results on f_{10}.

Different to f_1, there are many local optimums on the landscape of f_{10}. Although larger step size mutations such as Cauchy mutation were still capable to generate offspring closer to the global optimum as what they did in minimizing f_1. However, the offspring closer to the global optimum might not be better than those points further away from the optimum. Such a fact led smaller step size operators like Gaussian and OneG to have higher chances to find fitter offspring than other three larger step size mutations in minimizing f_{10}. Therefore, Gaussian and OneG took control at the early stage. The reason that Gaussian had higher rates than OneG is because Gaussian could make larger improvement than OneG even though at slightly lower chances than OneG when the solutions are rather poor at the beginning. When the solutions turned better, only OneG could still manage to find better solutions sometimes while other mutations barely made any improvement.

4.3   Operator Adaptation in EPOAI

Fig. 2 shows the evolution process of operator adaptation in EPOAI for f_1 and f_{10}. For f_1, larger step size operators like Cauchy, LevyS, and LevyL were no longer beneficial from making larger improvement in operator adaptation as what they did in EPOAP. In EPOAI, not absolute fitness benefit but chances of the winning or improvement decided how the operators were adapted. OneG and Gaussian had higher improvement probabilities so that they were chosen more frequently at the beginning. However, Gaussian started shortly to fail more and more in finding better solutions, and its rates were reduced to the level of other 3 large step size operators. It is interesting to see the rates of OneG dropped gradually while the rates of other 4 operators increased slowly at the later evolution process. It is because other 4 operators were rewarded in operator adaptation when OneG was punished by failing more and more in finding better solutions. However, such rewarding from side effect of failure of other operators could not raise the rates of operators too much.

f_1 f_{10}

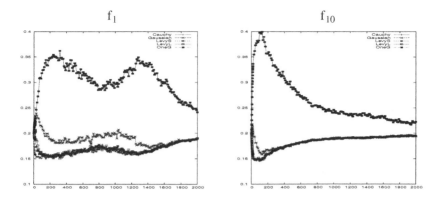

Fig. 2. Operator adaptation in EPOAI on minimizing f_1 and f_{10}. The vertical axis is the probabilities of five operators and the horizontal axis is the number of generations. The left figure show the results on f_1, and the right figure gives the results on f_{10}.

The operator adaptation had similar trend in minimizing f_{10} to what had happened in f_1 at the beginning. No matter in unimodal functions or multimodal functions, the mutations with smaller step sizes would usually have higher probabilities on finding better solutions than larger step size mutations although the improvement on the fitness might be small. However, not the amounts of improvement on fitness but the number of improvement had been adopted to modify the operator rates in EPOAI. It is the reason that the operator adaptation showed the similar trend in both f_1 and f_{10} at the beginning of the evolution.

Although other four mutations had similar adaptation in both f_1 and f_{10} through the whole evolution process, OneG was adapted differently after the initial generations. OneG was able to continuely find better solutions in f_1 since there is only one minimum. However, once the solutions fell in a local minimum in f_{10} with many local minimums, OneG was no longer able to find better solutions so that it was punished more and more while the other 4 mutations were therefore rewarded more frequently.

5   Conclusions

The behaviors of EP with operator adaptation at both population level and individual level have been studied in this paper. It has been found that a small-stepping operator could become a dominant operator when other operators have rather larger step sizes. Once most of the offspring were generated by such small-stepping operators, the evolution prcess could become rather slower.

It is worth pointing out that such small-stepping operators could even slow down the process in EP with mixed operators at the fixed rate as what had happened in EPNOA. Such behaviour was not found when the step sizes of mixed operators are not too small like OneG[3,2].

By comparing operator adaptation to self-adaptation of strategy parameter, there is a similar adaptation behavior between two. Without control, the values of strategy parameter could become extremely small in self-adaptation. It is the reason that lower-bound or adaptive bound should be placed in self-adaptation of strategy parameter. Such similar adaptation suggests that operators should not be adapted freely by themselves without proper control.

References

1. Meyer-Nieberg, S., Beyer, H.-G.: Self-Adaptation in Evolutionary Algorithms. In: Lobo, F.G., Lima, C.F., Michalewicz, Z. (eds.) Parameter Setting in Evolutionary Algorithms, Springer, Heidelberg (2007)
2. Dong, H., He, J., Huang, H., Hou, W.: Evolutionary Programming Using a Mixed Mutation Strategy. Information Sciences 177(1), 312–327 (2007)
3. Lee, C.Y., Yao, X.: Evolutionary programming using the mutations based on on the Lévy probability distribution. IEEE Transactions on Evolutionary Computation 8(1), 1–13 (2004)
4. Fogel, D.B.: System Identification Through Simulated Evolution: A Machine Learning Approach to Modeling. Ginn Press, Needham Heights, MA 02194 (1991)
5. Angeline, P.J.: Adaptive and Self-Adaptive Evolutionary Computations. In: Palaniswami, M., Attikiouzel, Y. (eds.) Computational Intelligence: A Dynamic Systems Perspective, IEEE Press, NJ, New York (1995)
6. Yao, X., Liu, Y., Lin, G.: Evolutionary programming made faster. IEEE Transactions on Evolutionary Computation 3(2), 82–102 (1999)
7. Bäck, T., Schwefel, H.-P.: An overview of evolutionary algorithms for parameter optimization. Evolutionary Computation 1(1), 1–23 (1993)
8. Igel, C., Kreutz, M.: Using Fitness Distributions to Improve the Evolution of Learning Structures. In: Angeline, P.J., Michalewicz, Z., Schoenauer, M., Yao, X., Zalzala, A. (eds.) Proceedings of the Congress on Evolutionary Computation, vol. 3, pp. 1902–1909. IEEE Press, New York (1999)
9. Igel, C., Chellapilla, K.: Fitness Distributions: Tools for Designing Efficient Evolutionary Computations. In: Spector, L., Langdon, W.B., O'Reilly, U.-M., Angeline, P.J. (eds.) Advances in Genetic Programming 3, ch. 9, pp. 191–216. MIT Press, Cambridge, MA, USA (1999)

A Comparison of GAs Using Penalizing Infeasible Solutions and Repairing Infeasible Solutions on Average Capacity Knapsack

Jun He[1] and Yuren Zhou[2]

[1] CERCIA, School of Computer Science, University of Birmingham, Edgbaston
Birmingham B15 2TT, UK
jun.he@ieee.org
[2] School of Computer Science and Engineering, South China University of Technology
Guangzhou 510640, China
yrzhou@scut.edu.cn

Abstract. Different constraint handling techniques have been incorporated with genetic algorithms (GAs), however most of current studies are based on computer experiments. The paper makes an theoretical analysis of GAs using penalizing infeasible solutions and repairing infeasible solutions on average knapsack problem. It is shown that GAs using the repair method is more efficient than GAs using the penalty method on average capacity knapsack problems.

Keywords: Genetic Algorithms, Constrained Optimization, Knapsack problem, Computation Time, Performance Analysis.

1 Introduction

In the real world, almost every optimization problem contains more or less constraints. Different constraint handling techniques have been incorporated with Genetic Algorithms (GAs) [1,2,3,4]. However, most of current studies are based on computer experiments, few theoretical contributions have been make to this research issue. Through computer experiments, Michalewicz has compared GAs using different constraint handling techniques [5], including the penalty function method, repair method and decode method. The following phenomenon was observed in the experiments [5]: for the average capacity knapsack, the penalty function algorithm $A_p[1]$ is the best among five GAs which are proposed in Section 4.5 of [5]. This paper tries to give a theoretical analysis to these experimental results.

There were a few theoretical analysis of running times of GAs for solving the knapsack problem. In [6], the constraint is handled by a multi-objective optimization technique. Two multi-objective evolutionary algorithms, SEMO and FEMO, are applied for a simple instance of the multi-objective 0/1 knapsack problem and the expected running time is analyzed. In [7], the constraints are also transformed into two objectives. A multi-objective evolutionary algorithm,

L. Kang, Y. Liu, and S. Zeng (Eds.): ISICA 2007, LNCS 4683, pp. 100–109, 2007.

REMO, is proposed to solve the knapsack problem. The paper formalizes a $(1+\epsilon)$-approximate set of the knapsack problem and presents a rigorous running time analysis to obtain the formalized set. The running time on a special bi-objective linear function is analyzed. The paper [8] has analyzed the role of penalty coefficients in GAs. Different from the previous research, this paper focuses on a comparison between two constraint-handling techniques.

The paper is organized as follows: Section 2 introduces the knapsack problem and describes GAs for solving the 0-1 knapsack problem. Section 3 compares GAs for average capacity problems. Section 4 reports supplemental experimental results. Section 5 gives conclusions.

2   Knapsack Problem and Genetic Algorithms

The 0-1 knapsack problem [9,5] is to

$$\text{maximize} \quad \sum_{i=1}^{n} p_i x_i,$$

$$\text{subject to} \quad \sum_{i=1}^{n} w_i x_i \leq C, \quad x_i = 0 \text{ or } 1, \quad i = 1, \cdots, n,$$

where

$$x_i = \begin{cases} 1 \text{ if item } i \text{ is selected,} \\ 0 \text{ else,} \end{cases}$$

and p_i is the profit of item i, w_i the weight of item i, and C the capacity of the knapsack.

Due to the limitation of paper length, this paper mainly discusses the uncorrelated average capacity knapsack and leaves other problems in future.

1. *uncorrelated*: p_i and w_i uniformly random in $[1, \nu]$. In this paper, the parameters ν is set to be $\nu = \frac{n}{20}$.
2. *average capacity knapsack:* the capacity of the knapsack is large,

$$C_2 = 0.5 \sum_{i=1}^{n} w_i.$$

The GA is described in [5]. The encoding is a binary representation $(x_1 \cdots x_n)$, where $x_i = 1$ if the i-th item is chosen in the knapsack, $x_i = 0$ if the i-th item is not in the knapsack. The genetic operators used in the GA are bitwise mutation with a mutation rate $1/n$ and an elitist selection. Crossover is not considered in the GA since the analysis of a crossover is too complex.

The fitness of an individual is dependent on the technique used for handling constraints. A general introduction of constraint-handling techniques incorporated with GAs can be found in many references, e.g. [1,5,4].

In this paper, two of them are considered: the methods of penalizing infeasible solutions and repairing infeasible solutions.

For the method using a penalty function, the fitness of an individual consists of two parts:

$$f(x) = \sum_{i=1}^{n} x_i p_i - g(x),$$

where the penalty term $g(x)$ is set to 0 for all feasible solutions, otherwise it is assigned a positive value. Michalewicz [5] used three types of penalty functions in experiments (algorithms $A_p[1]$, $A_p[2]$ and $A_p[3]$ respectively), given as follows:

$$g_1(x) = \ln(1 + \rho(\sum_{i=1}^{n} x_i w_i - C)) \tag{1}$$

$$g_2(x) = \rho(\sum_{i=1}^{n} x_i w_i - C) \tag{2}$$

$$g_3(x) = \rho^2(\sum_{i=1}^{n} x_i w_i - C))^2 \tag{3}$$

where

$$\rho := \max_{i=1,\cdots,n} \{\frac{p_i}{w_i}\}.$$

GAs using the repair method is almost the same as GAs using the penalty method. The only difference is at one point: if an infeasible individual is generated, then it will be repaired and then become a feasible solution.

The fitness function is determined as follows [5]:

$$f(x) = \sum_{i=1}^{n} x_i' p_i,$$

where x' is a feasible individual through repairing the infeasible individual x.

The repairing procedure can be found in [5]. Two repairing methods are used in [5] which are describe as follows:

1. $A_r[1]$: the algorithm takes a random repairing approach, i.e. the **select** procedure is to choose an item from the knapsack randomly.
2. $A_r[2]$: the algorithm takes a greedy repairing approach, i.e first the items are sorted according to the order of p_i/w_i, then the **select** procedure is to choose the smallest item.

Lemma 1. *For algorithm* $A_p[1]$, *denote* $x^* = (x_1^* \cdots x_n^*)$ *to be a global optimal solution and* $I(x^*)$ *the set of items in the optimal knapsack, and* $J(x^*)$ *the items outside the optimal knapsack.*

1. if the following condition holds,

$$\exp(\sum_{i \in J(x^*)} p_i) > 1 + 0.5\rho(\sum_{i=1}^{n} w_i) \tag{4}$$

then there is an infeasible solution y such that

$$f(y) \geq f_{\max} := \max\{f(x), x \text{ is feasible}\}.$$

2. *if the above condition doesn't hold, then for all infeasible solutions y, $f(y) < f_{\max}$.*

Proof: The conclusion can be drawn from the following fact:

$$f(\mathbf{1}) - f(x^*) = \sum_{i=1}^{n} p_i - \sum_{i \in I(x^*)} p_i - \ln(1 + \rho(\sum_{i=1}^{n} w_i - C_1))$$

$$= \sum_{i \notin I(x^*)} p_i - \ln(1 + \rho(\sum_{i=1}^{n} w_i - C_1)).$$

Corollary 1. *If $\mid J(x^*) \mid \geq 3 \ln n$, then the fitness from $g_1(x)$ satisfies 4.*

3 A Comparison Between Repairing Infeasible Solutions and Penalizing Infeasible Solutions

Given an average capacity knapsack, the genetic operators and the fitness function of algorithms A_p and A_r are the same in the feasible solution area. So the behaviors of algorithms A_p and A_r are different only in the infeasible solution area. In the following we only consider infeasible solutions.

Let's investigate the event of algorithms A_p and A_r generating an infeasible solution. Given an individual x, denote items in the knapsack by

$$I(x) = \{i \mid x_i = 1\}$$

and items outside the knapsack by

$$J(x) = \{j \mid x_j = 0\}.$$

The individual x is mutated into an infeasible solution only when at least one item from $J(x)$ is added, (where the added items are denoted by $A(x)$), several (or null) items are removed from $I(x)$ (where the removed items are denoted by $R(x)$), and the constraint is violated:

$$\sum_{i \in I(x)} w_i - \sum_{i' \in R(x)} w_{i'} + \sum_{j \in A(x)} w_j > C_2.$$

In practice, the maximum generation T is always fixed, for example, $T = 500$ in Michalewicz's experiment when $n = 100, 250, 500$ [5]. Therefor it is reasonable to assume that $T \leq 5n$ in the paper.

Let k be the number of items in $J(x)$ added into x simultaneously. The probability of adding K items is no more than

$$\binom{n}{k} \left(\frac{1}{n}\right)^k$$

If k is beyond a constant, for example, $k = \ln n$, then the above probability is very small, so it may seldom happen if the maximum generations $T \leq 5n$.

The analysis of a population-based GA is usually very complex. Here a simple idea is followed: if a better offspring x'' appears after a few generations, then the individual should come from some parent x (due to no crossover). It is necessary to trace how the individual x generates x. The individual x is mutated to an infeasible solution x' first.

- And then in the penalty method, x' survives in the selection and enters the next generation with some probability. In the next generation, it is possibly mutated into the individual x'' with a better fitness $f(x'') > f(x)$, or in more longer generations, x' is mutated to x''.
  From the above analysis, it is seen that at least two generations are needed in the penalty method to generate a better offspring: first an infeasible individual is generated, and then in the next generation or longer, the infeasible individual is mutated into a feasible solution.
- In the repair method, x' is repaired to generate a feasible solution x'' within the same generation.

This idea show it is possible to analyze the behavior of populations through analyzing that of individuals.

Let's start the discussion from adding an item into the individual x.

Event 1. *Given an individual x, one item j from $J(x)$ is added into x (x' is infeasible solution); then an item i is removed from x', and a better offspring x'' is generated.*

Analysis of algorithm $A_r[2]$ let's see when the above event will happen: let the item $i_{\min} \in I(x)$ be the item such that

$$\arg \min_{i \in I(x)} \left\{ \frac{p_i}{w_i} \right\}.$$

There exists some item $j \in J(x)$ which satisfies,

$$p_j > p_{i_{\min}}, \tag{5}$$
$$\frac{p_j}{w_j} \geq \frac{p_{i_{\min}}}{w_{i_{\min}}}. \tag{6}$$

Denote $J_a(x)$ to be the set of all items $j \in J(x)$ satisfying the above conditions.

Starting from the individual x, the repair method $A_r[2]$ chooses an item j from $J_a(x)$ with a probability not less than

$$c \binom{|J_a(x)|}{1} \frac{1}{n} = \Omega \left(\frac{|J_a(x)|}{n} \right).$$

where c is a constant. Then the repair method removes the minimum item $i_{\min}$ and finds a better solution with probability 1.

So if $J_a(x)$ is not empty the probability for $A_r[2]$ to producing a better solution is at least

$$\Omega\left(\frac{\mid J_a(x)\mid}{n}\right). \tag{7}$$

Note: if the set $J_a(x)$ is empty, the above probability is 0, and the repair procedure plays no role in handling infeasible solution.

However since the discussion here is restricted to the early search phase, i.e. $T \leq 5n$, the event of $J_a(x)$ being empty may seldom happen, especially in the initial population or very early phase population.

From the above analysis, Algorithm $A_r[1]$ can find a better solution x quickly until the condition holds: for all $j \in J(x)$,

$$\frac{p_j}{w_j} < \frac{p_{i_{\min}}}{w_{i_{\min}}}.$$

Analysis of algorithm $A_r[1]$ let's see when Event 1 will happen: there exists some item $j \in J(x)$, and j satisfies for some $i \in I(x)$,

$$p_j > p_i \tag{8}$$

Denote $J_b(x)$ to be the set of all item $j \in J(x)$ satisfying the above condition. It is obvious that $J_a(x) \subset J_b(x)$.

Given an item $j \in J_b(x)$, denote the items i with $p_i \leq p_j$ by $I_b(x,j)$.

Start from individual x, Algorithm $A_r[1]$ first adds one item j from $J_b(x)$ with some probability, and then an infeasible solution is generated. The probability of such event happening is not less than

$$c\binom{\mid J_b(x)\mid}{1}\frac{1}{n} = \Omega\left(\frac{\mid J_b(x)\mid}{n}\right),$$

where c is a constant.

Assume one item $j \in J_b(x)$ is added, the algorithm repairs the infeasible solution through removing one item from $I(x)$ at random, it will remove an item $i \in I_b(x,j)$ with a probability

$$\frac{\mid I_b(x,j)\mid}{\mid I(x)\mid}$$

So the probability of generating a better solution is

$$\Omega\left(\sum_{j\in J_b(x)}\frac{1}{n}\frac{\mid I_0(x,j)\mid}{\mid I(x)\mid}\right) \tag{9}$$

Analysis of algorithm A_p it is similar to the analysis of Algorithm $A_r[1]$. Let's see when Event 1 will happen: there exists some $j \in J(x)$, and j satisfies for some $i \in I(x)$,

$$p_j > p_i \tag{10}$$

Still denote $J_b(x)$ to be the same set of all item $j \in J(x)$ satisfying the above condition. Given an item $j \in J_b(x)$, denote the items i with $p_i \leq p_j$ by $I_b(x, j)$.

Starting from x, the probability for x generating an infeasible solution through adding one item $j \in J_b(x)$ is no more than

$$c \binom{|J_b(x)|}{1} \frac{1}{n} = O\left(\frac{|J_b(x)|}{n}\right).$$

And then the individual survives in the selection with some probability, denoted by $q(x')(q(x') < 1)$. This probability is dependent on the selection pressure and the fitness of other individuals in the population.

In the next generation, one item $i \in I_b(x, j)$ is removed and a better individual is generated with a probability no more than

$$c \binom{|I_b(x, j)|}{1} \frac{1}{n}.$$

If the above three events are considered together, the event of generating a better individual in two generations is

$$O\left(q \sum_{j \in J_b(x)} \frac{1}{n} \frac{|I_b(x, j)|}{n}\right). \tag{11}$$

However sometimes no better individual is generated in the second generation, so we should investigate the probability of generating a better individual in the third generation.

Similar to the above analysis, the probability is no more than

$$O\left(q^2 \sum_{j \in J_b(x)} \frac{1}{n} \frac{|I_b(x, j)|}{n}\right). \tag{12}$$

By deduction, the total probability for x to generate a better individual in up to T generations is

$$O\left((q + q + \cdots + q^T) \sum_{j \in J_b(x)} \frac{1}{n} \frac{|I_b(x, j)|}{n}\right), \tag{13}$$

$$= O\left(\frac{q(1 - q^{T+1})}{1 - q} \sum_{j \in J_b(x)} \frac{1}{n} \frac{|I_b(x, j)|}{n}\right), \tag{14}$$

This means the probability of algorithm $A_p[i], i = 1, 2, 3$ producing a better offspring is smaller than the algorithm $A_r[1]$.

From the above upper bound, it is seen that if the selection pressure is smaller, i.e. q is bigger, then the upper bound of the probability of generating a better individual may be larger under Event 1. This means among three algorithms

$A_p[1], A_p[2], A_p[3], A_p[1]$ may have the biggest probability to generate a better offspring.

However, this needs a precondition, i.e. the fitness of all feasible solutions is better that that of infeasible solution. From Lemma 1, for Algorithm $A_p[1]$, if the cardinality $\mid J(x^*) \mid \geq 3 \ln n$, then the problem to maximize the fitness function is not equivalent to the knapsack problem. Since the problem is an uncorrected averaged capacity knapsack, it holds $\mid J(x^*) \mid \geq 3 \ln n$ for many instances, and Algorithm $A_p[1]$ may find some infeasible solution with a large fitness rather a feasible solution. So it may be the worst among three penalty algorithms.

By using the same approach, we analyze other types of events, e.g.

Event 2. *Given an individual x, one item j from $J(x)$ is added into x, and the offspring x' is an infeasible solution. Then two items $i_1, i_1 \in I(x)$ are removed from x', then a better offspring x'' is generated.*

Event 3. *add two items $j_1, j_2 \in J(x)$ to generate a new infeasible individual x' first; and then remove one item i from $I(x)$ to generate a better individual.*
The following conditions hold

$$p_{j_1} + p_{j_2} > \min_{i \in I(x)} \{p_i\}$$

but

$$p_{j_1} < \min_{i \in I(x)} \{p_i\}$$

$$p_{j_2} < \min_{i \in I(x)} \{p_i\}$$

$$\sum_{i \in I(x)} w_i + w_{j_1} + w_{j_2} - \sum_{k \in R(x)} w_k < C_2$$

From the above analysis, it can be seen that starting from the same infeasible solution, Algorithm $A_r[2]$ has the largest probability to find a solution with a better fitness.

4   Experiments

The experiment setting is given as follows: the number of items is 500; the profits and weights in the knapsack is initialized at random; the initial population is generated at random. $\nu = r = n/20$. Bitwise mutation with rate $p_m = 0.05$, the same as in [5]. Crossover rate $p_c = 0.65$. The repairing algorithms $A_r[1]$ and $A_r[2]$ uses repairing rate 1. The maximum number of generation is set to 500.

Table 1 gives the experimental result with number of items being 500. The result is tested on 5 instances, and each is averaged over 10 independent runs. Table 2 gives results with different numbers of items for uncorrelated, weakly correlated and strongly correlated knapsack problems (their definitions may be referred to [9,5]).

From the tables, it is seen that the experimental results have supported the theoretical claim in the previous sections:

Table 1. The experiments results of 5 runs. The above table is for GAs without crossover, and the below table for GAs with crossover. "*" means no feasible solution is found during 500 generations.

experiment	no. of items	$A_p[1]$	$A_p[2]$	$A_p[3]$	$A_r[1]$	$A_r[2]$
1	500	1491.3	3602.3	3533.2	3703.9	5081.8
2	500	1475.5	3303.2	3294.9	3469.4	4746.9
3	500	1708.7	3289.3	3308.3	3524.2	4759.4
4	500	1559.4	3397.4	3372.7	3539.6	4905.0
5	500	1515.9	3529.6	3464.4	3633.3	5017.8
1	500	*	4003.0	3954.2	3812.8	5192.3
2	500	*	3899.0	3892.0	3704.9	5096.3
3	500	*	3894.5	3802.4	3604.5	5015.6
4	500	*	4013.4	3947.7	3701.4	5063.1
5	500	*	3869.2	3885.7	3641.2	5105.2

Table 2. The experiments results with different numbers of items. he above table is for GAs without crossover, and the below table for GAs with crossover. "*" means no feasible solution is found during 500 generations.

Correlation	No. of items	$A_p[1]$	$A_p[2]$	$A_p[3]$	$A_r[1]$	$A_r[2]$
	100	76.1	138.7	135.5	152.9	184.9
none	250	476.9	962.8	930.1	995.4	1289.9
	500	1491.3	3602.3	3533.2	3703.9	5081.8
	100	*	240.8	230.8	244.8	306.1
weak	250	1093.8	1308.3	1282.9	1370.6	1809.9
	500	2439.2	5219.2	5172.5	5473.9	7462.3
	100	403.4	411.0	406.7	409.6	441.9
strong	250	*	2316.9	2292.3	2355.9	2830.1
	500	6968.5	9658.7	9568.6	9836.9	11704.7
	100	129.3	157.8	159.5	165.5	194.5
none	250	221.9	929.0	922.9	978.9	1279.9
	500	3161.5	3661.3	3620.1	3755.2	5015.1
	100	103.4	207.3	194.0	212.7	262.6
weak	250	200.6	1245.8	1161.0	1335.5	1771.6
	500	3244.5	5185.1	5120.1	5391.1	7276.6
	100	155.6	359.4	335.8	389.2	444.0
strong	250	1506.7	2370.0	2368.8	2479.2	2857.6
	500	4881.0	9709.2	9588.6	9977.5	11709.4

- Algorithm $A_r[2]$ is the best.
- Algorithm $A_p[1]$ sometimes cannot find a feasible solution and is worst.
- Among three A_p, $A_p[2]$ is best.

5   Conclusions and Discussions

Michalewicz has made an experiment-based comparison among GAs using different constraint handling techniques on the 0-1 knapsack problem [5]. Such observations need a theoretical explanation. This paper has provided such a theoretical analysis for Michalewicz's experiments [5]. The main result of the paper is that GAs using the repair method is more efficient than GAs using the penalty method on both restrictive capacity and average capacity knapsack problems. The theoretical result about the average capacity knapsack is different from Michalewicz's experimental results. So supplemental experiments have been implemented to support the theoretical claim in the paper.

Since the repair method has used some kind of knowledge about the knapsack problem, this theoretical study has confirmed a general principle used in practice: a problem-specific constraint-handing technique (here the repair method) performs better than a general-purposed techniques (here the penalty method). In combinatorial optimization, the repair method is regarded as the best way in handling constraints [4].

Acknowledgment. The work of J. He was supported by the UK Engineering and Physical Research Council under Grant No. EP/C520696/1. The work of Y. Zhou was supported in part by the National Natural Science Foundation of China under Grant No.60673062, in part by the Natural Science Foundation of Guangdong Province of China under Grant No.06025686.

References

1. Michalewicz, Z., Schoenauer, M.: Evolutionary computation for constrainted parameter optimization problem. Evolutionary Computation 4(1), 1–32 (1996)
2. Michalewicz, Z., Janikow, C.Z.: Genocop: A genetic algorithm for numerical optimization problems with linear constraints. Commun. ACM 39(12es), 175–201 (1996)
3. Koziel, S., Michalewicz, Z.: Evolutionary algorithms, homomorphous mappings, and constrained parameter optimization. Evolutionary Computation 7(1), 19–44 (1999)
4. Coello Coello, C.A: Theoretical and numerical constraint-handling techniques used with evolutionary algorithms: a survey of the state of the art. Computer Methods in Applied Mechanics and Engineering 191(11), 1245–1287 (2002)
5. Michalewicz, Z.: Genetic Algorithms + Data Structure = Evolution Program. Springer, New York (1996)
6. Laumanns, M., Thiele, L., Zitzler, E.: Running time analysis of evolutionary algorithms on a simplified multiobjective knapsack problem. Natural Computing 3(1), 37–51 (2004)
7. Kumar, R., Banerjee, N.: Analysis of a multiobjective evolutionary algorithm on the 0-1 knapsack problem. Theor. Comput. Sci. 358(1), 104–120 (2006)
8. Zhou, Y., He, J.: A runtime analysis of evolutionary algorithms for constrained optimization problems. IEEE Transactions on Evolutionary Computation (2007) (in press)
9. Martello, S., Toth, P.: Knapsack Problems. John Wiley & Sons, Chichester (1990)

About the Limit Behaviors of the Transition Operators Associated with EAs

Lixin Ding[1,2] and Sanyou Zeng[1]

[1] School of Computer, China University of Geosciences,
Wuhan 430074, China
lxding@whu.edu.cn,
sanyou-zeng@263.net
[2] State Key Lab of Software Engineering,
Wuhan University, Wuhan 430072, China
lxding@whu.edu.cn

Abstract. This paper focuses on the limit behaviors of evolutionary algorithms based on finite search space by using the properties of Markov chains and Perron-Frobenius Theorem. Some convergence results of general square matrices are given, and some useful properties of homogeneous Markov chains with finite states are investigated. The geometric convergence rates of the transition operators, which is determined by the revised spectral of the corresponding transition matrix of a Markov chain associated with the EA considered here, are estimated. Some applications of the theoretical results in this paper are also discussed.

1 Introduction

Evolutionary algorithms(EAs for brevity) are a class of useful optimization methods based on a biological analogy with the natural mechanisms of evolution, and they are now a very popular tool for solving optimization problems. An EA is usually formalized as a Markov chain, so one can use the properties of Markov chains to describe the asymptotic behaviors of EAs, i.e., the probabilistic behaviors of EAs if never halted. Asymptotic behaviors of EAs has been investigated by many authors [1−12]. Due to the connection between Markov chains and EAs, a number of results about the convergence of EAs have been obtained by adopting the limit theorem of the corresponding Markov chin in the above works. In this paper, we will make further research on this topic, especially on convergence rate of EAs by using Perron-Frobenius Theorem and other analytic techniques.

The remaining parts of this paper are organized as follows. In section 2, we apply some basic matrix theory, such as Jordan Standard Form Theorem and Perron-Frobenius Theorem etc., to study the convergence of general square matrix A. We obtain that A^n converge with geometric convergence rate defined by the revised spectral of A. In section 3, we concern on homogeneous Markov chains with finite states. We give the relations among states classification, geometric convergence rate and eigenvalues of transition matrix. In section 4, we combine the results in section 2 and section 3 to investigate the limit behaviors of EAs. Under some mild

L. Kang, Y. Liu, and S. Zeng (Eds.): ISICA 2007, LNCS 4683, pp. 110–119, 2007.

conditions, we get that EAs converges to the optimal solution set with geometrical rate which is determined by the revised spectral of corresponding transition matrix of a Markov chain associated with the EA considered in this paper. In section 5, we give some applications of the theoretical results in this paper. Finally, we conclude this paper with a short discussion in section 6.

2   Preliminaries

In this section, we need to collect a number of definitions and elementary facts with respect to matrix classification, matrix decomposition and matrix convergence which will be useful throughout the whole paper. For a detailed reference on matrix theory, see the monograph by Steward[13]

Definition 1. *A $m \times m$ square matrix $\mathbf{A}$ is said to be*
(1) nonnegative($\mathbf{A} \geq 0$), if $a_{ij} \geq 0$ for all $i, j \in \{1, 2, \cdots, m\}$,
(2) positive($\mathbf{A} > 0$), if $a_{ij} > 0$ for all $i, j \in \{1, 2, \cdots, m\}$.

A nonnegative matrix $\mathbf{A} : m \times m$ is said to be
(3) primitive, if there exists a positive integer k such that $\mathbf{A}^k$ is positive,
(4) reducible, if there exists a permutation matrix $\mathbf{B}$ such that

$$\mathbf{BAB}^T = \begin{pmatrix} \mathbf{C} \ \mathbf{0} \\ \mathbf{R} \ \mathbf{T} \end{pmatrix},$$

where square matrix $\mathbf{C}$ and $\mathbf{T}$ are square matrices,
(5) irreducible, if it is not reducible,
(6) stochastic, if $\sum\limits_{j=1}^{m} a_{ij} = 1$ for all $i \in \{1, 2, \cdots, m\}$.
A $m \times m$ stochastic matrix $\mathbf{A}$ is said to be
(7) stable, if it has identical rows.

Definition 2. *For a square matrix $\mathbf{A} : m \times m$ with eigenvalues $\lambda_1, \cdots, \lambda_m$, its revised spectral gap is usually defined as $r(\mathbf{A}) = \max\{|\lambda_i| : |\lambda_i| \neq 1, i = 1, \cdots, m\}$, and its norm is defined as $||\mathbf{A}|| = \max\{|a_{ij}| : i, j = 1, \cdots, m\}$.*

The following two Lemmas are well-known and can be found in many literatures of matrix theory.

Lemma 1 (Jordan Standard Form Theorem). *Suppose that square matrix $\mathbf{A} : m \times m$ has r different eigenvalues $\lambda_1, \cdots, \lambda_r$. Then there exists an invertible matrix $\mathbf{B}$ such that*

$$\mathbf{B}^{-1}\mathbf{AB} = \mathbf{J} \equiv diag[\mathbf{J}(\lambda_1), \cdots, \mathbf{J}(\lambda_r)],$$

where

$$\mathbf{J}(\lambda_i) = \begin{pmatrix} \lambda_i \ 0 \ \cdots \ 0 \ 0 \\ 1 \ \lambda_i \ 0 \ \cdots \ 0 \\ \cdots\cdots\cdots\cdots\cdots \\ 0 \ \cdots \ 1 \ \lambda_i \ 0 \\ 0 \ 0 \ \cdots \ 1 \ \lambda_i \end{pmatrix}$$

$$\in \mathbf{C}^{n(\lambda_i) \times n(\lambda_i)}, 1 \le i \le r,$$

and $\sum\limits_{i=1}^{r} n(\lambda_i) = m$.

Lemma 2 (Perron-Frobenius Theorem). *For any nonnegative square matrix* $\mathbf{A} : m \times m$, *the following claims are true.*

(1) There exists a non-negative eigenvalue λ *such that there are no other eigenvalues of* $\mathbf{A}$ *with absolute values greater than* λ;

(2) $\min\limits_{i}(\sum\limits_{j=1}^{m} a_{ij}) \le \lambda \le \max\limits_{i}(\sum\limits_{j=1}^{m} a_{ij}).$

By using the above matrix theorems, we can get the following convergence results about $\mathbf{A}^n$ immediately.

Proposition 1. *Suppose that* 1 *is a simple eigenvalue of square matrix* $\mathbf{A} : m \times m$ *and all other eigenvalues have absolute values less than* 1. *Then* $\lim\limits_{n \to \infty} \mathbf{A}^n$ *exists and has geometric convergence rate.*

Proposition 2. *Suppose that square matrix* $\mathbf{A} : m \times m$ *has* m *linear independent eigenvectors and its eigenvalues except* 1 *have absolute values less than* 1. *Then* $\lim\limits_{n \to \infty} \mathbf{A}^n$ *exists and has geometric convergence rate determined by* $r(\mathbf{A})$.

3   Homogeneous Markov Chains with Finite States

Let $\mathbf{P}$ be the transition matrix associated with Markov Chain $\{X_n; n \ge 0\}$ defined on a finite state space $S = \{s_1, s_2, \cdots, s_m\}$. We will also classify the state space in the following.

Definition 3. *(1) a vector:* $v = (v_1, \cdots, v_m)$ *is called a probability vector if* $v_i \ge 0$ *and* $\sum\limits_{i=1}^{m} v_i = 1$,

(2) a probability vector v *is called an invariant probability measure(stationary distribution) of transition matrix* $\mathbf{P}$: *if* $v\mathbf{P} = v$.

The following notations are usually needed to classify the states of Markov chains.

$f_{ij}^n \doteq P\{X_0 = i, X_1 \ne j, \cdots, X_{n-1} \ne j, X_n = j\}$, is the probability that Markov chain starts at state s_i and reaches state s_j at time n for the first time;

$f_{ij}^* \doteq \sum\limits_{n=1}^{\infty} f_{ij}^n$, is the probability that Markov chain starts at s_i and reaches s_j after finite steps;

$m_{ii} \doteq \infty$, if $f_{ii}^* < 1$; otherwise $m_{ii} \doteq \sum\limits_{n=1}^{\infty} n f_{ii}^n$;

$d_i \doteq$ the biggest common divisor of $\{n : p_{ii}^n > 0\}$, is called the period of state s_i

Definition 4. *The state s_j is called a*
(1) transient state, if $f_{jj}^* < 1$;
(2) recurrent state, if $f_{jj}^* = 1$;
(3) positive recurrent, if $m_{jj} < \infty$;
(4) zero recurrent, if s_j is not a positive recurrent;
(5) aperiodic, if $d_i = 1$.

In the following, we will further describe the states classification of Markov chains. Let $N \subset S$ be the collection of all transient states of S, R^+ be the collection of all positive recurrent states, and R^0 be the collection of all zero recurrent states of S. Then $S = N \bigcup R^0 \bigcup R^+$. Furthermore, R^0 and R^+ can be divided into some irreducible sub-classes, that is, $R^0 = R_1^0 + \cdots + R_i^0$ and $R^+ = R_1^+ + \cdots + R_j^+$.

For Markov chain with finite states, it is well-known that

$$\lim_{k \to \infty} \frac{1}{k} \sum_{l=1}^{k} P_{ij}^l = \Pi_{ij}, \forall i, j \in S. \tag{1}$$

Researchers can refer to relative limit theorems, such as Proposition 3.3.1 in [14]. Moreover, since $\mathbf{P}$ is finite dimensional, hence the limit distribution Π is also a transition matrix on S.

Definition 5. *The subset $E \subset S$ is closed if $i \in E, j \notin E$, which implies that $p_{ij} = 0$, i.e., if $i \in E$ then $\sum_{j \in E} P_{ij} = 1$. The state space S is called reducible, if S have no-empty closed subset; otherwise, S is irreducible.*

In fact, S is reducible(irreducible) $\Leftrightarrow$ transition matrix $\mathbf{P}$ on state space S is reducible(irreducible).

We have another important fact that if every positive state of $\mathbf{P}$ is aperiodic, then $\lim_{k \to \infty} \mathbf{P}^k$ exists. Combining Proposition 1 and Proposition 2 as well as Theorem 16.0.1 and Theorem 16.0.2 in [14], we can get the following conclusion immediately.

Proposition 3. *Give a Markov chain with transition matrix $\mathbf{P} : m \times m$ on finite state space, for the following statements*
(1) $\mathbf{P}$ is aperiodic,
(2) $\mathbf{P}^k$ has geometric convergence rate,
(3) 1 is a simple eigenvalue and all other eigenvalues have absolute values less than 1,
(4) P has m linearly independent eigenvectors and and the eigenvalues except 1 have absolute values less than 1,
then the relations among them are that

$$(1) \Leftrightarrow (2); \quad (3) \Rightarrow (2); \quad (4) \Rightarrow (2).$$

For a reducible stochastic matrix, there is a very important convergence theorem given by M. Iosifescu[15], which is

Lemma 3. *Let* **P** *be a reducible stochastic matrix, where* **C** *is a primitive stochastic matrix and* **R**, **T** $\neq$ **0**. *Then*

$$P^\infty = \lim_{k\to\infty} P^k = \begin{pmatrix} \mathbf{C}^\infty & \mathbf{0} \\ \mathbf{R}_\infty & \mathbf{0} \end{pmatrix}$$

is a stable stochastic matrix.

In the following, Π is always defined as in Proposition 1 or Proposition 2. It is obvious that

$$\Pi\mathbf{P} = \mathbf{P}\Pi = \Pi = \Pi^2.$$

Thus, we have $(\mathbf{P} - \Pi)^k = \mathbf{P}^k - \Pi, \forall k \geq 1$. Moreover, by Proposition 1 and 2, **P** has geometric convergence rate, hence $\sum_{k=1}^\infty \|\mathbf{P}^k - \Pi\| < \infty$. Thus, if let $\mathbf{Z} = \mathbf{I} + \sum_{k\geq1} (\mathbf{P} - \Pi)^k = \frac{\mathbf{I}}{\mathbf{I} - \mathbf{P} + \Pi}$, then **Z** is **well-defined** and $\mathbf{Z} = (\mathbf{I} - \mathbf{P} + \Pi)^{-1}$. We can prove that **Z** has the following properties.

Proposition 4. *(1)* $(\mathbf{I} - \mathbf{P})\mathbf{Z} = \mathbf{Z}(\mathbf{I} - \mathbf{P}) = \mathbf{I} - \Pi$,
   (2)$\Pi\mathbf{Z} = \Pi, \mathbf{Z}\mathbf{1} = \mathbf{1}$,
   (3) all eigenvectors of **P** *are those of* **Z**; *moreover, if* $r_i(\neq 1)$ *is a eigenvalue of* **P**, *then* $\frac{1}{1-r_i}$ *is the eigenvalue of* **Z**.

It is easy to know from Perron- Frobenius theorem that if **P** is a transition matrix, then 1 is a eigenvalue of **P** and there is no other eigenvalues with absolute values greater than 1. This fact implies that $r(\mathbf{P}) \leq 1$.

4   Asymptotic Behaviors of Evolutionary Algorithms

In this section, we consider the following optimization problem: Given an objective function $f : S \to (-\infty, \infty)$, where $S = \{s_1, s_2, \cdots, s_M\}$ is a finite search space. A maximization problem is to find a $x^* \in S$ such that

$$f(x^*) = \max\{f(x) : x \in S\}. \tag{2}$$

We call x^* an **optimal solution** and write $f_{\max} = f(x^*)$ for convenience. If there are more than one optimal solution, then denote the set of all optimal solutions by S^* and call it an **optimal solution set**. Moreover, **optimal populations** refer to those which include at least an optimal solution and the **optimal population set** consists of all the optimal populations, write it by C^*.

An evolutionary algorithm with population size $N(\geq 1)$ for solving the optimization problem (2) can be generally described as follows:

   step 1.   initialize, either randomly or heuristically, an initial population of N individuals, denoted it by $\xi_0 = (\xi_0(1), \cdots, \xi_0(N))$, where $\xi_0(i) \in S, i = 1, \cdots, N$, and let $k = 0$.

step 2.   generate a new (intermediate) population by adopting genetic operators (or any other stochastic operators for generating offsprings), and denote it by $\xi_{k+1/2}$.

step 3.   select N individuals from populations $\xi_{k+1/2}$ and ξ_k according to certain select strategy , and obtain the next population ξ_{k+1}, then go to step 2.

For convenience, we write that

$$f(\xi_k) = \max\{f(\xi_k(i)) : 1 \leq i \leq N\}, \forall k = 0, 1, 2, \cdots,$$

which represents the maximum in populations $\xi_k, k = 0, 1, 2, \cdots$.

It is well-known that $\{\xi_k; k \geq 0\}$ is a Markov chain with the **state space** S^N because the states of the $(k+1) - th$ generation only depend on the $k - th$ generation. In this section, we assume that the stochastic process, $\{\xi_k; k \geq 0\}$, associated with an EA, is a **homogeneous Markov chain**, and denote its transition probability matrix by $\mathbf{P}$. It is easy to check the following results.

Remark 1. If the selection strategy in step 3 of the EA can lead to the fact that

$$f(\xi_k) \leq f(\xi_{k+1}), \tag{3}$$

then the corresponding transition matrix $\mathbf{P}$ is reducible.

The selection with the property of equation (3) is the so-called elitist selection, which insures that if the population has reached the optimal solution set, then the next generation population cannot reach any other states except those in the optimal population set. In practice, a lot of EAs have this kind of property. Hence, we always assume that EAs considered here possess the property of equation (3).

Remark 2. If population size $N = 1$ and the optimization problem has only one optimal solution, then

$$\Pi = \mathbf{P}^\infty = \begin{pmatrix} 1 & 0 & \cdots & 0 \\ 1 & 0 & \cdots & 0 \\ \cdots\cdots\cdots\cdots \\ 1 & 0 & \cdots & 0 \end{pmatrix}.$$

Remark 3. If population size $N \geq 1$ and the optimization problem has only one optimal solution, then

$$\Pi = \mathbf{P}^\infty = \begin{pmatrix} a_{11} & a_{12} & \cdots & a_{1m} & 0 & \cdots & 0 \\ a_{21} & a_{22} & \cdots & a_{2m} & 0 & \cdots & 0 \\ \cdots & \cdots & \cdots & \cdots & \cdots\cdots\cdots \\ a_{q1} & a_{q2} & \cdots & a_{qm} & 0 & \cdots & 0 \end{pmatrix},$$

where $q = M^N$, and the former m elements in matrix $\mathbf{P}$ exactly correspond to the m optimal states.

The remark 2 and 3 can be followed by Lemma 3 immediately.

Remark 4. For any initial distribution v_0, $v_k \doteq v_0 \mathbf{P}^k \to (b_1, b_2, \cdots, b_m, 0, \cdots, 0)$ $(k \to \infty)$, which implies that $P(\lim_{k\to\infty} \xi_k \in C^*) = 1$, that is, EAs converge to optimal population set in probability.

In the following, we will prove the main results in this paper.

Theorem 1. *Suppose the optimization problem has only one optimal solution* x^* *and the population size* $N = 1$. *If* $P\{\xi_1 = x^* | \xi_0 = s_j\} > 0$ *for all* $s_j \neq x^*$, *then*

(1) *all states except* x^* *are transient;*

(2)x^* *is positive recurrent and aperiodic;*

(3) $\mathbf{P}^k$ *converges, and if writing the limit by* Π, *then*

$$\Pi = \begin{pmatrix} 1 & 0 & \cdots & 0 \\ 1 & 0 & \cdots & 0 \\ \cdots & \cdots & \cdots & \cdots \\ 1 & 0 & \cdots & 0 \end{pmatrix}.$$

Proof. Note that $P(\xi_1 = x^* | \xi_0 = s_j) > 0$ and $P(\xi_1 = x^* | \xi_0 = x^*) = 1$. So, we have $f_{jj}^* < 1$ for all $s_j \neq x^*$, which means that $s_j(\neq x^*)$ is transient. This completes the proof of (1) of this theorem.

Since $\mathbf{P}$ is finite dimensional matrix, the positive recurrent states are not empty. Hence, x^* must be positive recurrent by (1) of this theorem. Combine the above fact and $P(\xi_1 = \xi^* | \xi_0 = \xi^*) = 1$, we get that x^* is aperiodic. This is (2) of this theorem.

By Remark 1, we know that $\lim_{k\to\infty} \mathbf{P}^k$ exists and the limit Π has the given form of (3) of this theorem. □

In order to deal with more complicate cases, such as f is not 1-1 and population size $N \geq 1$, we will introduce the following **analytic techniques**.

Denote the elements in image space of f by $I_f = \{y_1, \cdots, y_q\}$. For $i = 1, \cdots, q$, the **level sets** of original state space S^N are defined by

$$S_i = \{(x_1, \cdots, x_N) \in S^N : \max\{f(x_1), \cdots, f(x_N)\} = y_i\}.$$

Define new transition matrix $\overline{\mathbf{P}}(k)$ on new state space $\{S_1, S_2, \cdots, S_q\}$ by

$$\overline{p}_{ij}(k) = \frac{\sum_{x \in S_i, z \in S_j} P(\xi_{k+1} = z, \xi_k = x)}{\sum_{x \in S_i} P(\xi_k = x)}, \forall S_i, S_j.$$

We can check that $\overline{p}_{ij}(k) = \overline{p}_{ij}(1) \doteq \overline{p}_{ij}, \forall k \geq 1$, which means that $\overline{\mathbf{P}}(k)$ is homogenous. In particular, let $C^* = \{(s_1, \cdots, s_N) \in S^N : \max\{f(s_1), \cdots, f(s_N)\} = f_{max}\}$ be the optimal population set. Then

$$\overline{p}_{ij} = 0, \quad if \ S_i = C^*, S_j \neq C^*;$$

$$\overline{p}_{ii} = 1, \quad if \ S_i = C^*.$$

Consider new stochastic process $\{\bar{\xi}_k; k \geq 1\}$ defined on new state space $\bar{S} = \{S_1, \cdots, S_q\}$, the distribution of $\bar{\xi}_k$ is given by $P\{\bar{\xi}_k = S_i\} = P\{\xi_k \in S_i\}$. Obviously, $\{\bar{\xi}_k; k \geq 0\}$ is a homogenous Markov chain with transition matrix $\bar{\mathbf{P}}(k)$. We can get the following general results immediately.

Theorem 2. *If $P\{\bar{\xi}_1 = C^* | \bar{\xi}_0 = S_j\} > 0$ for all $S_j \neq C^*$, then transition matrix $\bar{\mathbf{P}}$ has the following properties*

(1) all states in new state space except C^* are transient;

(2) C^* is positive recurrent and aperiodic;

(3) $\lim\limits_{k \to \infty} \bar{\mathbf{P}}^k$ exists, and if writing the limit by $\bar{\Pi}$ then

$$
\bar{\Pi} = \begin{pmatrix} 1 & 0 & \cdots & 0 \\ 1 & 0 & \cdots & 0 \\ \cdots\cdots\cdots\cdots \\ 1 & 0 & \cdots & 0 \end{pmatrix}.
$$

Theorem 3. *If $P\{\bar{\xi}_1 = C^* | \bar{\xi}_0 = S_j\} > 0$ for all $S_j \neq C^*$, then transition matrix $\bar{\mathbf{P}}^k$ has geometric convergence rate determined by $r(\bar{\mathbf{P}})$.*

5 Applications

We consider the following linear function

$$
f(x) = w_0 + \sum_{i=1}^{n} w_i x_i, \tag{4}
$$

where $x = (x_1, x_2, \cdots, x_n)$ is a binary string, $w_n > w_{n-1} > \cdots > w_1 > 0, w_0 > 0$. In this example, there is only one global optimal solution $x^* = (1, \cdots, 1)$ for the problem (4) and $I_f = \{w_0, w_0 + w_1, w_0 + w_2, w_0 + w_1 + w_2, \cdots\}$.

We assume that a standard $(N + N)$−**EA** will be used to solve the problem (4).

First, we check the case that $n = 2, N = 2$. Note that, at this case, each bit in individual will remain unchangeable with probability $\frac{1}{2}$. The level set of S^N are

$$
S_0 = \{(00, 00)\},
$$
$$
S_1 = \{(00, 10), (10, 00), (10, 10)\},
$$
$$
S_2 = \{(00, 01), (01, 00), (01, 01), (01, 10), (10, 01)\},
$$
$$
S_3 = \{(00, 11), (11, 00), (11, 10), (10, 11), (11, 01), (01, 11), (11, 11)\},
$$

and we know that $C^* = S_3$. Denote the number of elements in set A by $\sharp A$. Then

$$
P\{\bar{\xi}_1 = C^* | \bar{\xi}_0 = S_j\} > 0
$$

for all $S_j \neq C^*$ and

$$
P\{\bar{\xi}_1 = S_0 | \bar{\xi}_0 = S_0\} = P\{\xi_1 = (00, 00) | \xi_0 = (00, 00)\} = \frac{1}{16},
$$

$$P\{\overline{\xi_1} = S_1|\overline{\xi_0} = S_1\} = \frac{P\{\overline{\xi_1} = S_1, \overline{\xi_0} = S_1\}}{P\{\overline{\xi_0} = S_1\}} = \frac{\sum\limits_{x,y \in S_1} P\{\xi_1 = y, \xi_0 = x\}}{\sum\limits_{x \in S_1} P\{\xi_0 = x\}}$$

$$= \frac{\sum\limits_{x,y \in S_1} P\{\xi_1 = y|\xi_0 = x\} \cdot P\{\xi_0 = x\}}{\frac{1}{4}\sharp S_1} = \frac{7}{24},$$

$$P\{\overline{\xi_1} = S_2|\overline{\xi_0} = S_2\} = \frac{3}{5}, P\{\overline{\xi_1} = S_3|\overline{\xi_0} = S_3\} = 1.$$

Thus, $r(\overline{P}) = \frac{3}{5}$. By Theorem 3, $\overline{P}$ has geometric convergence rate determined by $\frac{3}{5}$.

Second, we still consider optimal problem (4) with $n = 2, N = 2$. But in this time, the mutation probability $p = \frac{2}{3}$, that is each bit in individual will remain unchangeable with probability $\frac{1}{3}$. Using the same techniques as the above, we have

$$P(\overline{\xi_1} = S_0|\overline{\xi_0} = S_0) = \frac{1}{81},$$
$$P(\overline{\xi_1} = S_1|\overline{\xi_0} = S_1) = \frac{1}{3} \times \frac{31}{81},$$
$$P(\overline{\xi_1} = S_2|\overline{\xi_0} = S_2) = \frac{19}{27},$$
$$P(\overline{\xi_1} = S_3|\overline{\xi_0} = S_3) = 1.$$

Thus, $r(\overline{P}) = \frac{19}{27}$ and it is easy to check that $\overline{P}_{4\times4}$ has 4 linear independent eigenvectors. Hence by Theorem 3, $\overline{P}$ has geometric rate of convergence which determined by $\frac{19}{27}$.

From the above examples, we know that the practical convergence speed of the EA depends on parameters assignments strongly.

6   Concluding Remarks

This paper confirms mathematically some results on asymptotic behaviors of evolutionary algorithms. Several important facts of the asymptotic behaviors of evolutionary algorithms, which make us understand evolutionary algorithms better, are proved theoretically. In fact, there are still a number of open problems for the further investigation such as, what effect on asymptotic behaviors will be brought by selection strategy, genetic operators and population size, respectively; the question of non-asymptotic behaviors(when the number of iterations depends in some way of the population size); and others. Probably, one can think of many variants and generalization of the algorithm, but the results we obtained in this paper incite us to go on studying simplified models of evolutionary algorithms in order to improve our understanding of their asymptotic behaviors.

Acknowledgments. This work is supported in part by the National Natural Science Foundation of China(Grant no. 60204001), Chengguang Project of Science and Technology for the Young Scholar in Wuhan City (Grant no. 20025001002) and the Youthful Outstanding Scholars Foundation in Hubei Prov. (Grant no. 2005ABB017).

References

1. Agapie, A.: Theoretical analysis of mutation-adaptive evolutionary algorithms. Evolutionary Computation 9, 127–146 (2001)
2. Cerf, R.: Asympototic Convergence of Genetic Algorithms. Advances in Applied Probablity 30, 521–550 (1998)
3. He, J., Kang, L.: On the convergence rate of genetic algorithms. Theoretical Computer Science 229, 23–39 (1999)
4. Lozano, J.A., et al.: Genetic algorithms: bridging the convergence gap. Theoretical Computer Science 229, 11–22 (1999)
5. Nix, A.E., Vose, D.E.: Modeling genetic algorithms with Markov chains. Annals of Mathematics and Artificial Intelligence 5, 79–88 (1992)
6. Poli, R., Langdon, M.: Schema theory for genetic programming with one-point crossover and point mutation. Evolutionary Computation 6, 231–252 (1998)
7. Poli, R.: Exact schema theory for genetic programming variable-length genetic algorithms with one-point crossover. Genetic Programming and Evolvable Machines 2, 123–163 (2001)
8. Rudolph, G.: Convergence analysis of canonical genetic algorithms. IEEE Transactions on Neural Networks 5, 96–101 (1994)
9. Rudolph, G.: Convergence Properties of Evolutionary Algorithms. Verlag Dr. Kovac, Hamburg (1997)
10. Schmitt, L.M.: Theory of genetic algorithms. Theoretical Computer Science 259, 1–61 (2001)
11. Suzuki, J.: A further result on the Markov chain model of genetic algorithms and its application to a simulated annealing-like strategy. Man and Cybernetics-Part B 28, 95–102 (1998)
12. Vose, D.: The Simple Genetic Algorithms: Foundations and Theory. MIT Press, Cambridge (1999)
13. Steward, G.W.: Introduction to Matrix Computation. Academic Press, New York (1973)
14. Meyn, S.P., Tweedie, R.L.: Markov Chains and Stochastic Stability, 3rd edn. Springer, New York (1996)
15. Isoifescu, M.: Finite Markov Processes and Their Applications. Wiley, Chichester (1980)

Differential Evolution Algorithm Based on Simulated Annealing

Kunqi Liu[1,2], Xin Du[3,2], and Lishan Kang[1,3]

[1] School of Computer, China University of Geosciences, Wuhan, 430074, China
[2] Department of Computer Science, Shijiazhuang University of Economics, Shijiazhuang, 050031, China
[3] State Key Laboratory of Software Engineering, Wuhan University, Wuhan 430072, China
liu_kq@126.com

Abstract. Differential evolution algorithm is a simple stochastic global optimization algorithm.In this paper, the idea of simulated annealing is involved into original differential evolution algorithm and a simulated annealing-based differential evolution algorithm is proposed. It is almost as simple for implement as differential evolution algorithm, but it can improve the abilities of seeking the global excellent result and evolution speed. The experiment results demonstrate that the proposed algorithm is superior to original differential evolution algorithm.

1 Introduction

Differential Evolution Algorithm（DEA）is a kind of Evolution Algorithm, which solves the continuous global optimization problem basing on population difference. It was proposed by Rainer Storn and Kenneth Price in 1995 for solving Chebyshev multinomial. Although this algorithm surpassed all the others on the first session of evolutionary computation match at Nagoya in 1996, it aroused people's attention until 2000. The studies and application of this algorithm have become a new hot point at abroad, but the study of it at home is still very few at present.

However, the mutation operation of DEA amends the value of every individual based on difference vector of population. With the increase of evolution generation, the difference information of every individual is lessening gradually so that convergence speed becomes slow even sometimes falls into local optimal point. In order to overcome above defects, people put forward some improved DEA, for instance Lampinen proposed improved DEA in 2001, Huiyuan Fan and Lampinen proposed DEA based on triangle mutation in 2003, Thomsen proposed Niche DEA based on squeezing in 2004.GaoFei proposed DEA of population extinction based on space contraction in 2004.And Liu Guangming proposed improved EDA. These algorithms improve DEA from different aspects and improve the convergence speed and precision of DEA in different degree. In this paper, the idea of simulated annealing is involved into DEA and a differential evolution algorithm based on simulated annealing is proposed. This algorithm combines DEA with global optimization ability, simple realization and simulated annealing algorithm with

L. Kang, Y. Liu, and S. Zeng (Eds.): ISICA 2007, LNCS 4683, pp. 120–126, 2007.

stronger local search ability and skipping out of local optimal solution, so it avoids the defect of DEA with falling into local extreme point easily, improves the convergence speed and precision of algorithm at later period. It can be shown that DEA based on Simulated Annealing is better than original DEA from experimental results of some standard test functions.

2   Differential Evolution Algorithm

DEA is a kind of evolutionary algorithm based on real code. It gets middle population by applying the difference of current individual to recombine, and then lets middle population and its father population crossover directly and produces new population, lets new population and its father population compete directly and produces the next population. The whole structure of DEA is similar to genetic algorithm. Comparing with genetic algorithm, the main distinction is mutation operation. The mutation operation of DEA is based on differential vector of chromosome and the rest operation is similar to genetic algorithm. For explain it in detail, taking DEA/best/2/bin for example.

It produces N points as initial population in n dimension space randomly, evaluates their fitness. The best individual is recorded as Xbest.

(1) Mutation. First, it selects four individual: Xa,Xb,Xc,Xd randomly and gets Xabcd=Xa-Xb+Xc-Xd   ,which reflects differential degree of population. Then the noise is added on Xbest by Xabcd and gets the noise Vi=Xbest+F×Xabcd (F is the given contract and enlargement factor in advance).

(2) Crossover. For every individual Xi, Letting Vi crossover and produce individual XT and XT has at least a component which is got by corresponding component of Xi. As for other components, according to crossover factor pc , if rand(0,1)<pc, then they are got by corresponding components of Vi ,else they are got by corresponding components of Xi.

(3) Displace. If the fitness of XT is better than Xi's, Xi will be replaced by XT, else Xi will be kept. So the scale of population will not change after crossover.

Comparing with other evolutionary algorithm, DEA has some advantages, such as fast convergence speed, strong stability, simple operation and realization. But it has some defects, such as slow convergence speed at later stage of evolution and sometimes falling into local optimal point easily.

3   Simulated Annealing Algorithm

The basic idea of simulated annealing algorithm derives from the annealing principle of thermodynamics. It was first proposed by Metropolis etc in 1953. And in 1982, S.Kirkpatrick, C.D.Gelatt Jr. and M.P.Vecchi bring it into solving optimal problems and invent simulated annealing algorithm.

Simulated annealing algorithm is a kind of stochastic optimization algorithm based on the strategy of Monte Carlo iteration. It begins with a higher initial temperature T0, searches randomly in solution space using Metropolis sampling strategy with probability rush character and finds global optimal solution finally with temperature declining constantly. The procedure of this algorithm is as follows:

(1) Initialize the anneal temperature Tk(let k=0),produce initial stochastic solution x0 ;

(2) Repeat the following operation at the temperature Tk , until reaching the balance state of temperature Tk:

Produce new feasible solution x ´ in the field of solution x ;

Compute the difference value $\triangle$f between target function f(x ´) of x ´ and target function f(x) of x ;

Receive x ´ according to the probability of {1,exp(-$\triangle$f/Tk)}>random[0,1], random[0,1] is a random number among the region [0,1].

(3) Annealing operation: Tk+1←α×Tk, k←k+1,the annealing speedα∈(0,1). If it satisfies convergence criteria, then the procedure of annealing ends, else turns to (2).

Annealing temperature T controls solving procedure to progress towards the direction of optimal value and receives inferior solution with the probability of exp(-$\triangle$f/Tk).So it can skip out of local extreme value point. As long as it has high enough initial temperature and the annealing procedure is slow enough, it can find global optimal solution.

Simulated Annealing has the ability of stronger local searching and skipping out of local optimal solution. Although some researchers have combined simulated annealing and other optimized algorithm, they have not combined simulated annealing and DEA.

4   Differential Evolution Algorithm Based on Simulated Annealing

The distinction between this algorithm and differential evolution algorithm is selection operation. The selection operation of differential evolution algorithm is the pattern with "greed" selection, only when the function value of new individual is better, it can be kept in next population. Otherwise the father of new individual will be kept in next population. However the selection mode of differential evolution algorithm uses Metropolis criteria: Computes the fitness difference $\triangle$E between new and old individual. If $\triangle$E≤0, then the old individual will be replaced by new individual, else judging whether exp(-$\triangle$E/αT) is bigger than random[0,1] or not, if it is ,then the old individual will be replaced by it , else new individual will be abandoned and old individual will be kept.

The basic procedure of Differential Evolution Algorithm Based on Simulated Annealing is as follows:

(1) Produce initial population randomly and set initial temperature T, cooling speedα, crossover probability pc ;

(2)Evaluate every individual's fitness of population and compute the best individual ;

(3) Produce middle population according to mutation and crossover operation of differential evolution;

(4) Selection operation. For every individual, compute fitness difference $\triangle E$ between middle and old generation. If $\triangle E \leq 0$ or exp(-$\triangle E/\alpha T$) is larger than random[0,1], then the corresponding old individual will be replaced by the individual of middle generation and perform annealing operation: $T(t+1) \leftarrow \alpha * T(t)$, else new individual will be abandoned. Producing new population after this operation;

(5) Compute the best individual of new population and compare it with the best one of old population ,if it is prior to the old best individual, then the best individual of old population will be replaced by it ;

(6) If current generation <max generation, then turn to (3), else turn to (7) ;

(7) Output current best individual and then end.

5 Simulation Experiment

This sector will take solving minimum value of five standard test functions as examples and evaluate the performance of Differential Evolution Algorithm Based on Simulated Annealing according to computer simulation. The hard and software environment are : Petium(R) 4, CPU 2.20GHZ, 256MB memory and programming environment of VC++6.0.In DEA, popsize is 50, crossover probability pc is 0.3 and contract and enlargement factor F is 0.5. However in SADEA, popsize is 50, crossover probability pc is 0.3 and contract ,enlargement factor F is 0.5 ,initial annealing temperature T is 1000000 and annealing speed $\alpha$is 0.3.

(1) $f(x,y)=100*(x^2+y^2)^2+(1-x)^2,x,y\in[-2.048,2.048]$. This function has a minimum value 0 at point (1,1).

Average global minimum value, average minimum value and average evolution generation of DEA and SADEA running 50 times continuously are given by Table1.

Table 1. The result of function1

Function name	average global minimum value	average minimum value point	average evolution generation
DEA	0.000000000003	(0.999997,0.999996)	148
SADEA	0.000000000001	(0.999999,0.999992)	105

(2)$f(x,y)= (x^2+y^2) /4000-cos(x)*cos(y/\sqrt{2})+1$, $x,y \in [-600.0,600.0]$.This function has a minimum value 0 at point (0,0).

Average global minimum value, average minimum value and average evolution generation of DEA and SADEA running 50 times continuously are given by Table2.

Table 2. The result of function2

Function name	average global minimum value	average minimum value point	average evolution generation
DEA	0.000000000002	(-1.434464e-6, -2.192027e-6)	140
SADEA	0.000000000001	(5.64485e-7, 2.166715e-6)	105

(3) $f(x, y) = 0.5 + (\sin\sqrt{x*x + y*y} - 0.5)/(1 + 0.001*(x^2 + y^2))^2$, x,y
$\in$[-100.0,100.0]. This function has a minimum value 0 at point (0,0).

Average global minimum value, average minimum value and average evolution generation of DEA and SADEA running 50 times continuously are given by Table3.

Table 3. The result of function3

Function name	Average global minimum value	average minimum value point	average evolution generation
DEA	0.000071251250	(-4.02243e-2, 3.3344003e-3)	3000
SADEA	0.000002563571	(-4.0532e-7, -5.42977e-7)	3000

(4) $f(x,y)=x^2+y^2-10*\cos(2\pi x)-10*\cos(2\pi y)+20$,x$\in$[-100.0,100.0].This function has a minimum value 0 at point (0,0).

Average global minimum value, average minimum value and average evolution generation of DEA and SADEA running 50 times continuously are given by Table4.

Table 4. The result of function4

Function name	average global minimum value	average minimum value point	Average evolution generation
DEA	0.000000000001	(-3.3323e-8, $-$5.4096e-8)	113
SADEA	0.000000000001	($-$6.291e-9, 7.2983e-8)	98

(5) $f(x) = \sum x_i^2/40000 - \prod\cos(x_i/\sqrt{i}) + 1$, xi$\in$[-1000.0,1000.0].This

function has infinite extreme point and is hard to optimize. It is often used to evaluate evolution algorithm's exploration and development performance. This function has a minimum value 0 at point (0,0).

Average global minimum value, average minimum value and average evolution generation of DEA and SADEA running 50 times continuously are given by Table5.

Table 5. The result of function5

Function name	average global minimum value	average minimum value point	average evolution generation
DEA	0.000223349702	(0.590332031250,0.572021484375)	30000
SADEA	0.000000285500	(5.581112e-6,3.77762318e-4)	30000

Evolution curve comparison conditions of two algorithms are given by Figure1. Lateral axis denotes evolution generation and vertical axis denotes fitness of function.

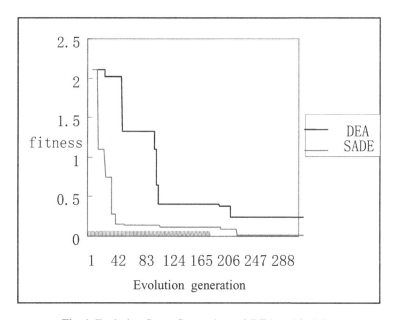

Fig. 1. Evolution Curve Comparison of DEA and SADEA

As can be seen from comparison results of five functions, the results of SADEA are all prior to DEA's and the convergence speed of SADEA is also prior to DEA's. As can be seen from Table5 and Figure1, SADEA can skip out of local optimal point in some degree and this is also prior to DEA.

6 Conclusion

DEA is a simple stochastic global optimization algorithm .In this paper, the idea of simulated annealing is involved into DEA.Receiving middle population after mutation and crossover according to simulated annealing mechanism and keeping good individual. The probability of receiving inferior is decreasing gradually with the decline of temperature and the progress of evolution process. So the convergence performance is improved. It is almost as simple for implement as differential evolution algorithm, but can improve the abilities of seeking the global excellent result, evolution speed and convergence precision. The experiment results demonstrate that the proposed algorithm is superior to original differential evolution algorithm.

Acknowledgments

This work is supported by the National Natural Science Foundation of China (Nos. 60473081 , 60473037), the National Natural Science Foundation of Education Department of Hebei Province of China under Grant (Nos.2004454, 2005338) and the High Science and Technology Research of Hebei Province of China under Grant (No.05213567, 06213562).

Reference

1. Stron, R., Price, K.: Differential evolution-a simple and efficient adapative scheme for global optimization over continuous spaces.Technical Report TR-95-012,ICSI (1995)
2. Babu, B.V., Jehan, M.M.L.: Differential evolution for multiobjective optimization [J]. Evolutionary Computation 4, 8–12 (2003)
3. Gao, F.: Differential evolution algorithms with extinction based on space contraction[J]. Complex systems and complexity science 19(1), 49–52 (2004)
4. Liu, G.: Differential evolution algorithms and modification[J]. Systems Engineering 23(2), 108–111 (2005)
5. Wang, F.S, Jang, H.J.: Parameter estimation of a bioreaction model by hybrid differential evolution [J]. Evolutionary Computation 1, 16–19 (2000)
6. Fan, H., Lampinen, J.: A trigonometric mutation operation to differential evolution [J]. Journal of Global Optimization 27, 105–129 (2003)
7. Thomsen, R.: Multimodaloptimization using crowding-based differential evolution. In: Proceedings of the 2004 Congress on Evolutionary Computation, vol. 2, pp. 1382–1389 (1389)
8. Hrstka, O., Kucerova, A.: Improvements of real coded genetic algorithms based on differential operators preventing premature convergence[J]. Engineering Software 35, 237–246 (2004)
9. Lampinen, J.: A constraint handling approach for the differential evolution algorithm [J]. Evolutionary Computation 2, 12–17 (2002)
10. Kang, L.: Non-numeric parallel algorithm (first volume) – Simulated annealing algorithm [M]. Science Publisher, Beijing (1997)
11. Wang, X., Wang, Y.: The combination of Simulated annealing algorithm and genetic algorithm[J]. Chinese Journal of Computers 20(4), 381–384 (1997)

A Novel Memetic Algorithm for Global Optimization Based on PSO and SFLA

Ziyang Zhen[1], Zhisheng Wang[1], Zhou Gu[2], and Yuanyuan Liu[1]

[1] College of Automation Engineering, Nanjing University of Aeronautics and Astronautics,
Nanjing, 210016, China
{zhenziyang, wangzhisheng, liuyuanyuan}@nuaa.edu.cn
[2] College of Power Engineering, Nanjing Normal University, Nanjing, 210042, China
guzhouok@yahoo.com.cn

Abstract. Memetic algorithms (MAs) which mimic culture evolution are population based heuristic searching approaches for the optimization problems. This paper presents a new memetic algorithm called shuffled particle swarm optimization (SPSO), which combines the learning strategy of particle swarm optimization (PSO) and the shuffle strategy of shuffled frog leaping algorithm (SFLA). In the proposed algorithm, the population is partitioned into several memeplexes according to the performance, and the memotypes in each memeplex evolve according to the self-learning and the learning from the best memotype of the memeplex. Furthermore, the memeplexes are shuffled and separated again to continue the evolutionary process. The combination approach contributes to the local exploration and the global exploration of SPSO. Experimental studies on the continuous parametric benchmark problems show the robustness and the global convergence property of the proposed memetic algorithm.

Keywords: Memetic algorithm; particle swarm optimization; shuffled frog leaping algorithm; global optimization.

1 Introduction

Global optimization problems, including combinatorial problems, constrained and unconstrained problems, arise from many practical applications, such as management science, engineering, computer science, and so on. General mathematical methods are incapable of searching the guaranteed optimum solutions to solve various global optimization problems in short time. In the field of global optimization, evolutionary algorithms (EAs) were investigated and proved to be powerful in solving these problems in the last decades. Evolutionary algorithms are stochastic search techniques with biological foundations. In addition to the genetic algorithm (GA) [1], other evolutionary algorithms with different basis of natural biological evolution have been developed, such as memetic algorithm (MA) [2], particle swarm optimization (PSO) [3-4], ant colony optimization (ACO) [5], shuffled frog leaping algorithm (SFLA) [6], and so on.

L. Kang, Y. Liu, and S. Zeng (Eds.): ISICA 2007, LNCS 4683, pp. 127–136, 2007.

Memetic algorithms is a special class of heuristic searching methods, which are considered as the evolutionary culture derived from the group behaviors of human and animals. Meme is considered as a unit of cultural evolution. Memetic algorithm is a marriage between the population based global search and the heuristic local search made by individuals [7-8]. The population consists of a number of memotypes carrying information about the culture ideas as the chromosomes carrying information about the genes in genetic algorithm (GA) [9]. The memetic algorithm was first put forward by Moscato in 1989, in which the evolutionary algorithms were combined with simulated annealing (SA) for the local search [2]. In recent years, memetic algorithms are used to solve the nondeterministic polynomial (NP) optimization problems such as multistage capacitated lot sizing problem [10], cell formation problem [11], channel assignment in wireless FDMA systems [12], and many others.

In this paper, a new memetic algorithm which combines the ideas of the particle swarm optimization and the shuffled frog leaping algorithm is investigated. After condensed descriptions of history, application and mathematical formulas of each evolutionary algorithm, comparison tests with four widely used benchmark functions are conducted in the end of the paper.

2 Particle Swarm Optimization

Particle swarm optimization, originally developed by Kennedy and Eberhart in 1995, is a population based stochastic optimization technique. PSO is inspired by the social behaviors of animals such as bird flocks and fish schools. Each individual named particle, adjusts its velocity associated with its own experience and the publicized knowledge, which contributes to making the flock rush toward the destination effectively. PSO can be easily and inexpensively computed, moreover, it has been proved to be an efficient method for many global optimization problems, not suffering the difficulties encountered by other evolutionary computation (EC) techniques [3-4].

In the population, each particle is treated as a point in a D-dimensional search space, and the best previous positions of itself and the population are represented as x_pbest and x_gbest respectively. The velocity v and the position x of the particle are calculated according to the following equations.

$$v^{k+1} = wv^k + c_1 r_1^k (x\_pbest - x^k) + c_2 r_2^k (x\_gbest - x^k) \tag{1}$$

$$x^{k+1} = x^k + v^{k+1} \tag{2}$$

where w is called inertia weight, c_1 and c_2 are the two positive constants, called cognitive and social coefficient respectively, r_1 and r_2 are the two random sequences in the range $[0,1]$, k is the iteration number.

The inertia weight w provides a balance between the global and the local exploration abilities, which is crucial for the convergence of PSO. Generally, the inertia weight w is set to be a large value to promote the ability of global search in initial iterations, and is decreased with time variant in order to get more refined solutions. Shi provided a linear decreasing weight approach in which the w is usually set around w_{max} initially and varies linearly to w_{min}, as expressed by following equation [13].

$$w = w_{max} - k(w_{max} - w_{min})/k_{max} \tag{3}$$

The parameters c_1 and c_2 determine the learning rate of the cognitive and the social components, and are often both set to be same level in experiments and applications [14].

The parameters r_1 and r_2, randomly distributed in the range $[0,1]$, impact the particle's learning degree from the experience of itself and the population. The random strategy improves the diversity of the particles' accelerations to increase the volume of the search space. A pseudo code for PSO procedure is summarized in following.

```
Begin;
    Initialize the values of the parameters c₁, c₂, w_max,
    w_min, k_max, and the population size P;
    Initialize the positions and the velocities of the
    particles randomly;
    For each particle;
       Calculate its fitness;
    Record the best previous positions: x_pbest and
    x_gbest;
    For each particle;
       Update the position and velocity according to (1),
       (2);
       Compute its fitness;
    Update the best previous positions: x_pbest and
    x_gbest;
    Repeat for iterations until the terminal conditions
    being satisfied;
End.
```

3 Shuffled Frog Leaping Algorithm

Shuffled frog leaping algorithm, developed by Eusuff and Lansey in 2000, is a population based memetic meta-heuristic for combinatorial optimization [6]. SFLA has been tested on several discrete combinatorial problems and found to be effective in searching the global solutions [15-16]. Eusuff and Lancey designed a computer model SFLANET, which combines SFLA and the hydraulic simulation software EPANET, to solve the water distribution network optimization problem which is a NP problem with high complexity [15].

One analogy for SFLA is based on the memes evolution carried by a culture population such as craftsmen, who are developing the ideas about the pottery through the exchange of information among the interactive individuals. The other analogy for SFLA is a group of frogs leaping in a swamp for searching a stone with maximum amount of food.

In the SFLA, the population is separated into several memeplexes, and in each memeplex there is a sub-memeplex including a number of frogs. Usually, the better the frog's position is, the higher the probability of the frog selected in the sub-memeplex will be. The position of the frog with worst performance in sub-memeplex is updated through learning from the best frog of the sun-memeplex or the population. Mixing of memeplexes is accomplished after a number of iterations for the local search happened in each memeplex. In view of there facts, the SFLA draws on the learning idea from PSO as a local search tool and the mixing idea from shuffled complex evolution (SCE-UA) algorithm as a global search tool [18].

Suppose that the search space is D-dimensional , and the population consists of F frogs which are sorted in order of decreasing performance value. According of this, the population is partitioned into m memeplexes, each containing n frogs. After that q $(q \leq n)$ frogs in the memeplex are selected to make up a sub-memeplex. Higher weight generated by a triangle probability distribution expressed in (4) is assigned to the frog with higher performance, according to which q frogs are selected out. And then the new position of the worst frog (q-th frog) in the sub-memeplex is calculated according to (5).

$$p_j = 2(n+1-j)/[n(n+1)] \tag{4}$$

$$x\_worst^{k+1} = x\_worst^k + r^k(x\_sbest - x\_worst^k) \tag{5}$$

where x_worst is the position of the worst frog in the sub-memeplex, x_sbest is the position of the best frog in the sub-memeplex, r is a random number in range $[0,1]$, k is the iteration number of the sub-memeplex, $j = 1,...,n$. If this process produces a better solution then replace the old position of the worst frog by the new one, otherwise, the worst frog learns from the best frog of the whole population, then the calculation of the new position can be expressed by

$$x\_worst^{k+1} = x\_worst^k + r^k(x\_gbest - x\_worst^k) \tag{6}$$

where x_gbest is the best position of the population having reached so far. If this action still cannot produce a better performance, then the new position of the worst frog is randomly produced. A pseudo code for SFLA procedure is summarized in following.

```
Begin;
    Initialize the values of the parameters F, m, n, q;
    Initialize the positions of the frogs;
    For each frog;
        Compute its fitness;
    Sort the frogs in order of decreasing fitness values;
    Record the global best position as x_gbest;
    Divide the population into m memeplexes;
    For each memeplex;
        Construct a sub-memeplex includeing n frogs by the
        selection strategy in (4);
```

```
    For the sub-memeplex;
        Record the position of the best frog as x_sbest;
        Update the position of the worst frog x_worst
        using (5);
        If it has better performance, replace it;
        Else update its position by (6);
            If it has better performance, replace it;
            Else randomly generating a new one;
        Update the memeplex and sort in descending order
        of performance;
    Repeat for a specific evolving iterations for
    memeplexes;
    Shuffle the population and sort in descending order
    of the performance;
    Repeat for generations until the terminal conditions
    being satisfied;
End.
```

4 Proposed Memetic Algorithm Based on PSO and SFLA

SFLA is a new member in the family of memetic algorithms [6]. In SFLA, the different memeplex, being considered as different culture, performs a local search. However, SFLA has a disadvantage of high computational cost, because both of the memeplexes and the population evolve for a number of generations. In PSO, particles have effective learning approach by introducing the concepts of velocity and accelerate.

In this paper, a new memetic algorithm called shuffled particle swarm optimization (SPSO) is presented, the basis of which is the theory of shuffle strategy in SFLA and adaptive learning strategy in PSO. In SPSO, the population consists of several memeplexs, and all the memotypes learn from the best one of the memeplex, after that, all the memeplexes are shuffled and separated to new memeplexes again. It should be mentioned that the dividing method for memeplexes is similar with that of SFLA. The learning actions of the memotypes in memeplex will help find better local solutions, and the self-learning actions will improve the convergence speed and help escape the local minima. Shuffling the memeplexes will promote the communication among the memotypes which enhances the global exploration. The update of the learning rate z and the culture level of each memotype x can be expressed by

$$z^{k+1} = c_1 r_1^k z^k + c_2 r_2^k (x\_mbest - x^k) \qquad (7)$$

$$x^{k+1} = x^k + z^{k+1} \qquad (8)$$

where x_mbest represents the best level in memeplex, c_1 and c_2 are the learning coefficients. r_1 and r_2 are random numbers which distribute in range $[0,1]$. It should be noted from (7) that the learning rate of the memotype consists of two parts, the first part is the self-learning rate which means that the memotype is capable of maintaining the inertia learning, the second part is the learning rate from the best memotype in the memeplex. It is obvious that there is one memeplex including the global best memotype of the population, which conduces the population to move toward the global best location. A pseudo code for PSO procedure is summarized as follows.

```
Begin;
    Initialize the learning rate coefficients c₁, c₂, and
    the level of each memotype in population with size of
    M randomly;
    For each memotype;
        Calculate its fitness value;
    Sort the population in descending order of fitness
    values;
    Record the global best culture level of population as
    x_gbest ;
    Divide the population into m memeplexes;
    For each memeplex;
        Record the best culture level of memeplex as
        x_mbest ;
        Update the learning rate and culture level of each
        memotype using (7), (8);
        For each memotype;
            calculate its fitness value;
        Sort memotypes in descending order of performance,
        and update the memeplex;
    Shuffle the memeplexes;
    Update the global best culture level x_gbest ;
    Repeat this evolving process until the terminal
    conditions are satisfied.
End;
```

5 Experiments and Results

5.1 Continuous Global Optimization

In this sub-section, the proposed memetic algorithm SPSO is evaluated on a set of four widely used benchmark functions with high complexity, as summarized in Table 1. Most of the test functions have several local minima, but the global minima of the functions are all equal to 0.

The parameters setting of PSO are performed: $P = 200$, $w_{max} = 1.2$, $w_{min} = 0.1$, $k_{max} = 200$, $c_1 = c_2 = 1.0$. The parameters setting of SFLA are performed: $F = 200$, $m = 10$, $n = 20$, $q = 10$, and the generation number for the sub-memeplex is assigned as 1 (SFLA1) and 5 (SFLA2). The parameters setting of SPSO are carried out: $M = 200$, $m = 10$, therefore the number of memotypes in each memeplex is 20, $c_1 = c_2 = 1.0$. Each function optimized by each algorithm is tested by 100 times, and each trial is terminated after 200 iterations reached. It is considered to be successful when the accuracy of 10^{-3} is achieved. For each function and each algorithm, the experimental results are recorded with successful times (ST), mean generation (MG) required for success, mean value (MV) of the optimized function, minimum (MIN) function value, and standard deviation (STD) of the function values by 100 times.

Table 1. Benchmark optimization test functions

Function	Function Expression
Sphere	$F(x) = \sum_{i=1}^{N} x_i^2, \quad x_i \in [-500, 500], \quad N = 30$
Rosenbrock	$F(x) = \sum_{i=1}^{N-1} [100(x_i^2 - x_{i+1})^2 + (x_i - 1)^2], \quad x_i \in [-5, 10], \quad N = 3$
Rastrigin	$F(x) = 10N + \sum_{i=1}^{N} (x_i^2 - 10\cos(2\pi x_i)), \quad x_i \in [-5.12, 5.12], \quad N = 5$
Ackley	$F(x) = 20 + e - 20\exp(-\frac{1}{5}\sqrt{\frac{1}{N}\sum_{i=1}^{N} x_i^2}) - \exp(-\frac{1}{N}\sum_{i=1}^{N} \cos(2\pi x_i)),$ $x_i \in [-15, 30], \quad N = 30$

5.2 Results and Discussions

Fig.1 shows the mean convergence process of the test functions. Table 2 summarizes the optimal results of the four test functions obtained by SFLA, PSO, SPSO. As can be seen from the experimental results, several points can be concluded: (1) PSO tends to trap in the local minima in initial stage, and obtain a near-optimum solution with fast convergence speed; (2) SFLA will have better performance of the global search when the generations of the memeplexes increase which makes it spend much more time for computation; (3) Comparing with PSO and SFLA, SPSO has the advantage of the fastest convergence speed and far away from the local minima. Therefore, it requires less time for searching the optimum or near-optimum solutions to the problem.

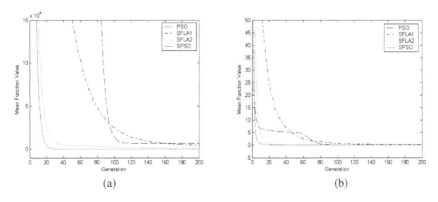

(a)   (b)

Fig. 1. Mean convergence curves of the global optimization functions by SFLA, PSO, SPSO: (a) Sphere; (b) Rosenbrock; (c) Rastrigin; (d) Ackley.

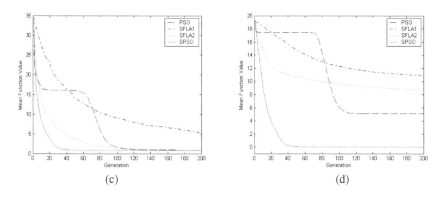

(c)   (d)

Fig. 1. (*continued*)

Table 2. Comparison results of PSO, SFLA, and SPSO

Function	Property	PSO	SFLA		SPSO
Sphere	ST	0	0	0	100
	MG	-	-	-	70.9100
	MV	6.7334×10^3	4.4556×10^3	839.3703	1.0846×10^{-19}
	MIN	4.9746×10^3	1.4086×10^3	215.0205	3.3811×10^{-21}
	STD	3.7859×10^3	1.5777×10^3	400.4633	2.6220×10^{-19}
Rosenbrock	ST	35	0	19	14
	MG	124.1143	-	82.1579	34
	MV	0.0100	0.1590	0.1590	0.0558
	MIN	3.3305×10^{-5}	0.0029	1.6699×10^{-8}	6.2475×10^{-12}
	STD	0.0215	0.1693	0.0728	0.0556
Rastrigin	ST	37	0	41	35
	MG	122.4324	-	128.6098	35.4000
	MV	0.8660	5.2932	0.7153	0.8955
	MIN	0	0.8857	7.1054×10^{-14}	0
	STD	0.8902	2.1795	0.8597	0.8544
Ackley	ST	0	0	0	99
	MG	-	-	-	73.0808
	MV	5.1123	10.8709	8.6868	0.0116
	MIN	3.5703	8.4299	6.8233	2.2995×10^{-12}
	STD	0.8943	0.6755	0.7288	0.1155

6 Conclusion

In this study, a new memetic algorithm named shuffled particle swarm optimization (SPSO) is investigated. SPSO draws the learning strategy from PSO and the memeplex shuffling strategy from SFLA. Comparisons among PSO, SFLA, and SPSO are made in the optimization of some continuous benchmark functions. Results of the experiments verify the effectiveness of the proposed memetic algorithm. In general, SFLA would have better ultimate performance if the evolving generations of sub-memeplex are increased, however, it requires much more processing time. PSO is a simplest evolutionary algorithm, but proved to be robust in solving the optimization problems. SPSO owns high convergence speed and strong capability of searching the global best solution of the multi-peak functions. Conclusively, the proposed memetic algorithm is characterized by robustness with small computational cost and high quality solution in the application of the global optimization problems.

References

1. Holland, J.: Adaptation in Natural and Artificial Systems. University of Michigan Press, Ann Arbor, MI (1975)
2. Moscato, P.: On Evolution, Search, Optimization, Genetic Algorithms and Martial Arts: Towards Memetic Algorithms. Caltech Concurrent Computation Program, Tech. Rep., California Institute of Technology, Pasadena, California, USA (1989)
3. Kennedy, J., Eberhart, R.: Particle Swarm Optimization. In: Proceedings of IEEE International Conference on Neural Networks, Perth, Australia, IEEE Service Center, Piscataway, NJ, pp. 1942-1948 (1995)
4. Eberhart, R., Kennedy, J.: A New Optimizer Using Particle Swarm Theory. In: Proceedings of the Sixth International Symposium on Micro Machine and Human Science, Nagoya, Japan, pp. 39–43 (1995)
5. Dorigo, M., Maniezzo, V., Colorni, A.: Ant System: Optimization by a Colony of Cooperating Cgents. IEEE Transactions Systems, Man and Cybernetics 26(1), 29–41 (1996)
6. Eusuff, M.M., Lansey, K.E.: Water distribution network design using the shuffled frog leaping algorithm. World Water Congress (2001)
7. Moscato, P.: A Memetic Approach for the Traveling Salesman Problem Implementation of a Computational Ecology for Combinatorial Optimization on Message-Passing Systems. In: Valero, M., Onate, E., Jane, M., Larriba, J.L., Suarez, B. (eds.) Parallel Computing and Transputer Applications, pp. 176–177. IOS Press, Amsterdam, The Netherlands (1992)
8. Moscato, P., Cotta, C.: A Gentle Introduction to Memetic Algorithms. In: Handbook of Meta-heuristics, pp. 1–56. Kluwer, Dordrecht (1999)
9. Merz, P.: Memetic Algorithms for Combinatorial Optimization Problems: Fitness Landscapes and Effective Search Strategies. University of Siegen, Siegen (2000)
10. Berretta, R., Rodrigues, L.F.: A Memetic Algorithm for a Multistage Capacitated Lot Sizing Problem. International Journal of Production Economics 87, 67–81 (2004)
11. Muruganandam, A., Prabhaharan, G., Asokan, P., Baskaran, V.: A Memetic Algorithms Approach to the Cell Formation Problem. International Journal of Advanced Manufacturing Technology 25, 988–997
12. Kim, S.S., Smith, A.E., Lee, J.H.: A Memetic Algorithm for Channel Assignment in Wireless FDMA Systems. Computers and Operations Research 34(6), 1842–1856 (2007)
13. Shi, Y.H., Eberhart, R.C.: Parameter Selection in Particle Swarm Optimization. In: The 7th Annual Conference on Evolutionary Programming, San Diego, USA (1998)
14. Carlisle, A., Dozier, G.: An Off-The-Shelf PSO. In: Proceeding of the 2001 Workshop on Particles Swarm Optimization, Indianapolis, pp. 1–6 (2001)
15. Eusuff, M.M., Lansey, K.E.: Optimization of Water Distribution Network Design Using the Shuffled Frog Leaping Algorithm. Journal of Water Resource Planning and Management (2003)

16. Eusuff, M.M., Lansey, K.E., Pasha, F.: Shuffled Frog-Leaping Algorithm: A Memetic Meta-heuristic for Discrete Optimization. Engineering and Technology, Mathematics and Optimization 38(2), 129–154 (2006)
17. Duan, Q., Gupta, V.K., Sorooshian, S.: A Shuffled Complex Evolution Approach for Effective and Efficient Global Minimization. Optimization Theory and Application 76(3), 501–521 (1993)
18. Duan, Q., Sorooshian, S., Gupta, V.K.: Effective and Efficient Global Optimization for Conceptual Rainfall-Runoff Models. Water Resources Research 28(4), 1031–1051 (1992)

Building on Success in Genetic Programming: Adaptive Variation and Developmental Evaluation

Tuan-Hao Hoang[1], Daryl Essam[1], Bob McKay[2], and Nguyen-Xuan Hoai[3]

[1] Australian Defence Force Academy, Canberra, Australia
[2] Seoul National University, Seoul, Korea
[3] VietNam Military Technical Academy, Hanoi, VietNam
hao_hth@yahoo.com,d.essam@adfa.edu.au,rim@cse.snu.ac.kr,nxhoai@gmail.com

Abstract. We investigate a developmental tree-adjoining grammar guided genetic programming system (DTAG3P$^+$), in which genetic operator application rates are adapted during evolution. We previously showed developmental evaluation could promote structured solutions and improve performance in symbolic regression problems. However testing on parity problems revealed an unanticipated problem, that good building blocks for early developmental stages might be lost in later stages of evolution. The adaptive variation rate in DTAG3P$^+$ preserves good building blocks found in early search for later stages. It gives both good performance on small k-parity problems, and good scaling to large problems.

Keywords: Genetic Programming, Developmental, Incremental Learning, Adaptive Mutation.

1 Introduction

Developmental tree adjoining grammar guided GP (DTAG3P) is a grammar guided GP system using L-systems [1] to encode tree adjoining grammar guided (TAG) derivation trees [2]. In [3,4], we introduced developmental evaluation, in which individuals are evaluated throughout development on problems of increasing difficulty. We demonstrated it could solve some difficult problems and in [5], that this good performance was associated with increased genotypic regularity. However we also noted that it might find, but subsequently lose, structures which had been successful in early developmental phases. This differs from natural evolution, in which archaic structures are highly conserved. To validate our hypothesis, that adaptive variation rates might ameliorate this, we compare DTAG3P$^+$ with both Koza-style GP [6] and the original TAG3P [2], on a family of k-parity problems of increasing difficulty. Also, we observe DTAG3P$^+$ results on a family of symbolic regression problems previously studied in [4], to provide comparison with the original DTAG3P representation.

The paper is organised as follows. The next section briefly mentions the literature on adaptive GP and surveys L-systems, concluding with k-parity problems, emphasising their scaling properties. Section 3 discusses the interaction of

L. Kang, Y. Liu, and S. Zeng (Eds.): ISICA 2007, LNCS 4683, pp. 137–146, 2007.

evolution, development and evaluation, and describes our L-system based developmental evolutionary system with adaptive variation. Experimental setups are described in section 4. Section 5 provides the results, with discussion in section 6. Conclusions and future work are laid out in the final section.

2   Background and Previous Work

2.1   Adaptive Evolutionary Parameters

We call an evolutionary algorithm "adaptive" if it modifies or updates some aspect of the method based on its behaviour in solving a problem. It has been an important theme in evolutionary and GP research over the years [7,8,9,10,11,12,13,14], generally involving either a heuristic algorithm for determining the adaptation based on previous behaviour, or incorporating parameters controlling the evolution into the evolutionary genotype (self-adaptation).

2.2   L-Systems and DOL-Systems

L-systems were introduced by Lindenmayer in 1968 [1], to simulate the developmental processes of natural organisms. For more details see [15]. We use the simplest form, a Deterministic L-system with O interactions (DOL-system), corresponding to Context Free Grammars (CFG). A DOL-system is an ordered triplet $G = (V, \omega, P)$ where:

- V is the alphabet of the system, V^* the set of all words over V.
- $\omega \in V^*$ is a nonempty word called the *axiom*.
- $P \subset V \times V^*$ is a finite *set of productions*. A production $(p, s) \in P$ is written $p \rightarrow s$; p and s are the *predecessor* and *successor* of this production.
- Whenever there is no explicit mapping for a symbol p, the identity mapping $p \rightarrow p$ is assumed.
- There is at most one production rule for each symbol $p \in V$.

Let $p = p_1 p_2 ... p_m$ be an arbitrary word over V. The word $s = s_1 s_2 ... s_m \in V^*$ is directly derived from p, denoted $p \Rightarrow s$, iff $p_i \rightarrow s_i$ for all $i \in \{1 ... m\}$. If there is a *developmental sequence* $p_0, p_1, \ldots, p_n$ with $p_0 = \omega$, $p_n = s$, and $p_0 \Rightarrow p_1 \ldots \Rightarrow p_n$, we say that G generates s in a derivation of length n.

2.3   Parity Problems

The developmental evaluation approach aims to handle scaling of GP problems by incrementally solving a family of problems of increasing difficulty (during the developmental process). The k-parity problems constitute a long-studied family of difficult GP benchmarks. The even (odd) task is to evolve a function returning 1 if an even (odd) number of the inputs evaluate to 1, and 0 otherwise. Langdon and Poli observed in [16] that the task is extremely sensitive to change in the value of its inputs, and that the commonly-used function set $OR, AND, NOR, NAND$ omits the usful XOR and EQ building blocks of this problem. Inspired by Poli and Page [17], we chose the function set AND, OR, XOR, NOT as a suitable compromise set for comparisons on the family of odd-k-parity problems – containing the XOR building block, unlike the first function set, but tougher than Poli and Page's set (which contained all binary Boolean functions).

3 Evolution, Development and Evaluation

3.1 Genotype and Development: TAG Based DOL-Systems

For our purposes, DOL systems must not only represent development, but must generate at each stage an individual which can be evaluated. We use TAG representation, as introduced in [2] [1] . Briefly, TAG representation consists of an α tree ('The cat sat on the mat') and instructions for adjoining β trees ('black' or 'which it liked') to form a more complex whole ('The black cat sat on the mat which it liked'). In our representation, the DOL triple $G = (V, \omega, P)$ is mapped to TAG representation by defining ω to consist of an α tree together with a predecessor from P, and each letter $\{L_1, L_2, L_3, \ldots\} \in V$ to be either a predecessor from P or a β tree. Thus the DOL-rewriting of ω corresponds to adjunction of successive β trees into the initial α tree. For example, assume an L-system $G' = (V', \omega', P')$ with $V' = \{L_1, L_2, L_3, L_4\}$, $\omega' = (\alpha_1, L_1)$, and P' being $P'_1: L_1 \rightarrow \beta_1\beta_2\beta_3L_4\beta_1L_2$, $P'_2: L_2 \rightarrow \beta_2\beta_1\beta_4\beta_3L_2L_3$, $P'_3: L_3 \rightarrow \beta_5\beta_6L_4\beta_7\beta_8L_1$, $P'_4: L_4 \rightarrow \beta_1L_2\beta_4\beta_6\beta_7L_3$. Figure 1 shows tree representations of these productions, together with three stages of the expansion of this system into a TAG derivation tree. The expansion starts with the TAG initial tree α_1 together with the predecessor L_1. In the stage 1 expansion, L_1 is replaced by its successor in the production rule P'_1. This successor has two predecessors L_2 and L_4. In the stage 2 expansion, these are replaced by their successors using the corresponding production rules P'_2 and P'_4. This leaves us with four available predecessors – two occurrences each of L_2 and L_3. This process continues until predefined limits on the number of stages are reached.

3.2 Developmental TAG GP (DTAG3P)

DTAG3P uses TAG-based DOL-systems to encode Tree Adjoining Grammars, so delimiting the language of the genetic programming system. It is a developmental form of the earlier TAG3P system, and shares many aspects. We describe these briefly, but refer readers to [2] for detail. We assume a lexicalised TAG (LTAG) grammar [2] G_{lex} defining the sets A of α trees, and B of β trees. We

---

[1] Any rooted subtree of a valid TAG derivation tree is also valid, as is any extension [2]. This property is not shared by other tree-based GP representations.

[2] Schabes [18] showed any CFG G generates an equivalent lexicalised TAG G_{lex}, and an inverse transformation converts a G_{lex} derivation tree to a G derivation tree. We use grammars G which generate the expression trees for the problem domain.

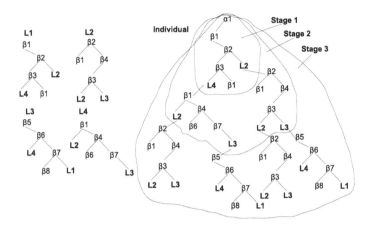

Fig. 1. TAG-Based L-Systems Example, Left: representations of individual rules, Right: stages of development

evolve DOL rulesets (each ruleset is an evolutionary individual). The ruleset specifies the development of the individual, generating a TAG derivation tree at each stage s of development. This tree is fitness-evaluated against the corresponding problem P_s from our target family of problems. Evaluation uses the standard conversion, first to the corresponding CFG derivation tree, and then to the expression tree [2]. We follow Koza's specification scheme [6], adapting it to incorporate developmental evaluation:

1. *Initialisation:* We randomly generate max_{pop} DOL-systems, each containing n_{rules} rules $R = \{R_1, R_2, \ldots, R_{n_{rules}}\}$. We denote the predecessors of these rules as $\Lambda = \{L_1, L_2, \ldots, L_{n_{rules}}\}$, so that $V = \Lambda \bigcup A \bigcup B$. We randomly select $\omega = (\alpha, L)$: $\alpha \in A$, $L \in \Lambda$. We construct the successor (RHS) of R_i by first randomly drawing β-trees from B and assigning them to the RHS of R_i, up to a random limit between $min_{betas}, \ldots max_{betas}$, and then randomly drawing num_{letter} predecessors from V and inserting them into the RHS.

2. *Development and Fitness Evaluation:* Each individual undergoes a fixed number max_{life} of developmental stages (corresponding to the size of the problem family) [3] . Each individual I is expanded through its development stages (see section 3.1). At stage s, this generates a TAG derivation tree I_s of G_{lex} and the corresponding CFG derivation tree [1] $CF(I_s)$ of G and expression $exp(I_s)$. We evaluate $exp(I_s)$ against the corresponding problem, P_s, to get a fitness value $fit(I_s)$.

3. *Selection* uses a developmental form of tournament selection of size $size_{tourn}$. We first compare the individuals on stage 1 fitness $fit(I_1)$. The fittest individuals are carried to stage 2. This is repeated as necessary with $I_2, \ldots, I_{max_{life}}$;

---

[3] Many fitness evaluations from later life stages are not used in selection. Lazy evaluation would eliminate their computational cost. For analysis, we perform these evaluations, but also report the computational cost had we avoided them.

if more than one reach max_{life}, we use random choice [3] . Two individuals I, J are considered equal iff $|fit(I_s) - fit(J_s)| \leq \delta|$. We use an elite of 1.

4. *Genetic operators:* individuals in the next generation are produced with probability p_X by *recombination* and $1 - p_X$ by *alteration*.

  – *Recombination* takes two individuals $\{P_1, P_2\}$ and creates two offspring $\{C_1, C_2\}$ by uniform crossover on rules: a rule in $C_1(C_2)$ is with probability p_{copy} copied from the corresponding rule of $P_1(P_2)$, otherwise from the corresponding rule of $P_2(P_1)$.

  – *Alteration* consists of three sub-operators [4] acting on RHSs of rules:

    • *internal crossover* [5] : subtree crossover is performed between rules
    • *subtree mutation*: subtrees are mutated by subtree mutation
    • *lexical mutation*: the symbol in a node is randomly substituted

  The probability of alteration uses an adaptive alteration rate p_{adapt}, initially set to a high value p_{bad}. When a rule is used in a developmental stage which was used to select the parent, it is reset to a lower value p_{good}. Thus the child is more likely to inherit this rule unchanged.

5. *Parameters:* The maximum number of generations max_{gen}, population size max_{pop}, and recombination (p_X, p_{copy}) and alteration (p_{bad}, p_{good}) rates specify the evolutionary system; the number of rules n_{rules} and min_{betas}, max_{betas}, num_{letter} – respectively minimum and maximum number of β trees and of predecessors in a rule RHS – together with the maximum lifetime max_{life} and minimum difference δ, specify the developmental system.

3.3   Domain Variables

Our previous work [3,4] used a family $\{F_1, \ldots, F_9\}$ of symbolic regression problems, with $F_i(X) = 1 + x + \ldots + x^i$. In these problems, $F_1, \ldots, F_9$ all have the same domain variable, x. By contrast, each new $k + 1$-parity problem introduces a new variable x_{k+1} not required for k-parity. But the learning system doesn't know this. How should it handle an expression with x_j when it is solving a problem $P_k, k < j$ where it has no meaning [6] ? We can see three different ways of dealing with such 'undefined' variables (denoted $undef$) during evaluation. We chose alternative 2 in these experiments after preliminary testing.

1. Always replace $undef$ with 1 (or equivalently, with 0)
2. $(undef$ OP $x_i | undef) = (x_i | undef$ OP $undef) = undef$; when an expression evaluates to $undef$, its error is calculated as 0.5
3. $undef$ OP x_i and x_i OP $undef$ depend on OP (for example, $undef$ AND 0 $= 0$; $undef$ AND 1 $= undef$), giving a Lukasiewicz logic [19].

---

[4] They are operators of the TAG3P GP system, described in full detail in [2].

[5] We emphasise that this is an exchange of information between components of an individual, not a recombination operator.

[6] The developmental process could prevent this, but we regard it as cheating.

4   Experimental Details

The first experiments compare the adaptive variation system DTAG3P$^+$ with the original DTAG3P on two benchmark problems: the structured family of symbolic regression problems $\{F_1, \ldots, F_9\}$, and the family of odd-k-parity problems $\{k = 2, \ldots, 10\}$. The context-free grammars G_1 and G_2 for symbolic regression and parity are given in table 1. The corresponding TAGs are $G_{ilex} = \{V_i, T_i, I_i, A_i\}$,

Table 1. Grammars for Solution Spaces: (top) Symbolic Regression, (bottom) Parity

$G_1 = (V_1, T_1, P_1, S_1)$	$P_1 = EXP \to EXP\,OP\,EXP \vert PRE\,EXP \vert VAR$
$S_1 = EXP$	$OP \to + \vert - \vert * \vert /$
$V_1 = \{EXP, PRE, OP, VAR\}$	$PRE \to sin \vert cos \vert lg \vert ep$
$T_1 = \{x, sin, cos, lg, ep, +, -, *, /\}$	$VAR \to x$
$G_2 = (V_2, T_2, P_2, S_2)$	$P_2 = BOOL \to BOOL\,OP\,BOOL$
$S_2 = BOOL$	$\vert PRE\,BOOL \vert VAR$
$V_2 = \{BOOL, PRE, OP, VAR\}$	$OP \to AND \vert OR \vert XOR \vert NOT$
$T_2 = \{AND \vert OR \vert XOR \vert NOT$	$PRE \to NOT$
$\vert x_i, i = 1, \ldots, k\}$	$VAR \to x_i, i = 1, \ldots, k$

$i = 1, \ldots 2$, with $I_1 \cup A_1$ for symbolic regression as in [3], and with $I_2 \cup A_2$ for parity as shown in figure 2. All individuals are composed of instances of these α-and β-trees. Detailed parameters for experiments on the two problem families

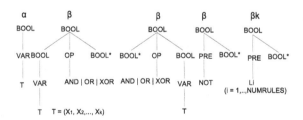

Fig. 2. Lexicalized TAG (G_{lex}) elementary trees for k-parity problem

are given in Table 2. For the parity experiments, three experimental settings have been used with different k-parity problems ($k = 8, 10, 12$). We start development with the 2-parity problem, giving lifetimes max_{life} of 7, 9 and 11.

The second set of experiments, addressed scaling issues, testing the performance of DTAG3P$^+$, TAG3P and GP on k-parity ($k = 8, 10, 12$). Counting generations gives a distorted view of DTAG3P$^+$'s computational cost: not all evaluations are used [3] , and the parsimony of DTAG3P$^+$ leads to cheaper evaluation. We use evaluations of nodes in the expression tree as the primary measure, with a budget of $1.25 * 10^8$ node evaluations, but also report function evaluations and generations in the results. Parameters were otherwise as in table 2.

Table 2. Evolutionary & Developmental Parameter Settings

Parameter	Symbolic Regression	Parity
Success Predicate	Error Sum $< \varepsilon = 0.01$	Zero Error
Fitness Cases	20 points in $[-1 \ldots + 1]$	All Boolean Combinationsdot
Fitness	Error Sum	$\sharp$ of Errors ($undef$ counts 0.5)
Genetic Operators	Tournament selection (3); Recombination,	
	internal crossover, subtree mutation, lexical mutation	
Elite Size	1	
$\sharp$ of Runs	100	30
max_{life}	8	7,9,11 [7] (1)
max_{gen}	51	100,100,100
max_{pop}	250	500
p_X	0.9	
p_{copy}	0.8	
p_{bad}	0.8	
p_{good}	0.05	
n_{rules}	15	20
$(min_{betas}, max_{betas})$	(1,2)	
num_{letter}	1	
δ	0.01	0

5   Results

Table 3 and the top row of figure 3 summarise the results of the first set of experiments, giving some impression of the overall performance of DTAG3P$^+$. The first part of the table shows the percentage of successful runs for GP, TAG3P, DTAG3P and DTAG3P$^+$ on F_9 from the symbolic regression problem family, with the second showing the same for DTAG3P and DTAG3P$^+$ on 6, 8 and 10-parity problems. The left figure shows the cumulative probability of success by generation for these systems on the symbolic regression, while the middle figure compares the median best fitnesses by generation for 6, 8 and 10-parity problems. In the latter, the combination of our treatment of "undefined" values, so that 0.5 fitness is easilty attained, and the use of elitism, probably explain the 'blocky' look of the plots. From the right figure we gain an impression of the scaling of DTAG3P$^+$ on the parity problems, showing as it does the cumulative probablity of success on each of the 6, 8, 10 and 12 parity problems.The bottom row of figure 3 gives a more detailed insight into the relative behaviour of GP, TAG3P and DTAG3P$^+$ on the difficult 12-parity problem. The left figure gives

Table 3. Percentage Success Rates, Left: Symbolic Regression, F_9, Right: Parity

GP	TAG	DTAG3P	DTAG3P$^+$
0%	8%	73%	100%

	6-Parity	8-Parity	10-Parity
DTAG3P	0%	0%	0%
DTAG3P$^+$	93%	80%	53%

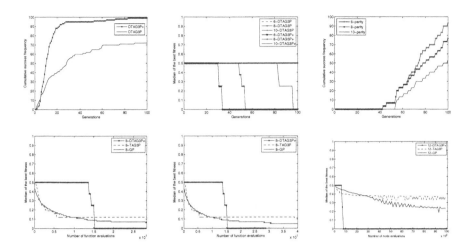

Fig. 3. Top: Overall Performance of DTAG3P$^+$: (Left: Cumulative Success Rate vs Generation (F_9, Symbolic Regression), Middle: Median Best Fitness vs Generation (8-parity), Right: Cumulative Success Rate vs Generation on family of k-parity problems). Bottom: Median of the best fitness of GP, TAG3P and DTAG3P$^+$ on 12-parity problem vs number of evaluations, left: all functions, middle: necessary functions, right: nodes.

Table 4. Success Rates after $1.25 * 10^8$ Node Evaluations: 8, 10 and 12 Parity, 30 runs

	GP	TAG	DTAG3P$^+$
8 Parity	23%	23%	100%
10 Parity	0%	0%	100%
12 Parity	0%	0%	100%

the conventional view of the relative computational complexity, using the number of function evaluations as X-axis. The middle figure takes into account lazy evaluation of fitnesses, and only counts the number of evaluations which would actually be necessary. The right figure gives the most accurate comparison of computational cost, showing the actual number of node evaluations (AND, OR, XOR, NOT operations) required. The Y axis, in all cases, is the median best fitness (calculated over 30 runs). Using this metric, table 4 shows the relative rates of success of GP, TAG3P and DTAG3P$^+$ on 8, 10 and 12 parity.

6 Discussion

The results in table 3 and figure 3(top) confirm that the good performance of DTAG3P on symbolic regression is not damaged by use of adaptive variation rates – it appears to be substantially enhanced – and that this performance improvement extends to parity problems. They show that DTAG3P scales well with the difficulty of this problem, and that reasonable computational resources can handle this difficult problem well. Figure 3(bottom) shows that the results further favour DTAG3P$^+$ when more realistic measures of computational cost are used: DTAG3P$^+$ is able to solve the difficult 12-parity problem reliably, with a computational effort equivalent to only 2-3 generations of standard GP – when the latter has made virtually no progress on the problem. Table 4 shows that, while GP and TAG3P scale very poorly on parity problems, DTAG3P$^+$ scales well. Finally we note that the developmental methods not only solve one problem (F_9 or 12-parity) in less computational cost than conventional approaches, but also solve the smaller problems (F_i, $i < 9$, k-parity, $k < 12$) for free.

7   Conclusions and Future Work

We demonstrated that developmental evaluation, combined with adaptive variation, both performs well in symbolic regression and parity problems, and scales well. We believe we have demonstrated a general purpose problem decomposition engine, applicable to a range of domains, and scaling well, for families of related problems. Future work will include

- replacing adaptive variation with self-adaptive variation, to reduce the number of system parameters and further improve scalability
- further investigating the role of replicated building blocks in DTAG3P$^+$ with compression methods
- more detailed investigation of the scalability of DTAG3P$^+$
- extension to a range of new problems
- investigating alternative ways to handle undefined variables

Acknowledgements. We gratefully acknowledge the benefit of discussions with Naoki Mori. This research was partially financially supported by a Seoul National University support grant for new faculty.

References

1. Lindenmayer, A.: Mathematical models for cellular interaction in development, parts i and ii. Journal of Theoretical Biology **18** (1968) 280–299 and 300–315
2. Hoai, N.X., McKay, R.I.B., Essam, D.: Representation and structural difficulty in genetic programming. IEEE Transactions on Evolutionary Computation **10**(2) (April 2006) 157–166
3. McKay, R.I., Hoang, T.H., Essam, D.L., Nguyen, X.H.: Developmental evaluation in genetic programming: the preliminary results. In Collet, P., Tomassini, M., Ebner, M., Gustafson, S., Ekárt, A., eds.: Proceedings of the 9th European Conference on Genetic Programming. Volume 3905 of Lecture Notes in Computer Science., Budapest, Hungary, Springer (10 - 12 April 2006) 280–289

4. Hao, H.T., Essam, D., McKay, R.I., Nguyen, X.H.: Developmental evaluation in genetic programming: A TAG-based framework. In Pham, T.L., Le, H.K., Nguyen, X.H., eds.: Proceedings of the Third Asian-Pacific workshop on Genetic Programming, Military Technical Academy, Hanoi, VietNam (2006) 86–97

5. Shin, J., Kang, M., McKay, R.I.B., Nguyen, X., Hoang, T.H., Mori, N., Essam, D.: Analysing the regularity of genomes using compression and expression simplification. In: Proceedings of the 10th European Conference on Genetic Programming (EuroGP2007, Valencia, Spain). Volume 4445 of Springer Lecture Notes in Computer Science. Springer-Verlag, Berlin, Germany (April 2007) 251–260

6. Koza, J.R.: Genetic Programming: On the Programming of Computers by Means of Natural Selection. MIT Press, Cambridge, MA, USA (1992)

7. Schwefel, H.P.: Numerical Optimization of Computer Models. John Wiley & Sons, Inc., New York, NY, USA (1981)

8. Bäck, T., Schwefel, H.P.: An overview of evolutionary algorithms for parameter optimization. Evol. Comput. 1(1) (1993) 1–23

9. Angeline, P.J., Pollack, J.B.: Coevolving high-level representations. In Langton, C.G., ed.: Artificial Life III. Volume XVII of SFI Studies in the Sciences of Complexity., Santa Fe, New Mexico, Addison-Wesley (15-19 June 1992 1994) 55–71

10. Rosca, J.P., Ballard, D.H.: Hierarchical self-organization in genetic programming. In: Proceedings of the Eleventh International Conference on Machine Learning, Morgan Kaufmann (1994)

11. Angeline, P.J.: Two self-adaptive crossover operators for genetic programming. In Angeline, P.J., Kinnear, Jr., K.E., eds.: Advances in Genetic Programming 2. MIT Press, Cambridge, MA, USA (1996) 89–110

12. Teller, A.: Evolving programmers: The co-evolution of intelligent recombination operators. In Angeline, P.J., Kinnear, Jr., K.E., eds.: Advances in Genetic Programming 2. MIT Press, Cambridge, MA, USA (1996) 45–68

13. Iba, H., de Garis, H.: Extending genetic programming with recombinative guidance. In Angeline, P.J., Kinnear, Jr., K.E., eds.: Advances in Genetic Programming 2. MIT Press, Cambridge, MA, USA (1996) 69–88

14. Angeline, P.J.: Adaptive and self-adaptive evolutionary computations. In Palaniswami, M., Attikiouzel, Y., eds.: Computational Intelligence: A Dynamic Systems Perspective. IEEE Press (1995) 152–163

15. Prusinkiewicz, P., Lindenmayer, A.: The algorithmic beauty of plants. Springer-Verlag New York, Inc., New York, NY, USA (1996)

16. Langdon, W.B., Poli, R.: Why "building blocks" don't work on parity problems. Technical Report CSRP-98-17, University of Birmingham, School of Computer Science (13 July 1998)

17. Poli, R., Page, J.: Solving high-order boolean parity problems with smooth uniform crossover, sub-machine code GP and demes. Genetic Programming and Evolvable Machines 1(1/2) (April 2000) 37–56

18. Schabes, Y., Waters, R.: Tree insertion grammar: A cubic-time parsable formalism that lexicalizes context-free grammar without changing the trees produced. Computational Linguistics 20(1) (1995) 479–513

19. Lukasiewicz, J. In: On Three-Valued Logic. Clarendon Press, Oxford, UK (1967) 16–18

A Granular Evolutionary Algorithm
Based on Cultural Evolution

Zuqiang Meng[1,2] and Zhongzhi Shi[1]

[1] Key Laboratory of Intelligent Information Processing, Institute of Computing Technology,
Chinese Academy of Sciences, Beijing 100080, China
[2] College of Computer, Electronics and Information, Guangxi University, Nanning 530004,
China
mengzuqiang@sohu.com

Abstract. Analogous to biological evolution, cultural evolution also is a kind of optimal mechanism of nature. Studying this mechanism might possibly provide a more efficient computation for solving complicated problems, such as knowledge acquisition in large data set. In this paper, an algorithm, granular evolutionary algorithm for data classification, simply written as GEA, is proposed based on cultural evolution and granular computing. The proposed algorithm is essentially a granular computation, which is characterized by computing with granules. Each granule consists of some individuals, which itself also is an evolutionary population. The algorithm is realized in PVM environment by agent technology, and the experimental results certify its validity. Further analysis can find that the proposed algorithm has relatively better performance from large data sets.

Keywords: GrC; cultural evolution; granular evolution; data classification.

1 Introduction

Granular Computing(GrC) presents a theoretical framework, which is regarded as an big umbrella covering the studies of granular theory, methodology, technology etc[1],[2],[3]. GrC is primarily characterized by computing with granules, and a granule is a clump of objects drawn together by indistinguishability, similarity, proximity or functionality[3]. In quotient space theory, an equivalence classe is used as a granule, and arbitrary concept may be represented by constructing upper approximation and lower approximation[4]. Furthermore, with complete foundations of mathematics of quotient space, it is quite convenient for us to investigate the relationship and transformation between different worlds consisting of equivalence classes. Therefore, quotient space theory is perfect from the theoretical perspective, and has been successfully applied to solving many of problems. However, real problem solving makes us find it difficult to satisfy the condition of equivalence relation in many situations, which limited the scope of its application. In fact, equivalence relation is usually required to be degenerated into other relations, such tolerance relation, fuzzy equivalence relation, etc. A tolerance relation can constitute coverage of universe.

L. Kang, Y. Liu, and S. Zeng (Eds.): ISICA 2007, LNCS 4683, pp. 147–156, 2007.

And computing with elements of the coverage is so called tolerance granular computing(TGrC)[5]. The major contribution to this kind of work is presented in [6],[7],[8],[9]. Paper[6] defined a granule as an element's neighbour. And with such granule, a concept can be approximately represented. The ultimate goal of this paper is to acquire decision rules in incomplete information system. Work in paper[7] focused on acquisition of certain rules with the similar principle of TGrC, which is a continuation of paper[6]. Although goal being identical to [6], paper[8] took on different forms of granules. Concretely, a maximal tolerance class is used as a granule, which is called maximal consistent block in [8]. In this way, a new approximation to an object set in incomplete information systems is formulated with higher level of accuracy, with which a more efficient computation for knowledge acquisition, especially in large incomplete systems, is provided. In paper[9], a granule was defined in more general, where a granule consisted of such objects that had same description. In other words, elements in a granule are drawn together only by the same description. It means that any two granules might joint, and again might not. Evidently, the former belongs to quotient space theory while the latter belongs to tolerance granular computing.

From what has been discussed above, it is easily found that flexible definition and employment of granules are required in problem solving. In fact, the idea and principle of granular computing have been applied under various names in many different fields. The major contribution of GrC is to open a new avenue to solving complicated problems, whose core idea is that problem's universe is partitioned into several subsets, regarded as granules; the complexity of problem is greatly reduced by using a granule as an arithmetic element and problem gets the settlement ultimately. In fact, this idea comes from simulation of human cognitive process in problem solving.

2 Cultural Evolution and Its Granular Model

In addition, the similar ideas also exist in many other mechanisms of social and biologic activities. For example, the simulation of biological evolution resulted in evolutionary computing, which is successfully applied in many fields. Recently, another evolutionary mechanism, cultural evolution, has drawn many scholars' attention[10],[11].

Cultural evolution is peculiar to human being. It refers to the development of human intelligence and relationship between human being and other biology, as well as the development of form of human society, the formation, diffusion, extinction of a new culture, etc. Cultural evolution began with the advent of human being, also can be considered as a particular evolution relating to human being and all activities of human being, including human intelligent evolution and social evolution, etc. Human being is only one species whose evolution is dependent on culture, although there is also social form in some other species to a certain degree. Culture has three features: culture is socially constructed; culture is all encompassing; and it functions to motivate individuals to behave in ways that benefit the group[12]. There is study suggesting that culture and genes coevolve and jointly influence human behavior and psychology[13]. It shows that human being's cultural evolution and its biological evolution are consistent. Therefore, human being's cultural evolution is incarnated in two aspects: one is social aspect, which is viewed as human macroscopic evolution, and the other is biological aspect, which is regarded as human microscopic evolution. Human microscopic evolution has been unraveled in many papers, and the following section will mainly introduce something about human macroscopic evolution.

Human macroscopic evolution, i.e. social aspect, is characterized by evolving with social populations, which composes new evolutionary population, called superpopulation. In other words, superpopulation consists of many populations, and each population is regarded as an individual with respect to superpopulation. To a great extent, cultural evolution incarnates superpopulation's evolution. In this sense, an individual is called superindividual, which is essentially a general population, for the convenience of discussion. Such a superpopulation is a special evolutionary population with evolutionary operations acting on superindividuals. For example, starting from hunting and foraging ages, human beings lived with the form of primitive tribe, and one primitive tribe struggled for survival against another and natural environment. In this way, one tribe might be extinguished or taken over by another, and then some stronger tribes are generated. At last, nations, special tribes, came into being, which had more reasonable structure, stronger fighting capacity and larger scale. Following this, human society evolved in a more rational and more reasonable manner by notions, as to emerge modern society in an advanced state of development. The fact that modern society consists of the United States, Germany, France, England, Italy etc. is one state of cultural evolution. We can believe that human society will further be optimized in this way. It is apparent that assimilation, differentiation, amalgamation, permutation and elimination are basic evolutionary operations in the process of cultural evolution. At the same time, as a unit with respect to cultural evolution, notion itself also is an evolutional entity. Obviously, there must exist a kind of efficient mechanism in cultural evolution, with which we might possibly provide a more efficient computation for solving complicated problems.

From the granular point of view, a superpopulation can be viewed as a granular space while superindividual as a granule, such as a tribe or a nation and so on, for superindividual is a set of individuals, which is consonant with the intension of granule. It is different from traditional granule that superindividual is "dynamic" while traditional granule is "static". In order to depict superindividual's dynamic behavior and model human microscopic evolution, we use an agent to represent a superindividual, whose software simulation also requires agent technology. And in this sense, it is quite natural that a superpopulation should be represented a multi-granular agent system(MGAS). Further, a $MGAS$ can be formalized as follows:

$$MGAS = (AS, R)$$

where $AS = \{GAgent_1, GAgent_2, \ldots, GAgent_n\}$, R is a set of relations on AS, relating to specific problem.

In social environment Γ, function $\zeta(MGAS, \Gamma)$ represents the fitness of $MGAS$ with respect to environment Γ.

In $MGAS$, assimilation, differentiation, amalgamation, permutation and elimination are represented with the following operators, respectively:

(1) Assimilation: $GAgent_1 = GAgent_2 \square GAgent_3$;
(2) Differentiation: $\Diamond GAgent_1 = \{GAgent_2, GAgent_3\}$;

(3) Amalgamation: $GAgent_1 \copyright GAgent_2 = GAgent_3$;
(4) Permutation: $GAgent_1 \otimes GAgent_2 = \{GAgent_1', GAgent_2' \}$;
(5) Elimination: $\circ GAgent_1 = \varnothing$.

Superoperator is the collective name for the above operators.

Suppose at moment t, a multi-granular agent system is represented as $MGAS_t$; after actions of superoperator, $MGAS_t$ is changed to another state, expressed in the form of $GV(MGAS_t, \square, \diamondsuit, \copyright, \otimes, \circ)$, written as $MGAS_{t+1}$, i.e.

$$MGAS_{t+1} = GV(MGAS_t, \square, \diamondsuit, \copyright, \otimes, \circ)$$

The function GV is commensurate with evolution of superpopulation. And together with evolution of $GAgent_i$, GV composes so-called granular evolution. Thus, in environment Γ, multi-agent based granular evolution can be recursively modeled as follows:

$$\max \zeta = \zeta(MGAS_{t+1}, \Gamma)$$
$$\text{s.t. } MGAS_{t+1} = GV(MGAS_t, \square, \diamondsuit, \copyright, \otimes, \circ)$$
$$GAgent_{t+1} = E(GAgent_t)$$

where E stands for additional evolution of population.

The above granular evolutionary model focuses on fitness of $MGAS$, evolution of population and superpopulation, which together characterize cultural evolution properly.

3 Granular Evolutionary Algorithm for Data Classification

In this section, application of granular evolution is considered. A recursive model-based algorithm for data classification is designed and its realization is achieved by using agent technology. Suppose decision system $<U, C \cup \{d\}>$ denotes a data set.

3.1 The Structure of Granular Agent and Its Evolution

As mentioned above, a superindividual is replaced by granular agent $GAgent$ during the simulation of granular evolution. A $GAgent$'s basic components include a set of chromosomes(individuals) consisting of genes, goal, knowledge database, which are represented as $Popu$, $Goal$, KD, respectively. Of course, an agent usually includes many components, such as sensor etc., but some of them are not essential to the simulation and then are omitted. Therefore, we represent a $GAgent$ with a triple $(Popu, KD, Goal)$, i.e.

$$GAgent = (Popu, KD, Goal)$$

Each chromosome in $Popu$ is encoded in the form as in Table 1 based on Pittsburgh's method[14], which consists of n genes, where $n = |C|$.

Table 1. Coding method of chromosomes

$gene_1$			$gene_2$			...		$gene_n$		
W_1	O_1	V_1	W_2	O_2	V_2		...	W_n	O_n	V_n

When vector $I=(gene_1, gene_2,..., gene_n)$ is used to stand for the chromosome that the above genes represent, parameter W_i, O_i, V_i are written as $I.W_i$, $I.O_i$, $I.V_i$, respectively, where $i =1,2,...,n$, $n=|C|$. Parameter $I.O_i$ and $I.V_i$ stand for operator and related value, respectively. If parameter $W_i \in [0,1]$ is less than a given threshold, $gene_i$ is removed from I, i.e., $gene_i$ exerts no effects on I, and $gene_i$ is called unefficient genes, otherwise $gene_i$ is called efficient genes. The set of efficient genes is equal to the set $\{gene_i \mid I.W_i \geq$ a threshold, $i = 1, 2,...,n\}$, written as $EGS(I)$.

Let vertor $g = (g_1,..., g_2, g_n)$, where $g_i \in [0,1]$, $i = 1,2,...,n$; $goal(g) = \{gene_i \mid gene_i \in I, g_i \in g$ and $g_i \geq 0.5$, $i = 1, 2, ...,n\}$. Then $goal$ can explained that $GAgent$ hopes that its $popu$ contains as many genes in $goal(g)$ as possible in evoluntionary processes.

Granule $m(I)$ is a set of these individuals(chromosomes) that satisfy all conditions configured by I, such as $I.W_i$ is large than a given threshold, $gene_i$'s value is also a given value, and so on. And then fitness of I is defined as follows:

$$fitness(I) = \frac{|goal(g) \cap EGS(I)|}{goal(g)} \times \frac{|m(I) \cap m(\varphi)|}{|m(I)|}$$

where φ is the consequent part of rule $I \rightarrow \varphi$, which is a logical expression constructed by class label attribute d and its value, and I is viewed as a logical expression constructed by itself.

The algorithm for $GAgent$'s evolution is written as GAgent_GA. It is similar in basic framework to genetic algorithm. So the detailed description for GAgent_GA is omitted here.

3.2   MGAS's Evolution and Evolutionary Operations

$MGAS$ is essentially a multi-agent system consisting of $GAgents$. But in this paper it is viewed as an updated granular space with each granule simulating a cultural group, which is introduced as $GAgent$(superindividual) above. And such a granular space forms an evolutionary population. As an evolutionary population, $MGAS$ have also fitness of its individuals, i.e. superindividuals, and evolutionary operations, which are discussed as follows.

Suppose $MGAS = (AS, R)$, for arbitrary superindividual $GAgent \in AS$, let $\psi(GAgent)$ denote fitness of superindividual $GAgent$, and $\psi(GAgent)$ is defined as classification accuracy on dataset U of rules constructed by I, where $I \in GAgent.Popu$.

Let $GAgent_1 = (Popu_1, KD_1, Goal_1)$ and $GAgent_2 = (Popu_2, KD_2, Goal_2)$, and vector goals of $goal_1$ and $goal_2$ is expressed as $g_1=(g_1', g_2',..., g_n')$ and $g_2=(g_1'', g_2'',..., g_n'')$, respectively. The similarity degree of $GAgent_1$ and $GAgent_2$ is defined as follows:

$$GoalSim(GAgent_1, GAgent_2) = \frac{|goal(g_1) \cap goal(g_2)|}{n}.$$

Suppose $U_t' \subseteq U$ is a data set that $GAgent$ can visited, the set of individuals which support I is called territory[15] in terms of I, written as $T(I)$, where $I \in GAgent.Popu$. Obviously, $T(I)=m(I) \cap U_t'$. In addition, let $T(Popu) = \cup \{T(I) \mid I \in Popu\}$.

For arbitrary $I_1 \in Popu_1$, $I_2 \in Popu_2$, if $T(I_1) \subseteq T(I_2)$, then let $Popu_1= Popu_1-\{I_1\}$; if $T(I_2) \subseteq T(I_1)$, then let $Popu_2= Popu_2-\{I_2\}$, and then let $Popu_3=Popu_1 \cup Popu_2$. The above operation is called combination of $Popu_1$ and $Popu_2$ with writing as $Popu_3 = \cup(Popu_1, Popu_2)$.

With above concepts, we can further construct the evolutionary operations acting on superindividual $GAgents$ as following, which are embodiments of evolutionary operation appearing in the above granular model.

(1) At one moment, $GAgent_1 = (Popu_1, KD_1, Goal_1)$ randomly chose another $GAgent_2 = (Popu_2, KD_2, Goal_2)$; if $GoalSim(GAgent_1, GAgent_2) \geq 2/3$, then let $GAgent =(\cup (Popu_1, Popu_2), KD_1 \odot KD_2, Goal_1 \ominus Goal_2)$, then make that $MGAS.AS = MGAS.AS - \{GAgent_1, GAgent_2\} \cup \{GAgent\}$, where $Goal_1 \ominus Goal_2$ means that $g = <min(g_1,g_2), min(g_1,g_2), ..., min(g_1,g_2)>$, while $KD_1 \odot KD_2$ will be defined in a following part. And the above operation is called *amalgamation* of $GAgent_1$ and $GAgent_2$, written as $GAgent = GAgent_1 \copyright GAgent_2$.

(2) At one moment, $GAgent=(Popu, KD, Goal)$ forms other two $GAgent$: $GAgent_1 = (Popu, KD, Goal_1)$ and $GAgent_2 = (Popu, KD, Goal_2)$, where $Goal_1$ and $Goal_2$ are constructed in the following way: randomly choose an element in $\{gene_i \mid i=1,2,...,n\} - goal(g)$ and add it to goal(g), then vector g will become a new one, written as g_1. With verctor g_1, $GAgent_1 = (Popu, KD, Goal_1)$ can be constructed. In the same way, another $GAgent_2 = (Popu, KD, Goal_2)$ can also be established. At last, let $MGAS.AS = MGAS.AS-\{GAgent\} \cup \{GAgent_1, GAgent_2\}$, and the above operation is called *differentiation* of $GAgent$, represented as $\diamondsuit GAgent = \{GAgent_1, GAgent_2\}$.

(3) For $GAgent =(Popu, KD, Goal)$, if there is a $GAgent' =(Popu', KD', Goal') \in MGAS.AS$, such that $T(Popu) \subseteq T(Popu')$, or $\psi(GAgent)$ is small enough, then let $MGAS.AS=MGAS.AS-\{GAgent\}$. The above operation is called *elimination* of $GAgent$, written as $\circ GAgent = \varnothing$.

(4) At one moment, $GAgent_1 = (Popu_1, KD_1, Goal_1)$ randomly chooses another $GAgent_2 = (Popu_2, KD_2, Goal_2)$; if $goal(g_2) \subset goal(g_1)$, then for all $gene_i \in goal(g_1)-goal(g_2)$ and $I \in AS.Popu_2$, let $I.gene_i.W_i$ be a stochastic real number that is greater than the given threshold, which results in the generation of a new population, $Popu'_2$. Then a new $GAgent = (Popu_1 \cup Popu'_2., KD_1, Goal_1)$ can be obtained, and let $MGAS.AS=MGAS.AS-\{GAgent_1, GAgent_2\} \cup \{GAgent\}$. The above operation is called *assimilation* of $GAgent_1$, written as $GAgent = GAgent_1 \square GAgent_2$.

(5) At one moment, $GAgent_1=(Popu_1, KD_1, Goal_1)$ randomly chooses another $GAgent_2 = (Popu_2, KD_2, Goal_2)$, and randomly exchanges p individuals with $GAgent_2$, where $p<min(|Popu_1|, |Popu_2|)$. As a result, $GAgent_1$ and $GAgent_2$ become two new $GAgents$ respectively, supposedly written as $GAgent'_1=(Popu'_1, KD_1, Goal_1)$ and $GAgent'_2=(Popu'_2, KD_2, Goal_2)$. Then let $MGAS.AS=MGAS.AS-\{GAgent_1, GAgent_2\} \cup \{GAgent'_1, GAgent'_2\}$. The above operation is called *permutation* of $GAgent_1, GAgent_2$, written as $GAgent_1 \otimes GAgent_2 = \{GAgent'_1, GAgent'_2\}$.

3.3 KD's Contents

When the above evolutionary operations being in action, KD as well as $Popu$ and $Goal$ of $GAgent$ changes, which is represented as $KD = KD_1 \odot KD_2$. KD consists of fact base and regular base, whose initial contents are introduced as follows.

- The fact base
i. Information about environment and resources, such as the maximal number of process that might be established in current environment, etc.
ii. The ratio of the number of punished individuals to the size of *Popu*, written as ξ.
- The regular base
i. *IF* there are more than τ rules on PC_i, *THEN* no *GAgent* can be established on PC_i.
ii. *IF* the ratio ξ exceeds 2/3 fine times running, *THEN* differentiation operation $\diamondsuit$ works.
iii. *IF* the ratio ξ does not exceeds 2/3 fine times running, *THEN* amalgamation operation $\copyright$ works.
iv. *IF* assimilation generator is activated(with probability $P_\square$), *THEN* assimilation operation $\square$ works.
v. *IF* permutation generator is activated(with probability $P_\otimes$), *THEN* permutation operation $\otimes$ works.
vi. *IF* elimination generator is activated(with probability $P_\circ$), *THEN* elimination operation $\circ$ works.

3.4 Granular Evolutionary Algorithm

In this section, granular evolutionary algorithm by using the above evolutionary operations for data classification is presented, which is based on agent technology.

The algorithm GEA(Granular Evolutionary Algorithm) is described as follows:

Input: training set U
Output: classifier consisting of classfication rules

Step1. establishing a PVM process, $GAgent_0$.
Step2. $GAgent_0$ initializes its population *Popu* and establishs its *KD* with initialization. And then establishs m GAgents: $GAgent_1$, $GAgent_2$, ..., $GAgent_m$ with reasonable distribution in all machines, where $m = |U/\{d\}|$.
Step3. $GAgent_0$ uses U to initialize $GAgent_1$, $GAgent_2$, ..., $GAgent_m$, including initializations of *Popu*, *KD* and *Goal*.
Step4. Each $GAgent_i$ establishs n_i local *GAgents*: $GAgent_{i,1}$, $GAgent_{i,2}$, ..., $GAgent_{i_{n_i}}$, and initializes these *GAgents* with information received from $GAgent_0$ and local machine.
Step5. Let finish = false.
Step6. par while(not finish)
{
   Step6.1 $GAgent_{i,j}$ evolves by following the above algorithm GAgent_GA. In this process, *KD* is updated after each generation; $GAgent_{i,j}$ decides whether to accept other *GAgents'* request operations(if receives) and send itself request operations to other *GAgents* or not, and takes necessary measures, such as assimilation $\square$, differentiation $\diamondsuit$, amalgamation $\copyright$, permutation $\otimes$; if receiving terminating signal, $GAgent_{i,j}$ ends itself.
   Step6.2 $GAgent_i$ randomly chooses another $GAgent_{i,j}$, and calculates its $\psi(GAgent_{i,j})$; if $\psi(GAgent_{i,j})$ is small enough, then eliminates $GAgent_{i,j}$ with larger probability of elimination, otherwise chooses some

individuals from $GAgent_{i,j}.Popu$ which have lager $fitness(I)$ and then submits them to $GAgent_0$.

Step6.3 $GAgent_0$ probes if there are submitted excellent individuals by way of non-block; if there are, then equal amount of individuals from $GAgent_0.Popu$ should be randomly chosen and replaces them with submitted individuals; $GAgent_0$ calculates classification accuracy of individuals from $GAgent_0.Popu$ on training set U; if the classification accuracy is larger than a given value or the number of evolutionary generations exceeds a given number, then $GAgent_0$ sends terminating signal to the other $GAgents$, and let finish = true.

}

Step7. GAgent0 constructs descriptions of concepts in U/{d}, with which related rules are established, and at last classifier generates naturally.

Step8. After outputting the classifier, GAgent0 ends itself.

4   Simulation and Experiment

In this section, performances of the proposed algorithm are evaluated by simulations. Four computers with configurations of 1.70GHz CPU、256MB memory and 40GB hard disk are used to establish a parallel virtual machines(PVM), which provides realization of the proposed algorithm with a soft environment.

In our experiments, three kinds of data set, Nursery, Adult and DataSet2, are used, where Nursery and Adult come from UCI Machine Learning Repository[16], and DataSet2 is randomly generated by a program. Some information about the two data sets is illustrated in Table 2.

Table 2. Some information about the two data sets

Data set	Instances	Discrete attributes	Continuous attributes	Decision attributes	Classes
Nursery	12,960	8	0	1	5.
Adult	45,222	8	6	1	2
DataSet2	200,000	24	0	1	5

Note: the instances in data set which contain miss values are directly deleted. Continuous attributes are dispersed by Equidistance method.

The proposed GEA is compared with traditional parallel genetic algorithm[17], written as TGA, with the ending condition being that classification accuracy on data set is equal or larger than 0.8. The result is shown in Table 3.

Table 3. Time comparison of algorithm TGA and GEA(sec.)

Data set	TGA	GEA
Nursery	396	371s
Adult	1,001s	896s
DataSet2	2,825s	1,762s

Note: not including the time for initialization.

The above results conclude that GEA has a distinct advantage over TGA in term of convergence rate. This advantage is not contingent. In fact, GEA has been keeping good convergence rate and stability in its running process. The advantage can be illuminated in Fig. 1, which gives the comparison between classification accuracies of GEA and TGA on DataSet2.

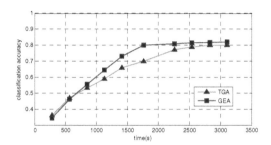

Fig. 1. The comparison between classification accuracies of GEA and TGA on DataSet2

After repeated experiments we can further find that GEA has relatively superior efficiency and classification accuracy on large data sets.

5   Conclusions and Further Work

This paper first makes a deeper analysis of the relationship between them and abstracts basic evolutionary operations from cultural evolution, then characterizes cultural evolution by establishing its granular model. Secondly, by using the basic principle, granular evolutionary algorithm for data classification, simply written as GEA, is presented based on cultural evolution's mechanism. GEA is described in detail by using agent technology. At last, GEA is realized in PVM environment, and the experimental results certify GEA's validity.

Granular evolutionary algorithm can be concluded as follows: (1) GEA can get relatively better performance from large data sets, while get relatively poor performance from small ones; (2) We lack a foundation for granular evolutionary theory, which makes it difficult to make a quantitative analysis of GEA.

In our next work, we will further study the mechanism of cultural evolution and try to establish a theoretical framework for GEA, as well as make more experimental study.

Acknowledgments. This work is supported by the National Natural Science Foundation of China under grant No. 60435010 and Scientific Research Fund of Guangxi Zhuang Autonomous Region's Education Department of China.

References

1. Zadeh, L.A.: Fuzzy Logic=Computing with Words. IEEE Transactions on Fuzzy Systems 4(1), 103–111 (1996)
2. Zadeh, L.A.: Towards a Theory of Fuzzy Information Granulation and its Centrality in Human Reasoning and Fuzzy Logic. Fuzzy Sets and Systems 19(1), 111–127 (1997)

3. Zadeh, L.A.: Some Reflections on Soft Computing, Granular Computing and their Roles in the Conception, Design and Utilization of Information/Intelligent Systems. Soft Computing 2(1), 23–25 (1998)

4. Pawlak, Z.: Rough Sets. International Journal of Information and Computer Science 11(5), 341–356 (1982)

5. Zheng, Z.: Tolerance Granular Space And Its Applications [Ph. D. dissertation]. Institute of Computing Technology, Chinese Academy of Sciences, Beijing (2006) (in Chinese)

6. Kryszkiewicz, M.: Rough Set Approach to Incomplete Information Systems. Information Sciences 112, 39–49 (1998)

7. Kryszkiewicz, M.: Rules in Incomplete Information Systems. Information Sciences 113, 271–292 (1999)

8. Leung, Y., Li, D.Y.: Maximal Consistent Block Technique for Rule Acquisition in Incomplete Information Systems. Information Sciences 153, 85–106 (2003)

9. Leung, Y., Wu, W.Z., Zhang, W.X.: Knowledge Acquisition in Incomplete Information Systems: A Rough Set Approach. European Journal of Operational Research 168, 164–180 (2006)

10. Meng, Z.Q., Cai, Z.X.: A New Computing: Granular Evolutionary Computing. Computer Engineering and application 42, 5–8 (2006) (in Chinese)

11. Zhang, J., Li, X.W.: Evolutionary Granular Computing Model and Applications, Advances in Natural Computation. In: Wang, L., Chen, K., Ong, Y.S. (eds.) ICNC 2005. LNCS, vol. 3612, pp. 309–312. Springer, Heidelberg (2005)

12. Philip, G., Chase: The Emergence of Culture. The Evolution of a Uniquely Human Way of Life, Springer, New York (2006)

13. Henrich, J., Henrich, N.: Culture, Evolution and the Puzzle of Human Cooperation. Cognitive Systems Research 7, 220–245 (2006)

14. Sen, S., Knight, L., Legg, K.: Prototype based Supervised Concept Learning Using Genetic Algorithms. In: Dasgupta, D., Michalewicz, Z. (eds.) Evolutionary Algorithms in Engineering Applications, pp. 223–239. Springer, Heidelberg (1997)

15. Tan, K.C., Tay, A., Lee, T.H., et al.: Mining Multiple Comprehensible Classification Rules Using Genetic Programming. In: Proceedings of the 2002 Congress on Evolutionary Computation CEC2002, pp. 1302–1307 (2002)

16. UCI Machine Learning Repository, http://www.ics.uci.edu/ mlearn/

17. Meng, Z.Q., Cai, Z.X.: A Method of Data Classification based on Parallel Genetic Algorithm. Computer Science 29(9s), 148–151 (2002) (in Chinese)

A Self-adaptive Mutations with Multi-parent Crossover Evolutionary Algorithm for Solving Function Optimization Problems

Guangming Lin [1,5,*], Lishan Kang [2,3], Yuping Chen[3], Bob McKay[4], and Ruhul Sarker[5]

[1] Shenzhen Inititute of Information Technology, Shenzhen 518000, China
lingm@sziit.com.cn
[2] School of Computer Science, China University of Geosciences (Wuhan) Wuhan, 430074, China
[3] State Key Laboratory of Software Engineering, Wuhan University, Wuhan 430072, China
[4] School of Computer Science & Engineering, College of Engineering, Seoul Natioinal University
[5] School of Information Technology and Electrical Engineering UNSW@ADFA, Australian Defence Force Academy, Northcott Drive, Canberra, ACT 2600, Australia

Abstract. In this paper, we introduce a new self-adaptive evolutionary algorithm for solving function optimization problems. The capabilities of the new algorithm include: a) self-adaptive choice of Gaussian or Cauchy mutation to balance the local and global search on the variable subspace, b) using multi-parent crossover to exchange global search information, c) using the best individual to take place the worst individual selection strategy to reduce the selection pressure and ensure to find a global optimization. These enhancements increase the capabilities of the algorithm to solve Shekel problems in a more robust and universal way. This paper will present some results of numerical experiments which show that the new algorithm is more robust and universal than its competitors.

Keywords: Shekel's family function problems; Self-adaptive Gaussian and Cauchy mutations; Multi-parent crossover;Exploration and exploitation.

1 Introduction

The allocation of responsibilities to the genetic operators is relatively well understood: Mutation introduces innovations into the population, recombination changes the context of already available useful information, and selection directs the search towards promising regions of the search space [8]. Acting together, mutation and recombination "explore" the search space while selection "exploits" the information represented within the population. Exploration is the creation of population diversity by exploring the search space; Exploitation is the reduction of the diversity by focusing on the

---

* Corresponding author.

L. Kang, Y. Liu, and S. Zeng (Eds.): ISICA 2007, LNCS 4683, pp. 157–168, 2007.
© Springer-Verlag Berlin Heidelberg 2007

individuals of higher fitness, in other words, exploiting the fitness information (or knowledge) represented within the population. The balance between exploration and exploitation or, in other words, between the creation of diversity and its reduction by focusing on the individuals of higher fitness, is critical in order to achieve a reasonable behavior of EAs for complicated optimization problems.

In this paper, we introduce a new self-adaptive evolutionary algorithm to solve global optimization. The basic idea of this algorithm can be traced back [1]-[4]. In [2, 3] we gave an analysis of the search capabilities of Gaussian and Cauchy mutations, and self-adaptive to choice Gaussian or Cauchy mutation to balance the local and global search on the variable subspace. We put the better mutation of Gaussian and Cauchy into [4] to form a new algorithm, and find it works very well. We get satisfactory results in solving a suite of benchmark problems. However the numerical experiments show that Classical Evolutionary Programming (CEP), Fast Evolutionary Programming (FEP) and Improved Fast Evolutionary Programming (IFEP) [3] are hard to solve Shekel's Family Functions. In this paper, we introduce multi-parent crossover with self-adaptive mutation together to solve Shekel Functions, it works very well. It is also an efficient general-purpose global optimization algorithm. This algorithm has rather coarse parallel granularity. So it is suitable for large-scale distributed parallel processes, in particular, for solving complex high-dimension non-linear programming problems in parallel.

The rest of this paper is organized as follows. Section 2 describes the problem to be solved. Section 3 describes related works. Section 4 describes a new self-adaptive evolutionary algorithm. Section 5 presents the experimental results. We use a suite of benchmark functions to test the new algorithm and compare the experimental results with other. Finally, Section 6 concludes with remarks and future research directions.

2   The Problems to Be Solved

The general non-linear programming (NLP) problem can be expressed in the following form:

$$\text{Minimize } f(X,Y) \tag{1}$$

$$s.t. \quad h_i(X,Y) = 0 \quad i = 1,2,...,k_1, \quad g_j(X,Y) \leqslant 0 \quad j=k_1+1, k_1+2,...,k$$

$$X^{lower} \leqslant X \leqslant X^{upper}, \quad Y^{lower} \leqslant Y \leqslant Y^{upper}$$

where $X \in R^p$, $Y \in N^q$, and the objective function $f(X,Y)$, the equality constraints $h_i(X,Y)$ and the inequality constraints $g_j(X,Y)$ are usually nonlinear functions which include both real variable vector X and integer variable vector Y.

Denoting the domain $D = \{(X,Y) \mid X^{lower} \leqslant X \leqslant X^{upper}, Y^{lower} \leqslant Y \leqslant Y^{upper}\}$, we introduce the concept of a subspace V of the domain D. m points $(X_j,Y_j), j=1,2,...,m$ in D are used to construct the subspace V, defined as :

$$V = \{ (X_V, Y_V) \in D \mid (X_V, Y_V) = \sum_{i=1}^{m} a_i (X_i, Y_i) \}$$

where a_i is subject to $\sum_{i=1}^{m} a_i = 1$, $-0.5 \le a_i \le 1.5$.

Because we deal mainly with optimization problems which have real variables and INequality constraints, we assume $k_I = 0$ and $q = 0$ in the expression (1).

Denoting $w_i(X) = \begin{cases} 0, & g_i(X) \le 0 \\ g_i(X), & \text{otherwise} \end{cases}$ and $W(X) = \sum_{i=1}^{k} W_i(X)$

Then problem (1) can be expressed as follows:

$$\text{Minimize } f(X) \qquad\qquad X \in D \qquad\qquad (2)$$

Subject to

$$W(X) = 0 \qquad\qquad X \in D$$

We define a Boolean function *"better"* as:

$$better\ (X_1, X_2) = \begin{cases} W(X_1) < W(X_2) & \text{TRUE} \\ W(X_1) > W(X_2) & \text{FALSE} \\ (W(X_1) = W(X_2)) \wedge (f(X_1) \le f(X_2)) & \text{TRUE} \\ (W(X_1) = W(X_2)) \wedge (f(X_1) > f(X_2)) & \text{FALSE} \end{cases}$$

If *better* (X_1, X_2) is TRUE, this means that the individual X_1 is "better" than the individual X_2.

3 Related Work

In 1999, Guo Tao proposed a multi-parent combinatorial search algorithm (GTA) for solving non-linear optimization problems in his PhD thesis [4]. Later it was developed as a kind of subspace stochastic search algorithm [5], that can be described as follows:

```
Guo Tao's Algorithm (GTA)

Begin

initialize popln P = {X1, X2,..., XN };  Xi ∈D since (q =
0 implies no integer variables)

generation count t := 0 ;
```

$$X \text{ best } = \text{arg } \underset{1 \le i \le N}{Min\ f(X_i)} \ ;$$

$$X \text{ worst } = \text{arg } \underset{1 \le i \le N}{Max\ f(X_i)} \ ;$$

```
while abs(f (X best)-f (X worst)) >ε do
```

```
select randomly m points X 1', X 2',..., X m'from P to form the
subspace V;
select randomly one point X'from V;

If  better (X', X worst) then  Xworst   := X';

t := t + 1;
```

$$Xbest = \arg \underset{1 \le i \le N}{Min\ f(X_i)} ;$$

$$Xworst = \arg \underset{1 \le i \le N}{Max\ f(X_i)} ;$$

```
end do

output  t , P ;

End
```

where N is the size of population P, $(m-1)$ is the dimension of the subspace V (if the m points (vectors) that construct the subspace V are linearly independent), t is the number of generations, ε is the accuracy of solution. $X_{best} = \arg \underset{1 \le i \le N}{Min\ f(X_i)}$ means that X_{best} is the variable (individual) in X_i ($i=1, 2,..., N$) that makes the function $f(X)$ have the smallest value.

The sub-population in GTA is families which reproduce sexually through the number of m individuals randomly selected from P. The best individual in the sub-population takes part in competition to replace the worst individual in P, therefore the pressure of elimination through selection is minimum. There is no mutation operator, only using multi-parents crossover in GTA.

4 A Self-adaptive Evolutionary Algorithm

Since Guo's algorithm deals mainly with continuous NLP problems with INequality constraints, to make it a truly universal and robust algorithm for solving general NLP problems, we extend Guo's algorithm by adding to it the following improvements:

(1) Guo selected randomly only one candidate solution from the current subspace V. Although he used the concept of a subspace to describe his algorithm, he did not really use a subspace search, but rather a multi-parent crossover. Because he selected randomly only one individual in the subspace, this action would tend to ignore better solutions in the subspace, and hence influence negatively the quality of the result and the efficiency of the search. If however, we select randomly several individuals from the subspace, and substitute the best one for the worst one in the current population, the search should be better. So we replace the instruction line in Guo's algorithm:

"select randomly one point $X'$from V; "

with the two instruction lines:

" select randomly s points X_1^*, X_2^*, ..., X_s^* from V;

$$X' = \arg \underset{1 \leq i \leq s}{Min} \; f(X_i^*) ;"$$

(2) The dimension m of the subspace in Guo's algorithm is fixed (i.e. m parents reproduce). The algorithm always selects a substitute solution in subspaces which have the same dimension, regardless of the characteristics of the solutions in the current population. Thus, when the population is close to the optimal value, the searching range is still large. This would apparently result in unnecessary computation, and affect the efficiency of the search. We can in fact reduce the search range, that is to say, the dimension of the subspaces. We therefore use subspaces with variable dimensions in the new algorithm, by adding the following instruction line to Guo's algorithm:

if $abs\,(f(X_{best}) - f(X_{worst})) \leq \eta$ **.and.** $m \geq 3$ **then** $m := m - 1$;

where η depends on the computation accuracyε, and $\eta > \varepsilon$. For example, if the computation accuracy$\varepsilon = 10^{-14}$, then we can set$\eta = 10^{-2}$ or 10^{-3}.

(3) We know in principle that Guo's algorithm can deal with problems containing EQuality constraints. For example, we can use the device of setting two INequality constraints $0 \leq h_i(X,Y)$ and $h_i(X,Y) \leq 0$ to replace the equality constraint $h_i(X,Y) = 0$, but the experimental results when employing this device are not ideal. However, equality constraints are likely to exist in real-world problems, so we should find methods to deal with them. One such method is to define a new function $W(X, Y)$.

Where $W(X, Y) = \sum_{i=1}^{k} W_i(X,Y)$

$$W_i(X,Y) = \begin{cases} \left\| h_i(X,Y) \right\|, & i = 1,2,\cdots,k_i \\ \max\{0, g_i(X,Y)\}, & i = k_1 + 1, k_1 + 2, \cdots, k. \end{cases}$$

(4) The penalty factor r is usually fixed. However, some people use it as a variable, such as Cello[7], who employed a self-adaptive penalty function, but his procedure was rather complex (using two populations). We also make r a variable namely $r = r(t)$, where t is the iteration count. It can self-adjust according to the reflection information, so we label it a "self-adaptive penalty operator". Since the constraints have been normalized, r is relative only to the range of the objective function, which ensures a balance between the errors of the fitness function and the objective function, in order of magnitude.

(5) Guo's algorithm can deal only with continuous optimization problems. It cannot deal directly with integer or mixed integer NLP problems. In our algorithm, when we are confronted with such problems, we need only replace the integer variables derived from the range of the float of the fitness function with "*integer function*" $int(Y^*)$, where $int(Y^*)$ is defined as the integer part of Y^*. No other changes to the algorithm are needed.

(6) The only genetic operator used in Guo's algorithm was crossover. However, we can add self –adaptive mutations in it, we introduce a better of Gaussian and Cauchy

mutation operator into the subspace search. For Gaussian density function f_G with expectation 0; and variance σ^2 is

$$f_G = \frac{1}{\sigma\sqrt{2\pi}} \; e^{-\frac{x^2}{2\sigma^2}} \; , \qquad -\infty < x < +\infty$$

For Cauchy density function f_C with scale parameter $t>0$ is,

$$f_C = \frac{1}{\pi} \frac{1}{t^2 + x^2} \; , \qquad -\infty < x < +\infty$$

Considering the above points, we introduce a new algorithm as follows:

Denoting $Z = (X, Y^*)$, where $Z \in D^*$, and

$D^* = \{(X, Y^*) | X^{lower} \leq X \leq X^{upper}, \; Y^{lower} \leq Y^* \leq Y^u, \; X \in R^p, \; Y^* \in R^q\}$, we define integer vector $Y = \text{int}(Y^*)$, where $Y^u = Y^{upper} + 0.999\cdots 9 I$

Denoting $W(Z) = W(X, \text{int}(Y^*))$,

we define the Boolean function "*better*" as follows:

$$better(Z_1, Z_2) = \begin{cases} W(X_1) < W(X_2) & \text{TRUE} \\ W(X_1) > W(X_2) & \text{FALSE} \\ (W(X_1) = W(X_2)) \wedge (f(X_1) \leq f(X_2)) & \text{TRUE} \\ (W(X_1) = W(X_2)) \wedge (f(X_1) > f(X_2)) & \text{FALSE} \end{cases}$$

The general NLP problem (1) can be expressed as follows:

$$\text{Minimize } f(X, \text{int}(Y^*)) \qquad \text{in } D^*$$

Subject to   (3)

$$W(Z) = 0, \qquad Z \in D^*$$

The new algorithm can now be described as follows:

```
Self-Adaptive  Mutations  with  Multi-Parent  Crossover
Evolutionary Algorithm  (SMMCEA)

Begin

initialize P = {Z1,Z2,…, ZN };    Zi ∈ D*;

t := 0;
```

$$\text{Zbest} = \underset{1 \leq i \leq N}{\arg Min\, f(Z_i)} \; ;$$

$$\text{Zworst} = \underset{1 \leq i \leq N}{\arg Max\, f(Z_i)} \; ;$$

```
while not abs ( F (Zbest) - F (Zworst)) ⩽ ε  do
```

```
select randomly M  points Z1', Z2',…, ZM'from P to form the
subspace V;
```

select s points randomly $Z_1^*, Z_2^* ... Z_s^*$ from V;

```
for i=1,…s  do
        for j=1,…p+q  do
```

$Z_{Gi}^*(j) := Z_i^*(j) + \sigma_i(j) N^j(0, 1)$

$Z_{Ci}^*(j) := Z_i^*(j) + \sigma_i(j) C^j(1)$

$\sigma_i(j) := \sigma_i(j) \exp(\tau N(0,1) + \tau' N_j(0,1))$

```
                                      endfor
```

if better(Z_{Gi}^*, Z_{Ci}^*) then $Z_i^{'*} := Z_{Gi}^*$ else $Z_i^{'*} := Z_{Ci}^*$;

```
endfor
```

$Z' = \arg \underset{1 \le i \le N}{Min} f(Z_i)$;

```
if  better (Z', Z worst)    then   Zworst   := Z';
t := t + 1;
```

Zbest $= \arg \underset{1 \le i \le N}{Min} f(Z_i)$;

Zworst $= \arg \underset{1 \le i \le N}{Max} f(Z_i)$;

```
if abs (f(Zbest)- f (Zworst)) ≤ η .and.  M ≥3  then
            M := M -1 ;
endwhile
output t , Zbest , f(Zbest) ;
end
```

Where $Z_{Gi}^*(j)$, $Z_{Ci}^*(j)$ and $\sigma_i(j)$ denote the j-th component of the vectors Z_{Gi}^*, Z_{Ci}^*, and σ_i, respectively. N(0,1) denotes a normally distributed one-dimensional random number with mean zero and standard deviation one. $N_j(0, 1)$ indicates that the Gaussian random number is generated anew for each value of j. $C_j(1)$ denotes a

Cauchy distributed one-dimensional random number with $t=1$. The factors τ and τ' have commonly set to $\left(\sqrt{2\sqrt{(p+q)}}\right)^{-1}$ and $\left(\sqrt{2(p+q)}\right)^{-1}$.

The new algorithm has the two important features:

(1) This algorithm is an ergodicity search. During the random search of the subspace, we employ a "non-convex combination" approach, that is, the coefficients a_i of $Z'=\sum_{i=1}^{m} a_i Z_i$ are random numbers in the interval [-0.5, 1.5] This ensures a non-zero probability that any point in the solution space is searched. This ergodicity of the algorithm ensures that the optimum is not ignored.

(2) The monotonic fitness decrease of the population (when the minimum is required). Each iteration ($t\rightarrow t+1$) of the algorithm discards only the individual having the worst fitness in the population. This ensures a monotonically decreasing trend of the values of objective function of the population, which ensures that each individual of the population will reach the optimum.

When we consider the population $P(0)$, $P(1)$, $P(2)$,..., $P(t)$,... as a Markov chain, we can prove the convergence of our new algorithm. See [9].

5 Numerical Experiments

5.1 SMMCEA for Solving Shekel Problems

Shekel's Family Functions

$$f(x) = -\sum_{i=1}^{m}[(x-a_i)(x-a_i)^T + c_i]^{-1}$$

with m=5, 7 and 10 for $f_{21}(x)$, $f_{22}(x)$ and $f_{23}(x)$, respectively, $0\le x_j \le 10$ [3]. These functions have 5, 7 and 10 local minima for $f_{21}(x)$, $f_{22}(x)$ and $f_{23}(x)$, respectively. $x_{local\_opt} \approx a_i, f(x_{local\_opt}) \approx 1/c_i$ for $1\le i \le m$, The coefficients are defined by Table 2.1.

In this section CEP, FEP, IFEP and SMMCEA are compared for solving Shekel problems. Empirical studies on 3 Shekel's family functions were carried out to compare the performance of CEP, FEP, IFEP and SMMCEA. All experiments were repeated for 50 runs. The initial population was generated uniformly at random in the ranges specified. The size of subspace is M=8. The population size is N=100. Comparing with CEP, FEP, IFEP and SMMCEA in **Table 5.2**, we can see SMMCEA performed better than CEP, FEP, IFEP. Therefore the multi-parent crossover operator is very importation in SMMCEA to overcome the EPs problems for solving Shekel's family functions; it is also a key factor to balance between exploration and exploitation.

The results are averaged over 50 runs, where "best" indicates the mean best function values found in the last generation of population.

To test the robustness and universality of our new algorithm, we chose another six different benchmark functions in the following section.

Table 5.1. Shekel Functions $f_{21}(x)$, $f_{22}(x)$ and $f_{23}(x)$ [3]

i	$a_{ij}, j = 1,...,4$				c_i
1	4	4	4	4	0.1
2	1	1	1	1	0.2
3	8	8	8	8	0.2
4	6	6	6	6	0.4
5	3	7	3	7	0.4
6	2	9	2	9	0.6
7	5	5	3	3	0.3
8	8	1	8	1	0.7
9	6	2	6	2	0.5
10	7	3.6	7	3.6	0.5

Table 5.2. Comparison between CEP, FEP, IFEP [3] and SMMCEA on Shekel's family functions

Functions	No of Gen.	CEP Mean Best	FEP Mean Best	IFEP Mean Best	SMMCEA Mean Best
Shekel-5	100	-6.43	-5.50	-6.46	-10.12
Shekel-7	100	-7.62	-5.73	-7.10	-10.38
Shekel-10	100	-8.86	-6.41	-7.80	-10.51

5.2 The Comparing of SMMCEA and GTA

In this section the Guo Tao's Algorithm (GTA) without mutation and the new algorithm are compared. Extensive empirical studies on 6 benchmark problems were carried out to compare the performance of GTA and the new algorithm. All experiments were repeated for 50 runs. The initial population was generated uniformly at random in the ranges specified in Table 1. The size of subspace is M=8. The population size is N=100. The dimension of test problem is n=2. The number of generations for each function was determined after some limited preliminary runs which showed that GTA and the new had converged (either prematurely or globally) after a certain number of generations.

Table 5.3. The 6 benchmark functions used in our experimental studies, f_{min} is the minimum of the function

Test function	z	f_{min}
$f_1(x) = \sum_{i=1}^{n} x_i^2$	$[-100,100]^2$	0
$f_2(x) = \sum_{i=1}^{n} \mid x_i^2 \mid + \prod_{i=1}^{n} \mid x_i \mid$	$[-10,10]^2$	0
$f_3(x) = \max\{\mid x_i \mid, 1 \le i \le n\}$	$[-100,100]^2$	0
$f_4(x) = \sum_{i=1}^{n} [\ x_i^2 - 10\cos(2\pi x_i) + 10)\ ]$	$[-5.12,5.12]^2$	0
$f_5(x) = -20\exp\left(-0.2\sqrt{\frac{1}{n}\sum_{i=1}^{n} x_i^2}\right) - \exp\left(\frac{1}{n}\sum_{i=1}^{n}\cos 2\pi x_i\right) + 20 + e$	$[-32,32]^2$	0
$f_6(x) = \left(x_2 - \frac{5.1}{4\pi^2} x_1^2 + \frac{5}{\pi} x_1 - 6\right)^2 + 10\left(1 - \frac{1}{8\pi}\right)\cos x_1 + 10$	$[-5,10] \times [0,15]$	0.398

Table 5.4. Comparison between SMMCEA and GTA on f_1-f_6. The results are averaged over 50 runs, where "best" indicates the mean best function values found in the last generation, and "ave." stands for the mean function values of population.

Function	No. of Gen.	SMMCEA best	SMMCEA ave.	GTA best	GTA ave.
f_1	100	2.0×10^{-8}	1.1×10^{-2}	3.2×10^{-7}	1.6×10^{-3}
f_2	100	1.4×10^{-4}	1.22×10^{-1}	7×10^{-4}	3×10^{-2}
f_3	100	1.31×10^{-4}	6.2×10^{-2}	5×10^{-4}	2.4×10^{-2}
f_4	100	1.51×10^{-6}	2.16×10^{0}	9.45×10^{-5}	2.91×10^{-1}
f_5	100	1.88×10^{-4}	7.3×10^{-1}	2.5×10^{-3}	1.1×10^{-1}
f_6	100	3.98×10^{-1}	1.226×10^{-0}	3.98×10^{-1}	4.05×10^{-1}

Comparing with SMMCEA and GTA in **TABLE 5.4**, we can see the best solution of SMMCEA performed better than that of GTA, and the value of average of SMMCEA is bigger than that of GTA, it means that there is more diversity of population in SMMCEA than that of in GTA. Therefore self-adaptive mutations operator are very important in SMMCEA. It is also a key factor to balance between exploration and exploitation.

6 Conclusion

We here proposed some self-adaptive methods to choose the results of Gaussian and Cauchy mutation, and the dimension of subspace. We used the better of Gaussian and Cauchy mutation to do local search in subspace, and used multi-parents crossover to exchange their information to do global search, and used the worst individual eliminated selection strategy to keep population more diversity.

Judging by the results obtained from the above numerical experiments, we conclude that our new algorithm is both universal and robust. It can be used to solve function optimization problems with complex constraints, such as NLP problems with inequality and (or) equality constraints, or without constraints. It can solve 0-1 NLP problems, integer NLP problems and mixed integer NLP problems. When confronted with different types of problems, we don't need to change our algorithm. All that is needed is to input the fitness function, the constraint expressions, and the upper and lower limits of the variables of the problem. Our algorithm usually finds the global optimal value.

Acknowledgements

This work was supported by the National Natural Science Foundation of China (No.60473081) and the Natural Science Foundation of Hubei Province (No. 2005ABA234). Thanks especially give to the anonymous reviewers for their valuable comments.

References

1. Lin, G., Yao, X.: Analyzing Crossover Operators by Search Step Size. In: ICEC'97. Proc. of 1997 IEEE International Conference on Evolutionary Computation, Indianapolis, U.S.A, 13-16 April, 1997, pp. 107–110. IEEE Computer Society Press, Los Alamitos (1997)
2. Yao, X., Lin, G., Liu, Y.: An analysis of evolutionary algorithms based on neighborhood and step sizes. In: Angeline, P.J., McDonnell, J.R., Reynolds, R.G., Eberhart, R. (eds.) Evolutionary Programming VI. LNCS, vol. 1213, pp. 297–307. Springer, Heidelberg (1997)
3. Yao, X., Liu, Y., Lin, G.: Evolutionary programming made faster. IEEE Trans. Evolutionary Computation 3(2), 82–102 (1999)
4. Guo, T.: Evolutionary Computation and Optimization. PhD thesis. Wuhan University, Wuhan (1999)

5. Guo, T., Kang, L.: A new evolutionary algorithm for function optimization. Wuhan University Journal of Nature Science 4(4), 409–414 (1999)
6. Deb, K., Gene, A.S.: A robust optimal design technique for mechanical component design. Evolutionary algorithm in engineering application, pp. 497–514. Springer, Heidelberg (1997)
7. Coello, C.A.: Self-adaptive penalties for GA-based optimization. In: Proceedings of the Congress on Evolutionary Computation, Washington, D.C USA, pp. 537–580. IEEE Press, NJ, New York (1999)
8. Bäck, T.: Selective pressure in evolutionary algorithms: A characterization of selection mechanisms. In: Michalewicz, Z. (ed.) Proceedings of the First IEEE Conference on Evolutionary Computation, vol. 1, pp. 57–62. IEEE Neural Networks Council, Institute of Electrical and Electronics Engineers (1994)
9. He, J., Kang, L.: On the convergence rates of genetic algorithms. Theoretical Computer Science 229, 23–29 (1999)

A Quantum Genetic Simulated
Annealing Algorithm for Task Scheduling

Wanneng Shu and Bingjiao He

College of Computer Science, South-Central University for Nationalities, Wuhan 430074
shuwanneng@yahoo.com.cn

Abstract. Based on quantum computing, a Quantum Genetic Simulated Annealing Algorithm (QGSAA) is proposed. With the condition of preserving Quantum Genetic Algorithm (QGA) advantages, QGSAA takes advantage of simulated annealing algorithm so as to avoid premature convergence. When the temperature in the process of cooling no longer falls, the result is the optimal solution on the whole. Comparing experiments have been conducted on task scheduling in grid computing. Experimental results have shown that QGSAA is superior to QGA and Genetic Algorithm (GA) on performance.

Keywords: Quantum genetic simulated annealing algorithm, Genetic algorithm, Quantum genetic algorithm, Task scheduling, Grid computing.

1 Introduction

Quantum computing is based on the concepts of qubits and superposition of states of quantum mechanics. So far, many efforts on quantum computing have progressed actively due to its superiority to traditional optimization method on various specialized problems. During the past two decades, evolutionary algorithm have gained much attention and wide applications, which are essentially stochastic search methods based on the principles of natural biological evolution[1].since later 1990s, research on merging evolutionary computing and quantum computing has been started and gained attention both in physics, and computer science fields.

Recently, some quantum genetic algorithm (QGA) have been proposed for some combinatorial optimization problems, such as traveling salesman problem [2], Knapsack problem[3,4].QGA is based on the concept and principles of quantum computing such as qubits and superposition of states. By adopting qubit chromosome as a representation, QGA can represent a linear superposition of solutions due to its probabilistic representation .However, the performance of QGA is easy to be trapped in local optimal so as to be premature convergence. In other words, the qubit search with quantum mechanism and genetic search with evolution mechanism need to be well coordinated and the ability of exploration and exploitation need to be well balanced as well.

In this paper, a Quantum Genetic Simulated Annealing Algorithm based on quantum computing, called QGSAA, is studied for task scheduling in grid computing.

L. Kang, Y. Liu, and S. Zeng (Eds.): ISICA 2007, LNCS 4683, pp. 169–176, 2007.
© Springer-Verlag Berlin Heidelberg 2007

The experimental results show the superiority of the QGSAA in terms of optimization quality, efficiency. The organization of the remaining content is as follows; in section 2 QGSAA is proposed, in section 3 simulations on task scheduling problem are carried out to investigate the effectiveness of the QGSAA. Finally we end with some conclusions in section 4.

2 The Structure and Description of QGSAA

2.1 Qubit Chromosomes

QGSAA is based on the concepts of qubits and superposition of states of quantum mechanics .the smallest unit of information stored in a two-state quantum computer is called a quantum bit or qubit. A qubit may be in the state "1", the state "0", or in any superposition of them. the state of a qubit can be represented as

$$|\psi\rangle = \alpha |0\rangle + \beta |1\rangle \tag{1}$$

Where α and β are complex numbers that specify the probability amplitudes of the corresponding states. Gives the probability that qubit will be found in state "0" and gives the probability that qubit will be found in state "1".Normalization of the state to unity guarantees

$$|\alpha|^2 + |\beta|^2 = 1 \tag{2}$$

QGSAA uses a new representation that is based on the concept of qubits. One qubit is define with a pair of complex numbers (α, β), which is characterized by Eq.(1) and Eq.(2). An m_qubits representation is defined as

$$q_j^t = \begin{bmatrix} \alpha_1^t & \alpha_2^t & \cdots & \alpha_m^t \\ \beta_1^t & \beta_2^t & \cdots & \beta_m^t \end{bmatrix} \tag{3}$$

Where $|\alpha_i^t|^2 + |\beta_i^t|^2 = 1, i = 1, 2 \cdots, m$ .m is the length of chromosomes, and t is the generation.

2.2 Selection Operation

Selection is the transmission of personal information from the parent individual to the offspring individuals. We are adopting roulette selection strategy; each individual selection probability is proportion to fitness value.

Step 1: Computing the selection probability of all the individuals

$$P(i) = f(i) / \sum_{i=1}^{M} f(i)$$

Step 2: Generate a random number r between [0,1];

Step 3:If $P(0)+ P(1)+...+ P(i-1)<r< P(0)+ P(1)+...+ P(i)$,the individual i is selected into the next generation

In step 1, $f(i)$ denotes the fitness value of individual i.

2.3 Crossover Operation

Crossover is the substitution between two individuals of the parent generation that is to generating new individual. In QGSAA, one point crossover operator is used for qubit, which is illustrated as follows. In particular, one crossover position is randomly determined (e.g. position i), and then the qubit of the parents before position i are reserved while the qubits after position i are exchanged.

$$\begin{bmatrix} \alpha_{1,1} & \alpha_{1,2} & \cdots & \alpha_{1,i} & \alpha_{1,i+1} & \cdots & \alpha_{1,m} \\ \beta_{1,1} & \beta_{1,2} & \cdots & \beta_{1,i} & \beta_{1,i+1} & \cdots & \beta_{1,m} \end{bmatrix} \Rightarrow \begin{bmatrix} \alpha_{1,1} & \alpha_{1,2} & \cdots & \alpha_{1,i} & \alpha_{2,i+1} & \cdots & \alpha_{2,m} \\ \beta_{1,1} & \beta_{1,2} & \cdots & \beta_{1,i} & \beta_{2,i+1} & \cdots & \beta_{2,m} \end{bmatrix}$$

$$\begin{bmatrix} \alpha_{2,1} & \alpha_{2,2} & \cdots & \alpha_{2,i} & \alpha_{1,i+1} & \cdots & \alpha_{1,m} \\ \beta_{2,1} & \beta_{2,2} & \cdots & \beta_{2,i} & \beta_{1,i+1} & \cdots & \beta_{1,m} \end{bmatrix} \Rightarrow \begin{bmatrix} \alpha_{2,1} & \alpha_{2,2} & \cdots & \alpha_{2,i} & \alpha_{2,i+1} & \cdots & \alpha_{2,m} \\ \beta_{2,1} & \beta_{2,2} & \cdots & \beta_{2,i} & \beta_{2,i+1} & \cdots & \beta_{2,m} \end{bmatrix}$$

2.4 Mutation Operation

Mutation operation enables the whole population to maintain a certain variety through the abrupt change of the mutation operator when the local convergence occurs in the population. In QGSAA, mutation for qubit is illustrated as follows. In particular, one position is randomly determined (e.g. position i), and than the corresponding α_i and β_i are exchanged

$$\begin{bmatrix} \alpha_1 & \alpha_2 & \cdots & \alpha_i & \cdots & \alpha_m \\ \beta_1 & \beta_2 & \cdots & \beta_i & \cdots & \beta_m \end{bmatrix} \Rightarrow \begin{bmatrix} \alpha_1 & \alpha_2 & \cdots & \beta_i & \cdots & \alpha_m \\ \beta_1 & \beta_2 & \cdots & \alpha_i & \cdots & \beta_m \end{bmatrix}$$

2.5 Quantum Gate

In QGSAA, a quantum gate $U(\theta)$ is employed to update a qubit individual as a variation operator.

$$\begin{pmatrix} \alpha_i' \\ \beta_i' \end{pmatrix} = U(\theta) \begin{pmatrix} \alpha_i \\ \beta_i \end{pmatrix} = \begin{pmatrix} \cos(\theta_i) & -\sin(\theta_i) \\ \sin(\theta_i) & \cos(\theta_i) \end{pmatrix} \begin{pmatrix} \alpha_i \\ \beta_i \end{pmatrix} \tag{4}$$

Where θ_i is given as $s(\alpha_i \beta_i)\Delta\theta_i$. The parameters used are shown in Tab.1, where $f(x), f(b)$ are the profit, $s(\alpha_i \beta_i)$ is the sign of θ_i. b_i and x_i are the ith bits of the best solution **b** and the binary solution x respectively.

Table 1. Lookup table of θ_i

x_i	b_i	$f(x) \geq f(b)$	$\Delta\theta_i$	$s(\alpha_i\beta_i)$			
				$\alpha_i\beta_i > 0$	$\alpha_i\beta_i < 0$	$\alpha_i = 0$	$\beta_i = 0$
0	0	false	0	0	0	0	0
0	0	true	0	0	0	0	0
0	1	false	0	0	0	0	0
0	1	true	0.05π	-1	+1	± 1	0
1	0	false	0.01π	-1	+1	± 1	0
1	0	true	0.025π	+1	-1	0	± 1
1	1	false	0.005π	+1	-1	0	± 1
1	1	true	0.025π	+1	-1	0	± 1

2.6 Procedure

The solution process of QGSAA as follows:

Step 1: Generate an initial population $Q(t)$, T_0, t=0, M ;

Step 2: Conduct the operations as follows at the current temperature T_t :

Step 3: Make a set of binary solutions $P(t)$ by observing $Q(t)$ states;

Step 4: Record the best solution denoted by **b** among $P(t)$;

Step 5: if stopping condition is satisfied, then output the best result; otherwise go on following steps.

Step 6: Update $P(t)$ to generate $P(t)^{'}$ using quantum rotation gates $U(\theta)$;

Step7: Perform selection and quantum crossover and mutation for $P(t)^{'}$ to generate $P(t+1)$;

Step 8: let t=t+1, $T_{t+1} = T_t \times (1 - \dfrac{t}{M})$ and go back to step 3.

$Q(t) = \begin{bmatrix} q_1^t & q_2^t & \cdots & q_n^t \end{bmatrix}$, q_j^t denotes the jth individual in the ith generation is initialized with $1/\sqrt{2}$, it means that one qubit chromosome $q_j^t |_{t=0}$ represents the linear superposition of all possible states with the same probability. n is the size of population. T_0 is the initial temperature, M is the size of the population;

$P(t) = \begin{bmatrix} x_1^t & x_2^t & \cdots & x_n^t \end{bmatrix}$ is a set of binary solution in the tth generation. one binary solution $x_j^t, j = 1, 2 \cdots n$,is a binary of length of m, and is formed   by selecting each bit using the probability of qubit, either $|\alpha_i^t|^2$ or $|\beta_i^t|^2, i = 1, 2, \cdots m$ .Generate a random number r from the range[0,1].if r>$|\alpha_i^t|^2$, the bit of the binary string is set to 1,otherwise 0.Each solution x_j^t is evaluated to give some measure of its fitness.

3   Simulation Experiment and Result Analysis

Task scheduling in grid computing is used to investigate the performance of QGSAA. In the grid computing, task scheduling is essentially to distribute n interdependent tasks to m heterogeneous available resources to make the whole task fulfilled in the shorten time and the resources used as fully as possible [5]. Here is the general definition of this problem:

Definition 1. Supposing $\Psi = \{R, D, L\}$ represents the gird computing environment. $R = \{r_1, r_2, \ldots, r_m\}$ represents the set of m heterogeneous resources ; $D = \{d_1, d_2, \ldots, d_n\}$ represents the set of n interdependent tasks; $L = \{l_1, l_2, \ldots, l_m\}$ respectively represent the dynamic load weight of m resource nodes.

Definition 2. Supposing the load parameter of resource node r_i such as usage rate of CPU, usage rate of memory, current network flow, access rate of the disk I/O, process amount are respectively represented by $C(i)\%$, $M(i)\%$, $N(i)$, $Io(i)\%$, $P(i)$ .the dynamic load weight of resource node r_i can be formulated as:

$$l_i = \pi_1 \times C(i)\% + \pi_2 \times M(i)\% + \pi_3 \times N(i) + \pi_4 \times Io(i)\% + \pi_5 \times P(i) \quad (5)$$

In which$\Sigma\pi_i = 1$ (i=1, …,n) , π_i reflects the importance degree of the each load parameters[6,7].

Definition 3. Supposing$\Theta$is a m×n matrix, in which Θ_{ij} stands for the complete time of the task d_j at the resource node r_i in the ideal state (without considering the load of the resource nodes).$\Delta$is a m×n matrix , in which $\Delta_{ij}=1$ stands for the task d_j distributed to the resource node r_i, otherwise$\Delta_{ij} =0$.

Definition 4. Supposing X represents one scheme of all possible distributive scheme, then under such a strategy, the target function (complete time) of the task scheduling in grid computing is $F(x) = \sum_{i=1}^{m} \sum_{j=1}^{n} C \times \Delta_{ij} \times \Theta_{ij} \times \sqrt{l_i} / \sqrt{\sum_{i=1}^{m} l_i}$, in which C is a constant.

To check the research capability of the operator and its operational efficiency, such a simulation result is given compared with the GA and QGA. The major parameters employed in the algorithm such as the population scale M =100, the initial annealing temperature T_0 =1, the terminal annealing temperature is 0, the crossover and mutation probability are 0.85 and 0.05 respectively. The platform of the simulation experiment is a Dell power Edge2600 server (double Intel Xeon 1.8GHz CPU, 1G memory, RedHat Linux 9.0, Globus Toolkit3.0 as middle ware).

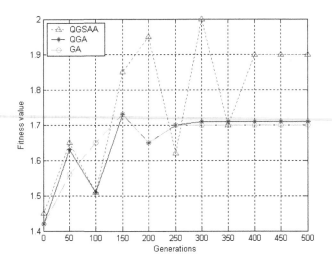

Fig. 1. The comparison of fitness value (50/3 problem)

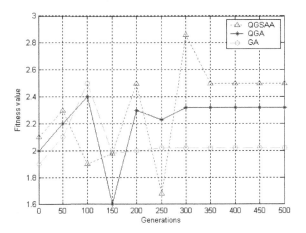

Fig. 2. The comparison of fitness value (50/5 problem)

Fig 1-2 are the change of fitness value on QGSAA, QGA and GA with the increase of evolution algebra in 50/3(50 tasks and 3 resource nodes), 50/5 problem. Figs 3-4 show the static quality of the algorithm and list the optimal solutions search by the algorithm in the different evolution algebra. They illustrate the simulation results of 50/3, 50/5 problem respectively.

In Figs 1-2, fitness values fluctuated acutely in the course of implement QGSAA, which indicate that QGSAA may keep father population diversity and avoid the phenomenon of prematurely convergence. In Figs 3-4, we can educe that QGSAA has a higher convergence speed and more reasonable selective scheme which guarantees the non-reduction performance of the optimal solution. Therefore, QGSAA is better than QGA and GA through the experimental results.

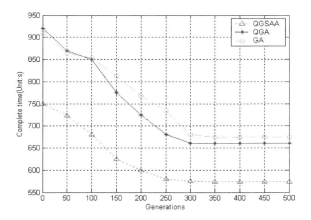

Fig. 3. The comparison of complete time (50/3 problem)

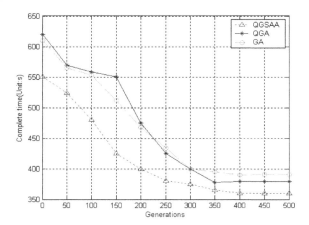

Fig. 4. The comparison of complete time (50/3 problem)

4 Conclusions

By combining quantum search and evolutionary search, a framework of QGSAA is proposed for task scheduling in grid computing. The experimental results of the task scheduling problem demonstrate the effectiveness and the applicability of QGSAA. The superiority of QGSAA is that combined with the advantages of QGA and GA so as to avoid premature convergence. Future research will investigate the appropriate quantum gates in compliance with the practical problem and explore other applications.

References

1. Wang, L.: Intelligent Optimization Algorithm with Application. Tsinghua University & Springer Press, Beijing (2001)
2. Narayanan, A., Moore, M.: Quantum inspired genetic algorithm. In: IEEE International Conference on Evolutionary Computation, Piscataway, pp. 61–66. IEEE Computer Society Press, Los Alamitos (1996)
3. Han, K.H., Perk, K.H., Lee, C.H., Kim, J.H.: Parallel quantum-inspired genetic algorithm for combinatorial optimization problem[A]. In: Proceedings of the 2001Congress on Evolutionary Computation[C], USA, IEEE Press, NJ, New York (2001)
4. Han, K.H., Moore, J.H.: Genetic quantum algorithm and its application to combinatorial optimization problem[A]. In: Proceedings of the 2000 IEEE Congress on Evolutionary Computation[C], USA, IEEE Press, NJ, New York (2000)
5. Shu, W., Zheng, S.: A Parallel Genetic Simulated Annealing Hybrid Algorithm for Task Scheduling. Wuhan University Journal of Natural Sciences 12(5) (2006)
6. Zheng, S., Shu, W., Chen, G.: A Load Balanced Method Based on Campus Grid. In: ISCIT 2005. International Symposium on Communications and Information Technologies, Beijing (October 12-14, 2005)
7. Shu, W., Zheng, S.: A Real-course-based Load Balanced Algorithm of VOD Cluster. In: ISCST 2005. International Symposium on Computer Science and Technology, Ningbo (October 20-24, 2005)

Optimized Research of Resource Constrained Project Scheduling Problem Based on Genetic Algorithms

Xiang Li, Lishan Kang, and Wei Tan

School of Computer, China University of Geosciences, Wuhan, China, 430074
lixiang@cug.edu.cn

Abstract. This paper mainly discusses the genetic algorithms in optimization of network planning project. It focuses on the study of multi-task scheduling (Resource-constrained project scheduling problem, RCPSP). For the problem of the resource-constrained task scheduling problem, the paper proposes a method to improve the genetic algorithms optimization of multi-task scheduling problem. From the aspects of the establishment of the optimized algorithms models, the algorithms design, the accomplishment of algorithms as well as the analysis of the result, we conducts this study in detail, and makes a comparison analysis with the other 11 heuristic methods. In this way, we testify our method and get the good results.

Keywords: Network planning Optimization, Genetic Algorithms, Multi-task.

1 Introduction

Network planning Optimization is very important in the project management, which determines the profit of project. In the long-term practical application, Network planning Optimization always adopts the network planning techniques of operations research and mathematical programming methods. However, these methods have many drawbacks. A genetic algorithms is a parallel global search and efficient solving method. It essentially deals with the discrete optimization search problem. It does not require the continuity of space issues and need the gradient information. Its Robust has been confirmed and made remarkable achievements in the large complex optimization problems. So compared with other methods, it has more matched advantages in solving large-scale multi-task problems.

2 Resource—Constrained Project Scheduling Model

Assuming there are several parallel tasks, and a shared resource pool containing a number of resources that can update (renewable resources), and all our resources which are supplied limitedly. In addition to sharing resources, tasks are independent with each other. To facilitate the description of the problem, the establishment of the mathematical model is as follows: RCPSP has P independent Multi-task scheduling

L. Kang, Y. Liu, and S. Zeng (Eds.): ISICA 2007, LNCS 4683, pp. 177–186, 2007.

tasks; the task k contains nk+ 1 works. The nk+ 1 task is for terminating work for virtual tasks, and it self occupies no resources and time. P missions' works are to share this kind of M renewable resources, the total for the mth resource is Rm. With Wi expressed the ith misson's work set, Wij expressed in the ith task jth work, its duration is dij, the requirement of the mth resources is rijm, the starting time is Sij, its all tight front duties formed the collection is Pij. The set composed of all tasks on time t is named I_t. Considering the importance of different tasks, ak is used for the weight k task. Taking these assumptions and symbols, a multi-task scheduling with resource constraints can be described as formula:

$$\min\sum_{k=1}^{P}\partial_k *(S_{k,n_k+1})\tag{1}$$

$$s.\ t.\qquad S_{i,j}\geq S_{i,h}+d_{i,h}\ ,\quad \forall h\in P_{ij},\forall i,j\tag{2}$$

$$\sum_{w_{i,j}\in I_t}r_{ijm}\leq R_m\ ,\quad \forall m,t.\tag{3}$$

$$I(t)=\left\{Wkj\in\bigcup_{k=1}^{P}Wk\mid S_{kj}\leq t\leq S_{kj}+d_{kj}\ \right\}\tag{4}$$

$$r_{ijm}\geq 0\tag{5}$$

$$\sum_{k=1}^{P}\partial_k=1\tag{6}$$

3 Algorithm Design

3.1 RCPSP Problems and the Solution

As a number of tasks share common resources, so the solution is different form single tasks'. In the way to solve this problem, some documents have proposed a bilevel decision method of multi-resource constraint and multi-task scheduling problem[4]. However, this method has two major flaws. First, the prerequisite is that the duration of shared resource in each task is determined by the distributed resource. The use of shared resources and the allocation of resources were inversely related to working time. (If two machine work six days, four machines will work three days) Working time = Workload + Resources. In real life, the workload is not an accumulation work. Working time = Workload /Resources. So the tasks do not accumulate in the work of the workload. The model can not be solved. The second defect is that the model adopts the specific resource allocation scheme between tasks. When resources are allocated to a particular task, it will always belong to the task; other tasks can no longer use the resources. This approach would make a larger allocation of resources wasted. Because such a distribution strategy, There is not exist mutual adjustment of flow of resources between tasks. So when a task in which there is a sort of free resources, other tasks will not use,and it is wasted.

The literature [6] also discussed the multi-task scheduling problem and its solution. However, the solution is to add more tasks network plan to merge into a virtual work of the network plan for the Solution. This method lowers the efficiency and flexibility of the solution.

This paper presents a method based on genetic algorithms to solve tasks of the type of work for more than an accumulation of the task. Between the various tasks and resources are dynamic to the circulation of each other. It will not waste the resources. Various tasks to multi-task and do not plan to conduct any merger of the network. Only needs to input all the tasks, resources and other information, we will be able to carry out all tasks scheduling.

3.2 Algorithm and Data Structure

Before the generation of initial population, it is first necessary to establish a data structure for storing information of the various tasks.

First, defining a task structure.

```
struct project
{ int n;   int TP; int TC;    int TS;   work *wMAX; entity
*EMAX;}*pro;
```

Among them, n is the number of tasks, TP is the task completion time. TC is the calculated completion time of the task, TS is the actual completion time of the task, and work is the definition of the structure. wMAX task is working at the point-array. Entity is the definition of the network structure. EMAX is the point-array related to the each task. Dynamic information to the importation of various tasks, Cross linked-list node is used store the relationship between the various tasks.

Cross linked-list node of the structure as follows:

```
typedef struct cnode
{ int hang; int lie; int value; struct cnode *down;
struct cnode *right;}crossnode;
typedef struct
{  int  n;  crossnode  *hcMAX;}clinknode;  clinknode
*Head=new clinknode;
```

'hang' and 'lie' in the structure crossnode mark the position in the linked-list. 'value' is the index of the work, 'down' is the eligible point of it and 'right' is the immediate point. 'In the 'clinknode' hcMAX' and 'Head' is the head note of cross linked-list.

In this paper, a key of the multi-task scheduling is for all of tasks using a cross linked-list. And the identification of various tasks is proposed in the linked-list. After the establishment of schedule, the activation of various tasks is based on the linked-list. This is an innovation of this paper.

3.3 The Structure and Coding of Chromosome

Resource-constrained multi-task scheduling problems can be also considered that the scheduling problem with certain constraints. The key to solving the problem is how to use genetic algorithms to find an appropriate sequencing of multitask. According to a certain sequence of chromosome, we code all of issues on the table. If there are M tasks, each task has n works, then there m*n works. The first expression Vk is Kth chromosome in current population. Suppose Pkij is a gene of chromosome. It expressed a work of the Pkth task at location j of chromosome.

Chromosome can be expressed as follows:

Vk=P1i1,..., P1ii,...,P1in,...,Pki1,..., Pkii,...Pkin,...,Pmi1,..., Pmii,...,Pmin
where Pkij bound by the random order with the constraint.

3.4 Initialization Chromosome

For the Chromosomes initialization, this paper is divided into two parts. One is randomly generating various chromosomes by the topological scheduling, the other is generating chromosome through a variety of heuristic methods.

Method 1: Randomly generated initial chromosome

Similar with the optimal scheduling method with the shortest Resource Constrained time of the single task, the only difference is that for every work. Firstly judge the task to which it belongs, then determine the constraint relationship. It is determined the first consider work by a Left to Right approach with a fixed sequence. At each stage of the work order has been maintained and assembled Scheduling can work and completely random from a pool selected to work in the pool. This process is repeated indefinitely until all the work was arranged.

In each iteration of the process all the work at one of the following three conditions:

1. The work has index: Part of the chromosome structure in the work.
2. The work can schedule: All immediate works are ordered.
3. Free work: All of work

v_k^t is a part of chromosome, including t activities. Q_t is in phase t with a corresponding arrangements can be made to the activities set, Pi is all the activities set, with a corresponding set of arrangements can be made activity set Q_t is defined as :

$$Q_t = \{ j \mid P_i \subset v_k^t \} \tag{7}$$

It includes specific arrangements in the end nodes, and it is all competition activity set. Scheduling changes which may work in the following manner: a. deleted Qt from the work j which has been selected. b. judging if there is work k in all immediate works or not, and it's all immediate works are ordered . c. if there is k, it can be added to the scheduling work set Qt.

Method 2: Initializing chromosome through various heuristic methods

Bai Sijun listed 31 heuristic methods. Based on the resources 11 more effective heuristic ones are picked out here6, 7. Application of this 11 heuristic methods generates a good number of chromosomes of relatively activation effects.

The11 heuristic criteria are as follows:

(1) Minimum total time difference priority standard (MinTS), according to the earliest starting time if they are same.
(2) Maximum total time difference priority standard (MinTS), according to the earliest starting time if they are same.

(3) Minimum criteria for priority period (SOF), according to the numbers of work if they are same.

(4) Maximum criteria for priority period (SOF), according to the numbers of work if they are same.

(5) LS priority minimum standard (MinLS), according to the numbers of work if they are same.

(6) EF priority minimum standard (MinEF), according to the numbers of work if they are same.

(7) LF priority minimum standard (MinLF), according to the numbers of work if they are same.

(8) ACTIM standard, the different between the length of the critical path and the latest starting time, the greater the value has the priority.

(9) Maximum priority greatest resource requirements standard (MaxR).

(10) Minimum priority greatest resource requirements standard (MinR).

(11) Most intended tight works priority criterion (MostIS).

The literature 9 also makes some heuristic criteria considered: First considering the work on the critical path. If there are conflicts among several key tasks, considering the following programs:

(1) The smallest total time priority. (2) First start priority. (3) First completion priorities. (4) Priority for the most intense work, the remaining processing time and delivery time ratio is smaller becomes tenser.

3.5 Genetic Operators Design

On the base of established cross linked-list, re-number various tasks in the work. Then apply the algorithms described in this paper. All tasks will be mixed with the work. After encoding into a chromosome, the problems become simple. Genetic chromosome operation with the single task is similar to the chromosomal genetic operations.

3.5.1 Crossover

PMX is used in literature [10] (Partly Matched Crossover), but the operator would have to generat) illegal entity. And it will need to repair chromosome, which will reduce the efficiency of the implementation process. Here it proposes a method that it is not used to repair chromosome according to the discrete crossover operator.

Participate in the two crossover operator for mothers M and fathers F. We choose a random integer q, $1<=q<=J$, J is chromosome length. M and F through radio spots overlapping operations produced two generations of daughters D and sons S. In the list D work, the former q tasks come from M.

$$j_i^D = j_i^M, i = 1, 2 \ldots q$$

i = q + 1,...J, come from the location of the F, And the relative position of the F resurveyed.

$$j_i^D = j_k^F, i = q+1, q+2, \ldots, J$$

Which $k = \min\{k \mid j_k^F \notin \{j_1^D, j_2^D, \cdots, j_{i-1}^D\}, k = 1, 2, \cdots, J\}$

S and D are similar to the formation of the linked-list, not going to repeat here.

3.5.2 Mutation

Mutation operator adopts centralized search strategy combined with the neighborhood technology to improve the offspring. D genes will not exceed the change in the scheduling of the neighborhood gathered as scheduling x (Figure 1). If x neighborhood than any other solution, then x is called scheduling optimization. Mutation process is as follows:

Individual i

A	B	C	D	F	E	H	G	I

The neighborhood of individual i

A	B	C	F	D	E	H	G	I
A	B	D	C	F	E	H	G	I
A	B	D	F	C	E	H	G	I
A	B	F	C	D	E	H	G	I
A	B	F	D	C	E	H	G	I

Fig. 1. Neighborhood maps chromosome

```
Begin
  If (rand () <pmutation)
  Take n continuous in a row as the chromosome genes
  Permutations and combinations of the n genes
     Check each gene sequence, the genome sequence will
  be its discarded if it unreasonable;
  Assessing all the neighborhood scheduling;
  Neighborhood as the best choice for future generations
End
```

In the process of mutation, two consecutive complete gene permutations and combinations. Of course, there has a violation of constrain sequence. Here can directly be discarded, but will not have an impact on the choice of the best neighborhood. This method avoids repairing the chromosome after mutation and improves the efficiency of the procedure.

3.6 Evaluation Function

Calculate the shortest duration of tasks, we must first decode chromosome. Calculate the earliest starting time, and then find the earliest completion time for each task. Finally, according to the weight ratio of the various tasks calculate the shortest weighted average duration. Reached the objective function formula (8):

$$\min \sum_{k=1}^{P} \partial_k * (S_{k,n_k} + 1) \tag{8}$$

Since the task is to minimize the total built-constrained Project Scheduling Problem issues. We change the original objective function to ensure that the individual is suited to meet the greatest value. The current population vk is located k chromosome; g (vk) is a fitness function. f (vk) is the original target value. fmax and fmin are the biggest target value and the minimum target value. Conversion methods such as formula (9)

$$g(v_k) = \frac{f_{max} - f(v_k) + \varepsilon}{f_{max} - f_{min} + \varepsilon} \qquad (9)$$

where ε is a positive number, often limited the open interval (0,1).

3.7 Choose Operator

The operation that winning individuals are chosen from the group and the bad individuals are eliminated is named selection.

Selection is based on the currently popular "breeding pool (Breeding Pool) choice", "To adapt to-value ratio of options", "ranking selection (Ranking Selection) ","Mechanism based on local competitive choices", and so on.Roulette choice (Roulette Wheel Selection) [10] is based on fitness than A choice of the most widely used methods. It is the first calculation of the relative fitness of individual fi/$\sum$fi, named Pi, and then chooses probability {pi. = 1,2,..., N) a disk cut into N copies, 2πPi fan angle of the center of Pi. In making the selection, rotating disk can be assumed that if a reference point to fall into the itch fan, then chose individual i. Generation a random number r in 0, 1, if Po+Pi+... +Pi-1 "r<Pl+P2... +Pi, i individual choice, the assumption here Po=0. It is easy to see that this was very similar to roulette choice of trouble. For sector bigger area, it has the greater the probability fall in and it was an opportunity to be chosen.

Thus, the structure of the gene was likely to pass it on to the next generation greater. This paper discusses the use of multi-robin scheduling optimization algorithms choice. Every generation of new groups in the use of optimal preservation strategy that will preserve the best, so far in the contemporary individual and overcome random sampling error.

4 Algorithm Implementation

4.1 Example Data

The examples of cases come from literature 6. This is a task of three parallel scheduling problems with the same tasks of the three network structure. The network structure is shown in Figure 2. Each task has 16 works, all 48 work share 9 resources. The resources are here alone mode. Every of them can only be used as a resource. Different tasks have different durations, and types of resource requirements and demand, as shown in Table 1. The type and total of resources, as shown in Table 2.

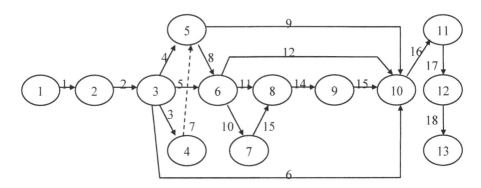

Fig. 2. Two mission codenamed network structure

Table 1. Task the resources required type and duration of work demand

Work Index	Begin Pot	Resources type	Mission 1		Mission 2		Mission 3	
			Resources	time	Resources	time	Resources	time
1	1,2	1	6	4	8	5	10	4
2	2,3	1	8	3	7	6	8	6
3	3,4	4	4	2	5	2	4	1
4	3,5	1	10	4	9	3	7	4
5	3,6	3	2	3	3	2	4	3
6	3,10	2	3	2	4	6	5	6
7	4,5	1	0	0	0	0	0	0
8	5,6	1	5	4	6	6	9	3
9	5,10	2	4	1	4	3	3	2
10	6,7	7	10	2	14	1	11	1
11	6,8	6	9	4	11	2	8	2
12	6,10	7	12	3	9	3	10	3
13	7,8	5	7	4	5	4	7	3
14	8,9	6	13	3	10	2	12	4
15	9,10	7	16	2	13	4	14	2
16	10,11	8	6	3	5	3	16	4
17	11,12	9	3	4	4	4	5	4
18	12,13	1	0	0	0	0	0	0

4.2 Genetic Algorithms Analysis and Conclusions

Set the weight coefficients of certain task are a1=0.3, a2=0.24, a3=0.36. The population size is 120, the max evolutional generation is 100, crossover probability is 0.8, mutation probability is 0.01.Mutation in the neighborhood length(λ) is 3. Set up the starting time of three tasks is same, without resource constraints. The completion time of tasks obtained by CPM is: 33, 38 and 35 days. The results are as follows: When the genetic algorithms optimization process optimization evolutionary adaptation to the 13th generation of convergence function, there chromosome best when the objective function value of 47.9 to produce the optimal chromosome

(36,38,40,42,39,43,0,44,44,45,1,2,6,5,4,3,8,7,18,46,47,48,9,11,10,19,23,21,20,24,22,
12,13,14,15,16,26,25,27,28,29.49,30,50,51,52,53,31,32,33,34,35,17).

Table 2. The type and count of resource

Index	Name	Total	Index	Name	Total
1	Designer	10	6	Machining Center	14
2	Buyers	6	7	CNC Machining Center	18
3	Craft	4	8	Assembly center	7
4	Programmer	5	9	D center	5
5	Heat Treatment Center	8			

Set the weight coefficients of certain task are a1=0.3,a2=0.24,a3=0.36. The population size is 150, the max evolutionary generation is 100, crossover probability is 0.8, mutation probability is 0.01.Mutation in the neighborhood length (λ) is 4. Set up the starting time of three tasks is same, without resource constraints. The completion time of tasks obtained by CPM is: 33, 38 and 35 days. The results are as follows: When the genetic algorithms optimization process optimization evolutionary adaptation to the 57th generation of convergence function, there chromosome best when the objective function value of 47.9 to produce the optimal chromosome (36, 37,38,40,42,39,41,43,0,44,1,2,6,45,5,4,3,7,8,18,46,47,48,9,19,11,22,10,23,21,20,12,1 3,14,15,16,24,49,50,17,26,25.28,27,29,51,30,31,32,52,53,33,34,35). One of the first tasks to be the beginning of time sequence followed (17,21,24,24,24,24,26,28, 28,32,32,34,34,38,41,43,46,50). The order of the first two tasks were to starting time (32,37,43,43,43,43,45,46,49. 52,52,53,53,57,59,63,66,70). The order of the first three tasks were starting time (0,4,10,10,10,16,11,14,14,17. 17,18,18,21,25,27,31,35).First mission is expected to be 50 to 70 hours to complete the first two tasks, task completion time of three to 35. In literature 6, this case optimized to achieve the best value for the objective function 59.44. In this paper the optimal objective function value of 47.9. Obviously, the methods used in this paper the optimization results better than the original. Heuristic Optimization compared with other methods such as Table 3.

Table 3. Scheduling algorithms for the different tasks

Algorithms	Task1	Task 2	Task3	Targeted days
GA In this paper	50	70	35	47.9
MinTS	76	64	70	70.36
MaxTS	67	76	72	71.46
SOF	37	70	55	53.2
MOF	70	66	76	71.8
MinLS	65	76	70	69.94
MinLF	62	74	68	67.64
MinEF	65	73	69	68.76

Table 3. (*continued*)

ACTIM	78	74	70	73.36
MaxR	80	66	60	67.44
MinR	67	76	72	64.2
MostIS	66	70	74	70.64
Critical of Path priorities (1)	76	64	70	70.36
Critical of Path priorities (2)	67	76	72	71.46
Critical of Path priorities (3)	76	46	70	66.04
Critical of Path priorities (4)	73	45	67	63.52

From Table 3 we can see, the effect of this optimized genetic algorithms is much better than other algorithms.

5 Conclusions

In this paper, resource-constrained multi-task scheduling problem, a new genetic algorithms approach is put forward. It overcomes certain deficiencies of the bilevel model proposed by Tan Ye[4]. Furthermore it does not need to add virtual network in the preparation of work plans for the merger of all tasks for a network. The method to solve large-scale multi-task problem is better than other methods.

References

1. Feng, C.-W., Liu, L., Burns, S.A.: Using Genetic Algorithms to Solve Construction Time-Cost Trade-Of Problems. Journal of Computingin Civil Engineering 11(3), 184–189 (1997)
2. Li, H., Love, P.: Using Improved Genetic Algorithms to Facilitate Time-Cost Optimization. Journal of Construction Engineering and management 123(3), 233–237 (1997)
3. Hegazy, T.: Optimization of Resource Allocation and Leveling Using Genetic Algorithms. Journal of Construction and Management 125(3), 167–175 (1999)
4. Yue, T., Zhong, W., Xu, N.: Two Decision-making Method of Multiple Resources Attributing Among Multi-projects. Systems Engineering Journal 14(3) (1997)
5. Liao, R., Chen, Q., Mao, N.: Genetic algorithms for resource - constrained project scheduling. Computer Integrated Manufacturing Systems 10(7) (2004)
6. Sijun, B.: Heuristic Mechod for Multiple Resource-Constrained in the Network. Systems Engineering Theory and Practice (7) (2004)
7. Sijun, B.: Evaluating Heuristics for Re source-constrained Activity Network(III), vol. 8(4) (December 1999)
8. Ren, L., Qingxin, C., Ning, M.: A Heuristic Algorithms for Resource–ConstrainedProject Scheduling, Journal of Engineering Management,supplement (2002)
9. Guangnan, X., Zunwei, C.: Genetic Algorithms and Engineering Design. Science publishing company (2000)
10. Xiaoping, W., Liming, C.: Genetic Arithmetic-theory, application and implement with sofrware. Press of Xi'an Jiaotong University (2002)

An Evolutionary Agent System for Mathematical Programming

Abu S.S.M. Barkat Ullah, Ruhul Sarker, and David Cornforth

School of Information Technology and Electrical Engineering,
University of New South Wales at the Australian Defence Force Academy,
Canberra 2600, Australia
{barkat, r.sarker, d.cornforth}@adfa.edu.au

Abstract. During the last decade, both evolutionary computation and multi-agent systems have been used for solving decision and optimization problems. This paper proposes a new evolutionary agent system by incorporating evolutionary process into agent concepts for solving mathematical programming models. Each of the agents represents a candidate solution of the problem, and able to sense and act on the society. The fitness of the agent improves through co-evolutionary adaptation of society with the individual learning of the agents. The performance of the proposed algorithm is tested on five new benchmark problems along with existing 13 well-known problems, and the experimental results show convincing performance.

1 Introduction

Evolutionary Computation (EC) have brought a tremendous advancement in the area of computer science and optimization with their ability to solve complex real world problems [1]. However, traditional EC is much simplified and lacks many important features of natural evolution process like dynamically changing environmental conditions [2], self learning of the individuals etc, and may suffer from premature convergence and so are not well suited for fine tuning search [3-6]. To improve the performance of EC, different hybridization of algorithms has been introduced in recent times e.g. memetic algorithms (MAs). One of the critical issues regarding the performance of MAs is the selection of appropriate local search (LS) while hybridizing LS with Genetic Algorithms (GAs). However if the selection of LS is not appropriate for a particular problem then MAs may not perform well; the performance may even be worse than GAs alone [7-9]. Again it is very difficult to know which type of LS is appropriate for a particular problem.

Recently some researches have been carried out to incorporate evolutionary processes into agent based systems [2, 6, 10-13]. Agent-based computation introduces a new paradigm for conceptualizing, designing and implementing intelligent systems, and has been widely used in many branches of computer science [14, 15]. The Agents are discrete individuals situated in an environment having a set of characteristics and rules to govern their behavior and interactions. They sense the environment and act on

L. Kang, Y. Liu, and S. Zeng (Eds.): ISICA 2007, LNCS 4683, pp. 187–196, 2007.
© Springer-Verlag Berlin Heidelberg 2007

it in pursuit of a set of goals or tasks for which they are designed [16]. For solving optimization problems some agent-based evolutionary algorithms have been introduced newly e.g. Dobrowolski et al [17] used evolutionary multi-agent system in solving multiobjective optimization, Siwik et al [11] used semi-elitist evolutionary multi-agent systems for multiobjective optimization, Zhong et al [6] used a multiagent genetic algorithm (MAGA) for global numerical optimization and Liu et al [12] used a multiagent evolutionary algorithm for constraint satisfaction problems. In MAGA [18], four evolutionary operators: neighborhood competition operator, neighborhood orthogonal crossover operator, mutation operator and self-learning operator are used to realize the behaviors of the agents.

Inspired by the performances of these integrated algorithms in this paper we have introduced an evolutionary agent system (EAS) for mathematical programming. Here an agent represents a candidate solution of the problems, goes through evolutionary processes and tries to find an improved solution. Agents carry out cooperative and competitive behaviors, and select an appropriate life span learning processes (LSLPs) adaptively. The main goal of an agent is to survive by improving its fitness i.e. to improve the objective function value of the problem otherwise leave room to the new potential agents i.e. it dies. Unlike the traditional evolutionary operators: selection, crossover, mutation in the proposed algorithm, the agents cooperate and compete through neighborhood orthogonal crossover [6, 19] and apply different types of life span learning process to solve a problem. Here we have designed the LSLPs based on several local and directed search procedures. The agent chooses a LSLP as a search operator adaptively. In traditional evolutionary algorithms (EAs), an individual in a generation produces offspring and the offspring may be mutated to change its genetic materials. However in reality, beside reproduction, an individual learns and gains experiences different ways during its life time which improves the individual's genetic materials. We have represented this process by LSLPs in this paper. An individual in the population of a certain generation lives for a certain period of time, and explores the environment and interacts with other individuals to enhance learning and then dies.

To test the performance of the algorithm, 13 well-known problems widely used [20-22] plus five new problems [23, 24] are solved and the results are compared with several existing well-known algorithms [20, 25-29]. The comparisons show that the results are of quite acceptable quality.

The rest of this paper is organized as follows. Section 2 describes the proposed evolutionary agent system. Section 3 describes the performance of the proposed approach on 18 benchmark problems and comparison of its results. Finally, Section 4 concludes the paper and provides future research direction.

2   Evolutionary Agent System

In natural systems the individuals of the population develop through the evolutionary adaptation of the population and individual learning of an individual [4, 30, 31]. Besides the evolutionary adaptation, certain individuals may develop themselves through self learning and exploiting their own potential. The agents may learn in different ways to improve their performances, and make their own decisions. This process is not centralized like traditional evolutionary algorithms. Following the natural adaptation model in the proposed EAS, the agents develop their fitness by different types of self learning processes with the evolutionary adaptation of the population.

In the environment E, the agents are arranged in a lattice-like form of size $M \times M$ (where M is always an integer). The agents communicate with their surrounding neighbor agents and interchange information with them through comparison and the crossover operator. Certain percentages of agents are allowed for different types LSLPs. As an evolutionary operator, when orthogonal crossover operator is applied on an agent, the agent senses its neighborhood and selects the best neighboring agent to mate. If the offspring is better than the parent then the parent dies to make room for the offspring. Every agent interacts with its neighborhood to exchange information, so the information is diffused to the whole agent lattice. A certain percentage of the agents are selected for different types of LSLPs for exploitation of the currently obtained solution. The agent tries to improve its present fitness during the learning process. Different types of LSLPs represent different type of local searches. As an appropriate choice of a LSLP is very important for the performance of the algorithm, here an agent selects one of the several LSLPs in an adaptive manner.

The main steps of the proposed algorithm are follows:

Step 1. Create a random population which consists of $M \times M$ agents.
Step 2. Arrange the agents in a lattice-like environment.
Step 3. Evaluate the agents individually.
Step 4. If the stopping criterion has been met, go to step 8; otherwise continue.
Step 5. For each agent examine its neighborhood.
         -Select an agent from its neighborhood and perform crossover.
Step 6. Select a certain percentage of agents.
         -Select a life span learning process for certain steps based on several rules.
Step 7. Go to step 3.
Step 8. Stop.

Details of the operators are discussed below.

2.1 Crossover Operator

We have used neighborhood orthogonal crossover operator which was proposed by Leung [19]. When this crossover operator is applied on an agent $A_{i,j}$, the agent searches for its best neighbor agent e.g. $B_{i,j}$ to mate. Then the best offspring among the uniformly generated individuals compares its fitness with $A_{i,j}$. If the new agent's fitness is better then it takes the place of $A_{i,j}$ and $A_{i,j}$ dies. The details of the crossover used here can be found in [6, 19].

Basically, the orthogonal crossover operator quantizes the search spaces defined by the two parents and generates new individuals. The search space is quantized into a finite number of points, and then orthogonal design is applied to select a small, but representative sample of points as the potential offspring. The reason of using this crossover is to search for a better offspring within the search space defined by the parents.

2.2 Different Types of Life Span Learning Processes

A certain percentage of agents (with P_L probability) of the population are selected for LSLP i.e. to apply local search after crossover. For EAS, we have designed four types of LSLPs. These LSLPs try to find better solutions by exploiting the present solution vectors of the agents in different ways.

The first and second LSLPs are random in nature. In the first LSLP, each of the variables of the solution vector is selected randomly. Then the selected variable is changed with a small perturbation Δ for several times. The changing a variable may be stopped if it decreases the fitness.

The second LSLP type is same as the first, except for the sequence of variable selection. Instead of selecting the variables randomly, the variables are selected in ascending order of the index i.e. a_1 will be selected first then a_2 and so on.

The third LSLP is like the gradient based search. Initially every variable is changed by Δ. Then based on the effect of changing each variable on the fitness, the variables are ranked. Variables are selected to be modified according to the rank. Suppose by changing a_1 in the positive direction, the fitness of the solution of the agent is improved by 10%, for a_2 the improvement is 50%, and for a_3 the improvement is 30%. Then the sequence of selection of the variables will be a_2, a_3, and then a_1.

The last type of LSLP is like directed search and tries to follow the best agent in the current generation. Here, the selected agent $A_{i,j}$ with solution vector $[a_1, a_2, ..., a_n]$ tries to reach nearer to the best agent of that generation with solution vector $[b_1, b_2, ..., b_n]$. It attempts to change each variable $a_1, a_2, ..., a_n$ by Δ to get closer to $b_1, b_2, ..., b_n$ by maximum m learning steps.

The third type of learning process attempts to change the solution vector in a directed way and the last type of LSLP leads the agents towards the current best solution. The last two type LSLPs try to make the agents converge; on the other hand the first two types maintain diversity. The details of the LSLPs can be found in our another paper [32].

2.3 Fitness Evaluation and Constrained Handling Techniques

The goal of the individual agent is to survive by minimizing the objective function value of the problem in hand with satisfying the constraints. The agents first apply crossover operators with their best neighbors. The best neighbor is found by using pair-wise comparison among the neighbors. The pair-wise comparison indirectly handles the constraints. The fitness of an agent is evaluated by comparing its corresponding objective function value, however for constrained optimization the scenario is different; when an agent violates the constraints i.e. an infeasible agent, it should have less fitness than an agent with a feasible solution. Like Deb [33], while comparing the fitness of two individual agents in minimization problems we have considered:

1. A feasible individual is always better than an infeasible individual.
2. If both of the individuals are feasible, then the individual with lower objective function value is better.
3. If both of them are infeasible, then the one with less constraint violation is better. The total constraint violation of an individual is considered here as the sum of absolute values by which the constraints are violated.

As such while comparing two agents, the infeasible agent is penalized and feasible agent is rewarded, so the constraints are handled indirectly.

2.4 Selection of Life Span Learning Processes

A certain percentage of the agents are allowed to apply different type of learning processes to exploit their present values and improve their fitness. For the selection of process of an appropriate LSLP, the agents apply an adaptive technique. Initially all the agents are assigned different types of LSLPs randomly with Improvement Index (I.I.) zero. Here, I.I. indicates the rate of fitness improvement made by a particular LSLP type assigned to an agent. A positive value of I.I. indicates that fitness is improved by the LSLP, while a negative value indicates deterioration of fitness. When selecting a LSLP, an agent checks the parents' type of LSLPs and I.I. values. The parents may have different type of LSLPs associated with I.I. values. The offspring will choose the LSLP which has better I.I. value.

While calculating the I.I., if an infeasible solution vector of an agent becomes feasible after applying a LSLP, the I.I. of that LSLP for that agent is assigned +1. If the situation is opposite, i.e. from a feasible to infeasible then I.I. is assigned -1. For minimization problems, as discussed in the previous subsection, we should consider the objective function values for agents with feasible solution vectors, and constraint violations for infeasible agents. If the solution remains feasible before and after the LSLP, the improvement index for feasible agent is based on the objective functions:

$$I.I. = \frac{Obj.\ func.\ value\ before\ LSLP - Obj.\ func.\ value\ after\ LSLP}{Obj.\ func.\ value\ before\ LSLP} \tag{1}$$

If an agent with an infeasible solution uses a LSLP which still results in an infeasible solution, the improvement index for infeasible agent is based on total constraint violations (CV):

$$I.I. = \frac{Total\ CV\ before\ LSLP - Total\ CV\ after\ LSLP}{Total\ CV\ before\ LSLP} \tag{2}$$

The value of I.I. is restricted in the range $-1 \leq I.I. \leq +1$ by assigning values outside this range to the boundary values.

3 Experimental Results and Discussions

We have used 18 benchmark problems to test the performance of the proposed algorithm. The first 13 problems are well-known in the literature (indicated as g1-g13), studied by Michaelwicz and Schoenauer [21], Koziel and Michalewicz [20], Michalewicz [22], and further studied by Runarsson and Yao [27], Chootinan and Chen [26], Elfeky et al [28] etc and the rest 5 problems (indicated here as p01-p05) are new and collected from the literature [23, 24]. The benchmark problems include different forms of objective function (linear, quadratic, cubic, polynomial, nonlinear) and different number of variables (n). The maximization problems are transformed into equivalent minimization problems.

The equality constraints of g03, g05, g11, g13, p01, and p02 $h_j(X)=0$ have been converted into inequality constraints $-\delta \leq h_j(X) \leq \delta$, where δ is a small tolerance value.

Initially δ is assigned 1 and after some generations it is divided by 10 up to 0.0001; which allows a large feasible search space at the starting and then reduces the search space after certain periods.

As the benchmark problems are completely different in number of variables, type of objective functions and characteristics of the constraints, for different types of problems we have used different parameters e.g. population size, probability of learning (P_L). Initially we have executed a single run for each of the problems varying the population size $(M{\times}M)$ from $M=9$ to 24 and P_L from 0.05 to 0.25 by varying 0.05. After this experimentation the population sizes (PS) and P_L are selected to give best results. The PS and the values of P_L used for the first 13 problems are given in the last column of Table 1. During the LSLP the agent is allowed at most $m=10$ steps. We have investigated the effect of learning steps: lower values of m delayed the convergence while larger values increase computational cost rather than performance. The maximum number of generations was 3500.

During the LSLP the variables are changed with $\pm\Delta$. The direction of Δ (add/ deduct) is selected by observing the modified fitness value of the agent. The value of Δ should be very small and decreasing with passing generations, as the solution vectors of the agents are improving gradually and need to be modified with high precision. We have considered $\Delta=absolute(G(0,1/g))$, where $G(0,1/g)$ is a Gaussian random number generator with zero mean and decreasing standard deviation $(1/g)$ as g is the present generation number.

We have defined the neighborhood of an agent in three ways, four neighbor approach (considering left, right, top and bottom agents), eight neighbor approach (considering left, right, top, bottom, left-top, right-top, left-bottom, and right-bottom agents), and combined approach (both four and eight neighbors approach alternatively) and experimented the performance of each of the approaches for the 13 benchmark problems in our previous paper [32]. The results show the performance of combined approach is better than the other two approaches as it posses both of their characteristics.

For the first 13 problems we have compared our algorithm (combined approach) with four previously published well-known algorithms [20] [26-28]. Runarsson and Yao (abbreviated as RY) [27] in their ES based algorithm used stochastic ranking to handle the constraints and solved all 13 problems. This algorithm is most well-known for its performance in these problems. Koziel and Michalewicz's [20] GA (abbreviated as KM) depends on a homomorphous mapping between an n dimensional cube and the feasible part of the search space. Its drawback is that requires an initial feasible solution. Chootinan and Chen [26](abbreviated as CC) used a repair procedure embedded into a simple GA as a special operator to solve the first 11 problems. Elfeky et al [28] used GAs with new ranking, selection, and triangular crossover methods (abbreviated as TC). TC has solved only nine problems which have only inequality constraints.

In Table 1 the best results of RY, KM, CC, TC, and proposed EAS from 30 runs are given. RY has found optimal results for eight problems, KM has found three optimal results, CC has found seven optimal results, and TC has found 5 optimal results. EAS has found optimal results for eight problems, including problem g02 which has not been solved by any of these other algorithms.

Table 1. Comparison of the best results of RY, KM, CC, and TC, and our algorithm (indicated as EAS) for the 13 benchmark problems and population size (*PS*) and probability of learning(P_L) used by EAS

| Prob | Optimal | Best results | | | | | PS/P_L |
		RY	KM	CC	TC	EAS	
g01	-15.000	**-15.000**	-14.786	**-15.000**	**-15.000**	**-15.000**	169/0.1
g02	-0.803619	-0.803515	-0.79953	-0.801119	-0.803616	**-0.803619**	361/0.05
g03	-1.000	**-1.000**	**-1.000**	1.000	-	**-1.000**	324/0.05
g04	-30665.539	**-30665.539**	-30664.5	-30665.538	**-30665.539**	**-30665.539**	225/0.25
g05	5126.498	5126.497	--	**5126.4981**	-	5126.5342	324/0.20
g06	-6961.814	**-6961.814**	-6952.100	-6961.814	**-6961.814**	**-6961.814**	144/0.20
g07	24.306	**24.307**	24.620	24.329	24.505	24.318	576/0.15
g08	-0.095825	**-0.095825**	**-0.095285**	**-0.095285**	**-0.095285**	**-0.095825**	100/0.15
g09	680.63	**680.630**	680.910	**680.630**	680.633	680.631	576/0.15
g10	7049.331	7054.316	7147.9	**7049.2607**	7049.474	7265.852	484/0.15
g11	0.750	**0.750**	**0.750**	**0.750**	-	**0.750**	196/0.15
g12	-1.000	**-1.000**	-0.999	--	**-1.000**	**-1.000**	256/0.20
g13	0.053950	0.053957	0.054000	--		**0.053954**	81/0.15

For stochastic search algorithms, sometimes the best results may be outlier by producing only a single best solution with other poor solutions. To detect this we need to perform statistical significance testing such as *Student's t-test. Student's t-test* compares the actual difference between two means in relation to the variation in the data. We have compared our EAS with RY using the *Student's t-test* in Table 2. We have considered the absolute value of t (indicated as t-C). If the calculated t value exceeds the tabulated t value, then it can be said that the means are significantly different with 95% confidence levels. If there is a significant difference we have indicated "Yes", otherwise "No", in Table 2. If the means are different then we have indicated the algorithm with lower mean (for minimization problems) as the better algorithm in the last column of Table 2. When there is no significant difference between the means, we have indicated Equal. The degrees of freedom is considered here 60.

Form Table 2 we can see for five problems EAS is significantly better than RY, for only two problems RY is better, and for the other six problems the performance of EAS and RY is Equal. These results suggest the superiority of EAS.

Table 2. *Student's t-test* between RY and EAS for 13 benchmark problems

| Fcn | Mean | | St. Dev. | | Significance Level | | Performance |
	RY	EAS	RY	EAS	t-C	95%	
g01	**-15.000**	**-15.000**	0.00E+00	2.39E-06	0.00	NO	Equal
g02	-0.781975	**-0.802032**	2.00E-02	1.34E-03	5.48	YES	**EAS**
g03	**-1.000**	**-1.000**	1.90E-04	4.68E-06	1.14	NO	Equal
g04	**-30665.539**	-30665.538	2.00E-05	1.45E-04	20.88	YES	**RY**
g05	**5128.000**	5130.259	3.50E+00	3.45E+00	2.52	YES	**RY**
g06	-6875.940	**-6961.811**	1.60E+02	1.24E-03	2.94	YES	**EAS**
g07	24.374	**24.348**	6.60E-02	2.23E-02	2.02	YES	**EAS**
g08	**-0.095825**	**-0.095825**	2.60E-17	9.51E-07	1.39	NO	Equal
g09	680.656	**680.638**	3.40E-02	3.73E-03	2.89	YES	**EAS**
g10	7559.192	**7411.024**	5.30E+02	6.99E+01	1.52	NO	Equal
g11	**0.750**	**0.750**	8.00E-05	1.86E-06	0.43	NO	Equal
g12	**-1.000**	**-1.000**	0.00E+00	0.00E+00	0.00	NO	Equal
g13	0.067543	**0.055283**	3.10E-02	1.22E-03	2.16	YES	**EAS**

To show the ability of our algorithm for solving new problems, we have also solved five more complex problems as shown in Table 3. Where n is the number of variables, LI is the number of Linear Inequality Constraints, NI is the number of Nonlinear inequality constraints, LE is the number of Linear Equality constraints, NE is the number of nonlinear equality constraints, a is the number of active constraints. The first four problems (p01-p04) are problem# 4a (page 396), #5 (page 397), #12 (page 407), and #16 (page 415) from [23] respectively and p05 is collected from [24]. We have compared the results of EAS with other two algorithms [25, 29] (with maximum FES 5×10^5) as these problems are not solved by any of these four algorithms [20, 26-28]. The multi-populated differential evolution algorithm [29] is indicated as MPDE in Table 3 and the modified differential-evolution based approach [25] indicated as MDE. The best, mean and standard deviations of the results achieved by these algorithms are given in Table 3. From the results given in Table 3 we can see that the results of EAS are very competitive with the other two algorithms.

Table 3. Definitions of the five new problems and comparison of results

Prob	n	Func. type	LI	NI	LE	NE	a	Optimal	Algo.	Best	Mean	St.Dev.
			Problem Definition									
p01	10	Nonlinear	0	0	3	0	3	**-47.765**	EAS	-46.778	-45.501	2.07E+00
									MPDE	-47.765	-47.511	2.87E-01
									MDE	-47.765	-47.765	0.00E+00
p02	3	Quadratic	0	0	1	1	2	**961.715**	EAS	962.941	965.234	1.28E-01
									MPDE	961.715	961.715	7.94E-15
									MDE	961.715	961.715	1.40E-05
p03	5	Nonlinear	4	34	0	0	4	**-1.905155**	EAS	-1.865617	-1.702509	2.08E+00
									MPDE	-1.905155	-1.905154	4.30E-06
									MDE	-1.905155	-1.905155	0.00E+00
p04	9	Quadratic	0	13	0	0	6	**-0.866025**	EAS	-0.866025	-0.865787	4.66E-05
									MPDE	-0.866025	-0.866025	0.00E+00
									MDE	-0.866025	-0.866025	0.00E+00
p05	2	Linear	0	2	0	0	2	**-5.508013**	EAS	-5.508013	-5.508013	0.00E+00
									MPDE	-5.508013	-5.508013	0.00E+00
									MDE	-5.508013	-5.508013	0.00E+00

5 Conclusions

This paper has proposed a novel evolutionary agent system by incorporating evolutionary process into agent concepts for solving mathematical programming models. The agents are engaged to solve the problems using evolutionary operator (crossover) and different types of learning processes. The results show the robustness of our proposed algorithm in its handling of both linear and nonlinear equality and inequality constraints and different types of functions. The equality constraints are efficiently handled by changing the constraints dynamically; the constraint handling techniques used here do not need any penalty functions or parameters. The performance of the proposed algorithms is compared with some well-known algorithms. The results show that the proposed approach gives mostly improved or comparable results to the algorithms.

In future work, we shall analyze the dynamics of different operators and parameters in solving different type of problems including multi-objective problems.

References

1. Sarker, R., Kamruzzaman, J., Newton, C.: Evolutionary optimization (EvOpt): a brief review and analysis. International Journal of Computational Intelligence and Applications 3, 311–330 (2003)
2. Kisiel-Dorohinicki, M.: Agent-Oriented Model of Simulated Evolution. In: Grosky, W.I., Plášil, F. (eds.) SOFSEM 2002. LNCS, vol. 2540, pp. 253–261. Springer, Heidelberg (2002)
3. Molina, D., Herrera, F., Lozano, M.: Adaptive local search parameters for real-coded memetic algorithms. The 2005 IEEE Congress on Evolutionary Computation 1, 881, 888–895 (2005)
4. Krasnogor, N., Smith, J.: A tutorial for competent memetic algorithms: model, taxonomy, and design issues. IEEE Transactions on Evolutionary Computation 9, 474–488 (2005)
5. Muruganandam, A., Prabhaharan, G., Asokan, P., Baskaran, V.: A memetic algorithm approach to the cell formation problem. The International Journal of Advanced Manufacturing Technology V25, 988–997 (2005)
6. Zhong, W., Liu, J., Xue, M., Jiao, L.: A multiagent genetic algorithm for global numerical optimization. IEEE Transactions on Systems, Man and Cybernetics, Part B 34, 1128–1141 (2004)
7. Ong, Y.S., Keane, A.J.: Meta-Lamarckian learning in memetic algorithms. IEEE Transactions on Evolutionary Computation 8, 99–110 (2004)
8. Davis, L.: Handbook of Genetic Algorithms. Van Nostrand Reinhold, New York (1991)
9. Hart, W.E.: Adaptive Global Optimization With Local Search. PhD thesis. Univ. California, San Diego, CA (1994)
10. Dreżewski, R., Marek, K.-D.: Maintaining Diversity in Agent-Based Evolutionary Computation. In: ICCS 2006. LNCS, pp. 908–911. Springer, Heidelberg (2006)
11. Siwik, L., Kisiel-Dorohinicki, M.: Semi-elitist Evolutionary Multi-agent System for Multiobjective Optimization. In: ICCS 2006. LNCS, Springer, Heidelberg (2006)
12. Liu, J., Zhong, W., Jiao, L.: A multiagent evolutionary algorithm for constraint satisfaction problems. IEEE Transactions on Systems, Man and Cybernetics, Part B 36, 54–73 (2006)
13. Deng, H., tan, Y.J., li, J.: The study of a new multiagent-based genetic algorithm. In: Proceedings of the First international conference on Machine learning and cybernetics, Beijing, vol. 3, pp. 1237–1240. IEEE Computer Society Press, Los Alamitos (2002)
14. Sycara, K.P.: Multiagent Systems. The American Association for Artificial Intelligence (1998)
15. Ferber, J.: Multiagent systems as introduction to distributed artificial intelligence. Addision-Wesley, London (1999)
16. Stan, F., Art, G.: Is It an agent, or just a program?: A taxonomy for autonomous agents. In: Intelligent Agents III Agent Theories, Architectures, and Languages. LNCS, vol. 3, pp. 21–35. Springer, Heidelberg (1997)
17. Dobrowolski, G., Kisiel-Dorohinicki, M., Nawarecki, E.: Evolutionary multiagent system in multiobjective optimisation. In: Proc. of the IASTED Int. Symp.: Applied Informatics, IASTED/ACTA Press (2001)

18. Zhong, W., Liu, J., Xue, M., Jiao, L.: A multiagent genetic algorithm for global numerical optimization. IEEE Transactions on Systems, Man and Cybernetics, Part B 34, 1128 (2004)

19. Leung, Y.-W.: An Orthogonal Genetic Algorithm with Quantization for Global Numerical Optimization Optimization. IEEE Transactions on Evolutionary Computation 5, 41–53 (2001)

20. Koziel, S., Michalewicz, Z.: Evolutionary algorithms, homomorphous mappings, and constrained parameter optimization. Evol.Comput. 7 (1999)

21. Michalewicz, Z., Schoenauer, M.: Evolutionary algorithms for constrained parameter optimization problems. Evolutionary Computation 4, 1–32 (1996)

22. Michalewicz, Z.: Genetic algorithms, numerical optimization and constraints. In: Proc. 6th Int. Conf. Genetic Algorithms, pp. 151–158 (1995)

23. Himmelblau, D.M.: Applied Nonlinear Programming. Mc-Graw-Hill, USA (1972)

24. Floudas, C.: Handbook of Test Problems in Local and Global Optimization. In: Nonconvex Optimization and its Applications, Kluwer Academic Publishers, The Netherlands (1999)

25. Mezura-Montes, E., Velazquez-Reyes, J., Coello Coello, C.A.: Modified Differential Evolution for Constrained Optimization, pp. 25–32 (2006)

26. Chootinan, P., Chen, A.: Constraint handling in genetic algorithms using a gradient-based repair method. Computers & Operations Research 33, 2263–2281 (2006)

27. Runarsson, T.P., Yao, X.: Stochastic ranking for constrained evolutionary optimization. IEEE Transactions on Evolutionary Computation 4, 284 (2000)

28. Elfeky, E.Z., Sarker, R.A., Essam, D.L.: A Simple Ranking and Selection for Constrained Evolutionary Optimization. In: Wang, T.-D., Li, X., Chen, S.-H., Wang, X., Abbass, H., Iba, H., Chen, G., Yao, X. (eds.) SEAL 2006. LNCS, vol. 4247, pp. 537–544. Springer, Heidelberg (2006)

29. Tasgetiren, M.F., Suganthan, P.N.: A Multi-Populated Differential Evolution Algorithm for Solving Constrained Optimization Problem, pp. 33–40 (2006)

30. Moscato, P.: On Evolution, Search, Optimization, Genetic Algorithms and Martial Arts Towards Memetic Algorithms. Caltech Concurrent Computation Program Report 826. Caltech Concurrent Computation Program Report 826,California Institute of Technology, Pasadena, CA, U.S.A (1989)

31. Krasnogor, N.: Studies on the Theory and Design Space of Memetic Algorithms. Ph.D. Thesis. University of the West of England (2002)

32. Barkat Ullah, A.S.S.M., Sarker, R., Cornforth, D., Lokan, C.: An Agent-based Memetic Algorithm (AMA) for Solving Constrained Optimization Problems. The 2007 IEEE Congress on Evolutionary Computation (2007)

33. Deb, K.: An efficient constraint handling method for genetic algorithms. Computer Methods in Applied Mechanics and Engineering 186, 311 (2000)

Agent-Based Coding GA and Application to Combat Modeling and Simulation*

Youming Yu, Guoying Zhang, and Jiandong Liu

Dept. of Computer, Beijing Institute of Petrochemical Technology, 102617 Beijing, China
yuyouming@bipt.edu.cn

Abstract. Agent-based coding genetic algorithms (AGA) is proposed by combining agent basic theory and encoding methods with agent attribute because simple genetic algorithms (SGA) cannot solve complex problem with good result or without reasonable solution. AGA algorithm is based on individual structure description. In the paper, AGA environment structure and agent structure and genetic operator target function are defined, and verified AGA with a test function and applied it to model combat situation and implement its simulation.

Keywords: agent coding, combat simulation, genetic algorithm.

1 Introduction

Genetic algorithms is an algorithm used to find an approximate solutions to difficult problems through application of the principles of genetics and evolutionary biology, which is applied broadly to auto control, computation science, pattern recognition, engineering design, intelligent fault diagnosis, management science and social science[1].Complex system is a research hotspot of the current management and engineering science field. Complexity research involves lots of advanced problems of many subjects[2][3].Complex system deals with some characteristics that are very universal[4].Complex system has administrative levels, independency, individual intelligence, adaptability, go-aheadism, parallelism via genetic algorithm has self-organization, self-adaptability and self-learning intelligence and genius parallelism[5]. Complex system and genetic algorithm have the same or close characteristics; genetic algorithm is good for improving the agent intelligence and adaptability in complex system model[6].

Some improvement on genetic algorithm can enhance its efficiency and ability to solve problems, but some problems cannot enough coded with strings, so here an agent-code based genetic algorithm is proposed.

---

* Supported by Excellence Person with Ability Training Special Item Outlay Imburse of Beijing under Grant No. 20042D0500508.

L. Kang, Y. Liu, and S. Zeng (Eds.): ISICA 2007, LNCS 4683, pp. 197–203, 2007.

2 Agent Based Coding Genetic Algorithms – AGA

The genetic algorithm based on agent coding, for short AGA, is based on genetic algorithm and coded with agents not coded with strings. An agent is a computation entity with the characteristics of station, reaction, sociality, go-aheadism and can exert function independently. The following gives some research on AGA for its environment, agent structure, genetic operator and target function, etc.

2.1 Environment

In AGA, all the operations of the agents carry through the environment. The environment is the space for agents to survive and exist. Agents can alternate and operate and communicate each other in the environment and the environment will affect and restrict agents' behavior. Environment factors include space location, space distance, obstacle and attribute, etc. Agent environment can be expressed by a 3D coordinate and k function value on it, such as formula (1):

$$E = (x, y, z, f_1(x, y, z), ..., f_k(x, y, z)) \qquad x \in R, y \in R, z \in R, k \in N . \tag{1}$$

2.2 Individual Agent Structure

In AGA, agent structure has great influence on GA function, problems especially complex system have to materialize their complexity, adaptability, intelligence by agent's structure, rules, behaviors and intelligence degree. In AGA, every agent is depicted by a six tuple group:

Agent=<mark ,type, knowledgebase, rule sets, attribute, parameter>

2.3 Genetic Operator

An agent is more complicated than a coding string, so AGA's genetic operator and evolutionary operator are more flexible than SGA's selection, crossover and mutation operators and have more kinds. In AGA, there are many operators for different usage: Selection: similar to SGA's selection operation to select appropriate agent to continue relative operation.

Wash out: to wash out some agents from the agent colony by agent performance evaluation value.

Propagate: to select the agent with excellent performance to reproduce more excellent offspring.

Inherit: the offspring can be produced by propagation with the characters of parents' or inherited from the parents'.

Others: according to the different demand, AGA can provide more operations, such as competition, combat, infection, correspond, assimilation, etc. It reflects the agility of the AGA.

2.4 Target Function

AGA has 3 kinds of target functions. They are based on agent level, or based on agent colony level and come-off type target function.

3 Application Example

Shaffer function is selected as the testing function, as function (2):

$$f(x,y)=(4-2.1x^2+1/3\ x^4)x^2+xy+(4\ y^2-4)\ y^2 \quad (-20<x,y<20) \tag{2}$$

In figure 1, the function has a maximum at point (0,0), around it is a group of circle with the one heart. Firstly, use the testing function to verify AGA's ability to solve problems, and then model the combat simulation.

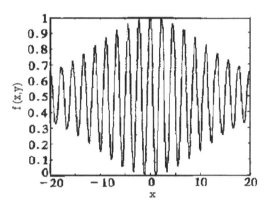

Fig. 1. Shaffer function curve

3.1 Environmental Setting

Here, let the environment to be (3):

$$E = (x, y, f(x, y)). \tag{3}$$

where -20<x<20, -20<y<20, and f(x, y) is the testing function. In simulating environment, it is corresponding to grids 80×80, environment is dispersed and zoom out by 2.Coordinate of the grid is corresponding to the point (x/2, y/2) in the environment. Every grid has a character value F, it equals to the value of testing function at the point.

3.2 The Agent Characters Setting

The agent here only has 3 tuples, they are rule sets, character and parameter, for the problem is easy to solve. The rule sets include mainly move rule and wash out rule, agent character is agent's visible distance value L, it denotes that an agent can discover and realize all grids character value F all around it. Here, L=1, and it denotes that an agent only can discover 4 or 8 grids value all around it.

3.3 Operation Rule

Move rule: an agent can find the grid with the maximum character value all around it and move to the location.

Wash out rule: if a grid has the character value less than the mean value where an agent stay, then the agent will be washed out at the probability P.

3.4 Simulation Result

At the simulation beginning=1, and the initial population is 400, agent can look around at 8 directions, and finally the agent converge the grid (0,0).

4 Combat Simulation Modeling

Combat simulation is to forecast or recall the combat process of military confronting two sides at a certain or existed situation, to simulate the operation of campaign equipment and personnel at battle field and to do statistics analysis [7].

4.1 The Environmental Setting

Environment is a logic map of many entities at battlefield from natural or social environment to model space. Agents and objects interact each other in environment, such as communicating between agents and controlling agents object. Agents interact with environment, such as agents obtain decision making information from the environment, the action result will influence the environment. The environment on the other side will restrict and influence the agents' all actions. Here the battlefield environment is defined in function (4):

$$E = (x, y, z, f_1(x, y, z), ..., f_k(x, y, z)) \tag{4}$$

Each $f_k(x, y, z)$ is an environment character function, can denote distance, terrain gradient, or threat degree function and other geography or situation information functions. In simulation world, E is a 3D map with grids, and the environment is dispersed, the grid can be magnified by need. The coordinate of the grid is corresponding to the point (x/n, y/n, z/n).Each grid has a character value F, it equals to the function value at the point[8].

4.2 Agent Characters Setting

Combat simulation model have 4 tuples to depict its information at least: {agents group, objects group, environment, model parameter}. In the corresponding agent, its 6 tuples structure { mark, type, knowledgebase, rule sets, attribute, parameter} should depict in detail. The mark tuple will be used to distinguish agents group. The type tuple to distinguish different arms of services. The knowledgebase tuple to form the sets of military rules and combat ordinances. The rule sets tuple to construct all behavior rules, which mean spirit action, scouting, communication, move neatly, aggression, recovery, and resuming, etc. rules. The attribute tuple to include space character, physics character and appended character, where space character means location, speed and acceleration ,physics character means sort,rank,size,health index, ammo repertory, and appended character means some agents have some special characters. The parameter tuple can be divided into control and observation parameters. Control parameters include all original conditions, action size, live or die conditions; observation parameters include all statistic analysis variables and some observing variables of the combat process for agents, objects and environment to change.

4.3 Operation Rules

Move rule: agent can move and find the grid with the maximum value, and move to the grid;

Wash out rule: computing the grid value of the agent staying, wash out the agent with the value less than the mean value;

Scout rule: agent obtains any kind of the battlefield information by some sensors;

Act rule: some agent which can make decision should have its target location. Considering combat aim, battlefield terrain, the current situation and their own factors, and then confirm their moving direction and path.

Combat rule: according to the obtained information, agents will select the targets to attack, and configure the weapons by aim, distance and fight environment factors.

Modify rule: After an action, modify all agents' characters and objects and environment status by destroying effect of theoretic computation.

Guarantee rule: agent maneuver and transportation, recruit transportation and allocation, the wounded salvage and cure, arm and equipment maintaining.[9]

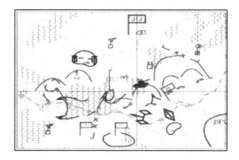

Fig. 2. Battlefield situation info. Simulation at a certain time

Study rule: to perfect and revise the action rule set. At the beginning of modeling, for the knowledge limited, agent action rule set may be not maturity and exists conflict between rules. With the simulating, agents abide by these rules and find some unsuitability and modify them, so agents can improve their adaptability by study rule.

4.4 Implementation of Agent

Agent is a logic concept of the model, how to realize it needs to represent its characters and structure and actions by an effective mode. Agent and object have many similar characters. Table 1 has the comparison result between them with composing, element partition, parameter and correlation and characteristic. The table shows that agent character can be defined by object's variable, and its action can be described by methods, and its arrangements can be realized by inheritance and nesting, so it is appropriate for agent to be realized by object-oriented methods.

Table 1. Comparison between agent and object

Sys.Tuple	Element	Parameter	Correlation	Characteristics		
Object	class	variable	method	inheritance	polymorphism	encapsulation
Agent	class	character	behavior	arrangement	intelligence	Go-aheadism

4.5 Combat Simulation Realization

Combined with military GIS system and object-oriented programming, using Visual C++6 and Oracle8i as programming tool and database environment, combat simulation system was realized。 Fig.2 is a diagram to simulate the combat situation information distribution of the battlefield at a certain time.

5 Conclusion

From the analysis of AGA and application examples, finished the combat simulation modeling, and realized the simulating of the battlefield scene and situation, incarnated that AGA has much more ability to describe and solve problems, and has more utility value for complex system modeling.

References

1. Srinivas, M.: Genetic algorithms: a survey. Computer 27(6), 17–26 (1994)
2. Hofbaur, M.W.: Hybrid Estimation of Complex Systems. Systems, IEEE Transactions on Man and Cybernetics, Part B 34(5), 2178–2191 (2004)
3. Shou-yun, W., Jing-yuan, Y.: The opened and complex huge system, pp. 32–66. ZheJiang Science Tech. Press, Zhe Jiang (1995)
4. Chen, S.-w.: Complex science and system engineering. Transaction on management science 2(2), 1–7 (1999)

5. Tan, Y.-j.: Space dynamic modeling of Complex economy system. System engineering theory and practice 10, 9–13 (1997)
6. Deng, H.-z.: Research on problem of complex system by agent based whole modeling simulation method. System engineering 18, 73–78 (2000)
7. Xu, X.-w., Wang, S.-y.: Modern combat simulation. Science Press, Beijing (2001)
8. Li, J.-w.: Agents path searching in real dynamic environment. Robot 26(1) (2004)
9. Xu, X.-z.: Distributed interaction scene simulation of battlefield situation info. System simulation transaction, 13.S (2001)

A Two-Stage Genetic Algorithm for the Multi-multicast Routing

Xuan Ma, Limin Sun, and Yalong Zhang

School of Automation and Information Enginerring, Xi'an University of Technology,
710048, Xi'an, China
maxuan@xaut.edu.cn

Abstract. The multi-multicast routing problem is to find a set of multicasts which allows transmission and routing of packets from many sources to many destinations. Due to the amount of residual bandwidth of a link, the problem becomes complicated. This paper presents an approach to solve the problem based on a two-stage genetic algorithm. The problem is decomposed into two objectives, the primary objective is to find a multicast group from a source to its destinations satisfied bandwidth and delay constraints, and the secondary objective is to find an optimal combinatorial solution in all multicasts. For the objectives, two evolving modules are designed. In the first module a method of encoding tree-like chromosome, and corresponding crossover and mutation operator are presented. The numerical simulation shows that the algorithm is efficient to solve the multi-multicast routing problem.

Keywords: genetic algorithm, multi-multicast routing, bandwidth and delay constraints, combinatorial optimization

1 Introduction

The modern network can support some of multimedia applications such as audio, live video and distance education. These multimedia applications require quality of service (QoS), such as delay, delay-jitter, bandwidth, and packet-loss rate metrics. Multicasting is an efficient method of supporting group communication that transmits multiple copies of a message from one source to a set of destinations in a network. Finding an optimal multicast is equivalent to the Steiner tree problem on graphs, which is a typical NP-hard problem[1]-[7].The Steiner tree problem and the multicast routing problem with QoS requirements are extensively studied, and many of methods have been proposed[8]-[11],[13-18].

Another case may occur in a communication network, that is, when many message packets are transmitted from different sources to its destinations, these packets experienced a link of the network at the same time should satisfy the QoS constraints. In this case, to employ fewer network resources finding a set of multicasts with

L. Kang, Y. Liu, and S. Zeng (Eds.): ISICA 2007, LNCS 4683, pp. 204–213, 2007.

minimal cost for all packets is necessary. We call it the multi-multicast routing (MMR) problem. Due to the constraint of the bandwidth capability of a link, competitions will occur while determining a minimal multicast routing of each source to its destinations. Thus, the MMR problem is more complex than the multicast routing problem.

Genetic Algorithm (GA) is a non-deterministic, stochastic-search adaptive method. It is not guaranteed to find a global solution, but it does find a global optimum in higher probability and is capable of finding near-optimal solutions in a moderate amount of time for a larger problem[4]. The one of characteristics of GA is that can find a number of different feasible solutions in a population. In this paper, for the MMR problem, we propose a GA with two stage evolving modules, based on the idea of decomposing the problem into two objectives. The primary objective is to find multiple multicasts of a source to its destinations, satisfied bandwidth and delay constraints. The secondary objective is to find an optimal combinatorial solution satisfied bandwidth capability in all multicasts corresponding to each source to its destinations.

In the case of multimedia applications, considering all the QoS criterions complicates the problem and makes it even intractable. Practically, a real-time multimedia application is delay sensitive and requires a certain amount of residual bandwidth. Thus, to simplify the issue, in this paper bandwidth and delay constraints are taken into account.

2 Problem Definition

The network is modeled as an undirected graph $G = (V, E)$. V represents a set of nodes, and E represents a set of edges. The cost is the mapping from E to a positive real value such that $C : E \rightarrow \Re^{+}$. $C(e)$ is the cost of an edge $e \in E$. On graph G there are a source set, $I = \{ s_k \mid k = 1, 2, \cdots, n \}$, and n destination sets $J = \{ S_k \mid k = 1, 2, \cdots, n \}$, $S_k = \{ D_i \mid i = 1, 2, \cdots, M_k \}$, $S_k \in V - \{ s_k \}$. A multicast tree $T_k(V_k, E_k) \subseteq G$, consists of a source s_k, and spans a set of destinations $S_k \in V - \{ s_k \}$. The cost $C(T_k)$ is calculated as follows:

$$C(T_k) = \sum_{e \in E_k} C(e) \tag{1}$$

The delay function $d(e) : E \rightarrow \Re^{+}$ assigns a positive real value to each edge in the network, and bandwidth function $B(e)$ is the same. $d(e)$ reflects the total delay that a packet experiences on edge e due to queuing delay, propagations delay, and processing delay. If $P_T(s, d)$ represents a path from source s to a destination d in a multicast tree T_k, then the total delay a packet experiences on $P_T(s, d)$ is given by:

$$D_{P_T(s,d)} = \sum_{e \in P_T(s,d)} d(e) \tag{2}$$

For a certain multicast tree T_k , it should satisfy the following constraints:

$$D_{P_T(s,d)} \leq \Delta, \ \forall d \in S_k \tag{3}$$

$$B(P_T(s,d)) \geq \beta, \quad \forall d \in S_k \tag{4}$$

where Δ is the delay constraint, and $B(e)$ is the bandwidth constraint, which represents the maximum rate that a link can offer to a new connection.

The MMR problem can now be defined as follows:

Find a set $P = \{T_k \mid k = 1, \cdots, n\}$, consists of the sources I , and the destinations J , compute a P , such that $C(P)$ is minimal:

$$C(P) = \sum_{k=1}^{n} C(T_k) \tag{5}$$

If b_k represents the bandwidth of the kth packet, while m packets experience an edge e on P at the same time, the aggregate bandwidth $b(e)$ is calculated as follows:

$$b(e) = \sum_{k=1}^{m} b_k \tag{6}$$

The P must satisfy the following constraints:

$$b(e) \geq \beta, \ e \in E \tag{7}$$

3   Description of the Algorithms

GA begins the optimization process by randomly creating a population of individuals, called chromosomes. Through selection, crossover, and mutation operations, the population evolves toward better regions of the space. The objective function delivers information about the quality of the search points, and the selection process favors individuals of greater fitness to reproduce more often than those of less fitness[12]. Based on the principle of GA, we propose the following algorithm.

3.1   Encoding

The chromosome representation is designed as follows. The locus, $L_1, L_2, L_3, \cdots, L_n$, is associated with the multicast tree sets, respectively. A gene value on each locus denotes one of trees of a multicast tree set. For example, the chromosome, as shown in Fig.1, implies that the gene value 5 denotes the tree t_{15} of the first multicast tree set, and the gene value 3 denotes the tree t_{23} of the second multicast tree set, and so on.

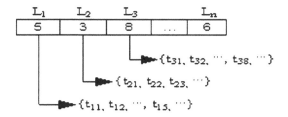

Fig. 1. The structure of the chromosome

According to the representation, finding n multicast tree sets is an elementary task. The algorithm consists of two evolving modules illustrated as follows.

3.2 Module for Finding Optimal Multicast Tree

For the multicast routing problem, how to encode a tree is critical for GA approach because the solution should be a tree. A few of encoding methods of GA for solving the Steiner tree problem and the multicast routing problem have been proposed. In ref. [4], [6], and [13], a binary string with fixed length is used to represent a tree. The nodes except the source node and the destination nodes, called mid nodes, are set 1 or 0. 1 denotes the connection node, and 0 denotes the non connection node. Since a chromosome may be non linked subgraph on a graph, the decoding becomes more complicated as the mid nodes are less. Ref. [5] and [6] used Prüfer encoding method, which uses a permutation of $n-2$ digits to uniquely represent a tree with n nodes where each digit is an integer between 1 and n inclusive. Since some links of a chromosome generated by crossover and mutation operations may not exist on a graph, the operations to modify link relations are needed. To a large original graph, the operation takes a long time, especially, as there are many non-leaf nodes. In ref. [16], one of paths from a source to a destination is used to compose the representation of a chromosome. Because crossover and mutation operations are implemented only in corresponding paths, the efficiency of searching an optimal tree is not high. In this paper we propose a coding method to create tree-like chromosome.

Tree-like chromosome representation. For a tree-like graph, as shown in Fig.2, the paths from source node s to destination nodes $\{D_1, D_2, D_3, D_4, D_5\}$ can be represented as $R = \{r_1, r_2, r_3, r_4, r_5\} = \{(s \rightarrow a \rightarrow D_1), (s \rightarrow a \rightarrow b \rightarrow D_2), (s \rightarrow D_3), (s \rightarrow D_3 \rightarrow c \rightarrow D_4), (s \rightarrow D_3 \rightarrow c \rightarrow D_5)\}$. Take the branch node as the initial node of each sub-path, the chromosome can be represented as $\{(s \rightarrow a \rightarrow D_1), (a \rightarrow b \rightarrow D_2), (s \rightarrow D_3 \rightarrow c \rightarrow D_4), (c \rightarrow D_5)\}$.

Generally, a weighted graph $T = r(s, D_1) \cup r(s, D_2) \cup \cdots \cup r(s, D_m)$ is formed by a set of paths from source node s to destination nodes. Because $r(s, D_i)$ is randomly generated, there may be same mid node in different paths.

It leads to create a loop in T , called tree loop here. A tree should be a non-loop subgraph. Thus, the loop must be eliminated. In fact, by comparing the two sub paths of the loop and getting rid of the longer, we can easily eliminate the generated tree loop.

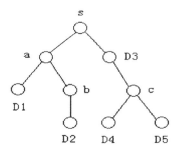

Fig. 2. Graph of tree structure

Crossover operator. Crossover operation means that the offspring pairs are created by exchanging some genes of the parent chromosomes. To tree-like chromosomes, the procedure of crossover is outlined as follows:

(1)   Select two parent chromosomes p_1 and p_2 from the population
(2)   Select a mid node j stochastically in p_1, then search the node j in p_2
(3)   Exchange the sub trees of rooted node j of p_1 and p_2

After crossover operation duplicate mid nodes may exist in one path, which leads to create another type of loop, called path loop here. The path loop and tree loop, if exist, should be eliminated from the chromosome.

Mutation operator. Select a node from the mid nodes of the chromosome to be mutated, and take the selected node as a root node of the sub tree in the chromosome. Another tree, with the same root node and leaf nodes, is stochastically generated on graph. Then, use it to replace the original sub tree. The operation of examining and eliminating the path loop and the tree loop also should be executed here.

Fitness function. Given $C(x)$ be the cost of chromosome x , i.e., the cost of multicast tree represented by x , the fitness is computed as a maximal problem, $F = C_{max} - C(x)$, C_{max} is a number greater than the maximum of $C(x)$ in population.

3.3   Module for Finding Optimal Combination

This evolving module is to find an optimal solution of the combination in all multicast tree sets. The chromosome of this GA is represented as an integer character string as described in Sect 3.1.

One-point crossover is taken in this module. The mutation operation is taken as follows: a gene locus of a chromosome is chosen randomly, and then the gene is replaced by a different gene selected from the character set. The fitness function in this module is designed as the same form as shown in Sect 3.2.

Due to the constraints of bandwidth capability, the chromosomes satisfied constraints are selected as the member of next generation population.

3.4 Main Procedure of GA for Multi-multicast Routing

The main procedure of the algorithm for MMR problem is outlined as follows:

Step1. Generate an initial population $P_k(t = 0)$, $k = 1$, in which a chromosome represents a multicast tree from a source s_k to its destinations, satisfied bandwidth and delay constraints.

Step2. Implement the crossover and mutation operations on $P_k(t)$.

Step3. Generate next generation population $P_k(t + 1)$ by selection strategy.

Step4. Return to Step 2 until stop criteria.

Step5. Output the set Q_k of excellent multicast trees.

Step6. Set $k = k + 1$, return to step 1, until $k = n + 1$, n denotes the number of the source nodes.

Step7. Generate an initial population $P(t = 0)$, the length of all chromosomes in it is n, and ordinal gene value is $x_k \in Q_k$.

Step8. Implement the crossover and mutation operations on $P(t)$.

Step9. Generate next generation population $P(t + 1)$ by selection strategy.

Step10. Return to Step 8 until stop criteria.

In the algorithm the pre-specified numbers of generations are taken as the stop criteria in both evolving modules.

4 Simulations

Up to now, we have not fond the paper on the multi-multicast routing problem. In order to illustrate the effectiveness of the proposed algorithm, we take the sum of the minimal Steiner trees as the lower limit for evaluation value in the problem of the MMR optimization.

4.1 Network

We use the graph shown in Fig.3 as the network for the experiment. It has 26 nodes and 65 edges with (cost, delay, bandwidth) on each edge. The value of cost, delay and bandwidth is generated randomly.

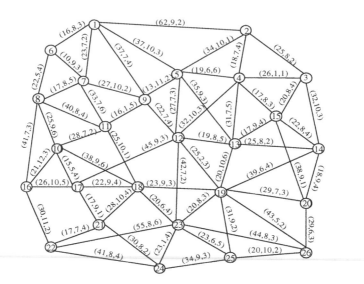

Fig. 3. A network graph

4.2 GA Parameters

The parameters are set as follows: in the module of finding an optimal multicast tree, the population size 100, the crossover rate 0.8, the mutation rate 0.02, and the number of generation 200. In the module of finding an optimal combination of multicast trees, the population size 200, the crossover rate 0.8, the mutation rate 0.05, and the number of generations 200. For the simple, $Q_k = 20, \quad k = 1,2,\cdots,n$, are taken.

4.3 Experiment Results

Firstly, we select node 5 and nodes {10, 25, 24, 20, 17} in the graph as the source and the destinations, respectively, to evaluate the performance of GA for finding multiple excellent multicast trees. The algorithm was executed 20 trials. The 6 excellent solutions including the optimal multicast tree, shown in Table 1, are obtained 18 times. Comparing with the corresponding minimal Steiner trees, we can see that, due to the constraints of bandwidth and delay the cost of the optimal multicast tree is greater than that of minimal Steiner tree. The results show that the GA proposed in this paper can efficiently search a number of excellent multicast trees satisfied with bandwidth and delay constraints.

Secondly, we randomly select 10 groups of nodes in the graph, respectively. Each group includes one source node and five destination nodes. Set the bandwidths of 10 packets as 1, 2, 2, 3, 2, 1, 1, 2, 1, 2, respectively. For the simple, the delay constraints for all packets are taken as the same value 20. The solution of the optimal multi-multicast routing is shown in Table 2. In the two tree sets, the tree in the left-hand column is the minimal Steiner tree of each source to its destinations. The tree in the right-hand column is the solution of the constrained MMR problem. Comparing the two columns, we can see that, the fifth tree, the sixth tree and the seventh tree in the right-hand column is the same to its corresponding minimal Steiner tree, respectively, while the cost of the others' is higher than that of corresponding minimal Steiner tree, respectively. The total cost of the left-hand column is 2004 and that of the right-hand column is 2123, which is close to the limit.

Table 1. The optimal multicast tree

No	minimal Steiner tree	cost	multicast tree bandwidth=3 and delay=20	cost
1	5 9 11 **10 17** 21 **24 25** 26 **20**	202	5 12 **17 10**,17 21 **24 25** 26 **20**	217
2	5 12 **17 10**, 12 19 **25 24**, 19 **20**	206	5 12 **17 10**,12 13 19 **25 24**, 19 **20**	220
3	5 12 **17 10**, 12 19 23 **25**, 19 **20**, 23 **24**	207	5 9 12 **17 10**,17 21 **24 25** 26 **20**	225
4	5 12 19 **25 24** 21 **17 10**, 19 **20**	208	5 12 **17 10**,17 18 19 **25 24**, 19 **20**	226
5	5 12 19 23 **24** 21 **17 10**, 19 **20**, 23 **25**	209	5 12 **17 10**,17 21 **24 25** 19 **20**	228
6	5 9 11 **10 17** 18 23 **25**, 23 19 **20**, 23 **24**	209	5 13 12 **17 10**,13 19 **25 24**, 19 **20**	228

Table 2. The optimal multi-multicast routing

No	source → destinations	minimal Steiner tree	cost	constrained multicast tree	cost
1	2 → 16, 19, 21, 23, 26	2 4 13 **19 23** 18 17 **16**,17 **21**,19 **26**	217	2 4 12 17 **16**, 12 **19**,17 **21**,19 23,19 **26**	226
2	3 → 8, 18, 20, 22, 24	3 14 **20** 19 **18** 17 10 **8**,17 21 **22**,21 **24**	228	3 15 **20** 19 **18** 11 **8**, 18 21 **22**,21 **24**	250
3	5 → 10, 17, 20, 24, 25	5 9 11 **10 17** 21 **24 25** 26 **20**	202	5 12 **17 10**, 17 21 **24 25** 26 **20**	217
4	8 → 14, 17, 19, 22, 25	8 10 **17** 18 **19 14**,17 21 **22**,19 **25**	189	8 16 **17** 18 **19 14**, 16 **22**, 19 **25**	212
5	12 → 6, 10, 13, 19, 23	12 9 7 **6**,7 8 **10**,12 **13** 19 23	160	12 9 7 **6**,7 8 **10**, 12 **13** 19 23	160
6	20 → 3, 4, 6, 7, 8	20 15 4 5 9 7 **6**, 15 **3**, 7 **8**	161	20 15 4 5 9 7 **6**, 15 **3**, 7 **8**	161
7	19 → 4, 6, 10, 21, 25	19 12 17 **10** 8 **6**,12 **4**,17 **21**,19 **25**	212	19 12 17 **10** 8 **6**,12 **4**,17 **21**,19 **25**	212
8	10 → 5, 13, 18, 22, 26	10 17 **18** 19 **13 5**, 17 21 **22**, 19 **26**	192	10 **18** 19 **13 5**, 18 21 **22** , 19 **26**	204
9	15 → 3, 7, 10, 22, 25	15 **3**,15 13 19 18 17 **10** 8 **7**,17 21 **22**,19 **25**	224	15 **3**, 15 13 12 9 **7**, 12 17 **10**,17 21 **22**,13 19 **25**	250
10	18 → 3, 8, 19, 22, 26	18 **19** 13 15 **3**, 18 17 10 **8**, 17 21 **22**, 19 **26**	219	18 **19** 13 15 **3**, 18 10 **8**, 18 21 **22**, 19 **26**	231

The algorithm was run through 20 separate trials on this problem. Fig.4 shows the generational changes for an average of 20 evaluation values for the elite chromosome in each generation and an average of all chromosomes of generational population. From the results, we can see that the convergence of the algorithm is fast. In the 20 trials the optimal solution 2123 can be reached 16 times. The results show that the algorithm has the good efficiency and the stability.

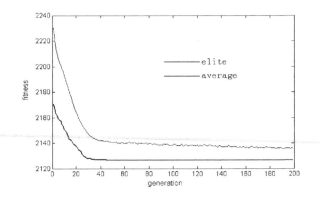

Fig. 4. Generational changes in the average evaluation value

The number of the multicast trees in Q_k is a critical parameter in the algorithm. It should be moderately set. If too large, it takes a long time to search the optimal solution. On the other hand, if too small, the solution satisfied the constraints in the space may inexistent. For our experiment, when set Q_k =5, the eligible chromosome in initial population can not be generated.

5 Conclusions

In this paper we have presented a genetic algorithm with two evolving modules for searching the optimal solution of the multi-multicast routing problem. The first module is to find a number of excellent multicast trees of each source to its destinations. The second module is to find an optimal combination of all multicast trees. For the first module, a new method of encoding a tree-like chromosome based on node connection for multicast tree, and corresponding crossover and mutation operators were also presented. The tree-like chromosome, which is different from that in former GA, need not be decoded. The results of numerical simulation show that the algorithm is effective to search the optimal solution of the multi-multicast routing problem with bandwidth and delay constraints.

Acknowledgments. This research is supported by the Department of Education of Shaanxi Province of China (05JK269).

References

1. Lee, K.J., Gersht, A., Friedman, A.: Multiple connection routing. Int. J. Digital Analog Commun. Syst. 3, 177–186 (1990)
2. Downsland, K.A.: Hill-climbing, simulated annealing and the Steiner problem in graphs. Eng.Optim. 17, 91–107 (1991)
3. Tanaka, Y., Huang, P.C.: Multiple destination routing algorithms. IEICE Trans. Commun. E76-B(5), 544–552 (1993)
4. Esbensen, H.: Computing Near-Optimal Solutions to the Steiner Problem in a Graph Using a Genetic Algorithm. Networks 26, 173–185 (1995)
5. Priwan, V., Aida, H., Saito, T.: The multicast tree based routing for the complete broadcast multipoint-to-multipoint communications. IEICE Trans. Commun. E78-B(5), 720–728 (1995)
6. Leun, Y., Li, G., Xu, Z.B.: A Genetic Algorithm for the Multiple Destination Routing Problems. IEEE Transactions on evolutionary computation 2(4), 150–161 (1998)
7. Yu, Y.P., Qiu, P.L.: An improved algorithm for Steiner tree. Journal of China Institute of Communications 23(11), 35–40 (2002)
8. Sun, Q., Langendoerfer, H.: Computation of costrained multicast trees using a genetic algorithm. Eur. Trans. Telecommun 10, 513–516 (1999)
9. Salah, A., Fawaz, A.: A Hybrid Evolutionary Algorithm for Multiple-destinations Routing Problem. International Journal of Computational Intelligence & Application 4(4), 337–353 (2004)
10. Chen, L., Yang, Z.Y., Xu, Z.Q.: A Degree-Delay-Constrained Genetic Algorithm for Multicast Routing Tree. In: Proc IEEE International Conference on Computer and Information Technology, IEEE Computer Society Press, Los Alamitos (2004)
11. He, X.Y., Fei, X., Luo, J.Z., Wu, J.Y.: A Scheme for QoS-Based Routing Using Genetic Algorithm in Internet. Chinese Journal of Computers 23(11), 1171–1178 (2000)
12. kim, J., Zeigler, B. P.: Hierarchical Distributed Genetic Algorithms: A Fuzzy Logic Controller Design Application, IEEE Expert, 76–84 (1996)
13. Liu, Y., Liu, S.Y.: Degree-Constrained Multicasting for Multimedia Communications. Chinese Journal of Computers 24(4), 367–372 (2001)
14. Chen, M., Li, Z.J.: A Real-Time Multicast Routing Algorithm Based on Genetic Algorithms. Journal of Software 12(5), 721–728 (2001)
15. Cao, Y.D., Cai, G.: Genetic Algorithm Study of Group Broadcast QoS Router. Computer Engineering 30(7), 80–82 (2004)
16. Sun, B.L., Li, L.Y.: A Multicast Routing Optimization Algorithm with Bandwidth and Delay-Constrained Based on Genetic Algorithm. Computer Engineering and Applications 11, 30–33 (2004)
17. Oliveira, C.A.S, Pardalos, P.M: A survey of combinatorial optimization problems in multicast routing. Comp.Oper.Res. 32(8), 1953–1981 (2005)
18. Oh, S., Ahn, C., Ramakrishna, R.S: A genetic-inspired multicast routing optimization algorithm with bandwidth and end-to-end delay constraints. In: King, I., Wang, J., Chan, L., Wang, D. (eds.) ICONIP 2006. LNCS, vol. 4234, pp. 807–816. Springer, Heidelberg (2006)

A Novel Lower-Dimensional-Search Algorithm for Numerical Optimization

Hui Shi[1,2], Sanyou Zeng[1,2], Hui Wang[1,2], Gang Liu[1,2], Guang Chen [1,2],
Hugo de Garis[1,3], and Lishan Kang [1,4]

[1] School of Computer Science, China University of Geosciences,
430074 Wuhan, China
[2] Research Center of Science and Technology, China University of Geosciences,
430074 Wuhan, China
[3] International School of Software, Wuhan University,
430072 Wuhan, China
[4] StateKey Laboratory of Software Engineering, Wuhan University,
430072 Wuhan, China
shihui0205@163.com, sanyou-zeng@263.net,
wanghui_cug@yahoo.com.cn, lg0061408@126.com

Abstract. In the present study, a novel strategy of Lower-dimensional-Search Algorithm (LDSA) is proposed for solving the complex numerical optimization problems. The crossover operator of the LDSA algorithm searches a lower-dimensional neighbor of the parent points where the neighbor center is the parents' barycenter, therefore, the new algorithm converges fast. The niche impaction operator and the offspring mutation operator preserve the diversity of the population. The proposed LDSA strategies are applied to 22 test problems. These functions are widely used as benchmark in numerical optimization. The experimental results are reported here show that the LDSA algorithm is an effective algorithm for the complex numerical optimization problems. What's more is that the LDSA algorithm is simple and easy to be implemented.

1 Introduction

At present, numerical optimization especially the multi-peaks and the nonlinear optimization problems have attract many researchers' interest. In general, many optimization techniques are difficult in dealing with the multi- peaks and the nonlinear numerical optimization problems. One of the main difficulties is that they can easily be entrapped in local minima. Moreover, many techniques cannot generate or even use the global information needed to find the global minimum for a function with multiple local minima.

A number of evolving algorithms have been used to solve problems in the complex numerical nonlinear optimization research fields [1][2].This success is due to EAs essentially are search algorithms based on the concepts of natural selection and survival of the fittest. They guide the evolution of a set of randomly selected individuals through a number of generations in approaching the global optimum solution. Besides that, the fact that these algorithms do not require previous considerations regarding the problem to be optimized and offers a high degree of parallelism.

L. Kang, Y. Liu, and S. Zeng (Eds.): ISICA 2007, LNCS 4683, pp. 214–223, 2007.
© Springer-Verlag Berlin Heidelberg 2007

However, at the present stage, EAs are basically limited to solving small-scale problems due to the constraint of computational efficiency. Some problems are computationally intensive regarding solution's evaluation, which makes the optimization by EA's slow for some situations. A complex simulator for an industrial plant often makes the evaluation functions time-consuming. These complex EAs slow down the performance of the algorithm to unacceptable levels.

More and more new algorithms such as Particle Swarm Optimization (PSO) [3][4][5] and Differential Evolution (DE) [6][7][8]are proposed to solve numerical optimization problems. Evolutionary programming (EP) [9] has also been applied with success to many numerical and combinatorial optimization problems in recent years. Xin Yao, Senior Member has proposed a "fast EP" (FEP) which uses a Cauchy instead of Gaussian mutation as the primary search [10].

PSO [3][4][5] is an extremely simple algorithm that seems to be effective for optimizing a wide range of functions. Unique to the concept of particle swarm optimization is flying potential solutions through hyperspace, accelerating toward "better" solutions. While other evolutionary computation schemes operate directly on potential solutions which are represented as locations in hyperspace. Much of the success of particle swarms seems to lie in the agents' tendency to hurtle past their target.

DE [6][7][8] is a practical approach to global numerical optimization that is easy to understand, simple to implement, reliable, and fast. Basically, DE adds the weighted difference between two population vectors to a third vector. This way no separate probability distribution has to be used which makes the scheme completely self-organizing. So DE is fairly robust.

In this paper, a lower-dimensional search evolutionary algorithm (LDSA) is proposed. LDSA makes use of the useful characteristics of PSO and DE. Like as DE, LDSA has also uses the idea of weighted difference, it adds the weighted difference between two population vectors to a third vector. Inspired by PSO, LDSA adds the third vector to preserves the information of the best parents. The special third vector leads LDSA convergence to the "better" solution through hyperspace to find the global optimization. In this way, LDSA makes good use of the parents' information. This strategy can seed up the constringency of LDSA.

Unique to the LDSA, the novel strategy of crossover probabilities makes the crossover operator searches a lower-dimensional neighbor of the parent points where the neighbor center is the parents' barycenter. Therefore, LDSA can converge fast especially for the multi-peaks and the nonlinear optimization problems. This is the key of LDSA. The weighted difference of LDSA makes it robust. What's more, the niche impaction operator and the offspring mutation operator preserve the diversity of the population.

The rest of this paper is organized as follows: in section 2, the Lower-dimensional search Algorithm (LDSA) is described detailedly. Section 3 describes the experiments and presents a discussion regarding the results achieved. In section 4, some conclusions and points toward future works are presented.

2   Lower-Dimensional-Search Algorithm (LDSA)

2.1   The Framework of the LDSA Algorithm

For the convenience of describing the LDSA algorithm, an extended explanation of the population of an evolutionary algorithm is given: unlike generic evolutionary algorithm, the population is not made up of individuals, but of niches [11] where individuals stay. Extremely, there is only one individual in a niche. The LDSA algorithm belongs to the extreme case. The fitness of the only individual in the niche is regarded as that of the niche. The LDSA algorithm generates a trial individual (offspring) for each niche to compete the old one in the niche. If the new offspring is better than the old individual in the niche, then the old one is replaced by the new one.

The framework of the LDSA algorithm is as follows:

Algorithm 1. The framework of LDSA

```
Step1. Randomly create a population P(0) of size N and
       each niche i has one individual x⃗ⁱ. Set the generation
       counter t = 0
REPEAT
FOR each niche i
   Step2. Execute offspring generation operator on niche i
          (Algorithm 2)
   Step3. Compare x⃗ˢⁱ with x⃗ⁱ in niche i. If x⃗ˢⁱ is better
          than x⃗ⁱ, then x⃗ⁱ is replaced by x⃗ˢⁱ for the niche i
END FOR
Step4. Generation t = t+1
UNTIL evolution ends
```

2.2 The Offspring Generation Operator

The offspring generation operator on niche i consists of three operators: parent crossover, niche impaction and offspring mutation. The details of the operator are as follows:

Algorithm 2. Offspring generation operator on niche i

```
Step1. Randomly choose three different
       individuals: x⃗ᵖ¹, x⃗ᵖ² and x⃗ᵖᵇᵉˢᵗ (the best parent) from
       population P(t). They are all different from x⃗ⁱ in
       niche i as well.
Step2. Execute crossover operator (Algorithm 3) on the
       above four chosen individuals which yields one
       offspring x⃗ˢⁱ
```

Step3. Environment impaction operator on $\overrightarrow{x}^{si}$ (Algorithm5)

Step4. Output $\overrightarrow{x}^{si}$

The crossover operator is executed on the chosen parents $\overrightarrow{x}^{p1}$, $\overrightarrow{x}^{p2}$ and $\overrightarrow{x}^{pbest}$ (the best parent). The details are as follows:

Algorithm 3. The parent crossover operator

Input: the chosen parents: $\overrightarrow{x}^{p1}$, $\overrightarrow{x}^{p2}$, $\overrightarrow{x}^{pbest}$

Step1. Randomly create a_1, a_2, a_3 (Algorithm 4)

with $a_1 + a_2 + a_3 = 1$ in the range $-1 \leqslant a_1$, a_2, $a_3 \leqslant 3$

Step2. Execute

$$\overrightarrow{x}^{si} = a_1 \overrightarrow{x}^{p1} + a_2 \overrightarrow{x}^{p2} + a_3 \overrightarrow{x}^{pbest} \tag{1}$$

Step3. Output offspring $\overrightarrow{x}^{si}$

A way to create the 3 random numbers a_1, a_2, a_3 is as follow:

Algorithm 4. Create 3 random numbers a_1, a_2, a_3

a₁=rand(); a₂=rand(); a₃=rand();

$a_1 = a_1^2$; $a_2 = a_2^2$; $a_3 = a_3^2$; //For diversity

sum = a₁+ a₂+ a₃;

a₁= a₁/ sum; a₂= a₂/ sum; a₃= a₃/ sum;

a₁=4a₁ - 1; a₂=4a₂ - 1; a₃=4a₃ - 1;

By Equation 1 and $a_1 + a_2 + a_3 = 1$, we have
$$a_1(\overrightarrow{x}^{si} - \overrightarrow{x}^{p1}) + a_2(\overrightarrow{x}^{si} - \overrightarrow{x}^{p2}) + a_3(\overrightarrow{x}^{si} - \overrightarrow{x}^{pbest}) = 0$$

Algorithm 3 is one of the most important in the algorithm. The comparison between figure 1 and figure 2 indicates that the operation of a1= a12; a2= a22; a3= a32 can make the distribution of the offspring quite even. Though Algorithm 3, the offspring $\overrightarrow{x}^{si}$ generated by the crossover stays in the linear space defined by the three parents $\overrightarrow{x}^{p1}$, $\overrightarrow{x}^{p2}$, $\overrightarrow{x}^{pbest}$ .The dimension of the space is smaller than 2. By the constraints: -1 $\leqslant a_1$, a_2, $a_3 \leqslant 3$, $\overrightarrow{x}^{si}$ has to stay in a neighborhoods of the three parents, and the neighbor center is the barycenter of the parents. Therefore, $\overrightarrow{x}^{si}$ stays in a smaller than 2 dimensional neighborhoods of the three parents no matter how many dimensions the decision space of the optimization problem is. Therefore, the LDSA algorithm converges fast especially for optimization problems with higher dimensions decision space. The $\overrightarrow{x}^{pbest}$ makes LDSA to convergence to the "better" solutions.

Offspring $\overrightarrow{x}^{si} = (x_1^{si}, x_2^{si}, ..., x_n^{si})$ generated by the crossover is impacted by niche i and global environment. Denote the only individual in niche i $\overrightarrow{x}^i = (x_1^i, x_2^i, ..., x_n^i)$

The environment impaction operator on $\overrightarrow{x}^{si}$ is as follow:

Algorithm 5. Environment impaction operator

```
Input: the above offspring x⃗ˢⁱ and the individual x⃗ⁱ in
  the niche i.

Step1. Randomly create a, −1⩽a⩽2

Step2. For each gene j of x⃗ˢⁱ , execute
```
$$x_j^{si} \Leftarrow ax_j^i + (1-a)random(l_j, u_j)$$
```
  where probability pᵢ. lⱼ is the lower bound of gene j,
  uⱼ upper bound.

Step3. Output x⃗ˢⁱ
```

The niche impaction operator and the offspring mutation operator preserve the diversity of the population.

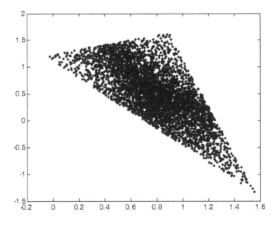

Fig. 1. The distribution of the offspring over the neighborhoods of the 3 parents. The 3 yellow 'o' points are the parents, and the red '*' point is the barycenter of the 3 parents. (Through the operation of $a_1 = a_1^2; a_2 = a_2^2; a_3 = a_3^2$).

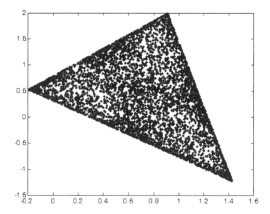

Fig. 2. The distribution of the offspring over the neighborhoods of the 3 parents. The 3 yellow 'o' points are the parents, and the red '*' point is the barycenter of the three parents. (Without the operation of $a_1 = a_1^2$; $a_2 = a_2^2$; $a_3 = a_3^2$).

3 Experimental Setup

3.1 The 22 Benchmark Functions

The current version of the LDSA algorithm is proposed to solve real-parameter problems. In this paper, it is used to solve the complex numerical problems. The benchmark functions used here is a set of 22 different functions selected from [10]. These functions are widely used as benchmark in numerical optimization. Functions f_1–f_{13} are high dimensional problems. Functions f_1–f_5 are unimodal. Function f_6 is the step function, which has one minimum and is discontinuous. Function f_7 is a noisy quartic function, where random[0, 1) is a uniformly distributed random variable in [0, 1). Functions f_8 –f_{13} are multimodal functions where the number of local minima increases exponentially with the problem dimension [12][13]. Functions f_{15}–f_{23} are low-dimensional functions which have only a few local minima [12].

In order to compare LDSA with "fast Evolutionary programming" (FEP) [10], the same parameter settings have been used in our experiments. We use fi (i=1,2,3,4,5,6,7,8,9,20,11,12,13,15,16,17,18,19,20,21,22,23)represent the benchmark functions. The order of the 22 benchmark functions f_i is the same as the 23 benchmark functions in [10]. (Note: We haven't use the benchmark function f_{14} in[10]) Due to the limit of the length of the paper, the detailed description of each function is not given here. A more detailed description of each function can be got from [10]. The22 benchmark functions are given in Table 1.

3.2 The Parameters Setting of LDSA

(1) The population size N in LDSA is 50.
(2) Environment impaction probability pI (cf. Algorithm 5)

Environment impaction probability pI is related to the dimensions of the decision space of the problems. For algorithm fast converging, the offspring generated by crossover operator is demanded not to be impacted by environment with probability q,

then we have $(1-p_I)^n = q$, that is $p_I = 1 - \sqrt[n]{q}$, where n is the dimension of the decision space of the problem. This paper sets q=0.5.

(3) The evaluations for each function of each experiment are the same as FEP. The evaluations are listed in table2.

3.3 Results Comparison and Discussion

The results of performing LDSA are listed in table 2. Table 2 shows best, worst, mean value found for each function, where each function is evaluated for independent 50 runs. The best vales of the 22 benchmark functions listed in table 2 are close to the minimum values listed in table 1.

Table 1. 22 benchmark functions used in our experimental. Where N is the dimension of the function, f_{min} is the minimum value of the function, and $S \in R^n$.

Function	N	S	f_{min}
f_1	30	$[-100,100]^n$	0
f_2	30	$[-10,10]^n$	0
f_3	30	$[-100,100]^n$	0
f_4	30	$[-100,100]^n$	0
f_5	30	$[-30,30]^n$	0
f_6	30	$[-100,100]^n$	0
f_7	30	$[-1.28,1.28]^n$	0
f_8	30	$[-500,500]^n$	-12596.5
f_9	30	$[-5.12,5.12]^n$	0
f_{10}	30	$[-32,32]^n$	0
f_{11}	30	$[-600,600]^n$	0
f_{12}	30	$[-50,50]^n$	0
f_{13}	30	$[-50,50]^n$	0
f_{15}	4	$[-5,5]^n$	0.0003075
f_{16}	2	$[-5,5]^n$	-1.0316285
f_{17}	2	$[-5,10]*[0,15]$	0.398
f_{18}	2	$[-2,2]^n$	3
f_{19}	4	$[0,1]^n$	-3.86
f_{20}	6	$[0,1]^n$	-3.32
f_{21}	4	$[0,10]^n$	-10
f_{22}	4	$[0,10]^n$	-10
f_{23}	4	$[0,10]^n$	-10

In this paper, the LDSA algorithm is compared with FEP [10], which uses a Cauchy instead of Gaussian mutation as the primary search operator. LDSA algorithm uses no gradient-based mutation and needs to set four parameters. Therefore, the LDSA algorithm does not demand the differentiability of the problem, easy to be implemented and converges fast. The results of this comparison are shown in tables 3 with respect to the mean best value found for each function and the standard deviations between LDSA and FEP. Although for f4, f5, f15, the results of LDSA is not better than FEP, it is apparent that FEP performs better than CEP for the rest part of the 22 benchmark functions.

4   Conclusions

This article introduced a lower-dimensional search algorithm (LDSA). The efficiency of the LDSA algorithm is mainly as the following points:

1) The LDSA algorithm searches a space with dimensions lower than 2 no matter how many dimensions the decision space of the optimization problem is. Therefore, the LDSA algorithm converges fast especially for optimization problems with higher dimensions.

Table 2. The best, worst, mean values found for each function. (Each function is evaluated for independent 50 runs.)

Function	Number of Evaluations	Best(LDSA)	Worst(LDSA)	Mean Best(LDSA)
f_1	1.5×10^4	4.54978×10^{-6}	4.01023×10^{-4}	1.00046×10^{-4}
f_2	2×10^5	1.99477×10^{-4}	5.78103×10^{-3}	2.89743×10^{-3}
f_3	5×10^5	3.40285×10^{-6}	2.56946×10^{-4}	4.33817×10^{-5}
f_4	5×10^5	0.70791	1.45839	1.06798
f_5	2×10^6	0.085692	84.8711	24.76520
f_6	1.5×10^4	0.00000	0.00000	0.00000
f_7	3×10^5	8×10^{-15}	9.70467×10^{-6}	6.65618×10^{-7}
f_8	9×10^5	-12569.486618	-12569.486404	-12569.48657
f_9	5×10^5	2.50230×10^{-6}	2.14194×10^{-4}	3.64700×10^{-5}
f_{10}	1.5×10^4	2.30957×10^{-3}	1.13518×10^{-2}	5.47740×10^{-3}
f_{11}	2×10^5	1.66379×10^{-4}	9.60723×10^{-2}	2.12613×10^{-2}
f_{12}	1.5×10^4	3.67936×10^{-7}	2.89860×10^{-5}	3.48339×10^{-6}
f_{13}	1.5×10^4	3.57966×10^{-6}	1.17130×10^{-2}	9.67833×10^{-4}
f_{15}	4×10^5	8.42182×10^{-4}	3.32085×10^{-2}	1.05434×10^{-2}
f_{16}	1×10^4	-1.031628453	-1.031628453	-1.031628453
f_{17}	1×10^4	0.3978873573	0.3978873573	0.3978873573
f_{18}	1×10^4	2.999999999999922	2.999999999999922	2.999999999999919
f_{19}	1×10^4	-3.862782147820755	-3.862782147820755	-3.862782147820753
f_{20}	2×10^4	-3.32199	-3.20310	-3.26017
f_{21}	1×10^4	-10.15319	-2.63047	-7.23625
f_{22}	1×10^4	-10.40294	-10.40294	-8.29288
f_{23}	1×10^4	-10.53640	-10.53640	-7.60918

2) Like as DE, LDSA has also uses the idea of weighted difference, it adds the weighted difference between two population vectors to a third vector. Inspired by PSO, LDSA adds the third vector to preserves the information of the best parents. The special third vector leads LDSA convergence to the "better" solution through hyperspace to find the global optimization.

3) The niche and the mutation operator preserves conserve the diversity, so the LDSA algorithm is not easy to be trapped in local optima as much as possible.

4) The LDSA algorithm does not demand the differentiability of the problem.

5) The LDSA algorithm is simple and easy to be implemented.

6) Another reason that LDSA is attractive is that there are few parameters to adjust. With slight variations, LDSA works well in constrained optimization problems.

Although some other algorithms such as Quantum-Inspired Evolutionary Algorithm [14] and Stochastic Genetic Algorithm [15] perform better than LDSA. However, LDSA proposed in this paper is originality, the characteristics listed above indicates that LDSA is a effective algorithm of great promise.

Table 3. Comparison between LDSA and FEP. Where "Funs", "Mean Beat" and " Std Dev" stand for the function, the mean best values and the standard deviation respectively.(Each function from LDSA and FEP is evaluated for independent 50 runs.)

Funs	Number of Evaluations	Mean Best Values		Std Dev	
		LDSA	FEP	LDSA	FEP
f_1	1.5×10^4	1.00×10^{-4}	5.7×10^{-4}	8.0×10^{-5}	1.3×10^{-4}
f_2	2×10^5	2.89×10^{-3}	8.1×10^{-3}	1.4×10^{-3}	7.7×10^{-4}
f_3	5×10^5	4.33×10^{-5}	1.6×10^{-2}	4.3×10^{-5}	1.4×10^{-2}
f_4	5×10^5	1.06	0.3	0.16	0.5
f_5	2×10^6	24.76	5.06	30.8	5.87
f_6	1.5×10^4	0	0	0	0
f_7	3×10^5	6.65×10^{-7}	7.6×10^{-3}	1.9×10^{-6}	2.6×10^{-3}
f_8	9×10^5	-12569.4	-12554.5	4.6×10^{-5}	52.6
f_9	5×10^5	3.64×10^{-5}	4.6×10^{-2}	3.7×10^{-5}	1.2×10^{-3}
f_{10}	1.5×10^4	5.47×10^{-3}	1.8×10^{-2}	1.9×10^{-3}	2.1×10^{-2}
f_{11}	2×10^5	2.12×10^{-2}	1.6×10^{-2}	2.2×10^{-2}	2.2×10^{-2}
f_{12}	1.5×10^4	3.48×10^{-6}	9.2×10^{-6}	5.0×10^{-6}	3.6×10^{-6}
f_{13}	1.5×10^4	9.67×10^{-4}	1.6×10^{-4}	2.8×10^{-3}	7.3×10^{-5}
f_{15}	4×10^5	1.05×10^{-2}	5.0×10^{-4}	6.9×10^{-3}	3.2×10^{-4}
f_{16}	1×10^4	-1.03	-1.03	0	4.9×10^{-7}
f_{17}	1×10^4	0.398	0.398	0	1.5×10^{-7}
f_{18}	1×10^4	2.99	3.02	3.0×10^{-15}	0.11
f_{19}	1×10^4	-3.86	-3.86	3.0×10^{-15}	1.4×10^{-5}
f_{20}	2×10^4	-3.26	-3.27	5.9×10^{-2}	5.9×10^{-2}
f_{21}	1×10^4	-7.23	-5.52	3.26	1.59
f_{22}	1×10^4	-8.29	-5.52	3.00	2.12
f_{23}	1×10^4	-7.61	-6.57	3.63	3.14

Acknowledgments

This work was supported by The National Natural Science Foundation of China (Nos. 60473037, 60473081, 40275034, 60204001, 60133010).

References

1. Back, T., Fogel, D.B., Michalewicz, Z. (eds.): Handbook of Evolutionary Computation. Institute of Physics Publishing (1997)
2. Michalewicz, Z.: Genetic algorithms + data structures = evolution programs, 2nd edn. Springer, New York (1994)
3. Clerc, M., Kennedy, J.: The particle swarm-explosion, stability, and convergence in a multidimensional complex space. IEEE Transactions on Evolutionary Computation 6(1), 58–73 (2002)
4. Parsopoulos, K.E., Vrahatis, M.N., N, M.: Recent approaches to global optimization problems through particle swarm optimization. Natural Computing 1(2-3), 235–306 (2002)
5. Andrews, P.S.: An Investigation into Mutation Operators for Particle Swarm Optimization. 2006 IEEE Congress on Evolutionary Computation, pp.3789–3796 (2006)
6. Storn, R., Price, K.V.: Differential evolution-A simple and Efficient Heuristic for Global Optimization over Continuous Spaces. Journal of Global Optimization 11, 341–359 (1997)
7. Lampinen, J., Zelinka, I.: Mixed Variable Non-Linear Optimization by Differential Evolution. In: Proceedings of Nostradamus'99,2nd International Prediction Conference, pp. 45–55. Zlin, Czech Republic. Technical University of Brno, Faculty of Technology Zlin, Department of Automatic Control (October 1999), ISBN 80-214-1424-3
8. Price, K., Storn, R., Lampinen, J.: Differential Evolution –A Practical Approach to Global Optimization. Springer, Heidelberg (2005)
9. Gehlhaar, D.K., Fogel, D.B.: Tuning evolutionary programming for conformationally flexible molecular docking. In: Fogel, L.J., Angeline, P.J., Bäck, T. (eds.) Evolutionary Programming V: Proc. of the Fifth Annual Conference on Evolutionary Programming, pp. 419–429. MIT Press, Cambridge, MA (1996)
10. Yao, X., Liu, Y., Lin, G.M.: Evolutionary programming made faster. IEEE Trans. Evolutionary Computation 3, 82–102 (1999)
11. Gustafson, S., Burke, E.K.: A Niche for Parallel Island Models: Outliers and Local Search. In: ICPPW'05. Proceedings of the 2005 International Conference on Parallel Processing Workshops, pp. 1530–2016 (2005)
12. Törn, A., Žilinskas, A.: Global Optimization. LNCS, vol. 350. Springer, Heidelberg (1989)
13. Schwefel, H.-P.: Evolution and Optimum Seeking. Wiley, New York (1995)
14. Tu, Z., Lu, Y.: A robust stochastic genetic algorithm (stga) for global numerical optimization. IEEE Trans. Evolutionary Computation 8(5), 456–470 (2004)
15. André, V., da Cruz, A., Vellasco, M.M.B.R.: Quantum-Inspired Evolutionary Algorithm for Numerical Optimization. 2006 IEEE Congress on Evolutionary Computation, pp.9181–9187 (2006)

Performance Evaluation of Three Kinds of Quantum Optimization

Bao Rong Chang[1,*] and Hsiu Fen Tsai[2]

[1] Department of Computer Science and Information Engineering
National Taitung University, Taitung, Taiwan 950,
Tel.: +886-89-318855 ext. 2607; Fax: +886-89-350214
brchang@nttu.edu.tw
[2] Department of International Business
Shu-Te University, Kaohsiung, Taiwan 824
soenfen@mail.stu.edu.tw

Abstract. Three kinds of quantum optimizations are introduced in this paper as follows: quantum minimization (QM), neuromprphic quantum-based optimization (NQO), and logarithmic search with quantum existence testing (LSQET). In order to compare their fitting ability among three quantum optimizations, the performance evaluation on these methods is implemented for the application of time series forecast. Finally, based on the predictive accuracy of time series forecast the concluding remark will be made to illustrate and discuss these three quantum optimizations.

Keywords: quantum minimization, neuromprphic quantum-based optimization, logarithmic search with quantum existence testing.

1 Introduction

Many computing and engineering problems can be tracked back to an optimization process which aims to find the extreme value (minimum or maximum point) of a so-called cost function or a database. From a quantum computing point view Grover algorithm [1] has been viewed as the most promising candidate. Unfortunately Grover-based solutions are efficient only in term of expected number of database quires. In order to tackle this main drawback we decide to introduce three algorithms to solve the problem of finding the extreme values. First, three kinds of quantum optimizations are introduced in this paper as follows: quantum minimization (QM) [2], neuromprphic quantum-based optimization (NQO) [3], and logarithmic search with quantum existence testing (LSQET) [4]. Next, in order to compare their ability among three quantum optimizations, the performance evaluation on these methods is implemented for optimizing an adaptive support vector regression (ASVR) [5] applied to the application of time series forecast. Finally, based on the predictive accuracy of time series the concluding remark will be made to illustrate and discuss these three quantum optimizations.

---

[*] Corresponding author.

L. Kang, Y. Liu, and S. Zeng (Eds.): ISICA 2007, LNCS 4683, pp. 224–233, 2007.
© Springer-Verlag Berlin Heidelberg 2007

2 Quantum Minimization

Quantum-based minimization that makes optimization task work out associated with probability of success at least 1/2 within an unsorted database is realized by quantum minimum searching algorithm [6]. A quantum exponential searching algorithm [7] is called by quantum minimum searching algorithm to be as a subroutine to serve a fast database searching engine.

2.1 Quantum Exponential Searching Algorithm

As reported in [7] where the searching problem is to find the index i such that $T[i] = x$, we are ready to describe the algorithm for finding a solution when the number t of solutions is known. For simplicity, we assume at first that $1 \leq t \leq 3N/4$.

Step 1: Initialize $m = 1$ and set $\lambda = 6/5$. (Any value of λ strictly between 1 and 4/3 would do.)

Step 2: Choose j uniformly at random among the nonnegative integers small than m.

Step 3: Apply j iterations of Grover's algorithm [1] starting from initial state

$$|\Psi_0\rangle = \sum_i \frac{1}{\sqrt{N}} |i\rangle .$$

Step 4: Observe the register: let i be the outcome.

Step 5: If $T[i] = x$, the problem is solved: exit.

Step 6: Otherwise, set m to $\min(\lambda m, \sqrt{N})$ and go back to step 2.

2.2 Quantum Minimum Searching Algorithm

We second give the minimum searching algorithm [6] in which the minimum searching problem is to find the index i such that $T[i]$ is minimum where $T[0,...,N-1]$ is to be an unsorted table of N items, each holding a value from an ordered set.

Step 1: Choose threshold index $0 \leq i \leq N-1$ uniformly at random.

Step 2: Repeat the following (2a, 2b, and 2c) and interrupt it when the total running time is more than $22.5\sqrt{N} + 1.4\lg^2 N$. Then go to stage 3.

(a) Initialize the memory as $\sum_j \frac{1}{\sqrt{N}} |j\rangle|i\rangle$. Mark every item j for which

$T[j] < T[i]$.

(b) Apply the quantum exponential searching algorithm of [7].

(c) Observe the first register: let i' be the outcome. If $T[i'] < T[i]$, then set threshold index i to i'.

Step 3: Return i

This process is repeated until the probability that the threshold index selects the minimum is sufficiently large.

3 Neuromorphic Quantum-Based Optimization

The synaptic weights w_{ijkl} are given in a Hopfield network [3].

$$E_{HN} = -\frac{1}{2}\sum_{ij}\sum_{kl}w_{ijkl}o_{ij}o_{kl} - \sum_{ij}h_{ij}o_{ij} \,,$$

(1)

where h_{ij} is the external bias for a neuron. The synaptic weights are obtained as

$$
\begin{aligned}
w_{ijkl} = &-2a\delta_{j,l}(1-\delta_{i,k}) - 2b\delta_{i,k}(1-\delta_{j,l}) \\
&- 2c\delta_{i+j,k+l}(1-\delta_{i,k}) - 2d\delta_{i-j,k-l}(1-\delta_{i,k}) \,,
\end{aligned}
$$

(2)

where δ_{ij} is the Kronecker delta. Let us consider that each qubit corresponds to each neuron of a Hopfield network. The state vector $|\psi\rangle$ of the whole system is given by the product of all qubit states.

The time evolution of the system is given by the following chr"odinger equation.

$$\left|\psi(t+1)\right\rangle = U(1)\left|\psi(t)\right\rangle = e^{-\frac{iH(t)}{h}}\left|\psi(t)\right\rangle \,,$$

(3)

Here, the operator $U(1)$ is given by the Pad´e approximation [8].

The quantum computation algorithm utilizing adiabatic Hamiltonian evolution has been proposed by Farhi et al. [9]. Adiabatic Hamiltonian evolution is given as

$$H(t) = \left(1-\frac{t}{T}\right)H_I + \frac{t}{T}H_F \,,$$

(4)

where H_I and H_F are the initial and final Hamiltonians, respectively. We assume that the quantum system starts at t = 0 in the ground state of H_I, so that all possible candidates are set in the initial state $|\psi(0)\rangle$. T denotes the period in which the Hamiltonian evolves and the quantum state changes, and we can control the speed of such changes to be suitable for finding the optimal solution among all candidates set in $|\psi(0)\rangle$. If a sufficiently large T is chosen, the evolution becomes adiabatic. The adiabatic theorem says that the quantum state will remain close to each ground state [10]. Therefore, the optimal solution can be found as the final state $|\psi(T)\rangle$. However, successful operation is not guaranteed in the case that there exists any degeneracy in energy levels or any energy crossing during the evolution [10]. The initial Hamiltonian H_I is chosen so that its ground state is given by the superposition of all states as

$$\left|\psi(0)\right\rangle = \frac{1}{\sqrt{2^n}}\sum_{i=0}^{2^n-1}\left|i\right\rangle \,,$$

(5)

where n is the number of qubits and $|i\rangle$ is the n-th eigenvector. The initial Hamiltonian H_I is given as

$$
\begin{aligned}
H_I = &\left(\sigma_x^{(0)} + \sigma_x^{(1)} + \cdots + \sigma_x^{(2^{16}-1)}\right) \\
= &(\sigma_x \otimes I \otimes \cdots \otimes I + I \otimes \sigma_x \otimes \cdots \otimes I \,, \\
&+ \cdots + I \otimes I \otimes \cdots \otimes \sigma_x)
\end{aligned}
$$

(6)

where σ_x is the x-component of the Pauli spin matrix. One can choose any other H_I which satisfies that its ground state is expressed by a linear combination of all states. For example, σ_x can be replaced by σ_y. It may be possible to solve optimization problems by composing a new Hamiltonian H_F considering the synaptic weights w_{ijkl}. The Hamiltonian H_F for the target problem is obtained as shown in the following equation. The eigenvalue ε_i of a basis state $|i\rangle$ should be obtained from the cost function in Eq. (1). Therefore, H_F has ε_i as the diagonal elements as

$$H_F = \begin{pmatrix} \varepsilon_0 & & & 0 \\ & \varepsilon_1 & & \\ & & \ddots & \\ 0 & & & \varepsilon_{2^n-1} \end{pmatrix} = \sum_{n=0}^{2^n-1} \varepsilon_i |i\rangle\langle i|, \tag{7}$$

By choosing a proper H_F, it is possible to solve the problems. However, the calculation cost of 2^n, where n is the number of qubits, is necessary in order to obtain the above-mentioned Hamiltonian H_F, and there is no difference when compared with a heuristic search. Therefore, the key of application is how we choose a more effective H_F with less calculation cost.

The Hopfield net [11] is a recurrent neural network as shown in Fig. 1 in which all connections are symmetric. Invented by John Hopfield (1982), this network has the property that its dynamics are guaranteed to converge. If the connections are trained using Hebbian learning then the Hopfield network can perform robust content-addressable memory, robust to connection alteration. Various functions of an artificial neural network (ANN) are realized by choosing suitable synaptic weights. A full connection neural network has n^2 synapses, where n is the number of neurons. In order to reduce the calculation cost of the Hamiltonian, we consider interactions between qubits and study a new method with a new H_F comprising nondiagonal elements considering its analogy to ANN. For convenience, we assume we have closely coupled 2-spin-1/2 qubits. The Hamiltonian of this quantum system is

$$H = J(\sigma_1, \sigma_2) = J_{12} \begin{pmatrix} 1 & 0 & 0 & 0 \\ 0 & -1 & 2 & 0 \\ 0 & 2 & -1 & 0 \\ 0 & 0 & 0 & 1 \end{pmatrix}, \tag{8}$$

where J_{12} is the magnitude of the interactions, and σ_i is the Pauli spin matrix. The eigenvalues and eigenvectors of this Hamiltonian when $J_{12}=1$ are shown as -3 for $|01\rangle - |10\rangle$ and 1 for $|00\rangle, |11\rangle, |01\rangle + |10\rangle$, respectively. The possible states to be measured are $|10\rangle$ or $|01\rangle$ if the system is in the ground state $|01\rangle - |10\rangle$. It can be said that the interaction of two neurons is inhibitory if we consider the analogy with an ANN model. Excitatory interaction is also possible with another Hamiltonian. From the above consideration, we can design a new Hamiltonian by converting the synaptic weights in Eq. (1) to the interactions of qubits.

The neuromorphic quantum -based optimization, as shown in Fig. 1, can apply w_{kj} in Eq. (1) to calculate final Hamiltonian H_F like an example as shown in Eq. (8) providing for adiabatic evolution algorithm in Eq. (3)-(4) to train the constrained optimization for an cost function expressed in Eq. (18)-(20).

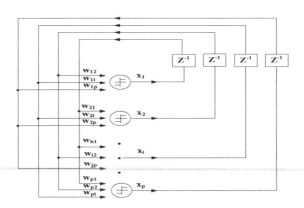

Fig. 1. A typical Hopfield network is exploited for neuronmorphic quantum-based optimization

4   Logarithmic Search with Quantum Existence Testing

From a quantum computing point of view Grover-based solutions [1] are efficient only in term of expected number of database quires. In order to overcome this major shortcoming we decide to use the quantum existence testing algorithm [4] as a core function. A well-known logarithmic search algorithm [12] with quantum existence testing is applied to finding the extreme values in an unsorted database.

4.1   Quantum Existence Testing

The classical accuracy c should be chosen such that in the case $M = 1$ the measured output of the IQFT contains at least one nonzero bit which allows distinguishing it from $|0>$. Let us assume again without loss of generality that we have a database $N = 2^n$ entry of size, therefore using $\frac{M}{N} = \sin(\frac{\Omega_\gamma}{2})$ we need

$$\min(\Omega_\gamma) = 2\arcsin\left(\sqrt{\frac{1}{N}}\right) \cong 2\sqrt{\frac{1}{N}} = 2^{(-n/2+1)} \geq 2^{-c}, \tag{9}$$

where we applied the well-known relation $\arcsin(y) = y$ if $y \ll 1$, from which we get

$$c = \left\lceil \frac{n}{2} \right\rceil - 1, \tag{10}$$

Of course we have to take care of quantum uncertainty of phase estimation as well, hence we need all together

$$n^* = \left\lceil \frac{n}{2} + ld(2\pi) + ld\left(3 + \frac{1}{\tilde{P}_{eP}}\right) \right\rceil - 2 \tag{11}$$

qbits, where $\breve{P}_{eP}$ stands for the allowed maximum quantum uncertainty and correction $ld(2\pi)$ is required because c refers to the accuracy of the estimated phase insteady of the phase ratio itself. Since the -2 term has marginal influence on the complexity we omit it during the further discussion, that is

$$n^* = \left\lceil \frac{n}{2} + ld(2\pi) + ld\left(3 + \frac{1}{\breve{P}_{eP}} \right) \right\rceil$$
(12)

Moreover if one gets $n^* < 1$ then n^* has to be set to 1. Only the the case $\Omega_\gamma \neq 0$ should be taken into consideration from a quantum error point of view when seeking the relationship between the required number of additional qbits and P_{eE}, where subscript E refers to existence testing. For the sake of controlling precisely this P_{eE} one needs p additional qbits to $\frac{n}{2} + ld(2\pi)$ qbits in the upper qregister to quarantee classical accuracy 2^{-c}. Provided that

$$P_{eE} \leq \frac{1}{8} \frac{1}{2^p} \frac{2^n - 1}{2^n}.$$
(13)

We have an engineering constraint $\breve{P}_e \geq P_{eE}$ one needs

$$p = ld\left(\frac{2^n - 1}{8 \cdot 2^n \breve{P}_e} \right) \leq ld\left(\frac{1}{8\breve{P}_e} \right),$$
(14)

qbits to fulfil it and total number of required qbits in the upper section of phase estimation is

$$n^* = \left\lceil \frac{n}{2} + ld(2\pi) + ld\left(\frac{1}{8\breve{P}_e} \right) \right\rceil.$$
(15)

4.2 Logarithmic Search

Let us assume that we have a function $y = g[x]$ which has integer input $x \in [0, N-1]$ and integer output $y \in [G_{\min 0}, G_{\max 0}]$. We are interested in y_{opt} such that $\min_x(g[x]) = g[x_{opt}] = y_{opt}$. To solve the above-mentioned problem we combine the well-known logarithmic search algorithm with quantum existence testing. In order to be more precise the proposed algorithm is now given in detail:

Step 1: We start with $s = 0$: $G_{\min 1} = G_{\min 0}$, $G_{\max 1} = G_{\max 0}$, and $\Delta G = G_{\max 0} - G_{\min 0}$

Step 2: $s = s+1$

Step 3: $G_{med\ s} = G_{\min s} + \left\lceil \frac{G_{\max s} - G_{\min s}}{2} \right\rceil$

Step 4: $flag = QET(G_{med 0})$

- if $flag = YES$ then $G_{\max s+1} = G_{med\ s}$, $G_{\min s+1} = G_{\min s}$
- else $G_{\max s+1} = G_{\max s}$, $G_{\min s+1} = G_{med\ s}$

Step 5: if $s < ld(\Delta G)$ then go to step 2 else stop and $y_{opt} = G_{med\ s}$

Already available quantum computing based solutions require $O(\sqrt{N} + ld^2(N))$ iterations (i.e. Grover operators) for an expected value. Conversely the propose new approach obtain optimal value y_{opt} using $O(ld(\Delta G)ld^3(\sqrt{N}))$ elementary steps originating from the computational complexity of the phase estimation, which means a fairly huge difference if $N \gg 1$.

5 Experimental Results and Discussions

Three kinds of quantum optimizations introduced in this paper are employed for optimizing a specified forecast model denoted by ASVR-BWGC/NGARCH [13]. As shown in Fig. 2 to Fig. 5 or Fig. 6 to Fig. 7, the predicted sequences indicate the predicted results for the following competing methods: (a) grey model (GM) [14], (b) auto-regressive moving-average (ARMA) [15], (c) radial basis function neural network (RBFNN) [16], (d) adaptive neuro-fuzzy inference system (ANFIS) [17], (e) support vector regression (SVR), (f) real-coded genetic algorithm [18] tuning BWGC/NGARCH model (RGA-BWGC/NGARCH), (g) quantum-minimized BWGC/NGARCH model (QM- BWGC/NGARCH), (h) neuromorphic quantum-based adaptive support vector regression tuning BWGC/NGARCH model (NQASVR-BWGC/NGARCH), and (i) logarithmic search with quantum existence testing for adaptive support vector regression tuning BWGC/NGARCH model (LSQETASVR-BWGC/NGARCH). In the experiments, the most recent four actual values is considered as a set of input data used for modeling to predict the next desired output. As the next desired value is obtained, the first value in the current input data set is discarded and joins the latest desired (observed) value to form a new input data set for the use of next prediction. The international stock price indexes prediction for four markets (New York Dow Jones Industrials Index, London Financial Time Index (FTSE-100), Tokyo Nikkei Index, and Hong Kong Hang Seng Index) [19] have been experimented where the monthly-closing index for 48 months from Jan. 2002 to Dec. 2005 are shown in Fig. 2 to Fig. 5. The accuracy of predicted result, including GM, ARMA, RBFNN, ANFIS, SVR, RGA-BWGC/NGARCH, QM-BWGC/NGARCH, and the proposed one, is also compared and the summary of first experiment is listed in Table 1. Criterion of mean square error for measuring the predictive accuracy is expressed below.

$$MSE = \sum_{t=1}^{l} \left(y_{t_c+t} - \hat{y}_{t_c+t} \right)^2 \Bigg/ l \qquad (16)$$

where l = the number of periods in forecasting, t_c = the current period, y_{t_c+t} = a desired value at the t_c+t th period and $\hat{y}_{t_c+t}$ = a predicted value at the t_c+t th period.

The goodness of fit on the first experiment is tested by Q-test successfully due to all of p-values greater than level of significance (0.05) [20]. London International Financial Futures and Options Exchange (LIFFE) [21] provide the indices of volumes of equity products on futures and options for 24 months from Jan. 2001 to Dec. 2002, and further their monthly-closing equity volume forecasts are shown in Fig. 6 and Fig. 7. Table 2 has listed the summary of forecasting accuracy for the comparison between methods. This is also tested by Q-test successfully due to all of p-values greater than level of significance (0.05).

Table 1. The comparison between different prediction models based on Mean Squared Error (MSE) on international stock price monthly indices for 48 months from Jan. 2002 to Dec. 2005. (unit=10^5)

Methods	NY-D.J. Industrials Index	London FTSE-100 Index	Tokyo Nikkei Index	HK Hang Seng Index	Average MSE
GM	1.9582	0.4006	3.2209	5.1384	2.6795
ARMA	1.8230	0.3883	2.9384	4.3407	2.3726
RBFNN	1.8319	0.3062	3.8234	3.1976	2.2898
ANFIS	1.4267	0.3974	3.5435	4.0279	2.3489
SVR	1.2744	0.3206	2.6498	3.1275	1.8431
RGA-BWGCNG	1.2642	0.2630	2.0175	3.1271	1.6680
QMASVR- BWGCNG	1.1723	0.2156	1.8656	3.0754	1.5822
NQASVR-BWGCNG	1.1694	0.2168	1.8671	3.0362	1.5724
LEASVR- BWGCNG	1.1714	0.2182	1.8669	3.0771	1.5834

Table 2. The comparison between different prediction models based on Mean Squared Error (MSE) on futures and options monthly-closing volumes indices of equity products for 24 months from Jan. 2001 to Dec. 2002

Methods	Futures Index of Equity Products	Options Index of Equity Products	Average MSE
GM	0.0945	0.0138	0.0542
ARMA	0.0547	0.0114	0.0331
RBFNN	0.0196	0.0087	0.0142
ANFIS	0.0112	0.0092	0.0102
SVR	0.0191	0.0085	0.0138
RGA-BWGCNG	0.0110	0.0082	0.0096
QMASVR-BWGCNG	0.0098	0.0072	0.0085
NQASVR-BWGCNG	0.0101	0.0070	0.0086
LEASVR- BWGCNG	0.0104	0.0071	0.0088

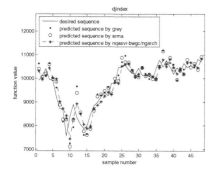 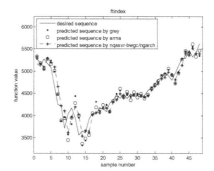

Fig. 2. Forecasts of N. Y. -D. J. Industrilas monthly-closing index

Fig. 3. Forecasts of London FTSE-100 monthly-closing index

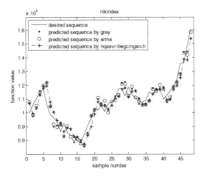

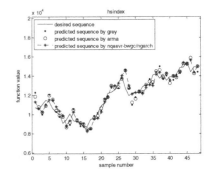

Fig. 4. Forecasts of Japan Nikkei monthly-closing index

Fig. 5. Forecasts of Hong Kong Hang-Seng monthly-closing index

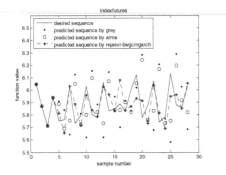

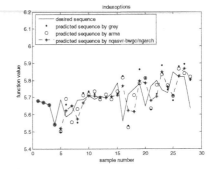

Fig. 6. Forecasts of futures monthly-closing volume index of equity product

Fig. 7. Forecasts of options monthly-closing volume index of equity product

6   Conclusions

The main objective of this paper is to explore the performance evaluation on three kinds of quantum optimization so as for understanding how good ability of fitting time series data is. In particularly these methods include quantum minimization (QM), neuromorphic quantum-based optimization (NQO), and logarithmic search with quantum existence testing (LSQET). Two experiments are tested by several models to verify their fitting ability under the time series forecast. Strictly speaking these quantum-based solutions are almost no difference between them and with the same fitting ability superior to the others. Hence from the performance point of view the quantum computing can be considered as a good approach to fit the time series analysis.

Acknowledgements

This work is fully supported by the National Science Council, Taiwan, Republic of China, under grant number **NSC 94-2218-E-143-001**.

References

1. Grover, L.K.: A Fast Quantum Mechanical Algorithm for Database Search. In: Proc. 28th Ann. ACM Symp, Theory of Comp., pp. 212–219. ACM Press, New York (1996)
2. Chang, B.R., Tsai, S.F.: Financial Prediction Applications Using Quantum-Minimized Composite Model ASVR and NGARCH. In: IJCNN06. Proc. International Joint Conference on Neural Network, pp. 7283–7290. Sheraton Vancouver Wall Centre Hotel, Vancouver, BC, Canada (2006)
3. Tank, D.W., Hopfield, J.J.: Simple 'neural' optimization networks: An A/D converter, signal decision circuit, and a linear programming circuit. IEEE Trans. Circuits Syst. 36, 533–541 (1986)
4. Imre, S., Balazs, F.: Quantum computing and Communication- An Engineering Approach, 3rd edn. Wiley, West Sussex, England (2005)
5. Chang, B.R.: Compensation and Regularization for Improving the Forecasting Accuracy by Adaptive Support Vector Regression. International Journal of Fuzzy System 7(3), 109–118 (2005)
6. Durr, C., Hoyer, P.: A Quantum Algorithm for Finding the Minimum, http://arxiv.org/abs/quant-ph/9607014
7. Boyer, M., Brassard, G., Hoyer, P., Tapp, A.: Tight Bounds on Quantum Searching. Fortschritte der Physik 46, 493–505 (1998)
8. Golub, G.H., van Loan, C.F.: Matrix Computations, 3rd edn. Johns Hopkins University Press, Baltimore (1996)
9. Farhi, E., Goldstone, J., Gutmann, S., Lapan, J., Lundgren, A., Preda, D.: A Quantum Adiabatic Evolution Algorithm Applied to Random Instances of an NP-Complete Problem. Science 292, 472–475 (2001)
10. Messiah, A.: Quantum Mechanics. Dover, New York (1999)
11. Hopfield, J.J.: Neural networks and physical systems with emergent collective computational abilities. In: Proceedings of the National Academy of Sciences of the USA, vol. 79(8), pp. 2554–2558 (1982)
12. Sandon, P.A.: Logarithmic Search in a Winner-Take-All Network. In: Proc. IEEE International Conference on Neural Network, pp. 454–459. IEEE Computer Society Press, Los Alamitos (1991)
13. Tsai, H.F., Chang, B.R.: Adaptive Support Vector Regression Tuning Composite Model of BWCG and NGARCH for Applications of Time-Series Prediction. Journal of Grey System 9(1), 1–7 (2006)
14. Deng, J.L.: Control Problems of Grey System. System and Control Letter 1(5), 288–294 (1982)
15. Box, G.E.P., Jenkins, G.M., Reinsel, G.C.: Time Series Analysis: Forecasting & Control. Prentice-Hall, New Jersey (1994)
16. Haykin, S.: Neural Networks, 2nd edn. Prentice Hall, New Jersey (1999)
17. Jang, J.-S.R.: ANFIS: Adaptive-Network-based Fuzzy Inference Systems. IEEE Transactions on Systems, Man, and Cybernetics 23(3), 665–685 (1993)
18. Ono, I., Kobayashi, S.: A Real-coded Genetic Algorithm for Function Optimization Using Unimodal Normal Distribution Crossover. In: Proc. 7th International Conf. on Genetic Algorithms, pp. 246–253 (1997)
19. International Federation of Stock Exchanges, FIBV FOCUS MONTHLY STATISTICS (2005), Available: http://www.fibv.com/WFE/home.Asp?nav=ie
20. Ljung, G.M., Box, G.E.P.: On a Measure of Lack of Fit in Time Series Models. Biometrika 65, 67–72 (1978)
21. London International Financial Futures and Options Exchange (LIFFE) (2002), Available: http://www.liffe.com/

An Efficient Multilevel Algorithm
for Inverse Scattering Problem

Jingzhi Li, Hongyu Liu, and Jun Zou

Department of Mathematics,
The Chinese University of Hong Kong,
Shatin, Hong Kong
{jzli, hyliu, zou}@math.cuhk.edu.hk

Abstract. In this paper we study an efficient multilevel algorithm for implementing the linear sampling method in inverse obstacle scattering problems. The algorithm performs a multilevel investigation of the concerned domain while avoiding the need for exploring the meshgrid point by point. The algorithm requires to solve only $O(n^{N-1})$ far-field equations for the concerned problem in R^N (N=2,3) instead of the original $O(n^N)$ equations. We present the results of numerical experiments to illustrate the efficiency and effectivenss of the algorithms.

Keywords: inverse problem, scattering.

1 Introduction

In the last one decade, the linear sampling method (LSM) has been successfully used for the solution of many inverse scattering problems arising from theoretical and application background, to name a few, 2D obstacle problems with Dirichlet, Neumann or impedance boundary conditions, and 2D medium problems with compact support for transverse magnetic or electric polarized incident waves [3, 6, 7], 3D isotropic obstacle problems [2], acoustic isotropic and orthotropic problems [8], and the detection of buried objects [1] or leukemia in human bone marrow [5]. The LSM is one of the most widely used methods for solving the inverse scattering problems in the resonance region. The major advantages of the LSM are as follows: 1) one no longer makes linear approximations (e.g. the Born approximation), which leads to some optimization problems that are highly nonlinear and computationally expensive; 2) its implementation only requires the solution of a linear integral equation, namely, the far field equation; 3) it can treat the cases whether the medium is penetrable or impenetrable and whatever the boundary conditions are in a uniform way; and 4) the LSM provides us the support of the scatter rather than the precise physical properties of the scatter, which is often enough in real world applications such as medical imaging and non-destructive testing and avoids the further computation for the physical values of the scatterer. One of the main weak points of LSM is its formidable computational cost. Thus, we think that the LSM is worth further refinement mainly for the following reasons.

L. Kang, Y. Liu, and S. Zeng (Eds.): ISICA 2007, LNCS 4683, pp. 234–242, 2007.
© Springer-Verlag Berlin Heidelberg 2007

Although it is fast for medium size problems compared with traditional nonlinear optimization methods, a few minutes as reported in the previous works, the computational expense is still intensive for in general one has to investigate $O(n^2)$ grid points in 2D or $O(n^3)$ grid points in 3D. To get a reliable resolution of the scatterer, one has to spend considerable time for even intermediate-sized n.

The purpose of this work is to address the computational complexity of the LSM. In particular, we propose a multilevel version of the LSM, which solves the linear integral equations at each test grid point in a multilevel way. Based on a heuristic idea, we ignore the region where $\|g\|_{L2(B)}$ is too large or too small, further investigation in those regions are in vain for we are sure that those regions are bound to lie away from the obstacles for the "too-large" case or inside the obstacles for the "too-small" case. We will show that the computational complexity is $O(n)$ in 2D and $O(n^2)$ in 3D when the boundary of the obstacle is of codimension one, which is optimal in the sense to locate the boundary of the obstacles of codimension one.

2   An Efficient Multilevel Algorithm

In this section we first introduce the linear sampling method (LSM) before further discussion.

For the sake of simplicity, let us consider the 2D case of a cylindrical scatterer represented by an impenetrable obstacle or an inhomogeneous object diffusing a fixed frequency electromagnetic field. If k is the wavenumber and y is a point inside the scatterer, then the support of the scatterer can be obtained by solving the far-field equation

$$\int_B u_\infty(x,d)g(d)ds(d) = \frac{e^{-i\pi/4}}{\sqrt{8\pi k}}e^{-ikx\cdot y} ,\qquad(1)$$

where $u_\infty(x, d)$ is the far field pattern of the scattered wave corresponding to incident angle d and observation angle x. It can be proven that $\|g\|_{L2(B)}$, where B denotes the unit circle in R^2, grows to infinity as y approaches the boundary of the scatterer from the inside. This "blow-up" property motivates the following simple idea to solve the inverse scattering problems (see Chap 4. 4).

LSM algorithm

For each grid point y on a grid containing the scatterer:

1. solve a one-parameter family of regularized solutions of the far field equation (1);
2. apply generalized Morozov discrepancy principle for determining an optimal value of the regularization parameter; and
3. plot the norm of the regularized solution $\|g\|_{L2(B)}$, determined by means of the previous two steps; or, alternatively, plot the value of the optimal regularization parameter (double phase visualization).

Note that the use of some regularization method for the solution of the far field equation is definitely crucial, owing to the severe ill-posedness of the problem (the norm of g over the unit circle may be of the order 10^{10-12}) by our experiments.

To reduce the computational complexity of the linear sampling method, we propose a multilevel linear sampling method (MLSM), a multilevel version of the original LSM, which is optimal due to the reason that we can determine the boundary of the obstacle in a n × n grid in 2D, which consists of O(n) points, by only investigating O(n) points. More precisely, the method can be generalized to the case of determining the boundary of the scatterer in ND within $O(n^{N-1})$ investigations.

The motivation of the MLSM is as follows. To locate the boundary in the grid as shown in Figure 1, we try to avoid the investigation of points far away or deeply inside the scatterer as much as possible. Obviously, there is no need to investigate regions that are clearly far away from or fully inside the obstacle. Let us denote the far-away region and the deeply inside region by **F-region** and **I-region** as shown in Figure 1, respectively. For those regions not being **F-regions** or **I-regions**, we denote them by **B-region,** or boundary region where the boundary curve of the scatterer may pass through. We can improve the efficiency of the LSM by taking advantage of labeling and removing the far away and inside regions to reduce the redundant computational cost. To get a high-resolution of the boundary of the scatterer, we can do the previous labeling and removing technique in a multilevel way, to be more precise, we trim larger regions in the coarser level to locate the profile of the scatterer boundary and remove smaller regions in the finer levels get the high-resolution boundary as shown in Figure 1, which achieves accuracy without sacrificing efficiency at the mean time.

Now let us formulate the main algorithm of the MLSM. Assume that we are given a nested sequence of grids T_k $(0 \leq k \leq L)$ on the test square domain $\Omega = [a, b] \times [a, b]$, with T_{k+1} $(n_{k+1} \times n_{k+1}$ mesh) being a refinement of T_k $(n_k \times n_k$ mesh), for $0 \leq k \leq L - 1$ and $n_{k+1} = 2n_k - 1$ and n_0 is some small constant and $n = n_L$. It is remarked that although the complexity analysis is done for the 2D case, the general result for the ND case can be easily obtained by a similar argument.

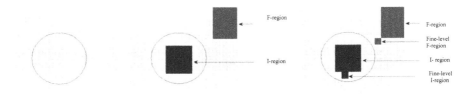

Fig. 1. Label-and-remove scheme

Suppose the boundary curve lies in $O(n_L)$ L-th level sub-squares in the finest grid, one can see that it also lies in $O(n_k)$ n_k-th level sub-squares in the k-th level grid. The reason behind this seemingly simple fact is not that easy. For a detailed proof of this assumption, please refer to [Lemma 2.1-2.6, 9].

MLSM algorithm

1. Set k=0 and label all the coarsest level sub-squares as B-region.
2. Apply the LSM scheme on the k-th level mesh only on B-region sub-squares to investigate the points exploited previously.

3. Label the k-th level sub-squares as three sets, far-away, boundary or inside, based on the heuristic principle of the LSM. Choose two cut-off values c_1 and c_2 with $c_1 < c_2$. Define the sub-squares as F-region if the norm of the regularized solution g on the four corners of the sub-square are all larger than c_2, and as I-region if all less than c_1. Otherwise the sub-squares are labeled as B-region.
4. Remove the F-region and I-region sub-squares.
5. Refine the mesh only on B-region sub-squares;
6. Set k := k + 1 and go to Step 2 until k > L, i.e., the finest grid has been be investigated.

This MLSM algorithm presented here improves the original one in [9] where $c_1 = c_2$, it is remarked that with this improvement one can have a better guess of the scatterer's boundary within a buffer layer.

To demonstrate that our new algorithm is asymptotically optimal, we will make some further assumptions on the boundary of the obstacle and the initial test grid in the algorithm.

Assumption A. The coarsest grid points should have points belongs to the "far-away" set and "inside" set. This excludes the extreme cases that the obstacle falls into the trap of the coarsest grid which stops the algorithm from initiating at the beginning.

Assumption B. On the finest grid, the boundary of the obstacle lies in O(n) sub-squares of the grid.

Let us first show a technical lemma for later use.

Lemma 2.1. *The boundary curve of the scatterer lies in at most M n_k sub-squares in the k-th level, where M is a constant independent of k.*

The idea behind this lemma is simple. Suppose the boundary curve of the scatterer lies in m sub-squares in the k-th level. It is proved in [9] that m $h_k^2 \leq L\, h_k$, then combined with the identity $n_k^2 h_k^2 = |\Omega|$ yields

$$m \leq \frac{L}{\sqrt{|\Omega|}} n_k,$$

which indicates the boundedness of the number of the sub-squares. For the detailed proof, please refer to [9]

With the above preparations, we have the following results, which can be proved in a similar manner as done in [9].

Theorem 2.1. *The MLSM algorithm only needs to investigate $O(n^{N-1})$ grid points in R^N and is asymptotically optimal.*

Remark 2.1. To avoid the obstacle falling in the trap in the coarsest mesh, we can take a meshgrid with some sub-squares both inside and outside the scatterer as the coarsest mesh.

3 Numerical Experiments and Analysis

In this section, we perform two tests to illustrate the effectiveness and efficiency of the proposed MLSM algorithm. All the programs in our experiments are written in Matlab and run on a Pentium 3GHz PC.

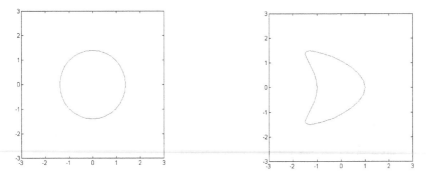

Fig. 2. Obstacle boundary curves in experiment 1(left) and 2(right)

The scatterer in system will be chosen to be a circle and a kite-shaped object which have been widely tested in the inverse scattering problems (see, e.g., [3], [6] and [8]). For these two tests, we referred them to be Circ and Kite, respectively. For the experiments Circ and Kite (see Figure 3), the scatterer D is composed of a circle of radius 1.4 and centered at the origin and a kite given by the following parameteric form:

$$x(t) = (\cos t + 0.65 \cos 2t - 0.65, 1.5 \sin t), \qquad 0 < t < 2\pi.$$

In the experiment Kite, we would add two percent random noise to the synthetic far field data, and in the experiment Circ, we add one percent random noise. The other parameters chosen for these experiments are listed in Table 1.

Table 1. Parameter setting in the experiments

	Test 1 (Circ)	Test 2 (Kite)
sampling domain Ω	$[-3, 3] \times [-3, 3]$	$[-3, 3] \times [-3, 3]$
incident wave number k	1	1
finest level n_L	129	129
upper threshold c_1	0.0380	0.0330
lower threshold c_2	0.0385	0.0335
noise level δ	0.01	0.02
no. of incident directions	32	32
no. of observation directions	32	32

It is worth noting that for our experiments, two positive constants c_1 and c_2 have been taken as the cut-off values, $c_1 < c_2$, instead of only one cut-off value c as in [9]. Since it is better to take a range of cut-off values, i.e., $[c_1, c_2]$, this enables us to get a buffer region to locate the boundary of the underlying object. Like in the original LSM, we label as inside corner points those points with the norm of distributed density g less than c_1, far-away corner points those points with the norm of distributed density g greater than c_2, and boundary corner points otherwise.

The synthetic far field data are generated by solving the layer potential operator equation using the Nyströme's method (see Sect. 3.5, Ch.3 in [4]). We compute the far field patterns at 32 equidistantly distributed observation points ($\cos t_j$, $\sin t_j$), $t_j = 2j\pi/32$, $j = 0, 1, \ldots, 31$, and 32 equidistantly distributed incident directions ($\cos \tau_j$, $\sin \tau_j$), $\tau_j = 2j\pi/32$, $j = 0, 1, \ldots, 31$. The far field patterns we obtained are subjected pointwise to uniform random noise. The uniform random noise is added according to the following formula,

$$u_\infty = u_\infty + \delta\, r_1\, |u_\infty|\, \exp(i\, \pi\, r_2),$$

where r_1 and r_2 are two uniform random numbers, both ranging from -1 to 1, and   is the noise level. For each mesh point z, the corresponding far-field equation (1) is solved by using Tikhonov regularization method.

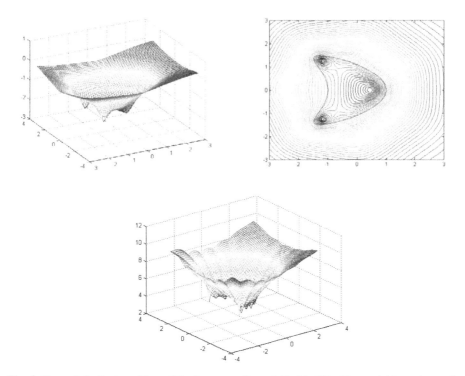

Fig. 3. Upper left: the logarithm of the L_2-norm of g_z plotted in 3D ; Upper right: contours of the logarithm of the L_2-norm of g_z; Bottom middle: the logarithm of the L_2-norm of g_z plotted in 3D without regularization in deriving g_z.

We now turn to the experiment Kite. First, we solve the far field equation (1) on the finest mesh (129 × 129) to find g_z with z being a sampling mesh point. In order to have a view of the behavior of this g_z over the sampling mesh, we plot the logarithm of its L_2-norm, namely log $\|g_z\|_{L2(B)}$, in a 3D graph (see the upper left picture of Fig. 4). The corresponding contours for log $\|g_z\|_{L2(B)}$ is also given in the upper right of Fig. 4 with a 2D view. Then, we can use the cut-off value principle to detect the kite and this gives the original LSM. And in this case, we would like to refer to [7] for a glance at the numerical outcome. We remark that the regularization is crucial in the numerical procedure. We have also plotted the logarithm of the L_2-norm of g_z which is obtained by solving the far-field equation without regularization, from which it can be seen that the reconstruction would be rather unsatisfactory; see the 3D display in Fig. 4 (bottom middle). This phenomenon was noted in [7] and a rigorous mathematical explanation was given in [1].

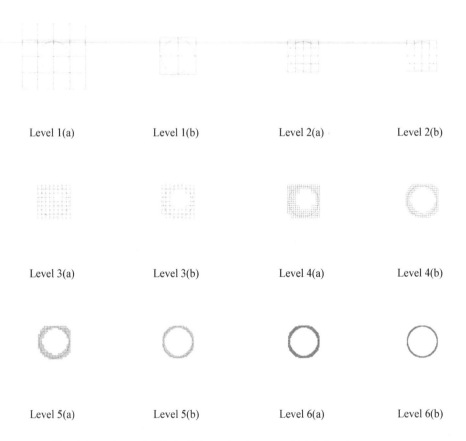

Level 1(a)	Level 1(b)	Level 2(a)	Level 2(b)
Level 3(a)	Level 3(b)	Level 4(a)	Level 4(b)
Level 5(a)	Level 5(b)	Level 6(a)	Level 6(b)

Fig. 4. Experiment 1 (Circ): (a) investigation step; (b) label-and-remove step

Now we apply our (6-level) MLSM to this problem with $n_L = n_6 = 129$ and plot the evolution of the detected boundary of the underlying object level by level. Figure 5 and 6 demonstrate that the boundary of the circle- and kite-shaped objects can be approximated in a clearly improving manner as we go from coarse to fine meshes, but the points examined are kept within the order $O(n_L)$.

For both experiments Circ and Kite, the MLSM performs as well as the original LSM method but the computational costs have reduced significantly (see Table 2). In fact, we have counted the number of points examined in the MLSM which is listed in Table 2; and for Test 1 and 2, it is roughly one twelfth of that for LSM which is 16641 ($= 129 \times 129$).

For comparison, we list in Table 2 the number of points examined at each level in the MLSM procedure and the total numbers points examined by MLSM and LSM for both tests. It can be seen from the table that the number of points examined at each level is about Mn_k with $M \approx 6$.

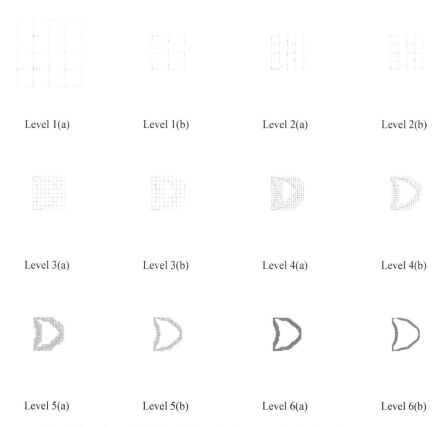

Level 1(a)	Level 1(b)	Level 2(a)	Level 2(b)
Level 3(a)	Level 3(b)	Level 4(a)	Level 4(b)
Level 5(a)	Level 5(b)	Level 6(a)	Level 6(b)

Fig. 5. Experiment 2 (Kite): (a) investigation step; (b) label-and-remove step

Table 2. Number of points checked by the MLSM at each level and LSM in the tests

	1	2	3	4	5	6	MLSM	LSM
Test 1(Circ).	25	25	80	190	326	638	1284	16641
Test 2(Kite).	25	25	81	169	334	751	1385	16641

Acknowledgments. The work was substantially supported by Hong Kong RGC grants (Project 404105 and Project 404606).

References

1. Colton, D., Coyle, J., Monk, P.: Recent developments in inverse acoustic scat- tering theory. SIAM Rev. 42(3), 369–414 (2000)
2. Colton, D., Giebermann, K., Monk, P.: A regularized sampling method for solving three-dimensional inverse scattering problems. SIAM J. Sci. Comput. 21(6), 2316–2330 (2000)
3. Colton, D., Kirsch, A.: A simple method for solving inverse scattering problems in the resonace region. Inverse Problems 12, 383–393 (1996)
4. Colton, D., Kress, R.: Inverse Acoustic and Electromagnetic Scattering Theory, 2nd edn. Springer-Verlag, Berlin (1998)
5. Colton, D., Monk, P.: A linear sampling method for the detection of leukemia using microwaves. SIAM J. Appl. Math. 58, 926–941 (1998)
6. Colton, D., Piana, M.: The simple method for solving the electromagnetic inverse scattering problem: the case of te polarized waves. Inverse Problems 14(3), 597–614 (1998)
7. Colton, D., Piana, M., Potthast, R.: A simple method using morozovs dis- crepancy principle for solving inverse scattering problems. Inverse Problems 13, 1477–1493 (1997)
8. Piana, M.: A simple regularization method for solving acoustical inverse scattering problems. J. Comput. Acoust. 9(2), 565–573 (2001)
9. Li, J., Liu, H., Zou, J.: Multilevel Linear Sampling Method for Inverse Scattering Problems. Technical report CUHK-2006-06(034). Department of Mathematics, the Chinese University of Hong Kong

A New Evolutionary Neural Network and Its Application for the Extraction of Vegetation Anomalies

Yan Guo[1], Lishan Kang[1,3], Fujiang Liu[2], and Linlu Mei[2]

[1] School of Computer, China University of Geosciences, Lumo Road 388,
430074 Wuhan, China
[2] Faculty of Information Engineering, China University of Geosciences, Lumo Road 388,
430074 Wuhan, China
[3] School of Computer Sciences, Wuhan University, Luojia Hill, 430072 Wuhan, China
guoyanwuhan@yahoo.com.cn, kang_whu@yahoo.com,
felixwuhan@yahoo.com.cn, meilinlu@163.com

Abstract. This paper proposes an evolutionary neural network (ENN). In this ENN model evolutionary algorithms (EAs) are adopted to train the multilayer perceptrons (MLPs) to overcome backpropagation (BP) algorithm shortcomings. The proposed ENN technique was used to the extraction of vegetation anomalies in remote sensing imagery compared against MLPs with BP algorithm. The experiments of extracting vegetation anomalies were carried out by ENN classifiers and BP classifiers in a 1241×1149 pixel Landsat-7 Enhanced Thematic Mapper plus (ETM+) high-resolution image of Zhaoyuan gold deposits, Shandong, China. We found that the use of EAs for finding the optimal weights of MLPs results mainly in improvements in overall accuracy of MLPs.

Keywords: Evolutionary neural networks, evolutionary algorithms, BP algorithm, vegetation anomalies, extraction, remote sensing.

1 Introduction

Satellite remote sensing has provided us an effective method for mineral surveying. Biogeochemical prospecting techniques have proven useful for mapping rock distributions, faults, and mineral anomalies [1]. Sources with vegetation anomalies are of particular interest and extraction vegetation anomalies become a key methodology within the mineral surveying field as they indicate the presence of mineral. Scientists currently identify the vegetation anomalies by Normalized Difference Vegetation Index (NDVI) and principal component analyses (PCA) [2], [3], which are adaptive to the region of small coverage area with strong anomalies information while failure to the area with weak anomalies information.

Our goal is to bring automation to the extraction of vegetation anomalies from remote sensing data using intelligent techniques, such as neural networks (NNs) to increasing classification accuracy of weak anomalies information. NNs were used in this research to determine the relationship of spectrum and vegetation anomalies in remote sensing imagery. Multilayer perceptrons (MLPs) are the most common type of

L. Kang, Y. Liu, and S. Zeng (Eds.): ISICA 2007, LNCS 4683, pp. 243–252, 2007.

NNs used for remote sensing studies and have proved to be effective in comparable studies [4].

But the training of the MLPs, normally featured with backpropagation (BP) algorithm or other gradient-descent-based algorithms, still faces certain drawbacks, e.g., very slow convergence, easily getting stuck in a local minimum, and inconsistent results due to random initial connection weights, etc.

One way to overcome gradient-descent-based training algorithms is evolutionary neural networks (ENNs) in which evolutionary algorithms (EAs) is adopted to formulate the training process as the evolution of parameters (including connection weights and bias) [5]. EAs refer to a class of population-based stochastic search algorithms that are developed from ideas and principles of natural evolution. One important feature of EAs is their population-based search strategy. Individuals in a population compete and exchange information with each other in order to perform certain tasks. EAs can be used effectively in the training of connection weights to evolve and find a near-optimal set of connection weights globally without computing gradient information. Unlike the case in gradient-descent-based training algorithms, the fitness (or error) function does not have to be differentiable or even continuous since EAs do not depend on gradient information. Moreover, another major advantage to the evolutionary approach over BP algorithm is the ability to escape local optima. More advantages include robustness and ability to adopt in a changing environment.

Our objective is to demonstrate that evolutionary algorithms can successfully address the training problems of MLPs, resulting in accurate networks with good generalization abilities.

This paper describes the MLPs with EAs-based training algorithm, called ENN classifiers, and their application to the identification of vegetation anomalies from remote sensing imagery. As an example, a land-cover classification experiment was carried out by ENN classifiers in a 1241×1149 pixel Landsat-7 Enhanced Thematic Mapper plus (ETM+) high-resolution image of Zhaoyuan in Shandong province in eastern China. Moreover the MLPs with BP algorithm, called BP classifiers, were employed to comparing the performance with ENN classifiers. The experiments showed that ENN classifiers produced significantly more accurate classification result than we could obtain by BP classifiers.

This paper is organized as follows: Section 2 outlines the problem of extracting vegetation anomalies in the remote sensing data, and provides details about the study area and data set. Section 3 describes the ENN algorithms in which EAs are adopted to train MLPs. Section 4 presents our experiments and reports the results. The paper concludes with our observations and plans for future work.

2   Study Area and Data Set

In vegetation area, as a result of that the plant absorb Au from rocks and soils, the contents of Au, pigment and water, the surface temperature, and the cells structure vary in the poisoned leaves, even the plant cause abnormality of biogeochemical effect, thus it appear abnormality in the spectral information of remote sensing image. For instances, the spectral reflectance of the poisoned leaves is 5%-30%lower than that in the normal areas in band 4 in spectrum curves of normal and anomalous vegetations as shown in Fig. 1. According to these feature and difference, the remote sensing image processing technologies can utilized to extract the anomalous features of biogeochemical effect [6]. This work focused on the extraction of vegetation anomalies from remote sensing imagery by applying the remote sensing image processing technologies and intelligent techniques. Such an analysis was performed by using PCA and ENN.

The study area was a 1433.18 km2 region of Zhaoyuan gold deposit in northwest Shandong Peninsular, China. It lies between 37°05′ and 37°33′ North Latitude and 120°08′ and 120°38′ East Longitude. Zhaoyuan is characterized by its rich gold resources; this area is one of the most famous in the eastern of China.

In July 2006, a field survey was undertaken in the study area to acquire ground data on vegetation anomalies, including the orientation of anomalous vegetation area and normal vegetation area using GPS facility.

A 1241×1149 pixel Landsat Enhanced Thematic Mapper plus (ETM+) image of the study area acquired on June 12 2000 (orbit number is 120/34) was used for the research as shown in Fig. 2. Here, the ETM+ data acquired in the six visible, near and short-wave infrared bands with a spatial resolution of 30m were selected for the study (the data acquired in ETM+ bands 6 and 8 were not used). These data were pre-processed carried out in ERDAS Imagine 8.7 by using RGB color composition, geometric correction, radiometric corrections and PCA.

PCA [7] is a linear transformation which decorrelates multivariate data by translating and/or rotating the axes of the original feature space, so that the data can be represented without correlation in a new component space. In our investigations, PCA was used as a data transform to enhance regions of vegetation anomalies in multi-spectral data. In this work PCA was done to reduce the six features to three principal features. The three bands are: Green band of wavelength (0.52–0.60 um), Red band of wavelength (0.63–0.68 um), and near infrared band of wavelength (0.76–0.90 um).After analyzing the principle component, supervised classification was performed by using ENN.

Six land-cover classes were obtained based on visual interpretation of the pre-processed image and the field survey. They are: (1) anomalous vegetation, (2) normal vegetation, (3) water, (4) urban, (5) farmland, and (6) road / bare ground.

3 Evolutionary Neural Networks

Evolutionary neural networks refer to a special class of neural networks in which evolution is another fundamental form of adaptation in addition to learning. Evolutionary algorithms have been used to train the weights of the networks, to select the most relevant input features of the training data, and to design the structure of the networks, and this section presents a brief review of previous work. The interested reader may consult the reviews by [8], [9], [10], [11], [12], [13] and [14].

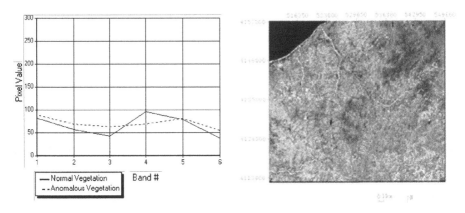

Fig. 1. Spectrum curves of normal and anomalous vegetation

Fig. 2. A sub-scene Landsat-7 ETM+ image of Zhaoyuan, Shandong, China

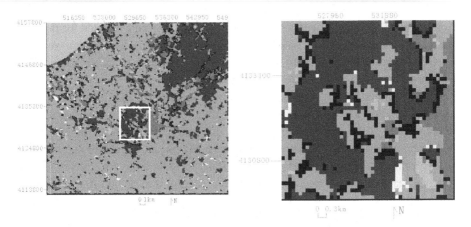

Fig. 3. Classified Image of the sub-scene in Fig.2 using ENN classifications

Fig. 4. The zooming-in image of the rectangle area within white boundary in Fig.3

Anomalous Vegetation	Water	Farmland
Normal Vegetation	Urban	Road and bare ground

In this study, EAs is adopted to train the MLPs, in which weights learning problems are involved. Training a NN is an optimization task with the goal of finding a set of weights that minimizes an error measure. The search space is high dimensional and, depending on the error measure. As we have shown in the previous section, the drawbacks of some form of the gradient search for neural networks training are numerous. So, we proposed a new approach for networks training based on EAs. The EAs process regarding to train the MLPs is described as follows.

3.1 Encoding Scheme for NNs

The architecture of the networks is fixed by the user prior to the experiment. It is well known that to solve non-linearly separable problems, the networks must have at least one hidden layer between the inputs and outputs; Therefore, three-layered MLPs networks were applied in this research and were trained with an evolutionary learning principle. The number of neurons in the hidden layer, H, also known as processing elements, is data dependent.

In this method, each individual in the EAs represents a real-number vector with all the weights of the networks. The most common approach for the genotype of the networks consists of the linearization of the connectivity matrix [15] (see Fig. 5). This codification is straightforward. However, one drawback is the so-called permutations problem [16]. It is caused by the many-to-one mapping from the representation (genotype) to the actual NN (phenotype) since two NNs whose hidden nodes just permuted will still be functionally the same though the representation of the weights in the chromosome would be different. Some permutations might be unsuitable for EAs because crossover might easily disrupt favorable combinations of weights.

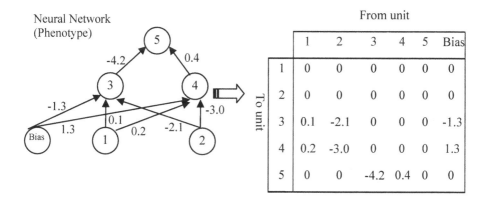

$0\ 0\ 0\ 0\ 0\ 0$ $0\ 0\ 0\ 0\ 0\ 0$ $0.1\ -2.1\ 0\ 0\ 0\ -1.3$ $0.2\ -3.0\ 0\ 0\ 0\ 1.3$ $0\ 0\ -4.2\ 0.4\ 0\ 0$ (Genotype)

Fig. 5. Standard real-number coding of a neural network

Another alternative is placing the incoming and outgoing weights of a hidden node next to each other [17]. The codification of the networks of Fig. 5 using this model is shown in Fig. 6. Thierens [18] presented a non-redundant coding that avoids different representations for exactly the same networks consequently solving permutation problem. In this codification all the $2^H H!$ equivalent networks, for a networks of H hidden nodes, can be transformed to a canonical form. Non-redundant coding was the encoding we used in this research.

Node 3: -1.3 0.1 -2.1 -4.2
Node 4:   1.3 0.2 -3.0 0.4

⟹   -1.3 0.1 -2.1 -4.2 1.3 0.2 -3.0 0.4

Genotype

Node codification
(Input and output nodes not codified)

Fig. 6. Non- redundant coding

3.2 Fitness Evaluation

The performance of each individual is measured according to a pre-defined fitness function, which is related to the problem to be solved. In this study, the objective (fitness) function was the same as that used in backpropagation algorithm: minimize the mean sum-squared error (*MSE*) function over all patterns S for G output neurons, that is

$$min\ MSE = \frac{1}{|G| \cdot |S|} \sum_{g \in G} \sum_{s \in S} (Y_g(s) - O_g(s))^2 \ . \tag{1}$$

where $Y_g(s)$ and $O_g(s)$ is the desired and actual outputs of training pixel s for class g. Equation (1) makes the error measure less dependent on the size of the training set and the number of output nodes.

3.3 Evolutionary Process

The evolution of the population of networks is as standard as possible. To obtain the next generation for a population of size N, the parents and the offspring is put together and the N best individual are selected. The best $P_e\%$ individuals of the population are replicated, the rest of the new population is obtained by multipoint crossover. The new population is subject to parametric mutation.

Multipoint crossover operator randomly chooses a pair of individuals and a number of crossover points along their chromosomes. Crossover exchanges segments of the two chromosomes delimited by the crossover points. Fig. 7 shows an example of two-point crossover.

Parametric mutation is carried out in order to fine-tune weights of the networks for the evolutionary process. In our model we considered two standard parametric mutation operations [15]:

(1) Backpropagation mutation. The networks are trained using a simple back-propagation algorithm for a fixed number of cycles. We use cross-validation and early-stopping. The validation set is taken randomly from the training set every time the back-propagation algorithm is run.

(2) Random mutation. A small quantity, normally distributed with zero mean and a small standard deviation, is added to all the weights of the networks.

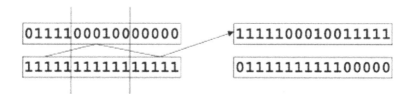

Fig. 7. An example of two-point crossover

The stop criterion, for the evolutionary algorithms used in this paper, is the stagnation of the fitness of the population. The evolution ends if, during the last 10 generations, the average fitness of the population does not improve above certain threshold. In the experiments this threshold is set to 1%, or after a limit of 100 iteration times.

4 Experiments

The proposed ENN algorithm was applied to real world classification tasks of extracting vegetation anomalies from remotely-sensed images. The MLPs classifiers with ENN algorithms were called ENN classifiers. The obtained optimal solutions of ENN classifiers were compared with BP classifiers, the MLPs classifiers with backpropagation algorithm. The programs and required analyses including accuracy assessment and kappa statistic were carried out in MATLAB Version 7.2.0.232 (R2006a). The experiments were executed on a single processor of a Microsoft Windows (XP Professional Service Pack 2) workstation with 2.4 GHz Intel Xeon processor and 512 Mb of memory.

In order to present the spectral variation of vegetation anomalies to NNs, sample sets for each class were selected from the 1241 by 1149 pixels image and a total of 6525 samples were generated. The input attributes used in this work were rescaled in the range [0, 1] and divided into ten non-overlapping splits, each one with 70% of the data for training while 30% is used for testing.

In order to make a fair comparison among the different methods for extracting vegetable anomalies, all the experiments were carried out using the same parameter set. According to the features of the input data used in the experiment, the MLPs networks had 3 inputs that correspond to each of the features in the input data, $H=14$ hidden nodes, and 6 outputs that correspond to each type of land cover. The transfer function of every neuron is the sigmoid hyperbolic tangent function. The range of possible weights of networks was [-10; 10]. The target training performance was set to 0.0100.

The comparisons among the performance of the different methods were made by means of standard t-tests with 95% confidence. A ten-fold crossvalidation trial was performed; that is, the algorithms were executed ten times, each time using a different split on the data.

We implemented the proposed method described in section 3 using EAs to train the MLPs networks. The population of the evolutionary algorithm was 20 MLPs with only one hidden layer and was evolved for 100 generations. Elitism is $P_e=0.1$, the probability of multipoint crossover is $P_c=0.8$, the probability of random and backpropagation mutation is $P_{bp}=0.25$ and $P_r=0.25$, respectively. For backpropagation algorithm learning rate $\eta=0.1$ and $\alpha=0.4$, and the number of cycles is 10. From each experiment the result of the evolution was the best of the population.

The second training method is using BP to train the MLPs. We implemented this method and used the same networks architecture and parameters as in the first experiment. Each networks was trained using maximum 1000 epochs of BP with a learning rate $\eta=0.1$ and $\alpha=0.4$.

A result of one run is shown in Fig.3. The Classified image of the sub-scene in Fig.2 by using an ENN classifier is illustrated in Fig. 3, in which the vegetation anomalies areas are yellow while normal vegetation areas are red. The rectangle area within white boundary in Fig.3 is zoomed in as shown in Fig.4.

Table 1. Comparison of average test accuracy: user's accuracy, kappa statistic and overall accuracy in % corresponding to the different type of land-cover for different methods (ten runs). The numbers in parenthesis are the standard deviations, and the results that are significantly different ($\alpha = 0.05$) from the BP classifier (first column) are highlighted in bold.

Land cover	User's accuracy (%)		Kappa statistic	
	BP	EAs	BP	EAs
Anomalous vegetation	78.01(2.78)	**87.57(0.97)**	0.7768	**0.8364**
Normal vegetation	87.76(3.02)	90.63(1.03)	0.8473	**0.8947**
Water	90.12(2.13)	92.74(0.83)	0.9214	0.9179
Urban	71.94(4.34)	**77.42(3.02)**	0.7346	**0.8166**
Road and bare ground	69.16(3.94)	**73.51(2.84)**	0.7186	**0.7931**
Farmland	91.36(2.36)	**86.7(0.95)**	0.9168	**0.8463**
Overall accuracy	85.63(2.47)	**90.58(1.39)**	0.8075	**0.8798**

Ten runs of experiments were conducted and the means of the overall performance and kappa statistic were reported. Here the classification results were compared with our interpretation results. These results are shown in Table 1.

The entries EAs and BP in Table 1 present the average accuracy of the best networks found in each run for two sets of experiments. The results highlighted in bold in the table are the better results significantly.

Table 1 shows accuracy measures for ENN classifiers and BP classifiers. It is observed from Table 1 the ENN classifiers performed better in terms of use's accuracy and kappa value than the BP classifiers for the classification of land cover. The overall accuracies for ENN classifiers and BP classifiers are 90.58% and 85.63% respectively.

The reason for better accuracy with ENN classifier than BP classifier is possibly because EAs could be used effectively in the training of connection weights to evolve and find a near-optimal combination of weights globally, which avoid to be trained several times with different NNs to prevent the networks becoming stuck into a local minimum but make the influence of NNs architecture and the initial values of the connection weights on the final classification accuracy least.

5 Conclusion and Future Work

This paper presented a comparison of MLPs with evolutionary training algorithms and with BP learning algorithm for the identification of vegetation anomalies in the remotely sensing data.

Our experiments suggest that, for this application, the EAs-based training algorithm can produce accurate classifiers that are competitive with networks trained by BP. For our application, we found significant differences among the ENN and BP networks. The consistently better method was to use the EAs to train the networks, which can find a near-optimal combination of weights globally.

There are several avenues to extend this work. In this paper we used a simple evolutionary algorithm to train neural networks. There are other combinations of EAs and NNs that we did not include in this study, but appear promising. For instance, a natural extension of this work would be to use evolutionary algorithms to select the most relevant input features of the training data, and to design the structure of the networks.

The major disadvantage to the EANN approach is it is computationally expensive. This can be an obstacle to applying these techniques to larger data sets, but there are numerous alternatives to improve the performance of EAs. For instance, we could exploit the inherently parallel nature of EAs using multiple processors.

Acknowledgements

This work was supported by the National Natural Science Foundation of China under Grant 60133010, the Research Foundation for Outstanding Young Teachers of China University of Geosciences (Wuhan) under Grant CUGQNL0628 and CUGQNL0640.

References

1. Nash, G.D., Moore, J.N., Sperry, T.: Vegetal-spectral Anomaly Detection at the Cove Fort-Sulphurdale Thermal Anomaly, Utah, USA: Implications for Use in Geothermal Exploration. Geothermics 32, 109–130 (2003)
2. Nash, G.D., Hernandez, M.W.: Cost-effective Vegetation Anomaly Mapping for Geothermal Exploration. In: Twenty-Sixth Workshop on Geothermal Reservoir Engineering, pp. 29–30. Stanford University, Stanford, California (2001)
3. Lasaponara, R.: On the Use of Principal Component Analysis (PCA) for Evaluating Interannual Vegetation Anomalies from SPOT/VEGETATION NDVI Temporal Series. Ecological Modeling 194, 429–434 (2006)
4. Foody, G.M., Cutler, M.E., McMorrow, J., Pelz, D., Tangki, H., et al.: Mapping the Biomass of Bornean Tropical Rain Forest from Remotely Sensed Data. Global Ecology and Biogeography 10, 379–387 (2001)

5. Yao, X., Liu, Y.: A New Evolutionary System for Evolving Artificial Neural Networks. IEEE Transaction on Neural Networks 8, 694–713 (1997)
6. Yue-liang, M.: Progress in Study on Methods of Prospecting for Gold Deposit Using Remote Sensing Biogeochemistry. Advance in Earth Sciences 17, 521–528 (2002)
7. Richards, J.A.: Remote sensing Digital Image Analysis, pp. 127–138. Springer, Kensington (1986)
8. Cantu-Paz, E., Kamath, C.: An Empirical Comparison of Combinations of Evolutionary Algorithms and Neural Networks for Classification Problems. IEEE Transactions on Systems, Man and Cybernetics. Part B: Cybernetics, 915–927 (2005)
9. Cantu-Paz, E., Kamath, C.: Evolving Neural Networks for the Classification of Galaxies. Neural Networks 16, 507–517 (2003)
10. Schaffer, J.D.: Combinations of Genetic Algorithms with Neural Networks or Fuzzy Systems. In: Zurada, J.M., Marks II, R.J., Robinson, C.J. (eds.) Computational Intelligence Imitating Life, pp. 371–382. IEEE Press, New York, NY (1994)
11. Lu, W.Z., Fan, H.Y., Lo, S.M.: Application of Evolutionary Neural Networks Method in Predicting Pollutant Levels in Downtown Area of Hong Kong. Neurocomputing 51, 387–400 (2003)
12. Yao, X.: Evolving Artificial Neural Networks. In: the Proceedings of IEEE, vol. 9(87), pp. 1423–1447. IEEE Computer Society Press, Los Alamitos (1999)
13. Guo, Y., Kang, L., Liu, F., Sun, H., Mei, L.: Evolutionary Neural Networks Applied to Land-cover Classification in Zhaoyuan, China. In: IEEE Symposium on Computational Intelligence and Data Mining, Hawaii, America, pp. 1423–1447. IEEE Computer Society Press, Los Alamitos (2007)
14. Liu, F., Wu, X., Guo, Y., Zhou, F., Mei, L.: A Pareto Evolutionary Artificial Neural Networks Approach for Remote-sensing Image Classification. In: The Proceedings of SPIE- Geoinformatics. vol.6419, pp. 64191L-1–64191L-7 (2006)
15. Garcia-Pedrajas, N., Ortiz-Boyer, D., Hervas-Martinez, C.: An Alternative Approach for Neural Network Evolution with a Genetic Algorithm: Crossover by Combination Optimization. Neural Networks 19, 514–528 (2006)
16. Radcliffe, N.J.: Genetic neural networks on MIMD computers. Doctoral dissertation, University of Edinburgh, Scotland (1990)
17. Thierens, D., Suykens, J., Vandewalle, J., Moor, B.D.: Genetic Weight Optimization of a Feedforward Neural Networks Controller. In: The Conference on Neural Nets and Genetic Algorithms, pp. 658–663. Springer, Berlin (1991)
18. Thierens, D.: Non-redundant Genetic Coding of Neural Networks. In: IEEE International Conference on Evolutionary Computation, pp. 571–575. IEEE Press, Piscataway, NJ (1996)

Dynamic System Evolutionary Modeling:
The Case of SARS in Beijing

Chenzhong Yang[1], Zhuo Kang[2,*], and Yan Li[2]

[1] School of Mathematics and Computer Science, Guizhou University for Nationalities,
Guiyang 550025, Guizhou, China
[2] Computation Center, Wuhan University, Wuhan 430072, Hubei, China
kang_zh@sohu.com

Abstract. In this paper a new evolutionary algorithm for automatically modeling of dynamic systems is proposed. The algorithm is based on a scalable multi-gene chromosome representation with fixed length, which is similar in form to gene expression programming (GEP) proposed by Ferreira. The complexity of the automatic programming of modeling is determined by length of chromosome, and the complexity of function set and terminal set used for modeling. For modeling dynamic systems, the systems of ordinary differential equations are used. The new algorithm is used to model the super-spreading events of severe acute respiratory syndrome (SARS) in Beijing, because Beijing experienced the largest outbreak of SARS, with >2500 cases reported between March and June, 2003. Two types of ODE models, systems of ordinary differential equations and higher order ordinary differential equations are automatically discovered by the new methodology from the reported data (http://www.Beijing.gov.cn/resource/Detail.asp?Resource ID=66070).

Keywords: Evolutionary modeling; gene expression programming; SARS models; System of ordinary differinary differential equations.

1 Introduction

In [1], Ferreira developed the basic ideas of gene expression programming (GEP) in 1999 almost unaware of their uniqueness. It may have great significance and deep influence on the research of automatic programming in the future.

In this paper a new evolutionary algorithm for automatically modeling of dynamic systems is proposed. The algorithm is based on a scalable multi-gene chromosome representation with fixed length, which is similar in form to gene expression programming. The new algorithm in this paper is used to model the super-spreading events of severe acute respiratory syndrome (SARS) in Beijing[6][7], because Beijing experienced the largest outbreak of SARS, with >2500 cases reported between March and June, 2003. The systems of ordinary differential equations is automatically

---

* Corresponding author.

L. Kang, Y. Liu, and S. Zeng (Eds.): ISICA 2007, LNCS 4683, pp. 253–261, 2007.

discovered by the new methodology from the reported data (http://www.Beijing. gov.cn/resource/ Detail.asp?Resource ID=66070).

The rest of this paper is organized as follows: In section 2, the function set F and the terminal set T in automatic programming is introduced. In section 3, multi-gene chromosome structure is discussed; Section 4 is the automatic modeling algorithm. Finally, numerical experiment results for evolutionary modeling of the super-spreading events of SARS in Beijing are given in section 5.

2 Function Set and Terminal Set in GEP

The formula of automatic programming [2] is described as follows:

    Automatic Programming = Evolutionary Algorithm + Program Structure

where Program Structure =Function Set $\odot$ Terminal Set [3]

$$= F \odot T$$

For example, the following expression:

$$\sqrt{(a-b)\times(c+?)}$$

Its operator (function) set $F = \{ -, *, +, \sqrt{} \}$, where $-$, $*$ and $+$ are dualistic operators, $\sqrt{}$ is unitary operator; and its terminal (data) set $T = \{a, b, c, ?\}$, where a, b and c can be not only integer but also real, and "?" is a random variable. Its expression tree (ET) is as follows:

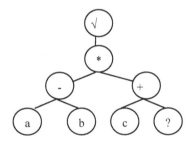

When F and T are determined, the program structure is determined. For a given problem, how to choose F and T, such that the program space is complete and without much redundancy is still an open problem. We call it problem-oriented automatic programming[2].

3 Multi-gene Chromosome Structure

In order to describe a system of ordinary differential equations, the multi-gene chromosome is introduced.

An m-gene chromosome g is composed of m genes:

$$G = (g_1, g_2, \cdots, g_m) \tag{1}$$

Where g_i, $i = 1, 2,\ldots,m$, are genes.

A gene is composed of three parts:

$$g = (\text{head, body, tail})$$

a head with length h, a body with length b and a tail with length t, and

$$t = (h + b)\times(n-1) + 1 \tag{2}$$

where n is the maximum number of variable arguments in function(operator). The head contains symbols from function (operator) set F , the tail contains symbols from terminal set T, and the body contains symbols from both F and T.

Given a function set $F = (+, -, *, /)$ and a terminal set $T = \{ a, b, x, y, z \}$ We want to construct a model of system of ordinary differential equations.

Example: The following chromosome with length $21 = 3\times7$ is composed of three genes g_1, g_2 and g_3 with length $7 = 1+2 + 4$:

$$\begin{Bmatrix} 1\ 2\ 3\ 4\ 5\ 6\ 7\ 1\ 2\ 3\ 4\ 5\ 6\ 7\ 1\ 2\ 3\ 4\ 5\ 6\ 7 \\ \underline{*\ -\ *\ a\ x\ b\ y}\ \underline{*\ x\ -\ y\ z\ y\ z}\ \underline{+\ /\ +\ y\ x\ z\ x} \\ \qquad g_1 \qquad\qquad g_2 \qquad\qquad g_3 \end{Bmatrix} \tag{3}$$

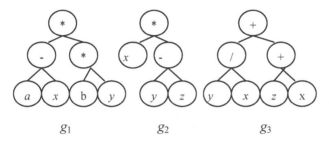

$$g_1 \qquad\qquad\qquad g_2 \qquad\qquad\qquad g_3$$

It can be used to express a system of non-linear ordinary differential equations[5] :

$$\begin{cases} \dfrac{dx}{dt} = (a-x)by \\ \dfrac{dy}{dt} = x(y-z) \\ \dfrac{dz}{dt} = \dfrac{y}{x} + z + x \end{cases}$$

All chromosomes form a model space denoted as G.

4 Automatic Programming Algorithm

Given observed data

$$(t_i, X_i),\ \ i = 1, 2, \ldots, m \tag{4}$$

where $X_i = (x_i, y_i, z_i) \in R^3$, and $t_{i+1} = t_i + \Delta t$.

Mathematical modeling problem[5]:

Find a system of non-linear ordinary differential equations

$$dX/dt = f(X, t) \tag{5}$$

such that

$$\min \sqrt{\sum_{i=1}^{n} \left[(x_i - x(t_i))^2 + (y_i - y(t_i))^2 + (z_i - z(t_i))^2 \right]}$$

where $x(t_i)$ is the numerical solution of ODEs models solved by Runge-Kutta method. The complexity of a model expressed as a system of ODEs is determined by the function set F, terminal set T and the length of chromosome which belongs to the model space $M^{[2]}$.

For solving the modeling problem (5) a function set F a terminal set T and a model space M must be given firstly. The frame of Evolutionary Modeling Algorithm (EMA) is like the frame of GA:

Procedure EMA

```
begin
      t := 0;
      initialize G (t) = {G₁,(t ), G₂,(t ), …, Gₙ(t )}, Gᵢ∈M;
              M is the model space
      Transformation T:   G (t )• P (t ) = {p₁(t),   p₂(t), …,
              pₙ(t)}, where   pᵢ∈P ;
      Evaluate   F(P(t)) = {  F(p₁),  F(p₂),  …, F(pₙ)};
      while not terminate do
            Gc (t ) = crossover  G(t )  with probability pc ;
            Gm (t ) = mutate  Gc(t )  with probability pm ;
            Transformation   T:   Gm (t ) •Pm(t) ;
            evaluate F(Pm(t) ) and choose a best model Gbest ;
            call GT to optimize parameters "?" of Gbest ;
            G(t +1) = select { Pm(t)∪P(t) } keeping elite(Gbest);
            t := t + 1;
      endwhile
      print pbest, F (pbest , t );
end
```

where F is the fitness function:

$$F(f) = \sqrt{\sum_{i=1}^{n} \left[(x_i - x(t_i))^2 + (y_i - y(t_i))^2 + (z_i - z(t_i))^2 \right]}$$

All crossover and mutation operators are chosen as in GEP[1].

GT is a subspace search method proposed by Guo Tao[4]. That is used for optimizing the randomly produced parameters in the best model G_{best}.

5 Evolutionary Modeling for Events of SARS in Beijing

SARS (severe acute respiratory syndrome) erupted in Guangdong Province in November 2002. In Chinese, the disease is called *feidianxing feiyan*, which means a-typical pneumonia, abbreviated to *feidian*. In Beijing, it was first detected in early March 2003. The outbreak of SARS constituted a serious danger for the population. Immediately, the authorities swung into action. After prompt measures were introduced in Beijing in response to the outbreak, opportunities for super-spreading were greatly reduced. Thus there may have been many other patients with host or viral characteristics conducive to super-spreading later in the Beijing outbreak, but successful infection control prevented these occurrences. A propaganda campaign unfurled. At the height of the epidemic in April and May, more than one million people left Beijing. Many residents were isolated in their living quarters out of fear of having been infected. Beijing experienced the largest outbreak of SARS, with >2,500 cases reported between March and June 2003. Some people [6] investigated probable and suspected cases reported from hospitals in Beijing to understand their relationship to each other, determine the incubation period between exposure and symptom onset, and describe clinical features at the time of symptom onset. They identified and followed close contacts of SARS patients to monitor their progress. They sought clinical data for patients associated with super-spreading. The chi-square statistic and where appropriate, Fisher exact test, were used to compare proportions. Some people used statistical models and systems of ordinary differential equations to model the super-spreading procedure of the largest outbreak of SARS in Beijing under some hypothesizes. We use a new evolutionary algorithm for automatically modeling the dynamics of super-spreading procedure of SARS from the data which describe the daily SARS situation from 20th of April to 23rd of June,2003 in Beijing as in Table 1, where the cumulative number 0f SARS patients at time t is denoted by $x(t)$, the cumulative number of persons died of SARS is denoted by $y(t)$ and the cumulative number of persons recovered from SARS is denoted by $z(t)$.

Consider the following three cases:

(1) The data $x(t)$ is regarded as an one dimensional time series which satisfies an ordinary differential equation:

$$\frac{dx}{dt} = f(x,t)$$
$$x(x_0) = x$$

Table 1.

t	$x(t)$	$y(t)$	$z(t)$	t	$x(t)$	$y(t)$	$z(t)$
1	339	18	33	34	2465	160	5
2	482	25	43	35	2490	163	6
3	588	28	46	36	2499	167	7
4	693	35	55	37	2504	168	7
5	774	39	64	38	2512	172	8
6	877	42	73	39	2514	175	8
7	988	48	76	40	2517	176	9
8	1114	56	78	41	2520	177	1
9	1199	59	78	42	2521	181	1
10	1347	66	83	43	2522	181	1
11	1440	75	90	44	2522	181	1
12	1553	82	10	45	2522	181	1
13	1636	91	10	46	2522	181	1
14	1741	96	11	47	2522	181	1
15	1803	100	11	48	2522	183	1
16	1897	103	12	49	2523	183	1
17	1960	107	13	50	2522	184	1
18	2049	110	14	51	2522	184	1
19	2136	112	15	52	2522	186	1
20	2177	114	16	53	2523	186	1
21	2227	116	17	54	2523	187	1
22	226.	120	18	55	2522	187	1
23	2304	129	20	56	2522	189	1
24	2347	134	24	57	2522	189	2
25	2370	139	25	58	2521	190	2
26	2388	140	25	59	2521	190	2
27	2405	141	27	60	2521	191	2
28	2420	145	30	61	2521	191	2
29	2434	147	33	62	2521	191	2
30	2437	150	34	63	2521	191	2
31	2444	154	49	64	2521	191	2
32	2444	156	44	65	2521	191	2
33	2456	158	52				

(2) The data $x(t)$ and $y(t)$ *are* regarded as a two dimensional time series which satisfies a system of ordinary differential equations :

$$\begin{cases} \dfrac{dx}{dt} = f_1(x, y, t) \\ \dfrac{dy}{dt} = f_1(x, y, t) \end{cases}$$

initial conditions: $x(t_0) = x_0, \ x(t_0) = y_0.$

(3) The data $x(t)$, $y(t)$ *and* $z(t)$ *are* regarded as a three dimensional time series which satisfies a system of ordinary differential equations :

$$\begin{cases} \dfrac{dx}{dt} = f_1(x, y, z, t) \\ \dfrac{dy}{dt} = f_1(x, y, z, t) \\ \dfrac{dz}{dt} = f_1(x, y, z, t) \end{cases}$$

initial conditions: $\begin{cases} x(t_0) = x_0 \\ y(t_0) = y_0 \\ z(t_0) = z_0 \end{cases}$

The fitness functions of three cases are

$$F(f) = \sqrt{\sum_{i=1}^{n}(x_i - x(t_i))^2}$$

$$F(f) = \sqrt{\sum_{i=1}^{n}\left[(x_i - x(t_i))^2 + (y_i - y(t_i))^2\right]}$$

$$F(f) = \sqrt{\sum_{i=1}^{n}\left[(x_i - x(t_i))^2 + (y_i - y(t_i))^2 + (z_i - z(t_i))^2\right]}$$

Where f denotes the model, $x(t_i)$, $y(t_i)$ and $z(t_i)$ are solutions of the systems of ODEs got by Runge-Kutta method with step-size $\Delta t = 0.01$. x_i 、 y_i and z_i are data from Table 1. The problem is to find a model $f \in M$, such that

$$\min_{f \in M} F(f),$$

where M is the model space.

The function set $F = \{+, -, *, /\}$ and the terminal set $T = \{x, y, z, t, ?\}$.
The length of head $h : 2 \le h \le 7$ and the length of body $b : 1 \le b \le 8$.
Using our new algorithm, we get the models as follows.

(1) One dimensional ODE model(See Fig. 1):
$dx/dt = (2533.368896 - x)(6.393551 + 23.809883t)$
initial condition: $x_0 = 339$

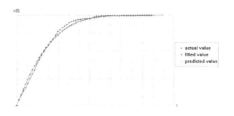

Fig. 1.

(2) One dimensional ODE model (See Fig. 2.):

$$\frac{dy}{dt} = 607.892749 - 1068.580646\ t$$

initial condition: $x_0 = 18$

Fig. 2.

(3) Two dimensional system of ODEs model(See Fig. 3.):

$$\begin{cases} \dfrac{dx}{dt} = 0.136719(252.058441 - x)y - 10.775696t + 3.231812 \\ \dfrac{dy}{dt} = 719.571045 - 2y - (y + 327.866882)t \end{cases}$$

initial condition: $x_0 = 33.9$, $y_0 = 18$

(4) Three dimensional system of ODEs model (See Fig. 4.):

$$\begin{cases} \dfrac{dx}{dt} = 184.261169(202.251507 - x) - 0.262329(x + z) \\ \dfrac{dy}{dt} = 2x - y - 2.368469z + 59.686375 \\ \dfrac{dz}{dt} = (1 + 4.784546t)y - 0.039063x - 0.643005 \end{cases}$$

initial condition: $x_0 = 33.9$, $y_0 = 18$, $z_0 = 3.3$

One may find that the models for modeling the super-spreading of SARS in Beijing discovered by our new evolutionary modeling algorithm can comparative with the models found by human beings.

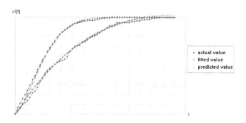

Fig. 3.

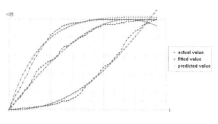

Fig. 4.

Acknowledgement

This work was supported by the National Natural Science Foundation of China (Nos.60473081, 60133010) and the Natural Science Foundation of Hubei Province (No. 2005ABA234).

References

1. Ferreira, C.: Gene Expression Programming: Mathematical Modeling by an Artificial Intelligence. Angra do Heroismo Portugal (2002)
2. Kang, L., Li, Y.,Chen, Y.: A Tentative research an complexity of automatic programming. Wuhan University Journal of Natural Sciences 6(1-2), 59–62 (2001)
3. Koza, J.R.: Genetic Programming: On the Programming of Computers by Means of Natural Selection. MIT Press, Cambridge, MA (1992)
4. Guo, T.: Evolutionary Computation and Optimization, Ph.D. Thesis, Wuhan University (1999)
5. Kang, L., Cao, H., Chen, Y.:The evolutionary modeling algorithm for system of ordinary differential equations. Chinese J. Computers 22(8), 871–876 (1999)
6. Liang, W., Zhu, Z., Guo, J., Liu, Z., He, X., Zhou, W., et al.: Severe acute respiratory syndrome, Beijing (2003). Emerg Infect Dis, vol. 10, pp. 25--31 (2004)
7. Lipsitch, I., Cohen, T., Cooper, B., et al.: Transmission dynamics and control of severe acute respiratory syndrome [J]. Science 10(1126), 1086616 (2003)

An Tableau Automated Theorem Proving Method Using Logical Reinforcement Learning

Quan Liu[1,2,3], Yang Gao[2], ZhiMing Cui[1], WangShu Yao[2], and ZhongWen Chen[3]

[1] Institute of Computer Science and Technology, Soochow University, Soochow 215006
[2] State Key Laboratory for Novel Software Technology, Nanjing University,
Nanjing 210093
[3] Institute of Mathematics, Soochow University, Soochow 215006
Quanliu@suda.edu.cn

Abstract. Logical reinforcement learning (LORRL) is presented with the combination of reinforcement learning and logic programming. Tableau method based on logic reinforcement learning is provided according to the real problem of tableau automated theorem proving method that need to extend for different logic formulae and it will influence the automated theorem proving efficiency. This method takes the combination of logic formulae and expansion result as abstract state, expansion rules as actions, node closes as the aim and receives a reward. On the one hand the method is suitable for a lot of types of tableau automated theorem proving and the blindness of reasoning is reduced. On the other hand simple automated theorem proving result can be used in complicated automated theorem proving and efficiency is raised.

Keywords: logical reinforcement learning; tableau automated theorem proving; LOMDP.

1 Introduction

Tableau method is a kind of automated theorem proving methods provided by Beth in 1955, the essence of which is displaying the connection of the semantic structure to contact the proving theory and semantics. In different logical systems, we just use the same tableau rule and extend the construct set of formulae to make it be closer to the corresponding logical system. From 1960's, for the universal and the intuition of tableau method, it has improved the interest of computer scientists, such as Smullyan, Fitting. As well as resolution, tableau method is regard as one of the important automated reasoning. Recently, tableau method has aroused a more widespread interest and became the key automated theorem proving method in automated theorem proving systems like semantic WEB[1]、 natural language under standing[2]、 database modify[3] and model checking[4].

For automated reasoning, an important guide line to check the efficiency in automated theorem proving is the time and space. The same problem also lies in tableau automated theorem proving. For example, we should limit the frequency of γ-formulae instantiation, and the function symbols and free variables in δ-formulae. If these controls can not be handled correctly, we'll make a simple proving be very

L. Kang, Y. Liu, and S. Zeng (Eds.): ISICA 2007, LNCS 4683, pp. 262–270, 2007.
© Springer-Verlag Berlin Heidelberg 2007

complex, delay the close time of tableau and even can not gain the proving process[5]. In addition, because of the symmetry of equality, the search space in tableau including equality will expand and even lead to an endless loop. In expansion from classical logic to non-classical logic, although the corresponding rules and methods of classical logic can be used in non-classical logic directly, a lot of redundant branches will be produced, which decrease the processing efficiency of computers. Nowadays, aiming at the problem of tableau automated theorem proving efficiency, a great deal of achievements which researching in tableau automated theorem proving techniques, policies and methods have been made[6][7]. But many of them are brought forward for especial logic formulae, so they are lack of universality and do no good to computer implement.

This paper proposes a logic reinforcement learning for automated reasoning on the basis of reinforcement learning. It considers the combination of logic formulae and expansion result as states, extend rules as actions and node closes as the aim to receive a reward. This method can find a best method in tableau expansion at the level of reinforcement learning, and it also can be applied in many kinds of tableau automated theorem proving. All these play an important role in decreasing the blindness and uncertainty of tableau extend.

2 Preliminary

2.1 Logic and Tableau Automated Theorem Proving

A first-order alphabet Σ is composed by a set of relation symbols P with arity $m \geq 0$ and constant C. If the arity m of relation symbol $p \in P$ is 0, then p is called proposition. Item is a variable X or a constant C. p is the predication sign with n arity, $t_1,\ldots,t_n$ are items, then $p(t_1,\ldots,t_n)$ is an atom, it is written At_Σ. A substitution is a finite set like $\{t_1/v_1,\ldots,t_n/v_n\}$, in which v_i is a variable assign, t_i is a item different from v_i, and in this set there is no two elements include the same variable sign behind bias sign, t_i is called the numerator of the substitution, and v_i is called the denominator of the substitution. When $t_1,\ldots,t_n$ are the ground items, the substitution is called ground. A conjunction A is an atom set. A conjunction A is said to be θ-subsumed a conjunction B, denoted by $A \leq_\theta B$, there exists a substitution θ such that $B\theta \subseteq A$. An item, atom and clause E is called ground when it contains no variables. The unifier of expression set $\{E_1,\ldots,E_k\}\sigma$ is called the most general unifier(MGU). If and only if for every unifier θ in this set there is a substitution λ, which makes $\theta=\sigma.\lambda$, then the most general unifier of atoms a and b is written mgu(a,b). The Herbrand base of Σ written as HB_Σ is a set of all ground atoms constructed by all the predicate and function symbols in the alphabet Σ.

Let $\{A_1,A_2,\ldots,A_n\}$ be a finite set of formulae. The following one branch tree is a tableau for formulae $\{A_1,A_2,\ldots,A_n\}$:

$$A_1$$
$$A_2$$
$$\vdots$$
$$A_n$$

If T is a tableau for $\{A_1, A_2, \ldots, A_n\}$, and T^* is a result of T by the application of a tableau expansion rules, then T^* is a tableau of $\{A_1, A_2, \ldots, A_n\}$.

Let's prove the theorem using tableau method, and divide the formulae to be proved into four types, these are α-, β-, δ- γ-formulae. The rules corresponding each formula are as follow:

α	β	γ	δ
α_1	$\beta_1 \mid \cdots \mid \beta_n$	$\gamma_1(y)$	$\delta_1(f(x_1,\ldots,x_n))$
$\vdots$			
α_n		y is a free variable	f is a new Skolem function symbol,
			$x_1,\ldots,x_n$ are free variables in branch

2.2 Reinforcement Learning

Reinforcement learning is a kind of learning that map from environment states to actions[9]. The main problem resolved is: how a autonomy agent who can perceive the environment though learning choose a optimization action that can reach the goal. The similar problems are lying in the area of automatic control, working procedure optimization and playing chess. When the agent takes an action in the environment, it will get a reward or a punishment to express whether the state result is right or not. The task of agent is learning from the indirectly delayed reward and making sure that the following actions will produce the biggest accumulative reward[10]. The standard framework of agent reinforcement learning is shown in Fig. 1.

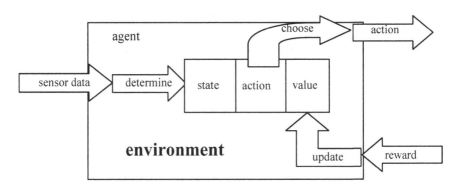

Fig. 1. The standard framework of agent reinforcement learning

Agent learning can be described as a Markov decision process (MDP). MDP is a tuple M=(S,A,T,R), in which S is a set of system states and A is a finite set of actions. To each state $s \in S$, the agent can gain a finite set of actions $A(s) \in A$. T is a transition, and for each s, $s' \in S$ and $a \in A(s)$ there is a transition T which makes state s transits into s', described as $s \xrightarrow{p:r:a} s'$. The p represents the probability of state s transits into s' aroused by action a, and for each $s \in S$, $a \in A(s)$: $\sum_{s' \in S} p(s,a,s') = 1$. For each transition, agent can achieve an immediate reward R(s, a, s') =r. In this case, reward function R just depends on current state and action. If R is probabilistic, then it is called nondeterministic, otherwise deterministic. A policy π: S$\rightarrow$A will choose the next action a_t on the basis of state s_t observed currently, that is $\pi(s_t) = a_t$. Therefore, the task of a agent is to find a policy π to maximize the value function $V^\pi(s_t)$ for all the $s_t \in S$. $V^\pi(s_t)$ are three primary forms as follow:

The infinite horizon discounted reward: $V^\pi(s_t) \equiv \sum_{i=0}^{\infty} \gamma^i r_{t+i}$

The finite horizon reward: $V^\pi(s_t) \equiv \sum_{i=0}^{\infty} r_{t+i}$

The average reward: $V^\pi(s_t) \equiv \lim_{h \to \infty} \frac{1}{h} \sum_{i=0}^{h} r_{t+i}$

In order to save space, please consult [7-8] to understand the symbols and concept that have not been explained here.

3   Logical Reinforcement Learning

The logic component of MDP corresponds to a finite state automaton. This is essentially a prepositional representation because the state and action symbols are flat, i.e. not structured. We can use logical symbols to replace the similar states and actions through logic Markov decision programs (LOMDPs) and reduce states and actions farthest.

Definition 1. Let Λ be a logic and Γ be a theory in Λ. Let P be a set of predicates in Λ, C be a set of constants in Λ and $\breve{A}$ is a set of special action predicates in Λ. A logic Markov decision process (LOMDP) is defined as $M_{LO}=(S_{LO},A_{LO},T_{LO},R)$, $S_{LO} \equiv \{s \in HB^P \cup {}^C |s \models \Gamma\}$, $A_{LO} \equiv \{a \in HB^{\breve{A}} \cup {}^C | a \models \Gamma\}$, $T_{LO} : S_{LO} \times A_{LO} \times S_{LO} \to [0,1]$.

Definition 2. An abstract state is a conjunction of logic atoms, i.e. logical query. In case of an empty conjunction, we write $\varnothing$.

Abstract states represent sets of states, state S_{LO} is a finite conjunction of ground facts over the alphabet, i.e. a logical interpretation of a subset of the Herbrand base. In the logic formulae$\{(\forall x)((g(x) \approx f(x)) \lor \neg(x \approx a)),(\forall x)(g(f(x)) \approx x),b \approx c,P(g(g(a)),b), \neg P(a,c)\}$ including equality, one possible abstract state S_{LO} is $g(f(x_1)) \approx x_1$, $g(x_2) \approx f(x_2)$, $P(g(f(a)),b)$. If there is a complementary couple in a possible state, then it is called close state. Abstract state $\neg P(a,c)$, $g(f(x_1)) \approx x_1$, $g(x_2) \approx f(x_2)$, $P(g(f(a)),b)$, $P(a,b)$, $P(x_2,c)$ is a close state under the substitution $\{a/x_2\}$.

Definition 3. A abstract transition T_{LO} is an expression of the form $B_{LO} \xrightarrow{p:r:\alpha} H_{LO}$ where $P(T_{LO}):=p \in [0,1]$, $R(T_{LO}):=r \in [0,1]$, a is an abstract action, $body(T_{LO}):= B_{LO}$, and $head(T_{LO}):= H_{LO}$ are abstract states.

We assume T_{LO} to be range-restricted, i.e. $vars(H_{LO}) \subseteq vars(B_{LO})$ and $vars(a) \subseteq vars(B_{LO})$, so that an abstract transition relies on the information encoded in the current state only. The semantics of an abstract transition are:

If agent is in state Z, such that $B \leq_\theta Z$, then it will go to the state Z' $:=[Z\backslash B\theta] \cup H\theta$ with probability p when performing action a and receive an expected next reward of r.

In tableau automated theorem proving, the unique reason for state transition is the using of expansion rules, that is to say, α-rule extends the conjunction elements inside the abstract state, β-rule divides one abstract state into many states, γ- and δ- rules import free variables and function symbols separately.

To explain this problem, consider the following abstract transition. The above formulae including equality transits from one state to another: $(\forall x)((g(x) \approx f(x)) \lor \neg(x \approx a))$, $(\forall x)(g(f(x)) \approx x), b \approx c$, $p(g(g(a)), b)$, $\neg p(a, c) \xrightarrow{0.9:-1:rl\_alfa} (g(x_2) \approx f(x_2)) \lor \neg(x2 \approx a)$, $g(f(x_1)) \approx x_1, b \approx c$, $p(g(g(a)), b)$, $\neg p(a,c)$

Applied it to state Exp:

$(\forall x)((g(x) \approx f(x)) \lor \neg(x \approx a))$, $(\forall x)(g(f(x)) \approx x), b \approx c$, $p(g(g(a)), b)$, $\neg p(a,c)$

the abstract transition tells us that when we execute rl_alfa, the successor state will be:

$(g(x_2) \approx f(x_2)) \lor \neg(x2 \approx a)$, $g(f(x_1)) \approx x_1, b \approx c$, $p(g(g(a)), b)$, $\neg p(a,c)$

with probability 0.9 gaining a reward of -1.

The following is tableau on the basis of logical reinforcement learning.

Algorithm 1

    Initialize Q_0, and assign all the (s,a) as 0

    e:=0

    do forever

    e:=e+1

    i:=0

    generate a random state s_0

    while not close(s_i) do   /* estimate whether s_i is a close state */

        select tableau rule a_i and execute it

        receive an immediate reward $r_i=r(s_i,a_i)$

        observe a new state s_{i+1}

        i:=i+1

    endwhile

    for j=i-1 to 0 do

        generate example $x=(s_j, a_j, q_j)$, $q_j:=r_j+max_{a'} \cdot Q_e(s_{j+1}, a')$

        if there is (s_j, a_j, q_{old}) in the example then use x to replace it, else insert x into it.

For a simple example $(p(x) \lor q(x)) \land p(a) \vdash q(a)$, we can apply algorithm 1 to generate a tableau process. The Result is shown in Table 1.

Table 1. The tableau process of formulae $(p(x) \lor q(x)) \land p(a) \vdash q(a)$ on the basis of logical reinforcement learning

Example 1	Example 2	Example 3	Example 4
Qvalue(0.81)	Qvalue(0.9)	Qvalue(1.0)	Qvalue(1.0)
rl_alfa	rl_alfa	rl_beta	rl_beta
$(p(x) \lor q(x)) \land \neg p(a)$	$p(x) \lor q(x)$	$p(x)$	$q(a)$
$\neg q(a)$	$\neg p(a)$	$\neg p(a)$	$\neg p(a)$
	$\neg q(a)$	$\neg q(a)$	$\neg q(a)$
		/*The state is closed in substitution $\{x/a\}$*/	/* The state is closed */

Theorem 1. Each LOMDP $M_{LO}=(S_{LO},A_{LO},T_{LO},\ R)$ is corresponding to a discrete $M=(S,A,T,R)$.

Proof

Let $HB^S_\Sigma \subseteq HB_\Sigma$ be the set of all ground atoms built over abstract state predicates, and let $HB^a_\Sigma \subseteq HB_\Sigma$ be the set of all ground atoms built over abstract action names. The countable state set S_{LO} consists of all finite subsets of HB^S_Σ, active set a is included in the finite set HB^a_Σ. A abstract transition T_{LO} is a form of $B_{LO} \xrightarrow{p:r:\alpha} H_{LO}$, among which both B_{LO} and H_{LO} are abstract transition, a is a abstract action, and the R in M is similar to the R in M_{LO}. So we can conclude that four truples in M and M_{LO} have a correlative relationship, each LOMDP $M_{LO}=(S_{LO},A_{LO},T_{LO},\ R)$ is corresponding to a discrete $M=(S,A,T,R)$.

Lemma 1. Given a tableau abstract state sequence $(S_j)_{0 \leq j \leq n}$, if tableau $S_j(0 \leq j \leq n)$ is Γ-satisfiable then tableau S_{j+1} is also Γ-satisfiable.

Proof

Let B be a branch from abstract state S_i to S_{i+1}, which formed by applying classical expansion rules or theory expansion rules. We can use theory close rules to delete close branches from abstract state. Let $M=<D,I>$ be a Γ-structure satisfactory to S_i.

β-rule: Let v be a random variable. There must be a branch B' such that $(M,v) \models B'$ in S_i. If B' is different from B, then $B' \in S_{i+1}$.

In addition, if B'=B, then $(M,v) \models B$. Let β be a formulae coming from B by β-rule. According to the property of β-rule, $(M,v) \models B$ must lead to $(M,v) \models \beta_1$ or $(M,v) \models \beta_2$. Therefore, there will be $(M,v) \models (B \backslash \{\beta\}) \cup \{\beta_1\}$ or $(M,v) \models (B \backslash \{\beta\}) \cup \{\beta_2\}$. As a result, $(B \backslash \{\beta\}) \cup \{\beta_1\}$ and $(M,v) \models (B \backslash \{\beta\}) \cup \{\beta_2\}$ are in S_{i+1}.

The proof of α- and γ-rules are similar to the β-rule.

δ-rule: Let δ be a δ_formulae coming from S_i to S_{i+1} by the δ-rule, where $\delta_1(f(x_1,...,x_m))$ is a formula added to the branch(f is a new skolem function symbol and $x_1,...,x_m$ are free variables in δ). Define a structure $M'=<D,I'>$ different from M, except that the new function symbol f is explained by I', others are defined as follow: for each element set $d_1,...,d_m$ in extent D, if there is a element d such that $(M,\xi) \models \delta_1(x)$, where $\xi(x_j)=d_j(1\leq j\leq m)$ and $\xi(x)=d$, then $f'(d_1,...,d_m)=d$. If there is an element d, then we can choose one from them, else we can choose any element from the extent. For all the variable assigns v: if $(M,v) \models \delta$, then $(M',v) \models \delta_1(f(x_1,...,x_m))$. Because f do not appear in Γ, M' is Γ-structure.

Let v be a random variable assign. There must be a branch B' in S_i such that $(M,v) \models B'$, if B' is different from B, then $(M',v) \models B'$(f is not in B'), and $B' \in S_{i+1}$.

When $\delta \in B'=B$, there must be $(M,v) \models \delta$, $(M',v) \models \delta_1(f(x_1,...,x_m))$. Thus $(B \setminus \{\delta\}) \cup \{\delta_1(f(x_1,...,x_m))\}$ is in a branch of S_{i+1} and satisfied by M'.

Theory expansion rule: let disprove $<\sigma,\{\rho_1,...,\rho_k\}>$ be used in expansion tableau, for S_i is $\Gamma\_$ satisfiable, then $S_i\sigma$ is Γ- satisfiable as well. Let M be a Γ- structure and satisfy $S_i\sigma$ and v is random variable assign, and then there must be $B' \in S_i\sigma$ and $(M,v) \models B'$.

According to the definition of γ-formulae, there is $B \models_\Gamma^\circ \{ (\forall x_1^i)\cdots(\forall x_{m_i}^i)\ \varphi_j | 1\leq j\leq p\}$, such that $B\sigma \models_\Gamma^\circ \{ (\forall x_1^i)\cdots(\forall x_{m_i}^i)\ \varphi_j\sigma$ i.e. $(M,v) \models \Phi\sigma$ here $\Phi=\{ (\forall x_1^i)\cdots(\forall x_{m_i}^i)\ \varphi_j | 1\leq j\leq p\}$. Because $<\sigma,\{\rho_1,...,\rho_k\}>$ is the disprove of Φ, such that $\Phi\sigma \models_\Gamma^\circ \rho_1 \vee ... \vee \rho_k$ for some $j \in \{1,...,k\}$ there must be $(M,v) \models \rho_j$ such that M satisfy the branch $B\sigma \cup \{\rho_j\}$ in S_{i+1}.

Theorem2. If there is a tableau abstract state proving sequence $\{\{\neg\varphi\}\}=S_0,S_1,...,S_{n-1}$, $S_n=\varnothing$ $(n\geq 0)$ in formulae Φ, then Φ is a Γ-tautology.

Proof

Φ is Γ-unsatisfiable. Then by lemma 1, T_n is Γ- unsatisfiable as well. So that $S_{n-1, ...,} S_1$, S_0 are Γ- unsatisfiable too. Then the first tableau $\{\{\neg\varphi\}\}$ is also Γ- unsatisfiable. That is to say $\neg\varphi$ is Γ-unsatisfiable. In other words, φ is Γ- unsatisfiable .

4 The Experiment Result

In the environment of windows, we apply the tableau method based on logical reinforcement learning to TableauTAP system. We implement the system with the SWI-PROLOG language and use this system to prove more than 400 problems in the TPTP of automated reasoning testing library. We compared this system to the old one, the results is shown in Table 2.

Table 2. Comparison of theorem proving results in TPTP

Algorithm	VarLim Limited	Testing close branches	Close branch	CPU time (second)	space (by te)
TableauTAP	982	1354	457	123. 243	9512
old TableauTAP	1560	10230	5020	845. 891	205422

From table 2, we can see that the performances of TableauTAP improved are better than those of old one, especially in CPU time. Because the TableauTAP has learning function, the subsequence proving can learn from the conclusion of anterior proving, the run speed is much higher than the old one. At the same time, for the generalization of this method, we do not need to consider too much in programming, which makes the program code much smaller than other systems, and it also be applied more simply to real application systems that use automated reasoning as automated theorem proving machine.

5 Conclusion

The same problem also lies in tableau automated theorem proving. For example, we should limit the frequency of γ-formulae instantiation, and the function symbols and free variables in δ-formulae. If these controls can not be handled correctly, we'll make a simple proving be very complex, delay the close time of tableau and even can not gain the proving process. In addition, because of the symmetry of equality, the search space in tableau including equality will expand and even lead to a endless loop. In the expansion from classical logic to non-classical logic, although the corresponding rules and methods of classical logic can be used in non-classical logic directly, a lot of redundant branches will be produced, which decrease the processing efficiency of computers. Aiming at the efficiency of tableau automated theorem proving, on the basis of reinforcement learning, this paper proposes logic reinforcement learning. It considers the combination of logic formulae and expansion result as states, expansion rules as actions and node closes as the aim to receive a reward. This method can find a best tableau expansion method from the lay of reinforcement learning, and it also can be applied in many kinds of tableau automated theorem proving. Experiments indicate that this method can make the tableau automated theorem proving be more approximate to the close result, and reduce the blindness and uncertainty of tableau expansion.

Acknowledgments. This paper is supported by National Natural Science Foundation of China (60673092, 60273080), Postdoctor Science Foundation of China (20060390919), Higher Education   Natural Science Foundation of Jiang Su, Postdoctor Science Foundation of Jiang Su (060211C) , Important Project of Ministry of Education(207040).

References

1. Shi, Z.Z., Dong, M.K., Jiang, Y.C., Zhang, H.J.: A logic foundation for the semantic web. Science in China, Series E 34(10), 1123–1138 (2004)
2. Blackburn, P., Bos, J.: Representation and inference for natural language. CSLI Publications, Stanford, CA (2005)
3. Bertossi, L., Schwind, C.: Analytic tableaux and database repairs. In: Eiter, T., Schewe, K.-D. (eds.) FoIKS 2002. LNCS, vol. 2284, Springer, Heidelberg (2002)
4. Su, K.L., Luo, X.Y., Lu, G.F.: Symbolic model checking for CTL. Chinese Journal of Computer 28(11), 1798–1806 (2005)

5. Liu, Q., Sun, J.G., Yu, W.J.: An improved method of δ-rule in free variable semantic tableau. Journal of Computer Research and Development 41(7), 1068–1073 (2004)
6. Paskevich, A.: Connection tableaux with lazy paramodulation. In: Furbach, U., Shankar, N. (eds.) IJCAR 2006. LNCS (LNAI), vol. 4130, pp. 112–124. Springer, Heidelberg (2006)
7. Liu, Q., Sun, J.G., Cui, Z.M.: An method of simplifying many-valued Generalized quantifiers tableau rule based on Boolean pruning. Chinese Journal of Computer 28(9), 1514–1518 (2005)
8. Horvitz, E.: Machine learning, Reasoning, and Intelligence in daily life: directions and challenges. In: Proceedings of ICML (1999)
9. Bryant, C., Muggleton, S., Oliver, S.: Combining inductive logic programming, active learning and robotics to discover the function of genes. Electronic Transactions in Artificial Intelligence 6(12) (2001)
10. Calzone, L., Chabrier-Rivier, N., Fages, F.: Machine learning biomolecular interactions from temporal logic properties. In: Plotkin, G. (ed.) Proceedings of CMSB (2005)
11. Teevan, J., Horvitz, E.: Personalizing search via automated analysis of interests and activities. In: Proceedings of SIGIR, Conference on Information Retrieval, Salvador, Brazi, pp. 449–456 (August 2005)
12. Fitting, M.: First-Order Logic and Automated Theorem Proving. Springer, New York (1996)
13. Gao, Y., Chen, S.F., Lu, X.: Research on reinforcement learning Technology: a review. Acta Automatica Sinica 30(1), 86–100 (2004)
14. Otterlo, M.: Reinforcement learning for relational MDPs. In: Nowé, A., Lenaerts, T., Steenhaut, K (eds.) BeNeLearn'04. Machine Learning Conference of Belgium and the Netherlands, pp. 138–145 (2004)
15. Liu, Q., Sun, J.G.: Theorem proving system based on tableau–TableauTAP. Computer Engineering 32(7), 38–45 (2006)

Gene Expression Programming with DAG Chromosome

Hui-yun Quan and Guangyi Yang

College of Mathematics and Computer Science, Hunan Normal University,
Changsha 410081, China
Hunan College of Information, Changsha 410200, China
quanhy@hunnu.edu.cn, yguangyi@gmail.com

Abstract. GEP(Gene Expression Programming) is applied to comprehensive fields such as Symbolic Regression,Parameter Optimization, Cellular Automate etc[2].With Kara-style chromosome,GEP can only express tree phynotype. This limits the expressiveness of the program that can be evolved. In this paper, a DAG(Directed Acyclic Graph) chromosome is integrated into GEP without increasing the computational complexity of fitness evaluation while improving the expressiveness of gene expression programming.

Keywords: gene expression programming,directed acyclic graph, symbolic regression.

1 Introduction

Linear representations of gene expression programming have been proven to be very amenable to fast evolution of solutions to complex problems. However what make this representation particularly evolvable is still not well understood. Linear representations have imitated many features seen in natural [2]. But the expressiveness of these is restricted on tree struct phynotype while many nontree structs existing in natural. So it is reasonable to explore some representations of nontree phynotype. In this work,we promote a method which can encode directed acyclic graph expressions into chromosomes.

Real coding is well suited to a large of programming languages and problems with a great number of variables [5]. In addition, more genetic operators can be used in gene expression programming with real coding representations.

This work considers some evolutionary algorithms which use biologically inspired multi-chromosomal to symbolic regression problem.

2 DAG Chromosome

2.1 Character Linear Chromosome

In gene expression programming, character linear representations are composed of genes which structurally organized in head and tail. These genes function as

L. Kang, Y. Liu, and S. Zeng (Eds.): ISICA 2007, LNCS 4683, pp. 271–275, 2007.

genome and are subjected to modification by means of genetic operator. This allows the algorithm to perform with high efficiency[4]. The calling relationship of functional alleles in character linear representations is implicated in their position. Every functional allele in this type of chromosome is succeed by their parameters. In another word,every allele is a parameter of its proceeder. The topology of terminals and functions is not explicitly encoded in the chromosome. So the topological structure of programs can not be evolved enough.

2.2 Directed Acyclic Graph Chromosome

In nature,an important feature which is seen in more complex biological systems is that of multiple chromosomes[6].Here, we present a multi chromosome representation with the ability of encoding the directed acyclic graph. We call this representation Directed Acyclic Graph Chromosome. A Directed Acyclic Graph Chromosome consists of main chromosome and topological chromosomes. The main chromosome contains functional and terminal alleles which is the same as the traditional character linear representation. Additional chromosomes consisting of numbers which indicate the DAG's topology are attached to the main chromosome. We call these chromosomes topological chromosome. The calling relationship of functional alleles is encoding in the topological chromosomes. For example the following chromosome is the genotype of DAG in figure 2.2

$$
\begin{array}{l}
\phantom{\text{main chromosome: }}\ 9\ 8\ 7\ 6\ 5\ 4\ 3\ 2\ 1\ 0 \\
\hline
\text{main chromosome: } +\ \text{-}\ *\ a\ /\ b\ b\ *\ a\ b \\
\text{topological chromosome: } 8\ \ 4\ 5\ 5\ 3\ 3\ 2\ 0\ 0\ 0 \\
\phantom{\text{topological chromosome: } 8\ \ }6\ 7\ 4\ 3\ 2\ 2\ 1\ 1\ 0\ 0
\end{array}
$$

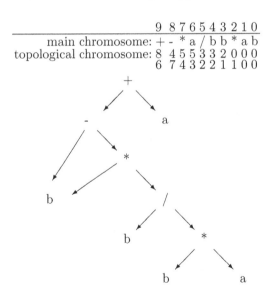

Fig. 2.2. A expression DAG example

The {8,6} under the {+} means that the parameters of plus is function {-} and terminal {a}. Through this representation,not only the component but also the topology of phenotype(expression DAG) is well exploited.

Except the last position all alleles can be functional or terminal symbol. One feature of this representation is that it can store multi solutions of a problem in a single chromosome. This is same as MEP[1].

3 Fitness Calculation

The translation of DAG chromosome to computer programs is obtained by parsing the chromosome right to left. A terminal allele specifies a single expression. A function allele specifies a complex expression which takes the expressions indicated by topological chromosome as the operands. For instance, there are ten expressions in previous example chromosome. These expressions are:

$$
\begin{array}{l}
\text{E0: b} \\
\text{E1: a} \\
\text{E2: a*b} \\
\text{E3: b} \\
\text{E4: b} \\
\text{E5: E3/E2} \\
\text{E6: a} \\
\text{E7: E4*E5} \\
\text{E8: E4-E7} \\
\text{E9: E8+E6}
\end{array}
$$

The fitness of an individual is equal to the lowest fitness of these expressions:

$$f(C) = \min_i f(E_i)$$

which $f(E_i)$ represents the fitness of expression E_i.

By using so-called dynamic programming technique[8] we can calculate the value of all expressions encoding in DAG chromosome with the same computing complexity as in linear representation GEP.

4 Genetic Operators

An additional genetic operator was introduced in order to modify the topology of programs(expressions). By exchanging the main and topological chromosome of parents we obtained two next generation individuals. The other operator used in this paper is the same as in traditional GEP. Also the gene transposition,mutation and crossover operator are applied to the topological chromosome.

5 Experiment

5.1 Test Problem

The aim of symbolic regression problem is to find a mathematical expression that satisfies a set of fitness cases. In this paper , the following problems are used for assessing the performance of DAG gene expression programming:

$$F_1(x) = 4.251x^2 + \ln(a^2) + 7.243e^a$$
$$F_2(x) = x^4 - x^3 + x^2 - x$$
$$F_3(x) = x^4 + 2 * x^3 + 3 * x^2 + 4 * x$$
$$F_4(x) = x^6 - 2 * x^4 + x^2$$

The metric of performance is the average best-of-run fitness over 100 runs. We randomly take ten points from (0,1) as the fitness case.

5.2 Parameter Settings

The parameters used per run are summarized in Table 5.2

Table 5.2. Settings used in the GEP(GEP) and DAG GEP (GEP^*)

parameter	GEP	GEP^*
Number of generations	500	500
Population size	50	50
Number of fitness cases	10	10
Function set	+,-,*,/,L,E,K, ,S,C	+,-,*,/,L,E,K, ,S,C
Head length	6	–
Number of genes	5	1
Chromosome Length	65	65
Mutation rate	0.044	0.044
One-point recombination rate	0.4	0.4
Two-point recombination rate	0.2	0.2
Gene transposition rate	0.1	0.1
IS transposition rate	0.1	0.1
IS elements length	1,2,3	1,2,3
RIS transposition rate	0.1	0.1
RIS elements length	1,2,3	1,2,3
Selection range	100	100
Precision	0.01	0.01

Table 5.3.

Average best-of-run fitness

Function	GEP	GEP^*
F_1	1952.3	1975.8
F_2	1982.7	1992.7
F_3	1975.2	1985.5
F_4	1972.7	1983.2

The meaning of {L,E,K, ,S,C} is same as in the Cândida Ferreira's book[2].

5.3 Experiment Results

In this subsection,we summarized our experiment results in Table 5.3.

6   Conclusions

A GEP chromosome representation capable of encoding expression DAG is promoted in this paper. The experiment results show that DAG chromosome can make evolutionary process more efficient. Furthermore this representation can also encode multi solutions of a problem in a single chromosome. The experiment results show that the DAG chromosome is competitive to the traditional character linear chromosome.

References

1. Mihai, O.: Solving Even-Parity Problems using Multi Expression Programming. In: The 7th Joint Conference on Information Science, September 26-30, pp. 315–318. Research Triangle Park, North Carolina (2003)
2. Ferreira, C.: Gene expression programming:Mathematical Modeling by an Artificial Intelligence. Angra do Heroismo, Portugal (2002)
3. Michalewicz, Z.: Genetic Algorithms + Data Structures = Evolution Programs, 3rd edn. Springer, Heidelberg (1996)
4. Kahrs, S.: Genetic programming with primitive recursion. In: GECCO 2006. Proceedings of the 8th annual conference on Genetic and evolutionary computation, Seattle, Washington, USA, vol. 1, pp. 941–942. ACM Press, New York (2006)
5. Leite, J.V., Avila, S.L., Batistela, N.J., Carpes, W.P., Sadowski, N., Kuo-Peng, P., Bastos, J.P.A.: Real coded genetic algorithm for Jiles-Atherton model parameters identification. IEEE Trans. Magn. 40(2), 888–891 (2004)
6. Cavill, R., Smith, S.L, Tyrrell, A.M: Multi-Chromosomal Genetic Programming. In: GECCO. Proceedings of Genetic and Evolutionary Computation, Washington (2005)
7. Baldonado, M., Chang, C.-C.K., Gravano, L., Paepcke, A.: The Stanford Digital Library Metadata Architecture. Int. J. Digit. Libr. 1, 108–121 (1997)
8. Mihai, O.: Improving the Search by Encoding Multiple Solutions in a Chromosome. In: Nedjah, N. (ed.) Evolutionary Machine Design, ch. 15, Nova Science Publisher, New-York
9. Koza, J.R.: Genetic Programming: On the Programming of Computers by Means of Natural Selection. MIT Press, Cambridge, MA (1992)

Zhang Neural Network for Online Solution of Time-Varying Sylvester Equation

Yunong Zhang, Zhengping Fan, and Zhonghua Li

Department of Electronics and Communication Engineering
Sun Yat-Sen University, Guangzhou 510275, China
ynzhang@ieee.org

Abstract. Different from gradient-based neural networks, a special kind of recurrent neural network has recently been proposed by Zhang *et al* for real-time solution of time-varying problems. In this paper, we generalize such a design method to solving online the time-varying Sylvester equation. In comparison with gradient-based neural networks, the resultant Zhang neural network for solving time-varying Sylvester equation is designed based on a matrix-valued error function, instead of a scalar-valued error function. It is depicted in an implicit dynamics, instead of an explicit dynamics. Furthermore, Zhang neural network globally exponentially converges to the exact solution of the time-varying Sylvester equation. Simulation results substantiate the theoretical analysis and demonstrate the efficacy of Zhang neural network on time-varying problem solving, especially when using a power-sigmoid activation function.

1   Introduction

The linear matrix equation, $AX - XB + C = 0$, known as the Sylvester equation, is closely related to the analysis and synthesis of dynamic systems; e.g., the design of feedback control systems via pole assignment [1][2]. Here, coefficients $A \in R^{m \times m}$, $B \in R^{n \times n}$, and $C \in R^{m \times n}$; and, $X \in R^{m \times n}$ is the unknown matrix to be obtained.

The dynamic-system approach, in the form of recurrent neural networks, is one of the important parallel-processing methods for solving optimization and equation problems [1]-[12]. For example, in recent years, many studies have been reported on real-time solutions of algebraic equations including matrix inversion and Sylvester equation [1]-[6][8]-[11]. Generally speaking, the methods reported in these references are generally related to the gradient descent method in optimization [13], which can be summarized as follows. Firstly, we construct an error function (e.g., $\|AX - XB + C\|^2/2$) such that its minimal point is the solution to the equation. Secondly, a recurrent neural network is developed to evolve along a descent direction of this error function until a minimum of the error function is reached. A typical descent direction is defined by the negative gradient.

However, if the coefficients of the equation are time-varying, then a gradient-based method [1][3]-[6] may not work well. Because of the effects of the time-varying coefficients, the negative gradient direction can no longer guarantee the

L. Kang, Y. Liu, and S. Zeng (Eds.): ISICA 2007, LNCS 4683, pp. 276–285, 2007.

decrease of such an error function. Usually, a recurrent neural network of much faster convergence in comparison to the time-varying coefficients is required for a real-time solution, if the gradient-based method is adopted. The shortcomings of applying such a method to time-varying cases are two-fold. 1) The much faster convergence is usually at cost of the precision or with stringent restrictions on design parameters. 2) Such a method is not applicable to the cases where the coefficients vary quickly or large-scale complex control systems are involved.

Different from gradient-based neural networks for solving Sylvester equation [1], a special kind of recurrent neural network has recently been proposed by Zhang *et al* [2][9][10] for real-time solution of time-varying problems including matrix inversion and Sylvester equation. That is, coefficient matrices A, B and C could be respectively $A(t)$, $B(t)$ and $C(t)$, time-varying ones. The design method of Zhang neural network (ZNN) is completely different from that of the gradient-based neural networks. In this paper, we generalize such a design method to the online solution of time-varying Sylvester equation, i.e., $A(t)X(t) - X(t)B(t) + C(t) = 0$ over time $t \geqslant 0$.

The remainder of this paper is organized as follows. Section 2 presents the problem formulation, the design method of Zhang neural network, and its online solution of the time-varying Sylvester equation. In Section 3, corresponding to different activation functions, convergence and robustness properties are studied for the resultant Zhang neural network. Section 4 presents illustrative examples of solving the time-varying Sylvester equation in real time t. Conclusions are finally given in Section 5.

2   Problem Formulation and Solvers

Consider a smoothly time-varying Sylvester equation:

$$A(t)X(t) - X(t)B(t) + C(t) = 0, \quad 0 \leqslant t < +\infty. \tag{1}$$

We are going to find $X(t)$ in real time and in an error-free manner such that the above equation holds true. Without loss of generality, $A(t)$, $B(t)$ and $C(t)$ are assumed to be known, while their time derivatives $\dot{A}(t)$, $\dot{B}(t)$ and $\dot{C}(t)$ are assumed at least to be measurable (if unknown). The existence of the theoretical solution $X^*(t)$ at any time instant t is also assumed for comparative purposes.

Here, it is worth mentioning that simplifications of Sylvester equation may result in various kinds of new problems, such as, the matrix-inverse problem and linear-equation problem. Specifically speaking,

- if $n = m$, $B(t) = 0$ and $C(t) = -I \in R^{n \times n}$, then the above time-varying Sylvester equation reduces to the matrix-inverse problem, $A(t)X(t) = I$; and
- if $n = 1$, $B(t) = 0$ and $C(t) = -c(t) \in R^m$, then $X(t) \in R^{m \times n}$ reduces to $x(t) \in R^m$ and the above time-varying Sylvester equation reduces to a set of linear time-varying equations, $A(t)x(t) = c(t)$.

The results achieved for time-varying Sylvester equation, for time-varying matrix inversion, and/or for linear time-varying equations could thus be tailored and used for one another.

2.1   Zhang Neural Network

For the time-varying problem solving, a special kind of recurrent neural network was proposed by Zhang *et al* [2][9][10]. By following such a design method, we can generalize a recurrent neural network to solving (1) in real time t as follows.

Step 1. To monitor the equation-solving procedure, the following matrix-valued error function is defined, instead of scalar-valued error functions.

$$E(t) := A(t)X(t) - X(t)B(t) + C(t) \in R^{m \times n}.$$

Step 2. The error-function time-derivative $\dot{E}(t) \in R^{m \times n}$ should be made such that every element $e_{ij}(t) \in R$ of $E(t) \in R^{m \times n}$ converges to zero, $\forall\, i = 1, \cdots, m$ and $j = 1, \cdots, n$. In mathematics, we need choose $\dot{E}(t)$ and/or $\dot{e}_{ij}(t)$ such that $\lim_{t \to +\infty} e_{ij}(t) = 0, \forall\, i, j$.

Step 3. A general form of $\dot{E}(t)$ could thus be

$$\frac{\mathrm{d}E(X(t), t)}{\mathrm{d}t} = -\Gamma \mathcal{F}\left(E(X(t), t)\right) \tag{2}$$

where design parameter Γ and activation-function mapping $\mathcal{F}(\cdot)$ are described as follows.

- $\Gamma \in R^{m \times m}$ is a positive-definite matrix used to scale the convergence rate of the neural-network solution. Γ could simply be γI, with scalar $\gamma > 0$ and I denoting the identity matrix. In addition, Γ (i.e., γI), being a set of inductance parameters or reciprocals of capacitive parameters, should be set as large as the hardware permits (e.g., in analog circuits or VLSI [12]) or selected appropriately for experimental/simulative purposes.
- $\mathcal{F}(\cdot) : R^{m \times n} \to R^{m \times n}$ denotes a matrix activation-function mapping (or termed, a matrix activation-function array) of neural networks. A simple example of activation-function mapping $\mathcal{F}(\cdot)$ is the linear one, i.e., $\mathcal{F}(E) = E$. Different choices of $\mathcal{F}$ will lead to different performance. In general, any monotonically-increasing odd activation function $f(\cdot)$, being the ijth element of matrix mapping $\mathcal{F}(\cdot) \in R^{m \times n}$, can be used for the construction of the neural network, where four types of activation functions $f(\cdot)$ are introduced here and shown in Fig. 1:

  - linear activation function $f(e_{ij}) = e_{ij}$,
  - sigmoid activation function $f(e_{ij}) = (1 - \exp(-\xi e_{ij}))/(1 + \exp(-\xi e_{ij}))$ with $\xi > 2$,
  - power activation function $f(e_{ij}) = e_{ij}^{p}$ with odd integer $p \geqslant 3$ (note that linear activation function $f(e_{ij}) = e_{ij}$ can be viewed as a special case of the power activation function with $p = 1$), and
  - power-sigmoid activation function

$$f(e_{ij}) = \begin{cases} e_{ij}^{p}, & \text{if } |e_{ij}| \geqslant 1 \\ \frac{1+\exp(-\xi)}{1-\exp(-\xi)} \cdot \frac{1-\exp(-\xi e_{ij})}{1+\exp(-\xi e_{ij})}, & \text{otherwise} \end{cases}$$

  with suitable design parameters $\xi > 2$ and $p \geqslant 3$.

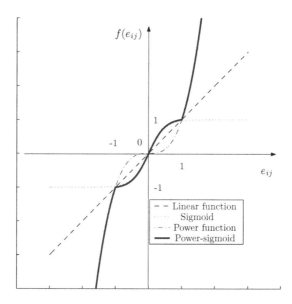

Fig. 1. Activation functions $f(\cdot)$ used in the neural networks

New activation-function variants $f(\cdot)$ can thus be generated readily based on the above four basic types.

Step 4. Expanding the design formula (2) leads to the following implicit dynamic equation of Zhang neural network for solving online the time-varying Sylvester equation, (1).

$$
\begin{aligned}
A(t)\dot{X}(t) - \dot{X}(t)B(t) = \\
- \dot{A}(t)X(t) + X(t)\dot{B}(t) - \dot{C}(t) - \gamma \mathcal{F}\left(A(t)X(t) - X(t)B(t) + C(t)\right)
\end{aligned}
\tag{3}
$$

where

- $X(t)$, staring from an initial condition $X(0) = X_0 \in R^{m \times n}$, is the activation state matrix corresponding to the theoretical solution of (1); and
- $\dot{A}(t)$, $\dot{B}(t)$ and $\dot{C}(t)$ denote the measurements or known analytical forms of the time derivatives of matrices $A(t)$, $B(t)$ and $C(t)$ [simply, $\dot{A}(t) := dA(t)/dt$, and so on].

It is worth mentioning that when using the linear activation function mapping $\mathcal{F}(E) = E$, Zhang neural network (3) reduces to the following classic linear one:

$$
\begin{aligned}
A(t)\dot{X}(t) - \dot{X}(t)B(t) = \\
- \left(\gamma A(t) + \dot{A}(t)\right)X(t) + X(t)\left(\gamma B(t) + \dot{B}(t)\right) - \left(\gamma C(t) + \dot{C}(t)\right).
\end{aligned}
$$

2.2   Gradient-Based Neural Network

For comparison, it is also worth mentioning that almost all the numerical algorithms and neural-dynamic computational schemes were designed intrinsically
for constant coefficient matrices A, B and C, rather than time-varying matrices
$A(t)$, $B(t)$ and $C(t)$. The aforementioned neural-dynamic computational schemes
[1][3]-[6] are in general related to the gradient-descent method in optimization,
which was described briefly in the second paragraph of Section 1. Its design
procedure is as follows.

- A scalar-valued error function, such as $\|AX - XB + C\|^2/2$, is firstly con-
  structed such that its minimum point is the solution of Sylvester equation
  $AX - XB + C = 0$.
- Then, the typical descent direction is the negative gradient of $\|AX - XB +
  C\|^2/2$; i.e.,

$$-\frac{\partial \left(\|AX - XB + C\|^2/2\right)}{\partial X} = -A^T \left(AX - XB + C\right) + \left(AX - XB + C\right) B^T.$$

By using the above design method of gradient-based neural networks [1][3]-[6],
we could have the classic linear gradient-based neural network

$$\dot{X}(t) = -\gamma A^T \left(AX - XB + C\right) + \gamma \left(AX - XB + C\right) B^T$$

and the general nonlinear form of gradient-based neural network

$$\dot{X}(t) = -\gamma A^T \mathcal{F} \left(AX - XB + C\right) + \gamma \mathcal{F} \left(AX - XB + C\right) B^T. \tag{4}$$

2.3   Method and Model Comparison

As compared to the above gradient-based recurrent neural network (4), the difference and novelty of Zhang neural network (3) for solving online the time-varying
Sylvester equation lie in the following facts.

- Firstly, the above Zhang neural network (3) is designed based on the elimination of every entry of the matrix-valued error function $E(t) = A(t)X(t) -
  X(t)B(t) + C(t)$. In contrast, the gradient-based recurrent neural network
  (4) was designed based on the elimination of the scalar-valued error function
  $\bar{E}(t) = \|AX(t) - X(t)B + C\|^2/2$ (note that here A, B and C could only be
  constant in the design of the gradient-based recurrent neural networks).
- Secondly, Zhang neural network (3) methodically and systematically exploits
  the time derivatives of coefficient matrices $A(t)$, $B(t)$ and $C(t)$ during the
  problem-solving procedure. This is the reason why Zhang neural network
  (3) globally exponentially converges to the exact solution of a time-varying
  problem. In contrast, the gradient-based recurrent neural network (4) has
  not exploited such important information.

– Thirdly, Zhang neural network (3) is depicted in an implicit dynamics, or in the concisely arranged form $M(t)\dot{y}(t) = P(t)y(t) + q(t)$ where $M(t)$, $P(t)$, $q(t)$ and $y(t)$ can be given by using Kronecker product and vectorization technique [1][2][10]. In contrast, the gradient-based recurrent neural network (4) is depicted in an explicit dynamics, or in the concisely arranged form $\dot{y}(t) = \bar{P}y(t) + \bar{q}$.

Before ending this section/subsection, it is worth pointing out that the implicit dynamic equations (or to say, implicit systems) frequently arise in analog electronic circuits and systems due to Kirchhoff's rules. Furthermore, implicit systems have higher abilities in representing dynamic systems, as compared to explicit systems. The implicit dynamic equations could preserve physical parameters in the coefficient/mass matrix, i.e., $M(t)$ in $M(t)\dot{y}(t) = P(t)y(t) + q(t)$. They could describe the usual and unusual parts of a dynamic system in the same form. In this sense, implicit systems are much superior to the systems represented by explicit dynamics. In addition, the implicit dynamic equations (or implicit systems) could mathematically be transformed to explicit dynamic equations (or explicit systems) if needed.

3   Theoretical Results

For Zhang neural network (3) solving the time-varying Sylvester equation (1), we have the following theorems on its global convergence and exponential convergence. However, the proof to the theorems is omitted due to space limitation.

Theorem 1. *Given time-varying matrices $A(t)$, $B(t)$ and $C(t)$, if a monotonically increasing odd activation function array $\mathcal{F}(\cdot)$ is used, then the state matrix $X(t)$ of Zhang neural network (3) starting from any initial state X_0 always converges to the time-varying theoretical solution $X^*(t)$ of Sylvester equation (1). In addition, the neural network (3) possesses the following properties.*

– *If the linear activation function is used, then the global exponential convergence with rate γ is achieved for (3).*
– *If the bipolar sigmoid activation-function is used, then the superior convergence can be achieved for (3) for error range $[-\epsilon, \epsilon]$, $\exists \epsilon > 0$, as compared to the case of using the linear function. This is because the error signal e_{ij} in (3) is amplified by the bipolar sigmoid function for such an error range.*
– *If the power activation function is used, then the superior convergence can be achieved for (3) for error ranges $(-\infty, -1]$ and $[1, +\infty)$, as compared to the linear case. This is because the error signal e_{ij} in (3) is amplified by the power activation function for error ranges $(-\infty, -1]$ and $[1, +\infty)$.*
– *If the power-sigmoid activation function is used, then the superior convergence can be achieved for (3) for the whole error range $(-\infty, +\infty)$, as compared to the linear case. This is in view of the above mentioned properties.*□

In the analog implementation or simulation of recurrent neural networks, we usually assume it to be under ideal conditions. However, there are always some

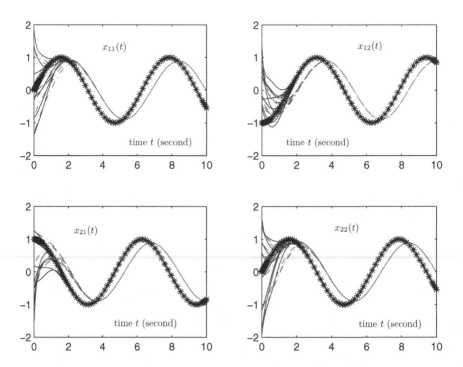

Fig. 2. Online solution of time-varying Sylvester equation (1) by Zhang neural network (3) and gradient-based neural network (4), where theoretical solution is denoted by asterisks in black, ZNN solution is denoted by solid lines in blue, and gradient-based neural solution is denoted by dash-dotted lines in red

realization errors. The implementation errors of matrices $A(t)$ through $C(t)$, the differentiation errors of matrices $A(t)$ through $C(t)$, and the model-implementation error appear most frequently in hardware realization. For these realization errors possibly appearing in Zhang neural network (3), we have the following analysis results.

Theorem 2. *Consider the perturbed ZNN model with imprecise matrix implementation, $\hat{A} = A + \Delta_A$, $\hat{B} = B + \Delta_B$ and $\hat{C} = C + \Delta_C$:*

$$\hat{A}(t)\dot{X}(t) - \dot{X}(t)\hat{B}(t) =$$
$$- \dot{\hat{A}}(t)X(t) + X(t)\dot{\hat{B}}(t) - \dot{\hat{C}}(t) - \gamma\mathcal{F}\left(\hat{A}(t)X(t) - X(t)\hat{B}(t) + \hat{C}(t)\right).$$

If $\|\Delta_A(t)\|$, $\|\Delta_B(t)\|$ and $\|\Delta_C(t)\|$ are uniformly upper bounded by some positive scalars, then the computational error $\|X(t) - X^*(t)\|$ over time t and the steady-state residual error $\lim_{t\to\infty}\|X(t) - X^*(t)\|$ are uniformly upper bounded by some positive scalars, provided that the theoretical solution $\hat{X}^*(t)$ (corresponding to the imprecise matrix implementation $\hat{A}$, $\hat{B}$ and $\hat{C}$) still exists. □

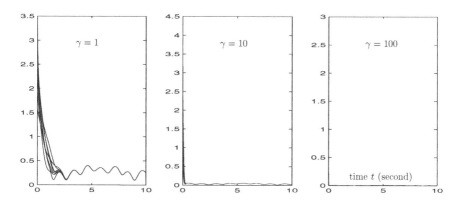

Fig. 3. Computational error $\|X(t) - X^*(t)\|$ of imprecisely-implemented ZNN (5)

For the differentiation errors and model-implementation error, the following ZNN dynamics is considered, as compared to the original dynamic equation (3).

$$A\dot{X} - \dot{X}B = -\left(\dot{A} + \Delta_{\dot{A}}(t)\right)X + X\left(\dot{B} + \Delta_{\dot{B}}(t)\right) - \dot{C}$$
$$- \gamma\mathcal{F}\left(AX - XB + C\right) + \Delta_R(t) \tag{5}$$

where $\Delta_{\dot{A}}(t)$ and $\Delta_{\dot{B}}(t)$ denote respectively the differentiation errors of matrices $A(t)$ and $B(t)$, while $\Delta_R(t)$ denotes the model-implementation error (including the differentiation error of matrix C as a part). These errors may result from truncation/round-off errors in digital realization and/or high-order residual errors of circuit components in analog realization.

Theorem 3. *Consider the ZNN model with implementation errors $\Delta_{\dot{A}}$, $\Delta_{\dot{B}}$ and Δ_R, which is finally depicted in equation (5). If $\|\Delta_{\dot{A}}(t)\|$, $\|\Delta_{\dot{B}}(t)\|$ and $\|\Delta_R(t)\|$ are uniformly upper bounded by some positive scalars, then the computational error $\|X(t) - X^*(t)\|$ over time t and the steady-state residual error $\lim_{t\to\infty}\|X(t) - X^*(t)\|$ are uniformly upper bounded by some positive scalars, provided that a so-called design-parameter requirement holds true. The design-parameter requirement means that γ should be large enough. More importantly, as γ tends to positive infinity, the steady-state residual error vanishes to zero.$\square$*

Theorem 4. *In addition to the general robustness results in Theorems 2 and 3, the imprecisely-constructed Zhang neural network (5) possesses the following properties.*

- *If the linear activation function is used, then the maximum element residual error $\lim_{t\to\infty}[X(t) - X^*(t)]_{ij}$ could be written out as a positive scalar under the design-parameter requirement.*
- *If the sigmoid activation function is used, then the steady-state residual error $\lim_{t\to\infty}[X(t) - X^*(t)]$ can be made smaller by increasing γ or ξ under the design-parameter requirement, as compared to the case of using the linear*

activation function. In addition, if the sigmoid activation function is used, then superior convergence and robustness properties exist for error range $|e_{ij}(t)| \leqslant \epsilon, \exists \epsilon > 0$, as compared to the case of using the linear function.

- *If the power activation function is used, then the design-parameter requirement always holds true for any $\gamma > 0$ and thus it is removed in this case. In addition, if the power activation function is used, then superior convergence/robustness properties exist for error range $|e_{ij}(t)| > 1$, as compared to the case of using the linear activation function.*

- *If the power-sigmoid activation function is used, then the design-parameter requirement also always holds true and is thus removable in this case. Moreover, if the power-sigmoid activation function is used, then the steady-state residual error $\lim_{t \to \infty}[X(t) - X^*(t)]$ can be made smaller by increasing γ or ξ, and superior convergence/robustness properties exist for the whole error range $e_{ij}(t) \in (-\infty, +\infty)$, as compared to the case of using the linear activation function.*   □

4   Illustrative Examples

For illustrative and comparative purposes, for the time-varying Sylvester equation (1), we assume that $n = m$, $B(t) = 0$, $C(t) = -I$ and $A(t) = [\sin t, \cos t; -\cos t, \sin t] \in R^{2 \times 2}$. Then, with $\gamma = 1$, we apply Zhang neural network (3) and gradient-based neural network (4) to solving such a time-varying Sylvester equation. Fig. 2 shows that, starting from any initial state, the state matrix $X(t)$ of Zhang neural network (3) all converges to theoretical solution $X^*(t)$. The convergence can also be expedited by increasing γ. For example, if $\gamma = 10^3$, the convergence time is within 2 milliseconds; and, if $\gamma = 10^6$, the convergence time is within 2 microseconds. In contrast, the steady-state error of gradient-based neural network (4) is considerably large, always lag behind $X^*(t)$.

To show the robustness property, some implementation errors are added to Zhang neural network (3) in higher-frequency sinusoidal forms: $\Delta_{\dot{A}}(t) = 0.5[\cos 3t, -\sin 3t; \sin 3t, \cos 3t]$ and $\Delta_R(t) = 0.5[\sin 3t, 0; 0, \cos 3t]$. Fig. 3 shows that, even with such large implementation errors, the ZNN computational error $\|X(t) - X^*(t)\|$ is still bounded and very small. Moreover, as the design parameter γ increases from 1 to 10 and finally to 100, such a steady-state computational error is decreased from around 0.41 to 0.057 and finally to 0.0064 (inappreciably small in the rightmost subplot of Fig. 3).

In general, Zhang neural network is much more efficient and effective than classic recurrent neural networks on the issue of time-varying problem solving.

5   Conclusions

By following the new design approach on recurrent neural networks, a neural network has been generalized to solving online time-varying Sylvester equation. Such a neural network could globally exponentially converge to the exact solution of time-varying problems. Simulation examples have substantiated the theoretical analysis and efficacy of the neural network. Future research directions may lie in its hardware-implementation and engineering-applications.

Acknowledgements. Before joining Sun Yat-Sen University in 2006, the corresponding author, Yunong Zhang, had been with National University of Ireland, University of Strahclyde, National University of Singapore, Chinese University of Hong Kong, since 1999. Continuing the line of this research, he has been supported by various research sponsors. This work is funded by National Science Foundation of China under Grant 60643004 and by the Science and Technology Office of Sun Yat-Sen University.

References

1. Zhang, Y.: Analysis and Design of Recurrent Neural Networks and Their Applications to Control and Robotic Systems. Ph.D. Thesis, Chinese University of Hong Kong (2002)
2. Zhang, Y., Jiang, D., Wang, J.: A Recurrent Neural Network for Solving Sylvester Equation with Time-Varying Coefficients. IEEE Transactions on Neural Networks 13, 1053–1063 (2002)
3. Manherz, R.K., Jordan, B.W., Hakimi, S.L.: Analog Methods for Computation of the Generalized Inverse. IEEE Transactions on Automatic Control 13, 582–585 (1968)
4. Jang, J., Lee, S., Shin, S.: An Optimization Network for Matrix Inversion. Neural Information Processing Systems, pp. 397–401, American Institute of Physics, New York (1988)
5. Wang, J.: A Recurrent Neural Network for Real-Time Matrix Inversion. Applied Mathematics and Computation 55, 89–100 (1993)
6. Zhang, Y.: Revisit the Analog Computer and Gradient-Based Neural System for Matrix Inversion. In: Proceedings of IEEE International Symposium on Intelligent Control, pp. 1411–1416. IEEE Computer Society Press, Los Alamitos (2005)
7. Zhang, Y.: Towards Piecewise-Linear Primal Neural Networks for Optimization and Redundant Robotics. In: Proceedings of IEEE International Conference on Networking, Sensing and Control, pp. 374–379. IEEE Computer Society Press, Los Alamitos (2006)
8. Zhang, Y.: A Set of Nonlinear Equations and Inequalities Arising in Robotics and its Online Solution via a Primal Neural Network. Neurocomputing 70, 513–524 (2006)
9. Zhang, Y., Ge, S.S.: A General Recurrent Neural Network Model for Time-Varying Matrix Inversion. In: Proceedings of the 42nd IEEE Conference on Decision and Control, pp. 6169–6174. IEEE Computer Society Press, Los Alamitos (2003)
10. Zhang, Y., Ge, S.S.: Design and Analysis of a General Recurrent Neural Network Model for Time-Varying Matrix Inversion. IEEE Transactions on Neural Networks 16, 1477–1490 (2005)
11. Steriti, R.J., Fiddy, M.A.: Regularized Image Reconstruction Using SVD and a Neural Network Method for Matrix Inversion. IEEE Transactions on Signal Processing 41, 3074–3077 (1993)
12. Mead, C.: Analog VLSI and Neural Systems. Addison-Wesley, Reading, MA (1989)
13. Bazaraa, M.S., Sherali, H.D., Shetty, C.M.: Nonlinear Programming – Theory and Algorithms. Wiley, New York (1993)

A Global Optimization Algorithm Based on Novel Interval Analysis for Training Neural Networks*

Hongru Li, Hailong Li, and Yina Du

Key Laboratory of Integrated Automation of Process Industry(Northeastern University), Ministry of Education
Shenyang, 110004, China
lihongru@ise.neu.edu.cn

Abstract. A global optimal algorithm based on novel interval analysis was proposed for Feedforward neural networks (FNN). When FNN are trained with BP algorithm, there exists some local minimal points in error function, which make FNN training failed. In that case, interval analysis was took into FNN to work out the global minimal point. For interval FNN algorithm, an interval extension model was presented, which creates a narrower interval domain. And more, in the FNN training, hybrid strategy was employed in discard methods to accelerate the algorithm's convergence. In the proposed algorithm, the objective function gradient was utilized sufficiently to reduce the training time in both interval extension and discard methods procedure. At last, simulation experiments show the new interval FNN algorithm's availability.

1 Introduction

Artificial neural networks have been developing in many years. Because of their excellent capability of self-learning and self-adapting, neural networks are very important in many fields[1,2]. In the numerous models of neural networks, the multilayer Feedforward neural networks(FFN) are widely used. The FFN model has powerful ability in approximation for any continuous functions, therefore, this ability provides basic theory for its modelling and controlling in nonlinear system[3].And the Back-Propagation(BP) algorithm or Quasi-Newton(QN) method etc, which are based on gradient descent algorithm, make the FNN more easily to adjust the weights vector. However, the gradient descent algorithm has the natural drawbacks: those training algorithm would often stick the error function in a local minimal point and stop searching the optimal weights for goal error.

For the weakness of those FFN algorithms, the randomness global optimization algorithms ,such as genetic algorithms, simulated annealing algorithm, randomness searching,were used into FNN[17,18]. These algorithms really brought

---

* This work is supported by national natural science foundation of P. R. China Grant ♯60674063, by national postdoctoral science foundation of P. R. China Grant ♯ 2005037755, by natural science foundation of Liaoning Province Grant ♯20062024.

L. Kang, Y. Liu, and S. Zeng (Eds.): ISICA 2007, LNCS 4683, pp. 286–295, 2007.

some exciting results. Nevertheless, for the large-scale continuous system optimization problem, those algorithms were too slow. In that theory, their random property would make sure that they will find the final solution in the enough time. But in fact, those algorithms would waste too much time in practice projects[4]. In sum, there are still some shortcomings in those randomness algorithms.

In contrast to randomness global optimization algorithms, deterministic global optimization algorithms were also developing quickly in recent years, for example, Interval Algorithm, Filled-Function, Tunnelling method, Downstairs method etc.[5,6,7,11]. Deterministic global optimization algorithms have the faster searching speed and now have been an important branch in optimization algorithm field[4].

Among the deterministic algorithms, the interval algorithm has a better quality. Interval algorithm has been used for neural networks to solve local minimum problem successfully [2]. In interval algorithm, there are two important iteration steps, which are interval extension and discard method. The neural networks interval extension without gradient had the rough interval domains, so they need more computing time. Both interval extensions and discard methods are key point to influent the speed of interval algorithm. A new interval extension method is presented for FNN interval algorithm in the paper. Also, some new discard methods in paper [4] are brought to reduce the calculated capacity. As a consequence, the new interval FNN algorithm could train the error function for neural networks efficiently and have a faster convergence speed.

The remainder of this paper is organized as follows. In Section 2, some preliminaries on interval algorithm are presented. In Section 3, two new important algorithm steps are proposed and proved. Global optimization neural networks algorithm is developed in Section 4. The illustrative examples are given in Section 5. Finally, we make concluding remarks in Section 6.

2   Basic Interval Algorithms

2.1   Interval Analysis and Interval Algorithm

R. E. Moore presented Interval analysis firstly in fifties, twenty century to check the computing results[5,6,7]. Later, he utilized the idea of interval analysis into optimization problem, so here came the interval algorithm. Interval analysis' main idea is to use an interval instead of the real number in the computing.

In recent years, interval algorithm was applied in global optimization problem and many authors have been studying how to improve interval algorithm [3,4,6,11]. Some new interval extension and discard method were presented in paper [3, 4, 15]. And some authors studied new multi-section technique to accelerate the searching speed in paper[10, 13]. Some different properties of objective functions were under considered. In paper [16], the authors have studied the X^0 domains is unbounded. Others also consider that the objective function is non-smooth or non-differentiable,such as in paper [8, 9]. And constrained optimization case is also under consideration in paper [14]. For FNN with sigmoid activation function, in paper [5, 6], authors have proved that interval algorithm is able to solve this problem. Before improving algorithm, we would introduce the basic knowledge firstly.

2.2   Interval Numbers and Interval Notations

An interval number $X = [a, b]$ is such a closed interval that $X = x : a \leq x \leq b$. The set of real numbers between and including the endpoint a and b . While the standard mutation for real quantities is a lowercase, we shall generally use a capital to indicate the interval. The domain X is named as **BOX** .

2.3   Basic Interval Arithmetic

Let II be the set of all real compact intervals in $[a, b], a, b \in R$ and $II(D)$ be the set of all real compact intervals in D . Let $+, -, \times, \div$ denote the operation of addition, subtractions, multiplication and division, respectively. This definition produces the following generating the endpoint of X and Y from two intervals $X = [a, b]$ and $Y = [c, d]$

$$X + Y = [a + c, b + d] \tag{1}$$

$$X - Y = [a - c, b - d] \tag{2}$$

$$X \times Y = [\min(ac, bc, ad, bd), \max(ac, bc, ad, bd)] \tag{3}$$

$$1/Y = [1/d, 1/c](0 \notin Y) \tag{4}$$

3   Improved Interval Algorithm

3.1   Interval Extension

There are some different existing interval extension forms, such as Meanvalue forms and Taylor forms. For the properties of FNN and need for algorithm speed, we propose a new interval extension. At first, note that some useful definitions as following:

Definition 1. *the midpoint of an interval* $X = [a, b]$ *is* $m(X) = 0.5(a + b)$

Definition 2. *the width of X is* $w(X) = b - a$

Definition 3. *the interval function F is an interval extension of f , if we have* $F(x_1, \cdots, x_n) = f(x_1, \cdots, x_n)$ *, for all* $x_i(i = 1, \cdots, n)$ *, then* $F(X_1, \cdots, X_n)$ *contains the range of values of* $f(x_1, \cdots, x_n)$ *for all* $X_i(i = 1, \cdots, n)$ *.*

Definition 4. *the interval function F is a natural interval extension of f , if F is interval extension of f and satisfy* $F = f(X)$ *with basic interval arithmetic.*

Definition 5. *There exists a* $X = [a, b]$ *, we denote that* $a = X^L$ *,* $b = X^R$

In this section, we would deal with the following optimization problem:

$$\min f(x) \tag{5}$$

Definition 6. *If f is a one-dimension smooth function, the derivative of f on point x is*

$$g(x) = f'(x) \tag{6}$$

Lemma 1. *Suppose that f is a one-dimension smooth function, there exists a real d satisfying $d \in g(x) \mid x \in (a,b)$ such that $f(b) - f(a) = d(b-a)$*

Theorem 1. *Suppose that f is a smooth function defined on box X , then let*

$$F(X) = \big[f(A) + \inf\{(X-A)^T G(X)\},$$

$$f(A) + \sup\{(X-A)^T G(X)\}\big] \tag{7}$$

Where A is any point in domain X, $G(X)$ is natural interval extension of derivative $g(x)$ on box X , $G(X) = [\underline{g}, \overline{g}]$. Then we call $F(X)$ is the interval extension of $f(x)$.

Proof. For any box $X \in I(X^0)$, choose any $A \in X$, the partial derivative of function f on point x along direction x_j is $g_i(x)$, whose natural interval extension is $G_j(X) = [\underline{g}_j, \overline{g}_j]$, from Lemma 1

$$f(x) - f(A) = \sum_{j=1}^{n} (f(A_1, \cdots, A_{j-1}, x_j, x_{j+1}, \cdots, x_n)$$

$$-f(A_1, \cdots, A_{j-1}, A_j, x_{j+1}, \cdots, x_n))$$

$$\in \sum_{j=1}^{n} G_j(A_1, \cdots, A_{j-1}, X_j, x_{j+1}, \cdots, x_n)(x_j - A_j) \subseteq \sum_{j=1}^{n} G_j(X)(x_j - A_j) \tag{8}$$

$$f(A) + \inf\{(X-A)^T G(X)\} \leq f(x) \leq f(A) + \sup\{(X-A)^T G(X)\} \tag{9}$$

Therefore

$$f(x) \in F(X) = \big[f(A) + \inf\{(X-A)^T G(X)\},$$

$$f(A) + \sup\{(X-A)^T G(X)\}\big] \tag{10}$$

Thus, we have proved the Theorem.

Choosing A: A is any point in the domain X. There is no limitation in choosing A. In the real project, we could choose A in many different ways, such as $A = m(X)$. Namely, we could let A be the middle point of **BOX** X.

3.2 Discard Method

Discard method is the accelerating device in the algorithm. It is to delete the non-minimum involved box. So the studying of discard methods has been an important part in interval analysis field.

Some different discard methods have been proposed before. The discard methods in FNN training were used, such as midpoint test, monotone test, interval Newton test, concavity test and so on. Some of them with simple principles are easy to use, but can not delete enough interval domains. Some of them need to calculate Hessian Matrix so the calculated capacity is huge. For the analysis above, we only apply discard method using derivative. With this hybrid strategy, the algorithm's speed would be increased efficiently.

Discard Method 1
Based on the paper [3], we apply a new discard method into neural networks training. For any box $X = (X_j)_{n \times 1} \in I(X^0)$, suppose that there exist i, j make $G_{ij}(X)$ satisfy $\underline{g}_{ij} < 0 < \overline{g}_{ij}$. Choose $C = (C_j)_{n \times 1} \in X$ to make $f(c) \geq \overline{f}$, where $\overline{f}$ is the current minimal point of function upper bound. Then from Theorem 1

$$f(x) - f(c) \subseteq \sum_{j=1}^{n} G_j(X)(x_j - c_j) \tag{11}$$

We would like to delete the domains $x = (x_j)_{n \times 1} \in X$ such that

$$f(c) + \sum_{j=1}^{n} G_j(X)(x_j - c_j) > \overline{f} \tag{12}$$

From the inequality above, we have

$$x_k \in X_k \cap (c_k + \frac{\overline{d}_k}{\overline{g}_k}, c_k + \frac{\overline{d}_k}{\underline{g}_k}), (k = 1, \cdots, n) \tag{13}$$

Where $\overline{d}_k = \overline{f} - f(c) - \sum_{j=1, j \neq k}^{n} \min(\overline{g}_j(a_j - c_j), \overline{g}_j(b_j - c_j), \underline{g}_j(a_j - c_j), \underline{g}_j(b_j - c_j))$

Where k is iteration epochs. Let $\underline{t}_k = c_k + \frac{\overline{d}_k}{\overline{g}_k}, \overline{t}_k = c_k + \frac{\overline{d}_k}{\underline{g}_k}$, then there are two different situations:

(1) If there exists a certain k such that $x_k \in X_k \cap (\underline{t}_k, \overline{t}_k) = X_k$,then delete the box X_k
(2) If there exists a certain k such that $x_k \in X_k \cap (\underline{t}_k, \overline{t}_k) = (\underline{t}_k, \overline{t}_k)$,then delete the box $(\underline{t}_k, \overline{t}_k)$

Discard Method 2(Monotone Test)
Let $I_X = \{i \in \{1, \cdots, m\}| \sup F_i(X) \geq \inf F(X)\}$, $\underline{g}_j = \min_{i \in I_X}\{\underline{g}_{ij}\}$, $\overline{g}_j = \max_{i \in I_X}\{\overline{g}_{ij}\}$, then use monotone test for interval X :

If $\underline{g}_j > 0$ then degrade X as $(X_1, \cdots, X_{j-1}, a_j, X_{j+1}, \cdots, X_n)^T$;
If $\overline{g}_j < 0$,then degrade X as $(X_1, \cdots, X_{j-1}, b_j, X_{j+1}, \cdots, X_n)^T$.

Theorem 2. *Suppose* $L(X) = \max\limits_{i \in I_X}\{\|G_i(X)\|_\infty\}; \overline{f}$ *is upper bound of* $\min\limits_{x \in X} f(x)$,
when for $x \in X \in I(X^0)$, *there exists* $f(x) > \overline{f}$,*let*

$$\alpha = \frac{f(x) - \overline{f}}{L(X)}, X_\alpha = ([x_1 \pm \alpha], \cdots, [x_n \pm \alpha])^T \tag{14}$$

Then there is no minimum in interval X_α . *From Theorem 2, we could attain Discard Method 2.*

Discard Method 3(Lipschitz test)
Let $L(X) = \max\limits_{i \in I_X}\{\|G_i(X)\|_\infty\}$,$\overline{f}$ is the upper bound of $\min\limits_{x \in X} f(x)$. Suppose that
$\alpha = \frac{f(m(x)) - \overline{f}}{L(X)}$, if α is larger than the second large branch , then delete the
domains $X_\alpha = ([m(x_1) \pm \alpha], \cdots, [m(x_n) \pm \alpha]$ from X .

4   The Global Optimization Algorithm for FNN

We define the FNN error function as:

$$e(w, \theta) \tag{15}$$

where w is weights and θ is bias . For the nets with n training sample data and q output neurons, the error function and sum error function of the kth sample are defined as follows:

$$e(k, w, \theta) = \frac{1}{2} \sum_{i=1}^{q} (y_i(k) - y_i(k, w, \theta))^2 \tag{16}$$

$$e(w, \theta) = \frac{1}{n} \sum_{k=1}^{n} e(k, w, \theta) \tag{17}$$

Where $y_i(k)$ is the output of the networks and $y_i(k, w)$ is the expected output, and w is the weight matrix. The bias could be viewed as weights as the input is -1, so the error function could be transformed to be the function only with argument w . Training the net is a procedure that finds proper weights to minimize the error function $e(w)$. So in this section, we consider:

$$\min e(w) \tag{18}$$

Where w is the weights in FNN and $e(w)$ is transformed to be $E(W)$ in interval function. Our presented training algorithm for FNN is as follows:

1. Set the initial box $W^0 \subset R^n$.
2. Choose a proper interval extension
   $E(W) = \left[E(m(W)) + \inf\{(W - m(W))^T G(W)\}, \right.$
   $\left. E(m(W)) + \sup\{(W - m(W))^T G(W)\}\right]$
3. Calculate $E(W^0) = [E^L(W^0), E^R(W^0)]$, $\bar{e} = E^R(W^0)$, $\underline{e} = E^L(W^0)$ and $\tilde{e} = E^R(c)$ where $c = m(W^0)$
4. Initialize list $L = (W^0, \underline{e})$
5. Choose a coordinate direction k parallel to which W^i , $i = 0, 1, \cdots, n$ in the list L has an edge of maximum length.
6. Bisect W^i along direction k getting boxes W^{i1} and W^{i2} such that $W^i = W^{i1} \bigcup W^{i2}$.
7. Calculate $E(W^{ij}) = [E^L(W^{ij}), E^R(W^{ij})]$
   $j = 0, 1, \cdots, n, \bar{e} = \min(\bar{e}, \min_{i=1}^{n} E^R(W^{ij}))$, $\underline{e}_i = E^L(W^{ij}), \tilde{e}_i = E^R(c_i)$ where $c_i = m(W^i)$.
8. Enter $(W^{i1}, \underline{e}_{i1})$, $(W^{i2}, \underline{e}_{i2})$ at the end of the list and delete $(W^i, \underline{e}_i)$ from the list.
9. If $w(W^{ij}) < \varepsilon$ where $w(W^{ij}) = W^{ijR} - W^{ijL}$ and ε is a number close to 0 enough, then go to Step 11
10. Utilize discard methods 1-3 to test the boxes in the list and delete the useless parts, arrange the boxes left in the sequence i and go to Step 5
11. If $\underline{e}_{ij} < \varepsilon$ where ε is a number close to 0 enough, then end. Or go to Step 1 and make $W^0 = 2W^0$

5   Simulation

Some simulation experiments were made with presented algorithms. In this paper, three experiments are chosen to show presented algorithm's availability.

5.1   Process Modeling

To test the validity of the new algorithm for real problem, we made a simulation about a project's process modeling. The object is fuel furnace in a certain steel company. The input is gas flux and the output is thickness of CO_2. 1000 sets of input-output data were taken to experiments from the project. The goal of modeling is to construct the nonlinear dynamic mapping relationship between input and output. The simulation used BP algorithm and new FNN Interval algorithm to training the net, respectively.

   In the simulation experiments, we chosen a set of initial weights which could lead BP algorithm to be stuck in local minimum. With the same initial weights, interval algorithm can still train the FNN to goal precision. It showed us interval algorithm is a global optimal algorithm for FNN.

5.2   Nonlinear System Recognition

Some simulation experiments were also made on recognizing nonlinear system with our new algorithms, and the results show us the proposed algorithm's advantage compared to BP algorithm. We also use a set of initial which could make

Table 1. Process Modeling Simulation Results

Algorithm	Time(s)	Epoch	Training Error
BP algorithm	129.5	50000	0.06521
FNN Interval algorithm	86.1	6012	0.00009786

BP algorithm to fall into local minimal point as last experiment. Assume that the nonlinear system model is

$$y(k) = \frac{y(k-2)y(k-3)u(k-2) + u(k-1)[y(k-1)-1]}{1 + y^2(k-3)} \tag{19}$$

The neural networks structure is 2-7-1. There are two inputs $u(k)$ and $y(k-1)$. The output is $y(k)$. The input is $u(k) = 0.7\sin(2k\pi/100) + 0.42\sin(2k\pi/25)$. The training results are shown as Table 2. The comparisons of different algorithms could demonstrate proposed algorithm is independent of choosing initial weights.

Table 2. Nonlinear System Recognition Simulation Results

Algorithm	Time(s)	Epoch	Training Error
BP algorithm	96.8	50000	0.01547
FNN interval algorithm	22.2	2250	0.00009973

Table 3. Iris Classification Simulation Results

Algorithm	Time(s)	Epoch	Training Error
FNN interval algorithm	25.8	2319	0.00009836
Genetic algorithm	32.7	3128	0.00009926

5.3   Iris Classification

Iris Classification is a typical classification problem: we will use 4 characters (Sepal length, Sepal width, Petal length and Petal width) to label the objective plant into one of three species(Setosa, Versicolor, Virginia). In this experiment, there are 150 sets of data. The FNN model is 4-5-3.

Besides of our new interval algorithm, we use genetic algorithm in the experiments, so that the performance of different algorithms could be compared to get a fair conclusion. The results show us the proposed algorithm's advantage in convergence speed.

In Genetic Algorithm, the coding is 8-bit binary system. We had 50 initial colonies and the range of initial weights is [-2,2]. We will choose 7 of them to be optimal individuals. The training results are shown as Table 3.The comparisons of different algorithms could demonstrate proposed algorithm has a faster speed than Genetic Algorithm.

6   Conclusions

In this paper, the FNN global optimization algorithm is researched . After introduction of basic interval algorithm, a new interval extension model was proposed to get a more precise interval domains of objective function . Furthermore, three discard methods are combined together into FNN training. This hybrid strategy could accelerate the algorithm speed. At last, the algorithm simulation results show that the presented global FNN interval algorithm could train FNN efficiently and has a fast rate of convergence.

At the same time, training time is still a problem. For real project, larger and larger neural networks structures are needed. For large scale neural networks, interval algorithm convergence speed is far away from the our satisfaction, so the writers are still doing some research in interval algorithm for feedforward neural networks.

References

1. Xue, J., Li, Y., Chen, D.: A Neural Network's Learning Algorithm Based on Interval Optimization[J]. Computer Engineering 32, 192–194 (2006)
2. Liu, B.: A Neural Network Global Optimization Algorithm Based on Interval Optimization[J]. Computer Engineering and Application 23, 90–92 (2005)
3. Shen, Z.H., Huang, Z.Y., Wolfe, M.A.: An interval maximum entropy method for a discrete minimax problem[J]. Applied Math And Compute. 87, 49–68 (1997)
4. Cao, D.X., Huang, Z.Y.: An interval algorithm for a discrete minimax problem[J]. Journal of Nanjing University Mathematical Biquarterly 14, 74–82 (1997)
5. Moore, R., Yang, C.: Interval Analysis [J]. Technical Document Lockheed Missiles and Space Division 12, 87–95 (1959)
6. Moore, R.: Methods and Applications of Interval Analysis [M]. Society for Industrial and Applied Mathematics, pp. 100–133 (1979)
7. Asaithambi, N.S., Shen, M.R.E.: On computing the range of values[J]. Computing 28, 225–237 (1982)
8. Corliss, G., Kearfott, B.: Rogorous global search industrial applications [J]. Reliable Computing 2, 7–16 (1999)
9. Shen, P.P., Zhang, K.C.: An interval method for non-smooth global optimization problem[J]. OR Transactions 6, 9–18 (2000)
10. Casado, L.G., Garcia, I., Martinez, J.A.: Experiments with a new selection criterion in a fast interval optimization algorithm[J]. Journal of Global Optimization 19, 1247–1264 (2001)
11. Ming, G., Liu, X., Li, S.: A New Global Optimization BP Neural Networks[J]. Jouenal of Binzhou Teachers College 20, 37–41 (2004)
12. Li, h.-q., Wan, b.-w.: A New Global Optimization Algorithm for Training Feedforward Neural Networks and Its Application[J]. System Engineering Theory and Practice 8, 42–47 (2003)

13. Casado, L.G., Garcia, I., Csendes, T.: A new multi-section technique in interval methods for global optimization[J]. Computing 65, 263–269 (2000)
14. Wolfe, M.A.: An interval algorithm for constrained globle optimization[J]. Journal of Computational and Applied Mathematics 50, 605–612 (1994)
15. Kearfott, R.B.: Interval extensions of non-smooth function for global optimization and nonlinear systems solvers[J]. Computing 57, 149–162 (1996)
16. Ratschek, H., Voller, R.L.: Global Optimization Over Unbounded Donmains. SIAM [J]. Control and Optimization 28, 528–539 (1990)
17. David, J.J., Frenzel, J.F.: Training Product Unit Neural Networks with Genetic Algorithms[J]. IEEE Expert 8, 26–33 (1993)
18. Kirkpatrick, S., Gelatt, C.D., Vecchi, M.P.: Optimization by simulated annealing[J]. Science 220, 371–380 (1993)

Approximate Interpolation by Neural Networks with the Inverse Multiquadric Functions

Xuli Han

School of Mathematical Sciences and Computing Technology,
Central South University, 410083 Changsha, China
xlhan@mail.csu.edu.cn

Abstract. For approximate interpolation, a type of single-hidden layer feedforward neural networks with the inverse multiquadric activation function is presented in this paper. We give a new and quantitative proof of the fact that a single layer neural networks with $n + 1$ hidden neurons can learn $n + 1$ distinct samples with zero error. Based on this result, approximate interpolants are given. They can approximate interpolate, with arbitrary precision, any set of distinct data in one or several dimensions. They can uniformly approximate any C^1 continuous function of one variable.

1 Introduction

Consider the following set of functions

$$\mathcal{N}_{n+1,\phi}^d = \left\{ N(\mathbf{x}) = \sum_{i=0}^{n} c_i \phi(\mathbf{w}_i . \mathbf{x} + b_i) : \mathbf{w}_i \in \mathbb{R}^d, c_i, \ b_i \in \mathbb{R} \right\},$$

where $\mathbf{w}.\mathbf{x}$ denotes the usual inner product of $\mathbb{R}^d$ and ϕ is a function from $\mathbb{R}$ to $\mathbb{R}$. We call single layer feedforward networks to the elements of $\mathcal{N}_{n+1,\phi}^d$. It is well known that single layer feedforward networks with at most $n + 1$ addends(elements of $\mathcal{N}_{n+1,\phi}^d$) can learn $n + 1$ distinct samples $(\mathbf{x}_i, f_i)$ with zero error (exact interpolants) and the inner weights ($\mathbf{w}_i$ and b_i) can be chosen "almost" arbitrarily, see [1], [2] and [3].

With regard to the problem of finding the weights of the network, all the algebraic proofs fix the inner weights in a more or less "arbitrary way" and the outer weights (c_i) are obtained by solving an $(n + 1) \times (n + 1)$ linear system, see [4] and [5]. Thus the methods can be scarcely effective when the number of neurons is large. Even the most simple (algebraic) methods require to invert an $(n+1) \times (n+1)$ dense matrix. Therefore, we are interested in finding the weights of an exact neural interpolant without training.

Approximate interpolants are used as a tool for studying exact interpolation, see [6]. Approximation capabilities of neural networks are discussed in [7] and [8]. Most of the methods of approximation by neural networks are reduced to the search of approximate interpolants. These are usually obtained by local optimization techniques or global optimization procedures.

L. Kang, Y. Liu, and S. Zeng (Eds.): ISICA 2007, LNCS 4683, pp. 296–304, 2007.
© Springer-Verlag Berlin Heidelberg 2007

In this paper we give a method for approximate interpolation, that is, for the interpolation problem and for any precision ε, we show a constructive method for obtaining a family of approximate interpolation in $\mathcal{N}_{n+1,\phi}^d$. Here we say that the function $g : \mathbb{R}^d \to \mathbb{R}$ is an approximate interpolant if

$$|g(\mathbf{x}_i) - f_i| < \varepsilon, \, i = 0, 1, \cdots, n.$$

In [9], [10] and [11], some methods are discussed for the approximation of a function with a neural network whose activation function is sigmoidal. Our approach shows that some simple explicit expressions are given and the given precision of approximating a function can be easily obtained.

This paper is organized as follows. Section 2 gives a new and quantitative proof of the fact that $n + 1$ hidden neurons can learn $n + 1$ distinct samples with zero error. Based on this result, Section 3 introduces the approximate interpolation in the uni-dimensional case. These networks do not require training and can approximately interpolate an arbitrary set of distinct samples. Section 4 gives a constructive proof of the uniform convergence of approximate interpolation to any C^1 continuous function of one variable. Section 5 extends the concept of approximate interpolation to multidimensional domains. Section 6 provides some concluding remarks.

2   Exact Neural Interpolants

Let $\Delta : a = x_0 < x_1 < \cdots < x_n = b$ be any partition of the finite interval $[a, b] \subset \mathbb{R}$. We want to find a neural net N that exactly interpolates the data (x_i, f_i), $i = 0, 1, \cdots, n$, where $f_i \in \mathbb{R}$, that is

$$N(x_i) = f_i, \, i = 0, 1, \cdots, n.$$

We consider an interpolant of type

$$N(x) := \sum_{i=0}^{n} c_i \varphi(w_i x + b_i), \tag{1}$$

such that

$$\sum_{i=0}^{n} c_i \varphi(w_i x_j + b_i) = f_j, \, j = 0, 1, \cdots, n, \tag{2}$$

where $\varphi(x) = 1/(1 + x^2)$ is the inverse multiquadric activation function.

The inner weights of the network are determined by the following procedure: Let A be a positive real variable. For $i = 0, 1, \cdots n$, let

$$z_i = \begin{cases} x_{i-1}, \, |x_i - x_{i-1}| < |x_{i+1} - x_i|, \\ x_{i+1}, \, |x_i - x_{i-1}| \geq |x_{i+1} - x_i|, \end{cases}$$

$$w_i = \frac{A}{z_i - x_i}, \quad b_i = \frac{x_i A}{x_i - z_i},$$

where $x_{-1} \le 2x_0 - x_1$, $x_{n+1} \ge 2x_n - x_{n-1}$, that is $z_0 = x_1$, $z_n = x_{n-1}$. Thus, we have

$$w_i x_i + b_i = 0, \quad w_i z_i + b_i = A,$$

for $i = 0, 1, \cdots, n$. That is, for each neuron i, we can choose its value in x_i equal to c_i and its value in $x_j \ne x_i$ near 0 when A is large enough.

The system (2) can be written as an $(n+1) \times (n+1)$ linear system in vectorial form

$$Mc = f,$$

where

$$M = (\varphi(w_j x_i + b_j))_{(n+1) \times (n+1)},$$
$$c = (c_0, c_1, \cdots, c_n)^T, \quad f = (f_0, f_1, \cdots, f_n)^T.$$

If the matrix M is invertible, then the outer weights are obtained by solving the liner system.

Theorem 1. If $A > \sqrt{n-1}$, then the matrix M is invertible.

Proof. The entries of the matrix M are

$$m_{ij} = \varphi(w_j x_i + b_j) = \varphi\left(\frac{x_i - x_j}{z_j - x_j} A\right).$$

Since $\varphi(x) = 1/(1 + x^2)$, we obtain $m_{ii} = 1$, $m_{ij} \le \varphi(A)$ ($|i - j| = 1$), $m_{ij} < \varphi(A)$ ($|i - j| > 1$). Thus

$$\sum_{j \ne i} |m_{ij}| = \sum_{j \ne i} m_{ij} \le n\varphi(A) = \frac{n}{1 + A^2}.$$

If $A > \sqrt{n-1}$, then

$$\sum_{j \ne i} |m_{ij}| < |m_{ii}|, \ i = 0, 1, \cdots, n.$$

This implies that matrix M is a strictly diagonally dominant matrix, and then matrix M is invertible. $\qquad\square$

Obviously, the matrix M trends to an unit matrix as A trends to positive infinity. Thus, we need to consider the favorable property of reproducing constant functions. A modified interpolant is given as follows

$$N_e(x, A) := \sum_{i=0}^{n} c_i^* \sigma(w_i x + b_i), \tag{3}$$

where

$$\sigma(w_i x + b_i) = \frac{\varphi(w_i x + b_i)}{\sum_{j=0}^{n} \varphi(w_j x + b_j)}, \tag{4}$$

for $i = 0, 1, \cdots, n$. Thus, interpolation conditions

$$N_e(x_j, A) = f_j, \ j = 0, 1, \cdots, n, \tag{5}$$

can be written as

$$Uc^* = f,$$

where $U = DM$,

$$D^{-1} = diag\left(\sum_{j=0}^{n} m_{0,j}, \sum_{j=0}^{n} m_{1,j}, \cdots, \sum_{j=0}^{n} m_{n,j}\right).$$

Corollary 1. If $A > \sqrt{n-1}$, then the matrix U is invertible.

3   Unidimensional Neural Approximate Interpolants

Consider the interpolation problem for the data $(x_i, f_i) \in \mathbb{R}^2$, $i = 0, 1, \cdots, n$. Corresponding to the exact interpolant $N(x) = N(x, A)$ given by (1), we define the following neural network

$$N_0(x, A) := \sum_{i=0}^{n} f_i \varphi(w_i x + b_i). \tag{6}$$

We note that $N_0(x, A)$ and $N(x, A)$ differ only in the outer weights. We will prove that $N_0(x, A)$ is an approximate interpolant that is arbitrarily near of the corresponding $N(x, A)$ when the parameter A increases.

Theorem 2. If $A > \sqrt{n-1}$, then

$$|N(x, A) - N_0(x, A)| < \frac{n}{A^2 + 1 - n} \sum_{i=0}^{n} |f_i| \tag{7}$$

for all $x \in [a, b]$.

Proof. For $A > \sqrt{n-1}$, the matrix M is invertible. We can get c by solving the linear system (2). Thus, we have

$$c - f = (I - M)c = (I - M)(c - f) + (I - M)f,$$

where I is an unit matrix. Therefore

$$\|c - f\|_1 \leq \frac{\|I - M\|_1}{1 - \|I - M\|_1}\|f\|_1.$$

From the explicit values of M_{ij}, we have

$$\|I - M\|_1 = \max_{0 \leq i \leq n} \sum_{j \neq i} m_{ij} \leq n\varphi(A).$$

For $A > \sqrt{n-1}$, we have $n\varphi(A) < 1$. Then, considering that the function $g(x) = x/(1-x)$ is strictly increasing on $(-\infty, 1)$, it follows that

$$\|c - f\|_1 < \frac{n\varphi(A)}{1 - n\varphi(A)}\|f\|_1.$$

On the other hand

$$|N(x, A) - N_0(x, A)| = |\sum_{i=0}^{n}(c_i - f_i)\varphi(w_i x + b_i)|$$

$$\leq \sum_{i=0}^{n}|c_i - f_i| = \|c - f\|_1.$$

From this we obtain (7) immediately.   □

Corresponding to the interpolant (3), we would rather give a modified approximate interpolant as follows:

$$N_a(x, A) := \sum_{i=0}^{n} f_i \sigma(w_i x + b_i). \tag{8}$$

Theorem 3. If $A > \sqrt{n-1}$, then

$$|N_e(x, A) - N_a(x, A)| < \frac{2n}{A^2 + 1 - n}\sum_{i=0}^{n}|f_i| \tag{9}$$

for all $x \in [a, b]$.

Proof. For $A > \sqrt{n-1}$, the matrix U is invertible. We can get c^* by solving the linear system (5). In the same way as the proof of Theorem 2, we have

$$\|c^* - f\|_1 \leq \frac{\|I - U\|_1}{1 - \|I - U\|_1}\|f\|_1.$$

From (5) , we have

$$\|I - U\|_1 = \max_{0\leq i\leq n} 2\left(1 - \frac{1}{\sum_{j=0}^{n} m_{ij}}\right)$$

$$= \max_{0\leq i\leq n} \frac{2\sum_{j\neq i} m_{ij}}{1 + \sum_{j\neq i} m_{ij}} \leq \frac{2n\varphi(A)}{1 + n\varphi(A)}.$$

For $A > \sqrt{n-1}$, we have

$$\frac{2n\varphi(A)}{1 + n\varphi(A)} < 1.$$

Therefore

$$\|c^* - f\|_1 \leq \frac{2n\varphi(A)}{1 - n\varphi(A)}\|f\|_1.$$

On the other hand

$$\|N_e(x, A) - N_a(x, A)\| = |\sum_{i=0}^{n}(c^* - f_i)\sigma(w_i x + b_i)|$$

$$\leq \sum_{i=0}^{n}|c_i^* - f_i| = \|c^* - f\|_1.$$

From this we obtain (9) immediately.   □

4   Uniform Approximation Properties

In this section we discuss the uniform approximation of continuous functions for the given interpolant (3) and approximate interpolant (8). We consider the uniform approximation with the following features.

(1) The partition of $[a, b]$ is uniform, that is, $x_i = a+(b-a)i/n$, $i = 0, 1, \cdots, n$.
(2) The real numbers f_i are the images of x_i under a function f, that is $f_i = f(x_i)$, $i = 0, 1, \cdots, n$.

Theorem 4. Let $f \in C^1[a, b]$. For each $\varepsilon > 0$, there exists a natural number N such that

$$|f(x) - N_a(x, A)| < \varepsilon$$

for all $n > N$ and for all $x \in [a, b]$.

Proof. Let $x \in [a, b]$, and suppose that $x \in [x_j, x_{j+1}]$, then

$$|f(x) - N_a(x, A)| = |\sum_{i=0}^{n}(f(x) - f(x_i))\sigma(w_i x + b_i)|$$

$$\leq \max\{|f(x) - f(x_j)| + |f(x) - f(x_{j+1})|\}$$

$$+ \|f'\|_\infty \sum_{i \neq j, j+1} |x - x_i|\sigma(w_i x + b_i)$$

$$\leq \left[h + \sum_{i \neq j, j+1} |x - x_i|\sigma(w_i x + b_i)\right]\|f'\|_\infty,$$

where $h = (b - a)/n$, and $\|f'\|_\infty := \max_{a \leq x \leq b}|f'(x)|$.
   A straightforward analysis gives that

$$\sum_{i \neq j, j+1} |x - x_i|\varphi(w_i x + b_i) < \frac{2h}{A}[1 - \ln 2 + \ln(n - 1)],$$

for $n \geq 3$, and

$$\sum_{i=0}^{n} n\varphi(w_i x + b_i) \geq \frac{8}{4 + A^2}.$$

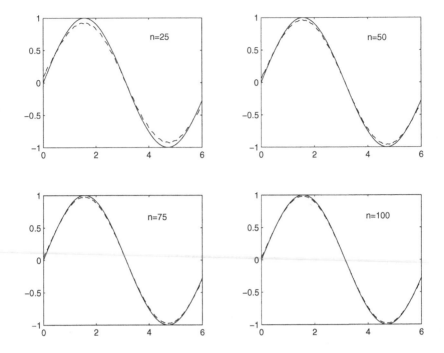

Fig. 1. Approximate interpolation curves

Thus, we have

$$\sum_{i \neq j, j+1} |x - x_i| \sigma(w_i x + b_i) \leq \frac{(b-a)(4+A^2)}{4nA^2} \big[1 - \ln 2 + \ln(n-1)\big].$$

From this the result follows.   □

Corollary 2. Let $f \in C^1[a, b]$. For each $\varepsilon > 0$, there exists a function $A(n)$ and a natural number N such that

$$|f(x) - N_e(x, A(n))| < \varepsilon$$

for all $n > N$ and for all $x \in [a, b]$.

Proof. Since

$$|f(x) - N_e(x, A(n))| \leq |f(x) - N_a(x, A(n))| \\ + |N_a(x, A(n)) - N_e(x, A(n))|,$$

the result follows by Theorem 3 and Theorem 4.   □

Figure 1 shows the uniform convergence of the interpolant $N_a(x, f)$ for $f(x) = \sin(x)$ when n increases. Here we have considered $A = 2$.

5   Multidimensional Neural Approximate Interpolants

Let the function $f : \mathbb{R}^d \to \mathbb{R}$ and let $S = \{\mathbf{x}_0, \mathbf{x}_1, \cdots, \mathbf{x}_n\} \subset \mathbb{R}^d$ be a set of distinct vectors. If $\{f_i : i = 0, 1, \cdots, n\}$ is a set of real numbers, we search for a neural network N such that

$$N(x_i) = f_i, \; i = 0, 1, \cdots, n.$$

Lemma.[1] Let $S = \{\mathbf{x}_0, \mathbf{x}_1, \cdots, \mathbf{x}_n\} \subset \mathbb{R}^d$ be a set of distinct vectors. Then there exists a vector $\mathbf{v}$ such that the $n + 1$ inner products $\mathbf{v}.\mathbf{x}_i (i = 0, 1, \cdots, n)$ are different from each other.

Let $\mathbf{v}$ denote a fixed vector that verifies the condition in the Lemma. Then we reorder the points of S in such a way that

$$\mathbf{v}.\mathbf{x}_0 < \mathbf{v}.\mathbf{x}_1 < \cdots < \mathbf{v}.\mathbf{x}_n.$$

Let $\mathbf{v}.\mathbf{x} = t$, $\mathbf{v}.\mathbf{x}_i = t_i$,

$$s_i = \begin{cases} t_{i-1}, & |t_i - t_{i-1}| < |t_{i+1} - t_i|, \\ t_{i+1}, & |t_i - t_{i-1}| \geq |t_{i+1} - t_i|, \end{cases}$$

$$w_i = \frac{A}{s_i - t_i}, \quad b_i = \frac{t_i A}{t_i - s_i}$$

where $t_{-1} = 2t_0 - t_1$, $t_{n+1} = 2t_n - t_{n-1}$, $i = 0, 1, \cdots, n$.

Based on the expressions (3) and (8), we can obtain the expressions of exact interpolant and approximate interpolant respectively as follows:

$$N_e(\mathbf{x}, \mathbf{v}, A) = \sum_{i=0}^{n} c_i \sigma(w_i t + b_i),$$

$$N_a(\mathbf{x}, \mathbf{v}, A) = \sum_{i=0}^{n} f_i \sigma(w_i t + b_i).$$

A multidimensional version of Theorem 3 can be stated using arguments similar to the ones in Section 3. We have the following result.

Theorem 5. If $A > \sqrt{n-1}$, then

$$|f_i - N_a(\mathbf{x}_i, \; \mathbf{v}, A)| < \frac{2n}{A^2 + 1 - n} \sum_{i=0}^{n} |f_i|.$$

As an numerical example, we consider the multidimensional approximate interpolation corresponding to the function $f(x,y) = \sin(\sqrt{x^2 + y^2})/\sqrt{x^2 + y^2}$ on .the 64 points $S = \{(i, j) : \; -8 \leq i \leq 8, \; -8 \leq j \leq 8\}$. We have set $v = (1, 10)$, $e = \max_{\mathbf{x} \in S} |f(\mathbf{x}) - N(\mathbf{x}, \mathbf{v}, A)|$. The results are given in Table 1.

Table 1. The results of the approximate interpolation

A	1	5	10	20	40
e	0.3290	4.96×10^{-2}	1.37×10^{-2}	3.50×10^{-3}	8.85×10^{-4}

6   Conclusions

With the inverse multiquadric activation function, we give a new and quantitative proof of the fact that a single layer neural network with $n+1$ hidden neurons can learn $n+1$ distinct samples with zero error. Based on this result we introduce a family of neural networks for approximating interpolation of real functions. These networks can approximately interpolate $n+1$ samples in any dimension with arbitrary precision and without training. We give a rigorous bound of the interpolation error. The given approximate interpolant can uniformly approximate any C^1 continuous function of one variable and can be used for constructing uniform approximents of continuous functions of several variables. Our approach shows that obtaining the number of neurons n as a function of the precision ε is an easy task. The given approximants are simple explicit expressions and the given precision of approximating a function can be easily obtained.

Compare our results with other results of exact interpolation by neural networks, our method do not need to solve a linear system for the outer weights. Compare our results with other results of approximate interpolation by neural networks, our method do not need local optimization techniques or global optimization procedures.

References

1. Huang, G.B., Babri, H.A.: Feedforward neural networks with arbitrary bounded nonlinear activation functions. IEEE Trans. Neural Networks 9(1), 224–229 (1998)
2. Sartori, M.A., Antsaklis, P.J.: A simple method to derive bounds on the size and to train multilayer neural networks. IEEE Trans. Neural Networks 2(4), 467–471 (1991)
3. Tamura, S., Tateishi, M.: Capabilities of a four-layered feedforward neural network. IEEE Trans. Neural Networks 8(2), 251–255 (1997)
4. Barhen, J., Cogswell, R., Protopopescu, V.: Single iteration training algorithm for multilayer feedforward neural networks. Neural Process. Lett. 11, 113–129 (2000)
5. Li, X.: Interpolation by ridge polynomials and its application in neural networks. J. Comput. Appl. Math. 144, 197–209 (2002)
6. Sontag, E.D.: Feedforward nets for interpolation and classification. J. Comp. Syst. Sci. 45, 20–48 (1992)
7. Chui, C.K., Li, X., Mhaskar, H.N.: Neural networks for localized approximation. Mathematics of Computation 63, 607–623 (1994)
8. Chui, C.K., Li, X., Mhaskar, H.N.: Limitations of the approximation capabilities of neural networks with one hidden layer. Adv. Comput. Math. 5, 233–243 (1996)
9. Debao, C.: Degree of approximation by superpositions of a sigmoidal function. Approx. Theory & its Appl. 9, 17–28 (1993)
10. Mhaskar, H.N., Michelli, C.A.: Approximation by superposition of sigmoidal and radial basis functions. Adv. Appl. Math. 13, 350–373 (1992)
11. Lianas, B., Sainz, F.J.: Constructive approximate interpolation by neural networks. J. Comput. Appl. Math. 188, 283–308 (2006)

Decomposition Mixed Pixel of Remote Sensing Image Based on Tray Neural Network Model

Zhenghai Wang, Guangdao Hu, and Shuzheng Yao

Institute of Mathematical Geo. & Remote sensing ,
China University of Geosciences, Wuhan, 430074
wzhyy@tom.com, hugd@cug.edu.cn

Abstract. Aiming at the characteristics of remote sensing image classification, the mixed pixel problem is one of the main factors that affect the improvement of classification precision in remote sensing classification. In this paper ,A tray neural network model was established to solve the mixed pixel classification problems. ETM+ data of DaLi in May 2001 is selected in the study. The results show that this method effectively solves mixed pixel classification problem, improves network learning speed and classification accuracy, so it is one kind of effective remote sensing imagery classifying method.

Keywords: remote sensing image; Decomposition mixed pixel; tray neural network.

1 Introduction

Remote sensing map takes pixel as a measurement unit in the ground detection. Since the limited precision of satellite remote sensing detection, sometimes there are several kinds of ground objects in the range of one pixel, which is named mixed pixel. On the contrary, the pixel that consists of one kind of ground object is described as typical pixel. Generally the spectrum features of different ground objects are different. The spectrum of mixed pixels is a mixture of poly-type spectrum of ground objects according to some proportion. The existence of mixed pixel is one of the main factors that affect classifying precision in image recognition. Especially it affects the classification and recognition of linear ground objects and petty ground objects fairly obviously, such as TM and SPOT data. A lot of mixed pixels are bound to appear at the edge of ground objects. The key to resolve the problem is to find out the proportion of different objects.

There are five methods mostly for the decomposition of mixed pixel as follows : linear discriminate analysis, probabilistic, geometric-optical, stochastic geometric, fuzzy and Artificial Neural Network (ANN) models. ANN is an important branch of artificial intelligence science rapidly developed in recent years. ANN is a system of highly connected but simple processing units called neurons, which is inspired by the architecture of the human brain. Calibration of an ANN is done by repeatedly presenting a set of training examples to it, which has the advantage that the class distributions do not have to be known nor need to comply to some parametric model.

L. Kang, Y. Liu, and S. Zeng (Eds.): ISICA 2007, LNCS 4683, pp. 305–309, 2007.

Another advantage of ANNs compared to classic statistical methods is that multiple data sources can easily be integrated. In recent years, ANNs have demonstrated great potentials a method to decompose mixed pixels as well. It can deal with this problem: can not described by algorithms but have large numbers of sample to study. It has restrict the application of ANN for the affirmatory numerical value of input and output samples. In this paper , we propose a Tray Neural Network model combine the grey system and neural network to pixel immixing of Landsat 7 (ETM) image.

2 The Tray Neural Network Model

Although many types of ANNs exist, for castigation or decomposition of image data mostly feed-forward networks such as the multi-layer perception (MLP) are used. A MLPconsists of units spread over an input layer, possibly some hidden layers, and an output layer see (Figure 1). Each neuron in the input layer is associated with one of the n spectral bands in which a pixel's reflectance is measured, while the output layer contains m units, one for each of the estimated class proportions; the number of hidden layers (one or two) and the number of neurons each of these layers contains should be based on the characteristics of the remote sensing data. BPNN adopts error feedback learning propagation. Its learning process composes two parts: positive spread (network is calculating) and reverse spread (error feedback).In the positive spread process, the input information through the implicit unit is handled and spread directly to the output layer. If the output layer care t get the expectation export, it then turns into the reverse spread process. The error between the value predicted(by the network) and the value actually observed(known data) is then measured and propagated back w ands along the feeds forty and connect dons. The weights of links between units and node thresholds are modified to various extents. If the MSE(mean square error) exceeds some small predetermined value, a new'epoch' (cycle of presentation of all training inputs) is started after termination of the current epoch, till it is smaller than the predetermined value .

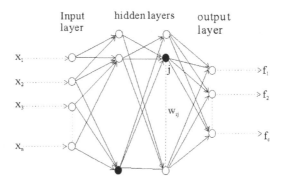

Fig. 1. A multi-layer perception designed for the decomposition of mixed pixels

If a number has no definitude value but a assumable bound we define it tray number. tray number is a set or a bound of number, Signed $\odot$. The gray of pixel of Remote Sensing Image is only a assumable bound but don't know it's definitude value, so The gray of pixel of Remote Sensing Image is a tray number. Supports A is a range, a is a number, if tray number $\odot$ belong to A, then a is a whiting of $\odot$; $\odot$ is a common tray number, $\odot(a)$ is a tray number with whiting a., we can express the input sample with tray number, then get the method of pixel immixing based on Tray Neural Network Model. Defines the sample set $\{x_1, d_1; x_2, d_2; \cdots, x_n, d_n;\}$. There into, $xi = (\odot(x_i), \odot(x_j), \cdots, \odot(x_n)T), I = 1, 2, \cdots, N$, x_i express the input sample; $d_i = (d_1, d_2, \cdots, d_c)T$ $i = 1, 2, \cdots, N$, d express the anticipated output sample.

Using monopole polarity S compressed function as activation function:

$$f(\gamma) = \frac{1}{1+\exp(-\lambda v)}$$

λ, the gradient fact of the shape of function. The Tray Neural Network algorithm is divided into four steps:

1) Initialize all of the weights using random number . choose the step parameter η > 0, $\iota \leftarrow 1$, used to compute the iterative times , the maximal iterative time is L; $p \leftarrow \iota$, p used to compute the times of sample; e $\leftarrow 0$, e used to compute the accumulative error, define the maximal error Emax.

2) input a sample x_p and its anticipated output d_p

$$\odot(o_i) \leftarrow \odot(x_i), c \; ; \qquad d \leftarrow d_p$$

First compute the output node in hidden lay(j):

$$\odot(y_j) = \sum_{i=1}^{n+1} (\odot(\omega ji) \cdot \odot(o_i)), o_{n+1} = 1 \; ; \quad \odot(o_j) = f(\odot(y_j)), j = 1, 2, \cdots, h$$

Then compute the output node in hidden lay(k):

$$\odot(y_k) = \sum_{j=1}^{h+1} (\odot(\omega k_j) \cdot \odot(o_j)), o_{h+1} = 1 \; ; \quad \odot(o_k) = f(\odot(y_k)), k = 1, 2, \cdots, c$$

3) accumulative error

$$\odot(e) \leftarrow \odot(e) + \sum_{k=1}^{c} (dk - \odot(o_k))$$

4) correct the weights coefficient
begin the output lays, first correct ωkj

$$\odot(\delta_k) = (d_k - \odot(o_k) \cdot \odot(o_k)) \cdot (1 - \odot(o_k)) \; ;$$
$$\odot(\omega_{kj}) \leftarrow \odot(\omega_{kj}) + \eta \cdot \odot(\delta_k) \cdot \odot(o_j), o_{h+1} = 1,$$
$k = 1, 2, \cdots, c, j = 1, 2, \cdots, h + 1$
correct ω_{kj} again

$$\odot(\delta_j) = \odot(o_j) \cdot (1 - \odot(o_j)) \cdot \sum_{k=1}^{c} \odot(\omega_{kj}) \cdot \odot(\delta_j) \; ;$$

$\odot(\omega_{ji}) \leftarrow \odot(\omega_{ji}) + \eta \cdot \odot(\delta_j) \cdot \odot(o_i)$

$o_{n+1} = 1, \; j = 1, 2, \cdots, c, \; i = 1, 2, \cdots, n+1$

5) if $p < N$ then $p \leftarrow p+1$, turn to $2 =$; otherwise turn to $6 =$

6) $1 \leftarrow 1+1$ if $1 > L$, then iterative end

if $\odot(e) < $ Emax iterative end

if $\odot(e) \geq $ Emax then $\odot(e) \leftarrow 0, p \leftarrow 1$, turn to 2), begin the next iterative.

3   Tray Neural Network Model Applied Decomposition of Mixed Pixel

We design an experiment to validate the feasibility of tray neural network model applied decomposition of mixed pixel .The experimental data is ETM satellite data of May 18, 2001. The research region is 1000×800 pixels (30 m resolution). The band range is the 1th ~5th bands and the 7th band. The decision class of training set is 5 because the serial number is given to 5 classes of ground overlay type.

1) Select a region of ETM image, reduce the resolution ,construct a ETM image with mixed pixels.

2) Assume there are no mixed pixels in the originality image , exactly Classification and get the proportion of the different objects(table 1).

3) Decompose mixed pixels for the image has been reduced resolution and get the acreage proportion of the different objects(fig2、 tabe 1).

Table 1. The proportion of the different objects base on different classification method

Proportion	bushes	farmland	grass	urban	road
supervised	22.20	32.02	12.01	14.63	21.14
decompose	19.09	34.16	13.90	13.00	19.85
exactly Classification	19.06	35.84	13.10	12.85	19.15

4)Analyze the experiment result

Define the standard precision fact e

$$e = \begin{cases} \dfrac{Pt}{ps} & Pt < ps \\ \dfrac{2\,ps - Pt}{ps} & Pt \geq ps \end{cases}$$

Pt : test data , ps : originality data , e : precision.

The compare precision between the tray neural network model and supervised method (table 2).

Table 2. The precision of the different objects base on different classification method

precision	bushes	farmland	grass	urban	road	average
supervised	86.36	89.34	91.67	86.15	89.60	88.62
decompose	99.83	95.32	93.89	98.85	93.09	96.19

4   Conclusion and Discussion

Through comparing the classification result and tested data, the average precision increased from 88.62 to 96.19 , it is shown that its reasonable and has practicality decompose mixed pixels based on the tray neural network model . its difficult to select special pixels exactly, By define the gray of special pixels as " tray" , the tray neural network model can increased average precision of classification.

Acknowledgement

This work is supported by the Key Brainstorm Project of the Ministry of Land and Resources of China (No. 20010305). The authors are grateful for the anonymous reviewers who made constructive comments.

References

1. Pawlak, Z.: Rough sets. International Journal of Computer and Information Sciences 11, 341–356 (1982)
2. Simpson, J.J., Mcintir, J.T.: A recurrent neural network classifier for improved retrievals of areal extent of snow cover. IEEE Transaction on Geoscience and Remote Sensing 39(10), 2135–2147 (2001)
3. Bagan, H., Ma, J., Li, Q., et al.: The self-organizingfeature map neural networks classification of the ASTER Data based on wavelet Fusion. Science in China, Series D 47(7), 651–658 (2004)
4. Daijin, K.: Data classification based on tolerant rough set. Pattern Recognition 34, 1613–1624 (2001)
5. Bagan, H., Ma, J., Zhou, Z.: Study on the artificial neural network method used for weather and AVHRR Thermal Data Classification. Journal of The Graduate School of The Chinese Academy of Sciences 20(3), 328–333 (2003)
6. Adams, J.B., Sabol, D.E., Kapos, V., Filho, R.A., Roberts, D.A., Smith, M.O., Gillespie, A.R.: Classification of multi-spectral images based on fractions of end members: Application to land-cover change in the Brazilian Amazon. Remote Sensing of Environment

A Novel Kernel Clustering Algorithm Based Selective Neural Network Ensemble Model for Economic Forecasting

Jian Lin[1] and Bangzhu Zhu[1, 2]

[1] School of Economics and Management, Beijing University of Aeronautics and Astronautics, Beijing 100083, China
[2] Institute of System Science and Technology, Wuyi University, Jiangmen 529020 Guangdong, China
Wpzbz@126.com

Abstract. In this study, a novel kernel clustering algorithm based selective neural network ensemble method, i.e. KCASNNE, is proposed. In this model, on the basis of different training subsets generated by bagging algorithm, the feature extraction technique, kernel principal component analysis (KPCA), is used to extract their data features to train individual networks. Then kernel clustering algorithm (KCA) is used to select the appropriate number of ensemble members from the available networks. Finally, the selected members are aggregated into a linear ensemble model with simple average. For illustration and testing purposes, the proposed ensemble model is applied for economic forecasting.

1 Introduction

Economic forecasting has been a common research problem in the past few years. Much attention has been paid to it by many scholars. Various approaches have been adopted to solve the problem. However, it is not easy to predict economic values due to its inherent high complexity and noise, which has encouraged us to developed more predictable models. As a result, artificial neural network (ANN), especially back-propagation neural network (BPNN), is considered as a good method. ANN has a good learning ability and generalization with its simple structure. Although a large number of successful applications have showed ANN can be a very useful tool for economic forecasting, some studies, however, showed that ANN also had some limitations such as the difficulty in determining the number of neural cells of the hidden layer, its initialization and easily dropping local minima and so on, which could influence its generalization and make the forecasting results not very good [1].

In order to overcome the limitations of single ANN, in 1990 a novel ensemble model, i.e. neural network ensemble [2], was developed. Because of combining multiple neural networks, neural network ensemble can remarkably enhance the forecasting ability and outperform any individual neural network. However, in the previous work on ensemble models, a few problems can be found, among which almost every research combines all the available networks to constitute an ensemble model. However, not all the circumstances are satisfied with the rule of "the more, the best" [3].

L. Kang, Y. Liu, and S. Zeng (Eds.): ISICA 2007, LNCS 4683, pp. 310–315, 2007.

In recent years, an idea of selective ensemble was proposed by Professor Zhou Zhi-hua, et al [4] to increase the diversity among the individual neural networks. In the selective ensemble methods, clustering algorithm can effectively eliminate the redundant individuals [5]. Therefore, in this paper a novel kernel clustering algorithm based selective neural network ensemble model (KCASNNE) is proposed and used to predict the increasing rates of Gross Domestic Product (GDP) of Jiangmen, Guangdong to illustrate its validity and feasibility.

The main aim of this study is to show how to carry out economic forecasting using the proposed KCASNNE. Therefore, this paper mainly describes the building process of the proposed ensemble model and the application of the ensemble approach in forecasting the increasing rates of GDP of Jiangmen, Guangdong, China.

The rest of the paper is organized as follows. The next section describes the building process of the proposed ensemble model in detail. In order to verify the effectiveness of the proposed model, empirical analysis of the increasing rates of GDP is reported in section 3. Finally, a few conclusions are contained in section 4.

2   The Building Processes of the Proposed Ensemble Model

In this section, a three-phase KCASNNE is described step by step and in detail. First multiple BPNNs are generated. Then an appropriate number of BPNNs are selected from the candidate predictors generated by the former phase. Finally, the selected BPNNs are aggregated into an ensemble predictor in a linear way.

2.1   Generating Individual BPNN Predictors

According to the bias-variance trade-off principle [6], in order to build a strong ensemble model, the individual models should be with high accuracy as well as high diversity. So how to generate the diverse models is a crucial factor. As for neural network ensemble, some methods have been put forward for generation of ensemble members, most of which depend on varying the parameters of neural networks.

In this study, we adopt bagging algorithm to generate different training subsets and kernel principal component analysis (KPCA) [7] to extract data features for training the individual networks.

2.2   Selecting an Appropriate Number of Ensemble Members

After training, each individual network can generate its own forecasting results. According to the selective ensemble learning theory [4], assembling many of the available neural networks may be better than assembling all of those networks. If there are a great number of available individual neural networks, it is very necessary to select a subset of representatives in order to improve ensemble efficiency.

In order to show that those "many" neural networks can be efficiently selected from all of the available ones. In this paper kernel clustering algorithm (KCA) [8] is presented to select appropriate ensemble members. This clustering approach can select the networks with little similarity to constitute a selective ensemble model.

2.3 Combining the Selected Members

Based on the previous two-phase work, some appropriate ensemble members can be picked out. The next task is to combine these members into an aggregated predictor in an appropriate ensemble strategy. Generally, there are two ensemble strategies: linear ensemble and nonlinear ensemble. Linear ensemble includes two approaches: simple average or weighted average, which has been widely applied in the existing literatures. Similar to the previous work, in this study we adopt simple average to combine the selected members into a linear ensemble model.

By solving the earlier three problems, KCASNNE has been built-up.

To summarize, the proposed selective neural network ensemble model can be described in detail as follows. At the beginning, for a given training set S, a series of training subsets S_1, S_2, ..., S_n can be determined by bagging method. Then, the key attributes F_1, F_2, ..., F_m, extracted by KPCA on the respective training subsets, can be taken as the inputs to create and train n individual neural networks. Next, KCA is applied to select k ensemble members. Finally, simple average is used to combine the selected individual neural networks into a linear ensemble model. The basic flow diagram can be shown in Fig.1.

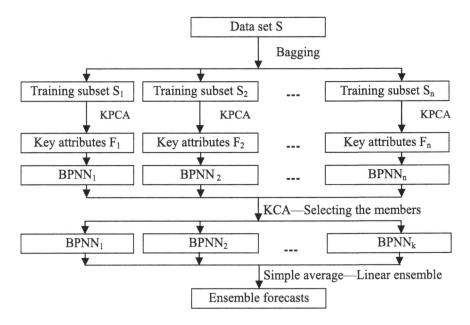

Fig. 1. A flow diagram of the proposed selective ensemble forecasting model

3 Empirical Analysis

The proposed selective ensemble model is applied to forecast the increasing rates of GDP of Jiangmen, China to illustrate its validity and feasibility. In this study, multiple BPNNs are used to build the KCASNNE.

3.1 Data Description

The economic forecasting data used in this paper are yearly and are obtained from Jiangmen Statistical Yearbooks from 1982 to 2004. This paper selects 8 forecasting indexes as follows [9]: x_1: GDP of Jiangmen; x_2: GDP of China; x_3: GDP of Guangdong; x_4: Government expenditures; x_5: Total exports; x_6: Total investment in fixed assets; x_7: Total sales of consumer goods; x_8: Foreign capital actually utilized.

While carrying on the importation of the input variables, this paper uses the increasing rate of the index in order to get rid of the long-term growth trend influence, price fluctuation influence and the difference of index scale. That means:

$$x_i(t)' = \frac{x_i(t)/w(t)}{x_i(t-1)/w(t-1)} - 1, i = 1,2,\cdots,8; t = 1982,1983,\cdots,2004$$

where $x_i(t)'$ is the increasing rate of indicator i in t year, $x_i(t)$ is the real value of indicator i in t year, $w(t)$ is the price index in t year and $w(t-1)$ is the price index in t -1 year. In addition, for convenience, let $x_i(t)'$ still be denoted as $x_i(t)$. In order to save space, the original data is not listed here, and detailed data can be obtained from the yearbooks or from the authors.

We take the data from1983 to 1999 corresponding to the increasing rates of GDP from 1984-2000 as the training set, that is, the number of the training samples is 17, i.e. $\|s\| = 17$. We also take the data from 2000 to 2003 corresponding to the increasing rates of GDP from 2001 to 2004 as the test set. In addition, we use the data in 2004 to forecast the increasing rate of GDP in 2005.

On the previous work, twenty training subsets, namely S_1, S_2, $\cdots$, S_{20} are bootstrap sampling from all the available data, and $\|S_i\| \geq 13, i = 1,2,\cdots,20$.

KPCA is used for dimension reduction and feature extraction of each training subsets. Three principal factors denoted as F_1, F_2 and F_3 of the above twenty training subsets can be respectively extracted. The accumulative total variances all achieve above 90%, so to a certain extent, these three factors can be used to describe the main development information of Jiangmen economic in past years, which can be taken as the 3 input neural cells of individual neural networks. Likewise, we also adopt KPCA to extract the date features from 2000 to 2004 and obtain another F_i (i=1, 2, 3).

3.2 Empirical Results

Matlab 7.01 neural network toolbox has been used to generate and train the individual neural networks. In this paper, we use tansig function as the hidden layer's threshold function, purelin as the output layer's threshold function and trainlm as the training function. Twenty neural networks, named BPNN$_i$ (i=1, 2,…, 20), are not trained until they all satisfy the request of error. F_i (i=1, 2, 3) trains BPNN$_i$ (i=1, 2,…, 20).These twenty neural networks are all 3-5-1 BPNNs with training epochs of 5000, learning rate of 0.001, target error of 0.000001 and initial weights of the random values among [-1,1].When all the training results satisfy the request of error, all the component neural networks have been trained well.

After training, KCA is adopted to cluster the available BPNNs into five groups, among which one representative BPNN is selected respectively. Hence, five BPNNs with more diversity can be used to constitute the selective ensemble model, and simple average is used to combine their forecasting results.

For the purpose of comparison, we have also built another ensemble forecasting model, called KPCANNE, which linearly combines all the available neural networks into an ensemble model. In addition, the root mean squared error (RMSE) is used as the evaluation criteria over the two different ensemble models and corresponding results are reported in Table 1.

Table 1. A comparison of results of ensemble forecasting models

Ensemble models	RMSE	Rank	2005
KCASNNE	0.0051	1	0.11775
KPCANNE	0.0077	2	0.11056

As are shown in the Table 1, we can conclude that (1) all the forecasting results are close to their real values, even though the number of the training set is only 17; (2) the forecasting ability of neural network ensemble is generally much higher than that of any individual network; (3) the forecasting ability of KCASNNE is slightly superior to that of KPCANNE on the whole; (4) Jiangmen increasing rate of GDP in 2005 is above 11% and (6) KCASNNE is feasible and effective for economic forecasting.

4   Conclusions

All the above is an attempt to presenting a novel kernel clustering algorithm based selective neural network ensemble forecasting model, and applying it for economic forecasting. The experiment results reported in this paper demonstrate the effectiveness of the proposed selective ensemble model, implying the proposed kernel clustering algorithm based selective neural network ensemble forecasting model can provide a new way for economic forecasting under small samples. On the whole, the research into the neural network ensemble in practical economic forecasting is in prospect, particularly under competitive market economic environment at present.

Acknowledgment

This work is partially supported by the National Natural Science Foundation of China under Grant 70471074 and Guangdong Provincial Department of Science and Technology under Grant 2004B36001051 to Lin Jian.

References

1. Lin, J., Zhu, B.Z.: Improved principal component analysis and neural network ensemble based economic forecasting. In: Huang, D.-S., Li, K., Irwin, G.W. (eds.) ICIC 2006. LNCS, vol. 4113, pp. 135–145. Springer, Heidelberg (2006)
2. Hansen, L.K., Salamon, P.: Neural network ensembles. IEEE Trans. Pattern Analysis and Machine Intelligence 12(10), 993–1001 (1990)

3. Yu, L.A., Wang, S.Y., Lai, K.K.: A novel nonlinear ensemble forecasting model incorporating GLAR and ANN for foreign exchange rates. Computer &Operation Research 32, 2523–2541 (2005)
4. Zhou, Z.H., Wu, J.X., Tang, W.: Ensemble neural networks: many could be batter than all. Artificial Intelligence 137(1-2), 239–263 (2002)
5. Pabitra, M., Murthy, C.A., Pal, S.K.: Unsupervised feature selection using feature similarity. IEEE Transactions on Pattern Analysis and Machine Intelligence 24(3), 301–312 (2002)
6. Yu, L.A., Wang, S.Y., Lai, K.K., Huang, W.: A bias-variance-complexity trade-off framework for complex system modeling. In: Gavrilova, M., Gervasi, O., Kumar, V., Tan, C.J.K., Taniar, D., Laganà, A., Mun, Y., Choo, H. (eds.) ICCSA 2006. LNCS, vol. 3980, pp. 518–527. Springer, Heidelberg (2006)
7. Zhu, B.Z.: Real estate investment environment evaluation based on kernel principal composition analysis. In: Proceedings of ICCREM 2006, Orlando, Florida, USA, October 5-6, 2006, pp. 355–357 (2006)
8. Ji, Q.Y., Lin, J.: Clustering algorithm based on kernel method and its application. Journal of Beijing University of Aeronautics and Astronautics 32(6), 747–750 (2006)
9. Xiao, J.H.: Intelligent forecasting for the regional economic. Mathematics in Economics 22(1), 57–63 (2005)

An Evolutionary Neural Network Based Tracking Control of a Human Arm in the Sagittal Plane

Shan Liu[1], Yongji Wang[1], and Quanmin Zhu[2]

[1] Department of Control Science and Engineering,
Key Laboratory of Image Processing and Intelligent Control,
Huazhong University of Science and Technology, Wuhan, Hubei, 430074, China
Shan_liu_susan@hotmail.com
[2] Faculty of Computing, Engineering and Mathematical Sciences (CEMS),
University of the West of England (UWE), Frenchay Campus,
Coldharbour Lane, Bristol, BS16 1QY, UK

Abstract. In this paper trajectory tracking control of a human arm moving in sagittal plane is investigated. The arm is described by a musculoskeletal model with two degrees of freedom and six muscles, and the control signal is applied directly in muscle space. To design the intelligent controller, an evolutionary diagonal recurrent neural network (EDRNN) is integrated with proper performance indices, which a genetic algorithm (GA) and evolutionary program (EP) strategy are effectively combined with the diagonal neural network (DRNN). The hybrid GA with EP strategy is applied to optimize the DRNN structure and a dynamic back-propagation algorithm (DBP) is used for training the network weights. The effectiveness of the control scheme is demonstrated through a simulated case study.

Keywords: trajectory tracking control, evolutionary diagonal recurrent neural network, genetic algorithm, evolutionary program, dynamic back-propagation, musculoskeletal model.

1 Introduction

Understanding the process and mechanisms of performing human moving activities can provide significant insight into the neural mechanisms of motor learning and adaptation, development of interactions between robots and human operators, and wide potential applications in computer animation of human motion and neuromotor rehabilitation. It may be advantageous to use humanlike control strategies for improving robot control, in particular with regard to safety.

There have been widespread research activities on modeling and controlling human arms movement in the last five decades. Hierarchical model [10] [15], neural network [10]-[13], optimal control [1] [14], adaptive control theory [5] [11], iterative learning and combination of these methods [16] have been applied in variable movement controls of the human arm. So far most of the studies have focused on arm movement control in the horizontal plane. However, the human arm movement in sagittal plane, without gravity compensation is more natural and realistic. Some experimental

L. Kang, Y. Liu, and S. Zeng (Eds.): ISICA 2007, LNCS 4683, pp. 316–325, 2007.

findings do suggest that gravity forces, at least for vertical arm movements under the gravity constraint, have an important role on the arm movement planning process [17] [18]. And gravitational force acting to the arm even is dependent on the arm posture [19]. Accordingly proper research efforts should be contributed in this field to provide a new angle to understand the operation mechanism of normal human arms.

It has been witnessed that neural networks have been increasingly adopted in biological motor control systems because such inductive approaches do not require detailed knowledge of plant models. A major problem in designing neural networks is the proper choice of the network architecture especially for the networks controlling multi-input multi-output nonlinear systems. In recent years, the combination of evolutionary algorithm and neural networks has attracted a great deal of attention [3] [7].

In this paper, we develop a new controller for the human arm by integrating diagonal recurrent neural network (DRNN) with a genetic algorithm (GA) and evolutionary program (EP) strategy. This paper is organized as follows. In section 2, a realistic musculoskeletal model of the human arm moving in sagittal plane and the control system structure are presented. In section 3, the DRNN based controller is constructed. Subsequently the GA and EP approaches for optimizing the DRNN structure and dynamic back-propagation algorithm (DBP) training algorithm for determining the DRNN weights are presented. In section 4, the feasibility of the proposed neural control system is demonstrated by numerical simulation studies. Finally in section 5, the conclusions and future works are provided.

2   Problem Description

2.1   Human Arm Model

The plant is a simulated, two degrees of freedom, arm model in the sagittal plane. A schematic picture of the arm model is shown in Fig.1. In this model there are two joints (i.e. shoulder and elbow) driven by three pairs of opposing muscles; two pairs of muscles individually actuate the shoulder and elbow joints, while the third pair actuates both joints simultaneously. The movement range of the shoulder is from $80°$ to $90°$ and that of the elbow is from $0°$ to $150°$. For simplify, it is assumed that the reference trajectory exists in joint space, mass effect of each muscle is not considered, and the external force only involves gravity.

The inverse dynamics of the arm model is

$$M(\theta)\ddot{\theta} + C(\theta,\dot{\theta}) + B\dot{\theta} + G(\theta) = \tau_m \ . \tag{1}$$

In Eq. (1), $\theta \in R^2$ is the joint angle vector (shoulder: θ_1, elbow: θ_2), $M(\theta) \in R^{2\times2}$ is the position definite symmetric joint inertia matrix, $C(\theta,\dot{\theta}) \in R^2$ is the vector of centrifugal of Coriolis torques, $B \in R^{2\times2}$ is the joint friction matrix, $G(\theta) \in R^2$ is the vector of gravity torques, and $\tau_m \in R^2$ is the joint torque produced by six muscles.

The muscle model is a simplified version of the model proposed by [2] [14]. It represents combined activation a, length l and velocity v dependent contractile forces. The tension produced by a muscle is represented as

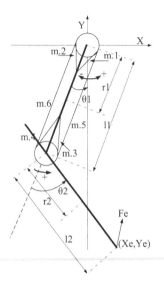

Fig. 1. The schematic picture of the arm model

$$T(a,\ l,\ v)=a(F_L(l)F_V(l,\ v)+F_P(l))\ .\tag{2}$$

and the muscle force is

$$F(a\ ,l,\ v)=F_{0max}T(a,\ l,\ v)\ .\tag{3}$$

Where F_{0max} is the maximum isometric force, and it is set by 31.8N.

The moment arm is defined as the perpendicular distance from the muscle's line of action to the joint's center of rotation, which is equal to the transpose of the Jacobian matrix from the joint space to the muscle space. Here, the relation between muscle forces and joint torques can be expressed by using the principle of virtual work as follows:

$$\tau_m + W(\theta)F(a,\ l,\ v)=0\ .\tag{4}$$

where $W(\theta)\in R^{2\times 6}$ is the moment arm.

2.2 Control System

The block diagram of the control system is illustrated in Fig.2. The system is composed of a DRNN controller and the plant. Inputs to the controller involve errors of the position and velocity, the next desired position in the joint space, the last input of the plant, and the integral information of the position error. Therefore, given a reference trajectory of movement in the joint space, the controller can fully take advantage of the useful input information to process through the neural network to generate desired control output commands to activate the muscles. The network structure is optimized by GA and EP; and the weights are trained by DBP.

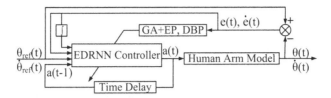

Fig. 2. The block diagram of the human arm control system

The reference trajectory equation is expressed as

$$\begin{cases} \theta_{1ref} = \frac{\pi}{3}\sin t \\ \theta_{2ref} = \frac{\pi}{3} - \frac{\pi}{3}\cos t \\ \dot{\theta}_{1ref} = \frac{\pi}{3}\cos t \\ \dot{\theta}_{2ref} = \frac{\pi}{3}\sin t \end{cases} \quad . \tag{5}$$

3   Evolutionary Diagonal Recurrent Neural Network Controller

3.1   Controller Structure

DRNN is a modified model of the fully connected recurrent neural network with one hidden layer, and the hidden layer is comprised of self-recurrent neuron [4]. In this paper, the input of DRNN controller is composed of the last errors of the position and the velocity ($\theta_{ref}(k-1) - \theta(k-1)$, $\dot{\theta}_{ref}(k-1) - \dot{\theta}(k-1)$, k is the current simulation step number), the current desired position ($\theta_{ref}(k)$), the last activation of muscles ($a(k-1)$), and the integral terms of the errors of the position. We use the integral terms because they can eliminate the steady state position error [8]. This network has fourteen neurons in the input layer and six neurons in the output layer. The number of hidden neurons for the network is determined by the hybrid GA and EP method.

3.2   Optimization of the DRNN Structure

The DRNN structure is determined by hybrid GA and EP method [7] because the approach can provide fast convergence rate. The goal of the controller is to minimize the cost function J (Eq. (6), i.e. the sum-squared of the tracking error (SSE)), with T the number of simulation epoch. At every generation, all networks in the population are trained to track the desire trajectories. The population is divided into three different groups, namely elitist, strong and weak group according to their cost function.

$$J = \frac{1}{2}\sum_{t=1}^{T}\sum_{j=1}^{2}[\theta_{jref}(t) - \theta_j(t)]^2 \quad . \tag{6}$$

Binary string representation is used to represent the structure of the DRNN. Fig.3 shows an example of the chromosome representation in a binary string, which means that there are 28 neurons and a log sigmoid transfer function in the hidden layer. It is assumed that the maximum of neuron in the hidden layer is 32, and a type of transfer function is tan sigmoid, log sigmoid or tangent.

The process of the evolution procedure is as follows:

1) Generate m number of networks in binary representation (m=13). Each network has different hidden neuron numbers (20-32). Weight matrices of every network are initialized to be uniformly distributed random matrices.
2) Train each network using DBP algorithm and calculate the cost function (i.e. each network is used to control the human arm model, and then the control performance is monitored.)
3) Divide the networks into three groups through the rank based selection method. If one network can make the behavior converge satisfactorily to the desired target, the procedure ends. If not, continues.
4) Apply GA operator onto the individuals in the weak group because GA can explore wider area of search in shorter number of evolution. The multipoint crossover is employed, and the simple flip over mutation method is used. Here we let crossover probability P_c=0.9 and mutation probability P_m=0.03. Then use these offspring networks to control the arm model, monitor the SSE, and compare their cost function with that of their parents. If the offspring is better, replace the parent. If not, the parent stays alive.
5) Apply the EP operator onto each individual in the strong group for EP can maintain the behavioral link of the selected offspring. Delete three neurons from each network, utilize the new network for controlling the plant and calculate its cost function. If the offspring is better, the offspring replaces the parent. If not, add three neurons into the hidden layer, use the new network for controlling and calculate again. If the offspring is better, the parent is replaced. If not, the parent stays alive.
6) Combine the networks obtained from step 4 and step 5 and the elitist to get new generation.
7) Repeat step 3 to 6.

3.3 DRNN Weight Training

The DBP method is used for training DRNN weight matrix, which can catch the dynamic information contained in the plant [4]. The error function is defined as

$$E = \tfrac{1}{2}\sum_{j=1}^{2}[\theta_{jref}(t) - \theta_j(t)]^2 \ . \tag{7}$$

The gradient of error in Eq. (7) with respect to an arbitrary weight matrix $W \in R^{n \times n}$ is represented by

$$\frac{\partial E}{\partial W} = -e(k)\frac{\partial \theta(k)}{\partial W} = -e(k)\theta_a(k)\frac{\partial a(k)}{\partial W} \ . \tag{8}$$

where $e(k) = \theta_{ref}(k) - \theta(k)$ is the error between the desired and output responses of the plant, and the factor $\theta_a(k) \equiv \partial \theta(k)/\partial a(k) \approx \frac{\theta(k)-\theta(k-1)}{a(k)-a(k-1)}$ represents the sensitivity of the plant with respect to its input.

The weights can be adjusted following a gradient method, i.e., the update rule of the weights becomes

$$W(k+1) = W(k) + \eta(-\partial E / \partial W) \ . \tag{9}$$

where η is the learning rate.

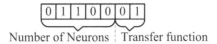

Number of Neurons ¦ Transfer function

Fig. 3. Binary string representation of the DRNN structure

4 Simulation Results

The controller is programmed in MATLAB. The plant parameters are from [14]. The reference trajectory is listed in section 2. SSE is monitored to study the performance of the proposed control scheme.

4.1 Evolution Process

The results of the evolution over thirty generations are shown in Fig.4. The graph shows the best cost function of each population decreasing sharply in the first six generations, and converging after 20 generations. By and large, the number of neuron in the hidden layer is converged to 26 to 30. The transfer function converges into log sigmoid. Judging from the cost function and the runtime shown in Table 1, twenty eight neurons is the optimum number of neuron.

4.2 Control Result

The DRNN is trained for variable epochs; when error function value is less than the set value (the desire error is set as 4), training stops; if error function value still is larger than the set value after network has been trained for 700 epochs, process stops. And learning rates η^o, η^D, η^I are chosen to be 0.03, 0.05, 0.005, respectively. The neural network with the same number of hidden neurons and the same activation functions, but with different initial weight matrices, will give different tracking error. In this paper, randomly generated weight matrices are used as the initial weights. So, the whole program has been run for several times until good results are gained.

We added pulse signals into the inputs of the arm model as disturbance. Each input has the same disturbance signal. The pulse amplitude is set to be 0.3, period is set to be 7 second, pulse width equals to 0.1% of period. The best results are shown in Fig.5 and Fig.6. It is observed that the DRNN controller is capable of controlling the human arm model to follow the desired joint reference with tracking errors in tolerance.

Table 1. Comparison of DRNN with different number of neuron

No. of neuron	26	27	28	29	30
Runtime(Sec)	179.21	181.3	198.62	236.40	300.25
Cost Function / T	0.04	0.034	0.02	0.02	0.02

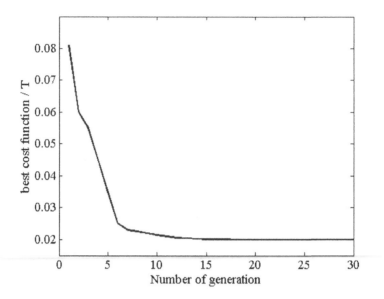

Fig. 4. Result of the evolution process

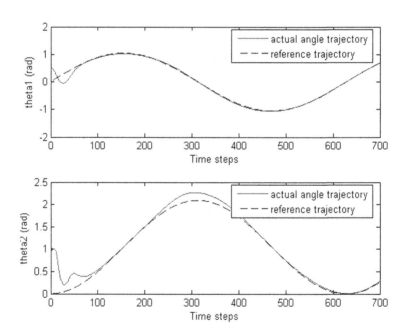

Fig. 5. Trajectory of the joint angles with pulse disturbance signals

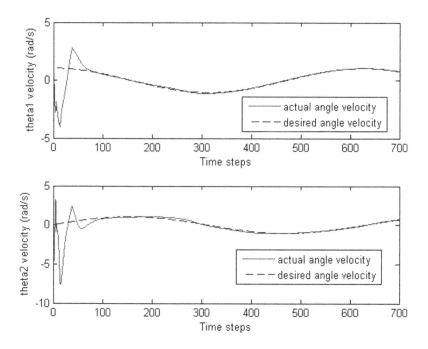

Fig. 6. Graph of the joint angle velocities with pulse disturbance signals

5   Conclusions

In the first stage of concept development and system design, a number of cutting edge techniques have been properly tailed into an interdisciplinary approach for trajectory tracking control of a human arm model in Sagittal plane. For initial feasibility study, simulation results are provided to show that GA and EP can find out suitable structure of the DRNN, and the controlled arm is well behaved to follow up those requested trajectories. It should be noticed that the proposed control procedure for the human arm movement can be further expanded for motor control, rehabilitation robot control, and various related applications.

With its favorable dynamical network structure, the DRNN has played a key role in the tracking control in this study. The authors have tried to use a multilayer perceptron network in the trajectory tracking, but unfortunately the outcomes are unacceptable, which are not included in this paper. Compared with pure GA searching method, the hybrid GA and EP strategy has faster converging rate and more effective.

In the second stage, further investigations will be expanded to the design of controller for generating trajectory tracking in the Cartesian coordinate rather than in joint space. To improve computational efficiency is also targeted for following development. Currently the simulation still takes a considerably long time due to the optimization of DRNN structure and training procedure precede simultaneously, and all controllers are trained in every generation. To reduce the runtime some efficient algorithms should be developed while a fast real-time control is request.

Acknowledgments. This work is supported by grants from the National Nature Science Foundation of China, No 60274020, No 60340420431 and No 60674105, and the Ph.D. Programs Foundation of Ministry of Education of China, No 20050487013. The authors would like to thank Prof. Jiping He for the guidance regarding muscular control and providing the arm model in Simulink.

References

1. Schouten, A.C., de Vlugt, E., van der Helm, F.C.T., Brouwn, G.G.: Optimal Posture Control of a Musculo-Skeletal Arm Model. Biol. Cybern. 84, 143–152 (2001)
2. Brown, I.E., Cheng, E.J., Leob, G.: Measured and Modeled Properties of Mammalian Skeletal Muscle. II. the Effects of Stimulus Frequency on Force-Length and Force-Velocity Relationships. J. of Muscle Research and Cell Motility 20, 627–643 (1999)
3. Chaiyaratana, N., Zalzala, A.M.S.: Hybridisations of Neural Networks and Genetic Algorithms for Time-Optimal Control. In: Angeline, P.J. (ed.) Proceedings of the 1999 Congress on Evolutionary Computation, vol. 1, pp. 389–396. IEEE Press, New Jersey (1999)
4. Ku, C.-C., Lee, K.Y.: Diagonal Recurrent Neural Networks for Dynamic System Control. IEEE Trans. on Neural Networks 6, 144–156 (1995)
5. Franklin, D.W., Milner, T.E.: Adaptive Control of Stiffness to Stabilize Hand Position with Large Loads. Exp. Brain Res. 152, 211–220 (2003)
6. Cheng, E.J., Brown, I.E., Loeb, G.E.: Virtual Muscle: a Computational Approach to Understanding the Effects of Muscle Properties on Motor Control. Journal of Neuroscience Methods 101, 117–130 (2000)
7. Hussenin, S.B., Jamaluddin, H., Mailah, M., Zalzala, A.M.S.: A Hybrid Intelligent Active Force Controller for Robot Arms Using Evolutionary Neural Networks. In: Angeline, P.J. (ed.) Proceedings of the 2000 Congress on Evolutionary Computation, vol. 1, pp. 117–130. IEEE Press, New Jersey (2000)
8. Nakazono, K., Katagiri, M., Kinjo, H., Yamamoto, T.: Force and Position Control of Robot Manipulator Using Neurocontroller with GA based Training. In: Nakauchi, Y. (ed.) Proceedings of 2003 IEEE International Symposium on Computational Intelligence in Robotics and Automation, vol. 3, pp. 1354–1357. IEEE Press, New Jersey (2003)
9. Katayama, M., Inoue, S., Kawato, M.: A Strategy of Motor Learning Using Adjustable Parameters for Arm Movement. In: Chang, H.K., Zhang, Y.T. (eds.) Proceeding of 20th IEEE EMBS. IEEE Engineering in Medicine and Biology Society, vol. 20, pp. 2370–2373. IEEE Press, New Jersey (1998)
10. Lan, N.: Analysis of an Optimal Control Model of Multi-Joint Arm Movements. Biol. Cybern. 207–117 (1997)
11. Sanner, R.M., Kosha, M.: A Mathematical Model of the Adaptive Control of Human Arm Motions. Biol. Cybern. 80, 369–382 (1999)
12. liu, S., Wang, Y., Huang, J.: Neural network based posture control of a human arm model in the sagittal plane. In: Wang, J., Yi, Z., Zurada, J.M., Lu, B.-L., Yin, H. (eds.) ISNN 2006. LNCS, vol. 3973, pp. 792–798. Springer, Heidelberg (2006)
13. Stroeve, S.H.: Impedance Characteristics of a Neuromusculoskeletal Model of the Human Arm. I. Posture Control. Biol. Cybern. 81, 475–494 (1999)
14. Li, W., TOdorov, E.: Iterative Quadratic Regulator Design for Nonliear Biological Movement system. In: Filipe, J. (ed.) Proceedings of the 1st International Conference on Informatica in Control, Automation and Robotics, vol. 1, pp. 222–229. Kluwer Academic Publisher, Netherlands (2004)

15. Li, W., Todorov, E., Pan, X.: Hierarchical Optimal Control of Redundant Biomechanical Systems. In: Hudson, D., Liang, Z.P. (eds.) Proceeding of 26th IEEE EMBS. IEEE Engineering in Medicine and Biology Society, vol. 26, pp. 4618–4621. IEEE Press, New Jersey (2004)
16. Todorov, E., Li, W.: A Generalized Iterative LQG Method for Locally-optiml Feedback Control of Constrained Nonlinear Stochastic Systems. In: Suhada, J., Balakrishnan, S.N. (eds.) Proceeding of 2005 ACC. American Control Conference, vol. 1, pp. 300–306. IEEE Press, New Jersey (2005)
17. Papaxanthis, C., Pozzo, T., McIntyre, J.: Arm End-point Trajectories Under Normal and Micro-gravity Environments. Acta Astronautica 43, 153–161 (1998)
18. Papaxanthis, C., Pozzo, T., Schieppati, M.: Trajectories of arm pointing movements on the sagittal plane vary with both direction and speed. Exp. Brain Res. 148, 498–503 (2003)
19. Kambara, H., Kim, K., Shin, D., et al.: Motor Control-Learning Model for Reaching Movements. In: 2006 International Joint Conference on Neural Networks, pp. 555–562 (2006)

New Ant Colony Optimization for Optimum Multiuser Detection Problem in DS-CDMA Systems

Shaowei Wang and Xiaoyong Ji

Department of Electronics Science and Engineering, Nanjing University, Nanjing, Jiangsu, 210093, P.R. China

Abstract. This paper presents a new ant colony optimization (ACO) method to solve the optimum multiuser detection (OMD) problem in direct-sequence code-division multiple-access (DS-CDMA) systems. The idea is to use artificial ants to keep track of promising areas of the search space by laying trails of pheromone which guides the search of the ACO, as a heuristic for choosing values to be assigned to variables. An effective local search is performed after each generation of the ACO to improve the quality of solutions. Simulation results show the proposed ACO multiuser detection scheme combined with local search can converge very rapidly to the (near) optimum solutions. The bit error rate (BER) performance of the proposed algorithm is close to the OMD bound for large scale DS-CDMA systems and the computational complexity is polynomial in the number of active users.

1 Introduction

The optimum multiuser detection (OMD) scheme for direct-sequence code division multiple access (DS-CDMA) systems searches exhaustively for all possible combinations of the users' entire transmitted bit sequence that maximizes the so-called log-likelihood function derived from the maximum likelihood sequence estimation rule [1]. Suppose that a DS-CDMA system has K active users and each user transmits an M-bit sequence, the number of the possible bit combinations is 2^{MK}, which the OMD must consider when evaluating the log-likelihood function. The computational complexity of the OMD can be reduced to 2^K by exploiting the Viterbi algorithm [1]. But it still increases exponentially with the number of active users.

From a combinatorial optimization viewpoint, the OMD problem is NP-complete [2]. Randomized search heuristics, such as evolutionary algorithms, local search, are useful methods for such problems [3][4]. Earlier works on applying heuristics to the OMD problem can be found in [5][6][7][8][9][10]. They show good performance while the number of active users is small. But the performances in large scale capacity communication systems are not discussed in these literatures. For a combinatorial optimization problem, searching for the global optimum becomes more and more difficult as the problem scale increases.

L. Kang, Y. Liu, and S. Zeng (Eds.): ISICA 2007, LNCS 4683, pp. 326–333, 2007.

In this paper, we propose a method to solve the OMD problem by using an ant colony optimization (ACO) algorithm. An effective local search method is used to improve the quality of solutions. The rest of this paper is organized as follows. In Sect.2, we present the general DS-CDMA communication model and state the OMD problem. The ACO based multiuser detection algorithm is described in Sect.3. In Sect.4, numerical simulation results are presented. Conclusion is found in Sect.5.

2 System Model and Optimum Multiuser Detection Problem

Assume a binary phase shift keying (BPSK) transmission through an additive-white-Gaussian-noise (AWGN) channel shared by K active users with packet size M in a DS-CDMA system. The baseband received signal can be expressed as

$$r(t) = \sum_{i=1}^{K} A_K \sum_{m=1}^{M} b_k(m) s_k(t - mT_b - \tau_k) + n(t) \tag{1}$$

where A_k is the signal amplitude of the kth user, $b_k(m)$ is the transmitted bit of the kth user, $s_k(t)$ is the normalized signature waveform of the kth user, T_b is the bit duration, $\tau_k \in [0, T_b]$ is the transmission delay of the kth user, $n(t)$ is the white Gaussian noise with power spectral density σ^2. Without loss of generality, the transmission delay τ is assumed to satisfy $0 = \tau_1 < \tau_2 < \ldots < \tau_K$ and $\tau_k - \tau(k-1) = T_c$, where T_c is chip duration time. The sufficient statistics for demodulation of the transmitted bits $\mathbf{b}$ are given by the length vector generated by matched filter banks [11]

$$\mathbf{y} = \mathbf{RAb} + \mathbf{n} \tag{2}$$

where $\mathbf{y} = [y_1(1), y_2(1), ..., y_K(1), ..., y_1(M), y_2(M), ..., y_K(M)]^T$, $\mathbf{b} = [b_1(1), b_2(1), ..., b_K(1), ..., b_1(M), b_2(M), ..., b_K(M)]^T$. $\mathbf{A}$ is the $MK \times MK$ diagonal matrix whose $k+iK$ diagonal element is the kth user's signal amplitude A_k and $i = 1, 2, ..., M$. $\mathbf{R}$ is the signature correlation matrix and can be written as

$$\mathbf{R} = \begin{bmatrix} \mathbf{R}(0) & \mathbf{R}^T(1) & \mathbf{0} & \ldots & \mathbf{0} & \mathbf{0} \\ \mathbf{R}(1) & \mathbf{R}(0) & \mathbf{R}^T(1) & \ldots & \mathbf{0} & \mathbf{0} \\ \mathbf{0} & \mathbf{R}(1) & \mathbf{R}(0) & \ldots & \mathbf{0} & \mathbf{0} \\ \vdots & \vdots & \vdots & \vdots & \vdots & \vdots \\ \mathbf{0} & \mathbf{0} & \mathbf{0} & \ldots & \mathbf{R}(1) & \mathbf{R}(0) \end{bmatrix} \tag{3}$$

where $\mathbf{R}(0)$ and $\mathbf{R}(1)$ are $K \times K$ matrices defined by

$$R_{jk}(0) = \begin{cases} 1, & \text{if } j = k; \\ \rho_{jk}, & \text{if } j < k; \\ \rho_{kj}, & \text{if } j > k; \end{cases} \tag{4}$$

$$R_{jk}(1) = \begin{cases} 0, & \text{if } j \geq k; \\ \rho_{jk}, & \text{if } j < k; \end{cases} \qquad (5)$$

ρ_{jk} denotes the partial cross-correlation coefficient between the jth user and the kth user. $\mathbf{n}$ is a real-valued zero-mean Gaussian random vector with a covariance $\sigma^2 \mathbf{H}$, $\mathbf{H} = \mathbf{ARA}$.

The optimum multiuser detection problem is to generate an estimation sequence $\hat{\mathbf{b}} = [\hat{b}_1(1), \hat{b}_2(1), \ldots, \hat{b}_K(1), \ldots, \hat{b}_1(M), \hat{b}_2(M), \ldots, \hat{b}_K(M)]^T$ to maximize the objective function

$$f(\mathbf{b}) = 2\mathbf{y}^T \mathbf{Ab} - \mathbf{b}^T \mathbf{Hb} \qquad (6)$$

It means to search 2^{MK} possible bit sequences exhaustively and is proven to be NP-complete[2].

A synchronous DS-CDMA system can be seen as a special case of an asynchronous one (for the case $M = 1$). On the other hand, an asynchronous DS-CDMA system can be interpreted as an equivalent synchronous system too [12]. In the following we only consider the synchronous case to simplify analysis without loss of generality.

3   Ant Colony Optimization for the OMD Problem

Ant colony optimization (ACO) is a well known population-based metaheuristic for combinatorial optimization problems [13] [14]. The inspiring idea behind ACO is the foraging behavior of real ants and the way they communicate. Consider an ant colony and a source of food. The ants manage to find the shortest path between the food and the colony. The optimization is the result of the collective work of all the ants in the colony. All ants move randomly in the search for food at the beginning and deposit a pheromone trail while they are walking. The higher the pheromone value on a path, the higher the probability the ant chooses that path. Due to evaporation, the old trails which are long and were not reinforced by new ants will eventually vanish. The characteristics of ACO include:

1. A method to construct solutions which balances pheromone trails (characteristics of past solutions) with a problem-specific heuristic;
2. A method to both reinforce and evaporate pheromone;
3. Local (neighborhood) search to improve solutions.

For the OMD problem, the estimation sequence is already in binary form (+1 or -1) and requires no additional effort for solutions encoding. So our efforts are concentrated on the desirability function, pheromone updating strategies and local search algorithm.

3.1   The Desirability Function

Each element of a solution vector in the OMD problem takes one out of N possible values, where N is the constellation size. For BPSK modulation synchronous DS-CDMA systems, $N = 2$ (The possible value is +1 or -1). Assume

$\mathbf{b} \in \{+1, -1\}^K$ is an initial solution vector, we can construct a $2 \times K$ table whose first row is the elements of $\mathbf{b}$, and the second row is the complement of the first. The desirability function for an ant starting at the ith element of $\mathbf{b}$ is defined as following

$$f_d(j) = \frac{1}{|y(i)| + C} \tag{7}$$

where $y(i)$ is the ith user's matched filter output, C is the absolute value of the average matched filter output, $C = \frac{1}{K}|\sum_{i=1}^{K} y(i)|$. As the ant moves along the elements of the solution vector, the desirability function at the $(i + j)$th stage can be redefined as follows

$$f_d(i + j) = \frac{1}{|y(i + j) + \sum_{l \in S} y(l)| + C} \tag{8}$$

where S is a set of positions where the ant has previously selected the complement elements of $\mathbf{b}$.

3.2 Pheromone Trail Intensity Update

To design ACO based multiuser detection algorithm, a meaningful pheromone deposition mechanism. First the pheromone are deposited in a $2 \times K$ table where the first row corresponds to the elements of $\mathbf{b}$, and the second row corresponds to the complement elements of $\mathbf{b}$. At the beginning of the search process, the pheromone table has equal amounts (unity) of pheromone in all of its entries. During the search process, pheromones are deposited and evaporated based on the path traversed by the ants.

The deposition rate p_d and evaporation rate p_e are parameters of the ACO. Here we adopt a similar strategy which is proposed in [8] to set the p_d and p_e. In our algorithm, p_d and p_e are inversely related to the number of iterations G. At any stage during the search, the higher the pheromone value in an entry (in the pheromone table), the greater is the probability of selecting the corresponding element value from $\mathbf{b}$ or its complement elements. An elitism philosophy is used in our pheromone deposition mechanism, only the ants that find good paths are allowed to deposit pheromones and the best solution produce by ACO is performed local search after each iteration.

3.3 Local Search Algorithm

An effective local search algorithm is used to improve the quality of each solution after a colony is generated. Denote the best solution of the tth generation by $\mathbf{b}_0^t = [b_1^t, b_2^t, \ldots, b_n^t]$, where $n = K$, T is the number of generation of the local search. $\mathbf{b}_j^t$ is a solution with only the jth bit different from $\mathbf{b}_0^t$, then $\mathbf{b}_j^t = [b_1^t, \ldots, -b_j^t, \ldots, b_n^t]$. The associated gain g_j^t from $\mathbf{b}_j^{t-1}$ to $\mathbf{b}_j^t$ is

$$g_j^t = f(\mathbf{b}_j^t) - f(\mathbf{b}_j^t), \qquad 1 \le j \le n \tag{9}$$

By flipping b_1^t of the $\mathbf{b}_0^t$, the greedy local search begins and the first current solution $\mathbf{b}_1^t$ is created, $\mathbf{b}_1^t = [-b_1^t, b_2^t, \ldots, b_n^t]$. During the algorithm runs, the current solution $\mathbf{b}_j^t$ is updated as

$$\mathbf{b}_j^t = \begin{cases} \mathbf{b}_j^t, & g_j^t > 0 \\ \mathbf{b}_{j-1}^t, & otherwise \end{cases}, \quad 1 \le j \le n \tag{10}$$

When the nth associated gain is calculated, the local optimal solution $\mathbf{b}_n^t$ is produced and taken as the current solution of the next generation, $\mathbf{b}_0^{t+1} = \mathbf{b}_n^t$. The local search terminates until there is no associated gain after flips in a generation. The pseudo code of the local search algorithm is illustrated in Table 1.

Table 1. Pseudo code of the local search algorithm in ACO

Initialization: $\mathbf{b} = sign(y) \in \{-1, +1\}^n$
Loop:
    $\mathbf{b}_0^t := \mathbf{b} = [b_1^t, b_2^t, \ldots, b_n^t]^T$;
    for $i = 1, 2, \ldots, n$
        Let $\mathbf{b}_i^t = [b_1^t, \ldots, -b_i^t, \ldots, b_n^t]^T$;
        Calculate gain:
            $g_i^t := f(\mathbf{b}_i^t) - f(\mathbf{b}_{i-1}^t)$;
        if $g_i^t > 0$
            $\mathbf{b}_i^t := \mathbf{b}_i^t$;
        else
            $\mathbf{b}_i^t := \mathbf{b}_{i-1}^t$;
        endif
    endfor
    $\mathbf{b} := \mathbf{b}_n^t$;
Until no associated gain after n flips;
Return: $\mathbf{b}$.

4   Simulation Results

Consider a synchronous DS-CDMA system with perfect power control and random binary sequences with length $L = 127$ are employed as spreading sequences. The outputs of the conventional detector are used to initialize the start solutions of all ants. The number of ants of ACO is set to $\lfloor \frac{K}{4} \rfloor$, where $\lfloor \frac{K}{4} \rfloor$ represents the maximum integer not more than $\frac{K}{4}$. $p_d = \frac{5}{G}$, $p_e = \frac{8}{G}$, where G is the number of iteration. A 20-user, 30-user and 40-user system which uses a BPSK transmission through an AWGN channel is considered respectively.

First we compare the convergence rate of the proposed new ACO based multiuser detector (NACO) with the other ACO (OACO) detector presented in [8]. Fig.1 shows the convergence curves of NACO and OACO as a function of number of iterations. Here the signal to noise rate (SNR) is set to 6. The number of users is 20 and 30. From Fig.1, it can be seen that in the two instances the NACO converges rapidly to (near) the OMD bound with less iterations than

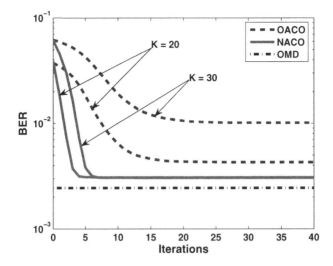

Fig. 1. The convergence curves of the NACO and the OACO in the perfect-power-control DS-CDMA system

that of the OACO. Local search in the new ACO detector plays an important role in improving the convergence rate.

The bit error rate (BER) performance of the conventional detector (CD), evolutionary programming detector EPD [6], $1 - opt$ [7], OACO [8] and the proposed NACO is illustrated in Fig.2 by the curves of BER versus SNR. The number of users is 30 and 40 ($K = 30, 40$) in Fig.2 (a) and (b) respectively. From Fig.2 we can see that in both cases the proposed NACO outperforms the CD,$1 - opt$, EPD, and OACO obviously. The local search in the ACO is the key to the improvement of the solution quality.

To simplify the analysis of the computational complexity of the NACO, we only consider the number of fitness function evaluations. Assume that the NACO has N ants and converges at the Gth generation. In each generation, there are N fitness function evaluations for selecting maximum fitness and K evaluations for local search. The total number of fitness function evaluations is K+N. Thus the number of fitness function evaluations is . Generally, the number of generations G is much smaller than K. At the same time N is set to $\frac{K}{4}$ in all cases discussed above. Then the computational complexity of the NACO is about $O(K^3)$, which is polynomial in the number of active users.

While the simulation results show a reduction in the number of iterations required for convergence when using the NACO algorithm, the amount of computation time increases for a single iteration increases because of the local search embedded in the algorithm. Following is the computation time estimated by curve fitting techniques (MATLAB programming environment, 2.66 GHz Pentium CPU and 512MB of RAM). The number of users is 40. We can see the computation time (in CPU time) of the NACO is less than the OACO.

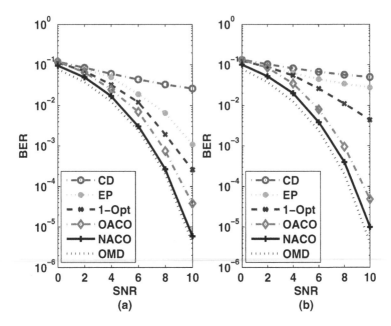

Fig. 2. BER as a function of SNR for the 30-user and 40-user DS-CDMA system with equal power. (a)$K = 30$ and (b) $K = 40$.

$$C_{NACO} = 4.76 \times 10^{-4}K^4 \tag{11}$$

$$C_{OACO} = 6.55 \times 10^{-3}K^4 \tag{12}$$

$$C_{OMD} = 3.20 \times 10^{-3}2^K \tag{13}$$

5   Conclusions

A new ant colony optimization (NACO) based multiuser detector for DS-CDMA systems has been proposed in this paper. The ACO is followed by a local search algorithm, which improves the quality of solution and the convergence rate dramatically. The proposed NACO detector can approach the BER bound of the optimum multiuser detection in the condition of small number of users. When the number of users is large, the proposed algorithm can also provide rather good performance while other heuristics based multiuser detectors perform poorly.

References

1. Verdu, S.: Minimum Probability of Error for Asynchronous Gaussian Multiple-access Channels. IEEE Transactions on Information Theory 32(1), 85–96 (1986)
2. Verdu, S.: Computational Complexity of Optimal Multiuser Detection. Algorithmica 4, 303–312 (1989)

3. Hoos, H.H., Stützle, T.: Stochastic Local Search: Foundations and Applications. Morgan Kaufmann Publishers/Elsevier, San Fransisco, USA (2004)
4. Fogel, D.B.: Evolutionary Computation, Toward a New Philosophy of Machine Intelligence, 2nd edn. IEEE Press, Piscataway, NJ (2000)
5. Wang, S., Zhu, Q., Kang, L. $(1+\lambda)$ Evolution Strategy Method for Asynchronous DS-CDMA Multiuser Detection. IEEE Communications Letters 10(6), 423–425 (2006)
6. Lim, H.S., Rao, M.V.C.: Multiuser Detection for DS-CDMA Systems Using Evolutionary Programming. IEEE Communications Letters 7(3), 101–103 (2003)
7. Hu, J., Blum, R.S., Gradient, A.: Guided Search Algorithm for Multiuser Detection. IEEE Communications Letters 4(11), 340–342 (2000)
8. Hijazi, S.L., Natarajan, B.: Novel Low-Complexity DS-CDMA Mutiuser Detector Based on Ant Colony Optimization. In: IEEE Vehicular Technology Conference. VTC2004-Fall, vol. 3, pp. 1939–1943 (September 2004)
9. Wang, S., Zhu, Q., Kang, L.: Landscape Properties and Hybrid Evolutionary Algorithm for Optimum Multiuser Detection Problem. In: Alexandrov, V.N., van Albada, G.D., Sloot, P.M.A., Dongarra, J.J. (eds.) ICCS 2006. LNCS, vol. 3991, pp. 340–347. Springer, Heidelberg (2006)
10. Wang, S., Zhu, Q., Kang, L.: Local Optima Properties and Iterated Local Search Algorithm for Optimum Multiuser Detection Problem. In: Huang, D.-S., Li, K., Irwin, G.W. (eds.) ICIC 2006. LNCS, vol. 4113, pp. 913–918. Springer, Heidelberg (2006)
11. Verdu, S.: Multiuser Detection. Cambridge University Press, Cambridge, UK (1998)
12. Proakis, J.G.: Digital Communications, 4th edn. McGraw-Hill, New York (2001)
13. Dorigo, M., Gambardella, L.: Ant colony system: a cooperative learning approach to the traveling salesman problem. IEEE Transactions on Evolutionary Computation 1(2), 53–66 (1997)
14. Dorigo, M.: Optimization, Learning and Natural Algorithms. Ph.D. Thesis, Politecnico di Milano, Italy (1992)

A Fast Particle Swarm Optimization Algorithm with Cauchy Mutation and Natural Selection Strategy

Changhe Li[1], Yong Liu[2], Aimin Zhou[3], Lishan Kang[1], and Hui Wang[1]

[1] China University of Geosciences, School of Computer, Wuhan, P.R. China, 430074
[2] The University of Aizu, Aizu-Wakamatsu, Fukushima, Japan, 965-8580
[3] Department of Computer Science, University of Essex Wivenhoe Park,
Colchester, CO4 3SQ
lch_wfx@yahoo.com.cn, yliu@u-aizu.ac.jp, azhou@essex.ac.uk,
kang_whu@yahoo.com, wanghui_cug@yahoo.com.cn

Abstract. The standard Particle Swarm Optimization (PSO) algorithm is a novel evolutionary algorithm in which each particle studies its own previous best solution and the group's previous best to optimize problems. One problem exists in PSO is its tendency of trapping into local optima. In this paper, a fast particle swarm optimization (FPSO) algorithm is proposed by combining PSO and the Cauchy mutation and an evolutionary selection strategy. The idea is to introduce the Cauchy mutation into PSO in the hope of preventing PSO from trapping into a local optimum through long jumps made by the Cauchy mutation. FPSO has been compared with another improved PSO called AMPSO [12] on a set of benchmark functions. The results show that FPSO is much faster than AMPSO on all the test functions.

Keywords: Particle swarm optimization, Cauchy mutation, swarm intelligence.

1 Introduction

Particle Swarm Optimization (PSO) was first introduced by Kennedy and Eberhart in 1995 [1,2]. PSO is motivated from the social behavior of organisms, such as bird flocking and fish schooling. Particles "fly" through the search space by following the previous best positions of their neighbors and their own previous best positions. Each particle is represented by a position and a velocity which are updated as follows:

$$X_{id}^{'} = X_{id} + V_{id}^{'} \tag{1}$$

$$V_{id}^{'} = \omega V_{id} + \eta_1 rand\ ()(P_{id} - X_{id}) + \eta_2 rand\ ()(P_{gd} - X_{id}) \tag{2}$$

where $X_{id}{'}$ and X_{id} represent the current and the previous positions of idth particle, V_{id} and $V_{id}{'}$ are the previous and the current velocity of idth particle, P_{id} and P_{gd} are the individual's best position and the best position found in the whole swarm so far respectively. $0 \le \omega < 1$ is an inertia weight which determines how much the previous velocity is preserved, η_1 and η_2 are acceleration constants, $rand()$ generates random number from interval [0,1].

L. Kang, Y. Liu, and S. Zeng (Eds.): ISICA 2007, LNCS 4683, pp. 334–343, 2007.
© Springer-Verlag Berlin Heidelberg 2007

In PSO, each particle shares the information with its neighbors. The updating equations (1) and (2) show that PSO combines the cognition component of each particle with the social component of all the particles in a group. The social component suggests that individuals ignore their own experience and adjust their behavior according to the previous best particle in the neighborhood of the group. On the other hand, the cognition component treats individuals as isolated beings and adjusts their behavior only according to their own experience.

Although the speed of convergence is very fast, many experiments have shown that once PSO traps into local optimum, it is difficult for PSO to jump out of the local optimum. Ratnaweera et.al.[3] state that lack of population diversity in PSO algorithms is understood to be a factor in their convergence on local optima. Therefore, the addition of a mutation operator to PSO should enhance its global search capacity and thus improve its performance. A first attempt to model particle swarms using the quantum model(QPSO) was carried out by Sun et.al. [4]. In a quantum model, particles are described by a wave function instead of the standard position and velocity. The quantum Delta potential well model and quantum harmonic oscillators are commonly used in particle physics to describe the stochastic nature of particles. In their studies[5], the variable of *gbest* (the global best particle) and *mbest* (the mean value of all particles' previous best position) is mutated with Cauchy distribution respectively, and the results show that QPSO with *gbest* and *mbest* mutation both performs better than PSO. The work of R. A. Krohling et.al.[6][7] showed that how Gaussian and Cauchy probability distribution can improve the performance of the standard PSO. Recently, evolutionary programming with exponential mutation has also been proposed [8].

In order to prevent PSO from falling in a local optimum, a fast PSO (FPSO) is proposed by introducing a Cauchy mutation operator in this paper. Because the expectation of Cauchy distribution does not exist, the variance of Cauchy distribution is infinite. Some researches [9][10] have indicated that the Cauchy mutation operator is good at the global search for its long jump ability. This paper shows that the Cauchy mutation is helpful in PSO as well. Besides the Cauchy mutation, FPSO chooses the natural selection strategy of evolutionary algorithms as the basic elimination strategy of particles. FPSO combines PSO with Cauchy mutation and evolutionary selection strategy. It has the fast convergence speed characteristic of PSO, and greatly overcomes the tendency of trapping into local optima of PSO.

The rest of this paper is organized as follows. Section 2 gives the analysis of PSO. The detail of FPSO is introduced in Section 3. Section 4 describes the experiment setup and presents the experiment results. Finally, Section 5 concludes the paper with a brief summary.

2 Fast Particle Swarm Optimization Algorithm with Cauchy Mutation and Natural Selection Strategy

2.1 Cauchy Mutation

From the mathematic theoretical analysis of the trajectory of a PSO particle [11], the trajectory of a particle X_{id} converges to a weighted mean of P_{id} and P_{gd}. Whenever the particle converges, it will "fly" to the personal best position and the global best particle's position. According to the update equation, the personal best position of the particle will gradually move closer to the global best position. Therefore, all the particles will converge onto the global best particle's position. This information sharing mechanism makes PSO have a very fast speed of convergence. Meanwhile, because of this mechanism, PSO can't guarantee to find the global minimal value of a function. In fact, the particles usually converge to local optima. Without loss of generality, only function minimization is discussed here. Once the particles trap into a local optimum, in which P_{id} can be assumed to be the same as P_{gd}, all the particles converge on P_{gd}. At this condition, the velocity update equation becomes:

$$V_{id}' = \omega V_{id} \tag{3}$$

When the iteration in the equation (3) goes to infinite, the velocity of the particle V_{id} will be close to 0 because of $0 \leq \omega < 1$. After that, the position of the particle X_{id} will not change, so that PSO has no capability of jumping out of the local optimum. It is the reason that PSO often fails on finding the global minimal value.

To overcome the weakness of PSO discussed at the beginning of this section, the Cauchy mutation is incorporated into PSO algorithm. The basic idea is that, the velocity and position of a particle are updated not only according to (1) and (2), but also according to Cauchy mutation as follows:

$$V_{id}' = V_{id} \exp(\delta) \tag{4}$$

$$X_{id}' = X_{id} + V_{id}' \delta_{id} \tag{5}$$

where δ and δ_{id} denote Cauchy random numbers

Since the expectation of Cauchy distribution doesn't exist, the variance of Cauchy distribution is infinite so that Cauchy mutation could make a particle have a long jump. By adding the update equations of (4) and (5), FPSO greatly increases the probability of escaping from the local optimum. In standard PSO, the position of a particle is updated according to equations (1) and (2). That is, for each particle, there is nowhere to move but following the direction of the best particle, and the flying direction is nearly determinate through the generation. From the above analysis of PSO, the particles incline to converge on a local optimum.

2.2 Natural Selection Strategy

In the standard PSO, all particles are directly updated by their offspring no matter whether they are improved. If a particle moves to a better position, it can be replaced by the updated. However if it moves to a worse position, it is still replaced by its offspring. In fact, the most particles fly to worse positions for most cases, therefore the whole swarm will converge on local optima. Like evolutionary algorithms, FPSO introduces an evolutionary selection strategy in which each particle survives according to a natural selection rule. Therefore, the particle's position at the next step is not only due to the position update but also the evolutionary selection. Such strategy could greatly reduce the probability of trapping into local optimum.

The evolutionary selection strategy is carried out as follows. Assume the size of the swarm is m, pair-wise comparison over the union of parents and offspring $(1,2,...2m)$ is made. For each particle, q opponents are randomly chosen from all parents and offspring with equal probability. If the fitness of particle i is less then its opponent , it will receive a "win". Then select m particles that have the more winnings to be the next generation.

The detail of the selection framework are as follows:

Step1: For each particle of parent and offspring, assign $win[i]=0$.

Step2: Randomly select q particles (opponents) for each particle in parent and offspring.

Step3: For each particle, compare it with its q opponents. For particle i, if the fitness of its opponent j is larger than particle i , then $win[i]++$.

Step4: Select m particles that have the more winnings to be the next generation.

2.3 Algorithm Framework

The major steps of FPSO are as follows:

Step1: Generate the initial particles by randomly generating the position and velocity for each particle.

Step2: Evaluate each particle's fitness.

Step3: For each particle, if its fitness is smaller than its previous best(P_{id}) fitness, update P_{id}.

Step4: For each particle, if its fitness is smaller than the best one (P_{gd}) of all the particles, update P_{gd}.

Step5: For each particle ,do

  1) Generate a new particle t according to the formula (1) and (2).

  2) Generate a new particle t' according to the formula (4) and (5).

  3) Compare t with t' ,chose the one with smaller fitness to be the offspring.

Step6: Generate the next generation according to the above evolutionary selection strategy.

Step7: if the stop criterion is satisfied, then stop, else goto Step 3.

3 Experiments and Results

Twelve benchmark functions $(f_1$-$f_{12})$ are used in this paper. Function f_1-f_9 are chosen from [12], and function f_{10}-f_{12} from[9]. Functions f_1-f_4 are unimodal functions while functions f_5-f_{12} have many local optima. Generally speaking, multimodal functions are often regarded as the most difficult in function optimization. Table 1 gives the details of these functions.

Algorithm parameters are as follows: acceleration constants of η_1 and η_2 are both set to be 1.496180, and inertia weight $\omega=0.729844$ as suggested by den Bergh [13]. In FPSO, a particle will be evaluated two times each generation, in order to be the same number of function evaluations to AMPSO [12] and PSO at each generation, the swarm size of FPSO is half of the size of AMPSO and PSO for all experiments. And the other parameters are given in the following experiments.

Table 1. Details of test functions, where n is the dimension of the function, fmin is the minimum value of the function, and S $\subseteq$ Rn

Test function	n	S	f_min
$f_1(x) = \sum_{i=1}^{n} x_i^2$	3	(-5.12,5.12)	0
$f_2(x) = 100(x_1^2 - x_2)^2 + (x_1 - 1)^2$	2	(-2.048,2.048)	0
$f_3(x) = 6 \cdot \sum_{i=1}^{5} \lfloor x_i \rfloor$	5	(-5.12, 5.12)	0
$f_4(x) = \sum_{i=1}^{n} i \cdot x_i^4 + U(0,1)$	30	(-1.28,1.28)	0
$f_5(x) = \dfrac{(\sin^2 \sqrt{x^2 + y^2}) - 0.5}{(1.0 + 0.001(x^2 + y^2))^2} + 0.5$	2	(-100.0,100.0)	0
$f_6(x) = \dfrac{1}{4000} \sum_{i=1}^{n} (x_i - 100)^2 - \prod_{i=1}^{n} \cos(\dfrac{x_i - 100}{\sqrt{i}}) + 1$	30	(-300.0,300.0)	0
$f_7(x) = -20 \cdot \exp(-0.2\sqrt{\dfrac{1}{n} \cdot \sum_{i=1}^{n} x_i^2}) - \exp(\dfrac{1}{n} \cdot \sum_{i=1}^{n} \cos(2\pi \cdot x_i)) + 20 + e$	30	(-30.0,30.0)	0
$f_8(x) = \sum_{i=1}^{n} 100((x_{i+1} - x_i^2)^2 + (x_i - 1)^2)$	30	(-2.048,2.048)	0
$f_9(x) = \sum_{i=1}^{n} (x_i^2 - 10\cos(2\pi x_i) + 10)$	30	(-5.12,5.12)	0
$f_{10}(x) = \sum_{i=1}^{n} - x_i \sin(\sqrt{|x_i|})$	30	(-500,500)	-12569.5
$f_{11}(x) = 4x_1^2 - 2.1x_1^4 + \frac{1}{3}x_1^6 + x_1 x_2 - 4x_2^2 + 4x_2^4$	2	(-5,5)	-1.0316285
$f_{12}(x) = [1 + (x_1 + x_2 + 1)^2 (19 - 14 x_1 + 3x_1^2 - 14 x_2 + 6x_1 x_2 + 3x_2^2)] \times [30 + (2x_1 - 3x_2)^2 (18 - 32 x_1 + 12 x_1^2 + 48 x_2 - 36 x_1 x_2 + 27 x_2^2)]$	2	(-2.0,2.0)	3

Two groups of experiments are conducted in this section.

Firstly, the proposed algorithm FPSO is compared with another improved PSO called AMPSO [12] on nine problems. In the experiments, the number of particles is 20 (40 in AMPSO) and other parameters are the same as in [12]. For AMPSO, it stops when no improvement can be made. And then, the mean fitness obtained by

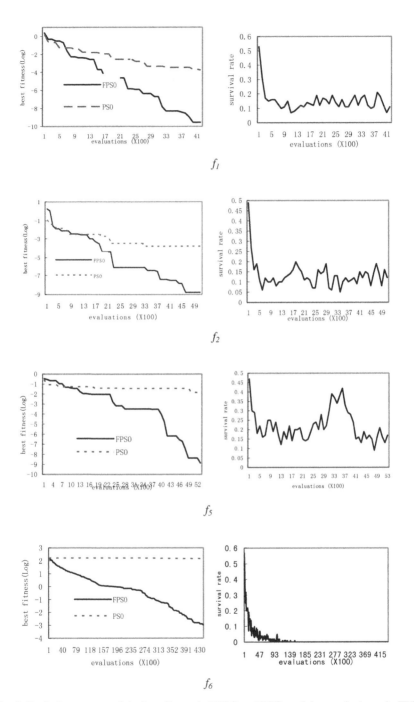

Fig. 1. Evolution process of the best fitness in FPSO and PSO, and the survival rate in FPSO

AMPSO is calculated as the target value of FPSO, and FPSO will stop when it surpasses this target value. The mean and standard deviation of fitness values achieved and generations spent are shown in Table 2 based on 30 independent runs. The results show that FPSO could find the target values achieved by AMPSO in shorter function evaluations for all test functions except for f_3. The function f_3 is an easy function with integer object function values. Both FPSO and AMPSO could find the global minimum in one or a few generations. It indicates that FPSO has the global search capability while AMPSO could only found the global minimum for functions f_1, f_2, and f_3, but failed on reaching the global minimum for the rest of functions including all multimodal functions. It can be known that FPSO not only has faster convergence, but also has better global search.

Secondly, the dynamics of FPSO and PSO are discussed, and the contribution of Cauchy mutation in FPSO is studied as well. For all experiments in the second group, the number of runs is *50* times, the particles size *m* is *50* for FPSO (100 for PSO), and the tournament size ($q=10$) is chosen for FPSO. The left side in Fig 1 and Fig 2 shows how the best fitness of the particle evolves in the search process. It suggests that PSO could only improve the best fitness for simple unimodal functions but hardly decrease the best fitness for the multimodal functions. The results of $f_3, f_4, f_7, f_8, f_{11}$ and f_{12} are not provided here due to space limitation. Table 3 shows the mean best values and the standard deviation got by FPSO and PSO for 50 runs.

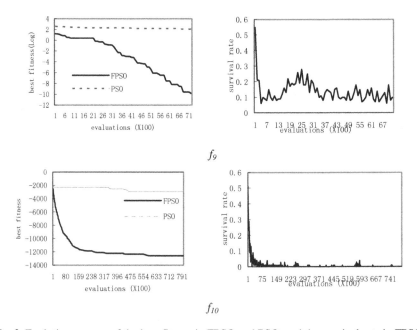

Fig. 2. Evolution process of the best fitness in FPSO and PSO, and the survival rate in FPSO

Table 2. Comparison between FPSO and AMPSO, all results have been averaged over 30 runs, where f is the test function

f	Target Value		Mean Evaluations	
	FPSO	**AMPSO**	**FPSO**	**AMPSO**
f_1	0.00017748±1.6323e-5	0.000300±0	656	1088.8
f_2	2.83656e-5±2.5674e-6	0.000049±0.000107	858.4	64471.2
f_3	0.0±0.0	0.0±0.0	124	40
f_4	16.9737±0.57268	41.469003±1.922524	1801.2	32499.6
f_5	0.01501±0.00079774	0.024139±0.004212	446.4	312049
f_6	63.3212±0.634179	106.118084±4.398166	660	79524.4
f_7	19.4785±0.0242021	19.660458±0.056733	1444	33084.8
f_8	2058.51±24.6575	2198.368745±86.302836	2804	22173.2
f_9	214.475±1.5299	224.765413±35.262816	2632	18287.2

The right side in Fig 1 and Fig 2 shows the survival rate of Cauchy mutation in FPSO through the evolution process. The survival rate is the ratio of the number of successful Cauchy mutation to the total number of Cauchy mutations. The higher survival rate, the more particles generated from Cauchy mutation survive. The survival rate remains at a certain level through the whole search process for some test functions, while it decreased for other test functions after the objective values of these functions close to the global minimum had been found. It can be explained that when the particles are far away from the global minimum, long jump made by Cauchy mutation was helpful to find the particles with smaller objective values. Once the particles were getting closer to the global optimum, FPSO would have to depend on the small search steps decided by the equations (1) and (2) rather than the long jump made by Cauchy mutation in order to fine tune the solutions. By viewing the results

Table 3. Comparison between FPSO and PSO on f1-f12, where "Mean Best" is mean best function values found in the last run, and "Std Dev" indicates the standard deviation

f	Number of evaluations	**FPSO**		**PSO**	
		Mean Best	Std Dev	Mean Best	Std Dev
f_1	1.5e4	2.299e-17	9.784e-18	4224.77	201.038
f_2	2.5e3	0	0	1.145e-13	6.509e-14
f_3	300	0	0	0	0
f_4	1.5e4	3.668e-3	2.072e-4	4.63e-3	2.100e-4
f_5	1.5e4	0	0	0.005	6.848e-4
f_6	1.5e4	0.082	0.031	101.41	1.523
f_7	1.5e4	1.356e-09	4.141e-10	1.306	0.148
f_8	1.5e4	7.53016e-14	4.53682e-14	4.59643e-26	1.83991e-26
f_9	1.5e4	0	0	0	0
f_{10}	1.5e4	-12563.6	3.437	-4005.02	54.012
f_{11}	4000	-1.03163	0	-1.03162	4.05433e-6
f_{12}	4000	3	5.01747e-10	3.00004	6.87691e-6

of Table 3, we can easily see that the mean best values got by FPSO are obvious better than PSO for function $f_1, f_2, f_5, f_6, f_7, f_{10}, f_{11}, f_{12}$, and slightly better for function f_4. For function f_3 and f_9, the performance of FPSO are nearly the same to PSO. Only for function f_8, the convergence speed of FPSO is slower than PSO, however if given more number of generation, FPSO will converge on the global optima. For most test cases, FPSO is more efficient than the standard PSO.

4 Conclusions

By analyzing the advantage and disadvantage of the standard PSO, FPSO based on Cauchy mutation and evolutionary selection strategy is proposed in this paper. Although PSO has a fast convergence rate, it is likely to trap into the local optimum, and can't guarantee converge to the global optimum. FPSO introduces Cauchy mutations into the position and velocity update equations in order to increase the probability of jumping out of the local optimum. FPSO was tested on *12* benchmark functions. From the experimental results of these functions, it can be seen that the FPSO performed much better than AMPSO on the selected problems. Although FPSO needs more time to perform Cauchy mutation, it decreases the fitness evaluations remarkably comparing to AMPSO.

Only functions with the dimension less than 30 were tested in this paper. Further research will focus on testing the performance of FPSO on higher dimensional problems in order to find whether FPSO would scale up well for the large function optimization problems.

References

1. Kennedy, J., Eberhart, R.C.: Particle Swarm Optimization. In: IEEE International Conference on Neural Networks, pp. 1942–1948. IEEE Computer Society Press, Los Alamitos (1995)
2. Eberhart, R.C., Kennedy, J., New, A.: Optimizer Using Particle Swarm Theory. In: Proceedings of the 6th International Symposium on Micro Machine and Human Science, pp. 39–43 (1995)
3. Ratnaweera, A., Halgamuge, S.K., Watson, H.C.: Self-organizing hierarchical particle swarm optimizer with time-varying acceleration coefficients. IEEE Transactions on Evolutionary Computation 8(3), 240–255 (2004)
4. Sun, J., Feng, B., Xu, W.: Particle swarm optimization with particles having quantum behavior. In: Proceedings of the IEEE Congress on Evolutionary Computation, Portland, Oregon USA, pp. 325–331. IEEE Computer Society Press, Los Alamitos (2004)
5. Liu, J., Xu, W., Sun, J.: Quantum-behaved particle swarm optimization with mutation operator. In: Proceedings of the 17th IEEE International Conference on Tools with Artificial Intelligence Pages, pp. 237–240. IEEE Computer Society Press, Los Alamitos (2005)
6. Krohling, R.A.: Gaussian particle swarm with jumps. In: Proceedings of the IEEE Congress on Evolutionary Computation, Edinburgh, UK, pp. 1226–1231. IEEE Computer Society Press, Los Alamitos (2005)

7. Krohling, R.A., dos Santos Coelho, L.: PSO-E: Particle Swarm with Exponential Distribution. In: Proceedings of the IEEE Congress on Evolutionary Computation, July 2006, pp. 1428–1433. IEEE Computer Society Press, Los Alamitos (2006)
8. Narihisa, H., Taniguchi, T., Ohta, M., Katayama, K.: Evolutionary Programming with Exponential Mutation. In: Proceedings of the IASTED Artificial Intelligence and soft Computing, Benidorm, Spain, pp. 55–50 (2005)
9. Yao, X., Liu, Y.: Fast evolutionary programming. In: Proc. of the Fifth Annual Conference on Evolutionary Programming (EP'96), San Diego, CA, USA, February 29-March 3, 1996, pp. 451–460. MIT Press, Cambridge (1996)
10. Yao, X., Liu, Y., Lin, G.: Evolutionary programming made faster. IEEE Trans. Evolutionary Computation, 82–102 (1999)
11. Clerc, M., Kennedy, J.: The Particle Swarm: Explosion, Stability and Convergence in a Multi-Dimensional Complex Space. IEEE Trans. on Evolutionary Computation 6, 58–73 (2002)
12. Pampara, G., Franken, N., Engelbrecht, A.P.: Combining Particle Swarm Optimisation with angle modulation to solve binary problems. In: Proceedings of the IEEE Congress on Evolutionary Computation, September 2005, pp. 89–96. IEEE Computer Society Press, Los Alamitos (2005)
13. van den Bergh, F.: An Analysis of Particle Swarm Optimizers. PhD thesis, Department of Computer Science, University of Pretoria, South Africa (2002)

Fast Multi-swarm Optimization with Cauchy Mutation and Crossover Operation

Qing Zhang[1,2], Changhe Li[1], Yong Liu[3], and Lishan Kang[1]

[1] China University of Geosciences, School of Computer, Wuhan, P.R. China, 430074
[2] Huanggang Normal University
[3] The University of Aizu, Aizu-Wakamatsu, Fukushima, Japan, 965-8580
zhangqing@hgnc.net, lch_wfx@yahoo.com.cn,
yliu@u-aizu.ac.jp, kang_whu@yahoo.com

Abstract. The standard Particle Swarm Optimization (PSO) algorithm is a novel evolutionary algorithm in which each particle studies its own previous best solution and the group's previous best to optimize problems. One problem exists in PSO is its tendency of trapping into local optima. In this paper, a multiple swarms technique(FMSO) based on fast particle swarm optimization(FPSO) algorithm is proposed by bringing crossover operation. FPSO is a global search algorithm witch can prevent PSO from trapping into local optima by introducing Cauchy mutation. Though it can get high optimizing precision, the convergence rate is not satisfied, FMSO not only can find satisfied solutions, but also speeds up the search. By proposing a new information exchanging and sharing mechanism among swarms. By comparing the results on a set of benchmark test functions, FMSO shows a competitive performance with the improved convergence speed and high optimizing precision.

Keywords: Particle swarm optimization, Cauchy mutation, swarm intelligence.

1 Introduction

Particle Swarm Optimization (PSO) was first introduced by Kennedy and Eberhart in 1995 [1,2]. It is motivated from the social behavior of organisms, such as bird flocking and fish schooling. Particles "fly" through the search space by following the previous best positions of their neighbors and their own previous best positions. Each particle is represented by a position and a velocity which are updated as follows:

$$X'_{id} = X_{id} + V'_{id} \tag{1}$$

$$V'_{id} = \omega V_{id} + \eta_1 rand\,()(P_{id} - X_{id}) \\ + \eta_2 rand\,()(P_{gd} - X_{id}) \tag{2}$$

where X_{id}' and X_{id} represent the current and the previous positions of idth particle, V_{id} and V_{id}' are the previous and the current velocity of idth particle, P_{id} and P_{gd} are the individual's best position and the best position found in the whole swarm so far respectively. $0 \leq \omega < 1$ is an inertia weight which determines how much the previous velocity is preserved, η_1 and η_2 are acceleration constants, $rand()$ generates random number from interval [0,1].

L. Kang, Y. Liu, and S. Zeng (Eds.): ISICA 2007, LNCS 4683, pp. 344–352, 2007.
© Springer-Verlag Berlin Heidelberg 2007

In PSO, each particle shares the information with its neighbors. PSO combines the cognition component of each particle with the social component of all the particles in a group. Although the speed of convergence is very fast, Once PSO traps into local optimum, it is difficult to jump out of local optimum. Ratnaweera et.al.[3] state that lack of population diversity in PSO algorithms is understood to be a factor in their convergence on local optima. Therefore, the addition of a mutation operator to PSO should enhance its global search capacity and thus improve its performance. A first attempt to model particle swarms using the quantum model(QPSO) was carried out by Sun et.al. [4]. In a quantum model, particles are described by a wave function instead of the standard position and velocity. The quantum Delta potential well model and quantum harmonic oscillators are commonly used in particle physics to describe the stochastic nature of particles. In their studies[5], the variable of *gbest* (the global best particle) and *mbest* (the mean value of all particles' previous best position) is mutated with Cauchy distribution respectively, and the results show that QPSO with *gbest* and *mbest* mutation both performs better than PSO. The work of R. A. Krohling et.al.[6][7] showed that how Gaussian and Cauchy probability distribution can improve the performance of the standard PSO. Recently, evolutionary programming with exponential mutation has also been proposed [8].

In order to prevent PSO from falling in a local optimum, a fast PSO (FPSO) is proposed by introducing a Cauchy mutation operator in this paper. Besides the Cauchy mutation, FPSO chooses the natural selection strategy of evolutionary algorithms as the basic elimination strategy of particles. Although FPSO greatly overcomes the tendency of trapping into local optima of PSO, the convergence rate isn't satisfied. Like distributed genetic algorithm, multiple swarms idea is a very useful for speeding up the search. In this paper, a multiple swarms algorithm(FMSO) based on FPSO is proposed by introducing a crossover operation, the new information exchanging and sharing mechanism of FMSO make it converge fast on the global optimum.

The rest of this paper is organized as follows. Section 2 gives the analysis of PSO, and the detail of FPSO. Section 3 gives a brief review of the multi-population technique and then describe FMSO explicitly. Section 4 describes the experiment setup and presents the experiment results. Finally, Section 5 concludes the paper with a brief summary.

2   Fast Multi-swarm Optimization with Cauchy Mutation and Crossover Operatio

2.1 Cauchy Mutation

From the mathematic theoretical analysis of the trajectory of a PSO particle [9], the trajectory of a particle X_{id} converges to a weighted mean of P_{id} and P_{gd}. Whenever the particle converges, it will "fly" to the personal best position and the global best particle's position. This information sharing mechanism makes PSO have a very fast speed of convergence. Meanwhile, because of this mechanism, PSO can't guarantee to find the global minimal value of a function. In fact, the particles usually converge to local optima. Without loss of generality, only function minimization is discussed here. Once the particles trap into a local optimum, in which P_{id} can be assumed to be the same as P_{gd}, all the particles converge on P_{gd}. At this condition, the velocity update equation becomes:

$$V_{id}^{'} = \omega V_{id} \tag{3}$$

When the iteration in the equation (3) goes to infinite, the velocity of the particle V_{id} will be close to 0 because of $0 \leq \omega < 1$. After that, the position of the particle X_{id} will not change, so that PSO has no capability of jumping out of the local optimum. It is the reason that PSO often fails on finding the global minimal value.

To overcome the weakness of PSO discussed at the beginning of this section, the Cauchy mutation is incorporated into PSO algorithm. The basic idea is that, the velocity and position of a particle are updated not only according to (1) and (2), but also according to Cauchy mutation as follows:

$$V_{id}^{'} = V_{id} \exp(\delta) \tag{4}$$

$$X_{id}^{'} = X_{id} + V_{id}^{'} \delta_{id} \tag{5}$$

where δ and δ_{id} denote Cauchy random numbers

Since the expectation of Cauchy distribution doesn't exist, the variance of Cauchy distribution is infinite so that Cauchy mutation could make a particle have a long jump. By adding the update equations of (4) and (5), FPSO greatly increases the probability of escaping from the local optimum.

2.2 Natural selection strategy

In the standard PSO, all particles are directly updated by their offspring no matter whether they are improved. If a particle moves to a better position, it can be replaced by the updated. However if it moves to a worse position, it is still replaced by its offspring. In fact, the most particles fly to worse positions for most cases, therefore the whole swarm will converge on local optima. Like evolutionary algorithms, FPSO introduces an evolutionary selection strategy in which each particle survives according to a natural selection rule. Therefore, the particle's position at the next step is not only due to the position update but also the evolutionary selection. Such strategy could greatly reduce the probability of trapping into local optimum.

The evolutionary selection strategy is carried out as follows. Assume the size of the swarm is m, pair-wise comparison over the union of parents and offspring $(1,2,...2m)$ is made. For each particle, q opponents are randomly chosen from all parents and offspring with equal probability. If the fitness of particle i is less then its opponent , it will receive a "win". Then select m particles that have the more winnings to be the next generation.

The major steps of FPSO are as follows:

Step1: Generate the initial particles by randomly generating the position and velocity for each particle.

Step2: Evaluate each particle's fitness.

Step3: For each particle, if its fitness is smaller than its previous best(P_{id}), update P_{id}.

Step4: For each particle, if its fitness is smaller than the best one (P_{gd}) of all particles, update P_{gd}.

Step5: For each particle ,do

1) Generate a new particle t according to the formula (1) and (2).
2) Generate a new particle $t^{'}$ according to the formula (4) and (5).
3) Compare t with $t^{'}$,chose the one with smaller fitness to be the offspring.

Step6: Generate next generation according to the above evolutionary selection strategy.
Step7: if the stop criterion is satisfied, then stop, else goto Step 3.

3 Multiple Swarms Optimization Technique

In order to escape from the local optima and avoid premature convergence, the search for global optimum should be diverse. Many researchers have improved the performance of the PSO by enhancing its ability with a more diverse search. Specifically, some have introduced using multiple swarms, and then exchange information among them. The fast converging behavior of the PSO makes this issue so critical for multimodal problems. Al-Kazemi and Mohan [10] divided the population into two sets to achieve a more diverse, one set moving to the gbest while another moving in opposite direction. After some generations, if the gbest would not improve, the particles would switch their group. Two cooperating swarms was used by Baskar and Suganthan [11] to search concurrently for a solution along with sharing the gbest information of two swarms. The two swarms track the *gbest* if it improves. Each swarm using different update equation: One uses the standard PSO while the other uses the Fitness-to-Distance ratio PSO [12]. Their approach improved the performance in solving single objective optimization problems. Then an improved algorithm was proposed by El-Abd and Kamel [13] through adding a twoway flow of information between two swarms. After running a fixed generations, if the best p particles improve, then they willl replace the worst p particles in the other swarm. This guarantees exchanging new information from the other swarm's experience for the two swarms.

In this study, A new learning mechanism is introduced among swarms. At each iteration, the particles not only update themselves according to the best particle of their own swarm, but also learn information from the best particle of other swarms. The information sharing and learning mechanism make swarm extend their search space and speedup the convergence speed. The information sharing and learning mechanism that we call it crossover operation is described as follows:

Step1: for each particle of swarm k, randomly select a best particle p' from a random swarm.

Strp2: for each dimension i of particle p's position $px[i]$ and velocity $pv[i]$, if rand()$<q_c$,

crossover particle p with p' as follows:
$$px[i]=(1-\alpha)*px[i]+ \alpha*p'x[i].$$
$$pv[i]=rand()*(p'x[i]-px[i]).$$

Step3: if all particles of swarm k are updated, end the operator, else go to Step 1.

where q_c is crossover rate , α is a random number of (0,1).

4 Experiments and Results

Eight benchmark functions (f_1-f_8) are used in this paper. Table 1 gives the details of these functions. Algorithm parameters are as follows for all experiments: acceleration constants of η_1 and η_2 are both set to be 1.496180, and inertia weight $\omega=0.729844$ as suggested by den Bergh [14], crossover rate q_c is 0.8, running time is 50. In order to be the same number of function evaluations to PSO, a particle will be evaluated two times each generation in FPSO and FMSO. The other parameters are given in the following experiments.

Two groups of experiments are carried out in this section.
Firstly, FMSO is compared with standard PSO and FPSO to show the performance of FMSO algorithm on 10 problems. In the experiments, the number of population is 60 ,

Table 1. Details of test functions, where n is the dimension of the function, fmin is the minimum value of the function, $S \subseteq Rn$

Test function	n	S	f_{min}
$f_1(x) = \sum_{i=1}^{n} x_i^2$	30	(-5.12,5.12)	0
$f_2(x) = \sum_{i=1}^{n} -x_i \sin(\sqrt{\|x_i\|})$	30	(-500,500)	-12569.5
$f_3(x) = 6 \cdot \sum_{i=1}^{5} \lfloor x_i \rfloor$	30	(-5.12, 5.12)	0
$f_4(x) = \sum_{i=1}^{n} i \cdot x_i^4 + U(0,1)$	30	(-1.28,1.28)	0
$f_5(x) = \dfrac{(\sin^2 \sqrt{x^2 + y^2}) - 0.5}{(1.0 + 0.001(x^2 + y^2))^2} + 0.5$	2	(-100.0,100.0)	0
$f_6(x) = \dfrac{1}{4000} \sum_{i=1}^{n} (x_i - 100)^2$ $- \prod_{i=1}^{n} \cos(\dfrac{x_i - 100}{\sqrt{i}}) + 1$	30	(-300.0,300.0)	0
$f_7(x) = -20 \cdot \exp(-0.2\sqrt{\dfrac{1}{n} \cdot \sum_{i=1}^{n} x_i^2})$ $- \exp(\dfrac{1}{n} \cdot \sum_{i=1}^{n} \cos(2\pi \cdot x_i)) + 20 + e$	30	(-30.0,30.0)	0
$f_8(x) = \sum_{i=1}^{n} 100((x_{i+1} - x_i^2)^2 + (x_i - 1)^2)$	30	(-2.048,2.048)	0

Table 2. The value of swarms(m) and swarm size(n),q is the tournament size

	Set 1	Set 2	Set 3	Set 4
m	2	4	8	16
n(q)	40(6)	20(5)	10(4)	5(3)

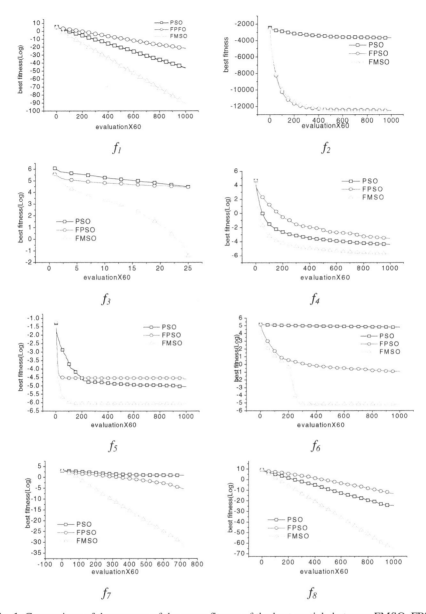

Fig. 1. Comparison of the process of the mean fitness of the best particle between FMSO, FPSO and PSO for 50 runs, the vertical axis is the function value and the horizontal axis is the number of evaluations

the tournament size ($q=10$) is chosen for FPSO. 3 swarms are used in FMSO, swarm size is 20, the tournament size ($q=5$) is chosen and crossover rate q_c is 0.6 for FMSO. The other parameters are the same as the above. Fig 1 shows the comparison results of

Table 3. The maximum(*Max*) ,minimum(*Min*) and average(*Avg*) best fitness over *50* runs. *Std* is the standard deviation

F	evaluations		PSO	FPSO	FMSO
f_1	$6*10^4$	Max	3.42138E-20	8.8762e-009	3.87449e-039
		Min	1.24857E-23	5.53311e-014	1.08821e-042
		Avg	5.06623E-21	3.73898e-010	3.4003e-040
		std	8.87525e-021	1.48328e-009	6.55003e-040
f_2	$6*10^4$	Max	-2793.79	-12086.3	-12569.5
		Min	-4836.19	-12569.5	-12569.5
		Avg	-3682.48	-12506.2	-12569.5
		std	457.426	101.935	0
f_3	$6*250$	Max	0	0	0
		Min	0	0	0
		Avg	0	0	0
		std	0	0	0
f_4	$6*10^4$	Max	0.024044	0.0575288	0.009357
		Min	0.00349205	0.0101007	0.000910975
		Avg	0.0120778	0.0284553	0.00321394
		std	0.00527977	0.0109074	0.00156956
f_5	$6*10^4$	Max	0.00971591	0.126991	0.00971591
		Min	0	0	0
		Avg	0.00641252	0.0100854	0.00233185
		std	0.00460249	0.0177373	0.00414948
f_6	$6*10^4$	Max	140.958	2.16826	0.0245904
		Min	87.6869	2.68284e-011	0
		Avg	120.464	0.39966	0.00580988
		std	11.4045	0.528499	0.00774408
f_7	$6*10^4$	Max	6.25549	0.000532368	0
		Min	3.12093e-007	5.90058e-007	0
		Avg	2.25332	5.85617e-005	0
		std	1.07449	0.000107675	0
f_8	$6*10^4$	Max	9.60947e-010	1.31399e-005	4.25245e-028
		Min	5.7947e-015	2.24269e-009	0
		Avg	2.07824e-011	1.36425e-006	7.15891e-029
		std	1.34386e-010	2.76779e-006	8.37964e-029

the process with the same evaluations and table 3 show the statistical results for all test problems over 50 runs. By viewing the results of Fig 1 and table 3, we can easily see that FPSO show better performance than PSO on function f_2, f_3, f_6 and f_7. All the results of FMSO are better than PSO. The results of Table 3 demonstrate that FMSO finds the global optima for function f_2, f_3, f_5, f_6, f_7 and f_8, especially for function f_2, f_3 and f_7, the global optima of them are fond for each run over 50 runs. From the comparison results, we can know that Cauchy mutation is helpful for some problems, and the multiple swarms technique works for all test problems. It indicates that FMSO not only speeds up the search, but also improves the optimizing precision.

Secondly, the aim of group 2 is to analyze the effect of different swarms and swarm size($m*n$) for a same population size 80. Function f_1, f_4, f_6, f_8 are chosen to test. the maximum generation is 1000, m and n are respectively the number of swarms and swarm size. 4 sets experiments are conducted, Table 2 shows the value of m and n. All the other parameters are the same as above. By viewing the comparison results of Table 4, the more number swarms is better for function f_4 and f_6 ,however , it isn't the case for function f_1 and f_8. That is to say, although multiple swarms can speed up the convergence rate, not the more the better. it is hard to find a optimal swarm size and number of swarms for general problems. Actually, the optimal swarm size and swarm numbers depend on the distributions of optimal solutions and number of optimal solutions. For functions with a few optimum, small number of swarms might be enough. However, for functions with a lot of optimum, large swarms might be needed.

5 Conclusions

By analyzing the advantage and disadvantage of the standard PSO, FPSO based on Cauchy mutation and evolutionary selection strategy is proposed in this paper. Although FPSO greatly overcomes the tendency of trapping into local optima of PSO, the convergence rate isn't satisfied, so a multiple swarms algorithm(FMSO) based on FPSO is proposed by introducing a crossover operation, FMSO is tested on 8 benchmark functions. From the experimental results of these functions, it can be seen that the FMSO performed much better than PSO and FPSO on the selected problems.

Table 4. Comparison with different swarms and swarm size for FMSO over 50 runs, The maximum(Max) ,minimum(Min) and average(Avg) best fitness over 50 runs. Std is the standard deviation

F		2 swarms*40	4 warms*20	8 swarms*10	16swarm*5
f_1	Max	1.97774e-39	4.37499e-40	3.97464e-40	5.18094e-34
	Min	3.68037e-44	1.63423e-42	8.54154e-44	6.24203e-36
	Avg	1.36448e-40	4.47269e-41	2.01959e-41	1.11419e-34
	std	3.89216e-40	8.05011e-41	5.88923e-41	1.04846e-34
f_4	Max	0.00573433	0.0050845	0.0050845	0.00284168
	Min	0.000839845	0.000748565	0.000748565	0.000767533
	Avg	0.00304075	0.00184794	0.00184794	0.00168235
	std	0.00110113	0.000686807	0.000686807	0.000493363
f_6	Max	0.0513256	0.0319228	0.0343841	0.00739604
	Min	0	0	0	0
	Avg	0.00934373	0.00285618	0.00708914	0.000443762
	std	0.011266	0.00607201	0.00913191	0.00175646
f_8	Max	1.57772e-028	3.59918e-028	6.40949e-29	4.47899e-21
	Min	0	0	0	1.95173e-23
	Avg	8.94864e-30	9.30856e-29	3.57452e-30	3.94318e-22
	std	2.75515e-29	7.01503e-29	1.22276e-29	6.52812e-22

Only functions with the dimension less than *30* were tested in this paper. Further research will focus on testing the performance of FMSO on higher dimensional problems in order to find whether FMSO would scale up well for the large function optimization problems.

References

1. Kennedy, J., Eberhart, R.C.: Particle Swarm Optimization. In: IEEE International Conference on Neural Networks, pp. 1942–1948. IEEE Computer Society Press, Los Alamitos (1995)
2. Eberhart, R.C., Kennedy, J., New, A.: Optimizer Using Particle Swarm Theory. In: Proceedings of the 6th International Symposium on Micro Machine and Human Science, pp. 39–43 (1995)
3. Ratnaweera, A., Halgamuge, S.K., Watson, H.C.: Self-organizing hierarchical particle swarm optimizer with time-varying acceleration coefficients. IEEE Transactions on Evolutionary Computation 8(3), 240–255 (2004)
4. Sun, J., Feng, B., Xu, W.: Particle swarm optimization with particles having quantum behavior. In: Proceedings of the IEEE Congress on Evolutionary Computation, Portland, Oregon USA, pp. 325–331. IEEE Computer Society Press, Los Alamitos (2004)
5. Liu, J., Xu, W., Sun, J.: Quantum-behaved particle swarm optimization with mutation operator. In: Proceedings of the 17th IEEE International Conference on Tools with Artificial Intelligence Pages, pp. 237–240. IEEE Computer Society Press, Los Alamitos (2005)
6. Krohling, R.A.: Gaussian particle swarm with jumps. In: Proceedings of the IEEE Congress on Evolutionary Computation, Edinburgh, UK, pp. 1226–1231. IEEE Computer Society Press, Los Alamitos (2005)
7. Krohling, R.A., dos Santos Coelho, L.: PSO-E: Particle Swarm with Exponential Distribution. In: Proceedings of the IEEE Congress on Evolutionary Computation, July 2006, pp. 1428–1433. IEEE Computer Society Press, Los Alamitos (2006)
8. Narihisa, H., Taniguchi, T., Ohta, M., Katayama, K.: Evolutionary Programming with Exponential Mutation. In: Proceedings of the IASTED Artificial Intelligence and soft Computing, Benidorm, Spain, pp. 50–55 (2005)
9. Clerc, M., Kennedy, J.: The Particle Swarm: Explosion, Stability and Convergence in a Multi-Dimensional Complex Space. IEEE Trans. on Evolutionary Computation 6, 58–73 (2002)
10. Al-Kazemi, B., Mohan, C.K.: Multi-phase discrete particle swarm optimization. In: Proc. of 4th Int. Workshop on Frontiers on Evolut. Alg., Research Triangle Park, NC (2002)
11. Baskar, S., Suganthan, P.N.: A novel concurrent particle swarm optimization. In: Proc. of Cong. on Evolut. Comput., Portland, OR, pp. 792–796 (2004)
12. Peram, T., Veeramachaneni, K., Mohan, C.K.: Fitness-distanceratio based particle swarm optimization. In: Proc. of IEEE Swarm Intell. Symp., Indianapolis, IN, pp. 88–94. IEEE Computer Society Press, Los Alamitos (2003)
13. El-Abd, M., Kamel, M.: Information exchange in multiple cooperating swarms. In: Proc. of Cong. on Evolut. Comput., Edinburgh, UK, pp. 138–142 (2005)
14. van den Bergh, F.: An Analysis of Particle Swarm Optimizers. PhD thesis, Department of Computer Science, University of Pretoria, South Africa (2002)

Particle Swarm Optimization Using *Lévy* Probability Distribution

Xingjuan Cai[1], Jianchao Zeng[1], Zhihua Cui[2,1], and Ying Tan[1]

[1]Division of system simulation and computer application, Taiyuan University of
Science and Technology, Shanxi, P.R. China, 030024
[2] State Key Laboratory for Manufacturing Systems Engineering
Xi'an Jiaotong University, Xi'an, P.R. China, 710049
cai_xing_juan@sohu.com,
zengjianchao@263.net,
cuizhihua@gmail.com

Abstract. Velocity threshold is an important parameter to affect the
performance of particle swarm optimization. In this paper, a novel velocity threshold automation strategy is proposed by incorporated with
Lévy probability distribution. Different from Gaussian and Cauchy distribution, it has an infinite second moment and is likely to generate an
offspring that is far away from its parent. Therefore, this method employs
a larger capability of the global exploration by providing a large velocity
scale for each particle. Simulation results show the proposed strategy is
effective and efficient.

1   Introduction

Particle swarm optimization (PSO) is a new swarm intelligent strategy firstly
proposed by J.Kennedy and R.C.Eberhart[1][2]. Because of the simple concepts
and the ease of implementation, many researchers have worked on it to improve
performance[3][4][5].

The PSO algorithm emulates the social behaviors among swarm animals such
as flocking of birds, schooling of fish, and herding of insects. Each individual
(called particle) searches food by collaborative and competitive manner. In this
paper, we only consider the following unconstrained problem:

$$\min \quad f(\overrightarrow{x}) \quad x \in [L, U]^D \subseteq R^D \tag{1}$$

In PSO algorithm, each particle represents one potential solution in the prob-
lem space. Suppose j^{th} particle locates in position $\overrightarrow{x}_j(t)$ at time t, then in the
next time it will fly to the position $\overrightarrow{x}_j(t+1)$ with the velocity $\overrightarrow{v}_j(t+1)$:

$$\overrightarrow{x}_j(t+1) = \overrightarrow{x}_j(t) + \overrightarrow{v}_j(t+1) \tag{2}$$

where velocity vector $\overrightarrow{v}_j(t+1)$ is defined by:

$$\overrightarrow{v}_j(t+1) = w\overrightarrow{v}_j(t) + c_1 r_1(\overrightarrow{p}_j(t) - \overrightarrow{x}_j(t)) + c_2 r_2(\overrightarrow{p}_g(t) - \overrightarrow{x}_j(t)) \tag{3}$$

L. Kang, Y. Liu, and S. Zeng (Eds.): ISICA 2007, LNCS 4683, pp. 353–361, 2007.
© Springer-Verlag Berlin Heidelberg 2007

where $\overrightarrow{p}_j(t)$ denotes the best location found by particle j up to time t, whereas $\overrightarrow{p}_g(t)$ represents the best position found by the entire swarm. Inertia weight w falls between 0 and 1, cognitive parameter c_1 and social parameter c_2 are known as acceleration coefficients, r_1 and r_2 are two random numbers generated with uniform distribution within $(0, 1)$.

The velocity vector of each particle consists of three parts: momentum part, cognitive part and social part while the performance of a PSO is determined with balancing among these parts. The momentum part provides the necessary momentum for particles to roam across the search space. The cognitive part represents the personal thinking of each particle while the social part represents the collaborative effect of the particles. Because many variants have been proposed to improve the performance of PSO, there are still some work need to do. In this paper, we propose a novel *Lévy* velocity threshold to improve the performance on multi-modal problems.

The rest of paper is organized as follows. Section 2 gives a brief description of the previous methods for velocity threshold. In Section 3, we discuss the definition and characteristics of the *Lévy* probability distribution and the algorithm for generating random variables owning the *Lévy* probability distribution. Section 4 discusses the details of the proposed method –particle swarm optimization using *Lévy* probability distribution (PSO-LPD), in which the velocity threshold v_{max} is a *Lévy* random number. Finally, the simulation results show the proposed algorithm can adjust the exploration and exploitation capabilities dynamically, and improve the performance effectively.

2   Previous Variants for Velocity Threshold of PSO

In 1998, Shi and Eberhart[6] found that the velocity threshold can be eliminated at the expense of a greater number of function evaluations for a fixed velocity threshold setting.

Fourie et al.[7][8] proposed a velocity threshold setting to improve the performance. If the best location of particle j does not change within a certain iterations h, the velocity threshold will reduce by the fraction β as follows:

$$v_{max}^j = \beta v_{max}^j, \quad if \ f(\overrightarrow{p}_j(t+h)) = f(\overrightarrow{p}_j(t)) \tag{4}$$

where v_{max}^j represents the velocity threshold of particle j.

Inspired by the evolutionary programming[9], Z.H.Cui et al.[10] proposed a stochastic velocity threshold. In evolutionary programming, the individual is a pair of real-valued vector (x_j, η_j), where x_j is a position vector while η_j is a standard deviation vector, and the offspring (x_j', η_j') is computed by

$$\eta_j'(k) = \eta_j(k)exp(\tau' N(0,1) + \tau N_k(0,1)) \tag{5}$$

$$x_j'(k) = x_j(k) + \eta_j(k)N_k(0,1) \tag{6}$$

where $x_j(k)$ is the kth variable of individual x_j (k=1,2,...,n), $N_k(0,1)$ and $N(0,1)$ are the two random numbers generated with mean zero and standard deviation one while $N_k(0,1)$ is renewed for different dimension. The factors τ and τ' are commonly set to $(\sqrt{2\sqrt{n}})^{-1}$ and $(\sqrt{2n})^{-1}$ respectively[11]. In [10], the uniform random number $N_k(0,1)$ is replaced by a Cauchy random number.

3 *Lévy* Probability Distribution

P.Lévy, in the 1930s, discovered a class of probability distributions having an infinite second moment and governing the sum of these random variable[12][13]. Such a probability distribution is called the *Lévy* probability distribution and has the following form:

$$L_{\alpha,\gamma} = \frac{1}{\pi} \int_0^\infty e^{-\gamma q^\alpha} \cos(qy) dq, \quad y \in R \tag{7}$$

The distribution is symmetric with respect to $y = 0$ and has two parameters γ and α. γ is the scaling factor satisfying $\gamma > 0$, and α controls the shape of the distribution, requiring $0 < \alpha < 2$. The analytic form of the integral is unknown for general α, but is known for a few cases.

Therefore, a numerical algorithm for generating *Lévy* random numbers was introduced in [14]. This algorithm utilizes two independent random variables x and y from standard Gaussian distribution and performs a nonlinear transformation:

$$v = \frac{x}{|y|^{\frac{1}{\alpha}}} \tag{8}$$

With this, it was shown that the sum of these variables with an appropriate normalization

$$z_n = \frac{1}{n^{\frac{1}{\alpha}}} \sum_{j=1}^n v_k \tag{9}$$

converges to the *Lévy* probability distribution with larger n.

A variable in the *Lévy* distribution, w, can then be generated using the nonlinear transfer

$$w = \{[K(\alpha) - 1]\exp(\frac{-v}{C(\alpha)}) + 1\}v \tag{10}$$

To obtain *Lévy* distributions with a scale factor γ other than 1, the linear transformation

$$z = \gamma^{\frac{1}{\alpha}} w \tag{11}$$

is applied. The values of σ_x, $K(\alpha)$ and $C(\alpha)$ for specified α are given in [14]. In this paper, the random number with *Lévy* distribution is always computing with the above manner for $\alpha = 0.8$.

4   Introduction of PSO-LPD Algorithm

Since many foragers and wandering animals have been shown to follow a *Lévy* distribution of steps, several researchers incorporates this distribution into the optimization algorithm[15][16].

Velocity threshold v_{max} is a parameter used to control the scale of velocity information. It can provide a chance to escape from local optima. Large v_{max} increases the search region, enhancing the global search capability, as well as small v_{max} decreases the search region, adjusting the search direction of each particle frequently. Since then, a proportional threshold v_{max} selection principle can balance the exploitation and exploration capability of PSO, utilizing more information about search directions to avoid premature convergence.

However, due to the development of scientific technology, the optimization problems from industry areas become very complex, high-dimensional, non-differential, and multi-modal. Therefore, a problem-based velocity threshold is hard to design. Since then, this paper introduces a stochastic velocity threshold automation strategy. It assigns the different velocity threshold for each particle based on a random number with *Lévy* probability distribution. That is, some particles with the same performance may acquire the different velocity threshold values. From this manner, a random character can reduce the hardness to predict the unknown optimization model, and obtain some average performance for most cases.

The detailed steps of PSO-LPD are listed as follows.

Step1. Generate the initiate population with m particles, and set the velocity threshold value of dimension k of particle j as v_0. The position vector $\vec{x}$ of each particle is selected within the domain region as well as velocity vector of each particle is chosen within the interval $[0, v_0]$ uniformly.

Step2. Update the velocity vector of each particle with formula (3).

Step3. Suppose $v_{max}^j(t)$ denotes the velocity threshold value of $j'th$ particle at time t. Then, it is generated with the *Lévy* probability distribution with formula (8) and (10), if the $v_{max}^j(t)$ is out of the problem space, we provide a reflection treatment.

Step4. Rearrage the velocity vector of each particle with the corresponding velocity threshold.

Step5. Update the position vector of each particle with formula (2).

Step6. Update the best locations found by each particle and the entire swarm.

Step7. If the stop criterion is satisfied, output the final best solution of the swarm. Otherwise, goto Step2.

5   Simulation Results

In order to certify the effectiveness of proposed particle swarm optimization using *Lévy* probability distribution(PSO-LPD), we select three famous benchmark functions: Schwefel Problem 2.26, Ackley and Penalized Function, to test the performance, and compare PSO-LPD and standard PSO (SPSO) and PSO with time-varying velocity threshold(PSO-TVV)[7]. The details of these four benchmark functions are listed as follows:

Schwefel Problem 2.26:

$$f_1(x) = -\sum_{j=1}^{n}(x_j \sin(\sqrt{|x_j|}))$$

where $|x_j| \leq 500.0$, and

$$f_1(x^*) = f_1(420.9687, 420.9687, ..., 420.9687) \approx -418.9829n$$

Ackely:

$$f_2(x) = -20exp(-0.2\sqrt{\frac{1}{n}\sum_{j=1}^{n}x_j^2}) - exp(\frac{1}{n}\sum_{k=1}^{n}\cos 2\pi x_k) + 20 + e$$

where $|x_j| \leq 32.0$, and

$$f_2(x^*) = f_2(0, 0, ..., 0) = 0.0$$

Penalized Function:

$$f_3(x) = 0.1\{\sin^2(3\pi x_1) + \sum_{i=1}^{n-1}(x_i - 1)^2[1 + \sin^2(3\pi x_{i+1})] + (x_n - 1)^2[1 + \sin^2(2\pi x_n)]\}$$

$$+ \sum_{i=1}^{n}u(x_i, 5, 100, 4)$$

where $|x_j| \leq 50.0$, and

$$u(x_i, a, k, m) = \begin{cases} k(x_i - a)^m, & \text{if } x_i > a \\ 0, & \text{if } -a \leq x_i \leq a \\ k(-x_i - a)^m, & \text{if } x_i < -a \end{cases}$$

$$y_i = 1 + \frac{1}{4}(x_i + 1)$$

$$f_3(x^*) = f_3(1, 1, ..., 1) = 0.0$$

The coefficients are set as follows: the inertia weight w is decreased linearly from 0.9 to 0.4. Acceleration coefficients c_1 and c_2 are both set to 2.0. Total individual is 100, and the dimensionality is 30 and 100, respectively. In PSO-TVV, the velocity threshold v_{max} is set by formula (4) where $\beta = 0.95$, $h = 3$[7]. In each experiment,the simulation runs 30 times, while each time the largest evolutionary generations are $50 \times dimension$.

Table 1. The Comparison Results for Dimension 30

Function	Algorithm	Mean Value	Std Value
f_1	SPSO	-6.611115600258316e+003	9.449995477386343e+002
	PSO-TVV	-7.367697070496831e+003	9.378599211266609e+002
	PSO-LPD	-6.450481842676059e+003	6.394216188263249e+002
f_2	SPSO	6.958953254354583e-006	6.106325746998641e-006
	PSO-TVV	1.942643379493347e-010	9.438609252675202e-011
	PSO-LPD	3.178116347868354e-008	2.767697292014740e-008
f_3	SPSO	1.641873384803907e-007	4.493491018898719e-007
	PSO-TVV	8.090579225340944e-020	7.874412586522213e-020
	PSO-LPD	1.312180846127977e-015	2.851915251256295e-015

Table 2. The Comparison Results for Dimension 100

Function	Algorithm	Mean Value	Std Value
f_1	SPSO	-1.953860453466208e+004	3.694232670594450e+002
	PSO-TVV	-2.112565906292736e+004	1.220455250477929e+003
	PSO-LPD	-2.446616249288609e+004	1.007920264274735e+003
f_2	SPSO	1.619210344196327e-001	2.426820287574869e-001
	PSO-TVV	1.155502862732416e+000	6.488721050939146e-001
	PSO-LPD	1.783418102352385e-005	6.998277108195449e-006
f_3	SPSO	3.959959448930896e+001	2.315927992071475e+001
	PSO-TVV	2.197473167176759e-003	4.632680203588346e-003
	PSO-LPD	2.584389645059405e-010	8.762152159124651e-011

Schwefel Problem 2.26, Ackley and Penalized Function are all multi-modal with many local optima. For a small dimension 30, the algorithm PSO-TVV is superior to other two algorithms: PSO-LPD and SPSO, whereas the performance of PSO-LPD is always better than SPSO except for Schwefel Problem 2.26. For Schwefel Problem 2.26, the performance of PSO-LPD and SPSO are nearly equivalent.

However, for a high dimension case:100, Table 2 shows that the performance of PSO-LPD outperforms PSO-TVV and SPSO significantly especially for penalized function. Because this penalized function is very complex to optimize, several variants of PSO are not stable[17].

Fig.1 and Fig.2 are the comprison results for Schwefel Problem 2.26 with different dimensions. In the first period, PSO-TVV is easy to find some better solutions than other two algorithm, however, this tendecy can not persist on a long time. From the 500 iterations, PSO-TVV is trapped into some unforeseen local optima. However, PSO-LPD still maintains a powerful global search ability especially for a large dimension. This phenomenon can be explained by Fig.2. The same phenomenon is always true for other four figures. In one word, PSO-LPD is good at high-dimensional multi-modal problems with many local optima.

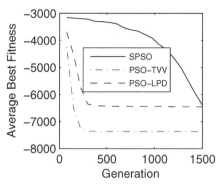

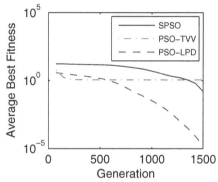

Fig. 1. Dynamic Comparison of f_1 for Dimension 30

Fig. 2. Dynamic Comparison of f_1 for Dimension 100

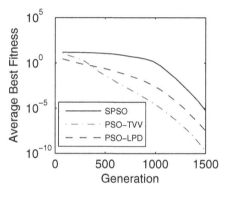

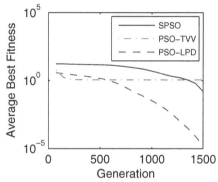

Fig. 3. Dynamic Comparison of f_2 for Dimension 30

Fig. 4. Dynamic Comparison of f_2 for Dimension 100

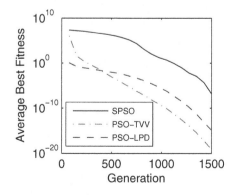

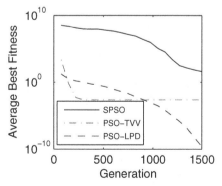

Fig. 5. Dynamic Comparison of f_3 for Dimension 30

Fig. 6. Dynamic Comparison of f_3 for Dimension 100

6   Conclusion

This paper introduces a novel velocity threshold automation strategy. Different other previous strategies, it is not determinstic one. To provide an average performance for most cases, it selects a stochastic strategy to enhance the escaping ability from a local optima. Three famous benchmar functions are used to compare. Simulation results show PSO-LPD is less than PSO-LVV in low dimension, however, it outperforms PSO-LVV in high dimension significantly. The reason is partly because of the introduction of the *Lévy* probability distribution. Since this distribution has a more chance to search further points than Gaussian probability distribution. Further research is how to apply other distributions into the PSO methology.

Acknowledgement

This paper was supported by National Natural Science Foundation under Grant No.60674104.

References

1. Kennedy, J., Eberhart, R.C.: Particle swarm optimization. In: Proceedings of IEEE International Conference on Neural Networks, pp. 1942–1948. IEEE Computer Society Press, Los Alamitos (1995)
2. Eberhart, R.C., Kennedy, J.: A new optimizer using particle swarm theory. In: Proceedings of 6th International Symposium on Micro Machine and Human Science, pp. 39–43 (1995)
3. Cui, Z.H., Zeng, J.C., Sun, G.J.: A fast particle swarm optimization. International Journal of Innovative Computing, Information and Control 2, 1365–1380 (2006)
4. Monson, C.K., Seppi, K.D.: The Kalman swarm: a new approach to particle motion in swarm optimization. In: Proc. of the Genetic and Evolutionary Computation Conference, pp. 140–150 (2004)
5. Liang, J.J., Qin, A.K., Suganthan, P.N., Baskar, S.: Comprehensive learning particle swarm optimizer for global optimization of multimodal functions. IEEE Transactions on Evolutionary Computation 10, 281–295 (2006)
6. Shi, Y., Eberhart, R.: Parameter selection in particle swarm optimization. In: Porto, V.W., Waagen, D. (eds.) Evolutionary Programming VII. LNCS, vol. 1447, pp. 591–600. Springer, Heidelberg (1998)
7. Fourie, P.C., Groenwold, A.A.: The particle swarm optimization algorithm in size and shape optimization. Structure Multidisciplinary Optimization 23, 259–267 (2002)
8. Schutte, J.: Particle swarms in sizing and global optimization. Master's thesis, University of Pretoria, Department of Mechanical Engineering (2002)
9. Fogel, L.J., Owens, A.J., Walsh, M.J.: Artificial intelligence through simulated evolution, New York, Wiley (1966)
10. Cui, Z.H., Zeng, J.C., Sun, G.J.: Adaptive velocity threshold particle swarm optimization. In: Proceedings of the First International Conference on Rough Sets and Knowledge Technology, China, pp. 327–332 (2006)

11. Fogel, D.B.: An introduction to simulated evolutionary optimization. IEEE Transactions on Neural Networks 5, 3–14 (1994)
12. Lévy, P.: Theorie de l'addition des veriables aleatoires, Guathier-Villars, Paris, France (1937)
13. Gnedenko, B., Kolmogorov, A.: Limit distribution for sums of independent random variables. Addition-Wesley, Cambirdge, MA (1954)
14. Mantegna, R.: Fast,accurate algorithm for numerical simulation of Lévy stable stochastic process. Physics Review E 49, 4677–4683 (1994)
15. Lee, C.Y., Yao, X.: Evolutionary programming using mutations based on the Lévy probability distribution. IEEE Transactions on Evolutinary Computation 8, 1–13 (2004)
16. Richer, T.J., Blackwell, T.M.: The Lévy particle swarm. In: Proceedings of IEEE Congress on Evolutionary Computation, Canada, pp. 808–815. IEEE Computer Society Press, Los Alamitos (2006)
17. Ratnaweera, A., Halgamuge, S.K., Watson, H.C.: Self-Organizing Hierarchical Particle Swarm Opitmizer with Time-Varying Acceleration Coefficients. IEEE Transactions on Evolutionary Computation 8, 240–255 (2004)

Re-diversification Based Particle Swarm Algorithm with Cauchy Mutation

Hui Wang[1,2], Sanyou Zeng[1,2], Yong Liu[3], Wenjun Wang[4],
Hui Shi[1,2], and Gang Liu[1,2]

[1] School of Computer, China University of Geosciences, Wuhan, 430074 China
[2] Research Center of Science and Technology, China University of Geosciences,
430074 Wuhan, China
[3] University of Aizu, Tsuruga, Ikki-machi, Aizu-Wakamatsu, Fukushima 965-8580 Japan
[4] Applied psychology Institution, China University of Geosciences, Wuhan, 430074 China
wanghui_cug@yahoo.com.cn, sanyou-zeng@263.net,
yliu@u-aizu.ac.jp, wangwenjun881@126.com

Abstract. Particle Swarm Optimization (PSO) has shown its fast search speed in many complicated optimization and search problems. However, PSO could often easily fall into local optima. This paper presents a hybrid PSO algorithm called RPSO by applying a new re-diversification mechanism and a dynamic Cauchy mutation operator to accelerate the convergence of PSO and avoid premature convergence. Experimental results on many well-known benchmark optimization problems have shown that the RPSO could successfully deal with those difficult multimodal functions while maintaining fast search speed on those simple unimodal functions in the function optimization.

1 Introduction

Particle Swarm Optimization (PSO) was firstly introduced by Kennedy and Eberhart in 1995 [1]. It is a simple evolutionary algorithm which differs from other evolutionary algorithms in which it is motivated the simulation of social behavior. PSO has shown good performance in finding good solutions to optimization problems [2], and turned out to be another powerful tool besides other evolutionary algorithms such as genetic algorithms [3].

Like other evolutionary algorithms, PSO is also a population-based search algorithm and starts with an initial population of randomly generated solutions called particles [4]. Each particle in PSO has a position and a velocity. PSO remembers both the best position found by all particles and the best positions found by each particle in the search process. For a search problem in an n-dimensional space, a potential solution is represented by a particle that adjusts its position and velocity according to Eqs. (1) and (2):

$$V_i^{(t+1)} = w * V_i^{(t)} + c_1 * rand_1() * (pbest_i - X_i^{(t)})$$
$$+ c_2 * rand_2() * (gbest - X_i^{(t)}) \tag{1}$$

L. Kang, Y. Liu, and S. Zeng (Eds.): ISICA 2007, LNCS 4683, pp. 362–371, 2007.
© Springer-Verlag Berlin Heidelberg 2007

$$X_i^{(t+1)} = X_i^{(t)} + V_i^{(t+1)} \qquad (2)$$

where X_i and V_i are the position and velocity of particle i, $pbest_i$ and $gbest$ are previous best particle for the ith particle and the global best particle found by all particles so far respectively, and w is an inertia factor proposed by Shi and Eberhart [5], and $rand_1()$ and $rand_2()$ are two random numbers independently generated within the range of [0,1], and c_1 and c_2 are two learning factors which control the influence of the social and cognitive components.

One problem found in the standard PSO is that it could easily fall into local optima in many optimization problems. Some researches have been done to tackle this problem [6-9]. The standard PSO was inspired by the social and cognitive behavior of swarm. According to equation (1), particles are largely influenced by its previous best particles and the global best particle. Once the best particle has no change in a local optimum, all the rest particles will quickly converge to the position of the best particle. At such situation, the diversity of population is almost 0, and the particles are very difficult to find other better positions. This paper presents a hybrid PSO algorithm called RPSO by applying a new re-diversification mechanism [10] and a dynamic Cauchy mutation operator with local search [11]. The re-diversification mechanism will keep high diversity in the population to get a large search space in the evolution process. It is to hope that the long jump from Cauchy mutation [11] on the global best particle could get the best position out of the local optima where it has fallen. RPSO has been tested on both unimodal and multimodal function optimization problems. Comparison bas been conducted between RPSO and standard PSO.

The rest of the paper is organized as follows: Section 2 describes the proposed RPSO algorithm. Section 3 defines the benchmark continuous optimization problems used in the experiments, and gives the experimental settings. Section 4 presents and discusses the experimental results. Finally, Section 5 concludes with a summary and a few remarks.

2   RPSO Algorithm

2.1   Re-diversification

Tim Blackwell [12] used swarm diameter as a measure of swarm diversity. The swarm diameter is defined as the largest distance, along any axis, between any two particles. It is computed as follows:

$$Diversity = \underset{i,j=1,2,\dots,PopSize}{Max} \underset{i \neq j}{\{Dist(p_i, p_j)\}} \qquad (3)$$

Where $Dis(p_i, p_j)$ is the distance between two different particles p_i and p_j.

Many researches [12-17] have shown PSO has a problem of diversity loss. Loss of diversity arises when swarm is converging to a position. If the swarm falls into local optima, the diversity of swarm will decrease to 0. At such situation, the average velocity of swarm is almost 0, and all particles have the same position. So the swarm will hardly achieve a good solution. There are four principle mechanisms for either re-diversification or diversity maintenance: randomization [10], repulsion [13], dynamic networks [14-15] and multi-populations [16-17].

To avoid premature convergence, a new random re-diversification mechanism is employed. Choose m worst particles in the swarm, and randomly re-initialize it. This happens every a fixed number of generations. In order to speed up the convergence, the m worst particles are randomly initialized in a dynamic interval $[a^p_i, b^p_i]$, where a^p_i and b^p_i are the minimum and maximum values of each dimension in current population respectively.

2.2 Dynamic Cauchy Mutation Operator

Some theoretical results have shown that the particle in PSO will oscillate between their pervious best particle and the global best particle found by all particles so far before it converges [18-19]. If the searching neighbors of the global best particle would be added in each generation, it would extend the search space of the best particle. It is helpful for the whole particles to move to the better positions. This can be accomplished by having a Cauchy mutation [11] on the global best particle in every generation. The one-dimensional Cauchy density function centered at the origin is defined by:

$$f(x) = \frac{1}{\pi} \frac{t}{t^2 + x^2} , \quad -\infty < x < \infty \tag{4}$$

where $t > 0$ is a scale parameter [20]. The Cauchy distributed function is

$$F_t(x) = \frac{1}{2} + \frac{1}{\pi} \arctan\left(\frac{x}{t}\right) \tag{5}$$

The reason for using such a mutation operator is to increase the probability of escaping from a local optimum [21]. The Cauchy mutation operator used in RPSO is described as follows:

$$W(i) = \left(\sum_{j=1}^{PopSize} V[j][i]\right) / PopSize \tag{6}$$

where $V[j][i]$ is the ith velocity vector of the jth particle in the population, PopSize is the population size. $W(i)$ is a weight vector within $[-W_{max}, W_{max}]$, and W_{max} is set to 1 in this paper.

$$gbest'(i) = gbest(i) + w(i) * N(X_{min}, X_{max}) \tag{7}$$

where N is a Cauchy distributed function with the scale parameter $t = 1$, and $N(X_{min}, X_{max})$ is a random number within $[X_{min}, X_{max}]$, which is a defined domain of a test function. The main steps of RPSO algorithm are as follows:

Table 1. The main steps of RPSO algorithm

Begin
  n = population size;
  P = current population;
  m = number of worst particles;
  $nGen$ = number of generation with a initial value 0;
  Num = a constant number;
  $x^p_j \in [a^p_j, b^p_j]$; //interval boundaries of the jth dimension in current population
  $best\_fitness\_value\_so\_far$ = the fitness value of the best particle found by all particles so far;
  VTR = value to reach;
  NFC = number of function calls;
  Max_{NFC} = maximum number of function calls (NFC);

  while ($best\_fitness\_value\_so\_far$ >= VTR and NFC <= Max_{NFC})
    if (($nGen$!= 0) && ($nGen$ % Num == 0))
      Update the interval boundaries $[a^p_j, b^p_j]$ in current population;
      Select m worst particles in current population;
      Randomly re-initialize the m worst particles in $[a^p_j, b^p_j]$;
      Update $pbest_i$, $gbest$ if needed;
    If end
    for each particle P_i
      Calculate particle velocity according to equation (1);
      Update particle position according to equation (2);
      Calculate fitness value of particle P_i;
      Update $pbest_i$, $gbest$ if needed;
    for end
    for i = 1 to n
      Update W[i] according to equation (6)
      if fabs(W[i] > W_{max}) W[i] = W_{max}
      If end
    for end
    Mutate $gbest$ according to equation (7);
    if the fitness value of $gbest'$ is better than $gbest$
      $gbest = gbest'$
    if end
    $nGen$++;
  while end
End

---

3   Experiments

3.1   Benchmark Problems

8 well-known test functions used in [21-22] have been chosen in our experimental studies. They are high-dimensional problems, in which functions f_1 to f_4 are unimodal functions, and functions f_5 to f_8 are multimodal functions. All the functions used in this paper are to be minimized.

3.2   Experimental Setup

The algorithms used for comparison were PSO and RPSO. Each algorithm was tested with all of the numerical benchmarks shown in Table 2. For each algorithm, the maximum number of function calls (MAX_{NFC}) allowed was set to 100,000. If the fitness value of the best particle found by all particles so far (*best_fitness_value_so_far*) is reach to a approximate value VTR (-12569.5 for f_5 and 10^{-25} for the rest functions), we consider the current population has converged to the optimum, and the algorithm is terminated. All the experiments were conducted 50 times with different random seeds, and the average fitness of the best particles throughout the optimization run was recorded. The results below 10^{-25} will be reported as 0.

Table 2. The 8 test functions used in our experimental studies, where n is the dimension of the functions, f_{min} is the minimum values of the function, and $X \subseteq R_n$ is the search space

Test Function	n	X	f_{min}
$f_1(x) = \sum_{i=1}^{n} x_i^2$	30	$[-5.12, 5.12]$	0
$f_2(x) = \sum_{i=1}^{n} i * x_i^2$	30	$[-5.12, 5.12]$	0
$f_3 = \sum_{i=1}^{n} i * x_i^4 + random[0,1)$	30	$[-1.28, 1.28]$	0
$f_4(x) = \sum_{i=1}^{n}[100(x_{i+1} - x_i^2)^2 + (1 - x_i^2)^2]$	30	$[-30, 30]$	0
$f_5 = \sum_{i=1}^{n} -x_i * \sin(-\sqrt{|x_i|})$	30	$[-500, 500]$	-12569.5
$f_6 = \sum_{i=1}^{n}[x_i^2 - 10\cos(2\pi x_i) + 10]$	30	$[-5.12, 5.12]$	0
$f_7 = -20 * \exp\left(-0.2 * \sqrt{\frac{1}{n}\sum_{i=1}^{n} x_i^2}\right) - \exp\left(\frac{1}{n}\sum_{i=1}^{n}\cos(2\pi x_i)\right) + 20 + e$	30	$[-32, 32]$	0
$f_8 = \frac{1}{4000}\sum_{i=1}^{n} x_i^2 - \prod_{i=1}^{n}\cos(\frac{x_i}{\sqrt{i}}) + 1$	30	$[-600, 600]$	0

3.3 Parameter Settings

The selection of the parameters w, c_1, c_2 of Eq. (1) is very important. It can greatly influence the performance of PSO algorithms and its variations. By following the suggestions given in [23], $c_1 = c_2 = 1.49618$, $w = 0.72984$ and the maximum velocity V_{max} was set to the half range of the search space on each dimension [24]. To evaluate the performance of convergence, the average number of function calls (NFC) was employed. All common parameters of PSO and RPSO are set the same to have a fair comparison. The specific parameter settings are listed as follows.

Table 3. The specific parameter settings

PopSize	Maximum number of function calls (MAX_{NFC})	W_{max}	m	M
10	100,000	1	2	100

4   Experimental Results

4.1   Comparisons Between PSO and RPSO

Table 4 shows the comparison between PSO and RPSO for function f_1 to f_8, where "Mean Best" indicates the mean best function values found in the last generation, and "Std Dev" stands for the standard deviation ,and "Best" and "Worst" shows the best value and the worst value achieved by different algorithms over 50 trials respectively. It is obvious that RPSO performs better than standard PSO. Figure 1 shows performance comparison between standard PSO and RPSO.

Table 4. The results achieved for f_1 to f_8 using different algorithms

Function	PSO				RPSO			
	Mean Best	Std Dev	Best	Worst	Mean Best	Std Dev	Best	Worst
f_1	6.06e-7	2.07e-6	1.98e-19	8.95e-6	0	0	0	0
f_2	3.50e-5	1.60e-4	1.26e-18	8.66e-4	0	0	0	0
f_3	9.75e-2	7.03e-2	1.33e-2	0.282	5.56e-3	3.36e-3	7.42e-4	8.83e-3
f_4	2.87	6.31	3.02e-4	21.46	8.67e-19	1.25e-18	9.46e-21	2.99e-18
f_5	-6705.1	631.2	-8621.4	-5303.6	-11057.3	269.6	-11503.1	-10792.9
f_6	62.22	12.70	34.82	83.58	1.02e-4	1.77e-4	0	3.06e-4
f_7	7.49	2.08	4.17	12.83	2.04e-14	2.8e-14	0	5.11e-14
f_8	0.659	1.15e-4	0.896	3.82	2.22e-16	2.94e-16	0	5.15e-16

4.2   Average Number of Evaluations

However, the Cauchy mutation operator employed in the RPSO will increase the computational complexity. So investigating the average number of function calls for each algorithm will be very meaningful for evaluating the performance of PSO and RPSO. All the results are given in Table 5. By contrast standard PSO, RPSO does not only cost fewer evaluations, but also achieves better solutions on f_1, f_2, f_6, f_7 and f_8. On function f_3, f_4 and f_5, though RPSO and PSO have the same average number of evaluations, RPSO gets better solutions than PSO.

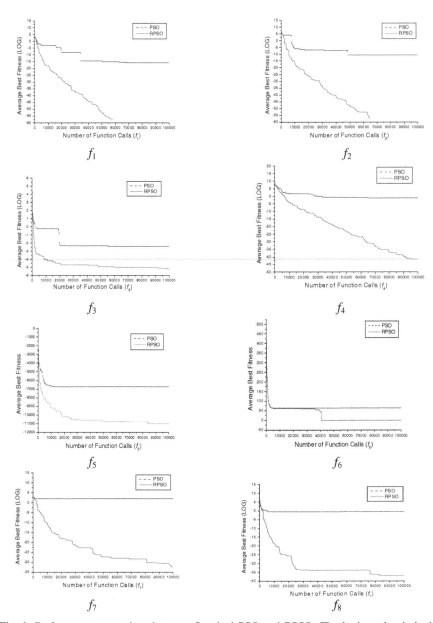

Fig. 1. Performance comparison between Standard PSO and RPSO. The horizontal axis is the average number of function calls and the vertical axis is the average best fitness value over 50 trials.

4.3 Diversity of Population

To investigate the effects of the re-diversification mechanism used in the RPSO, the diversity of population was calculated according to equation (3). Figure 2 shows the comparison results between PSO and RPSO. Because of the limited number of

Table 5. The average number of function calls (NFC)

Function	f_1	f_2	f_3	f_4	f_5	f_6	f_7	f_8
NFC (PSO)	100,000	100,000	100,000	100,000	100,000	100,000	100,000	100,000
NFC (RPSO)	57,895	65,101	100,005	100,005	100,005	94,884	90,874	86,149

pages, only the evolution process of the diversity of population on functions f_5 and f_8 is shown in Figure 2. It is obvious that the diversity of population in PSO is almost 0 after a few evaluations, because the particles have converged to local optima. However, the diversity of population in RPSO is much higher than PSO. That's to say, the search space of RPSO in the whole evolution process is large enough to find a better position. So the RPSO could search better solutions.

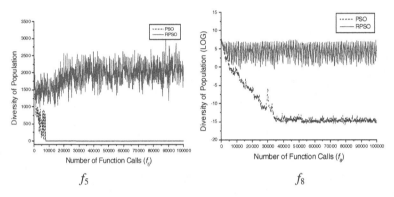

Fig. 2. Comparison of diversity of population

The significant improvement achieved by RPSO can be contributed to the new re-diversification mechanism and the search ability of dynamic Cauchy mutation operator. The re-diversification mechanism keeps a high diversity of population in the evolution process, and the Cauchy mutation operator extends the search space of the best particles. Such high diversity and extended neighbor search space will greatly help particles move to better positions. Therefore, RPSO had reached better solutions than the standard PSO.

5   Conclusions

The idea of RPSO is to use a new re-diversification method and a dynamic Cauchy mutation operator to help avoid local optima and accelerate the convergence of PSO. By randomly re-initializing some worst particles every a few generations, and applying a Cauchy mutaiton on the best particle found by all particles so far in each generation, RPSO could find better solutions than PSO.

RPSO has been compared with the standard PSO on both 4 unimodal functions and 4 multimodal functions. The results have shown that RPSO could have faster convergence on those simple unimodal functions, and better global search ability on those multimodal functions compared to the standard PSO. However, there are still fewer cases where RPSO had fallen in the local optima as what had happened on RPSO for the function f_5. It suggests that the proposed mthod might not be enough to prevent the search from falling in the local optima. Further study will foucs on how to improve the efficiency of the re-diversification mechansim and the Cauchy mutation operator.

Acknowledgments

This paper was supported by the Postgraduate's Innovation Foundation of China University of Geosciences (Wuhan) (No. CUGYJS0732).

References

1. Kennedy, J., Eberhart, R.: Particle Swarm Optimization. In: IEEE International Conference on Neural Networks, Perth, Australia, IEEE Computer Society Press, Los Alamitos (1995)
2. Parsopoulos, K.E., Plagianakos, V.P., Magoulas, G.D., Vrahatis, M.N.: Objective Function "stretching" to Alleviate Convergence to Local Minima. Nonlinear Analysis TMA 47, 3419–3424 (2001)
3. Eberhart, R., Shi, Y.: Comparison between Genetic Algorithms and Particle Swarm Optimization. In: The 7th Annual Conference on Evolutionary Programming, San Diego, USA, pp. 69–73 (1998)
4. Hu, X., Shi, Y., Eberhart, R.: Recentt Advenes in Particle Swarm. In: Congress on Evolutionary Computation, Portland, Oregon, June 19-23, 2004. June 19-23, pp. 90–97 (2004)
5. Shi, Y., Eberhart, R.: A Modified Partilce Swarm Optimzer. In: Proceedings of the IEEE Congress on Evolutionary Computation (CEC 1998), Piscataway, NJ, pp. 69–73. IEEE Computer Society Press, Los Alamitos (1998)
6. van den Bergh, F., Engelbrecht, A.P.: Cooperative Learning in Neural Networks using Particle Swarm Optimization. South African Computer Journal, 84–90 (November 2000)
7. Xie, X., Zhang, W., Yang, Z.: Hybrid Particle Swarm Optimizer with Mass Extinction. In: International Conf. on Communication, Circuits and Systems (ICCCAS), Chengdu, China, pp. 1170–1174 (2002)
8. Lovbjerg, M., Krink, T.: Extending Particle Swarm Optimisers with Self-Organized Criticality. Proceedings of Fourth Congress on Evolutionary Computation 2, 1588–1593 (2002)
9. Coelho, L.S., Krohling, R.A.: Predictive controller tuning using modified particle swarm optimization based on Cauchy and Gaussian distributions. In: Proceedings of the 8th On-Line World Conference on Soft Computing in Industrial Applications. WSC8 (2003)
10. Hu, X., Eberhart, R.C.: Adaptive particle swarm optimization: detection and response to dynamic systems. In: Proc. Congress on Evolutionary Computation, pp. 1666–1670 (2002)
11. Wang, H., Liu, Y., Li, C.H., Zeng, S.Y.: A hybrid particle swarm algorithm with Cauchy mutation. In: IEEE Swarm Intelligence Symposimu 2007. SIS 2007, Honolulu, Hawaii, USA (in press)

12. Blackwell, T.M.: Particle swarm and population diversity I: Analysis. Dynamic optimization problems, pp. 9–13 (2003)
13. Blackwell, T.M., Bentley, P.J.: Dynamic search with charged swarms. In: Langdon, W.B., et al. (eds.) Genetic and Evolutionary Computation Conference, pp. 19–26. Morgan Kaufmann, San Francisco (2002)
14. Janson, S., Middendorf, M.: A hierachical particle swarm optimizer for dynamic optimization problems. In: Raidl, G.R., Cagnoni, S., Branke, J., Corne, D.W., Drechsler, R., Jin, Y., Johnson, C.G., Machado, P., Marchiori, E., Rothlauf, F., Smith, G.D., Squillero, G. (eds.) EvoWorkshops 2004. LNCS, vol. 3005, pp. 513–524. Springer, Heidelberg (2004)
15. Li, X., Dam, K.H.: Comparing particle swarms for tracking extrema in dynamic environments. In: Congress on Evolutionary Computation, pp. 1772–1779 (2003)
16. Blackwell, T.M., Branke, J.: Multi-swarm optimization in dynamic environments. In: Raidl, G.R., Cagnoni, S., Branke, J., Corne, D.W., Drechsler, R., Jin, Y., Johnson, C.G., Machado, P., Marchiori, E., Rothlauf, F., Smith, G.D., Squillero, G. (eds.) EvoWorkshops 2004. LNCS, vol. 3005, pp. 489–500. Springer, Heidelberg (2004)
17. Parrott, D., Li, X.: A particle swarm model for tracking multiple peaks in a dynamic environment using speciation. In: Congress on Evolutionary Computation, pp. 98–103 (2004)
18. Ozcan, E., Mohan, C.K.: Particle Swarm Optimization: Surfing the Waves. In: Proceedings of Congress on Evolutionary Computation (CEC1999), Washington DC, pp. 1939–1944 (1999)
19. van den Bergh, F., Engelbrecht, A.P.: Effect of Swarm Size on Cooperative Particle Swarm Optimizers. In: Genetic and Evolutionary Computation Conference, San Francisco, USA, pp. 892–899 (2001)
20. Feller, W.: An Introduction to Probability Theory and Its Applications, 2nd edn., vol. 2. John Wiley & Sons, Inc., Chichester (1971)
21. Yao, X., Liu, Y., Lin, G.: Evolutionary Programing Made Faster. IEEE Transacations on Evolutionary Computation 3, 82–102 (1999)
22. Veeramachaneni, K., Peram, T., Mohan, C., Osadciw, L.A.: Optimization Using Particle Swarms with Near Neighbor Interactions. In: Cantú-Paz, E., Foster, J.A., Deb, K., Davis, L., Roy, R., O'Reilly, U.-M., Beyer, H.-G., Kendall, G., Wilson, S.W., Harman, M., Wegener, J., Dasgupta, D., Potter, M.A., Schultz, A., Dowsland, K.A., Jonoska, N., Miller, J., Standish, R.K. (eds.) GECCO 2003. LNCS, vol. 2723, pp. 110–121. Springer, Heidelberg (2003)
23. van den Bergh, F.: An Analysis of Particle Swarm Optimizers. PhD thesis, Department of Computer Science, University of Pretoria, South Africa (2002)
24. Zhang, W., Xie, X.: DEPSO: Hybrid particle swarm with differential evolution operator. In: IEEE Int. Conf. on System, Man & Cybernetics (SMCC), Washington, USA, pp. 3816–3821 (2003)

An Improved Multi-Objective Particle Swarm Optimization Algorithm

Qiuming Zhang and Siqing Xue

School of Computer, China University of Geosciences, Wuhan, P.R. China
qmzhang@cug.edu.cn, xue.xues@gmail.com

Abstract. This paper proposes a dynamic sub-swarms multi-objective particle swarm optimization algorithm (DSMOPSO). Based on solution distribution of multi-objective optimization problems, it separates particles into multi sub-swarms, each of which adopts an improved clustering archiving technique, and operates PSO in a comparably independent way. Clustering eventually enhances the distribution quality of solutions. The selection of the closest particle to the gbest from archiving set and the developed pbest select mechanism increase the choice pressure. In the meantime, the dynamic set particle inertial weight, namely, particle inertial weight being relevant to the number of dominating particles, effectively keeps the balance between the global search in the preliminary stage and the local search in the later stage. Experiments show that this strategy yields good convergence and strong capacity to conserve the distribution of solutions, specially for the problems with non-continuous Pareto-optimal front.

Keywords: Particle swarm optimization, multi-objective optimization, dynamic sub-swarms, clustering ,Inertial weight.

1 Introduction

Particle swarm optimization (PSO) is firstly proposed by James Kennedy and R.C.Eberhart on the ground of swarm intelligence. Its advantages of single objective function stimulates the efforts by many researchers for the further study in its multi-objective applications. In the recent years, better algorithms for multi-objective particle swarm optimization keep pouring out.The most influential ones are embodied by the follows: weighted categorization MOPSO [1](by Parsopoulos and Vrahatis), NSGA2[2](by Kalyanmoy Deb), Xiadong Li improves NSGA2[3] , Gregorio Toscano-Pulido and Carlos A. Coello adopt Clustering technique in MOPSO[4]. The characteristic of Fieldsend's MOPSO[5] is to take advantage of the newest data structure in the storage of the essence among each generation. The last but not the least, Thomas Bartz-Beielstein and the like make experiment analysis of multi-objective particle swarm optimization's archiving mechanism[6] and Sanaz Mostaghim makes a thorough summary as a whole[7].

This essay submits an dynamic sub-swarms multi-objective particle swarm optimization algorithm(DSMOPSO), which introduces dynamic sub-swarms strategy

L. Kang, Y. Liu, and S. Zeng (Eds.): ISICA 2007, LNCS 4683, pp. 372–381, 2007.

into the search of differential sections by setting multiple sub-swarms in order to improve its application abilities and the capacities to dispose of Pareto front non continuous multi-objective optimization problems. Each sub-swarm use improved clustering archiving technique, operate PSO process in a comparably independent way, on the purpose of gaining solution set the most approaching Pareto front, simultaneously producing more properly distributed Pareto optimal set. Experiments have been proved that this algorithm achieves better results.

2 Dynamic Sub-swarm Multi-Objective Particle Swarm Optimization

2.1 Multi-swarm Strategy

There have been some studies on the existing multi-swarm strategy in the utilization of the multi-objective optimization. The adoption of multi-swarm is mainly aimed at two-sided of development: 1) to retain parallels, enhance the quality and extension of the algorithm, and ultimately making use of high-quality parallel computer to solve complicated multi-objective problems; 2)to classify search space, to enable each sub-swarm to detect a smaller section, and realizing the utmost precision for non-dominated set toward true Pareto front. Multi-swarm strategy (AMOPSO) advocated by Toscano is typical of the trial in this field[8]. This algorithm sets a fix number of sub swarm, with given number of particles for each, and carry out an independent PSO process . Mostaghim establishes another multi-swarm strategy[9]. Begin with the implementation of a single swarm multi-objective particle optimization algorithm, the algorithm cultivates optimal set. Next, in the light of the established optimal set, a group of sub-swarms are created. Finally, the particles appearing in the first stage will be categorized into relevant swarms to help the sub-swarms conduct independent operation. To better the application ability of multi-swarms, dynamic sub-swarm strategy is put forward in its number and search section installation. In this event, not only individuals in the swarm can try out dynamic adjustment directing at their own experience as well as their neighbors', the swarm in itself can also make dynamic improvement in contrasting cases. When it comes to specific optimization problems, along with swarm particle distribution, the algorithm makes its decision about the necessity of the division on the swarm's side. The above serves as a brief description of the strategy adopted by the dynamic sub-swarms multi-objective optimization algorithm (DSMOPSO).

2.2 DSMOPSO Algorithm

DSMOPSO replaces single swarms with multi-swarms, whereas its differentiation from fix multi-swarm strategy lies in the fact that its multi-swarm installation is done determined by the distribution in decision making space on the part on optimal set during the implementation of the algorithm, and confirmed by Clustering technique.

The algorithm starts with the set of an initial swarm, after running a certain generations, cluster the handy optimal set, and then isolates the swarm into multiple sub-swarms. The particles in the initial swarm are allotted to the nearest sub-swarms in line with the distance to clustering center. Afterwards, each sub-swarm carries out PSO process independently. The same split occurs to each sub-swarm after some generations. Those that can't satisfy the demand for split drop out. At the same time, those that meet the need of elimination will also be crossed out after checked. The same procedure is repeated until the algorithm comes to an end.

1) Using the standard PSO initial process, initialize the swarm, with all the particles belonging to the same swarm.

2) Circularization evolution

   2a) Sub-swarms implement PSO process. Operate individually.

   2b) Check whether the existing sub-swarm need split. Do it if necessary.

   2c) Check whether the sub-swarms satisfy the requirement for elimination. Do it if so.

3) Meet the closing requirement for the algorithm. End it.

4) Output the optimal set.

DSMOPSO uses the exterior storage optimal set technique, focusing on the maintenance of the global optimal set ----- GnodominatedSet, together with the preservation of all the optimal set found in the sub-swarms. This optimal set is regarded as the output set after the operation of the algorithm. In every sub-swarm, interior optimal sets are set simultaneously ----- LNodominatedSet☐used for the storage of the optimal set found in the sub-swarm. After sub-swarms independently run for some generations, all the optimal set in sub-swarms will merge into the global optimal set according to the Pareto dominating relationship and eliminate the dominated solution in the global optimal set.

2.3 Process of Sub-swarm

In DSMOPSO, the sub-swarms utilize improved Clustering archiveing technique and conduct the PSO process in a comparable independent way. The detailed algorithm is as follows:

1) Initialize swarm P, and create an exterior optimal set ------ NodominateSet:

2) Select non-dominated particles in P, in accordance with their space information make a copy in NodomiateSet;

3) Delete the dominated particles in NodominateSet;

4) If the number of particles exceeds the set number in NodominateSet, cut it down with Clustering technique;

5) Renew individual optimality (PBest) of particles in swarm P decided by Pareto dominating relationship;

6) Take the non-dominated particles with the smallest distance from the particles in NodominateSet as the partial optimality (Lbest) in P;

7) Calculate the new particle speed and position via particle swarm optimization updating formula;

8) Judge the ending conditions. If not satisfying, then go to Step 2.

The concentration of particle swarm optimization algorithm lies in the choice of individual optimality (pbest) and global optimality (gbest). In single objective problems, it could be very spontaneous to hit the target from the objective function. In contrast, in multi-objective problems, it would be much more complex to decide which particle should take the first priority., which might result in a tough choice.

The algorithm chooses particle guide (gbest) from the archived optimal set, calculates the distance of solution between particles and archived optimal set in each generation. Then choose the optimal solution with the smallest distance as its gbest. The measurement of the distance is of European standard. Similar to neighboring method for distance choice, the most approaching optimal solution follows.

Nowadays, the existing multi-objective particle swarm optimization algorithms mainly focus on the improvement of the choice of gbest. Generally speaking, for pbest's updating, it only happens when the new position dominates pbest, otherwise it remains unchanged. Due to the reason that in the closing period of algorithm, the present particles and Pbest are frequently non-dominated, Pbest is rarely renewed, and its function is as a result decreased.

Here, an improvement for the choice of individual optimality is provided: Update Pbest with the current particles if Pbest is dominated by the latter. Retain Pbest in case Pbest dominate the current particles. But when the current particle and Pbest are nondominated by each other, first calculate the particle number in the swarm dominated by it and then choose the particle with the greatest number of dominating particles as Pbest.

2.4 Clustering Archiving Technique

The design of multi-objective optimization algorithm not only requires the good convergence quality, but also demands the appropriate distribution quality of the founded solution in the whole Pareto front space. The technique of Clustering can be used to enhance the distribution quality; what's more, if combined with gbest choice, it can extend the varieties of the swarm.

Allowing for the time limit for the algorithm, less complex Single Linkage Method appears. This clustering algorithm needs to calculate the distance between all the particles, with the standard of European distance measurement. The specific process is shown below:

1) Create an initial C; put every individual into a different cluster, and calculate the distance between each pair;
2) If C is smaller than the set N, go to the end; otherwise, go to Step 3;
3) Choose the two clusters with the smallest distance and incorporate them. Take the central point of the ongoing two as the position of the new cluster;
4) Renew distance matrix, and then go to Step 2;

2.5 Dynamic Setting of Inertial Weight

Inertial weight value W plays a crucial role in the convergence quality of particle swarm optimization algorithm. It takes control of the effect from the historic speed on the present one, and balance the global research and the partial one. The larger inertial weight is beneficial for the global research; in comparison, the smaller one is preferential for the partial. In dynamic installation of inertial weight, to calculate each particle's inertial weight, the following formula is used:

Inertial weight value w= swarm particle number N/ (swarm particle number N + particle number dominated by the particle)

It can be concluded from the above formula that inertial weight w is [0.5,1], particle inertial weight in the earlier stage tends to be 1; in the later stage it is liable to be 0.5. Superior particles seem to have smaller inertial weight value. According to the experiment ,a more balanced global and partial search can be achieved, which ultimately realizes the higher convergence speed.

2.6 Swarm Division Mechanism and Sub-swarm Elimination Mechanism

The main advantage of DSMOPSO is to adopt dynamic swarm division strategy, which is accomplished by Clustering technique. Swarm number is considered as parameter. Every time sub-swarm finishes independent PSO process, the first thing is to check the necessity of division. If necessary, cluster sub-swarm optimal set LnodomiatedSet. Assign the particles in sub-swarm to each sub-cluster in line with the distance to the central point of the new cluster. The adopted Clustering algorithm is the same as that mentioned above.

The criterion to decide whether sub-swarm satisfies the requirement of division is the average distance in sub-swarm smaller than the parameter Dmindistance. The average distance within a swarm is of European standard. The set of Dmindistance depends on specific problems. Another simple method is that when sub-swarm number reaches the largest number, no more division is allowed.

In addition to swarm division operation, the algorithm will also clear out the sub-swarm where no more new global optimal set can be found at anytime possible. After the accomplishment of each independent implementation of PSO by sub-swarm, if all the solution in sub-swarm optimal set LnodomiatedSet is dominated by global optimal set GnodominatedSet, sub-swarm will contribute nothing to the problem solution and then die. In this way, the search in some invalid sections can be reduced to the minimum, as a result which validity can been greatly improved.

The DSMOPSO algorithm concentrates on its swarm division mechanism and elimination mechanism. Here we use Deb2 function to do the experiment in order to

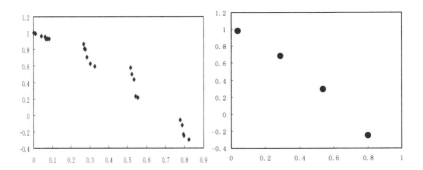

Fig. 1. First Swarm Division Optimal Set   **Fig. 2.** Central Points in Sub-swarm

prove the validity of this method Pareto front of Deb2 is non-continuous. The parameters in the experiment are as follows: swarm individual number 100, generation 100, and limited sub-swarm number 10.

1) After the 25th generation, the situation prior to the first swarm division is to find 23 optimalities. The specific distribution is shown by the Fig.1:
2) Clustering archiving optimal set to determine the central points of sub-swarm when swarm divides. Four central points of sub-swarm are retained as shown in Fig 2.
3) Carry out swarm division operation, and establish 4 sub-swarms. After 25th generation, the distribution of the 4 sub-swarms in search space and decision-making space is shown by Fig. 3 and Fig. 4.

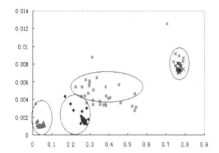

Fig. 3. Particle Distribution of Sub-swarms

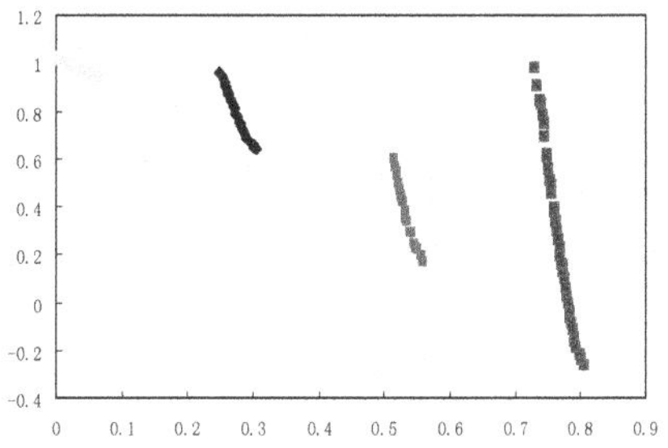

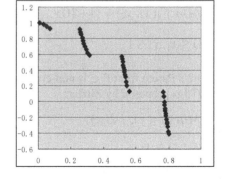

Fig. 4. Corresponding Solution Distribution of Sub-swarm

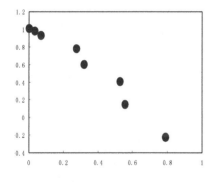

Fig. 5. 8 Central Points in Sub-swarms

Fig. 6. Deb2 Optimal Set

4) Proceed with swarm division process in Step 2, sub-swarms split once again, and gain 8 central points this time, shown by Fig. 5:
5) Algorithm process progresses, individual swarms independently transform, and finally Deb2 optimal set comes into existence, as shown in Fig. 6.

Table 1. Deb2 Swarm Division Process

generation	Sub swarm number	eliminated sub-swarm number	general swarm number
1	1	0	1
25	4	0	0
50	10	2	8
75	12	2	10
100	12	2	10

Table 1 give out the division swarm number and eliminated swarm number.From the above table, the swarm division process demonstrates the expected outcome of sub-swarm division mechanism and elimination strategy.

By means of dynamic sub-swarm strategy,DSMOPSO will divide particles into multiple sub swarms. Although DSMOPSO does not apparently separate decision-making space, it can still confine clustering split sub-swarm in the neighboring areas around the clustering center. That is why this algorithm can well dispose of such non-continuous problems.

3   Experimental Outcome and Analysis

We choose 2 Pareto front non-continuous testing function Deb2 and ZDT3 to test DSMOPSO: They are defined as follows:

Deb2 is a double objective problem proposed by Deb:

$$Min\ f_1(x_1, x_2) = x_1,$$

$$Min\ f_2(x_1, x_2) = (1+10x_2) * \left[1 - (\frac{x_1}{1+10x_2})^2 - \frac{x_1}{1+10x_2} * \sin(2\pi 4x_2)\right]$$

$$0 \le x_1, x_2 \le 1$$

Deb2's Pareto front Fig. is non-continuous.
ZDT3 is proposed by Zitzler:

$$Min\ (f_1(\overline{x}), f_2(\overline{x}))$$

$$f_1(\overline{x}) = x_1$$

$$f_2(\overline{x}) = g(\overline{x})h(f_1, g)$$

$$g(\overline{x}) = 1 + 9\sum_{i=2}^{m} x_i / (m-1)$$

$$h(f_1, g) = 1 - \sqrt{f_1/g} - (f_1/g)\sin(10\pi f_1)$$

$$0 \le x_1 \le 1, m = 30$$

ZDT3's Pareto front consists of a couple of non-continuous convex surfaces.

In DSMOPSO,swarm individual number is 100; general generation is 100, and confined sub-swarm number is 5. Exterior NodominateSet size as 80,   w=0.5,c1=1, c2=1. In the light of DSMOPSO algorithm, Fig. 6 illustrates Deb2 Pareto front; Fig. 7 reveals ZDt3 Pareto front.

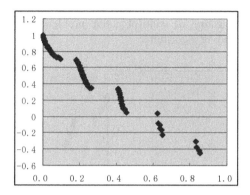

Fig. 7. ZDT3 Pareto Front

From Fig. 6 and 7, we can infer that DSMOPSO has got the ideal Pareto front in the two testing problems above.

To further confirm the validity of DSMOPSO algorithm, this essay chooses ZDT3 function for contrasting testing. Compare DSMOPSO algorithm with SPEA, NSGA, and CoelloMOPSO, the experimental data for SPEA and NSGA originates from data testing website maintained by Aitzler: http://www.tik.ee.ethz.ch/~zitzler/testdata.html

The data for CoelloMOPSO results from the literature review for its algorithm program. In the following experiment, the parameters for CoelloMOPSo include swarm number 100, w=1, c1=1,c2=1, section division number 25, generation   100, and archived optimal set size 80.

Fig. 8 illustrates the experimental outcome. Taking the uncertainty of particle algorithm into consideration, all the method takes the best as the output among the 10 operations. Table 2 makes a list of the number of optimality gained from the four algorithms, the number of dominated optimality by DSMOPSO and the number of dominating optimality in DSMOPSO optimal set.

Table 2. Optimal Set Number

	SPEA	**NSGA**	**CoelloMOPSO**	**DSMOPSO**
Optimality Number	72	58	80	80
Dominated Number	72	58	64	80
Dominating Number	48	42	44	80

From the results of ZDT3 experiment, we can see obviously that DSMOPSO are overwhelming advantageous over SPEA and NSGA. CoelloMOPSO and this algorithm has the same optimality num,with part of each dominating each other , with metric value C(DSMOPSO, CoelloMOPSO) 80% and C(CoelloMOPSO ,DSMOPSO) 55% respectively, but DSMOPSO still gets the upper hand as a whole.

Via individual and contrasting experiments, the algorithm in this essay is effective for the disposal of the testing of multi-objective problems, and brings about better result than the corresponding algorithms.

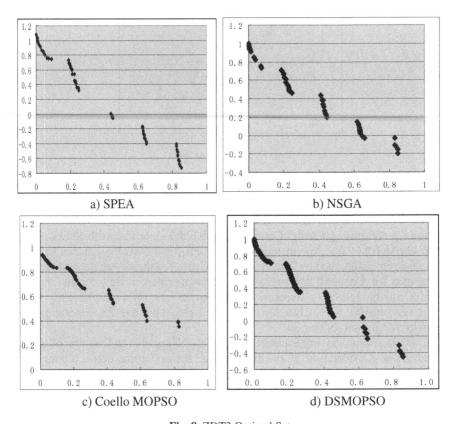

a) SPEA

b) NSGA

c) Coello MOPSO

d) DSMOPSO

Fig. 8. ZDT3 Optimal Set

4   Conclusion

Dynamic sub-swarm multi-objective particle swarm optimization algorithm DSMOPSO utilizes dynamic sub-swarm strategy and clustering archiving technique in order to enhance the distribution quality of solution. It increases the choice pressure through archiving method and selecting the nearest gbest and improved pbest choice mechanism. Meanwhile, the actively set particle inertial weight effectively keeps the balance between the global search in the preliminary stage and the partial search in the later stage. It has been proved by the available experiments that this strategy processes better convergence and stronger capacity to sustain the distribution of solutions. In addition, it can more effectively dispose of experimental multi-objective optimization problems, in particular in accordance with the non-continuous Pareto front problems, DSMOPSO achieves better results.

References

[1] Parsopoulos, K.E., Vrahatis, M.N.: Particle Swarm Optimization Method in Multiobjective Problems. In: Nyberg, K., Heys, H.M. (eds.) SAC 2002. LNCS, vol. 2595, pp. 603–607. Springer, Heidelberg (2003)
[2] Deb, K.: A fast and elitist multiobjective genetic algorithm:NSGA-II. IEEE Transactions on Evolutionary Computation, 182–197 (2002)
[3] Li, X.: A Non-dominated Sorting Particle Swarm Optimizer for Multiobject Optimization. In: Cantú-Paz, E., Foster, J.A., Deb, K., Davis, L., Roy, R., O'Reilly, U.-M., Beyer, H.-G., Kendall, G., Wilson, S.W., Harman, M., Wegener, J., Dasgupta, D., Potter, M.A., Schultz, A., Dowsland, K.A., Jonoska, N., Miller, J., Standish, R.K. (eds.) GECCO 2003. LNCS, vol. 2723, Springer, Heidelberg (2003)
[4] Coello Coello, C.A., Lechunga, M.S., MOPSO,: A Proposal for Multiple Objective Particle Swarm Optimization. In: Proceedings of the 2002 Congess on Evolutionary Computation, pp. 1051–1056 (2002)
[5] Fieldsend, J.E., Singh, S., Multi-Objective, A.: Algorithm based upon Particle Swarm Optimisation, an effcient Data Structure and Turbulence. In: Proceedings of UK Workshop on Computational Intelligence (UKCI'02), Birmingham,UK, pp. 37–44 (2002)
[6] Bartz-Beielstein, T., Limbourg, P., Konstantinos, E., et al.: Particle Swarm Optimizers for Pareto Optimization with Enhanced Archiving Techniques. In: Proceedings of the 2003 Congress on Evolutionary Computation (CEC'2003), Canberra, Australia, vol. 3, pp. 1780–1787 (2003)
[7] Mostaghim, S., Teich, J.: Strategies for Finding Good Local Guides in Multi-objective Particle Swarm Optimization (MOPSO). In: 2003 IEEE Swarm Intelligence Symposium Proceedings, Indiana,USA, pp. 26–33 (2003)
[8] Toscano, G., Coello, C.: Using Clustering Techniques to Improve the Performance of a Multi-Objective Particle Swarm Optimizer. In: Deb, K., et al. (eds.) GECCO 2004. LNCS, vol. 3102, pp. 225–237. Springer, Heidelberg (2004)
[9] Mostaghim, S., Teich, J.: Covering Pareto-optimal Fronts by Subswarms in Multi-objective Particle Swarm Optimization. In: Congress on Evolutionary Computation (CEC2004), Oregon, USA, vol. 2, pp. 1404–1411 (2004)

Dynamic Population Size Based Particle Swarm Optimization

ShiYu Sun, GangQiang Ye, Yan Liang, Yong Liu, and Quan Pan

College of Automation, Northwestern Polytechnical University, Xi'an 710072, China
{ssy425, ygqml}@163.com, liangyan@nwpu.edu.cn, dudu2077@163.com
quanpan@nwpu.edu.cn

Abstract. This paper is the first attempt to introduce a new concept of the birth and death of particles via time variant particle population size to improve the adaptation of Particle Swarm Optimization (PSO). Here a dynamic particle population based PSO algorithm (DPPSO) is proposed based on a time-variant particle population function which contains the attenuation item and undulate item. The attenuation item makes the population decrease gradually in order to reduce the computational cost because the particles have the tendency of convergence as time passes. The undulate item consists of periodical phases of ascending and descending. In the ascending phase, new particles are randomly produced to avoid the particle swarm being trapped in the local optimal point, while in the descending phase, particles with lower ability gradually die so that the optimization efficiency is improved. The test on four benchmark functions shows that the proposed algorithm effectively reduces the computational cost and greatly improves the global search ability.

Keywords: Particle Swarm Optimization; Dynamic Population Size;Population; Swarm Diversity.

1 Introduction

Particle swarm optimization（PSO）[1] is a kind of swarm searching based evolutionary computation first introduced by Eberhart and Kennedy in 1995. The algorithm is motivated by the regulations of animal social behaviors such as fish schooling, bird flocking, etc. The swarm search tendency is combined with the individual (particle) search process. PSO randomly creates an initial particle swarm and sets every particle a random velocity. While "flying", the velocity of a particle is dynamically adjusted by the experiences of itself and companies. Therefore, through cooperation among individuals in the swarm, PSO finally gets the best solutions to the problem. PSO is now widely used in many domains such as image processing, data mining, target optimizing etc. because of its simplicity, easy realization and fewer parameters needed to adjust.

Although the PSO have many advantages, it is easy to become premature, and the population diversity is easy to be destroyed with the time elapse. For this question, there has been tremendous work done in improving its performance. Shi and Eberhart [2] introduce the inertia weight on the base of canonical PSO which makes the

L. Kang, Y. Liu, and S. Zeng (Eds.): ISICA 2007, LNCS 4683, pp. 382–392, 2007.
© Springer-Verlag Berlin Heidelberg 2007

algorithm have a balanced searching ability on both global and local scale. He Ran [3] introduces an adaptive escape PSO algorithm (AEPSO) that can adaptively change the velocity of particles and has a better global optimization. Based on the self-organizing algorithm, Ratnaweera [4] introduces the self-organizing hierarchical PSO (HPSO) with time-varying mutation operations in order to strengthen the search ability. Ratnaweera also provides the way in which the adaptive parameter is selected. But the inertia part is removed from the velocity equation and the condition of the mutation happening is that the velocity is at zero, which makes the particle unable to escape the local optima quickly and efficiently. Keiichiro [5] introduces an algorithm dynamically changing the max velocity of the particles which has a good convergence speed and local convergence capability.

We analyse the effect of PSO in different population size, and proposed DPPSO algorithm which through changing population size to completing the optimization. DPPSO algorithm contains the attenuation item and undulates item. The attenuation item makes the population decrease gradually in order to reduce the computational cost because the particles have the tendency of convergence as time pass. The undulate item consists of periodical phases of ascending and descending. In the ascending phase, new particles are randomly produced to avoid the particle swarm being trapped in the local optimal point; while in the descending phase, particles with lower ability gradually die so that the optimization efficiency is improved.

2 Analyze the Changing of Population Size

In the canonical PSO, each individual in PSO flies in the search space with a velocity which is dynamically adjusted according to its own flying experience and its companions' flying experience. Each individual is treated as a volume-less particle (a point) in the D-dimensional search space. The ith particle is represented as $x_i = (x_{i1}, x_{i2} ..... x_{iD})$. The best previous position (the position giving the best fitness value) of the ith particle is recorded and represented as $P_i = (p_{i1}, p_{i2} ..... p_{iD})$. The index of the best particle among all the particles in the population is represented by the symbol g. The rate of the position change (velocity) for particle i is represented as $v_i = (v_{i1}, v_{i2} ..... v_{iD})$. The particles are manipulated according to the following equation:

Where c_1 and c_2 are acceleration coefficients, they are two positive constants, and $rand_1()$ and $rand_2()$ are two random functions in the range [0, 1].

$$v_{id}(t+1) = wv_{id}(t) + c_1 rand_1(t)(p_{id}(t) - x_{id}(t)) + c_2 rand_2(t)(p_{gd}(t) - x_{id}(t)) \qquad (1)$$

$$x_{id}(t+1) = x_{id}(t) + v_{id}(t+1) \qquad (2)$$

Function (1) and (2) constitute the canonical PSO. However, it is easy to premature and falls into local extremum. [8] analyses the effect of different population size for PSO. The result shows it is easy trap into local extremum using fewer particles. Moreover, it is also increase computing cost and decrease convergent velocity using more particles. This paper research on the effect of convergence velocity and precision in difference population size based on [8].The experiment is process in canonical PSO for f3.Where the experimental parameter is w =0.6 ; c_1=2 ; c_2=2, V_{max} =100, generation is set to 3000.

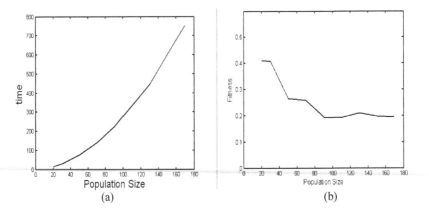

Population Size
(a)   (b)

Fig. 1. (a) illustrates the effect of population size on convergence velocity. It shows that the computing cost goes up fast while increase of population size. (b) shows the effect of population size on convergence precision. This figure shows that the convergence ability can not be improved because the prematurely of PSO, but it will be stabilization when the particles number reach certain numbers. The reason is that the population diversity is destroyed when the gradually centralize of particles. So using fewer particles to improve the searching ability is the main goal of DPPSO.

3 Dynamic Population Size Based Particle Swarm Optimization, DPPSO

It is shown in the above analysis that the fewer particles will result in the decrease of computing cost, but will increase the probability of trapping local extremum. And more particles will result in the increase of computing cost, but the optimization ability will not be improved. So the population can change its size according the state of searching.

Equation (3) is the designed the changing function of population size.

$$N(t) = N_d(t) + N_l(t) \tag{3}$$

Where $N(t)$ is the total particle number in time t, N_d is the attenuation item, N_l is the period fluctuate item. In the attenuation item, the population size will be smaller and smaller because of the die of redundancy and inefficient particles with centralizing of the particles, this can decrease computation cost and have little effect to particles' searching ability. In the ascending phase of period fluctuate item, new particles are randomly produced to avoid the particle swarm being trapped in the local extremum; while in the descending phase of period fluctuate item, particles with lower ability gradually die so that the optimization efficiency is improved.

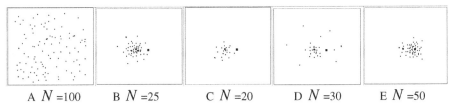

A N =100   B N =25   C N =20   D N =30   E N =50

Fig. 2. Shows the process and character of DPPSO. The population size form 100 decreases to 20 and then increase to 50 for f_1 (shows in figure 1) and the black pane is the global extremum (Represented as P); the big black point is local extremum (Represented as P'); other black point is the working particles. Figure A is the initial state of particle. The particles are randomly placed. Fig. B and Fig. C show the number of particles is decreased continually with centralizing of population, but the population lost the global search ability because of the population trap in P'. Fig. D shows it produces some new particle to improve the diversity of population and searching ability. As can b seen form Fig. e, particles are centralized form P' to P continually.

The flow chart of DPPSO as fellows:

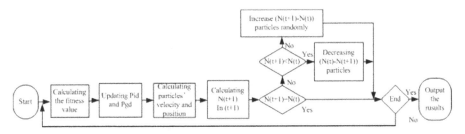

Fig. 3. Flow chart of DPPSO Algorithm

- Initialize the population $P(t)$ and velocity V_0 ;
- Calculating the particle fitness $F(t)$;
- Updating every particle's local best position P_{id} and global best position P_{gd} ;
- Calculating the particles' velocity and position according equation (1) and (2);
- Calculating population size of next generation;
- Compare current population size to the next generation population size, if N(t+1)=N(t),then N(t+1) is not change; If N(t+1)<N(t),then decrease some ineffective particles, If N(t+1)>N(t),then increase(N(t+1)-N(t)) particles randomly.
- If the iteration is end, then output the result, otherwise jump to step 2.

4 Experimental Results

4.1 Experiment Designing

For validate the DPPSO, we design a simple population size change function $N(t)$, it denotes the total particles in time t (Equation (4))

$$N(t) = \max(\min(N_d(t) + N_l(t), N_{max}), N_{min})$$

(4)

Where N_{max} is the maximum population size, N_{min} is the minimum population size; According to the analytical results, more or fewer particles will all result in the negative effect, so we design the maximum and minimum population size for DPPSO algorithm. To verify the robust of DPPSO, we only consider the time and optimize target factors and let the $N(t)$ change with the change of time and target. For the two items of $N(t)$, we let $N_l = 0$ and $N_d = 0$ respectively, and construct the line change strategy, Sine change strategy and Sine attenuation change strategy. Moreover, we analyze the algorithm in different population size.

4.1.1 The Damping Function (Equation (5))

Equation (5) is the population size function when $N_l = 0$.

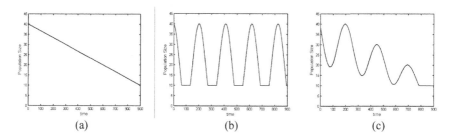

(a)   (b)   (c)

Fig. 4. Three strategies of different Population Size (a) Line change strategy. (b) Sine change strategy. (c)Sine attenuation change strategy.

The abbreviation of this PSO algorithm is DPPSO-1. The main property of this strategy is fully using larger population size at the beginning to achieve quick convergence and then gradually decreasing the particles to decrease computational cost.

$$N1 = (N_{max} - N_{min}) \cdot \frac{T - k}{T} + N_{min}$$

(5)

Where T is the total iterative times, k is the current iterative times.

4.1.2 The Sine Function (Equation (6))

Equation (6) is the population size function when $N_l = 0$.

This strategy is called DPPSO-2. Its property is the amount of particles fluctuates periodically and the advantages are mainly lie in following three aspects:

- When the curve goes down, the amount of particles decreases, which reduces the computational cost.
- When the curve goes up, part of particles need to be initialized again, which improves the swarm diversity and the global convergence ability.

- There is a platform at the bottom of the curve which makes the particles around the minimal point convergent smoothly.

$$N2 = m\,ax(N_{max}/2 \cdot \sin(k/A) + N_{max}/2, N_{min})\tag{6}$$

Where A is the factor of periods which is determined by the convergent precision and the particle's velocity.

4.1.3 Sine Attenuation Function (Equation (7))
The shortening of this strategy is DPPSO-3, which is constructed considering the advantages of DPPSO-1 and DPPSO-2. It can not only make the amount of particles decreases linearly but also oscillates periodically, which reduces the computational cost without losing global convergence ability.

$$N3 = m\,ax(N1/3 \cdot \sin(k/A) + 2N1/3, N_{min})\tag{7}$$

4.2 Experimental Design and Analysis

The experiments include two parts: the first part is the analysis of the computation cost and search ability of three strategies of DPPSO qualitatively. In the second part we analyses the search ability of DPPSO quantitatively. They are all compare to PSO, AEPSO, MPSO and HPSO in four benchmark functions.

4.2.1 Four Benchmark Function Used in Experiment (Table 1)
f_1 and f_3 are multimode functions which having many local extremum around the global extremum. Therefore, the search is very easy to fall into local extreme. f_2 and f_4 are single mode functions. Its main function is validating the convergent precision.

Table 1. Four Benchmark test Functions

Function	Name	Dim	$[v_{min}, v_{max}]$	Minimum value	Best position
$f_1 = \dfrac{\sin^2\sqrt{x_1^2 + x_2^2} - 0.5}{[1 + 0.001(x_1^2 + x_2^2)]^2} + 0.5$	Schaffer	20	[-100,100]	0	(0, 0)
$f_2 = \sum\limits_{i=1}^{n}(x_i - i)^2$	Sphere	20	[-200,200]	0	(1,2....n)
$f_3 = \dfrac{1}{4000}\sum\limits_{i=1}^{n}x_i^2 - \prod\limits_{i=1}^{n}\cos\dfrac{x_i}{\sqrt{i}} + 1$	Griewank	20	[-100,100]	0	(0,0....0)
$f_4 = \sum[100(x_{i+1} - x_i^2)^2 + (x_i - 1)^2$	Rosenbrock	20	[-300,300]	0	(1,1....1)

4.2.2 Comparing DPPSO with Other PSO in Single Mode Functions
To analyses the different characteristics of DPPSO in single mode functions, the experiment compares canonical PSO, AEPSO, HPSO and MPSO to three strategies of

DPPSO for f_1. The generations is set to 1000, N_{max} and N_{min} refer to [2] [8], other parameters refer to table 2. All the functions are set to 20 dimensions in this paper. The computer used in the experiment is P4 1.7G, 256M Memory.

Table 2. Experimental parameters of DPPSO and other Algorithms

Algorithms	w	C1	C2	N
PSO, AEPSO, MPSO	0.6	2	2	40
HPSO	0	[2.5 0.5]	[0.5 2.5]	40
DPPSO-1, DPPSO-2, DPPSO-3	0.6	2	2	[50 10]

Fig.5 shows the convergent precision and velocity of all algorithms are identical nearly, almost stable in 200~250 generations. The difference is only that the convergent precision of DPPSO are good than other algorithms.

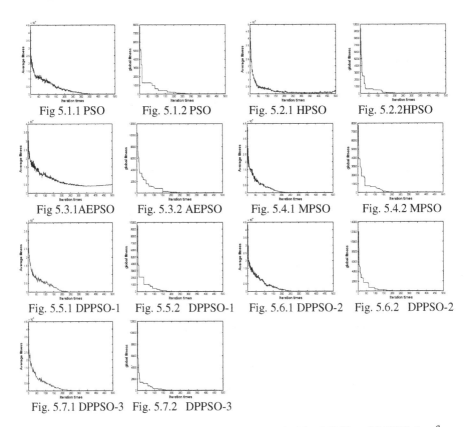

Fig 5.1.1 PSO   Fig. 5.1.2 PSO   Fig. 5.2.1 HPSO   Fig. 5.2.2HPSO

Fig 5.3.1AEPSO   Fig. 5.3.2 AEPSO   Fig. 5.4.1 MPSO   Fig. 5.4.2 MPSO

Fig. 5.5.1 DPPSO-1   Fig. 5.5.2 DPPSO-1   Fig. 5.6.1 DPPSO-2   Fig. 5.6.2 DPPSO-2

Fig. 5.7.1 DPPSO-3   Fig. 5.7.2 DPPSO-3

Fig. 5. The illustration of canonical PSO、HPSO、AEPSO、MPSO and DPPSO for f_1

4.2.3 Comparing DPPSO with Other PSO in Multimode Functions

To analyses the different characteristic of DPPSO in multimode functions, this paper compares canonical PSO, AEPSO, HPSO and MPSO to three strategies of DPPSO for f_1. The generation is set to 3000; other simulation parameters refer to 4.2.2.

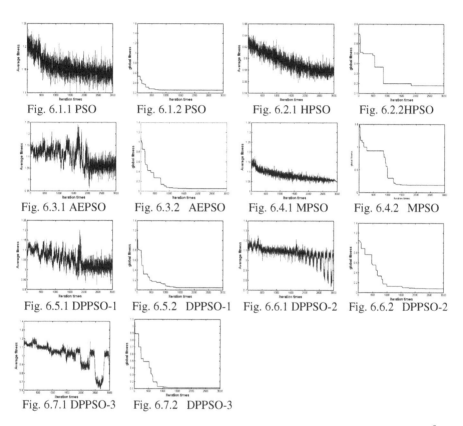

Fig. 6.1.1 PSO   Fig. 6.1.2 PSO   Fig. 6.2.1 HPSO   Fig. 6.2.2 HPSO

Fig. 6.3.1 AEPSO   Fig. 6.3.2 AEPSO   Fig. 6.4.1 MPSO   Fig. 6.4.2 MPSO

Fig. 6.5.1 DPPSO-1   Fig. 6.5.2 DPPSO-1   Fig. 6.6.1 DPPSO-2   Fig. 6.6.2 DPPSO-2

Fig. 6.7.1 DPPSO-3   Fig. 6.7.2 DPPSO-3

Fig. 6. The Simulation of canonical PSO、HPSO、AEPSO、MPSO and DPPSO for f_3

We analyze and compare the average fitness value and global best particle's fitness value of f_3 in every generation. Figure 6.1.1shows that the average fitness value of canonical PSO changes little and the best fitness is almost stable after 500 generations which implies that the swarm convergent to a minimum, the swarm diversity is damaged and the global search ability becomes weak. But in figure 6.6.1 and 6.6.2 the average fitness value of swarm undulates greatly and the global best fitness value decreases as sidesteps under the situation that the population size changes in the form of sine function. The best fitness value keeps unchanged for a very long period in figure 6.6.2, but this is mostly because of getting into a local minimum. As the diversity of swarm changes, after 1500 generations it begins to decrease again, which definitely proves the advantage of DPPSO. The analysis on figure 6.7.1 and 6.7.2 is the same with figure 6.6 while it shows a better performance. Moreover, Fig. 6.6.1 and Fig. 6.7.1 show the average fitness has a severe fluctuate, which indicates the search ability is improved with the centralization and diffusion of particles. So the results show that DPPSO have the strong global search ability, especially for the multimode functions, strategy 2 and 3 are more effective. Although AEPSO, HPSO and MPSO can also jump out the local best point in different degree, but the convergent precision and computation cost are not well balanced as DPPSO.

4.2.4 The Experimental Results of Comparing DPPSO with PSO、 AEPSO、 HPSO and MPSO

To analyze the effect of DPPSO in different population size changing strategies, this paper compares canonical PSO, AEPSO, HPSO and MPSO with three strategies of DPPSO in four functions. The generations is set as 500, 600, 2000 and 2000 for f_1 , f_2 f_3 and f_4 respectively, A total of 50 runs for each experimental setting are conducted. Other simulation parameters refer to table 2.

Analysis of the experimental results for f_1 and f_3: Table 3 shows DPPSO-2 and DPPSO-3 finding the global optimum probability are higher than other algorithms. The reason is that the population diversity of DPPSO algorithm is destroyed with the particles centralizes gradually. Moreover, AEPSO, HPSO and DPPSO-3 have a good search ability that is because those algorithms maintained the population diversity. Especially for DPPSO-3, it can produce new particles to improve population diversity. And the particles are decrease gradually, so the computational cost is lower than other algorithms. For f_3, we can find that the computational cost of DPPSO-2 and DPPSO-3 are comparatively low meanwhile the convergence precision is high.

Analysis of the experimental results for f_2 and f_4 : Table 3 shows that the convergent precision of all algorithms are almost identical. The experimental results for f_2 and f_4 are better than f_1 and f_3, that is because the single mode functions are not need maintain the population diversity. We can also see from table 3 that DPPSO-1 decrease computational cost while improving the convergent precision. Moreover, the computational cost of DPPSO-2 and DPPSO-3 are all lower than other algorithm while in a better convergent precision.

In the whole experiments, for the single mode functions such as f_2 and f_4, PSO、 AEPSO、 HPSO and DPPSO all have a good effect. And for multimode functions such as f_1 and f_3, AEPSO, HPSO and DPPSO can all avoid the prematurity. Especially, DPPSO has a higher convergence precision and lower computational cost than other algorithms.

Table 3. Simulation results of DPPSO and other Algorithms

Function	Name	Minimum value	Maximum value	Average fitness	Run time
f_1	PSO	2.3654e-28(68%)	7.3250e-14	5.2621e-20	45.9
	AEPSO	6.2154e-29(84%)	5.3621e-15	1.2367e-22	49.2
	HPSO	4.2102e-29(80%)	2.0021e-15	9.2356e-21	43.9
	MPSO	3.2015e-28(70%)	5.2154e-15	4.2314e-21	46.2
	DPPSO-1	6.8448e-29(72%)	5.9854e-16	1.0025e-20	43.2
	DPPSO-2	7.0250e-30(82%)	4.3201e-14	7.6215e-21	44.2
	DPPSO-3	6.2154e-30(90%)	4.2015e-16	4.3268e-22	38.5
f_2	PSO	5.5568e-17	6.1564e-15	2.5641e-16	1.30
	AEPSO	1.2545e-21	8.2564e-19	5.5145e-20	1.45
	HPSO	1.9254e-21	6.5148e-16	3.2516e-19	1.31
	MPSO	2.8882e-17	1.2780e-14	1.5692e-16	1.44
	DPPSO-1	3.2223e-18	2.525e-15	5.2365e-17	1.42
	DPPSO-2	2.7848e-18	1.6575e-14	6.5244e-17	1.48
	DPPSO-3	2.6587e-21	9.2548e-18	9.2156e-20	1.29
f_3	PSO	0.1026	0.1952	0.2129	71.3
	AEPSO	0.0845	0.1525	0.1109	73.7
	HPSO	0.1025	0.1587	0.1337	70.4
	MPSO	0.1022	0.2055	0.2360	71.6
	DPPSO-1	0.1171	0.1765	0.1562	62.0
	DPPSO-2	0.0985	0.1342	0.1217	63.1
	DPPSO-3	0.0854	0.1025	0.1125	62.5
f_4	PSO	1.0002e-32	5.3265e-19	2.5148e-25	1.95
	AEPSO	2.1025e-33	8.2541e-20	4.2514e-24	2.10
	HPSO	2.3685e-38	3.0021e-20	6.2107e-24	2.04
	MPSO	6.2155e-33	2.3214e-20	7.2198e-25	2.03
	DPPSO-1	6.2150e-36	7.2010e-19	6.5214e-26	1.31
	DPPSO-2	1.2001e-39	5.2367e-23	3.2541e-25	1.48
	DPPSO-3	9.2105e-40	8.2541e-21	4.2158e-24	1.42

5 Conclusions

Based on the analysis of effects of population size on PSO, this paper introduces the PSO with the population size change strategy. Improving the global research ability and keeping the population diversity through changing particle population size. Particle population function contains the attenuation item and undulates item. The attenuation item makes the population decrease gradually in order to reduce the computational cost because the particles have the tendency of convergence as time pass. The undulate item consists of periodical phases of ascending and descending. In the ascending phase, new particles are randomly produced to avoid the particle swarm being trapped in the local optimal point; while in the descending phase, particles with lower ability gradually die so that the optimization efficiency is improved. We do the test with four benchmark functions and the results show that DPPSO performs better than canonical PSO, MPSO, HPSO and AEPSO in single mode problems but it is not that remarkable, for multimode functions both the global search ability and computational cost improve significantly.

References

1. Kennedy, J., Eberhart, R.: Particle swarm optimization [J]. Conf. Neural Networks 11, 1942–1948 (1995)
2. Shi, Y., Eberhart, R.: Empirical study of particle swarm optimization. In: International Conference on Evolutionary Computation [A], Washington, pp. 1945–1950 (1999)
3. Ran, H., Yong-Ji, W., Qing, W., Jin-Hui, Z., Chen-Yong, H.: An Improved Particle Swarm Optimization Based on Self-Adaptive Escape Velocity. Journal of Software 16(12) (2005)
4. Ratnaweera, A., Halgamuge, S.K., Watson, H.C.: Self-Organizing hierarchical particle swarm optimizer with time-varying acceleration coefficients [J]. Evolutionary Computation 8(3), 240–255 (2004)
5. Fan, H.Y.: A modification to particle swarm optimization algorithm [J]. Engineering Computation 19(8), 970–989 (2002)
6. van den Bergh, F., Engelbrecht, A.P.: A Cooperative Approach to Particle Swarm Optimization [J]. IEEE Transactions on Evolutionary 8(3) (2004)
7. Kennedy, J., Mendes, R.: Neighborhood Topologies in Full-Informed and Best-of-Neighborhood Particle Swarm [J]. In: Soft Computing in Industrial Application. Proceedings of the 2003 IEEE International Workshop (2003)
8. El-Gallad, A., El-Hawary, M., Sallam, A., Kalas, A.: Enhancing the Particle Swarm Optimizer Via Proper Parameters Selection [A]. In: Proceedings of the 2002 IEEE Canadian Conference on Electrical & Computer Engineering (2002)
9. Hu, X.H., Eberhart, R.C.: Adaptive particle swarm optimization: Detection and response to dynamic system [A]. In: Proc. Congress on Evolutionary Computation, pp. 1666–1670 (2002)
10. Clerc, M., Kennedy, J.: The particle swarm——Explosion, stability, and convergence in a multidimensional complex space[J]. IEEE Trans. on Evolutionary Computation 6(1), 58–73 (2002)
11. Eberhart, R.C., Shi, Y.: Comparing inertia weights and constriction factors in particle swarm optimization [A]. In: Proc. 2000 Congress Evolutionary Computation [C], pp. 84–88. IEEE Press, Los Alamitos (2000)
12. Shi, Y., Eberhart, R.C.: Parameter selection in particle swarm optimization [J]. Lecture Notes in Computer Science—Evolutionary Programming VII 14(47), 591–600 (1998)

An Improved Particle Swarm Optimization for Data Streams Scheduling on Heterogeneous Cluster

Tian Xia, Wenzhong Guo, and Guolong Chen

College of Mathematics and Computer Sciences, Fuzhou University, Fuzhou 350002, China
summer0360@163.com, {guowenzhong,cgl}@fzu.edu.cn

Abstract. An improved particle swarm optimization (PSO) algorithm for data streams scheduling on heterogeneous cluster is proposed in this paper, which adopts transgenic operator based on gene theory and correspondent good gene fragments depend on special problem to improve algorithm's ability of local solution. Furthermore, mutation operator of genetic algorithm is introduced to improve algorithm's ability of global exploration. Simulation tests show that the new algorithm can well balance local solution and global exploration and is more efficient in the data streams scheduling.

Keywords: heterogeneous cluster; data streams scheduling; particle swarm optimization; transgenic operator; mutation operator.

1 Introduction

In recent years, the flow of backbone network increase exponentially, and the router process rate has reached the level of 10Gbps, so it asks for higher performance of network services, such as fire wall, IDS, and so on. In order to improve system's capability, veracity and real-time performance, the system can be constructed with the architecture of cluster [1]. Some network device capture all data packets on network, then classify them by some rules, at last transfer them to corresponding machine in cluster to process. A better scheduling scheme of data streams will make full use of cluster parallel processing ability, and achieve low loss and high throughput. So the key is how to allocate data streams to machines reasonably and make scheduling length shortest.

Data streams scheduling is a NP-complete combination optimization problem like task scheduling problem, there are many heuristic algorithms to solve this kind problem, such as MMC(Min-Min complete Time), SMM(Segmented Min-Min) [2] and so on. With the development of evolutionary algorithm in recent years, Genetic Algorithm (GA) has been used to solve these problems [3], and the literature [4] provides a method to improve performance of GA by introducing immunity operator. In these algorithms, GA can get better solution, but it still have some disadvantages and the solution still can be optimized.

This paper puts forward a particle swarm optimization (PSO) algorithm to solve this problem. To improve the ability of GA in solving this problem, Zhang Yan-nian [5]

L. Kang, Y. Liu, and S. Zeng (Eds.): ISICA 2007, LNCS 4683, pp. 393–400, 2007.
© Springer-Verlag Berlin Heidelberg 2007

introduce a transgenic operator base on gene theory. So this paper also designs a transgenic operator and correspondent good gene fragments depend on special problem to improve the PSO's ability of local solution. Furthermore, Darwinian evolutionism points out that gene mutation are very important to keeping population diversity, so this paper introduce mutation operator to improve the algorithm's ability of global exploration.

2 Problem of Data Streams Scheduling

In order to discuss this problem expediently, we abstract it reasonably. On the assumption that the cluster have m machines and all data packets captured on network can be partitioned to n data streams by some strategies, then these n data streams would satisfy two conditions:

1. The executed time of any data stream on any machine can be computed in advance.
2. There is no priority or communication between any two data streams.

The aim is to allocate these n separate data streams to m machines reasonably and make total complete time shortest. Using a $n \times m$-matrix ETM(Executed Time Matrix) to express data streams' executed time on machines, the element etm_{ij} express data stream i's executed time on machine j. Using a $n \times m$-matrix SOL to store allocation scheme, if the data stream i is allocated to machine j, then the element sol_{ij} equals 1, otherwise equals 0. So one machine's complete time is the sum of all data streams allocated to this machine, namely

$$\text{exe}_j = \sum_{i=1}^{n} \text{mat}_{ij} \{ i \,|\, \text{sol}_{ij} = 1 \} \tag{1}$$

The task for allocating data streams is to make scheduling length shortest, also mean:

$$\text{Minimize} \left\{ \underset{i=1}{\overset{m}{\text{Max}}} (\text{exe}_i) \right\}$$

At the same time, machines should take as more average load as possible, it also mean that the balance of cluster load should be big. The balance of cluster load can be defined as:

$$\frac{1}{m} \sum_{i=1}^{m} \frac{\text{exe}_i}{\text{len}} \tag{2}$$

In this equation, len express scheduling length.

3 An Improved Particle Swarm Optimization (IPSO)

3.1 Components of IPSO

Firstly, a discrete PSO must be constructed to fit the problem, so the particle and the operators should be redefined.

1. position

Particle position can be encoded directly, which means that the position of one particle X expresses one allocation scheme. So X is an N-dimension vector, and data stream i is allocated to machine x_i:

$$\mathbf{X} = (x_1, x_2, \cdots, x_i, \cdots, x_N), \quad 1 \leq x_i \leq M \tag{3}$$

In this equation, M is the number of machines, and N is the number of data streams.

2. velocity

Particles velocity represents the change of position, so velocity V is also a N-dimension vector:

$$\mathbf{V} = (v_1, v_2, \cdots, v_i, \cdots, v_N), 1 \leq i \leq N, 0 \leq v_i \leq M \tag{4}$$

Like above, M is the number of machines, and N is the number of data streams. If v_i equals 0, then data stream i is still allocated to primary machine, otherwise reallocated to machine v_i.

3. position and speed addition

Particle position will be updated by the effect of velocity:

$$\mathbf{X}_2 = \mathbf{X}_1 + \mathbf{V} \tag{5}$$

Every dimension value of X can be confirmed as this:

$$x_{2i} = \begin{cases} x_{1i} & \text{if } v_i = 0 \\ v_i & \text{otherwise} \end{cases} \tag{6}$$

4. position subtraction

Velocity expresses the change of position, so its value can be confirmed by position subtraction:

$$\mathbf{V} = \mathbf{X}_2 - \mathbf{X}_1 \tag{7}$$

Every dimension value of V should be confirmed as this:

$$v_i = \begin{cases} 0 & \text{if } x_{1i} = x_{2i} \\ x_{2i} & \text{otherwise} \end{cases} \tag{8}$$

5. velocity multiplication

Velocity can be updated by multiplication, namely:

$$V^2 = \alpha \cdot V^1 \quad \alpha \in [0, 1] \tag{9}$$

Velocity multiplication is defined by probability, which means that for velocity V^2, the probability to equal V^1 is α.

6. velocity addition

Velocity can also be updated by addition, namely:

$$V = V^1 + V^2 \tag{10}$$

Velocity addition is defined by probability too, which means that for velocity V, it has the same probability, namely 50%, to equal V^1 or V^2.

7. particle motion equation

After redefining operators above, now we give particle motion equation as follow:

$$V^{t+1} = w \times V^t + c_1 \times (P^t_{pbest} - X^t) + c_2 \times (P^t_{gbest} - X^t) \tag{11}$$

$$X^{t+1} = X^t + V^t \tag{12}$$

3.2 Transgenic Operator

In standard PSO, particles share information by global best solution and local best solution. But in practical problem, they don't make full use of domain knowledge of special problem so that algorithm's ability of local solution is still poor. Transgenic operator can avoid this disadvantage.

In a better solution, there must be a lot of data streams allocated to their best or better machine, so can make use of this domain knowledge to get good gene fragments. Using a $n \times m$-matrix G to express good genes, namely for data stream i, the j-th good machine is G_{ij} ($1 <= j <= m$). It's easy to get matrix G by sorting row vectors of *ETM*.

When transgenic operator start up, it first intercept some genes from global best particle as good gene fragment, and copy them to the particle, then calculate its fitness, if fitness is improved, then break, otherwise do transgenic operation by bit.

When transgenic by bit, first find the machine whose executed time is the longest, called problem machine. If problem machine's executed time can be shorten, then scheduling length will be shorten too. So for each data stream i allocated to problem machine j, considering the best machine G_{i1} for data stream i, if i is not allocated to G_{i1}, then allocate i to G_{i1} and see whether the fitness is improved, if yes, then break, otherwise swap data stream i with any data stream originally allocated to G_{i1} and see whether the fitness is improved. However, it's impossible that all data streams are allocated to their best machines, a lot of them should be allocated to their secondary, tertiary best machines, so besides considering the best machine G_{i1} for data stream i, can also consider secondary best machine G_{i2}, tertiary best machine G_{i3} and more. Introduce the definition of transgenic depth: if depth equals 1, it means only considering the best machine, if depth equals 2, it means considering secondary machine when there is no improvement after considering the best machine, and the like. The bigger the depth is, the more possible to get improvement, but the more time is spent.

3.3 Mutation Operator

PSO is easy to fall into local optimum, so we introduce mutation operator to improve population diversity and the ability of global exploration [6].

First introduce the definition of particle similarity:

$$S_{ij} = \frac{1}{N} \sum_{k=1}^{N} (i_k == j_k ? 1 : 0) \tag{13}$$

S_{ij} express similarity between particle i and j, while k is the k-th gene, and N is the number of genes.

particle concentration:

$$C_i = num/s \qquad (14)$$

num is the number of particles which are similar to particle i(similarity is more than S_0), and s is population size. C_i express the percentage of particles which are similar to particle i.

population diversity:

$$D = 1 - \frac{1}{s}\sum_{i=1}^{s} C_i \qquad (15)$$

Like above, s is population size.

During standard PSO research, with particles converging towards global best solution, population diversity reduce rapidly and it restricts the ability of global exploration. An improved method is that by monitoring population diversity D, once D is less than D_0, then start up mutation operator on all particles which are similar to the particle i which has the most concentration. The concrete operation is: for every gene in these particles, change its value with probability C_i. By this way, population diversity can keep a high level during research and the ability of global exploration is strengthened.

3.4 Algorithm Procedure

In this section, we describe the process of the improved PSO to solve data streams scheduling as following:

Step 1: Generate initial population by randomized algorithm;
Step 2: Update global best particle and local best particle;
Step 3: Update particles base on equation (11) and (12);
Step 4: Apply transgenic operator on each particle;
Step 5: Monitor population diversity D, if it's less than threshold D_0, then apply mutation operator on the particles which has the greatest concentration.
Step 6: If the termination conditions are satisfied, then the algorithm terminate, otherwise jump to step 2.

4 Experiment Results

4.1 Generation of Test Data

Use the same way as paper[2] to generate test data, the generation includes 3 parameters: data streams heterogeneity φ_d, machines heterogeneity ϕ_m, and data consistency. If data streams heterogeneity is low, then φ_d equals 100, which means that the difference among data streams is small, otherwise φ_d equals 3000. Similarly, if machines heterogeneity is low, then ϕ_m equals 10, which means that the difference among machines is small, otherwise ϕ_m equals 1000. Firstly, generate a n-vector B randomly, and each element $b_i \in [1, \varphi_d-1]$, then generate a $n \times m$-matrix X randomly too, and each element $x_{ij} \in [1, \varphi_m-1]$, at last calculate estimated executed time, $etm_{ij}=b_i \times x_{ij}$, $etm_{ij} \in [1, \varphi_d \times \varphi_m-1]$. Considering data consistency, if consistent, then sort elements in ETM by row, which make sure that for any element, if $etm_{ik} < etm_{il}$ then $etm_{jk} < etm_{jl}$; if semi-consistent, then sort elements only in even columns by row separately, which make sure that for any element in even columns, if $etm_{ik} < etm_{il}$ then $etm_{jk} < etm_{jl}$; if inconsistent, then do nothing with ETM.

Name test data according to this format: x-y-z. x is data streams heterogeneity, while taking 'l' means low (φ_d equals 100), and 'h' means high (φ_d equals 3000). y is machines heterogeneity, similarly to x, y taking 'l' means low (φ_m equals 10), and 'h' means high (φ_m equals 1000). z is data consistency, taking 'c' means consistent, 's' means semi-consistent and 'd' means inconsistent.

4.2 Simulation Test

Considering 128 data streams, 16 machines, execute each test data for five times and take the best solution.

After a lot of tries, IPSO can get satisfied solution in short time(10 seconds level) when parameters are set in this way: iterations is 2000, population size is 20, transgenic depth is the number of machines when data is consistent, and 4 when semi-consistent and inconsistent, particle similarity threshold S_0 is 0.8, population diversity threshold D_0 is 0.2.

When data is semi-consistent or inconsistent, there are great differences between good genes and normal genes, so it's easy to find good genes when transgenic depth is only 4. But when data is consistent, good genes are similar to normal genes, so transgenic depth should be set a big value -the number of machines.

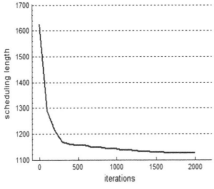

Fig. 1. Convergence process

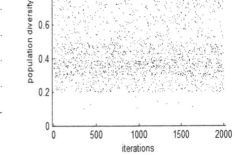

Fig. 2. Change of population diversity

Fig.1 and Fig.2 respectively show IPSO's convergence process and change of population diversity when test one l-l-c problem. It's proved that IPSO is of better convergence and its population diversity keeps a high level all the time.

Then test respectively by SMM, GA with seed (GAS). GAS's parameters are set in this way: iterations is 3000, crossover probability is 0.8, mutation probability is 0.2, use SMM's solution as seed.

Table 1. Contrast of scheduling length

problem type	SMM	GAS	IPSO
l-l-c	1292	1234	1126
l-h-c	78483	76812	69215
h-l-c	34644	33442	32291
h-h-c	2255566	2158983	2010580
l-l-s	748	713	678
l-h-s	47039	45619	39407
h-l-s	22164	21783	19964
h-h-s	1394428	1280191	1150713
l-l-d	599	587	542
l-h-d	36386	35525	28103
h-l-d	17042	16539	14979
h-h-d	1060335	941292	844154

Then compare balance of cluster load by Fig.3:

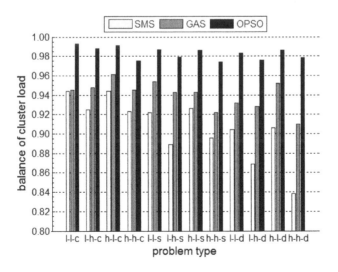

Fig. 3. Balance of cluster load

GAS use SMM's solution as seed, so its solution must be better than SMM's, but because of some disadvantages of GA in nature, the improvement is limited. From Table.1 and Fig.3, it is proved that transgenic operator and mutation operator can bring on a better data streams scheduling.

5 Conclusion

This paper construct a discrete PSO which is fit for data streams scheduling on heterogeneous cluster, base on gene theory, design transgenic operator and correspondent good gene fragments depend on special problem to improve PSO's ability of local solution, then introduce mutation operator of GA to improve PSO's ability of global exploration. We name the new algorithm IPSO. Simulation test shows that IPSO can get better solutions than other algorithms before in a short time.

Acknowledgement

This Work supported by the National Natural Science Foundation of China under Grant No.60673161, the Key Project of Chinese Ministry of Education under Grant No.206073, Fujian Provincial Natural Science Foundation of China under Grant No.A0610012.

References

1. Yu, R., Sun, Z., Chen, J., Mei, S.-l., Dai, Y.-q.: Traffic distributor for high-speed network intrusion detection system [J]. Journal of Tsinghua University(Science and Technology) 45(10), 1377–1380 (2005)
2. Wu, M.-Y., Shu, W., Zhang, H.: Segmented Min-Min: A Static Mapping Algorithm for Meta-Tasks on Heterogeneous Computing Systems[A]. In: 9th IEEE Heterogeneous Computing Workshop[C], pp. 375–385. IEEE Computer Society Press, Los Alamitos (2000)
3. Braun, T., Siegel, H., Netal, B.: A comparison study of static mapping heuristics for a class of meta-tasks on heterogeneous computing systems[A]. In: 8th IEEE Heterogeneous Computing Work shop[C], pp. 15–29. IEEE Computer Society Press, Los Alamitos (1999)
4. Zhong, Y.-w., Yang, J.-g.: A hybrid genetic algorithm for independent tasks scheduling in heterogeneous computing environments [J]. Journal of Beijing University of Aeronautics and Astronautics 30(11), 1080–1083 (2004)
5. Zhang, Y.-n., Liu, B., Dong, J.-k., Guo, P.-f.: Application of improved genetic algorithm in optimum design of building structures. Journal of Northeastern University (Natural Science) 25(7), 692–694 (2004)
6. Li, N., Sun, D.-b., Cen, Y.-g., Zou, T.: Particle Swarm Optimization with Mutation Operator. Computer Engineering and Applications 17, 12–15 (2004)

A Steepest Descent Evolution Immune Algorithm for Multimodal Function Optimization

Li Zhu [1,2], Zhishu Li [1], and Bin Sun [3]

[1] School of Computer, Sichuan University
[2] Network Center, Chengdu Sport University
[3]Engineering Development Center, Chengdu Aircraft Corporation
Chengdu, P.R. China
zhulicd@sina.com

Abstract. This paper presents an novel evolution and immune hybrid algorithm for multimodal function optimization. The algorithm constructs a multi-dimensional shape-space based on immune theory and approaches optima by steepest descent evolution strategy along each dimension, adjusts steps adaptively based on fitness in each iteration, as a result, gets steepest and surefooted ability approaching the optima. By suppressing close individuals in immune shape-space within a restraint radius and supplying new individuals to exploit new searching space, the algorithm obtains very good diversity. Experiments for multimodal functions show that the algorithm achieved global searching effect, obtained all the optima in shorter iterations and with lesser size of population compared with the GA, CSA and op-aiNet.

Keywords: Immune, evolution strategy, steepest descent, multimodal function optimization.

1 Introduction

The Evolution strategies (ESs) is a main branches of evolutionary algorithms. Schwefel [1] investigated the (1+1)-ES with binomially distributed mutations on a two dimensional parabolic ridge in 1965 and then introduced multi-member ES, the $(\mu+\lambda)$ - ES, and the (μ,λ) - ES. Evolution strategies have a important characteristic of self-adaptation in search direction and step size, but may end up in a local optimum or even exhibit so-called premature convergence. Cheng-Han [10] proposed a restricted evolution strategy for multimodal function optimization.

The concept of immune is derived from life sciences. Jerne (1975) put forward idiotype immune network theory. Farmer (1986) initiated artificial immunity system (AIS) by combining the immune with artificial intelligence (AI). Fukuda (1999) presented a immune learning algorithm [2] by incorporating immune network theory and genetic algorithm. The algorithm could search out multiple feasible-solution set

L. Kang, Y. Liu, and S. Zeng (Eds.): ISICA 2007, LNCS 4683, pp. 401–410, 2007.

and was used in multimodal function optimization. De Castro (1999) put forward a clonal selection algorithm (CSA) [3][4] which was used for pattern recognition and function optimization, then he presented an immune network theory (aiNet) [5] [6], used it in data classification and multimodal function optimization. Tieying [7] presented an improved AIS algorithm, analyzed its convergence and used it in multimodal function optimization. Carlos [8] solved a multi-objective optimization using artificial immune system. David [9] presented a migration strategy to increase the diversity of populations and suggested diversity as one selection criteria besides fitness for individuals.

Inspired by immune bio-mechanism, this paper presents a novel evolution strategies and immune hybrid algorithm - Steepest Descent Evolution Immune Algorithm (SDEIA) and makes the following contributions. First, proposes an evolution strategy - steepest descent with self-adaptive gradient which can obtain each local optima in high efficiency, the evolution directions and step sizes can be adjusted self-adaptively. Second, adopts immune shape-space ideas to promote the diversity of solutions and achieves global search effect. Third, the performance of the algorithms was studied in numerical experiments for multimodal functions optimization and got very good effect.

Remainder of this paper is organized as follows: section 1.1 presents the optimization model, Section 2 describes the SDEIA, Section 3 describes numerical experiments for multimodal functions, where the effect and efficiency of SDEIA were verified and compared with GA, CSA and op-aiNet. Some conclusions are presented in the last part.

1.1 Optimization Model and Shape-Space

Optimization Model
Let R^L is a L-dimension real space, $D \in R^L$, $f : D -> R^L$ is a objective vector, $h_i(x)$, $g_j(x)$ are constraint conditions in D, the multi-constraint optimization model is described as searching for a vector X, $X \in R^L$, which satisfies:

$$Min_{x \in D} f(X)$$
$$h_i(X) = 0 \qquad i, j = 1, 2 ... \tag{1}$$
$$g_j(X) \geq 0$$

Shape-Space
Shape-space describes complementary space between antibody and antigen of AIS in normalized L-dimension space. Suppose X is an antibody (or antigen), S^L is a L-dimension shape-space, then $X \in S^L \subseteq R^L$

Space between antibody and antigen in shape-space are measured in Hamming or Euclidean method. If an antibody is formed by $\{b_1, b_2, ..., b_L\}$, an antigen is presented by $\{g_1, g_2, ..., g_L\}$, then the Euclidean space between an antibody and an antigen is defined as:

$$d(b, g) = \sqrt{\sum_{k=1}^{L} (b_k - g_k)^2} \qquad (2)$$

The space between antibody x_i and antibody x_j is defined as:

$$d(x_i, x_j) = \sqrt{\sum_{k=1}^{L} (x_i^k - x_j^k)^2} \qquad (3)$$

Where x_i^k is the kth dimension value of antibody x_i.

2 SDEIA Description

SDEIA treats the optimized objective as antigen and solution as antibody, regards the optimization model as a continuous L-dimension mesh based on immune network theory. The mesh presents non-uniform distribution with different size sub-mesh and spreads smoothly. When approaching optima, mesh space may shrunk rapidly in order to converge to the extreme point. In each sub-mesh, a finite antibody set was permitted existing, depending on the restraint radius in immune shape-space. The smaller space between an antibody and an antigen, the bigger the affinity is. The smaller space between antibodies, the stronger immune repression is. First, SDEIA evolves antibodies by steepest descent evolution strategy in each iteration and then adjusts the evolution gradient self-adaptively by estimating variance of mesh space according to antibody's fitness. The lower the fitness, the shorter the evolution gradient it takes, then the better convergence it has. SDEIA adjusts antibody's step direction from time to time and makes them reaching the peaks stably through iterations.

SDEIA restrains same or close antibodies within a restraint radius in shape-space and complements new antibodies to improve the diversity of antibodies. By clonal selection and steepest descent evolution strategy, SDEIA achieved global random searching effect without using any memory and the experimental results show that SDEIA kept all the optimums and sub-optimums appeared without lost dominant solutions.

2.1 Antibody Restraint

SDEIA restrains antibodies too close each other in shape-space and measures in real number Euclidean space. Let σ presents restraint radius, $d(x_i, x_j)$、 $d_n(x_i, x_j)$ expresses space and normalized space between antibody x_i and x_j , if

$d_n(x_i, x_j) \le \sigma, \{x_i, x_j\} \in S^L$ and $\text{fit}(x_i) < \text{fit}(x_j)$, then x_i survives and x_j dies. Where $d(x_i, x_j)$ calculate by (3) and

$$d_n(x_i, x_j) = \frac{d(x_i, x_j) - \min(d)}{\max(d) - \min(d)} \tag{4}$$

Therein $\max(d) = \max(d(x_i, x_j))$, $\min(d) = \min(d(x_i, x_j))$, $i, j = 1,2,...N, i \ne j$.

The smaller space between antibodies, the more similarity is, whereas the larger the space, the better the diversity is.

The restraint radius in SDEIA is defined as $\sigma = N/\alpha$, where α is a positive constant, N is the size of antibody population.

2.2 Steepest Descent Evolution Strategy

Passed by restraint, SDEIA leads the predominant antibodies into the steepest descent evolving routine. Antibodies descend steeply along each of the L-dimension space.

Evolving Direction
Both positive and reverse directions in each dimension may exist more mature antibodies, let δ_i represents step size of antibody x_i, $x_i \in S^L$,

$i = 1,2,...N$, then the progress for SDEIA to evolve antibody x_i by descending steeply can be described as:

Along positive direction	Along opposite direction
While true	While true
$x_i' = x_i + \delta_i$	$x_i' = x_i - \delta_i$
If $fit(x_i') < fit(x_i)$	If $fit(x_i') < fit(x_i)$
$x_i = x_i'$, $fit(x_i) = fit(x_i')$	$x_i = x_i'$, $fit(x_i) = fit(x_i')$
Continue	Continue
Else	Else
Break	Break
End if	End if
End while	End while

If the fitness of antibody x_i decreases no more while descending, then a peak is approached and the step needs to be adjusted.

Evolving Step Size

Assuming *fthreld* is the threshold of share fitness, when $shfit(i) \geq fthreld$,

$$\delta_{i\,new} = \delta_i \cdot e^{-shfit(i)}, i = 1,2,...N \qquad (5)$$

$$\text{Whereas, } shfit(x_i) = \frac{\max(fit) - fit(x_i)}{\max(fit) - \min(fit)} \qquad (6)$$

According to (5) and (6), δ_i is defined by the fitness of antibody x_i. The lower the fitness or the larger the $shfit(x_i)$, the smaller $\delta_{i\,new}$ takes. This makes different antibody with different steps, thereby adapted to the mesh fluctuations. When the fitness takes maximum, i.e. $shfit = 0$, then δ_i keeps unchanged. The aim to set up fitness threshold is to make it impossible for antibodies with higher fitness reducing step size to a lower level. Let N represent the size of population, the scope of lth dimension space is given by $[l_2 - l_1]$, then the initial step size of lth dimension is defined as: $\delta_0 = (l_2 - l_1)/N$.

Steepest Descent Process

The process to evolving an antibody by steepest descent strategy refers to Fig.1, a random solution initiated from position 'A' with step size δ_i. First, SDEIA searches toward positive direction in step δ_i which leads the fitness dropped, then the solution transferred to position 'B'. Repeat the process till to position 'C'. But both the positive or reverse direction at 'C' in step δ_i couldn't make the fitness dropped any more, then SDEIA adjusts step size self-adaptively to $\delta_{i\,new}$ according to (5) and then ends current iteration. In the next iteration, evolution starts from position 'C' in the new step size $\delta_{i\,new}$, descends along positive direction first, then the opposite, the solution transfers to position D, the process continues until reaches the minimum.

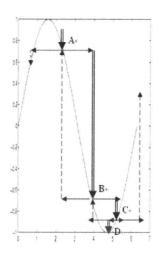

Fig. 1. Steepest descent evolving Evo lving from position 'A' to 'D'Solid line – try one step in a direction Double solid line – descends one step Dash line –descends fail in trying

2.3 Affinity Evaluation

Affinity indicates the binding capability between an antibody's basement and an antigen's epitope. The stronger the binding strength between them, the higher the affinity is. The affinity in SDEIA is defined as:

$$affinity(x_i) = -fit(x_i), i = 1,2,...N \qquad (7)$$

2.4 SDEIA Procedure Description

SDEIA algorithm was defined as 6 tuples: SDEIA = $(IT, N, \sigma, fthreld, LT, T)$

Where, IT represents maximum of iterations, N represents the size of population, σ is restraint radius $(0 \leqslant \sigma \leqslant 1)$, $fthreld$ is share fitness threshold, LT is a limitation vector with $2*L$ dimensions, T is terminate conditions, here takes the maximum of iterations.

The algorithm procedure is as follows:

Step1: Initialization, let $it = 0$, random generates initial population P with size N.
Step2: Evaluates affinity of population P according to (7), restrains antibody based on restraint radius σ. For antibodies satisfying: $d_n(x_i, x_j) \leq \sigma, \{x_i, x_j\} \in S^L$, reserves the one with higher affinity, eliminates the others, forms population $P1$ with size $N1$.
Step3: descends steeply in each dimension for antibodies in $P1$, adjusts steps, forms population C.
Step4: Generates recruit population T and combined with C forming population P, the size of P remains N.
Step5: while $T=true$ or $it>=IT$, then SDEIA end.
Step6: $it = it+1$, turn to step2.

3 Optimization Experiments

3.1 Multimodal Function

Three multimodal functions are selected in experiments, each has multiple minima and a lot of sub-minima.

Shubert Function
The Shubert function is defined as:

$$f(x_1, x_2) = \sum_{i=1}^{5} i \cos[(i+1)x_1 + i] * \sum_{i=1}^{5} i \cos[(i+1)x_2 + i] \qquad -10 \leq x_1, x_2 \leq 10$$

It has multi-thin peaks and lot of lower peaks. The maxma are about *210.482* and the minima are about -186.73. The function is shown in Fig. 2.(a).

Rastrigin Function
The experiments use reverse Rastrigin function which is defined as:

$$f(x) = -\sum_{i=1}^{n} (x_i^2 - 10\cos(2\pi x_i) + 10)$$

It consists of a lot of peaks, the minima are about *-80.707*, the maximum is *0*. There are about *10*n* lower peaks within the scope of $S = \{x_i \mid x_j \in (-5.12, 5.12)\}$,

$i = 1,2,...n$. The function vibrates intensely. The function with two dimensions is shown in Fig. 2.(b).

Roots Function

The experiments use reverse Roots function which is defined as:

$g(z) = -1/(1 + z^6 - 1)$,

$z \in C, z = x + iy, x, y \in [-2,2]$.

It presents a large plateau at the height -0.5, centered at (0, 0) and surrounded by six thin peaks. It takes minima at the height -1.0 in a complex plane[5].The Roots is shown in Fig. 2.(c).

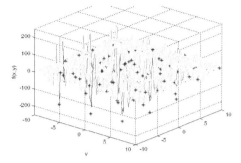

(a) Shubert function

3.2 Experimental Comparison

The experiments run on Matlib7.1 and aim to locate all the minima for the functions. The experiments compared the SDEIA with CSA, opt-aiNet and GA. In SDEIA, *fthreld* takes *0.1*, N and σ see the experimental tables. In CSA, *Pc* takes *0.5*, *Pm 0.01*, clonal multiple fat *0.1*. In GA, Pc *0.7*, *Pm 0.01*, generation gap *0.9*. For each situation, the experiments run 30 times above and takes the average and the standard deviation.

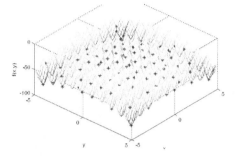

(b) Reverse Rastrigin function

Performance Comparison

The experimental results for searching minima in SDEIA are listed in table 1, table 2, and table 3. Therein, *Peaks* mean the number of peaks identified, *ItG* is the average number of iterations locate the first global minimum, *ItC* is the average number of iterations to locate all the global minima, *N* is the size of population.

For Shubert function (see table 1.), SDEIA found the first minimum after average *9.9* iterations and found the

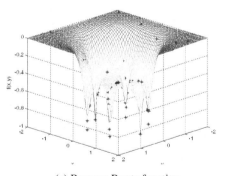

(c) Reverse Roots function

Fig. 2. Multimodal functions and optimization results in SDEIA

whole *9* minima after *150.1* iterations which located at *(-7.08,5.48)*, *(-7.08,-0.80)*, *(-0.80,-7.08)*, *(-0.80,-0.80)*, *(5.48,-0.80)*, *(-7.08,-7.08)*, *(-0.80,5.48)*, *(5.48,5.48)* and *(5.48,-7.08)* respectively. CSA and GA searched out only 1 - 2 minima.

Table 1. Performance comparison when applied to Shubert function

(σ=0.0167, N=100)

Algorithm	Peaks	ItG	ItC
SDEIA	9	9.9±3.9	150.1±30.2
CSA	2	20.4±3.3	-
GA	1	52.3+4.4	-

For Rastrigin (see table 2.), SDEIA found the first minimum in average 5.1 iterations, CSA took *20* iterations, GA took *35.3* iterations. SDEIA searched out the whole *4* minima in average *17.1* iterations, which located at *(4.52,-4.52), (-4.52,4.52), (-4.52,-4.52)* and *(4.52,4.52)*. CSA took *45.0* iterations, GA couldn't search out all the minima.

Table 2. Performance comparison when applied to Rastrigin function

(σ=0.0083, N=100)

Algorithm	Peaks	ItG	ItC
SDEIA	4	5.1±0.9	17.1±1.9
CSA	4	20.1±1.0	45.0±5.1
GA	1	35.3±3.9	-

For Roots (see table 3.), SDEIA found the first optimum in average *9.0* iterations, opt-aiNet took *86.9* iterations, CSA took *23.3* iterations with 5 times size of population in SDEIA. SDEIA searched out the whole 6 optimum in average *15.1* iterations, which located at: *(-0.50,0.866), (1.00,0.00), (-1.00,0.00), (0.50,-0.866), (0.50,0.866)* and *(-0.50,-0.866)* respectively. Opt-aiNet took average *295.0* iterations, CSA didn't search out all the optima.

Table 3. Performance comparison when applied to Roots function (σ=0.0033)

Algorithm	N	Peaks	ItG	ItC
SDEIA	20	6	9.0±1.2	15.1±3.5
opt-aiNet	20	6	86.9±34.3	295.0±129.7
CSA	100	6	23.3±5.5	-

Diversity Comparison

Fig.3. shows the relationship between the average affinity and the evolution generations and compared the SDEIA with CSA and GA, (a) for Shubert, (b) for Rastrigin. From Fig.3.(a) and (b), we can see that the average affinity of GA and CSA rose with the increasing of evolution generations, but the average affinity of SDEIA kept in almost horizontal level all the time which illustrates very good population diversity during the evolution progress.

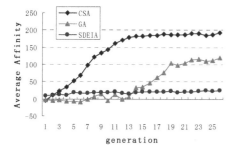

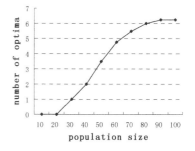

Fig. 3(a). Average affinity comparison between algorithms when applied to Shubert function, N=100

Fig. 3(b). Average affinity comparison between algorithms when applied to Rastrigin function, N=100

Size of Population and Searching Effect

Fig. 4. shows a relationship between the size of population and the number of minima. The number of minima grew while the size of population increasing. But the effect increased no more when the size reached some level. This means that it's enough for the size of population taking moderate size, oversize may result in repression and reduction in performance.

Experimental Results

The experimental results indicate that SDEIA trend toward achieving good diversity of population and got steady and

Fig. 4. Relationship between size of population and number of minima for SDEIA when applied to Shubert function

efficient effect on optimization at the same time. CSA put the population towards the optima at a cost of cloning with duplication to enlarge searching space which isn't suited for higher dimensional space. GA has diversity characteristic, but due to lack supervision and memorizing, GA is apt at lost dominant solutions even if they appear in the searching process. Opt-aiNet evolves too slow.

4   Conclusion

This paper presents a novel evolution and immune hybrid algorithm for multi-modal function optimization, combines the high efficiency of numerical analysis approach and the global search ability on bio-inspired immune. By descending steeply along each dimension in shape-space and tuning the step size self-adaptively, SDEIA gains steepest and steady ability to achieve the minima. By repressing close solutions within restraint radius in shape-space and supplies new individuals to exploit new search space, SDEIA obtains evolved population with wonderful diversity. By setting proper size of population, SDEIA is able to search out all the optima and a lot of sub-optima in fewer iterations. Experiments for multi-modal functions optimization show that SDEIA achieved global optimization effect by steepest descent evolution strategy combined with random searching on immune intelligence and is suitable for multi-dimension continuous space optimization. The experiments illustrate the effect and efficiency of SDEIA.

References

1. Beyer, H.-G., Schwefel, H.-P.: Evolution strategies. Natural Computing 1, 3–52 (2002)
2. Fukuda, T., Mcri, K., Tsukiama, M.: Parallel Search for Multi-Modal Function Optimization with Diversity and Learning of Immune Algorithm. In: Dasgupta, D. (ed.) Artificial Immune Systems and Their Applications, pp. 210–220. Springer, Heidelberg (1999)
3. de Castro, L.N., Von Zuben, F.J.: The Clonal Selection Algorithm with Engineering Applications. In: Workshop Proceedings of GECCO, USA. Artificial Immune Systems and Their Applications, vol. 7, pp. 36–37 (2000)
4. de Castro, L.N., Von Zuben, F.J.: Artificial Immune Systems: part I - Basic theory and applications. Technical Report, vol. 12 (1999)
5. De Castro, F.J.: An Evolution Immune Network for Data Clustering. In: Proc. Of the IEEE, vol. 11, pp. 84–89. IEEE Computer Society Press, Los Alamitos (2000)
6. de Castro, L.N., Timmis, J.: An Artificial Immune Network for Multimodal Function Optimization. In: IEEE2002, pp. 699–704. IEEE Computer Society Press, Los Alamitos (2002)
7. Tang, T., Qiu, J.: An Improved Multimodal Artificial Immune Algorithm and its Convergence Analysis. In: Proceedings of the 6th World Congress on Intelligent Control and Automation, Dalian, China, vol. 6, pp. 3335–3339. IEEE, Los Alamitos (2006)
8. Coello Coello, C.A.: Solving Multiobjective Optimization Problems Using an Artificial Immune System. Genetic Programming and Evolvable Machines 6, 163–190 (2005)
9. Power, D., Ryan, C., Azad, R.M.A.: Promoting diversity using migration strategies in distributed genetic algorithms. IEEE Transaction on Evolutionary Computation 5, 1831–1838 (2005)
10. Im, C.-H., Kim, H.-K., Jung, H.-K., Choi, K.: A novel algorithm for multimodal function optimization based on evolution strategy. IEEE Transaction Magnetics 40(2), 1224–1227 (2004)
11. Yao, J., Kharma, N., Grogono, P.: BMPGA: A Bi-Objective Multi-population Genetic Algorithm for Multi-modal Function Optimization. In: IEEE2005, pp. 816–823. IEEE, Los Alamitos (2005)
12. Singh, G., Deb, K.: Comparison of Multi-Modal Optimization Algorithms Based on Evolutionary Algorithms. In: GECCO '06. Proceedings of the 8th annual conference on Genetic and evolutionary computation, Washington, USA, vol. 7, pp. 1305–1312 (2006)
13. Corns, S.M., Ashlock, D.A., McCorkle, D.S., Bryden, K.M.: Improving Design Diversity Using Graph Based Evolutionary Algorithms. In: Congress on Evolutionary Computation, Vancouver, BC, Canada, vol. 7, pp. 333–339. IEEE, Los Alamitos (2006)
14. Yoshiaki, S., Noriyuki, T., Yukinobu, H., Katsuari, K.: A Fuzzy Clustering Based Selection Method to Maintain Diversity in Genetic Algorithms. In: Congress on Evolutionary Computation, Vancouver, BC, Canada, vol. 7, pp. 3007–3012. IEEE, Los Alamitos (2006)
15. Narihisa, H., Kohmoto, K., Taniguchi, T., Ohta, M., Katayama, K.: Evolutionary Programming With Only Using Exponential Mutation. In: Congress on Evolutionary Computations, Vancouver, BC, Canada, vol. 7, IEEE, Los Alamitos (2006)

A Hybrid Clonal Selection Algorithm Based on Multi-parent Crossover and Chaos Search

Siqing Xue, Qiuming Zhang, and Mailing Song

School of Computer Science, China University of Geosciences,
430074 Wuhan, China
xue.xues@google.com

Abstract. This paper proposes a novel hybrid immune clonal selection algorithm coupling with multi-parent crossover and chaos mutation (CSACC). CSACC takes advantages of the clonal selection mechanism and the learning capability of the clonal selection alorithm (CLONALG). By introducing the multi-parent crossover and neighbourhood mutation operator, CSACC achieves a dynamic balance between exploration and exploitation. And by using the characteristics of ergodicity and dynamic of chaos variables, the chaotic optimization mechanism is introduced into CLONALG to improve its search efficiency. The experimental results on function optimization show that the hybrid algorithm is more efficient than the clonal selection algorithm.

Keywords: clonal selection, multi-parent crossover, chaos search.

1 Introduction

Over the last few years, the study of artificial immune systems has become increasingly important due to the large number of its possible applications in fields of science and engineering [1], [2], [3]. Based on the learning and evolutionary principle in the adaptive immune response, De Castro developed CLONal selection ALGorithm (CLONALG) [4], which uses individual mutation to perform greed search and uses random search to explore the whole solution space. CLONALG emphasizes proportion selection and clone, which can hold the optimal solutions by set a memory cell. The essence of the clonal operator is producing a variation population around the parents according to their affinity, and then the searching area is enlarged. Compared with standard genetic algorithm (GAs) [5], CLONALG is convergent faster and the diversity is much better. But when the objective function becomes complex, the search ability of the algorithms becomes weak. Although the algorithms proposed in Ref. [4] try to simulate the characters of the diversity of population, the CLONALG emphasizes the more local search. On the other hand, the immune system is a typical complicated nonlinearity system, and recent study revealed that chaos also exists in the process of immune response [6], [7], so it is important to design the clonal mutation operator based on chaos theory.

This paper introduces multi-parent crossover and chaos search into CLONALG and constructs a novel hybrid immune Clonal Selection Algorithm coupling multi-parent Crossover and Chaos mutation (CSACC). In the next section, a brief review of

L. Kang, Y. Liu, and S. Zeng (Eds.): ISICA 2007, LNCS 4683, pp. 411–419, 2007.

CLONALG and GuoTao Algorithm is presented, and then a detailed description of the CSACC algorithm is put forward in Section 3. In Section 4, numerical experiments are performed on a suite of benchmark functions to investigate the effectiveness of the proposed algorithm. Finally, some conclusions are presented in Section 5.

2   The Clonal Selection Algorithm and GuoTao Algorithm

2.1   The Clonal Selection Algorithm

The clonal selection algorithm developed on the basis of clonal selection theory [4], which main idea is that only those immune cells that recognize the antigens are selected to proliferate. The antigens can selectively react to the antibodies, which are the natural production and spread on the cell surface in the form of peptides. The reaction leads to cell proliferating clonally and the clone has the same antibodies. Some clonal cells will become immune memory cells to boost the second immune response. The clonal selection is a dynamic process of the immune system self-adapting against antigen stimulation, which has some biologic characters such as learning, memory and genetic diversity.

The CLONALG proposed by De Castro can be described as follows: 1) randomly choose an antigen and present it to all antibodies in the repertoire, and calculate the affinity of each antibody; 2) select the n highest affinity antibodies to compose a new set; 3) clone the n selected antibodies independently and proportionally to their antigenic affinities, generate a repertoire set of clones; 4) submit the repertoire set to an affinity maturation process inversely proportional to the antigenic affinity. A maturated antibody population is generated; 5) re-select the highest affinity one from this set of mature clones to be a candidate to enter the set of memory antibodies set; 6) finally, replace the d lowest affinity antibodies by some random ones.

2.2   GuoTao Algorithm

Multi-parent crossover operators [8], [9], [10], where more than two parents are involved in generating offspring, are a more flexible version, generalizing the traditional two-parent crossover of nature. Guo proposed a new algorithm, named GuoTao algorithm (GTA) [11], which generates offspring by linear non-convex combination of m parents. Non-convexity tends to guide the population to search areas outside where it resides. Simplicity and high efficiency are the main advantages of GTA. GTA is as follows:

1) select randomly m individuals x^j in the population, j=1, 2 ,..., m.

2) generate m random real numbers α_j , j=1, 2 ,..., m,

subject to

$$\sum_{j=1}^{m} \alpha_j = 1, \text{ and } -\varepsilon \leq \alpha_j \leq 1+\varepsilon, \text{ j=1, 2 ,..., m.}$$

where ε is a positive parameter.

3) generate the offspring individual x^c as follows:

$$x^c = \sum_{j=1}^{m} \alpha_j x^j .$$ (1)

3 Hybrid Clonal Selection Algorithm

Clonal proliferation is the main feature of the clonal selection theory. In clonal proliferation, random hypermutation are introduced to the variable region genes, and occasionally one such change will lead to an increase in the affinity of the antibody [4]. In CLONALG, the antibody clone and mutation are employed to perform greed search and the global search is carried out by introducing new antibodies to replace low affinity antibodies. During optimization search, the appropriate balance between exploitation and exploration is the key issue [12]. So it is necessary to introduce some control mechanism to create a balance between local and global searches in optimization process.

Chaos is a universal phenomenon in nonlinear mechanical system [13], [14]. Chaotic variables seem to be a random change process, but chaotic variables can go through every state in a certain area according to their own regularity without repetition. Due to the ergodic and dynamic properties of chaos variables, chaos search is more capable of hill-climbing and escaping from local optima than random search, and thus has been applied to the area of optimization computation rules [6], [7], [13] [14], and [15].

Taking advantages of the simplicity and the high efficiency of GTA and the ergodic and dynamic properties of chaos system, we introduce the multi-parent crossover of GTA and the chaotic search mechanism of chaos optimization algorithm into the CLONALG to improve its search efficiency. In the algorithm proposed in this paper, i.e., CSACC, the chaotic neighbourhood mutation is employed to exploit local solution space, chaotic clonal mutation and chaotic multi-parent crossover operator are introduced to explore the global search space.

The chaos system used in this paper is the well-known Logistic mapping defined by

$$s_{i+1} = \mu s_i (1 - s_i), \text{ i=0,1,2,...,}$$ (2)

where μ is a control parameter, s_{i+1} is the value of the variable s at the *i*th iteration. If μ ∈ (3.56, 4.0), then the above system enters s_i into a chaos state and the chaotic variable s_{i+1} is produced. The chaos system has some special characteristics such as ergodicity, randomicity and extreme sensitivity to the initial conditions [14]. We can iterate the initial arbitrary value $s_0 \in [0, 1]$, and produce a cluster of chaos sequences s_1, s_2, s_3....

In this paper, the chaotic variables are generated by the following Logistic mappings:

$$s_{i+1}^j = \mu s_i^j (1 - s_i^j), \text{ i=0,1,2,...k and j=1,2,...,}$$ (3)

where i is the serial number of chaotic variables, and μ= 4, . Let j = 0, and given the chaotic variables initial values s_i^0 (i =0), then the values of the k chaotic variables

s_i^0 (i = 1,2,. . . , k) are produced by the Logistic equation (3). And by the same method, other N-1 number serials are produced.

Without the loss of universality, the proposed CSACC is used to solve the following continuous function optimization problem:

$$\begin{cases} \max & f(x_1, x_2..., x_r) \\ s.t. & x_i \in [a_i, b_i], \quad i = 1, 2, ..., r \end{cases}, \tag{4}$$

where $[a_i, b_i] \subset R$, f is a real-valued continuous function; r is the number of optimization variables.

3.1 Operators

Chaotic multi-parent Crossover

1) select m individuals x^j in the population P by the following:

$$i = \left\lfloor Ls^j \right\rfloor, \quad x^j = p_i, \tag{5}$$

where j=1, 2 ,..., m, s^j is the jth chaotic variable, and L is the size of the popular P, P_i is the ith (i=0, 1, 2, ..., L-1) individual of popular P.

2) generate m real number sequences by the following equation:

$$\begin{cases} \alpha^j = s^j(1+2\varepsilon) - \varepsilon, & where \ j = 1, 2, ..., m-1 \\ \alpha^j = 1 - \sum_{t=1}^{m-1} a^t, & where \ j = m \end{cases}, \tag{6}$$

where α^j is the jth sequence, $\alpha^j = (\alpha_1^j, \alpha_2^j, \cdots, \alpha_r^j)$, ε is a positive parameter, and s^j is the jth chaotic variable sequence. And the above equation should subject to

$$\sum_{j=1}^{m} \alpha_i^j = 1, \text{ and } -\varepsilon \leq \alpha_i^j \leq 1+\varepsilon,$$

where α_i^j is the ith real number of the jth equence, j=1, 2 ,..., m, and i=1,2,...,r.

3) generate the offspring individual x^c as follows:

$$x^c = \sum_{j=1}^{m} \alpha^j x^j, \tag{7}$$

where $x^j = (x_1^j, x_2^j, \cdots, x_r^j)^T$.

Different from GTA, in CSACC, c(c>1) new individuals will be generated to search the subspace spanned by m individuals.

Chaotic Neighbourhood Mutation

Suppose the individual x is expressed by a vector $x=(x_1,x_2,...,x_r)^T$, then the neighborhood of an individual x is defined as the following:

$$U(x, \delta)=\{y|y_i\in [a_i, b_i] \wedge |y_i-x_i|<\delta, i=1,2,...,r\}, \tag{8}$$

where δ is a positive parameter to control the range of the neighborhood and is dependent on the range of the variable.

A new individual x' by chaotic mutation in the neighbourhood of x is obtained by

$$x'=x+\delta|2s-1|, \tag{9}$$

where $s=(s_1, s_2,..., s_r)$ is a chaotic vector, and $s_i\in [0, 1]$ is the value of the ith chaotic variable. The vector $x' =(x_1', x_2',..., x_r')^T$ should be restricted by the following:

$$x_i' = \begin{cases} a_i & x_i' < a_i, \\ x_i' & x_i' \in [a_i,b_i], \\ b_i & x_i' > b_i. \end{cases} \tag{10}$$

Obviously, $x'\in U(x, \delta)$, and x' is in the chaotic neighbourhood of x.

Chaotic Mutation

A new individual x' cloned from the selected individual x by chaotic mutation is obtained by

$$x'=a+ds, \tag{11}$$

where $s=(s_1, s_2,..., s_r)$ is a chaotic vector, and $s_i\in [0, 1]$ is the value of the ith chaotic variable; $a=(a_1,a_2,...,a_r)^T, x' =(x_1', x_2',..., x_r')^T, d=(b_1-a_1,b_2-a_2,...,b_r-a_r)^T$.

3.2 Algorithm Description

In this paper, all chaos sequences are produced by the chaotic system described in formula (3). The steps of the CSACC can be described as follows:

step 1. generate N individuals by chaotic mutation operator composing the initial popular P(k), and calculate the affinity (fitness values) of the individuals.

step 2. sort the individuals in popular P(k) by affinity based on the logic function better() in Ref.[11].

step 3. perform chaotic multi-parent crossover c times to compose a popular named CP.

step 4. select the n (0<n<N) highest affinity individuals and perform clonal operator independently and proportionally to their affinities, generating n sub popular of clones.

step 5. perform hypermutation chaotic neighbourhood mutation in the former k sub popular, and performing hypermutation chaotic mutation in the rest sub popular. The highest affinity individual in each sub popular is selected to compose a new popular named MP. A new individual x' cloned from the selected individual x by hypermutation chaotic mutation is obtained by

$$x_i' = \begin{cases} x_i & s_i \leq p \\ a_i + s_i(b_i - a_i) & s_i > p \end{cases} \tag{12}$$

where $s_i \in [0, 1]$ is the value of the ith chaotic variable, and p is the mutation probability.

step 6. generate N-n-c new individuals to compose a new popular named NP, which will replace the lowest affinity individuals in the popular P.

step 7. Calculate the affinity of each individual in popular CP and NP; refresh popular P with CP, MP and NP composing a new popular P(k+1).

step 8. sort the individuals in popular P by fitness based on the logic function better() in Ref.[11]. If the affinity of the best individual in P(k+1) is lower than P(k), then the lowest affinity antibody in P(k+1) is replaced by the highest affinity antibody in P(k).

step 9. if the halt condition is satisfied, stop the algorithm; otherwise, go step 3.

4   Experiments and Results

In order to examine the performance of CSACC, the following complex functions are employed as test functions and compared with CLONALG. All of these functions have more than one local optimum, and some even have infinite local optima solutions.

f_1: Maximize: $f(x, y) = x \cdot \sin(4\pi x) - y \cdot \sin(4\pi y + \pi) + 1$, where $0 < x < 2$, $0 < y < 2$.

f_2: Maximize: $f(x, y) = 0.5 - \dfrac{(\sin(\sqrt{x^2 + y^2}))^2 - 0.5}{1 + 0.001*(x^2 + y^2)}$, where $-10 \leq x \leq 10$, $-10 \leq y \leq 10$.

f_3: Minimize: $f(x, y) = -[21.5 + x\sin(4\pi x) + y\sin(20\pi y)]$, where $12.1 \leq x \leq -3$, $5.8 \leq y \leq 4.1$.

f_4:   Minimize: $f_n(X) = -\left| \dfrac{(\sum\limits_{i=1}^{n} \cos^4(x_i) - 2\prod\limits_{i=1}^{n} \cos^2(x_i))}{\sqrt{\sum\limits_{i=1}^{n} i x_i^2}} \right|$,   where $\prod\limits_{i=1}^{n} x_i \geq 0.75$,

$\sum\limits_{i=1}^{n} x_i \leq 7.5n, 0 < x_i < 10$,   i=1, 2, ..., n.

The first function is a multimodal function used by De Castro et al. [4], which has many local optima peaks and one global optimum peak. The second is a benchmark for genetic algorithms in Davis [16], which is very difficult to optimize, as the highly discontinuous data surface features many local optima. And f_3 has a lot of local and global peaks [17]. f_4 is an n-dimensional function was designed to reproduce some of the features of practical engineering design optimization problems [18].

The main parameters of the CSACC were chosen as N=100, n=70, k=21, m=10, c=20, and δ= 0.3 according to experience. The parameters for CLONALG were N=100, n=70, k=0, and c=00. In order to study the affection of chaos multi-parent crossover and chaos neighborhood mutation on the performance of the algorithm, the other two groups of parameters were chosen as N=100, n=70, k=0, m=10, c=20 (named $CSACC_1$) and N=100, n=70, k=21, c=0, δ= 0.3 (named $CSACC_2$). The algorithms were written in Matlab language. Each algorithm optimized each function 10 times and the maximum number of iteration for algorithms is 6000.

The function f_4 (n=2) is considered to compare the performance of CLONALG, $CSACC_1$, $CSACC_2$, and CSACC. When n=2, the criterion value for success is -0.36497974587066 [19]. Table 1 shows the simulation results. In table 1, RS is rate of success, AFE is the average number of function evaluation, and AV is the average value of the optimization solution.

Table 1. Simulation results for f_4 (n=2) function optimization

	RS	AFE	AV
CLONALG	0	153834	-0.3643990293235738
$CSACC_1$	100%	221705	-0.3649797458706568
$CSACC_2$	0	200989	-0.3649703038028384
CSACC	100%	200683	-0.3649797458706569

It can be seen from Table 1 that CSACC and $CSACC_1$ outperform CLONALG and $CSACC_2$, since the function has a lot of local optimal solutions, the CLONALG and $CSACC_2$ is easy to stick at local optima and can not search high precision solution. Otherwise it can be observed that CSACC takes advantages of $CSACC_1$ and $CSACC_2$, that is, CSACC employed the chaotic neighborhood mutation and chaotic multi-parent crossover, which can balance the local search and global search in optimization process. Fig. 1 illustrates the process of obtaining optimal solution of function f_3.

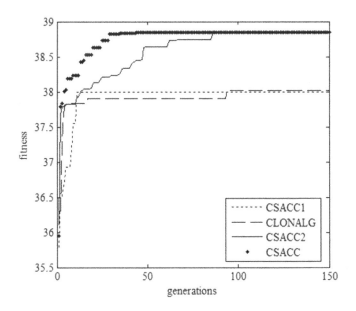

Fig. 1. The process of obtaining optimal solution of function f_3

Table 2 and table 3 shows the best optimization value obtained from simulations for optimization of the above test functions using CLONALG, CSACC$_1$, CSACC$_2$, and CSACC. From table 2 it can be seen that CSACC can obtain the better solution than the criterion value [17], [18], [19].

Table 2. Comparison of the results of the algorithms

	f_1	f_2
CLONALG	4.2538884433172273	0.9999999266444974
CSACC$_1$	4.2538884433172273	0.9999999925962908
CSACC$_2$	4.2538884411074545	0.9999999677547100
CSACC	4.2538884433172282	0.9999999999998260

Table 3. Comparison of the results of the algorithms

	f_3	$f_4 (n=5)$
CLONALG	-38.0501988630926680	-0.6187382161197592
CSACC$_1$	-38.0502502152185610	-0.6222810255678128
CSACC$_2$	-38.8502944753383870	-0.6222805900548480
CSACC	-38.850294479440151	-0.6344486869490337

5 Conclusions

In this paper, a mechanism for adapting the chaotic multi-parent crossover and chaotic mutation are introduced into clonal selection algorithm to enhance the efficiency. The simulation results show that CSACC has nice performance and the proposed method is quite promising. CSACC is expected to have the following features: 1) By introducing the multi-parent crossover and neighbourhood mutation operator, CSACC achieves a dynamic balance between exploration and exploitation; 2) By using the characteristics of ergodicity and dynamic of chaos variables, the chaotic optimization mechanism is introduced into CLONALG to improve its search efficiency; 3) CSACC takes advantages of the clonal selection mechanism and the learning capability of the CLONALG. Future work includes the application research of CSACC to other evolutionary model, such as Gene Expression Programming (GEP).

Acknowledgments. This work is supported by the Research Foundation for Outstanding Young Teachers of China University of Geosciences (CUGQNL0644).

References

1. Dasgupta, D., Ji, Z., Gonzlez, F.: Artificial Immune System (AIS) Research in the Last Five Years. In: CEC2003. Proceedings of the International Conference on Evolutionary Computation, pp. 123–130 (2003)
2. De Castro, L.N., Von Zuben, F.J.: Artificial immune systems: part I-basic theory and applications. FEEC/Unicamp, Campinas, SP, Tech. Rep. – RT DCA 01/99 (1999)

3. De Castro, L.N., Von Zuben, F.J.: Artificial Immune Systems: Part II – A Survey of Applications, Tech. Rep. – RT DCA 02/00 (2000)
4. De Castro, L.N.: Learning and optimization using the clonal selection principle. IEEE Transaction on evolutionary computation 6(3), 239–251 (2002)
5. Holland, J.H.: Genetic algorithms, Scientific American, pp. 66--72 (1992)
6. Canabarro, A.A., Gl_eria, I.M., Lyra, M.L.: Periodic solutions and chaos in a non-linear model for the delayed cellular immune response. Physica A 342, 234–241 (2004)
7. Zuo, X., Fan, Y.: A chaos search immune algorithm with its application to neuro-fuzzy controller design. Chaos, Solitons and Fractals 30, 94–109 (2006)
8. Li, J., Kang, L., Wu, Z.: A self-adaptive neighborhood-based multi-parent crossover operator for real-coded genetic algorithms. In: Proceedings of 2003 Congress on Evolutionary Computation, pp. 14–21 (2003)
9. Ting, C.K.: An Analysis of the Effectiveness of Multi-parent Crossover. In: Yao, X., Burke, E.K., Lozano, J.A., Smith, J., Merelo-Guervós, J.J., Bullinaria, J.A., Rowe, J.E., Tiňo, P., Kabán, A., Schwefel, H.-P. (eds.) Parallel Problem Solving from Nature - PPSN VIII. LNCS, vol. 3242, pp. 131–140. Springer, Heidelberg (2004)
10. Eiben, E.: Multiparent recombination in evolutionary computing. In: Ghosh, A., Tsutsui, S. (eds.) Advances in evolutionary computing. Natural computing series, pp. 175–192. Springer, Heidelberg (2002)
11. Tao, G., Kang, L., Li, Y.: A New Algorithm for Solving Function Optimization Problems with Inequality Constraints. Wuhan University Journal of Natural Sciences 45(5), 771–778 (1999)
12. Lin, H., Kang, L.: Balance between Exploration and Exploitation in Genetic Search. Wuhan University Journal of Natural Sciences 4(1), 28–32 (1999)
13. Lu, H., Zhang, H., Ma, L.: A new optimization algorithm based on chaos. Journal of Zhejiang University [Science A] 7(4), 539–542 (2006)
14. Porto, D.M.: Chaotic Dynamics with Fuzzy Systems. StudFuzz 187, 25–44 (2006)
15. Liu, R., Chen, L., Wang, S.: Immune Clonal Strategies Based on Three Mutation Methods. In: Jiao, L., Wang, L., Gao, X., Liu, J., Wu, F. (eds.) ICNC 2006. LNCS, vol. 4222, pp. 114–121. Springer, Heidelberg (2006)
16. Davis, L. (ed.): Handbook of Genetic Algorithms. Van Nostrand Reinhold, New York (1991)
17. Deb, K.: GeneAS: A robust Optimal Design Technique for mechanical component design. In: Dasgupta, D., Michalewicz, Z. (eds.) Evolutionary Algorithms in Engineering Applications, pp. 497–514. Springer, Heidelberg (1997)
18. E-Beltagy, M.A, Nair, P.B, Keane, A.J.: Metamodeling techniques for evolutionary optimization of computationally expensive problems: Promise and limits. In: Proc of Genetic and Evolutionary Computation, pp. 196–202 (1999)
19. kang, Z., Li, Y., Liu, P., Kang, L.: An All -Purpose Evolutionary Algorithm For Solving Nonlinear Programming Problems. Journal of Computer Research and Development 139(11), 1471–1477 (2002)

Spatial Clustering Method Based on Cloud Model and Data Field

Haijun Wang[1] and Yu Deng[1,2]

[1] School of Resources and Environment Science, Wuhan University, 129 Luoyu Road,
Wuhan 430079, China
[2] Institute of Geographic Sciences and Natural Resources Research, CAS,
Beijing 100101, China
landgiswhj@163.com

Abstract. The purpose of this paper is to set forth a spatial clustering method based on cloud model and data field, which can be widely applied to the research on classification and hierarchy in realm of spatial data mining and knowledge discovery.

1 Introduction

With rapid development of contemporary science and technology, the ability of data collection, manufacture, storing and manipulation has been increasing largely. Nevertheless, it is difficult to extract knowledge and principles hidden in mass data increasing steadily. In order to dig more valuable knowledge from mass data, there yield data mining and knowledge discovery which have become the focus of research and application throughout the world. Since spatial data acts as an important part of mass data, the process to dig knowledge form spatial data and to search those undefined and implicit knowledge, spatial relations or other modes- spatial data mining and knowledge discovery has been catching more and more attentions too. Spatial clustering is one of those important methods applying to spatial data mining and knowledge discovery. The spatial knowledge for spatial data mining mainly consists of principles including association, feature, classification, clustering and the like. This paper will discuss another method of spatial clustering, that is, spatial clustering method based on cloud model and data field.[1]

2 Date Field

Physics research is often involved in the distribution of certain regions of physical factor, commonly referred to as the "field", such as the gravitational field, electric field and so on. In the domain space, every data exposures its energy data on the entire domain space, in order to show their presence and thus data field. Data field has a common feature of field, but it can not simply be compared with the field model. The existence of field data must satisfy the independence, close at hand sex, ergodicity, superposition, decay, and anisotropy. We use the potential function to describe

L. Kang, Y. Liu, and S. Zeng (Eds.): ISICA 2007, LNCS 4683, pp. 420–427, 2007.

attributes of the data field. Potential function is the function of location or distance and can be superimposed. Therefore, in domain space every data object have contributed to the potential of any point field, and contributions is inversely proportional to the square of distance between them. [2] [3]

3 Cloud Theory

Professor Deyi Li puts forward the cloud theory that contains Cloud Model, Virtual Cloud, Cloud Operation, Cloud Transform, Reasoning under Uncertainty, and the like. Suppose that **U** is a quantitative domain represented by accurate numerical value, and **C** is a qualitative concept under **U**. If quantitative value **x**∈**U**, and **x** is a random implement of **C** in qualitative concept. The certainty degree of x to **C**, **u(x)** ∈[0,1], is a random number with stable tendency **μ**that satisfies that when **U**→[0,1] and **Px**∈ **U**, then **x** →**μ(x)**. Then, the distribution on discussed field **x** is cloud model, and is called cloud for short. Moreover, each **x** is called cloud drop, and represented as **C(Ex,En,He)**. [4] [5]

4 Spatial Clustering Based on Cloud Model and Date Field

4.1 Data Acquisition

Data used for spatial data mining and knowledge discovery usually includes spatial data and attribute data, which are fused under certain regulations and form a fusion combined index which is input as original data for classification. The formula to get combined index is as follows:

$$score = w_1 * x + w_2 * y + \sum_{k=3}^{n} w_k * a_i \qquad (1)$$

In the formula, $score$ is the value of fused index; w_1, w_2 ...... w_k represent weights; $w_1 + w_2 + \sum_{k=3}^{n} w_k = 1$, and **n** is the count of attribute data; **a** is attribute value; **x** and **y** are coordinates of spatial data.[6]

4.2 Potential Transformation of Data Field

Clustering by the attributes of spatial information, do not fully take into account the spatial location of the information. Data field origin in data radiation, is a description of the form of space of the radiation data, and is the process of abstract mathematics and assumptions on data radiation. The introduction of the market, well resolved the issue. Interactions or interactions of spatial data, is expressed through potential and attenuation function ,which is more accurately reflected on the different points of space radiation range. The potential function of any point in the field data is defined as the impact by all the data points. Given n data points,the distence D = (d1,... dn},thus ,the potential function is defined as :

$$F(x) = \sum_{i=1}^{n} \rho_i e^{-\frac{(d(x,di))^2}{2\delta^2}} \qquad (2)$$

d (x, di) express the distance between x and di,ρi is the value,σ is site light factor.

The transformation of the site light factor will directly affect the equipotential line spacing.the smaller σ,the smaller the individual data points of the scope,so equipotential lines more closely; the greater σ, the greater individual data points of the scope and equipotential lines more rare. As shown in figure1 and 2:

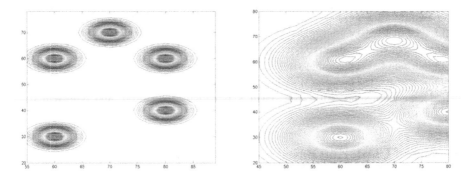

Fig. 1. Double dimension data field with differentσ

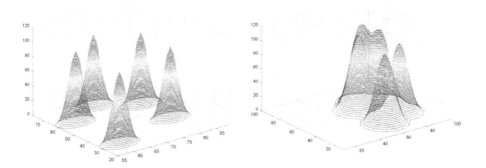

Fig. 2. Three dimension data field with differentσ（σ=2、σ=6）

Individual potential function is 1 at a distance of zero, and tending to zero for a space of infinite distance. In fact, according to the normal distribution of "three σ rules", when the distance is greater than or equal to 3σ, the potential function has almost zero value; Therefore, each of the radius can be approximated as follows : r≈3σ. Based on original data ρi, distance d (x, di) can be obtained between the two points. According to the formula (2). calculate the various points of all potential. Potential superposition is defined as the total potential F. [2]

4.3 Preprocessing Data

Before the data fusion, exception value in data should be eliminated and data standardization should finish. During collection and collation of data, it is impossible to omit errors or mistakes. Therefore, in order to ensure the reliability of data analysis, exception value should be found and extracted through experiences and other methods. There lies difference between spatial coordinate and attribute characteristic, and between attribute characteristics in aspects such as index unit and quantity degree. Sometime, the specific differences may be quite obvious. In purpose of eliminating effects brought by data unit and quantity degree, the values of coordinates and attribute characteristics should be changed into dimensionless. The relevant formula is as follows:

$$u_i = \frac{V_i - V_{\min}}{V_{\max} - V_{\min}} \times 100 \tag{3}$$

In the formula, $V_{\min}$ is the minimal value of vi; $V_{\max}$ is the minimal value of vi; u_i is the index value after being changed into dimensionless.

After so-called dimensionless processing, the value of u_i fluctuates form 0 to 100. And there exits horizontal comparability between attribute characteristics, and between attribute characteristic and spatial coordinate.

4.4 Cloud Model Establishment

Original Data Distribution
Applying rounding off method, data are changed into integral form. And then, appearance frequency of index value in this group of data. If integral index value is set as X-axis, and its relevant frequency is set as Y-axis, the demanded original distribution figure is got then. The formula is as follows:

$$f = n/N , \quad n = numel(round(a_i)) , \quad N = numel(Data) \tag{4}$$

In the formula, **n** represents the appearance time of a certain index value; **N** is the number of data in data group; **numel** is the number that a certain data appears; **round** is the function for rounding off process; a_i is the index value; Data is the original data group; and **f** is frequency.

Cloud Transform
Theory has proven that several normal functions can constitute any distribution of frequency[7]. By fitting the distribution of original data, it is feasible to superimpose clouds with different fineness and then realize the transform continuous variable into discrete variable. The peak point cloud transform is a cloud transform method widely applied[8]. From frequency figure, it is easy to figure out that the curve consists of peaks and troughs by turns. It is applicable to raise the operation accuracy through setting initial restriction (θ) of frequency. This method is implemented by deciding peak value satisfying conditions and fitting it with normal function. The specific implementation method is as follows. First of all, set maximal value as the expectation value of normal function. Secondly, search its neighboring nodes so that the difference between average value and expectation value is with the limit of acceptance ability. Then, calculate and get the variance of all points. Finally, fit normal

distribution curve based on the expectation value and variance, and then get series of sub-clouds, that is, leaf node in pan-concept-tree. The formula is as follows.

$$f_{wave\_crest} > \theta, round(x_i) = (E_x or E_x \pm 1 or E_x \pm 2), E_x = E_{xware\_crest}, E_n = Var(x_i) \quad (5)$$

In the formula, f is frequency satisfying limits; θ is limit that has been set; **round** is function for rounding off data; x_i is the neighboring node of Ex; V_{ar} is the variance of data group.

Climbing of Universal Conceptual Number

A series of sub-clouds got by fitting normal function are called atomic cloud vividly, of which the climbing of universal conceptual number rises in conceptual level on basis. The conceptual tree built by cloud model has property of uncertainty, and the boundary between concepts is vague. The attributions that index values offer to decide the concept vary a lot; concept fineness in lower level can climb to node in pan-concept-tree that is necessary to reach concept fineness in higher level. During application in reality, the number of types decides the number of root nodes. Therefore, it is necessary to upgrade concepts in various levels taking advantages of a set of cloud synthetic operators, and gain concept fineness in higher level. The atomic cloud can be described as: $C_1(Ex, En, He), C_2(Ex_2, En_2, He_2), \cdots, C_m(Ex_m, En_m, He_m)$

Applying Soft-Or method, cloud synthetic algorithm can be ameliorated and represented as follows[8] [9]: $Ex_1 < Ex_2$: $Ex_3 = (Ex_1 + Ex_2)/2 + (En_2 - En_1)/4$;

$$En_3 = (En_1 + En_2)/2 + (Ex_2 - Ex_1)/4 ; \quad He_3 = \max(He_1, He_2) \quad (6)$$

4.5 Determining Class

When the number of father clouds in the highest level reaches to the number of classification categories, pan-concept-tree based on cloud model has been generated, and the cloud synthesis finishes. Then, the relationship degree of each value to its relevant concept in higher level respectively is gained on basis of X-condition normal cloud model. Among all classes, the one with the highest relationship degree is defined as the relationship analysis result of its relevant object. The specific algorithm can be described in following steps: Step1: Compute P_i with formula $P_i = R_1(En, He)$, and get P_i as a random number under normal distribution with En as its expectation value and He as its standardization difference. Step2: Compute μ_i with formula $\mu_i = \exp[-0.5(x - Ex)^2 / Pi^2]$. Step3: Compute Max μ_i, and the object that is being studied now falls into the relevant Class-i .[7]

5 Application Examples

In this paper, the spatial clustering method based on cloud model and data field is testified by setting the classification of residential land as an instance. In order to comprehensively reflect the classification based on attributes and spatial, obtain the superposition potential of all the points by data field. From figure3, we can see that attribute role in rigged as a "potential value" ,which is a good integration of spatial information.

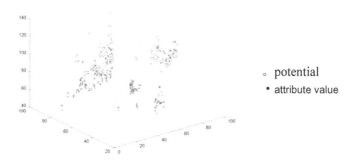

○ potential
● attribute value

Fig. 3. The comparison of potential and attribute value

Figure 4 shows the distribution of potential after data fusion, and Figure 5 shows the distribution of original data with equal spacing frequency. It is obvious that the later one does not have effective result.

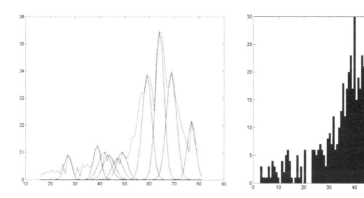

Fig. 4. Distribution of primary data

Fig. 5. Distribution of frequency by the equivalent distance

According to the method advanced in this paper, the procedure is as follows. First of all, get the peak limit values (limit value θ>0.006) by fitting normal function respectively, set peak value as expectation value（Ex）, and find its neighboring nodes. Secondly, get expectation value（Ex）and variance（En）regarding those points' data as basic data group, and get basic fitting Figure (See Figure6). Finally, draw the integral fitting curve based on development tendency of function and the basic fitting Figure[9]. (See Figure7)

In Figure 7, it is obvious that normal functions fit data quite well, and reflect the basic distribution of data perfectly. The error analysis after fitting proves that observable residual fluctuates within the limits of acceptance ability (See Figure8).

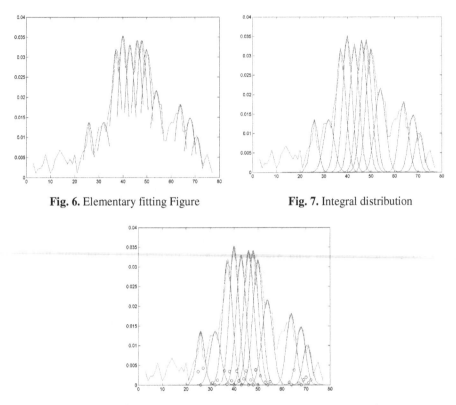

Fig. 6. Elementary fitting Figure   **Fig. 7.** Integral distribution

Fig. 8. Fitting spoil difference Figure("o" is spoil difference point)

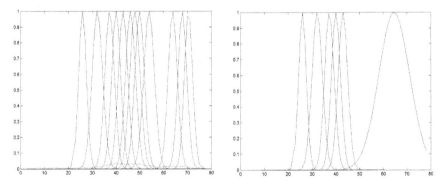

Fig. 9. The leaf node of pan-concept-tree   **Fig. 10.** The final picture of classification

When the accuracy satisfies the standard, the pan-concept-tree based on cloud model is built and Figure 9 shows its leaf nodes distribution. Then, realize the climbing of those nodes according to cloud computing principle so as to get concept in higher level. Finally, stop the second step when the number of concepts equals to the number of classification grades. (See Figure 10).

The last process is to compute the relationship degree of each a region to five-grade land, and process soft segmentation to plots based on the principle of highest relationship degree. The basic result is as follows.

Table 1. The chart of dwelling fields classification

Land grade	scale	member of plot
1	>49	98
2	42~49	201
3	36~42	204
4	32~36	120
5	<28	80

6 Conclusion

Spatial clustering is one of the most important ways of Spatial Data Mining and Knowledge Discovery. The introduction of data field can be an effective integration of attribute data and location information, While retaining the attributes of the object characteristics, the other more objective reflection of the interaction between each other. On based of the potential ,use cloud model clustering method to realize the true meaning of "soft segmentation" ,which is more consistent with the actual situation. How to choose a reasonable potential function parameters, how to control transmission errors and to improve the overall algorithm, need further study.

References

1. Li, D., Wang, S., Shi, W., Wang, X.: On Spatial Data Mining and Knowledge Discovery(SDMKD). Geomatics and Information Science of Wuhan University 26(5), 491–499 (2001)
2. Dai, X., Liu, C.: Application of Data Field in Information Token. Journal of Fudan University (Natural Science) 43(5), 933–937 (2004)
3. Gan, W., Li, D.: An Hierarchical Clustering Method Based on Data Fields. Acta Electronica Sinica 34(2), 258–262 (2006)
4. Di, K., Li, D., Li, D.: Cloud Theory and Its Applications in Spatial Data Mining and Knowledge Discovery. Journal of Image and Graphics 4A(11), 930–935 (1999)
5. Li, D., Meng, H., Shi, X.: Membership Clouds and Membership Cloud Generators. Computer R&D 32(6), 15–20 (1995)
6. Li, X., Zheng, X., Yan, H.: On Spatial Clustering of Combination of Coordinate and Attribute. Geography and Geo-Information Science 20(2), 38–40 (2004)
7. Qin, K., Li, D., Xu, k.: Image Segmentation Based on Cloud Model. Journal of Geomatics 31(5), 3–5 (2006)
8. Li, D., Liu, C., Du, Y., Han, X.: Artificial Intelligence with Uncertainty. Journal of Software 15(11), 1583–1594 (2004)
9. Jiang, R., Li, D., Fan, J.: Automatic Generation of Pan-Concept-Tree on Numerical Data. Chinese Journal of Computers 23(5), 470–476 (2000)

Diversity Analysis of Information Pattern and Information Clustering Algorithm

Shifei Ding[1,2], Wei Ning[3], and Zhongzhi Shi[2]

[1] School of Computer Science and Technology, China University of Mining and Technoogy,
Xuzhou 221008 P.R. China
[2] Key Laboratory of Intelligent Information Processing, Institute of Computing Technology,
Chinese Academy of Sciences, Beijing 100080 P.R. China
[3] School of Computer Science and Technology, Xuzhou Normal University,
Xuzhou 221116 P.R. China
dingsf@cumt.edu.cn

Abstract. According to information theory, a basic concept of the measure of diversity is defined, and a basic inequation of the measure of diversity is discussed and proved, then a concept of increment of diversity is given. The diversity of the information pattern is carried out the analysis. On the basis of theses discussions, the information coefficient measure (ICM) is defined, and a new information clustering algorithm is built up according to the ICM, and then carried out the information clustering analysis for soil fertility data processing in land. Compared with Hierarchical Clustering Algorithm (HCA) traditionally, the result of simulated application shows that the algorithm presented here is feasible and effective.

Keywords: diversity analysis; increment of diversity; information coefficient measure (ICM); information clustering; soil fertility.

1 Introduction

The classic clustering method is the Hierarchical Clustering Algorithm (HCA). The method is the most extensive method in clustering analysis. HCA with the distance matrix for foundation, according to a certain clustering method gradual combines the type, and needs to be saving distance matrix in processing. When the sample quantity is very big, this method is requested to have enough big memory space, and so it is inconvenient for us to be used to some degree. Another type called the Dynamic Clustering Algorithm, (DCA), resolved this problem from the technique, in which the C-means (CM) and fuzzy C-means (FCM) are the most common use. In these methods, we don't carry on optimization for features of the samples, and make use of features of samples to carry on clustering directly. So the efficiency of these algorithms is decided by the sample distribution to a large extent. Only the natural distribution of the class is the globosity or neared to the globosity, that is to say, the variance in each class is near equality, is there a better clustering result. Moreover the method of CM or FCM, adjusting a category of sample each time, the mean of each kind of sample has to calculate, and so they are called again step-wise sample revising method, while

L. Kang, Y. Liu, and S. Zeng (Eds.): ISICA 2007, LNCS 4683, pp. 428–434, 2007.
© Springer-Verlag Berlin Heidelberg 2007

the Iterative Self- Organizing Data Analysis Techniques Algorithm, (ISODATA) is called batch the sample correction method. The ISODATA algorithm is one of the better methods with characteristics to reason logically, analyze the direct and control clustering structure and person's machines to hand over with each other etc., is more of gather one of a methods[1-3].

The algorithms mentioned above can reach different clustering category under the different distribution and structure for the data. But their algorithms have no been analyzed and studied from the angle of information. We know that Shannon, C.E., the American electrical engineer, put forward the concept of the information entropy for the very first time in 1948, which is a measure of uncertainty of a random variable. [4,5]. So it is necessary for us to combine the information theory and clustering algorithm. In this paper, we discuss basic concept and its basic properties of the measure of diversity[6,7]. On this foundation, we build up the information coefficient measure (ICM), and point out that the ICM is a kind of diversity index. According to the index sign of the ICM, a kind of information clustering algorithm is constructed based on ICM, and applied it to soil fertility data processing.

2 Measure of Diversity

2.1 Definition of Measure of Diversity

For the state space $X = \{x_1, x_2, \cdots, x_s\}$, where $N = \sum_{i=1}^{s} n_i$, and n_i denotes the times of x_i appeared $(i = 1, 2, \cdots, s)$. We say that the state space constitutes the source of diversity, means for

$$[X \cdot N]: \begin{cases} X : x_1, x_2, \cdots, x_s \\ N : n_1, n_2, \cdots, n_s \end{cases} \tag{1}$$

where $N = \sum_{i=1}^{s} n_i$.

From the angle of information, the quantity of the measured diversity is called measure of diversity, denoted by $D(n_1, n_2, \cdots, n_s)$. It is defined as follows

Definition 1. Let X be a discrete random variable with the diversity source space $[X \cdot N]$, then the measure of diversity of X is defined as

$$D(X) = -\sum_{i=1}^{s} n_i \log \frac{n_i}{N} = N \log N - \sum_{i=1}^{s} n_i \log n_i \tag{2}$$

It can be seen by the formula (2):

① when N is very big, we have $p(x_i) \approx n_i / N$, according to the definition of information entropy , combining the formula (2), we get

$$H(X) = -\sum_{i=1}^{s} p(x_i) \log p(x_i) \approx -\sum_{i=1}^{s} \frac{n_i}{N} \log \frac{n_i}{N} = \frac{1}{N} D(X) \tag{3}$$

so we have

$$D(X) \approx NH(X) \tag{4}$$

where $H(X)$ denotes the information entropy. From formula (4), $D(X)$, measure of diversity, means the total information content of all individuals in X.

2.2 An Inequation of Measure of Diversity

Let Y be a discrete random with diversity source space $[Y \cdot M]$, i.e.

$$[Y \cdot M]: \begin{cases} Y: y_1, y_2, \cdots, y_s \\ M: m_1, m_2, \cdots, m_s \end{cases} \tag{5}$$

where $M = \sum_{i=1}^{s} m_i$.

While the sum of two diversity sources called combination diversity source space, can be expressed as

$$[X + Y \cdot M + N]: \begin{cases} X + Y: x_1 + y_1, \cdots, x_s + y_s \\ N + M: n_1 + m_1, \cdots, n_s + m_s \end{cases} \tag{6}$$

where $M + N = \sum_{i=1}^{s} (n_i + m_i)$.

The measure of diversity satisfies following an important inequation, i.e.

Theorem 1. Let $X, Y, X + Y$ have the diversity source space $[X \cdot N], [Y \cdot M]$ and $[X + Y, M + N]$, then we have

$$D(X + Y) \geq D(X) + D(Y) \tag{7}$$

Proof. Let

$$x_1 = \frac{n_i}{N}, \ x_2 = \frac{m_i}{M}; \ \lambda_1 = \frac{N}{M + N}, \ \lambda_2 = \frac{M}{M + N}$$

Because the function $f(x) = x \log x$ is a convex function, according to the Jensen inequation, we have

$$(\lambda_1 x_1 + \lambda_2 x_2) \log(\lambda_1 x_1 + \lambda_2 x_2) \leq \lambda_1 x_1 \log x_1 + \lambda_2 x_2 \log x_2$$

i.e.

$$\left(\frac{N}{M + N} \frac{n_i}{N} + \frac{M}{M + N} \frac{m_i}{M} \right) \times \log\left(\frac{N}{M + N} \frac{n_i}{N} + \frac{M}{M + N} \frac{m_i}{M} \right)$$

$$\leq \frac{N}{M + N} \frac{n_i}{N} \log \frac{n_i}{N} + \frac{M}{M + N} \frac{m_i}{M} \log \frac{m_i}{M}$$

so

$$(n_i + m_i) \log \frac{n_i + m_i}{M + N} \leq n_i \log \frac{n_i}{N} + m_i \log \frac{m_i}{M}$$

therefore

$$-\sum_{i=1}^{s}(n_i+m_i)\log\frac{n_i+m_i}{M+N} \geq -\sum_{i=1}^{s}n_i\log\frac{n_i}{N}-\sum_{i=1}^{s}m_i\log\frac{m_i}{M}$$

i.e.

$$D(X+Y) \geq D(X)+D(Y)$$

the theorem 1 is finished.

Theorem 1 shows that the combination measure of diversity is no smaller than the sum of every measure of diversity.

3 Information Clustering Algorithm

3.1 Definition of Increment of Diversity

According to the formula (7), we get that the combination measure of diversity is no smaller than the sum of every measure of diversity.

Definition 2. Let $X, Y, X+Y$ have the diversity source space $[X \cdot N], [Y \cdot M]$ and $[X+Y, M+N]$, then the increment of diversity is given by

$$D(X+Y)-D(X)-D(Y) \tag{8}$$

denoted by $\Delta(X,Y)$.

Obviously, the increment of diversity satisfies nonnegativity and symmetry, therefore, and so it is a kind of semi-metric or generalized distance measure.

3.2 Diversity Analysis of the Information Pattern

Suppose the pattern i contains n states, and x_{ik} represents numbers of individuals in the ith pattern and the kth state, where $N_i = \sum_{k=1}^{n} x_{ik}$. From here we get the measure of diversity which the pattern i contained, namely the total information content of pattern i is

$$D_i = N_i \log N_i - \sum_{k=1}^{n} x_{ik} \log x_{ik} \tag{9}$$

the total information content of the pattern j is

$$D_j = N_j \log N_j - \sum_{k=1}^{n} x_{jk} \log x_{jk} \tag{10}$$

the total information content after the two patterns merged is

$$D_{i+j} = (N_i + N_j)\log(N_i + N_j) - \sum_{k=1}^{n}(x_{ik}+x_{jk})\log(x_{ik}+x_{jk}) \tag{11}$$

the increment of diversity after the two patterns merged is

$$\Delta_{ij} = D_{i+j} - D_i - D_j \tag{12}$$

suppose that the pattern i and the pattern j are combined into a new pattern k, and then the increment of diversity of the new pattern k and the pattern r is

$$\Delta_{kr} = D_{k+r} - D_k - D_r \tag{13}$$

where

$$D_{k+r} = (N_i + N_j + N_r)\log(N_i + N_j + N_r) - \sum_{l=1}^{n}(x_{il} + x_{jl} + x_{rl})\log(x_{il} + x_{jl} + x_{rl}) \tag{14}$$

$$D_k = (N_i + N_j)\log(N_i + N_j) - \sum_{l=1}^{n}(x_{il} + x_{jl})\log(x_{il} + x_{jl}) = D_{i+j} \tag{15}$$

$$D_r = N_r \log N_r - \sum_{l=1}^{n} x_{rl} \log x_{rl} \tag{16}$$

where N_i, N_j and N_r are the numbers of individuals the pattern i, j and pattern r contained respectively.

3.3 Information Coefficient Measure

According to the increment of diversity and its properties, we can give the definition of information coefficient measure (ICM).

Definition 3. Suppose two diversity source spaces $[X \cdot N]$ and $[Y \cdot M]$ are given and shown as formulae (1) and (5), then the $\Delta(X, Y)$, the increment of diversity after the two diversity sources merged, is called ICM.

Obviously the ICM is a dissimilitude index. If the value of ICM is smaller, the two diversity sources is more alike, whereas more dissimilitude.

3.4 Information Clustering Algorithm

According to the discussion above, we construct following information clustering algorithm based on ICM as follows.

Step 1. Compute increment of diversity Δ_{ij}. Its relevant calculation formulae see formulae (9), (10), (11) and (12), because of symmetry, so let $i = 1, 2, \cdots, m$; $j > i \geq 1$.

Step 2. Merge the pattern. Choosing the least value from the Δ_{ij}, we merge the pattern i with the pattern j into the new pattern k. At this time $D_k = D_{i+j}$.

Step 3. Compute again increment of diversity. Let $i = k, j = r$, return the Step 1, compute Δ_{kr} again. At this time, the relevant formula is (13), (14), (15) and (16).

Step 4. Compute repeatedly. Repeated Step 1 and Step 2, until all patterns were merged.

4 An Example

Given 12 sets of data with soil fertility now with 12 type patterns of land, write P1, P2, …,P12 respectively. We select 7 influence factors with closely-related soil fertility, such as the geography type, the organic quality content(g/kg), soil capacity(g/cm^3), soil to change the quantity(mol/kg), soil total nitrogen(g/kg), soil rapidly available phosphate (mg/kg) and soil rapidly available potassiumsoon (mg/kg) etc. as the information characteristic index.

Computed by the DPS data processing system software [8], we get the information content and increment of diversity of each pattern. Information clustering reflects directly the similarity and the difference of 12 sets of soil fertility. according to Fig. 1, we can divide the 12 sets of soils into 4 classes, i.e. four kinds of types with soil fertility. Class 1 includes the pattern P1, P11; Class 2 includes P2, P6, P5, P9 and P10; Class 3 includes the pattern P3, P7 and P12; Class 4 includes the pattern P4, P8.

In order to compare the effects of information clustering algorithm [9], we carried out the system clustering again (the data carries on the standardization; clustering distance uses Europe distance; clustering method adopts sum of squares of deviations). Its clustering result is similar with information clustering here.

5 Conclusions

According to the information theories, we built up the information coefficient measure (ICM), designed the information clustering algorithm based on ICM, applied it to identify the soil fertility evaluation classification, and obtained satisfactory result of clustering. If the pattern dimensionality is higher, we may consider compressing the sample space firstly, and then carrying on the information clustering in the characteristic space of the low dimensionality. Thus, we will receive better result of clustering.

Acknowledgements

This work is supported by the National Science Foundation of China (No. 60435010, 90604017, 60675010, 40574001, 50674086), the 863 National High-Tech Program (No.2006AA01Z128), the National Basic Research Priorities Program (No. 2003CB317004), the Doctoral Foundation of Chinese Education Ministry (No. 20060290508), the Nature Science Foundation of Beijing (No.4052025) and the Science Foundation of China University of Mining and Technology.

References

1. Duda, R.O., Hart, P.E. (eds.): Pattern Classification and Scene Analysis. Wiley, New York (1973)
2. Bian, Z.Q., Zhang, X.G. (eds.): Pattern Recognition. Tsinghua University Press, Beijing (2000)
3. Sun, J.X. (ed.): Modern Pattern Recognition. The Defence University of Science and Technology Publishing House, Changsha (2002)

4. Shannon, C.E.: A Mathematical Theory of Communication. Bell Sys. Tech. Journal 27, 379–423 (1948)
5. Cover, T.M., Thomas, J.A. (eds.): Elements of Information Theory. Wiley, New York (1991)
6. Laxton, R.R.: The Measure of Diversity. J. Theor. Biol. 71, 51–67 (1978)
7. Yuan, Z.F., Zhou, J.Y. (eds.): Multivariate Statistical Analysis. Science Press, Beijing (2003)
8. Tang, Q.Y., Feng, M.G. (eds.): Practical Statistics and DPS Data Processing System. China Agricultural Press, Beijing (1997)
9. Hartigan, J.A. (ed.): Clustering Algorithm. Wiley, New York (1975)

Instant Message Clustering Based on Extended Vector Space Model[*]

Le Wang, Yan Jia, and Weihong Han

Computer School, National University of Defense Technology, Changsha, China
wanglelemail@163.com,
jiayanjy@vip.sina.com,
hanweihong@gmail.com

Abstract. Instant intercommunion techniques such as Instant Messaging (IM) are widely popularized. Aiming at such kind of large scale mass-communication media, clustering on its text content is a practical method to analyze the characteristic of text content in instant messages, and find or track the social hot topics. However, key words in one instant message usually are few, even latent; moreover, single message can not describe the conversational context. This is very different from general document and makes common clustering algorithms unsuitable. A novel method called *WR-KMeans* is proposed, which synthesizes related instant messages as a conversation and enriches conversation's vector by words which are not included in this conversation but are closely related with existing words in this conversation. *WR-KMeans* performs clustering like k-means on this extended vector space of conversations. Experiments on the public datasets show that *WR-KMeans* outperforms the traditional *k*-means and bisecting *k*-means algorithms.

Keywords: instant messages clustering, k-means, Vector Space Model.

1 Introduction

With the rapid development of internet and communication technology, Instant Messaging (IM, e.g. E-mails, SMS, chats through MSN or ICQ, etc.) on internet or mobile network is widely popularized[1], e.g. more than 160 million short text messages were sent over the National Week holiday in Beijing[2]. Considering economic interest or public security, corporations and governments, which provide this service, have stored instant messages in text database for further analytical and mining applications [1]. Instant message clustering is very useful for analyzing its content characteristic or establishing other mining application.

The most common text processing approach is to represent the documents with vectors. This is so-called vector space model, in which a vector corresponds to one

---

[*] This project is sponsored by national 863 high technology development foundation (No. 2006AA01Z451, No.2006AA10Z237).

[1] http://www.instantmessagingplanet.com/
[2] http://english.cri.cn//3100/2006/10/10/63@148745.htm

L. Kang, Y. Liu, and S. Zeng (Eds.): ISICA 2007, LNCS 4683, pp. 435–443, 2007.

document and the dimensions correspond to words in this document. Once the high-dimensional vectors are derived, the major challenge left for document clustering is how to deal with these high dimensional data. However, instant message is extremely shorter than the common document. There are usually only several key words in one instant message, and key words about the message topic are even latent sometimes. The sparsity of key words makes word-frequency based methods inappropriate to measure the similarity among instant messages. Table 1 is an example to demonstrate above problem, where instant messages are all about sports and have considerable similarity with each other. Although IM-1 is similar to IM-2 and IM-3 with 0.71 and 0.58 degree respectively, no similarity exists between IM-2 and IM-3, which conflicts with the reality. So the bag-of-word model and term frequency-based measure are not applicable in instant messages mining.

Table 1. Example to illustrate bag-of-word model vectors and similarity between instant messages according to vector inner product

	ball	basketball	football	foot		IM1	IM2	IM3
IM-1	0	1	1	0		-	0.71	0.58
IM-2	0	2	0	0			-	0
IM-3	1	0	2	1				-

This paper proposes two methods to enhance the description of instant messages to response the problem of sparse key words when clustering on instant messages.

Firstly, we notice that instant message is a kind of semi-structured data, which has source and destination addresses with time stamp. Instant messages sent back and forth among specific persons during some specific time intervals form a conversation, which groups these instant messages into a specific topic. So we combine these messages as one conversation. It is obvious that conversation has more key works and more integral context information than simply single message. Then clustering is performed toward conversations instead of messages.

Secondly, we enhance the content description of a conversation with words, which are not in the conversation but are closely related with existing words in this conversation. Fox example, words 'ball' and 'football' are added to IM-2, which are not appear in IM-2 but have obvious correlation with the word, 'basketball', in IM-2.

In this paper, we propose an instant message clustering method called WR-KMeans, which can automatically scan instant message corpora, construct conversations and enhance traditional TF-IDF model by adding relevant words in conversations. WR-KMeans performs clustering on this evolved model of conversations like k-means [2].

WR-Kmeans method is evaluated and compared with two other well-known text clustering methods which is based on traditional TF-IDF model. HowNet knowledge base is used to quantify the relation strengths between words in conversations during the experiments. Experimental evidence shows that WR-KMeans is significantly outperformed. Furthermore, HowNet is Chinese-English bilingual linguistics, so WR-KMeans and its components can be smoothly transformed to process Chinese [3].

The rest of the paper is organized as follows. We present related works in Section 2 and present WR-KMeans method in Section 3. The experimental results are reported in Section 4. Finally, we conclude the paper and discuss future works in section 5.

2   Related Works

Various methods can be used to confirm the boundary of conversation for different type of instant messages. Conversations can be easily captured by the posting threads in Usenet due to its inherent threaded nature [4]. Methods based on certain relevant patterns are used by Faisal M. Khan [5] to identify chat thread starts in chat-room medium flows. These patterns are made up of several sentences, such as "hi, hey" or "how are you", which are developed by human experts through observing chat conversations. It is an effective method for medium with very strong interaction like chat-room. Marti A. Hearst utilizes TextTiling algorithm to locate topic boundaries within expository text [6]. This algorithm is designed to separate expository text into paragraphs, and uses lexical analyses based on TF-IDF model to determine topic starting point. Methods mentioned above are mainly based on the content of corpora.

The relationships among words have been widely studied in fields of nature language processing, text mining and information retrieval, etc. One method is Latent Semantic Indexing (LSI) [7], which automatically discovers latent relationships among corpora through Singular Vector Decomposition. However, the method is time-consuming when applied to a large corpus. Kenneth Ward Church proposed 'association ratio' based on the notion of mutual information to estimate word association norms by their co-occurrence probability [8]. It is not appropriate to refer to words co-occurrence for instant messages because of the key words sparsity. Satoru Ikehara proposed a vector space model based on semantic attributes of words, which uses the Semantic Attribute System [9]. This method aims to reduce the vector dimension using upper-lower relations between semantic attributes of words and achieves good efficiency in processing Japanese.

3   WR-KMeans Method

3.1   Synthesizing Conversation

Message Database (MDB) is a message set, which store messages in a form which facilitates accessing a group of messages. T_1 and T_2 are used to denote the starting and end time of a specific period, respectively. s_i and d_i stand for source and destination addresses. c_i is the text content of instant message. Then the concept of conversation can be formalized as:

Definition 1 (Conversation). $M \in MDB$, $\|M\| = n$, If $\forall m_i = < t_i, d_i, s_i, c_i >$ and $\forall m_j = < t_j, d_j, s_j, c_j >$, $m_i \in M$ and $m_j \in M$, $0 < i, j \leq n$, if m_i and m_j satisfy $T_1 < t_i < T_2$, $T_1 < t_j < T_2$, $s_i = d_j$ or $d_i = s_j$, then the string, $c_1 | c_2 | ... | c_n$, synthesizes a conversation between two persons at the interval of T_1 and T_2 .   ∎

Further observation into the instant message data set[3] from the reality mining project of MIT allows us to find that the frequency of IMs transmission before and after one

---

[3] http://reality.media.mit.edu/dataset.php

conversation is usually lower than that during the conversation, which is shown in Figure 1. So the boundaries of conversations can be determined in some sense by the concaves of frequency. This can be explained as following. From the point of temporal view, peoples tend to densely communicate with each other about the same topic. In other words, instant messages which are produced by one pair of persons and are related to the same topic could be aggregated together approximatively according to their generation time, i.e. the communicating requency.

We define the following rules for synchronizing instant messages into conversations on the basis of above analysis. $V_{i,i+1}$ is used to denotes the time interval between two adjoint instant messages, m_i and m_{i+1}. We assume that if $V_{i,i+1} < \alpha$ and $V_{i,i+1} < V_{i+1,i+2}$, then m_{i+1} and m_i belongs to the same conversation; otherwise, m_{i+1} is the starting of next conversation. Where α is a statistic constant which describes the biggest interval between two adjoint instant messages that belong to the same conversation. *WR-KMeans* orderly compares the intervals of adjoint IMs between two specific persons and synthesizes conversations for each pair of all persons.

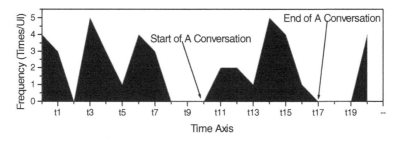

Fig. 1. Frequency Change of IM transmission between two persons in a specific time interval

3.2   Enhancing the Representation of Conversation Through Relevant Words

Assume that there are m conversations, which consist of n different words totally. We calculate the relevant strength of each pair of words according to *HowNet*. Given a conversation C_l, the word, t_j, which is not in conversation C_l, is used to enhance the vector representation of C_l if the relevant strength ($\delta_{i,j}$) between t_j and t_i, which is originally in C_l, is beyond one threshold of relevant strength.

In addition, the weights of the words in corpora are not equal to each other. The important word in conversations is usually the one which defines few conversations. That is the more irregular word which is the more important for distinguish the conversations. This relies on the information entropy of the word, which is defined as $E_i = -p_i \cdot \log_2 p_i, (0 < i \leq n)$ and $p_i = \lambda_i / m$. λ_i is the sum of conversations that include term t_i. Then the weight of word t_i in C_l is defined as formula (1) where $num(C_l)$ is the total num of words in conversation C_l.

$$\beta_i = \frac{E_i}{\sum_{k=1}^{num(C_l)} E_k}, (0 < i \le num(C_l)) \qquad (1)$$

Then the value in vector delegating word t_j, which is added to enhance the vector representation of the conversation, can be determined according to formula (2) where t_k is the value of k-th word in the vector of C_l according to TF-IDF model.

$$t_j = \sum_{k=1}^{k=num(C_l)} \beta_k \cdot t_k \cdot \delta_{k,j} \qquad (2)$$

For example, in table 2, a word-by-conversation matrix is constructed from 3 conversations (C1, C2, and C3) and 7 words. Only relevant strengths that are beyond 0.4 are considered and set $\delta_{i,j}$ equal to $\delta_{j,i}$. For T1 in C1, the value in TF-IDF is $2*\log_{10}(3/1)=0.9542$. The word, T4, which is not in C1, has a relevant strength beyond 0.4 with T1. So T4 should be added into the vector of C1. The value of T4 in vector is $0.9542*0.5283*0.62=0.3126$, where $-1/3*\log_2(1/3)=0.5283$ is the weight of T1.

WR-KMeans is developed as an instant messages clustering method, which is a variant on standard k-means algorithm. This algorithm preprocesses the instant messages, synthesizes conversations and extends the vectors of conversations before performing clustering on enhanced TF-IDF model. WR-KMeans measures the similarity between conversations according to a cosine measure.

The sum of terms is in a finite bound, preprocesses relate mainly with the volume of instant message set. So WR-KMeans is an extensible method.

Table 2. Example of extending word-by-conversation matrix and word relevant strengths obtained through querying of *HowNet*

	Original word Frequency			word relevant strengths							TF-IDF model enhanced by WR		
	C1	C2	C3	T1	T2	T3	T4	T5	T6	T7	C1	C2	C3
T1	2	0	0	N/A	-	-	0.62	-	-	-	0.9542	0.1277	0.0852
T2	1	1	0		N/A	-	-	0.43	-	-	0.1761	0.1761	0.4336
T3	0	1	1			N/A	-	-	-	0.42	0.2117	0.1761	0.1761
T4	0	3	2				N/A	-	-	-	0.3126	0.5283	0.3522
T5	0	0	4					N/A	-	-	0.0295	0.0295	1.9085
T6	0	2	0						N/A	-	0	0.9542	0
T7	2	0	0							N/A	0.9542	0.0288	0.0288

4   Experimental Evaluations

Three different algorithms, *WR-KMeans*, Bisecting k-means and standard k-means, are implemented and compared. All these experiments are performed on two public datasets with manually predefined categorizations.

4.1 Evaluation Criteria

Two cluster validation methods, Silhouette Coefficient (SC) [11] and normalized mutual information (NMI) [12], are used to evaluate the clustering performance because both of them are independent from the number of clusters, k.

Suppose that instant messages are synchronized into m conversations, which have l classes. $C_M = \{\overline{C_1}, \overline{C_2}, \cdots, \overline{C_k}\}$ defines a clustering result.

Silhouette Coefficient (SC)

$S(C_i, \overline{C_j})$ is the similarity of a conversation C_i to a cluster $\overline{C_j}$, which equals to the average similarity between C_i and each conversation in $\overline{C_j}$. Let $f(C_i, C_M) = S(C_i, \overline{C_j})$ be the similarity between the conversation C_i and its cluster $\overline{C_j}(C_i \in \overline{C_j})$, and $g(C_i, C_M) = \max_{\overline{C_j} \in C_M, C_i \notin \overline{C_j}} S(C_i, \overline{C_j})$, the similarity between C_i and the nearest neighboring cluster. The silhouette of C_i is defined as:

$$SC(C_i, C_M) = \frac{f(C_i, C_M) - g(C_i, C_M)}{\max\{f(C_i, C_M), g(C_i, C_M)\}} \tag{3}$$

The silhouette coefficient is defined as formula (4). Its value is usually between 0 and 1. Values beyond 0.5 indicate that clustering results are separable clearly. If they fall below 0.25, it becomes very difficult to find practically significant clusters.

$$SC(C_M) = \frac{\sum_{C_i \in MDB} SC(C_i, C_M)}{|MDB|} \tag{4}$$

Normalized Mutual Information (NMI)

Let m_i, m_j be the numbers of conversations in the i-th class M_i and j-th cluster $\overline{C_j}$ respectively, m_{ij} is the number of conversations of M_i that are assigned to $\overline{C_j}$, m is the total number of conversations in MDB. NMI is then defined as formula (5), which equals to 1 when clustering results perfectly match the external category labels and is close to 0 for a random partitioning. NMI measures the consistent level between the clustering result and the original classification in data set, the later is usually provided by human experts. The bigger value of NMI is the more ideal consistency with the outcome of human beings, which illuminates a perfect clustering method.

$$NMI(C_M) = \frac{\sum_{i=1}^{l} \sum_{j=1}^{k} m_{ij} \cdot \log(\frac{m \cdot m_{ij}}{m_i \cdot m_j})}{\sqrt{(\sum_{i=1}^{l} m_i \cdot \log\frac{m_i}{m})(\sum_{j=1}^{k} m_j \cdot \log\frac{m_j}{m})}} \tag{5}$$

4.2 Experimental Setting

We use two data sets: (1) the Reuters-21578 corpus[4], (2) 20-newsgroups data[5]. These two datasets comprise priori categorizations of documents, and their domains are broad enough to be as realistic as conversations. We preprocess the raw datasets mentioned above using the Bow toolkit[6] and Porter stemming function [13].

HowNet system version 2000, a free edition of this software, is used to quantify the mutuality between terms. One HP unit with 4 Itanium II 1.6G processors and 48 GB memory is used as hardware platform.

The word-by-conversation matrixes of two datasets are preprocessed according to TF-IDF model (for standard *k*-means and Bisecting *k*-means algorithms) and enhanced TF-IDF model (for *WR-KMeans* method) respectively.

4.3 Experimental Results

We use a maximum number of iterations of 20 (to make a fair comparison) for all these three algorithms. Each experiment is running ten times. We set the threshold of relevant strength between two words to 0.4.

Table 3. NMI results on 20-Newsgroup

k	5	15	20	25
Std k-means	.23±.03	.24±.02	.26±.03	.25±.02
Bis k-means	.37±.02	.40±.02	.42±.01	.41±.03
WR-KMeans	**.54±.03**	**.68±.02**	**.79±.01**	**.72±.02**

Table 4. NMI results on Reuters-21578

k	40	60	80	100
Std k-means	.21±.03	.22±.01	.25±.02	.24±.03
Bis k-means	.32±.03	.36±.02	.40±.01	.38±.02
WR-KMeans	**.46±.03**	**.52±.01**	**.64±.02**	**.58±.02**

Table 3 and Table 4 report the the effect of k on NMI results on NG20 and Reuters-21578, respectively. NMI measures the degree of consistency between clustering results and manually predefined categorizations, i.e. the superposition of clusters and classes. *WR-KMeans* has clear better clustering results, which is indicated by Table 3 and Table 4. The reason is that the extended vector space model has more enriched semantic information than traditional TF-IDF model, and the strengthened vector represents the real theme of text content. The vectors, in which only correlated terms are added, are used to compute the similarities. This approach magnificently avoids the warp resulted from the sparsity of key words when measuring the similarities of text and then achieve the better effectiveness than primal representation method of TF-IDF model.

---

[4] http://www.daviddlewis.com/resources/testcollections/reuters21578/
[5] http://kdd.ics.uci.edu/databases/20newsgroups/20newsgroups.html
[6] http://www.cs.cmu.edu/mccallum/bow

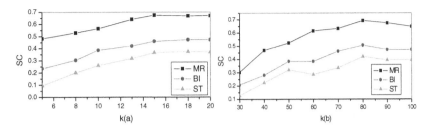

Fig. 1. Comparing the best SC results on NG20(a) and on Reuters-21578(b)

We perform experiments on NG20 and Reuters-21578 to study the effect of k on SC, and the result is shown in Figure 1. The SC studies the divisibility of clusters. At the point of original num of classes in dataset, *WR-KMeans* can get clear partitions of corpora, which can be induced according to SC in figure 1 (NG20 0.67 when k=20, Reuters-21578 0.69 when k=80). This is a reasonable result.

We can draw the conclusion from above results that the extended vector space model, which is combined with term mutual information, has more linguistic knowledge than TF-IDF model. It takes context information to distinguish the category of documents (conversations).

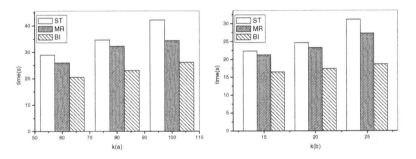

Fig. 2. Comparing the average time on NG20(a) and on Reuters-21578(b)

Figure 2 illuminates the running time of three algorithms on two datasets. We can see that *WR-KMeans* would comparatively needs more time than Bisecting k-means, but a little faster than standard k-means. The reason is that WR-KMeans is optimized only in preprocessing and text representing, not including the clustering process. Although WR-KMeans achieve better effectiveness than other two algorithms, it has not too much advantage in efficiency.

5 Conclusion and Future Works

In this paper, we focus on the instant messages clustering and propose *WR-Kmeans* method to solve the sparsity of key words arising from it. *WR-KMeans* automatically synthesizes instant messages into conversation, which has more key words and more integral context information than simply single message, and extends traditional TF-IDF model of conversations by relevant words by the aid of *HowNet*. Experimental evidence shows that *WR-KMeans* is significantly outperformed against other two method based on traditional TF-IDF model.

We plan to improve the speed of *WR-KMeans* when performing clustering by optimizing its initial partitions. In addition, we want analyze the network of IM by social network analysis in the future works.

References

1. Resig, J., Teredesai, A.: A framework for mining instant messaging services. In: Proceedings of the 2004 SIAM Lake Buena Vista, Florida (2004)
2. MacQueen, J.: Some methods for classification and analysis of multivariate observations. In: proceedings of 5th berkeley SMSP, pp. 281–297 (1967)
3. Guan, Y., et al.: Quantifying Semantic Similarity of Chinese Words from Hownet. In: IEEE Proceedings of ICMLC02, Beijing, vol. 1, pp. 234–239. IEEE Computer Society Press, Los Alamitos (2002)
4. Sack, et al.: A Content-Based Usenet Newsgroup Browser. In: Proceedings of the international conference on Intelligent user interfaces, New Orleans, Louisianna, pp. 233–240 (2000)
5. Khan, F.M., Fisher, T.A., Shuler, L., Wu, T., Pottenger, W.M.: Mining chat-room conversations for social and semantic interactions (2002)
6. Hearst, M.A.: TextTiling: A Quantitative Approach to Discourse Segmentation, Technical Report UCB: S2K-93-24 (1993)
7. Deerwester, S., et al.: Indexing by latent semantic analysis. Journal of the American Society of Information Science 41(6), 391–407 (1990)
8. Ding, C.H.Q.: A probabilistic model for dimensionality reduction in information retrieval and filtering. In: Proc. of the 1st SIAM, Raleigh, NC (2000)
9. Ikehara, S., et al.: Vector space model based on semantic attributes of words. In: Proc. of the Pacific Association for Computational Linguistics (PACLING), Kitakyushu, Japan (2001)
10. Daemi, A., et al.: From Ontologies to Trust through Entropy. In: Proceedings of the International Conference on Advances in Intelligent System, Luxembourg (2004)
11. Hotho, A., et al.: Ontology-based Text Document Clustering. KI 16(4), 48–54 (2002)
12. Strehl, A., Ghosh, J.: Cluster ensembles - a knowledge reuse framework for combining partitions. Journal of Machine Learning Research 3, 583–617 (2002)
13. Porter, M.F.: An algorithm for suffix stripping. Program 14(3), 130–137 (1980)

Continuous K-Nearest Neighbor Queries for Moving Objects

Hui Xiao[1,3], Qingquan Li[2], and Qinghong Sheng[3]

[1] Transportation Research Center, Wuhan University,129 Luoyu Road, Wuhan, China, 430079,
Tel.: 86-27-68778222-8101, Fax: 86-27-68778035-8113
[2] State Key Laboratory of Information Engineering in Surveying , Mapping and Remote
Sensing ,Wuhan University,129 Luoyu Road, Wuhan, China, 430079
[3] Institute of Remote Sensing and Information Engineering, Wuhan University,
129 Luoyu Road, Wuhan, China, 430079
xylogen@163.com, qqli@whu.edu.cn, sqh96.student@sina.com

Abstract. Continuous k-nearest neighbors (CKNN) search has been in the core
of spatiotemporal database research during the last decade. It is interested in
continuously finding the k closest objects to a predefined query object q during
a time interval. Existing methods are either computationally intensive
performing repetitive queries to the database or restrictive with respect to the
application settings. In this paper we develop an efficient method in order to
computes CKNN queries for a query point during a specified time interval. The
basic advantage of the proposed approach is that only one query is issued per
time interval. The R-tree structure is used to index the static objects. An
extensive performance evaluation shows that all the techniques we presented
boost the query performance greatly.

Keywords: moving objects databases, continuous k-nearest neighbor queries,
R-tree.

1 Introduction

With rapid advances in electronics miniaturization, wireless communications, and
positioning technologies, the acquisition and transmission of spatiotemporal data
using mobile devices are becoming pervasive. This meets the requirements for
location based services (LBS)[10], [11].

Unlike traditional databases, moving objects have some distinctive characteristics.
Spatio-temporal queries are continuous in nature. In contrast to snapshot queries,
which are invoked only once, continuous evaluation as the result becomes invalid
after a short period of time. An important query, which is definitely useful for LBS
processing, called as continuous k-nearest neighbors queries. An example of this type
of query is "find the nearest gas stations to the taxi in the next 10 minutes". It is
interested in continuously finding the k closest objects to a predefined query object q
during a time interval. Existing methods are either computationally intensive
performing repetitive queries to the database or restrictive with respect to the
application settings.

L. Kang, Y. Liu, and S. Zeng (Eds.): ISICA 2007, LNCS 4683, pp. 444–453, 2007.
© Springer-Verlag Berlin Heidelberg 2007

This paper proposes an algorithm that efficiently computes CKNN queries for a query point during a specified time interval. Suppose that the query points are continuously moving in a plane and the data points are static objects and the query time interval starts at or after the current time. The basic advantage of the proposed approach is that only one query is issued per time interval. The main contributions of this paper are as follows:

◆ an efficient query processing algorithm to perform CKNN search based on a R-tree structure[9]; and

◆ a comprehensive comparison with existing methods through considering several parameters that affect query processing performances.

The rest of this paper is arranged as follows. Following the introduction, related work is presented. Before the experimental studies are performed, the details of the algorithm are proposed. Finally, conclusions are drawn at the end of this paper.

2   Related Work

In the last decade, nearest neighbor (NN) queries have fueled the spatial and spatiotemporal database community with a series of interesting noteworthy research issues. An affluence of methods for the efficient processing of NN queries for static query points already exist, the most influential probably being the branch-and-bound R-tree traversal algorithm proposed by Roussopoulos et al. [7] for finding the nearest neighbor of a single stationary point. The algorithm utilizes two metrics, MINDIST and MINMAXDIST, in order to implement tree pruning and ordering.

Later, Cheung and Fu [8] proved that, given the MINDIST-based ordering, the pruning obtained by [7] can be preserved without the use of MINMAXDIST metric (the calculation of which is computational expensive).

We now pass on to moving objects. Kollios et al.[5] proposed an elegant solution for answering nearest neighbor queries for moving objects in a one-dimensional space. Their algorithm uses a duality transformation. However, this solution has difficulties to be straightforwardly extended to a 2-dimensional case, where the trajectories of the points become lines in 3-dimensional space.

Zheng and Lee [4] proposed a method for computing a single NN (k =1) for a moving query according to a R-tree indexed static points. The method is based on Voronoi diagrams and seems quite difficult to be extended for other values of k ($k>1$) and high space dimensions.

A method in Song and Roussopoulos [3] is presented to answer such queries on moving-queries, static-objects cases. Objects are indexing by an R-tree, and sampling is used to query the R-tree at specific points. However, the method may return incorrect results if a split point missed because of the nature of sampling.

The work most closely related to ours are by Tao et al.[12]. The authors propose an NN query processing algorithm for a geometric point of view, based on the concept of time-parameterized queries. On the other hand, our algorithm is based on a mathematical characteristic. Further we also provide performance analysis.

3 CKNN Query Processing

3.1 *CKNN* Algorithm

For the Euclidean distance between query point q and static object s, the distance $dq,s(t)$ can be estimated as a function of time. The function of time is given as follows:

$$d_{q,s}(t) = \sqrt{a*t^2 + b*t + c} \qquad (1)$$

where a, b, c are given by:

$$a = v_{qx}^2 + v_{qy}^2, \, b = 2*[v_{qx}*(q_x - s_x) + v_{qy}*(q_y - s_y)], \, c = (q_x - s_x)^2 + (q_y - s_y)^2$$

$(s_x, s_y),(q_x, q_y)$ are the positions of static object s and query point q respectively in the Euclidean plane, v_{qx} and v_{qy} are the velocities of the query point in each dimension. Since the calculations cost of square root is expensive, we assume that the distance is given by $(dq,s(t))^2$.

Let SS and q be a set of static objects and a moving query. The objects and the query are represented as points in a multi-dimensional space. Query point q is characterized by its reference position and velocity vector. Static objects are characterized by its reference fixed position. So we can define the distance function $(dq, s(t))^2$ for every object $s \in SS$. Although the proposed method is restricted to 2-D space for clarity and convenience, it can be used for arbitrary dimensionality.

Fig. 1 shows that we obtain the k NNs of the moving query point during the time interval $[t_s, t_e]$. The y-axis shows that the distances between the query point and static data objects, and the x-axis shows that query time interval. $p1$-$p5$ are the static data objects. For example, for $k = 2$ the NNs of q for the time interval are contained in the shaded area of Fig. 1. The pair of objects above each time point t_x declare the objects that have an intersection at t_x. These time points where a modification of the result is performed, are called split points. Note that not all intersection points are split points. For example, the intersection of objects $p3$ and $p4$ in Fig. 1 is not considered as a split point for $k = 2$, where it is a split point for $k = 3$.

The previous example demonstrates that the k NNs of a moving query can be determined by using the functions that represent the distance of each stationary object with respect to the moving query point. Based on the previous discussion, the subsequent part presents the design of an algorithm for K-Nearest Neighbor query processing (*CKNN*) which operates on moving objects. The algorithm report all split points with a single traversal.

Given a moving query q, a set SS of N static objects, a time interval $[t_s, t_e]$, and the k NNs of q are requested. The target is to partition the time interval into one or more sub-intervals, in which the list of NNs remains unchanged. Each time sub-interval is defined by two time split points, declaring the beginning and the end of the sub-interval. During the calculation, the set SS is partitioned into three sub-sets: 1) set $SS1$, which always contains k objects that are currently the NNs of q; 2) set $SS2$, which contain objects that are possible candidates for subsequent time points; and 3) set $SS3$, which contains rejected objects whose contribution to the answer is impossible for given time interval $[t_s, t_e]$.

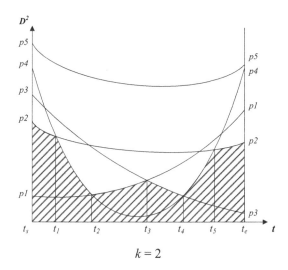

$k = 2$

Fig. 1. Relative distance between a moving query and static objects

Initially, $SS1 = \varnothing$, $SS2 = \varnothing$, and $SS3 = \varnothing$. The first step is to determine the k NNs for time point t_s. By inspecting Fig. 1 for $k = 2$ we get these objects are $p1$ and $p2$. Therefore, $SS1 = \{p1, p2\}$, $SS2 = \{p3, p4, p5\}$, and $SS3 = \varnothing$. Next, for each $p \in SS1$ the intersections with objects in $SS1 + SS2$ are determined. If there are any objects in $SS2$ that do not intersect any objects in $SS1$, they are removed from $SS2$ and are put in $SS3$, meaning that they will not be considered again. In our example, object $p5$ is removed from $SS2$ and we have $SS1 = \{p1, p2\}$, $SS2 = \{p3, p4\}$ and $SS3 = \{p5\}$. The currently determined intersections are kept in an ordered list, in increasing time order. Each intersection is represented as $(t_x, \{m, n\})$, where t_x is the time point of the intersection and $\{m, n\}$ is the objects that intersect at t_x.

Each intersection is defined by two objects, let's say m and n. The currently determined intersection points comprise the current list of time split points. According to the example, the split point list has as follows: $(t_1, \{p2, p3\})$, $(t_x, \{p2, p3\})$, $(t_2, \{p1, p4\})$, $(t_3, \{p1, p3\})$, $(t_4, \{p3, p4\})$, $(t_5, \{p2, p4\})$. For each intersection we distinguish between two cases: 1) $m \in SS1$ and $n \in SS1$; 2) $m \in SS1$ and $n \in SS2$. In the first case, the current set of NNs does not change. However, the order of the currently determined objects changes. Since two objects in $SS1$ intersect, they exchange their position in the ordered list of NNs. Therefore, objects m and n exchange their position. In the second case, object n is inserted into $SS2$ and therefore the list of NNs must be updated accordingly.

According to the currently determined split points, the first split point is t_1, where object $p2$ and $p4$ intersect. Since $p2 \in SS1$ and $p4 \in SS2$ $p2$ removed from $SS1$ and it is inserted into $SS2$. On the other hand, object $p4$ is removed from $SS2$ and it is inserted $SS1$ taking the position of $p2$. Up to this point concerning the sub-interval $[t_s, t_1)$ the nearest neighbor of q are $p1$ and $p2$. Moving to the next intersection, t_x, we see that this intersection is caused by objects $p2$ and $p3$. However, neither of these objects is

```
Algorithm CKNN(q, [ts, te], k, ss)
1  ss1 = 0 , ss2 = 0 , ss3 = 0
2  initialize a split-list slist: insert into slist points ts and te
3  initialize a k-list klist, find the k NNs of q at time point ts
4  For each (mi ∈ ss1) do:
       4.1 find intersections with n ∈ ss1 and n ∈ ss2
       4.2 update slist
       4.3 move irrelevant objects from ss2 to ss3
5  while (si ∈ slist)
       5.1 check next time split point tx(intersection)
       5.2 if(m∈ss1) and (n∈ss1), update clist, exchange positions in k-list
       5.3 if(m∈ss1)and (n∈ss2) ,
               5.3.1 move m from ss1 to ss2,
               5.3.2 move n from ss1 to ss3
               5.3.3 update klist
               5.3.4 update clist
               5.3.5 if(n participates for the first time in k-list),
                       5.3.5.1 determine intersections of n with objects in ss2
                       5.3.5.2 update slist
       5.4 if (m∈ss1)and (n∈ss1),  ignore split point tx
6  return clist
```

Fig. 2. A pseudocode representation of the CKNN algorithm

contained in SS2. Therefore, we ignore t_x and remove it from the list of time split points. We are now ready to check the next split point, which is t_2 where objects $p1$ and $p4$ intersect. Since both objects are contained in SS1, no new objects are inserted into SS1, and simply object $p1$ and $p4$ exchange their position. Up to this point concerning the sub-interval $[t_1, t_2)$ the nearest neighbor of q are $p1$ and $p4$. We are ready now to check the next split point, which is t_3 where objects $p1$ and $p3$ intersect. Since $p1 \in SS1$ and $p3 \in SS2$ object $p1$ is removed from SS1 and it is inserted into SS2. On the other hand, object $p3$ is removed from SS2 and it is inserted into SS1 taking the position of $p1$. Up to this point, another part of the answer has been determined, since in the sub-interval $[t_2, t_3)$ the NNs of q are $p4$ and $p1$. Since a new object $p3$ has been inserted into SS1, we check for new intersections between $p3$ and objects in SS1 and SS2. No new intersections are discovered, and therefore we move to the next split point t_4. Currently, for the time subinterval $[t_3, t_4)$ the NNs of q are $p4$ and $p3$. At t_4 objects $p4$ and $p3$ intersect, and this causes a position exchange. We move to the next split point t_5 where objects $p2$ and $p4$ intersect. Therefore, object $p4$ is removed from SS1 and it is inserted into SS2, whereas object $p2$ is removed from SS2 and it is inserted into SS1. Since $p2$ does not have any other intersections with objects in SS1 and SS2, the algorithm terminates. The outline the method is illustrated in Fig. 2

3.2 Query Processing with R-Trees

In the previous section, the CKNN algorithm is completed when $m \geq k$ objects have been collected from the dataset. In the section we are ready to elaborate the method is combined with the R-tree. The method is how to get the relevant m objects from the whole static dataset. These m static objects, i.e. *SS* (mentioned in the previous section), are used as input to the *CKNN* algorithm in order to determine the result from t_s to t_e. The R-tree can index static points in one, two, or three dimensions. Each R-tree entry r has the from <*r.MBR*, *r.pointer*> where *r.MBR* is the minimum bounding rectangle of r, and *r.pointer* to its child note at the lower level (for a non-leaf entry), or the actual record (for a leaf entry).

The order of entry accesses is very important for avoiding unnecessary visits. We employ branch-and-bound techniques to prune the search space. The proposed method can be applied to CKNN queries, but for simplicity we elaborate the method with single nearest neighbor query. Let E be an intermediate entry or a leaf entry and q be a query segment. *mindist(E,q)* is the shortest among six distances $(d1,d2,d3,d4,d5,d6)$ in Fig. 3. $MAXDIST = dist(q_s.NN, q_e)$, q_s is the start point of a query segment and q_e is the end point of a query sement. $q_s.NN$ is the nearest neighbor of q_s, such as $p1$ in Fig. 4. MAXDIST is the distance between the nearest neighbor of the start point and the end point. If *mindist(E,q)* < *MAXDIST* and E is an intermediate entry, the subtree of E can be visited. If *mindist(E,q)* < *MAXDIST* and E is a leaf node, E will be inserted into *SS*. If R intersects L, *midist* is set to zero. In the sequel, the heuristics prunes unnecessary node accesses.

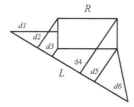

Fig. 3. *mindist* between a line segment and a rectangle

The proposed method can be applied with both the depth-first and best-first traversal. For simplicity, we elaborate the method using depth-first traversal for the query of Fig. 4. Initially the static set *SS* is null. $p1$ is found as the nearest neighbor of q_s and $MAXDIST = dist(p1, q_e)$. The root of the R-tree is retrieved and its entries are sorted by their distances to segment q. Since the *mindist* of *R1* are 0, its child node is visited, and the entries inside it are sorted(order *R4*, *R3*, *R5*, *R2*). The node of *R4* is accessed, all points in *R4* are processed. Since *mindist(p3, q)* < *MAXDIST*, *mindist(p4, q)* < *MAXDIST*, and *mindist(p5, q)* < *MAXDIST*, *p3* and *p4* and *p5* are inserted into *SS*. Now the algorithm backtracks to the upper level. The algorithm recurs. The nodes of *R3* and *R5* are accessed in turn. The heuristics prunes the node of *R2* for *mindist(R2, q)* > *MAXDIST* and the algorithm terminates. Fig. 3 shows that the final result is *SS* = {$p1, p5, p3, p4, p2, p9$}.

The above discussion can directly apply to other values of k ($k>1$) nearest neighbors queries except that $MAXDIST = dist(q_s.NN, q_e)$ should be replaced with the $MAXDIST = dist(q_s.kNN, q_e)$, $q_s.kNN$ is q_s'kth (i.e., farthest) NN.

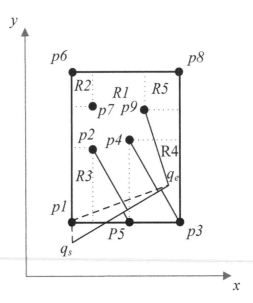

Fig. 4. Processing of *CKNN* with R-tree

4   Experimental Evaluation

We performed an extensive experiment to evaluate the efficiency of the proposed method. The experiments were performed in a PC running Microsoft Windows XP with an Inter Pentium 2.4GHZ processor and 512 MB RAM. The methods under consideration are (i) the *CKNN* algorithm described in the previous section, and (ii) the *CNN* algorithm which is proposed by Tao et al in 2002 [12]. Both algorithms as well as R-tree access method have been implemented in the C++ programming language.

The datasets used for the experimentation are synthetically generated using the uniform or the gauss distribution. The data space extends are 800,000 * 800,000 meters and the velocity vectors of the moving objects are uniformly generated, having speed values between 0 and 30m/sec.

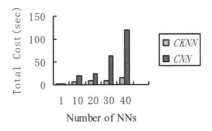

Fig. 5. Evaluation for different number of NNs

The first experiment is to evaluate the results of the methods for different values of the requested NNs. The corresponding results are shown in Fig. 5. The total cost includes CPU time and R-tree node accesses time. By increasing k, more split points are introduced for the *CKNN* method, whereas more influence calculations are needed by the *CNN* method. It is clear that the *CKNN* significantly outperforms the *CNN* method. Although both methods are highly affected by k, the performance of *CNN* degrades more rapidly.

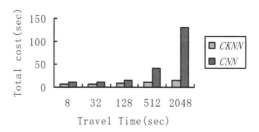

Fig. 6. Evaluation for different values of the travel time

The second experiment illustrates the impact of the travel time to the performance of the methods. The corresponding results are illustrated in Fig. 6 . The total cost includes CPU time and I/O time. Small travel times are favorable for both methods, because less CPU and I/O operations are required. However, *CKNN* performs much better for large travel times in contrast to *CNN* whose performance is significantly affected.

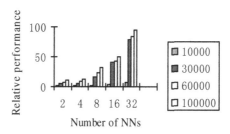

Fig. 7. Evaluation of different value of database size

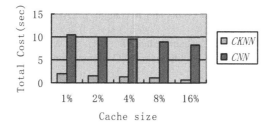

Fig. 8. Evaluation of different cache size

Fig. 7 shows the impact of database size in terms of query performance. By increasing the number of moving objects, more tree nodes are generated and more time is needed to search a R-tree. Moreover, by keeping the buffer capacity unchanged, the buffer hit ratio decreases and produces more page faults. As Fig. 7 illustrates, the performance ratio increases with the database size.

Finally, we evaluate performance under cache size by varying the cache size from 1% to 16% of the tree size. Fig. 8 demonstrates the total query time as a function of the cache size. *CKNN* receives larger improvement than *CNN* because its I/O time accounts for a higher percentage of the total cost.

To summarize, *CKNN* outperforms *CNN* significantly under all settings. The improvement is due to the fact that *CKNN* performs only a single traversal on the dataset to retrieve all split points and visits only the nodes necessary for obtaining the final result.

5 Summary and Future Work

An important query type in moving objects databases [2] is the continuous k-nearest neighbor query, which requires the determination of the k closest objects to the query for a given time interval $[t_s, t_e]$. The major difficulty is that query point changes positions continuously, and therefore the methods that solve the problem in spatial databases can not be applied directly.

In the paper, a study of efficient methods for CKNN query processing in moving object databases is performed, and several performance evaluation experiments are conducted to compare their efficiency. The main conclusion is that proposed algorithm outperforms significantly the repetitive approach.

As such, future work includes study the performance of the method to other access methods like the R*-tree [1]. A second research direction includes the pruning techniques may be revised.

Acknowledgments. This work was supported by the National Natural Science Foundation of China (Grant No. 40571134). We are also grateful to Dr Bisheng Yang for his suggestions.

References

1. Beckmann, N., Kriegel, H., Schneider, R., Seeger, B.: The R*-tree: An Efficient and Robust Access Method for Points and Rectangle. In: ACM SIGMOD, ACM Press, New York (1990)
2. Sistla, A.P., Wolfson, O., Chamberlain, S., Dao, S.: Modeling and querying moving objects. In: Proceedings of the International Conference on Data Engineering, pp. 422–432 (1997)
3. Song, Z., Roussopoulos, N.: k-nearest neighbor search for moving query point. In: Jensen, C.S., Schneider, M., Seeger, B., Tsotras, V.J. (eds.) SSTD 2001. LNCS, vol. 2121, pp. 79–96. Springer, Heidelberg (2001)

4. Zheng, B., Lee, D.: Semantic caching in location-dependent query processing. In: Jensen, C.S., Schneider, M., Seeger, B., Tsotras, V.J. (eds.) SSTD 2001. LNCS, vol. 2121, pp. 97–116. Springer, Heidelberg (2001)
5. Kollios, G., Gunopulos, D., Tsotras, V.J.: Nearest neighbor queries in a mobile environment. In: Böhlen, M.H., Jensen, C.S., Scholl, M.O. (eds.) STDBM 1999. LNCS, vol. 1678, pp. 119–134. Springer, Heidelberg (1999)
6. Tao, Y., Papadias, D.: Time-parameterized queries in spatio-temporal databases. In: Proceedings ACM SIGMOD Conference, pp. 334–345. ACM Press, New York (2002)
7. Roussopoulos, N., Kelley, S., Vincent, F.: Nearest Neighbor Queries. In: Proceedings of ACM SIGMOD, ACM Press, New York (1995)
8. Cheung, K.L., Fu, A.: Enhanced Nearest Neighbor Search on the R-tree. SIGMOD Record 27(3), 16–21 (1998)
9. Guttman, A.: R-Trees: A Dynamic Index Structure for Spatial Searching. In: ACM SIGMOD, ACM Press, New York (1984)
10. Benetis, R., Jensen, C., Karciauskas, G., Saltenis, S.: Nearest Neighbor and Reverse Nearest Neighbor Queries for Moving Objects. In: Proc. Int'l Symp. Database Eng. & Applications, pp. 44–53 (2002)
11. Xiong, X., Mokbel, M.F., Aref, W.G., Hambrusch, S.E., Prabhakar, S.: Scalable Spatio-Temporal Continuous Query Processing for Location-Aware Services. In: Proc. 16th Int'l Conf. Scientific and Statistical Database Management, pp. 317–326 (2004)
12. Tao, Y., Papadias, D., Shen, Q.: Continuous Nearest Neighbor Search. In: Bressan, S., Chaudhri, A.B., Lee, M.L., Yu, J.X., Lacroix, Z. (eds.) CAiSE 2002 and VLDB 2002. LNCS, vol. 2590, pp. 287–298. Springer, Heidelberg (2003)

Texture Classification of Aerial Image Based on Bayesian Networks with Hidden Nodes

Xin Yu[1], Zhaobao Zheng[1], Jiangwei Wu[1], Xubing Zhang[1], and Fang Wu[2]

[1] School of Remote Sensing Information Engineering, Wuhan University, 129 Luoyu Road,Wuhan 430079, China
china_yuxin@163.com
[2] China Aero Geophysical Survey & Remote Sensing Center for Land and Resources, Beijing 100083, China

Abstract. Bayesian networks have emerged in recent years as a powerful data mining technique for handling uncertainty in Artificial Intelligence community. However, researchers in the classification area were not interested in Bayesian networks until the simplest kind of Bayesian networks, Naive Bayes Classifiers (NBC), came forth. From that time on, their success led to a recent furry of algorithms for learning Bayesian networks from raw data and triggered experts to explore more deeply into Bayesian networks as classifiers. Although many of learners produce good results on some benchmark data sets, there are still several problems: nodes ordering requirement, computational complexity, lack of publicly available learning tools. Therefore, this paper puts up a new method, Bayesian networks with hidden nodes, which adds some hidden nodes between correlated feature variables to Bayesian networks based on the maximal covariance criterion. Experimental results demonstrate that the proposed method is efficient and effective, and outperforms NBC and Bayesian Network Augmented Naive Bayes (BAN).

Keywords: Bayesian networks, hidden nodes, classification, aerial images.

1 Introduction

Since 1988 Pearl et al. provided the concept of Bayesian networks for the first time, they have been popular in the Artificial Intelligence (AI) community. Bayesian networks are a powerful formalism for representing and reasoning under conditions of uncertainty, their success have led to a recent furry of algorithms for learning Bayesian networks from raw data [1]. Although many of these learners produce good results on some benchmark data sets, there are still several problems: nodes ordering requirement, computational complexity, lack of publicly available learning tools. However, Naive Bayesian network is the simplest kind of Bayesian networks and in the classification domain the performance of Naive Bayesian network is somewhat surprising although the independent assumption among feature variables is clearly unrealistic [2]. Since the "Naive" independent assumption in the Naive Bayesian network cannot be hold in many cases, researchers have wondered whether the performance will be better if we relax the strong independent assumption among

L. Kang, Y. Liu, and S. Zeng (Eds.): ISICA 2007, LNCS 4683, pp. 454–463, 2007.

feature variables. Therefore, this paper puts up a new method, Bayesian networks with hidden nodes (HBN), which adds some hidden nodes between correlated feature variables to Bayesian networks based on the maximal covariance criterion. On the one hand, the proposed method can almost hold the independent assumption, and on the other hand it can avoid nodes ordering requirement and cut down computational complexity greatly. In this paper, Bayesian networks with hidden nodes are applied in the texture classification of aerial images. Experimental results demonstrate that the proposed method is efficient and effective, and outperforms Naive Bayesian Classifier [3] and Bayesian Network Augmented Naive Bayes (BAN) [4].

This paper is organized as follows. In section 2, we introduce some basic concepts about Bayesian networks. Next, section 3 presents and explains our proposed method in detail. Then, section 4 describes how to apply the proposed approach to texture classification of aerial images. After that, in section 5 we present experimental results, which evaluate the performance of our technique based on the real world data (aerial images). Finally, in section 6, we draw conclusions and point to further work in the future.

2 Bayesian Networks

A Bayesian network is a directed acyclic graph (DAG) for representing probabilistic relationships among a set of random variables $X = \{X_1, X_2, \cdots, X_n\}$ [5]. Over the last decades, the Bayesian network has become a popular representation for encoding uncertain expert knowledge in expert system (Heckerman et al., 1995). Usually, Naive Bayesian network as one of Bayesian networks is accustomed to naming Naive Bayesian Classifier (NBC) in the classification domain. In fact, NBC also consists of two parts and can be represented by $NBC = \langle S, \Theta \rangle$ [3]. One part is the network structure S that encodes a set of conditional independent assumption about variable X_i in X. In the classification application, the nodes in S are in one-to-one correspondence with the feature variables X and each arc between nodes in S represents a probabilistic dependency between the associated nodes [5]. In addition, we use X_i to denote both the feature variable and its corresponding node, and use $Pa(X_i)$ to denote the parents of the node X_i in S as well as the variables corresponding to these parents. And another part is a set of local probability distribution associated with each variable, collectively represented by Θ, which quantifies how much a node depends on its parents. Together, these components define the joint probability distribution for X. The lack of possible arcs in S encodes conditional independencies. In particular, given structure S, the joint probability distribution for X is given by [6]

$$P(X_1, X_2, \ldots, X_n) = \prod_{i=1}^{n} P(X_i \mid Pa(X_i)) \qquad (1)$$

Figure 1 is an example of NBC applied in the classification. The node C denotes the class variable, and $X_1, \cdots, X_{n-1}$ and X_n denote the texture feature that are extracted from image classification unit. From the figure, we can see that the class variable C is the root node, $Pa(C) = \phi$ and each feature, which is assumed independent among each other, has the class variable C as its unique parent node, namely, $Pa(X_i) = C(1 \le i \le n)$.

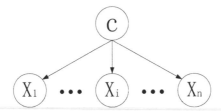

Fig. 1. Naive Bayes classifier

Due to the above Equation (1) and the Bayes's rule, we can get [7]

$$P(C \mid X_1, \ldots, X_n) = \alpha P(C) \prod_{i=1}^{n} P(X_i \mid C) \qquad (2)$$

where $\alpha = \prod_{i=1}^{n} P(X_i)$, and it is one constant [3]. The formula $P(C \mid X_1, \ldots, X_n)$ denotes the posterior probabilities of the class C. And $P(C)$ denotes the prior probabilities of C, assumed known. The formula $P(X_i \mid C)$ denotes the class-conditional density functions and it is also named the local probability distribution. In a general way, we can assume $X_i \sim N(\mu_i, \sigma_i)$, so we have

$$P(X_i \mid C) = \frac{1}{\sqrt{2\pi}\sigma_i} \exp(-\frac{(X_i - \mu_i)^2}{2\sigma_i^2}) \qquad (3)$$

where the sample mean μ_i and the sample covariance σ_i can be estimated based on the training samples by Maximum Likelihood Method.

3 Bayesian Networks with Hidden Nodes

Actually, researchers have provided some methods on hidden nodes [8,9]. For instance, Pearl proposed the use of a hidden node to satisfy the axioms of conditional independence and he created a star-decomposable structure. In his work, he dealt with star decomposition for binary-valued random variables, whereas in our application, feature random variables are not binary-valued but continuous random variables, for example covariance statistic. Hence, this paper proposes a new method, Bayesian networks with hidden nodes (HBN), which adds some hidden nodes between correlated feature variables to Bayesian network based on the maximal covariance criterion.

3.1 How to Add Hidden Nodes to Bayesian Networks

In the subsection, we introduce a technique for introducing a hidden node into an existing Bayesian Network. The new node will be inserted where it was found that several groups of child nodes were not conditionally independent given the states of their parent. Those origin feature variables can be grouped under some heads in terms of correlation among feature variables. That is, those feature variables with high correlation are grouped together. And then, according to the maximal covariance criterion the best projection direction can be found for each group and the corresponding feature variable is projected at the best projection direction. Consequently, this projection node is the hidden node and the projection value is the observed (virtual) value of the hidden node. After that, these hidden nodes are almost independent between each other, but there is a little correlation between those hidden nodes, which can be ignored and will not affect the performance of the mathematic model.

Given an observation vector $Y_{1\times n}^T = [Y_1, \cdots, Y_m]$, and suppose a training set is available that contains samples from M different classes. Consequently, the sample correlation matrix R can be unbiasedly estimated based on the training samples. If, in terms of R, there is high correlation among p random variables of Y containing m random variables ($m \geq p$), these p random variables will be grouped together. In practice, a random sample of n individuals is obtained on p variables. The training samples consist of a ($n \times p$) data matrix X, where rows represent individuals and columns represent feature variables. For the purpose of this derivation, we will assume that X is standardized and Σ denotes the corresponding sample covariance matrix. Our objective is to find the best projection direction u in order to maximize the variance of the elements of $Z = u^T X$.

A linear transformation that maps vectors of a given vector space onto a vector subspace is called a projection. From the matrix theory, the vector u that maximizes $u^T \Sigma u$, subject to the constraint that $u^T u = 1$, is the characteristic vector associated with the largest root of the eigenequation $|\Sigma - \lambda I| = 0$. Thus, the best projection direction is the eigenvector u corresponds to the maximal eigenvalue of the above eigenequation.

$$Z = u^T X = u_1 X_1 + \cdots + u_p X_p \tag{4}$$

The new variable Z can account for as much of the information contained in the original variables X as possible via the best projection because the covariance of Z exhibits the greatest possible variance.

3.2 Mathematic Inference Model of HBN

In the classification domain, nodes represent feature variables and feature variables have an influence upon hidden nodes [10]. In order to describe the mathematic inference model after adding hidden nodes to Bayesian networks further, in this paper we extracted seven texture features, which contained statistical texture features [11] and structural texture features [12]. They are skewness statistics, information entropy and inverse difference moment based on the gray co-occurrence matrix, which are denoted as X_1, X_2 and X_3 respectively. Others are the mean of LL sub-image, and standard deviation of LH sub-image and HL sub-image at the first decomposition level through the Symlets wavelet transform and fractal feature, which are denoted as X_4, X_5, X_6 and X_7 respectively.

In the light of section 3, we add the two hidden nodes X_8 and X_9. The original nodes X_1, X_4, X_6 and X_7 are replaced by X_8, and accordingly X_2 and X_3 are replaced by X_9. In the Figure 2, the nodes with shadow denote hidden nodes and broken lines denote the projection. Wherefore, based on the above network structure and the equation (1), we can get

$$
\begin{aligned}
P(X_1, X_2, \ldots, X_7, C) &= P(X_5, X_8, X_9, C) \\
&= P(C)P(X_5 \mid Pa(X_5))P(X_8 \mid Pa(X_8))P(X_9 \mid Pa(X_9)) \\
&= P(C)P(X_5 \mid C)P(X_8 \mid C)P(X_9 \mid C)
\end{aligned}
\tag{5}
$$

Then by Bayes' theorem, we have

$$
P(C \mid X_1, \ldots, X_7) = \frac{P(X_1, \ldots, X_7 \mid C)P(C)}{P(X_1, \ldots, X_7)} = \frac{P(X_1, \ldots, X_7, C)}{P(X_1, \ldots, X_7)}
\tag{6}
$$

Where the formula is one constant, which is independent of C. And then

$$
P(C \mid X_1, \ldots, X_7) \propto P(X_1, \ldots, X_7, C) = P(C)P(X_5 \mid C)P(X_8 \mid C)P(X_9 \mid C)
\tag{7}
$$

In the final, when posterior probability $P(C_i \mid X_1, X_2, \ldots, X_7)$ is maximal, we can decide the class C^* will be $\max_i \{P(C_i \mid X_1, X_2, \ldots, X_7)\}$.

3.3 HBN Applied in the Classification

In order to explain Bayesian networks with hidden nodes applied in the texture classification of aerial image in detail, the complete classification scheme is summarized below.

① Training samples (images) and testing samples of each class are randomly chosen from the whole database;

② Extract seven kinds of texture features from each classification unit;

③ In terms of the sample correlation matrix R, we grouped the original feature variables (nodes) under some heads, for example three heads in this paper;

④ Based on the maximal covariance criterion, we can choose the best projection for each group, and by the formula (4) we can get virtual observation value, for instance .

⑤ If we assume $P(X_i \mid C)$ follows a normal distribution, the mean and covariance for each class can be estimated by Maximum Likelihood Method;

⑥ In the light of the training results, we can compute the posterior probability of a new sample X belonging to each class C_i, $P(C_i \mid X_1, \ldots, X_n), 1 \leq i \leq n$, by virtue of the formula (3) and (7), where n denotes the number of class. In the end, the sample X is labeled as belonging to the class C^* if

$$C^* = \max_i \left\{ P(C_i \mid X_1, \cdots, X_n) \right\};$$

⑦ Statistical analysis of classification results.

4   Experiments

4.1   Data for Classification Experiments

To validate the feasibility and effectivity of HBN applied in the texture classification of aerial image, six pieces of 23cm×23cm aerial images about Australia and ten pieces of 23cm×23cm aerial images about Wuhan city in China are used in the experiments. In fact, these images are segmented into 465 small areas (i.e. image classification unit) and grouped into three classes (or groups). The first class, residential area (a), has 167 images, including images D1 through D167; the second class, paddy field (b), contains 144 images, including images D168 through D311; the third class, water area (c), contains 154 images, including images D312 through D465. Whereas, among them the maximal area is 40×40 pixels and the minimum area is 16×16 pixels. One sample image of each class is shown in Figure 2.

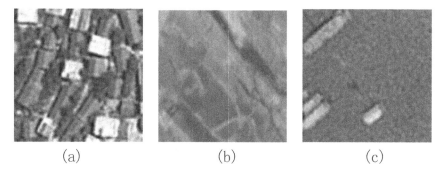

(a)   (b)   (c)

Fig. 2. A sample image of each class

4.2 Experimental Results

Table 1 is the correlation matrix R for $X_1, \cdots, X_7$, which was estimated based on the training sample. Obviously, the weakest is between X_5 and the other feature variables ($|r| < 0.04$), and the strongest is between X_1, X_4, X_6 and X_7 ($|r| > 0.78$). Nevertheless, the correlation coefficient between X_2 and X_3 is no more than 0.5. Thus, X_1, X_4, X_6 and X_7 are one group and X_2 and X_3 are another group. The corresponding Bayesian network with hidden nodes is shown in Figure 3.

Table 1. The correlation matrix of seven texture features

	X_1	X_2	X_3	X_4	X_5	X_6	X_7
X_1	1	-0.03	-0.13	0.80	0.03	0.80	0.78
X_2	-0.03	1	0.48	-0.16	-0.01	-0.16	-0.16
X_3	-0.13	0.48	1	-0.21	0.00	-0.21	-0.22
X_4	0.80	-0.16	-0.21	1	-0.01	0.89	0.89
X_5	0.03	-0.01	0.00	-0.01	1	0.04	0.02
X_6	0.80	-0.16	-0.21	0.89	0.04	1	0.89
X_7	0.78	-0.16	-0.22	0.89	0.02	0.89	1

And then, in terms of the maximal covariance criterion their best projection direction can be acquired and are as follows.

$$X_8 = -0.4800X_1 - 0.5054X_4 - 0.5065X_6 - 0.5076X_7$$

$$X_9 = -0.7083X_2 - 0.7059X_3$$

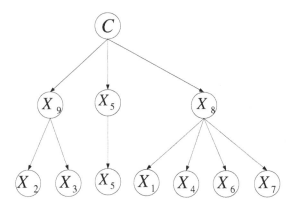

Fig. 3. The Bayesian network with hidden nodes

Table 2. The correlation matrix of X_5, X_8 and X_9

	X_8	X_9	X_5
X_8	1.0000	-0.1679	-0.0059
X_9	-0.1679	1.0000	0.0364
X_5	-0.0059	0.0364	1.0000

Table 2 is the sample correlation matrix after adding the hidden nodes X_8 and X_9. Distinctly, the correlation among variables will be extremely cut down and the maximal correlation coefficient is -0.1679, which can be ignored. The hidden nodes simplify the network structure and almost hold the independency among the hidden nodes. Thereby it can improve the learning efficiency and the performance of classification. In addition, Pearl compared the hidden node to the impresario visually, which expressed veritably that the hidden node played an important role in the inference model.

4.3 Comparison Experiments

In the end, the corresponding parameters were learned based on the above network structure and the training samples, and then the test samples were classified. The classification accuracy is calculated from the confusion matrix, which contains information about the correct classification and misclassification of all classes. To evaluate the efficiency of HBN, classification results were calculated based on HBN, NBC and BAN in terms of overall classification accuracy. The experimental results are shown in Table 3.

Table 3. The comparison results of three methods

N	35	40	45	50	μ	σ
HBN	87.31%	87.10%	86.88%	86.88%	86.80%	0.0032
BAN	85.38%	85.59%	88.17%	88.60%	86.56%	0.0122
NBC	66.24%	67.53%	67.10%	66.88%	66.48%	0.0154

Table 3 displays comparison of the accuracy among three methods in the condition of different training samples, and N denotes the number of training samples of each class. The best mean overall classification accuracy is 86.80% (HBN) and at the same time the fluctuation (std, standard deviation) of HBN is less than that of BAN and NBC. As expected, HBN gives better classification results than PCA-NBC and NBC. For the sake of intuitionistic vision effect, it is also depicted in Figure 4, where the horizontal axis denotes the number of training samples of each class and the vertical axis denotes the overall classification accuracy. Obviously, HBN is the best one and NBC is the worst one among them in the experiments. Whereas, when N is more than 45, the performance of HBN is not as good as that of BAN.

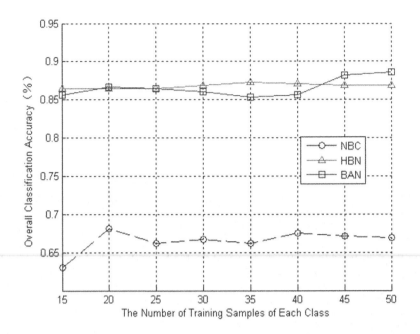

Fig. 4. The curve of the overall classification accuracy

5 Conclusions and Future work

A Bayesian network is a graphical model that encodes probabilistic relationships among variables of interest. When used in conjunction with statistical techniques, the graphical model has several advantages for data analysis. In this paper, Bayesian networks with hidden nodes, are used for texture classification of aerial images. Experimental results show that HBN outperforms than NBC and BAN. However, when N is more than forty-five in the experiments, the performance of HBN is not as good as that of BAN. In addition, other projection methods will be attempted and how to explain the factual sense of hidden nodes is difficult, which will be our future work.

Acknowledgments. This paper is financially supported by NSFC (No. 40571102 and 40271094). The authors wish to thank the anonymous reviewers for the comments and suggestions on this paper.

References

1. Cheng, J., Greiner, R., Kelly, J., Bell, D., Liu, W.: Learning Bayesian networks from data: An information-theory based approach. Artificial Intelligence 137, 43–90 (2002)
2. Friedman, N., Geiger, D., Goldszmidt, M.: Bayesian Network classifiers. Machine Learning 29, 131–163 (1997)

3. Yu, X., Zheng, Z., Li, L., Ye, Z.: Texture Classification of aerial image based on PCA-NBC. In: MIPPR 2005: Image Analysis Techniques. Proceedings of SPIE - The International Society for Optical Engineering, vol. 6044 (2005)

4. Cheng, J., Greiner, R.: Comparing Bayesian Network Classifiers. In: UAI-99

5. Cooper, G.F, Herskovits, E.: A Bayesian Method for the Induction of Probabilistic Networks from Data. Machine Learning 9, 309–347 (1992)

6. Heckerman, D., Geiger, D., Chickering, D.M.: Learning Bayesian networks: the combination of knowledge and statistical data. Machine Learning 20, 197–244 (1995)

7. Yu, X., Zheng, Z., Tang, L., Ye, Z.: Aerial Image Texture Classification Based on Naive Bayesian Network. Geomatics And Information Science of Wuhan University 31, 108–111 (2006)

8. Kwoh, C.K, Gillies, D.F: Using Hidden Nodes in Bayesian Networks. Artificial Intelligence 88, 1–38 (1996)

9. Croft, J., Smith, J.Q.: Discrete mixtures in Bayesian Networks with hidden variables: a latent time budget example. Computational Statistics & Data Analysis 41, 539–547 (2003)

10. Danwiche, A.: A Differential Approach to Inference in Bayesian Networks. Journal of the ACM 50, 280–305 (2004)

11. Heckerman, D.: Bayesian Networks for Data Mining. Data Mining and Knowledge Discovery 1, 79–119 (1997)

12. Yu, X., Zheng, Z., Ye, Z., Tian, L.: Texture Classification Based on Tree Augmented Naive Bayes Classifier. Geomatics And Information Science of Wuhan University 32, 287–289 (2007)

Human Motion Recognition Based on Hidden Markov Models

Jing Xiong and ZhiJing Liu

Computer Vision Laboratory, Xidian University,
Xi'an 710071, China
jxiong@mail.xidian.edu.cn, Liuprofessor@163.com

Abstract. Hidden Markov Models are widely used in forecasting the unknown sequence based on observation on outside system. In this paper, they are applied in Human Motion Recognition. With the human's silhouettes, the paper mainly deals with how to get the models of regular actions and combine them with HMM to recognize the motions of motive people. As for the localization on gray images of silhouettes, an algorithm combining silhouette contrasting and centroid tracking is put forward. The results show that the new algorithm has better performance.

Keywords: Motion models, HMM, Centroid, Silhouette.

1 Introduction

Nowadays, the transformation from observation of people to comprehension of them is becoming one of the most active subjects in the field of Computer Vision. Its core is to check, trace, recognize people's behaviors, and comprehend and describe them as well. Its important intension is to get rid of the traditional way of man-machine conservation (for example, the information import through mouse or keyboard), to make it possible for the computer to gain information from outside by itself and to give corresponding response to it. Ultimately, we can make our computer system more and more intelligent in this way.

Vision guard on motive object is a burgeoning subject in the field of Computer Vision. It distinguishes to the traditional monitor and control systems in its intelligence [2]. For its wide application foreground and great potential economic value, it is attracting more and more interest of famous research institutions and researchers all over the world. There are lots of methods about Motive Target Behavior Recognition, for example, Region Partition and Pivotal Point Trace [1]. However, it is very complicated to implement. In allusion to some suspicious behaviors in our daily life (for example, jumping, creeping, tussling, peeping) the author presents a method by combining the Hidden Markov Model with the Theory of Centroid. We can understand it by simple natural language and recognize these suspicious behaviors efficiently. Besides, it is comparatively expandable.

L. Kang, Y. Liu, and S. Zeng (Eds.): ISICA 2007, LNCS 4683, pp. 464–471, 2007.

2 Related Work

2.1 Prevalent Process About Motive Target Monitor

The work the author does belongs to motion judgment .We use Method of Template Matching to transform image sequence to a set of static modes, compare them with the standard behavior samples stored beforehand and judge human beings' motions[5].

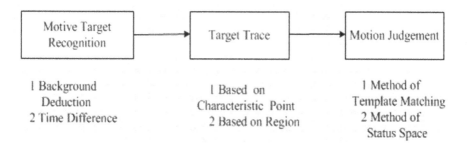

Motive Target Recognition		Target Trace		Motion Judgement

1 Background Deduction
2 Time Difference

1 Based on Characteristic Point
2 Based on Region

1 Method of Template Matching
2 Method of Status Space

Fig. 1. Prevalent process about Motive Target Monitor

2.2 Expression of Motive Human Beings' Image Sequence

We can divide motive human beings' actions into a set of image sequences[4], for example, walking, running and jumping .We use a set of continuous frames to express a cycle of canonical behaviors ,process every frame , distill people's contours, and establish libraries for them (Standard Behavior Library).

The picture below shows a set of image sequence of a walking man.

Fig. 2. A process of walking

2.3 Hidden Markov Models

Hidden Markov Models (HMM) are widely used to forecast the potential sequence of an event according to its samples [7]. It uses statistical methods to express transferring rate between every other status and describes a complex system .It is very efficient in learning and matching behavior models.

Suppose there is a sample sequence $O_1, O_2, \ldots O_t$ to every random event, which contains a potential sequence $X_1, X_2, \ldots X_t$. [3]We can presume:

(1)Markov Hypothesis:

$$P(X_i \mid X_{i-1}, X_{i-2}, \ldots X_1)=P(X_i \mid X_{i-1}) \tag{1}$$

(2) Immobility Hypothesis:
For every arbitrary i, j, the conclusion below is tenable.

$$P(X_{i+1} \mid X_i)=P(X_{j+1} \mid X_j) \tag{2}$$

(3) Output Independence Hypothesis:

$$P (O_1, O_2, \ldots O_t \mid X_1, X_2, \ldots X_t)= P(O_t \mid X_t) \tag{3}$$

There are three foundational questions on Hidden Markov Models, including evaluation, decoding and learning. Decoding means to work out the most possible state sequence for the present model and observation sequence. As we know, the Viterbi Arithmetic can settle this problem.

2.4 Comprehension of Human Motion Using Natural Language

Natural Language Comprehension is to research how to let the computer comprehend and give corresponding response to the natural language we use in our daily life. Technology of Natural Language Comprehension can be divided into several aspects including Machine Translation, Machine Comprehension and Man-Machine Conversation [8]. Among these, Machine Comprehension, by combining research production of linguistics with computer technology, is to realize comprehension of the meaning of the word. Computer used in this system can comprehend simple status of motive target by similarity contrast, and describe it in natural language, for example, a walking man with nothing in his hand, a walking man with a box, a jogging man .etc.

3   The System Realization and Its Ameliorated Algorithm

In this intelligent system, motions that we define as "regular behaviors" include walking, jogging, etc. Besides, it is also licit if he is walking or jogging with something in his hand (idealized as a box).While other behaviors, for example, creeping, jumping, peeping, tussling are identified as anomalistic ones. Therefore, we can establish our "Standard Behavior Library". As the contour of a jogging man is similar to that of a walking man, we classify them into one library.(Library A),define as "A walking or jogging person"; while behavior with a box into another one,(Library B) define as "A person walking with a box" .The process is like this:

3.1 Process of This Algorithm

(1)In order to ensure the integrality of the contour, after we have got the person's contour using Background Deduction and Time Difference Method, we use Region

Inosculating combined with the Theory of Morphologic Erosion and Expansion [6] to process it.

(2)Establish a coordinate, using circular search, find out the point which has the maximal abscissa and y-axis and the point which has the minimal horizontal coordinate and vertical coordinate. The line connected with the two points is the rectangle's diagonal that contains the person's contour and the rectangle is the minimal one.

(3)To solve the problem when people enter or leave the scene the contour is not integrity, we cut off the frames of the process. We detect the motion while the target has entered the scene completely.

(4) Action judgment combined with HMM and standard action exemplar.

Describe the HMM using the model which has five factors and is expressed as $\lambda = (N, M, \Pi, A, B)$. The factor N represents the amount of motion bases, M is the deputy of the pictures' total number in each motion base, Π is the probability in which a certain motion base is chosen, A is the transfer probability, B represents the image distribution in each base. The question is to get a correct state sequence S= $q_1, q_2, \ldots q_T$, for the observation sequence O= $o_1, o_2, \ldots o_T$ and the model λ, so that S will be the best explanation of the observation sequence O. As for decoding, we put forward The Viterbi Arithmetic, the details are as follows.
Definition:

$$\delta_t(i) = \max_{q_1, q_2, \ldots q_{t-1}} P[q_1, q_2, \ldots q_{t-1}, q_t = i, O_1, O_2, \ldots O_t \mid \lambda] \tag{4}$$

N is the amount of states and the T is the length of sequence. The sequence we are looking for is the one that the biggest $\delta_T(i)$ represents at time point T. Initialize at first:

$$\delta_1(i) = \pi_i b_i(O_1), 1 \leq i \leq N$$
$$\varphi_1(i) = 0, 1 \leq i \leq N \tag{5}$$

Recursion by the formula:

$$\varphi_t(j) = \arg\max_{1 \leq j \leq N} [\delta_{t-1}(i)a_{ij}], 2 \leq t \leq T, 1 \leq j \leq N \tag{6}$$

Then, we can work out the result:

$$P^* = \max_{1 \leq j \leq N} [\delta_T(i)] \qquad q_T^* = \arg\max_{1 \leq j \leq N} [\delta_T(i)] \tag{7}$$

And the sequence S we are looking for is:

$$q_t^* = \varphi_{t+1}(q_{t+1}^*), t = T-1, T-2, \ldots 1 \tag{8}$$

The sequence is the image sequence that describes the action in the motion base, including walking, jogging and so on. System can recognize more and more motions by extending motion bases. Contrarily, if we cannot find out the suitable state sequence for the observation sequence, the motion is judged as suspicious action and the system could give an alarm.

3.2 The Ameliorated Arithmetic

As the system recognizes the motions of people by contrasting the contour, there is a problem, for example, the shapes of his contours are almost consistent with the one mentioned when he is side-jumping and walking. It is difficult to distinguish them just by the method above. We put forward a method that combines the Hidden Markov Model with the Theory of Centroid to solve it. The details are as follows:

(1) The Absolute Distance

Work out the transformations of centroid in horizontal coordinate and vertical coordinate, compare the current frame with the former one, expressed as $\triangle X$ and $\triangle Y$, and we can divide the changing condition of centroid into several instances considering directions. Specially, if the horizontal coordinate of centroid changes while the vertical coordinate is almost unaltered when he is walking. We only calculate the changes in vertical coordinate, that is to say $\triangle X$ are ignored. However, when people are walking from far away to nearby or the case is opposite, we have to consider the changes both in horizontal coordinate and vertical coordinate. The picture below shows the sources of the absolute distance.

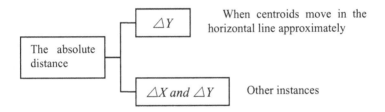

Fig. 3. The sources of the absolute distance

Adopt statistics principle through reduplicative experiments to simulate the critical value SA. Compare the absolute distance with SA, and we can gain a result.

The picture below shows a side-jumping man and his centroid varying from the time in the process of jumping. The blue points are centroids, the horizontal coordinate is the representation of time and the vertical coordinate is the representation of height (suppose the origin point is the point at the left bottom of screen).

(2) The Relative Distance

Calculate the changes between the current frames with the several frames before, and that is the relative distance. Then, we consider the absolute distance and relative distance synthetically.

Adopt statistics principle through reduplicative experiments considering directions to simulate the critical value SR. And then compare the relative distance with the critical value; we can gain the other result.

(3) Combine the two results, take the absolute distance and the relative distance into account, and we can recognize the motions of people combined with the consequence of HMM. System recognizes human motions only after the two conditions are satisfied.

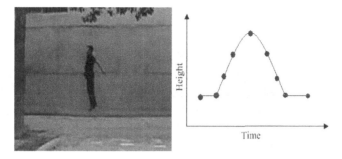

Fig. 4. A jumping man and his centroid varying

4   Realization and Performance Estimate of This System

4.1   Realization of This System

In order to check the performance of this system, the author exploits a Human Motion Recognition System based on natural language in C++ language in the VC++ environment.

The hardware configuration we used in this system is Pentium3.0G、 1G RAM, and the Software Operating System is Windows XP. Under this platform, we made a performance validation of this algorithm.

4.2   Result and Performance Estimate of This System

We made a test about this algorithm. After we defined walking and running as "regular behavior", other behaviors are identified by blue rectangles.

As we can see, the result can meet our needs. Fig. 5. shows the result of a single person:

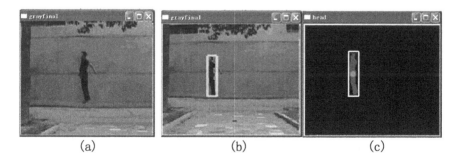

| (a) | (b) | (c) |

Fig. 5. Result of A Single Person. Both picture (a) and (b) show a side-jumping man, (a) is the result only use the silhouette contrasting, while (b) is the result of the new algorithm combining silhouette contrasting and centroid tracking. As we can see, the ameliorated arithmetic can recognize the suspicious action. The green point in picture (c) is the center of mass, and the write rectangle is the minimum one that includes the people's silhouettes.

The picture below shows the result of more than one people:

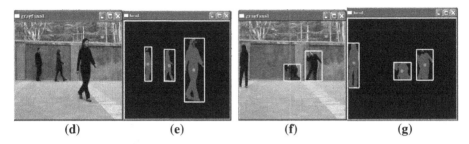

(d)   **(e)**   **(f)**   **(g)**

Fig. 6. Result of several people. Picture (d) and (f) show the motions of three people in different directions. All the people's actions in picture (d) are regular, so they are not indicated, but there are two anomalistic behaviors in picture (f), and the people who are located in the middle and on the right are identified by blue rectangles, while the left person is walking, so she is not indicated. The results can meet our need.

5 Conclusion and Discussion

We have described an approach to detect motions of people using HMM combined with the Theory of Centroid, which can recognize some simple behaviors of moving people .On the base of the result, we could build an   intelligent security system .In some special scene, after we have captured motive target by camera, we can trace it and give an alarm aiming at the suspicious actions. If we can do this, we could save money and manpower, and the more important is that we could prevent crime from happening .Thus, computer can not only be people's eyes, but also be the assistant, who can understand people and describe the actions of motive targets.

There are several directions towards which we are pursuing to improve the performance of the system. Aiming at the problems created by shadows which are too close to people, better results could be obtained if shadows are removed from silhouettes, perhaps based on stereo or motion. However, when several people enter the background at the same time, defilade will be a problem. In that case, we can screen it by several cameras from different directions, syncretize the data obtained, and then we could solve the problem ultimately.

References

1. Wren, C., Pentland, A.: Dynamic modeling of hu-man motion. In: Proc. of Third Face and Gesture Recognition Conf., pp. 22–27 (1998)
2. Lu, X.G.: The application of IPL and OpenCV in VC++ environment [J]. Microcomputer Applications 19(1) (2003)
3. Du, S.P, Li, H: The second order HMM and the applications in language computing [J], 41(4) (2004)
4. Zheng, N.N.: Computer Vision and Pattern Recognition [M]. National defence industry publishing company (1998)

5. Gonzalez, R., Woods, R.: Digital image managing [M]. electron industry publishing company (2002)
6. Intel Corporation: Open Source Computer Vision Reference Manual [EB/OL], http://www.intel.com/research/mrl/research/opencv/.2000-15-08
7. Birney, E.: Hidden Markov Models in Biological sequence analysis[J]. IBM Journal of Research and Development 45(3/4) (2001)
8. Bart, E., Ullman, S.: Class-based matching of object parts. In: VideoRegister04, p. 173 (2004)

Parameter Setting for Evolutionary Latent Class Clustering

Damien Tessier[1], Marc Schoenauer[1], Christophe Biernacki[2],
Gilles Celeux[3], and Gérard Govaert[4]

[1]Projet TAO and [4]Projet SELECT, INRIA Futurs, France
[2]Laboratoire de Mathématiques Paul Painlevé, USTL, France
[3]Heudiasyc, UTC, France

Abstract. The latent class model or multivariate multinomial mixture
is a powerful model for clustering discrete data. This model is expected
to be useful to represent non-homogeneous populations. It uses a conditional independence assumption given the latent class to which a statistical unit is belonging. However, it leads to a criterion that proves difficult
to optimise by the standard approach based on the EM algorithm. An
Evolutionary Algorithms is designed to tackle this discrete optimisation
problem, and an extensive parameter study on a large artificial dataset
allows to derive stable parameters. Those parameters are then validated
on other artificial datasets, as well as on some well-known real data: the
Evolutionary Algorithm performs repeatedly better than other standard
clustering techniques on the same data.

1 Introduction

When modeling an optimisation problem, all practitioners face similar dilemmas:
the most accurate models result in very difficult, if not intractable, optimisation
problems; And simplifying the model in order to obtain an optimisation problem
that is tractable by standard optimisation methods, with proven convergence,
might result in a poor fulfilment the original requirements for the problem at
hand, because of the weaknesses of the model itself.

Evolutionary Algorithms, on the other hand, can handle complex optimisation
problems because of their flexibility, that allows them to work well on non-
standard search spaces, non-regular objective functions, with many local optima
– at the cost of a high computational cost, and, sometimes, a poor fine-tuning
of the solution.

The issue is then whether it is better to obtain a very accurate answer to the
wrong question, or a possibly approximate answer to the correct question.

There exist many works describing situations where the second branch of the
alternative (using Evolutionary Computation to solve the exact model) does
give better solutions than working on some simplified problem, at least for some
instances of the problem at hand. Examples include many situations where there
is a choice between parametric and non-parametric models (e.g. in Structural
Mechanics, in Geophysical Inverse problems [10]).

L. Kang, Y. Liu, and S. Zeng (Eds.): ISICA 2007, LNCS 4683, pp. 472–484, 2007.

This paper is concerned with a similar situation in the context of model based cluster analysis for qualitative data. In this context the latent class model is a reference model (see for instance [7]). Usually the parameters of this model are estimated with the maximum likelihood methodology. But embedding the latent class model in a non informative Bayesian framework, it is possible to get a predictive clustering by integrating over the latent class model parameters. Such an approach is expected to be more stable, but it involves a difficult optimisation problem that is considered in this paper.

The paper is organised the following way: Section 2 introduces the latent class model, derives the resulting log-likelihood function to be maximised, it presents the predictive clustering approach derived from a Bayesian perspective and the Hill-Climbing method that had been used up to now to optimise the resulting criterion. Section 3 gives the details of the Evolutionary Algorithm. Section 4 presents the parametric study on an artificial data with a large number of examples, and comes up with a robust set of parameters. In section 5, the EA then is compared to the Hill-Climber algorithm, first using intensive experiments on smaller sets of examples drawn using the same artificial data generator, then on a well-known real problem, the so called *Toby* dataset. Finally, Section 7 discusses further works and concludes the paper.

2   Predictive Clustering with the Latent Class Model

2.1   The Latent Class Model

Observations to be classified are described with d discrete variables. Each variable j has m_j response levels. Data are $(\mathbf{x}_1, \ldots, \mathbf{x}_n)$ where $\mathbf{x}_i = (x_i^{jh}; j = 1, \ldots, d; h = 1, \ldots, m_j)$ with

$$\begin{cases} x_i^{jh} = 1 \text{ if } i \text{ has response level } h \text{ for variable } j \\ x_i^{jh} = 0 \text{ otherwise.} \end{cases}$$

In the standard latent class model, data are supposed to arise from a mixture of g multivariate multinomial distributions with probability distribution function (pdf)

$$f(\mathbf{x}_i; \boldsymbol{\theta}) = \sum_{k=1}^{g} p_k f_k(\mathbf{x}_i; \boldsymbol{\alpha}_k) = \sum_{k=1}^{g} p_k \prod_{j=1}^{d} \prod_{h=1}^{m_j} (\alpha_k^{jh})^{x_i^{jh}}$$

where α_k^{jh} is denoting the probability that variable $\mathbf{x}^j$ has level h if object i in cluster k, and $\boldsymbol{\alpha}_k = (\alpha_k^{jh}; j = 1, \ldots, p; h = 1, \ldots, m_j)$, $\mathbf{p} = (p_1, \ldots, p_g)$ is denoting the vector of mixing proportions of the g latent clusters, $\boldsymbol{\theta} = (p_k, \boldsymbol{\alpha}_k, k = 1, \ldots, g)$ denoting the vector parameter of the latent class model to be estimated. Latent class model is assuming that the variables are *conditionally independent* knowing the latent clusters.

Analysing multivariate categorical data is made difficult because of the curse of dimensionality. The standard latent class model which require

$(g-1)+g*\sum_j(m_j-1)$ parameters to be estimated is an answer to the dimensionality problem. It is much more parsimonious than the saturated log-linear model which requires $\prod_j m_j$ parameters. For instance, with $g = 5$, $d = 10$, $m_j = 4$ for all variables, the latent class models is characterised with 154 parameters whereas the saturated log-linear model requires about 10^6 parameters... Moreover, the latent class model can appear to produce a better fit than unsaturated log-linear models while demanding less parameters.

Maximum Likelihood Inference. Since the latent class model is a mixture model, the EM algorithm is a privileged tool to derive the ml estimates of the latent class model parameters (see [8]). Denoting $\mathbf{z} = (\mathbf{z}_1,\dots,\mathbf{z}_g)$ with $\mathbf{z}_k = (z_{1k},\dots,z_{nk})$ and $z_{ik} = 1$ if $\mathbf{x}_i$ arose from cluster k, $z_{ik} = 0$ otherwise, the unknown indicator vectors of the g clusters, the completed log-likelihood is

$$L(\boldsymbol{\theta};\mathbf{x},\mathbf{z}) = \sum_{i=1}^{n}\sum_{k=1}^{g} z_{ik} \log\left(p_k \prod_{j=1}^{d}\prod_{h=1}^{m_j}(\alpha_k^{jh})^{x_i^{jh}} \right).$$

From this completed log-likelihood, the equations of the EM algorithm are easily derived and this algorithm is as follows from an initial position $\boldsymbol{\theta}^{(0)} = (\mathbf{p}^{(0)},\boldsymbol{\alpha}^{(0)})$.

- E step: calculation of $\mathbf{t}^{(r)} = (t_{ik}^{(r)}, i = 1,\dots,n, k = 1,\dots,g)$ where $t_{ik}^{(r)}$ is the conditional probability that $\mathbf{x}_i$ arose from cluster k

$$t_{ik}^{(r)} = \frac{p_k^{(r)} f_k(\mathbf{x}_i;\boldsymbol{\alpha}_k^{(r)})}{\sum_{\ell=1}^{g} p_\ell^{(r)} f_\ell(\mathbf{x}_i;\boldsymbol{\alpha}_\ell^{(r)})}.$$

- M step: Updating of the mixture parameter estimates,

$$p_k^{(r+1)} = \frac{\sum_i t_{ik}^{(r)}}{n} \quad \text{and} \quad (\alpha_k^{jh})^{(r+1)} = \frac{\sum_{i=1}^{n} t_{ik}^{(r)} x_i^{jh}}{\sum_{i=1}^{n} t_{ik}^{(r)}}.$$

Bayesian Inference. Since the Jeffreys non informative prior distribution for a multinomial distribution $\mathcal{M}_r(q_1,\dots,q_r)$ is a conjugate Dirichlet distribution $\mathcal{D}(1/2,\dots,1/2)$ a fully non-informative Bayesian analysis is possible for latent class models contrary to Gaussian mixture models (see [9]). The prior distribution of the mixing weights is a Dirichlet $\mathcal{D}(1/2,\dots,1/2)$ distribution. Then, denoting $n_k = \#\{i : z_{ik} = 1\}$ and $n_k^{jh} = \#\{i : z_{ik} = 1, x^{jh} = 1\}$, the full conditional distribution of $(p_k, k = 1,\dots,g)$ is a Dirichlet distribution $\mathcal{D}(1/2+n_1,\dots,1/2+n_g)$. The conditional probabilities of the allocation variables are given, for $k = 1,\dots,g$ and $i = 1,\dots,n$, by

$$t_{ik} = \frac{p_k f_k(\mathbf{x}_i;\boldsymbol{\alpha}_k)}{\sum_{\ell=1}^{g} p_\ell f_\ell(\mathbf{x}_i;\boldsymbol{\alpha}_\ell)}.$$

In a similar way, the prior distribution of $(\alpha_k^{j1}, \ldots, \alpha_k^{jm_j})$ is a $\mathcal{D}(1/2, \ldots, 1/2)$ for $k = 1, \ldots, g$ and $j = 1, \ldots, d$ and the full conditional distribution for $\{\alpha_k^{jh}\}$, $(j = 1, \ldots, d; k = 1, \ldots, g)$ is

$$\alpha_k^{jh}|\ldots \sim \mathcal{D}(1/2 + n_k^{j1}, \ldots, 1/2 + n_g^{jm_j}).$$

The Gibbs sampling implementation of the fully non informative Bayesian inference can straightforwardly be deduced from those formulae.

2.2   Predictive Clustering

In a fully Bayesian perspective, it is possible to derive a classification of the data from the joint predictive distribution

$$f(\boldsymbol{x}, \boldsymbol{z}) = \int_{\Theta} f(\boldsymbol{x}, \boldsymbol{z}; \theta)\pi(\theta)d\theta.$$

Such an approach can involve various difficulties for general mixture models. But for the standard latent class model, it leads to a simple formulation. Assuming non informative Dirichlet prior distributions for the mixing proportions and the latent class parameters

$$\pi(\mathbf{p}) = \mathcal{D}(a, \ldots, a) \quad \text{et} \quad \pi(\boldsymbol{\alpha}_k^j) = \mathcal{D}(a, \ldots, a), \tag{1}$$

with $a = 1/2$ for a Jeffreys prior, we get using conjugate property of the Multinomial-Dirichlet distributions (see for instance [9])

$$f(\boldsymbol{x}, \boldsymbol{z}) = \frac{\Gamma(ga)}{\Gamma(a)^g} \frac{\prod_{k=1}^{g}\Gamma(n_k + a)}{\Gamma(n + ga)} \prod_{k=1}^{g}\prod_{j=1}^{d} \frac{\Gamma(m_j a)}{\Gamma(a)^{m_j}} \frac{\prod_{h=1}^{m_j}\Gamma\left(n_k^{jh} + a\right)}{\Gamma(n_k + m_j a)}.$$

The predictive clustering approach consists of maximising $f(\mathbf{z}|\mathbf{x})$. Since $f(\mathbf{z}|\mathbf{x}) \propto f(\mathbf{x}, \mathbf{z})$, it leads to maximise the criterion

$$C(\boldsymbol{z}) = \sum_{k=1}^{g} \log \Gamma(n_k + a) - \log \Gamma(n + ga) +$$
$$\sum_{k=1}^{g}\sum_{j=1}^{d}\left\{\sum_{h=1}^{m_j}\log \Gamma\left(n_k^{jh} + a\right) - \log \Gamma(n_k + m_j a)\right\}. \tag{2}$$

2.3   A Naive Algorithm

In this predictive approach, the unique and difficult task is to find $\boldsymbol{z}$ (vector of dimension n) optimising $C(\boldsymbol{z})$. A simple solution consists of iterating the optimisation of each dimension in turn: For $i = 1, \ldots, n$

$$z_i^+ = \arg\max_{z_i} C(z_1^+, \ldots, z_{i-1}^+, z_i, z_{i+1}^-, \ldots, z_n^-).$$

However, such local algorithm will be quite sensitive to its initial position. Thus, it is highly recommended to start from a reasonable z^0 vector. For instance, one can use the Maximum A Posteriori of the Maximum Likelihood estimate found by the EM algorithm.

The main advantage of such an algorithm is it simplicity and relative speed. But, since the space where z lies is of large dimension and discrete, this algorithm can be expected to be suboptimal.

3   The Evolutionary Algorithm

This section introduces the problem-specific parts of the Evolutionary Algorithm that has been used to tackle this optimisation problem, namely the genotype, the variation operators (crossover and mutation) and the initialisation procedure. All representation-independent parts will be briefly described in next section, together with the experimental results.

3.1   Representation

The genotype is the vector (z_i) giving for each example the cluster it is assigned to. It is of size n, the number of examples, and takes values $z_i \in [0, g-1]$, where g is the number of clusters. Because the latent-class predictive clustering technique tries different number of clusters to find the optimal number, the general integer representation has been chosen, even in the case where $g = 2$. However, though the values are represented as integers, they are treated as purely symbolic, as there is no notion of order or proximity among the different classes.

3.2   Representation-Specific Operators

The **initialisation** is straightforward: each component (z_i) is chosen uniformly in $[0, g-1]$. The variation operators are standard for vectors of symbolic values:

- **Uniform crossover** is analogous to the corresponding bitstring operator: the parents exchange the values for each example independently with probability 0.5. Note that 1-point crossover could also be used, randomly choosing a crossover point and swapping the values of all examples on the second half of the vector (similar to standard bitstring 1-point crossover).
- **Uniform mutation** applies some *gene-level mutation* to each position with a given probability $p_{mutGene}$. Here, the gene-level mutation amounts to choose for the class value of the given example a new value (i.e., different from its original value to ensure variation) chosen uniformly within all available values.

The algorithm described above has been implemented using the C++ Evolving Object library [6].

4   Parameter Study on Artificial Data

The first series of experiments was done on artificial data and aims at designing an Evolutionary Algorithm (together with its parameters) for the optimisation of the predictive criterion C (Equation 2). Indeed, parameter tuning can be considered as Achille's heel of Evolutionary Algorithms practitioners. Most parameters have to be fixed by the user based on her/his own experience, and/or on work on similar problems. Hence, since [5], the systematic trial-and-error remains the most widely used method to determine the best set of parameters, a noteworthy exception being the statistical approach proposed in [3].

The artificial data have been generated from a known mixture of $g = 2$ six-variate multinomial distributions ($d = 6$), with 4 response levels for the first four ones ($m_1 = \ldots = m_4 = 4$) and 6 for the last two ones ($m_5 = m_6 = 6$). The overlapping rate of the two components is about 10% (see [13] for the detailed values). A sample of size $n = 3200$ involving 2 components was used for parameter tuning.

4.1   Design of Experiments

Evolution Engine. The Evolution Engine describes the (representation-independent) way used to evolve a population of individuals, i.e., the selection and replacement procedures, as well as the population size and the number of generated offspring at each generation.

Though they can all be put into the same framework [2], some well-known general classes of selection/replacement procedures can be distinguished:

– Generational Genetic Algorithm (GGA): P selected parents give birth to P offspring, that in turn replace all P parents. The parameters are P (the population size), the selection mechanism, and its selective pressure. Tournament selection has been chosen here for its simplicity and robustness (it is insensitive to bad fitness scaling for instance). Three values for the selection pressure have been tried, namely 1.6 (the so-called "stochastic tournament" with parameter 0.8, that uniformly draws 2 individuals and returns the best one with probability 0.8), 2 and 4 (corresponding to "deterministic tournaments" of sizes 2 and 4 respectively).
– Steady-State Genetic Algorithm (SSGA): 1 offspring is generated from 1 or 2 selected parents (depending on whether crossover should be applied or not), and it is inserted back in the population by removing a low-fitness parent. The same parameters than for GGA apply for selection. An additional parameter comes from the replacement procedure: it was chosen to be either deterministic (the worst parent dies), or stochastic, involving a tournament of size 10 (the worst of 10 uniformly chosen parents dies).
– Evolution Strategies (ES): λ offspring are generated from the μ parents, without selection (i.e. all parents will give birth to the same number of offspring on average); two replacement procedures can be used, namely (μ, λ), where the best μ of the λ offspring become the new population, and $(\mu + \lambda)$,

where the μ best of the μ parents PLUS the λ offspring become the new population. This schema is borrowed from the historical Evolution Strategies (ES, see e.g. [1]), and admittedly good setting is to take $\lambda = 7 * \mu$. Note that the "population size" to be considered when comparing the ES engine to one of the GA engines is λ rather than μ, in connection with the required computing effort.

Population sizes of 50, 100 and 200 have been considered for the GGA and SSGA engines, while the values of μ for the ES engines have been chosen among 1, 10 and 30, to approximately obtain the same number of evaluations per generations (for 10 and 30) while testing the $(1 \dotplus 7) - ES$ (the GGA and SSGA engines generally require larger populations).

Variation Operators. Parameters related to variation operators are twofold.

At the population level, selected individuals first undergo crossover with probability p_{cross} and then mutation with probability p_{mut}. In order to limit the number of values, but to nevertheless test the extreme cases, values 0, 0.5 and 1.0 were considered (value 0 being excluded for mutation, known to be mandatory).

The only parameter at the individual level is here $p_{mutGene}$, the probability that a given example is given a different class in the mutation operator (see Section 3.2). It is well-known in the binary case [1] that the value $\frac{1}{n}$ (where n is the size of the vector) is a robust choice. An exploration of higher values was also done here, namely $\frac{2}{n}$ and $\frac{3}{n}$.

Table 1. Set of parameters used on the two latent class simulated data

	GGA	SSGA	ES
Pop size	50/100/200	50/100/200	01/10/30
Sel. Pressure	1.6/2/4	1.6/2/4	1.0
No offspring	100.00%	1	700.00%
Parent survive	0.00%	PopSize-1	0%/100%
Repl, "Pressure"	–	$+\infty$/10	–
p_{cross}	0/0.5/1	0/0.5/1	0/0.5/1
p_{mut}	0.5/1	0.5/1	0.5/1
$n * p_{mutGene}$	1/2/3	1/2/3	1/2/3

Experimental Settings. Table 1 summarises the different parameter values that have been used for those experiments. For each set of parameters, 11 runs were performed. The total number of runs are thus 11 times 162, 324, and 108 respectively for the GGA, SSGA and ES evolution engines, leading to a total of 6354 runs altogether.

All runs were given the same stopping criterion based on the number of fitness evaluations: a run stopped after a maximum of 500000 evaluations, or after 3000 evaluations without improvement, whichever came first. The average running time for a run was about 30mn, and overall computational cost for the 40-nodes cluster was about two weeks, including node crashes and global power failures.

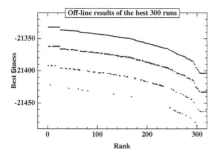

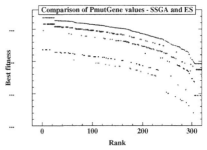

Comparing Evolution Engines: From top to bottom, all runs, SSGA, ES, and GGA.

Comparing values of $n * P_{mutGene}$, for SSGA and ES engines. From top to bottom: all values, (SSGA, 1), (SSGA,2), (SSGA,3). (ES.1), (ES.2) and (ES.3).

Fig. 1. Off-line results on the two classes artificial data. One dot corresponds to one run, the x-axis indicates its relative rank among all 6354 runs (only the 300 best runs are plotted), and the y-axis is the fitness reached at the end of the run: on both plots, the top (discrete) curve groups results for different parameters, and each curve below is an excerpt of the top one for one parameter, that has been artificially move downward to make it readable (hence the absence of values on the y-axis).

4.2   Comparing Evolution Engines

First of all, out of the 6354 runs, 3393 (53.4%) found a better solution than that found by the EM+HC algorithm described in Section 2.3. However, out of those, only 304 were using the GGA engine (17% of the GA runs), compared to the 2413 and 676 for the SSGA and ES engines (respectively 67.7% and 56.9% – remember that there were 3 times more runs for SSGA than for ES).

Moreover, the best fitness value (-21 332.2) was obtained 24 times altogether, but only once by a GGA engine, compared to respectively 18 and 5 times for SSGA and ES.

Figure 1-a displays the best 300 off-line results: The upper plot corresponds to the actual fitness, and represents all 300 runs. Each curve below is an excerpt of the top one, artificially translated downward to make it clearly readable. From top to bottom: SSGA, ES, and GGA. The latter is obviously outperformed, as witnessed by the sparseness of the plots.

At the other extreme (not plotted here, see [13]), though less significant, the worse 500 results have been obtained by GGA, and out of the worse 1000 off-line results, only 17 and 83 were using the SSGA and ES engines respectively.

From those results, it is clear that the GGA engine is outperformed by both SSGA and ES. Hence it was decided to abandon GGA from thereon.

4.3   Comparing Gene-Mutation Probabilities

Figure 1-b shows the off-line results, for the best 300 SSGA and ES runs. Three values were tested for $n * P_{mutGene}$, 1, 2 and 3 (see Section 4.1). Here again, the top line contains all runs (one point for each run), and the other scatter plots are artificially translated downward to make them readable: the first three plots downwards are the SSGA runs with respective values 1, 2 and 3 for $n * P_{mutGene}$, and the three other plots represents the results of the ES runs for the same values (1, 2, and 3) of $n * P_{mutGene}$. Be careful not to miss the plots with $n * P_{mutGene} = 3$, that only appear between ranks 240 and 300.

Again, a clear conclusion can be drawn here: the value 1 for $n * P_{mutGene}$ is far more efficient and robust that the value 2 and 3. Moreover, the worst results, at the other end of the plot, were all obtained with $n * P_{mutGene} = 3$. Hence all subsequent experiments will only consider $n * P_{mutGene} = 1$, together with SSGA and ES evolution engines (Note that similar conclusions w.r.t. $n * P_{mutGene}$ could also be drawn for the GGA engine).

4.4 Other Parameters

The situation however is not so clear when it comes to study the influence of the other parameters of Table 1.

As far as ES runs are concerned, the "plus" strategy seems more efficient to find high values of the criterion than the "comma" strategy, while a value of 1 for μ is worse than both values 10 and 30, both giving equivalent offline results. And, surprisingly, nothing can actually be deduced about the values of P_{cross} and P_{mut}.

As for the SSGA engine, again no very strong conclusion could be drawn. However, one could notice in the off-line results a slight advantage of the deterministic over the stochastic replacement, a slight advantage of the standard selective pressure (value 2) over the values 1.6 and 4, and a clearer disadvantage when no crossover was present ($P_{cross} = 0$) compared to both other values ($P_{cross} = 0.5$ and $P_{cross} = 1$). No influence of the mutation parameter (with possible values 0.5 or 1.0) could be identified.

4.5 A Robust Parameter Setting

From this parametric study, different sets of parameters for the EA seemed equivalently efficient and robust. One was nevertheless chosen for all following experiments: Steady State GA with population size 50, selection by tournament of size 2, deterministic replacement, crossover and mutation rate 1, gene-mutation rate $\frac{1}{n}$ (on average, one mutation per genotype). The stopping criterion remains the same, 500000 evaluations, or 3000 without improvement.

5 Results on Artificial and Real Data

5.1 Best Results on Artificial Data

The same artificial dataset of 3200 examples was then tested, using the hopefully robust setting described above, with fitnesses hypothesing more than 2 classes. The overall best results within 11 independent runs are presented in Table 2: whatever the number of classes in the fitness (between 2 and 5), the EA significantly outperforms the EM+HC algorithm. Moreover, the results of the EA suggest that the optimal number of classes is here 2, while EM+HC favours 3 classes.

Table 2. Best results on the artificial data obtained with fitnesses assuming different number of classes (best value reached in 11 independent runs)

# cl.	2	3	4	5
EM+HC	-21 479.2	-21 407.2	-21 849.3	-21 858.8
EA	-21 332.2	-21 361.9	-21 373.1	-21 469.9

In order to make a more extensive evaluation of both the optimisation strategy (ability to obtain the optimal value of C) and the predictive criterion (ability to detect the right number of classes), extensive simulations have been performed, with different samples of size $n = 200$ (to keep the computational cost low), obtained using the same data generator, but sampled from mixtures of different numbers of distributions, i.e. having different actual number of classes. Both the EM+HC algorithm and the EA with the robust parameters from Section 4.5 were run. For the EA, as usual, 11 independent runs were performed, in order to check the variability of the algorithm.

The results obtained for two classes confirm the tendency observed on the 3200-example dataset: out of 20 sets of 11 runs, 13 found a better value than the EM+HC algorithm, only one found a worse value, and the other 6 runs found exactly the same value. However, those 6 latter datasets correspond to easy problems: the EA found its best value much more often than on the other 6 datasets.

When running both algorithms on problems with 3 to 5 classes, the EA outperforms EM+HC on 17 datasets for 3 classes, on 14 datasets for 4 classes and on 16 datasets for 5 classes (out of 20 runs). Moreover, when the number of classes of the algorithm is larger than the actual one, the EA often finds its best results by removing one or more classes (i.e., not using one of the possible values for the cluster number): for instance, on 2-classes problems, 8 runs using the 3-classes fitness ended up with only 2 classes in their best result; when using the 4-classes fitness, 11 of the 20 best runs ended up with 3 classes, and 3 with only 2 classes; and for the 5-classes fitness, 9 results used only 4 classes, 5 only 3, and even 1 ended up using only 2 classes. This ability to use less classes than what is asked is unique to the evolutionary approach, and gives some clear indications about the actual optimal number of classes.

6   Results on Real-World Data

The data from Stouffer and Toby [11] have been used in many works devoted to latent class models (see for instance [4]). The dataset is made of 216 examples

involving four binary variables. The number of classes is unknown. For the anec-
dote, those data were gathered in a sociological questionnaire where the goal
was to find out whether you are more faithful to a friend than to the law ...

Table 3 presents the best results obtained with four different algorithms on those
data: the EM+HC algorithm described in Section 2.3, the Evolutionary Algorithm
described in Section 3, using the parameter setting from Section 4.5, and two algorithms from the well-known *WEKA* toolbox (http://www.cs.waikato.ac.nz/
ml/weka), EM and k-means. Beware that the two latter algorithms do **not** optimise the log-likelihood objective function described by equation 2 – the values
given in Table 3 have been computed a posteriori from the optimal clusters given
by the algorithms.

It is clear from the results of Table 3 that the Evolutionary Algorithms gives
the best results in all cases. Note that the results of the EA that are presented
in this Table are the best results out of 11 runs: 11 runs seem sufficient here to
outperform the EM+HC algorithm, though of course more runs might always
find better results, as the optimal values are not known. Those results also
suggest that the optimal number of classes is 2, a solution which makes sense
from the statistical viewpoint.

Table 3. Results on the Stouffer and Toby data

# cl.	EM + HC	EA	EM	K-means
2	-559.7498	**-553.4546**	-562.8296	-569.6086
3	-573.6737	**-563.0172**	-592.4112	-579.6012
4	-603.6050	**-576.1582**	-582.7882	-609.6562
5	-609.6562	**-593.2363**	-598.6893	-622.9583

7   Further Work

7.1   Links with Hill-Climbing

The method that was used was a standard hill-climbing, based on a simple
change of one example from one class to another. Hence it is easily trapped
into local optima for this move operator. Going back to the results presented
in section 4, the barrier at fitness level -21479.18 is easily seen on figure 1: it is
the fitness reached using the EM+HC algorithm, but it is also the best fitness
reached by some EA runs. On the same figure, other barriers can be seen: for
instance, running the HC algorithm from a solution obtained by a run that gave
a slightly better answer than -21479.18, say -21477.8, stops again at next barrier,
namely -21404. Again, starting the HC algorithm from a solution obtained in a
run that stopped at -21402.7 now gives the best answer that was ever obtained
at -21332.2, and no further improvement has ever been obtained using the hill-
climbing, neither from this stopping point, nor from any of its neighbours.

Those post-experiments strongly suggest to hybridise an Evolutionary Algorithm and the EM+HC procedure in order to obtain faster, if not better, results.

However, there are many possible hybridisation, and on-going investigation are dedicated to finding which one works best.

7.2 Inoculation

Inoculation has been proposed many years ago as a way to introduce domain knowledge into an evolutionary algorithm through a non-uniform initialisation [12]. Here, the EM algorithm can be a provider of a good starting point – as it does for the EM+HC algorithm. Indeed, preliminary experiments in that direction suggest that, again, initialising the population using some perturbations of the EM solution does greatly speed-up the convergence. Of course, this biases the search toward a specific region of the search space, while the global optimum might lie somewhere else, and uniform initialisation should always be used, at least partially, to ensure a global exploration. However, though further detailed experiments are required, our initial tests never showed that a better solution could be obtained by uniform initialisation and not by EM-based initialisation.

8 Conclusion

Overall, the results presented in this paper witness that indeed, EAs are a good choice to optimise the latent class criterion for qualitative clustering: the (naive) EM+HC algorithm was outperformed except for very few tests. Moreover, the results on real well-studied data confirmed the efficiency of the evolutionary approach.

Because parameter tuning is known to be one of the weaknesses of EAs, it is important to try to obtain a set of parameters that can be considered as robust – this is what we tried to achieve in section 4.1. However, only off-line results were considered here. Further experiments (see [13] demonstrated that while the crossover rate didn't seem to have any influence on the dynamics of the runs, a smaller mutation rate of 0.5 needed twice less fitness evaluations to reach the same final value. Another further improvement that would certainly speed up the convergence of the Evolutionary approach is a clever inoculation of some perturbed EM+HC solution in the initial population.

Nevertheless, we claim that those results are yet another success of the evolutionary approach in Machine Learning and Statistics, demonstrating that we should consider not only objective functions that can be solved by standard methods, with guaranteed convergence, but also measures of success that are more difficult to optimise, but yet give more accurate insight on the data at hand.

References

1. Báck, T.: Evolutionary Algorithms in Theory and Practice. Oxford University Press, New-York (1995)
2. Collet, P., Lutton, E., Schoenauer, M., Louchet, J.: Take it easea. In: Deb, K., Rudolph, G., Lutton, E., Merelo, J.J., Schoenauer, M., Schwefel, H.-P., Yao, X. (eds.) Parallel Problem Solving from Nature-PPSN VI. LNCS, vol. 1917, pp. 891–901. Springer, Heidelberg (2000), http://sourceforge.net/projects/easea

3. François, O., Lavergne, C.: Design of evolutionary algorithms: a statistical perspective. IEEE Transactions on Evolutionary Computation 5(2), 129–148 (2001)
4. Goodman, L.A.: Exploratory latent structure analysis using both identifiable and unidentifiable models. Biometrika 61, 215–231 (1974)
5. Grefenstette, J.J.: Optimization of control parameters for genetic algorithms. IEEE Trans. on Systems, Man and Cybernetics SMC-16 (1986)
6. Keijzer, M., Merelo, J.J., Romero, G., Schoenauer, M.: Evolving Objects. In: Collet, P., Fonlupt, C., Hao, J.-K., Lutton, E., Schoenauer, M. (eds.) EA 2001. LNCS, vol. 2310, pp. 229–241. Springer, Heidelberg (2002), http://eodev.sourceforge.net/
7. Magidson, J., Vermunt., J.: Latent class analysis. In: Kaplan, D. (ed.) The Sage Handbook of Quantitative Methodology for the Social Sciences, pp. 175–198. Sage Publications, Thousand Oakes (2000)
8. McLachlan, G.J., Peel, D.: Finite Mixture Models. Wiley, New York (2000)
9. Robert, C.P.: The Bayesian Choice, 2nd edn. Springer, Heidelberg (2001)
10. Schoenauer, M., Sebag, M.: Using Domain Knowledge in Evolutionary System Identification. In: Giannakoglou, K., et al. (eds.) Evolutionary Algorithms in Engineering and Computer Science, John Wiley, New York (2002)
11. Stouffer, S.A., Toby, J.: Role conflict and personality. American Journal of Sociology 56, 395–406 (1951)
12. Surry, P.D., Radcliffe, N.J.: Inoculation to initialize evolutionary search. In: Ebeling, W., Rechenberg, I., Voigt, H.-M., Schwefel, H.-P. (eds.) Parallel Problem Solving from Nature - PPSN IV. LNCS, vol. 1141, pp. 269–285. Springer, Heidelberg (1996)
13. Tessier, D., Schoenauer, M., Biernacki, C., Celeux, G., Govaert, G.: Evolutionary latent class clustering of qualitative data. INRIA Technical Report RR-6082 (2006)

Automatic Data Mining
by Asynchronous Parallel Evolutionary Algorithms

Yan Li[1], Zhuo Kang[1], and Hanping Gao[2]

[1] Computation Center, Wuhan University, Wuhan, 430072, Hubei, China
[2] School of Computer Science, China University of Geosciences, Wuhan, 430074, China
kang_whu@yahoo.com

Abstract. In this paper An Asynchronous Parallel Evolutionary Modeling Algorithm (APEMA) for automatically modeling of dynamic systems is proposed. The algorithm is based on a two –level hybrid evolutionary modeling algorithm HEMA].The APEMA is used to automatically discover knowledge modeled by higher order of ordinary differential equations from dynamic data by using different computing systems, especially, the MIMD computers. Two cases of modeling examples are used to demonstrate the potential of APEMA .One is for modeling the limit of the solutions of BUMP problem as its dimension n tending to infinity, another is for modeling the super-spreading events of severe acute respiratory syndrome (SORS) in Beijing, 2003.The results show that the dynamic models automatically discovered in data by computer sometimes can compare with the models discovered by human beings.

Keywords: Evolutionary modeling; asynchronous parallel algorithms, higher order of ordinary differential equations, dynamic data.

1 Introduction

In the field of data mining (DM) or knowledge discovery in databases (KDD), there have been major efforts in developing automatic methods to find significant and interesting models (or patterns) in complex data and forecast the future based on those data[1]. In general, however, the success of such efforts has been limited in the degree of automation during the process of data mining and in the level of the models discovered by data mining methods. There are some papers [2]-[4] using evolution algorithms for knowledge discovery in dynamic data. Our research focuses on discovering high-level knowledge in dynamic data modeled by ordinary differential equations (ODEs). An asynchronous parallel evolutionary modeling algorithm APEMA for suiting different computing systems, especially, the MIMD computers is proposed. Some numerical experiments were done to test its effectiveness. We run the APEMA for modeling the limit of the solutions of BUMP problem as its dimension n tending to infinity and the super-spreading events of severe acute respiratory syndrome(SARS) in Beijing,2003 to demonstrate the potential of APEMA for discovering the dynamic models in observed data automatically.

L. Kang, Y. Liu, and S. Zeng (Eds.): ISICA 2007, LNCS 4683, pp. 485–492, 2007.

The rest of the paper is organized as follows. Section 2 is the parallel algorithm for ODEs modeling problem. Section 3 gives an example that is to find the limit of solutions of the Bump problem as its dimension n tending to infinity.

2 Parallel Algorithm for ODEs Modeling Problem

Suppose that a series of observed data collected from one-dimensional dynamic system x(t) sampling at m time points can be written as

$$x(t_0), x(t_1), \cdots, x(t_m) \tag{1}$$

where $t_i = t_0 + i\Delta t, i = 0,1,\cdots,m$, t_0 denotes the starting observed time and Δt denotes the time step-size.

We want to discover a model of ordinary differential equation:

$$x^{(n)}(t) = f(t, x(t), x^{(1)}(t), \cdots, x^{(n-1)}(t))$$

with initial conditions

$$x^{(i)}(t_0) = y_{i+1}(t_0), \quad i = 0,1,\cdots,n-1 \tag{2}$$

Such that

$$\min_{f \in F} \sqrt{\sum_{i=0}^{m-1} (x^*(t_i) - x(t_i))^2} \tag{3}$$

where F is the model space and $x^*(t)$ is the solution of problem (2). The order of the ODE usually satisfies $n \le 4$.

Using the variable substitution:

$$y_{j+1} = \frac{d^j x}{dt^j} \qquad j = 0,1,\cdots,n-1 \tag{4}$$

The problem (2) can be converted into a system of first order ODEs as follows

$$\frac{dy_j(t)}{dt} = y_{j+1}(t), \qquad j = 1,2,\cdots,n-1$$

$$\frac{dy_n(t)}{dt} = f(t, y_1, y_2, \cdots, y_n) \tag{5}$$

With initial conditions:

$$y_{j+1}(t_0) = \frac{d^j x(t)}{dt^j} \bigg|_{t=t_0}, \quad j = 0,1,\cdots,n-1 \tag{6}$$

Denote $x(t_i)$ by x_i. Using the difference formula

$$\Delta^{i+1} x_k = \Delta^i x_k - \Delta^i x_{k+1}, \qquad i = 0,1,\ 2,3,$$

we can compute the difference table (Table 1):

Table 1. Difference table

i	x_i	Δx_i	$\Delta^2 x_i$	$\Delta^3 x_i$	$\Delta^4 x_i$
0	x_0	Δx_0	$\Delta^2 x_0$	$\Delta^3 x_0$	$\Delta^4 x_0$
1	x_1	Δx_1	$\Delta^2 x_1$	$\Delta^3 x_1$	$\Delta^4 x_1$
2	x_2	Δx_2	$\Delta^2 x_2$	$\Delta^3 x_2$	$\Delta^4 x_2$
3	x_3	Δx_3	$\Delta^2 x_3$	$\Delta^3 x_3$	$\Delta^4 x_3$
$\vdots$	$\vdots$	$\vdots$	$\vdots$	$\vdots$	$\vdots$
m-4	x_{m-4}	Δx_{m-4}	$\Delta^2 x_{m-4}$	$\Delta^3 x_{m-4}$	$\Delta^4 x_{m-4}$
m-3	x_{m-3}	Δx_{m-3}	$\Delta^2 x_{m-3}$	$\Delta^3 x_{m-3}$	
m-2	x_{m-2}	Δx_{m-2}	$\Delta^2 x_{m-2}$		
m-1	x_{m-1}	Δx_{m-1}			
m	x_m				

Denote $t_{s+i} = t_i + s\,\Delta t$, $i \le s \le i+4$.

The Newton-Gregory forward polynomial is

$x(t_{s+i}) = P_4(t_{s+i}) + \text{error}$

$$= x_i + s\,\Delta x_i + \frac{s(s-1)}{2!}\Delta^2 x_i + \frac{s(s-1)(s-2)}{3!}\Delta^3 x_i + \frac{s(s-1)(s-2)(s-3)}{4!}\Delta^4 x_i + \text{error} \quad (7)$$

where

$$\text{error} = \frac{s(s-1)\cdots(s-4)}{5!}(\Delta t)^5 x^{(5)}(\xi), \quad t_i \le \xi \le t_{i+4} \quad (8)$$

we have the approximate formula:

$$x^{(j)}(t_{s+i}) = \frac{d^j}{ds^j} P_4(t_{s+i}) \qquad j = 1, 2, 3 \quad (9)$$

Using (4) and (9), we can get the following data (matrix Y):

$$Y = \begin{bmatrix} y_1(t_0) & y_2(t_0) & y_3(t_0) & y_4(t_0) \\ y_1(t_1) & y_2(t_1) & y_3(t_1) & y_4(t_1) \\ \vdots & & & \\ y_1(t_m) & y_2(t_m) & y_3(t_m) & y_4(t_m) \end{bmatrix} \quad (10)$$

The modeling problem (1)-(3) then converts to the following modeling problem:
Given data Y, find (5) (6) such that

$$\min_{f \in F} \sqrt{\sum_{i=1}^{m} \sum_{j=1}^{n} (y_j(t_i) - y_j^*(t_i))^2} \tag{11}$$

where F is the model space, $n \le 4$, and $Y^*(t)$ is the solution of problem (5) and (6).

To make sure of the validity of models, we assume that system of ODEs implied in the observed data satisfies some degree of stability with respect to the initial condition. Namely, the small change of the initial condition will not give rise to the great change of the solution of ODEs.

The APEMA is the asynchronous parallel form of the HEMA (hybrid evolutionary modeling algorithm) to approach the task of automatic modeling of ODEs for dynamic system[5]. Its main idea is to embed a genetic algorithm (GA) in genetic programming (GP) where GP is employed to optimize the structure of a model, while a GA is employed to optimize the parameters of the model. It operates on two levels. One level is the evolutionary modeling process and the other one is the parameter optimization process.

Denote the population of ODEs by $P = \{p_1, p_2, \cdots, p_N\}$, where individual pi is an ordinary differential equation represented by a parse tree. We assume that a population of N individuals is assigned to each of K processors. Each processor executes the same program PROCEDURE APEMA to steer the asynchronous parallel computation.

PROCEDURE APEMA
```
Begin {
  t := 0;
  initialize the ODE model population P ( t ) ;
  evaluate  P ( t ) ;
  while    not (termination-criterion1)    do
  { evolutionary modeling process begins}
    simplify  P ( t );
    for   k := 1  to MAX  do {
    { MAX is the number of models chosen to optimize}
      choose p from P ( t )
      check out all the parameters contained in p;
      s := 0;
      initialize the parameter population  P*(s );
      evaluate  P*(s );
      while    not (termination-criterion2)    do
      { parameter optimization process begins }
        s := s + 1;
        select P*(s ) from P*(s - 1 );
        recombine P*(s ) by using genetic operators;
        evaluate P*(s );
      endwhile  { parameter optimization process ends }
  replace the parameters of p with the best individual in P*(s);
    endfor
    locate p_best and p_worst  by sorting P ( t );
    if ( t ≡ 0  (mod T )) then broadcast p_best to Q neighbors;
```

```
while   (any received message probed )   do
      if   ( recv - individual  better  than p_best )   then
         p_best := recv-individual
      else   p_worst :=   recv-individual;
      locate p_worst  by sorting P ( t );
   endwhile  }
   t: = t + 1;
   select P( t )   from P(t-1);
   recombine P ( t ) by using genetic operators;
   handle the same class of models in P( t ) by sharing
techniques;
      evaluate P( t );
   endwhile  { evolutionary modeling process  ends }
 end
```

where $t \equiv 0$ (mod T) denotes that t is congruent to zero with respect to modulus T.

Remark 1. The asynchronous communication between processors is implemented by calling **pvm-mcast** ( ), **pvm-prob**( ) and **pvm-nrecv** ( ) which are provided by PVM.

Remark 2. T determines the computational granularity of the algorithm, and together with Q, the number of neighbor processors to communicate with, control the cost of communication. That is why the granularity of the algorithm is scalable.

Once the best evolved model is obtained in one run, to check its effectiveness, we take the last line of observed   data as the initial conditions, advance the solution of the ODEs model by numerical integration using some numerical methods, such as the modified Euler method or Runge-Kutta method and get the predicted values for unknown data in the next time steps.   As for the representation, fitness evaluation and genetic operators of these two processes, interested readers can   refer to [5] to get more details.

3 Numerical Experiments

Example 1
In 1994, Keane [6] proposed the BUMP problem in optimum structural design as follows:

$$\text{Maximize } f_n(x) = \frac{\left| \sum_{i=1}^{n} \cos^4(x_i) - 2 \prod_{i=1}^{n} \cos^2(x_i) \right|}{\sqrt{\sum_{i=1}^{n} i x_i^2}}$$

subject to $0 < x_i < 10$, $i = 1, 2, \cdots, n$, $\prod_{i=1}^{n} x_i \geq 0.75$ and $\sum_{i=1}^{n} x_i \leq 7.5n$

The solutions of the BUMP problem are unknown.

According to this problem, Liu [7]proposed a challenge problem in his doctoral dissertation as follows:

$$\lim_{n \to \infty} Max\, f_n(x)$$

subject to $0 \le x_i \le 10, 1 \le i \le n$, $\prod_{i=1}^{n} x_i >= 0.75$ and $\sum_{i=1}^{n} x_i <= 7.5n$

and he got the best solutions of the BUMP problem for n = 2, 3, ... , 50 as follows (Table 1)

Table 2. Solution table

n	f_n	n	f_n	n	f_n	n	f_n	n	f_n
1		11	0.76105561	21	0.80464587	31	0.82210164	41	0.83148885
2	0.36497975	12	0.76256413	22	0.80833226	32	0.82442369	42	0.83226201
3	0.51578550	13	0.77333853	23	0.81003656	33	0.82390233	43	0.83226624
4	0.62228103	14	0.77726156	24	0.81182640	34	0.82635733	44	0.83323002
5	0.63444869	15	0.78244496	25	0.81399253	35	0.82743885	45	0.83285734
6	0.69386488	16	0.78787044	26	0.81446495	36	0.82783593	46	0.83397823
7	0.70495107	17	0.79150564	27	0.81694692	37	0.82915387	47	0.83443462
8	0.72762616	18	0.79717388	28	0.81648731	38	0.82896840	48	0.83455114
9	0.74126604	19	0.79800887	29	0.81918437	39	0.83047389	49	0.8318462
10	0.7473103	20	0.80361910	39	0.82188436	40	0.82983459	50	0.83523753

where $f_n = Max\, f_n(x)$. The best solutions are depicted in Figure 1.

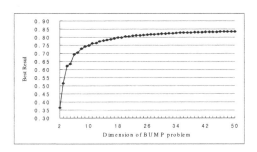

Fig. 1. Best results of f_2 to f_{50}

Using the method described in section 2, the following model is discovered by computer automatically:

$$d^2 f(t)/dt^2 = -15.658156\ df(t)/dt\,(t + df(t)/dt\,)$$
$$f(2) = 0.36497978$$
$$df(t)/dt\,|_{t=2} = 15.08058$$

This model fit the solutions of BUMP problem very well. We use it to predict the solution of the challenge problem by using Runge-Kutta method with $\Delta t = 0.01$ in 1000000 steps, and get $f(1000000)=0.849730$, which is nearly the same as Liu reported $f_{1000000}=0.8445861$ in [7].

Example 2

Evolutionary Modeling for Events of SARS (severe acute respiratory syndrome) in Beijing[8] because Beijing experienced the largest outbreak of SARS, with >2,500 cases reported between March and June 2003. We use a new evolutionary algorithm for automatically modeling the dynamics of super-spreading procedure of SARS from the data which describe the daily SARS situation from 20th of April to 23rd of June,2003 in Beijing as in Table 3, where the cumulative number Of SARS patients at date is denoted by $x(t)$ and the cumulative number of persons died of SARS is denoted by $y(t)$.

Table 3. Data of SARS

$x(t)$	$y(t)$	$x(t)$	$y(t)$	$x(t)$	$y(t)$	$x(t)$	$y(t)$	$x(t)$	$y(t)$	$x(t)$	$y(t)$
339	18	1553	82	2347	134	2490	163	2522	181	2522	189
482	25	1636	91	2370	139	2499	167	2522	181	2521	190
588	28	1741	96	2388	140	2504	168	2522	183	2521	190
693	35	1803	100	2405	141	2512	172	2523	183	2521	191
774	39	1897	103	2420	145	2514	175	2522	184	2521	191
877	42	1960	107	2434	147	2517	176	2522	184	2521	191
988	48	2049	110	2437	150	2520	177	2522	186	2521	191
1114	56	2136	112	2444	154	2521	181	2523	186	2521	191
1199	59	2177	114	2444	156	2522	181	2523	187	2521	191
1347	66	2227	116	2456	158	2522	181	2522	187		
1440	75	2304	129	2465	160	2522	181	2522	189		

The data of cumulative number Of SARS patients $x(t)$ or the cumulative number of persons died of SARS y(t) is regarded as an one dimensional time series which satisfies a higher order ordinary differential equation. The modeling results are shown as follows:

(1) Higher order ODE model for x(t) :

$$\frac{d^2x}{dt^2} = 8.259905[(1-t)\frac{dx}{dt} - 84.653069] - 140.021942t$$

initial condition: $xo = 339$, $dx / dt = 143$. (See Figure. 2)

(2) Higher order ODE model for $y(t)$:

$$\frac{d^2y}{dt^2} = 37707.141054 - y^2 - 56.440029\frac{dy}{dt}$$

initial condition: $x_o = 18$, $dy /dt = 7$.(See Figure.3)

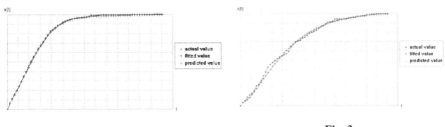

Fig. 2.   Fig. 3.

One may find that the models for modeling the super-spreading of SARS in Beijing discovered by computer can compare with the models found by human beings.

These numerical experiments are performed on a distributed parallel computer system (an MIMD machine). Because the algorithm is performed in an asynchronous parallel way, the processors need not to wait for each other, and the linear speed-up is obtained.

Acknowledgement. This work was supported by the National Natural Science Foundation of China (Nos.60473081, 60133010) and the Natural Science Foundation of Hubei Province (No. 2005ABA234).

References

1. Fayyad, U.M, Piatetsky-Shapiro, G., Smyth, P., Uthurusamy, R. (eds.): Advances in Knowledge Discovery and Data Mining. AAAI Press/The MIT Press (1966)
2. Iba, H., Sasaki, T.: Using genetic programming to predict financial data. In: Proceedings of the Congress on Evolutionary Computation, July 6-9, vol. 1, pp. 244–251 (1999)
3. Hafner, C., Frohlich, J.: Generalized function analysis using hybrid evolutionary algorithms. In: Proceedings of the Congress on Evolutionary Computation, July 6-9, vol. 1, pp. 287–294 (1999)
4. Yoshihara, I., Numata, M., Sugawara, K., Yamacla, S., Abe, K.: Time series prediction model building with BP-like parameter optimization. In: Proceedings of the Congress on Evolutionary Computation, July 6-9, vol. 1, pp. 295–301 (1999)
5. Cao, H.Q., Kang, L.S., Michalewicz, Z., Chen, Y.P.: A Hybrid evolutionary modeling algorithm for stem of ordinary differential equations. In: Neural, Parallel & Scientific Computations, June 1998, vol. 6(2), pp. 171–188. Dynamic Publishers, Atlanta (1998)
6. Keane, A.J.: Experiences with optimizers in structural design. In: Parmee, I.C., Plymouth (eds.) Proc.of the Conf. on Adaptive Computing in Engineering Design and Control 94, pp. 14–27 (1994)
7. Liu, P.: Evolutionary Algorithms and Their Parallelization. Doctoral Dissertation, Wuhan University (2000)
8. Liang, W., Zhu, Z., Guo, J., Liu, Z., He, X., Zhou, W., et al.: Severe acute respiratory syndrome, Beijing (2003). Emerg Infect Dis 10, 25–31(2004)

Texture Image Retrieval Based on Contourlet Coefficient Modeling with Generalized Gaussian Distribution*

Huaijing Qu[1,2], Yuhua Peng[1], and Weifeng Sun[1]

[1] School of Information Science and Engineering, Shandong University,
Jinan, Shandong, People's Republic of China
[2] School of Information & Electric Engineering, Shandong Jianzhu University,
Jinan, Shandong, People's Republic of China
huaijqu@sdu.edu.cn, pyuhua@sdu.edu.cn

Abstract. This paper presents a texture image retrieval scheme based on contourlet transform. In this scheme, the generalized Gaussian distribution (GGD) parameters are used to represent the detail subband features obtained by contourlet transform. To obtain these parameters, an improved maximum likelihood (ML) parameter estimation method is proposed, in which a new initial estimation value is exploited and a modified iterative algorithm is used. Compared with existing features used for the texture image retrieval, the use of the GGD parameters to represent the contourlet detail subbands provides richer information to improve the retrieval accuracy. The proposed retrieval scheme is demonstrated on the VisTex database of 640 texture images. Experimental results show that, compared with the current ML estimation and texture retrieval method, the proposed scheme can give more accurate estimates of the GGD parameters, and it improves more effectively the average retrieval rate from 76.05% to 78.09% with comparable computational complexity.

Keywords: Contourlet transform, Generalized Gaussian distribution, Modeling, Texture image retrieval.

1 Introduction

With the rapid advancement of camera technology, networking technology, and data storage technology, more and more information are represented, transferred, and stored in forms of digital images and videos. The content-based image retrieval (CBIR) provides an effective way to access the achieved images and videos [1] [2], and in CBIR, the representation of the visual image contents plays a key role in improving the retrieval performance. According to human texture perception study, texture can efficiently reflect the structural, directional, granularity, or regularity differences of diverse regions in a visual image [3]. Therefore, texture features become one of efficient and effective models used in CBIR systems.

---

* This project is sponsored by SRF for ROCS, SEM (2004.176.4) and NSF SD Province (Z2004G01) of China.

L. Kang, Y. Liu, and S. Zeng (Eds.): ISICA 2007, LNCS 4683, pp. 493–502, 2007.

To improve the retrieval performance, different texture feature extraction methods have been developed, which can be classified into the statistical methods, the model-based methods and the signal processing methods [4]. Among these methods, the signal processing methods have advantages in the characterization of the directional and scale features of textures. The early development of the signal processing methods focused on the use of discrete wavelet transform to express the texture features into some subband properties such as the subband energy. Later, Gabor wavelets and complex wavelets were developed to remove the limitation of the wavelet transform to capture more directional information in subbands. However, these methods suffer from heavy computational complexity. Thus, contourlet transform has recently received an increasing attention, but again, only the subband energies are used as the texture feature representations.

In wavelet domain, the detail subband coefficients of a texture image usually exhibit a striking non-Gaussian behavior, and their marginal distribution can be well characterized by the generalized Gaussian distribution (GGD). Thus, the parameters of GGD including the scale (or variance) and the shape parameter can be used as the features of the retrieval of texture images. They are estimated from the detail subband coefficients. The early works of estimating the GGD parameters were reviewed by Varanasi and Aazhang [5]. Several recent methods include the moment estimation method and the entropy matching and the maximum likelihood (ML) estimation. Comparatively, the ML estimation turned out to be a particularly efficient and unique solution, and can be assured [5]. In [6], Do and Vetterli applied ML method to estimate the GGD parameters by combining the Newton-Raphson iteration with an initial value from the moment estimation. However, this method may not deliver correct estimates by using small samples, particularly for small shape parameters. Recently in [7], Song proposed a globally convergent and consistent approach for the GGD parameter estimation. However, since the scale parameter estimation uses the result of ML method, it still suffers some shortcomings.

In this paper, the GGD parameters are used to represent the texture features obtained by the contourlet transform, and an improved algorithm is developed to estimate the GGD parameters of contourlet detail subband coefficients. Using these features, a contourlet-based texture image retrieval scheme is proposed. The effectiveness of the proposed retrieval scheme has been demonstrated through experiments.

2 Background

2.1 Contourlet Transform

The contourlet transform, developed by Do and Vetterli, is a directional multi-scale image representation [8]. It starts from the discrete-domain by using 2-D non-separable filter banks, and then converges to a continuous-domain expansion in a multi-resolution analysis framework. Taking advantage of the anisotropic scaling and the sufficient directional vanishing moments, contourlets can attain the optimal approximation rate for natural image with the 2-D piecewise smooth contours [9]. Moreover, they outperform the wavelets in capturing the directional information of natural images.

The contourlet transform has a pyramid directional filter bank structure. It consists of a Laplacian pyramid (LP) followed by a directional filter bank (DFB). The LP is first used to decompose the input image into multiple scale bandpass versions, and the DFB decomposes each scale bandpass version into different numbers of directional subbands. In order to obtain the minimal approximation error, the DFB decomposition levels are doubled at every other finer scale. Typically, for an $N \times N$ bandpass image and an L-level DFB, the first half of the 2^L directional subbands are of $N/2^{L-1} \times N/2$ in size, while the other half are of $N/2 \times N/2^{L-1}$ in size. In addition, the DFB is critically sampled, and the transform has only a redundancy factor of 4/3 due to the LP decomposition.

2.2 GGD Probability Density Function (PDF)

The PDF of a zero-mean generalized Gaussian distribution is defined as [10]

$$p(x;\ \sigma^2, \beta) = \frac{\beta \eta(\sigma, \beta)}{2\Gamma(1/\beta)} \exp\left\{ -\left[\eta(\sigma, \beta)|x| \right]^\beta \right\} \tag{1}$$

where σ^2 and β denote the variance and the shape parameter of the distribution respectively. Γ is the gamma function given by $\Gamma(x) = \int_0^\infty t^{x-1} e^{-t} dt$, $x > 0$, and

$$\log \Gamma(\mathrm{x}) = -\gamma\mathrm{x} - \log(\mathrm{x}) + \sum_{\mathrm{k}=1}^\infty \left[\frac{\mathrm{x}}{\mathrm{k}} - \log(1 + \frac{\mathrm{x}}{\mathrm{k}}) \right] \tag{2}$$

where $\gamma = 0.577$ denotes the Euler constant. In addition, $\eta(\sigma, \beta)$ is given by

$$\eta(\sigma, \beta) = \frac{1}{\sigma} \left[\frac{\Gamma(3/\beta)}{\Gamma(1/\beta)} \right]^{1/2} \tag{3}$$

Here, the smaller β corresponds to the sharper distribution, while the bigger β represents the flatter distribution.

3 GGD Modeling of Contourlet Direction Subband Coefficients

3.1 Estimation of GGD Parameters Using a Modified ML method

Given N samples $\mathbf{x} = (\mathrm{x}_1, \mathrm{x}_2, \cdots, \mathrm{x}_N)$ with zero mean PDF of GGD, The ML method gives the estimates of parameters σ and β as follows [6].

$$\hat{\sigma} = \left(\frac{\Gamma(3/\hat{\beta})}{\Gamma(1/\hat{\beta})}\right)^{1/2} \left(\frac{\hat{\beta}}{N}\sum_{i=1}^{N}|x_i|^{\hat{\beta}}\right)^{1/\hat{\beta}} \tag{4}$$

$$g(\hat{\beta}) = 1 + \frac{\Psi(1/\hat{\beta})}{\hat{\beta}} - \frac{\sum_{i=1}^{N}|x_i|^{\hat{\beta}}\log|x_i|}{\sum_{i=1}^{N}|x_i|^{\hat{\beta}}} + \frac{\log\left(\frac{\hat{\beta}}{N}\sum_{i=1}^{N}|x_i|^{\hat{\beta}}\right)}{\hat{\beta}} = 0 \tag{5}$$

where the digamma function is given by $\Psi(x) = \dfrac{d\log(\Gamma(x))}{dx}$. For the

transcendental equation (5), $\hat{\beta}$ can be solved numerically, and then $\hat{\sigma}$ is obtained

from (4). Therefore, it is sufficient to address the estimation of $\hat{\beta}$.

Do *et al* used the Newton-Raphson method with an initial value $E[|x|]/\sigma$ from the

moment estimation to estimate $\hat{\beta}$ [6]. For conveniences of citation in the subsequent
analysis, we denote this algorithm as ML1. Although ML1 can efficiently estimate the
GGD parameters, it has some drawbacks. First, since the convergence of the ML1
depends crucially on the choice of the initial shape parameter value, the method
cannot yield correct estimation for small samples. To see this, we consider a random

variable X of GGD. We denote $E_m(\sigma, \beta) = \int_{-\infty}^{\infty}|x|^m p(x; \sigma^2, \beta)dx$,

and $\log F(x) = \log\left(\dfrac{E_{m_1}(\sigma, x)}{\left(E_{m_2}(\sigma, x)\right)^{m_1/m_2}}\right)$. After some manipulations combining

with (2), we can obtain $F(x) < \dfrac{(1+m_2)^{m_1/m_2}}{1+m_1}$, for $m_2 > m_1$, and

$\lim_{x\to\infty} F(x) = \dfrac{(1+m_2)^{m_1/m_2}}{1+m_1}$. Let $m_1 = 1$ and $m_2 = 2$, then we

have $F(x) = \dfrac{E[|x|]}{\sigma} < \sqrt{3}/2$. If $\dfrac{E[|x|]}{\sigma}$ is greater than $\sqrt{3}/2$ for small samples, the

ML1 estimation result may not be correct [11]. The second drawback of the ML1
algorithm is that it gives some inaccurate estimation values when the shape parameter
is small (See section **3.2**). This is an intrinsic problem of the ML1 algorithm.

For improving the estimation performance of the ML1 algorithm, we modify the
ML method to estimate the shape parameter. The modified method is denoted as ML2
for short. It is given as follows.

Step1. Given initial value β_0 and a precision ε. Let $n = 0$.

Step2. If $\left|g(\beta_n)\right| < \varepsilon$, return β_n and stop.

Step3. Compute $d_n = -\dfrac{g(\beta_n)}{g'(\beta_n)}$, and let step factor $\lambda = 1$.

Step4. If $\left|g(\beta_n + \lambda d_n)\right| < \left|g(\beta_n)\right|$, $\beta_{n+1} = \beta_n + \lambda d_n$; else, $\lambda = \lambda/2$, Go to **Step4**.

Step5. Let $n = n+1$. Go to **Step2**.

In the ML2 algorithm, the initial value β_0 is determined from [7] as follows.

$$G(\beta_0) = \frac{1/N \sum_{i=1}^{N} \left|X_i\right|^{2\beta_0}}{\left(1/N \sum_{i=1}^{N} \left|X_i\right|^{\beta_0}\right)^2} - (\beta_0 + 1) \tag{6}$$

It can be proved that the equation $G(\beta_0) = 0$ is convergent when $N \to \infty$ [7].

3.2 Estimation Performance of the Modified ML Method

According to (4), $\hat{\sigma}$ depends on $\hat{\beta}$ for a set of given sample data. Therefore, the following experiments mainly test the power of estimation of the shape parameters. In order to measure the efficiency of our method, we first generate sets of samples obeying a zero-mean unity-variance GGD randomly from the standard random generator, where the shape parameters are chosen in a range given by $\beta \in [0.2, 2.0]$, and the size of samples is chosen as $N = 2000$. Then, our method is applied to obtain the estimation values of shape parameters, by taking the precision $\varepsilon = 10^{-6}$.

To evaluate the estimation performance, we use the relative mean square error (RMSE) as in [12], which is defined as

$$RMSE = \frac{1}{M} \sum_{i=1}^{M} \left(\frac{\hat{\beta}_i - \beta}{\beta}\right)^2 \tag{7}$$

where $M = 2500$ are the repeated experimental numbers, $\hat{\beta}_i$ is the estimate obtained in the *i-th* experiment and β is the true shape parameter value. The estimated results

are reported in Table1, where $\overline{\hat{\beta}} = 1/M \sum_{i=1}^{M} \hat{\beta}_i$, $\mathrm{Var} = 1/M \sum_{i=1}^{M} \left(\hat{\beta}_i - \beta\right)^2$.
The experimental results show that the proposed algorithm can effectively estimate the shape parameters.

Table 1. GGD shape parameter true and estimated by ML2 method

β_0	0.2	0.4	0.6	0.8	1.0	1.2	1.4	1.6	1.8	2.0
RMSE	0.0006	0.0013	0.0015	0.0017	0.0018	0.0019	0.0021	0.0022	0.0024	0.0026
$\overline{\hat{\beta}}$	0.2004	0.4006	0.6016	0.8017	1.0030	1.2039	1.4047	1.6039	1.8073	2.0060
Var	0.00002	0.00019	0.0005	0.0010	0.0018	0.0027	0.0040	0.0059	0.0077	0.0098

To compare the estimation performance of two methods while considering the contourlet subband sizes, we take the size of experimental samples $N = 500$. The estimation results are reported in Table 2. The data show that: (1) the estimation performance of ML2 method always outperforms that of ML1 method. (2) ML1 method gives some inaccurate estimation values when the shape parameters are small.

Table 2. Comparisons of the estimation performance of two methods (N=500)

β_0		0.2	0.4	0.6	0.8	1.0	1.2	1.4	1.6	1.8	2.0
RMSE	ML2	0.0026	0.0054	0.0062	0.0070	0.0075	0.0082	0.0087	0.0095	0.0103	0.0107
	ML1	0.8098	0.0332	0.0090	0.0075	0.0078	0.0085	0.0087	0.0096	0.0103	0.0107
$\overline{\hat{\beta}}$	ML2	0.2015	0.4026	0.6067	0.8096	1.0102	1.2125	1.4225	1.6247	1.8299	2.0250
	ML1	0.3570	0.4536	0.6265	0.8194	1.0163	1.2198	1.4250	1.6265	1.8307	2.0251
Var	ML2	0.0001	0.0009	0.0022	0.0045	0.0075	0.0119	0.0170	0.0244	0.0333	0.0429
	ML1	0.0324	0.0053	0.0032	0.0048	0.0078	0.0122	0.0170	0.0245	0.0334	0.0428

3.3   GGD Modeling of Contourlet Subband Coefficients

It has been shown that the marginal distributions of multiple subbands of a natural image are sufficient to characterize the image [13]. In this paper, we propose using a zero mean GGD model to characterize the contourlet subband coefficients.

Like the application of the GGD modeling method to the subband coefficients resulting from wavelet transform [6], we fit the distribution to the histogram of the subband images. Figure 1 shows some typical experimental results of contourlet coefficient modeling. In the contourlet transforms, the LP and DFB filters are chosen as the non-separable biorthogonal filters of support size (23,23) and (45,45) [14] respectively. These filters are called PKVA filters. The contourlet transforms of the texture images *Building*9 and *Flowers*5 (128×128 from VisTex database [15]) use 3 LP levels and 8 directions at the finest level of DFB, respectively. From Figure 1, it can be seen that the estimated GGD can effectively fit the actual sample distribution.

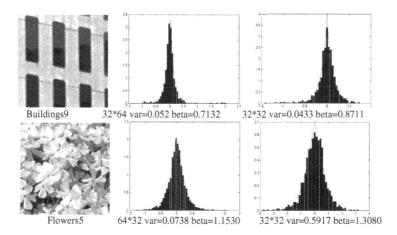

Fig. 1. Coefficient modeling of the contourlet transform detail subbands with different sizes. (var and beta denote the estimated variance and shape parameter, respectively.)

4 Contourlet-Based Texture Image Retrieval by Using GGD Parameters

Using the texture features developed above, we proposed a contourlet-based retrieval system. The retrieval task is to search the top N texture images that are similar to one query image within a large database of total M unlabeled texture images ($N \ll M$). In the proposed system, each texture image is first transformed into directional subband coefficients. Then, the GGD parameters, σ^2 and β, are extracted as image features using ML2 method. The Kullback-Leibler distance (KLD) between the query image and each database image is measured [6], and the top N database images that have the smallest KLD are retrieved.

To demonstrate the effectiveness of the proposed retrieval system, we use 40 texture classes obtained from VisTex database [6] [15]. A test database of 640 texture images is constructed by divided each 512×512 image into 16 non-overlapping 128×128 subimages, but only the gray-scale levels of the subimages are used. In retrieval experiments, the query image is any one of 640 subimages, and relevant candidate images are the other 15 subimges from the same class.

We examine the performance of the texture image retrieval by using the proposed texture features. The performance is first compared by considering different filter banks and decomposition levels used in the contourlet transform. The filter banks under consideration include the "9-7" biorthogonal filters [16], the PKVA filters and the Daubechies' filters [17]. The LP uses three and four level decompositions respectively, and the corresponding DFB uses (8, 4, 4) and (8, 8, 4, 4) directional decompositions from the finest scale to the coarser ones respectively. The experimental results are summarized in Table3.

Next we compare the retrieval performance of using the proposed texture features in the presence of different scales and directional subbands used in the contourlet transform. The filter banks under consideration only include the PKVA filters and the Daubechies' filters. The image retrievals are performed when the contourlet transform takes the scales and the directional decompositions as listed in Table4. The retrieval rates obtained by using the GGD feature representations for each scale and decomposition are also summarized respectively in Table4. The data in the tables indicate that the mid- and high-frequency directional subbands such as decompositions (8, 4, 4), (8, 8, 4), (8, 8, 4, 4) and (8, 8, 8, 4) are more important for characterizing textures in CBIR. Moreover, the better retrieval performance is obtained in two and three scale decompositions when the LP and the DFB are performed by using the PKVA filters, and the better performance is also obtained in four and more scale decompositions when the LP and the DFB are performed by the db8 filter and the PKVA filter, respectively. It is also worthy to note that the average retrieval rates of using the proposed texture features are dramatically reduced for the directional decomposition (8, 8, 8, 8). This is because the subband dimensions in the presence of coarse resolutions are of 8×16 or 16×8 which are too small to be used to obtain the accurate estimates of the GGD parameters. The data in the tables also show that the use of the PKVA filters to perform the LP and DFB decomposition can obtain good retrieval performance.

Table 3. Average retrieval rate (%) using GGD parameters for different filters

LP	DFB	(8 4 4)	(8 8 4 4)
PKVA	PKVA	78.09	76.45
9-7	PKVA	77.79	74.58
db8	PKVA	77.42	76.64

Table 4. Average retrieval rate (%) using GGD parameters for different scales and directional decompositions

LP	DFB	(4 4)	(8 4)	(8 8)	(4 4 4)	(8 4 4)	(8 8 4)
PKVA	PKVA	72.33	75.40	75.38	75.36	78.09	77.72
db8	PKVA	72.11	75.18	75.05	75.00	77.42	77.34
LP	DFB	(8 8 8)	(4 4 4 4)	(8 4 4 4)	(8 8 4 4)	(8 8 8 4)	(8 8 8 8)
PKVA	PKVA	76.54	72.69	75.66	76.45	76.05	50.62
db8	PKVA	76.36	73.36	76.15	76.64	76.04	53.87

To proceed, we investigate using 3 LP levels and 8 directions at the finest level in contourelet transform. Following the retrieval method of [6], the experimental results on 640 texture images of 40 classes indicate that the ML2 approach improves the average retrieval rate from 76.05% to 78.09% over the ML1 method. We also compare our retrieval results with one using DWT-based method [6]. The results of average retrieval rates are listed in Table 5.

Table 5. Average retrieval rates for the VisTex collection and for different methods and implementations

Method	DMT[6]	ML1	ML2
Average retrieval rate	76.57%	76.05%	78.09%

Lastly, we examine the retrieval time of the proposed contourlet-based retrieval system. The retrievals using ML1 method and using ML2 method are respectively carried out on the VisTex database which consists of 640 images. The retrievals are performed on a Pentium 4 computer with 2.40 GHz. Experimental results show that the average retrieval time per image is 1.05 seconds for using ML1 method and 1.14 seconds for using ML2 method. Thus, the proposed estimation method has a comparable average retrieval time to that of existing methods.

5   Conclusions

We have developed a new method of modeling the contourlet-based coefficients with GGD parameters. It improves existing results in the iterative algorithm and the selection of the initial value of the shape parameter, and it especially suits for estimating the small shape parameters (e.g. $\beta \leq 1.0$). Using the proposed GGD parameters to represent the texture features, a contourlet-based texture image retrieval system is developed. Compared with existing retrieval methods, the new approach could more effectively improve the average retrieval rate, which is demonstrated on a database of 640 texture images with comparable computational complexity.

References

1. Rui, Y., Huang, T.S.: Image retrieval: Current techniques, promising directions and open issues. J. Vis. Commun. Image Represent 10, 39–62 (1999)
2. Smeulders, A.W.M., Worring, M., Santini, S., Gupta, A., Jain, R.: Content-based image retrieval at the end of the early years. IEEE Trans. Pattern Anal. Mach. Intell. 22(12), 1349–1380 (2000)
3. Rao, A.R., Lohse, G.L.: Towards a texture naming system: identifying relevant dimensions of texture. In: Proc. IEEE Conf. Visualization, San Jose, Calif., 25-29 October, pp. 220–227. IEEE Computer Society Press, Los Alamitos (1993)
4. Randen, T., Husoy, J.H.: Filtering for texture classification: A comparative study. IEEE Transactions on Pattern Analysis and Machine Intelligence 21(4), 291–310 (1999)

5. Varanasi, M.K., Aazhang, B.: Parametric generalized Gaussian density estimation. J. Acoust. Soc. Amer. 86(4), 1404–1415 (1989)
6. Do, M.N, Vetterli, M.: Wavelet-based texture retrieval using generalized Gaussian density and Kullback-Leibler distance. IEEE Transactions on Image processing 11(2), 146–158 (2002)
7. Song, K.-S.: A globally convergent and consistent method for estimating the shape parameter of a generalized Gaussian distribution. IEEE Transactions on Information Theory 52(2), 510–527 (2006)
8. Do, M.N., Vetterli, M.: The contourlet transform: an efficient directional multiresolution image representation. IEEE Transactions on Image Processing 14(12), 2091–2106 (2005)
9. Do, M.N.: Contoulets and sparse image expansions. In: Unser, M.A., Aldroubi, A., Laine, A.F. (eds.) Proceedings of SPIE, Applications in Signal and Image Processing X, vol. 5207, pp. 560–570 (November 2003)
10. Aiazzi, B., Alparone, L., Baronti, S.: Estimation based on entropy matching for generalized Gaussian pdf modeling. IEEE Signal Processing Letters 6(6), 138–140 (1999)
11. Meignen, S., Meignen, H.: On the modeling of small sample distributions with generalized Gaussian density in a maximum likelihood framework. IEEE Transactions on Image Processing 15(6), 1647–1652 (2006)
12. Krupinski, R., Purczynski, J.: Approximated fast estimator for the shape parameter of generalized Gaussian distribution. Signal Processing 86(2), 205–211 (2006)
13. Liu, X., Wang, D.: Texture classification using spectral histograms. IEEE Transactions on Image Processing 12(6), 661–670 (2003)
14. Phoong, S.-M., Kim, C.W., Vaidyanathan, P.P., Ansari, R.: A new class of two-channel biorthogonal filter banks and wavelet bases. IEEE Transactions on Signal Processing 43(3), 649–665 (1995)
15. MIT Vision and Modeling Group. Vision Texture, [Online]. Available: http://vismod.www.medis.mit.edu
16. Vetterli, M., Herley, C.: Wavelet and filter banks: theory and design. IEEE Transactions on Signal Processing 40(9), 2207–2232 (1992)
17. Daubechies, I.: Ten lectures on wavelet. SIAM, Philadelphia, PA (1992)

Heterogeneous Spatial Data Mining Based on Grid

Yong Wang and Xincai Wu

Faculty of Information and Engineering, China University of Geosciences,
Wuhan, 430074, China
{giswy, xin_caiwu}@126.com

Abstract. The continuous increase of spatial data volumes available from many
sources raises new challenges for their effective understanding. Spatial
knowledge discovery in spatial data repositories involves processes and
activities that are heterogeneous, collaborative, and distributed in nature. The
Grid is a profitable infrastructure that can be effectively exploited for handling
distributed spatial data mining and knowledge discovery. We design a new
architecture of spatial knowledge grid, which is on top of computational grid
mechanisms. The spatial knowledge grid is a high-level framework providing
Grid-based knowledge discovery tools and services. All of these services are
currently being designed and implemented as Grid Services. This paper
highlights design aspects and implementation of spatial knowledge grid
services.

Keywords: Data Mining; Grid Services; Knowledge Grid; Spatial Data.

1 Introduction

Nowadays with the increase of GIS application, people have to face an overwhelming
amount of spatial data. Most spatial data is maintained in multiple disparate databases
and different in spatial data type, file formats, data schema, access mechanism, etc.
Spatial data is characterized to be heterogeneous, incomplete and usually involves a
huge amount of records. In order to discover spatial knowledge, spatial data must be
transformed in a set of patterns, rules or some kind of formalism, which helps to
understand the underlying information. Data mining is a widely used approach for the
transformation of data to useful patterns, aiding the comprehensive knowledge of the
concrete domain information. Data mining has been defined as the process of
discovering useful patterns in data. This process must be automatic, or at least semi-
automatic, since a manual process would be excessively costly and prone to error.
Nevertheless, traditional data mining techniques find difficulties in their application
on current scenarios, due to the complexity of spatial data. The participation of
several organizations in this process makes the assimilation of spatial data more
difficult.

Grid computing has emerged as a new technology, whose main challenge is the
complete integration of heterogeneous computing systems and data resources with the
aim of providing a global computing space. We consider that grid computing provides

L. Kang, Y. Liu, and S. Zeng (Eds.): ISICA 2007, LNCS 4683, pp. 503–510, 2007.

a new framework in which data mining applications can be successfully deployed. In this paper, we propose a new spatial data mining grid architecture, which defines a set of grid services for spatial data mining and usage patterns. Data mining process is allowed to be deployed in a grid environment, in which data and services resources are geographically distributed, belong to several virtual organizations and the security can be flexibly solved.

The outline of the paper is as follows. Section 2 describes Grid, OGSA and Spatial Information Grid. Section 3 discusses architecture of spatial data mining based on grid. Section 4 outlines the implementation of spatial knowledge services and shows an example of spatial data mining. Section 5 concludes the paper and introduces the future work.

2   Related Work

2.1   The Grid and OGSA

Grids are geographically distributed platforms for computation, composed of a set of heterogeneous machines accessible to their users via a single interface. Grid computing has been proposed as an important computational mode, distinguished from conventional distributed computing by its focus on large-scale resource sharing, innovative applications, high-performance orientation. The main original application area was advanced science and engineering. Grid technologies allow existing distributed computing to build dynamic cross organizational applications and provide a standard base for new concepts in distributed utility computing and autonomic computing promoted by big computer vendors like IBM, HP and Sun. Recently, grid computing is emerging as an effective paradigm for coordinated resource sharing and problem solving in dynamic, multi-institutional virtual organizations operating in the industry and business area. A set of individuals and/or institutions defined by such sharing rules form what we call a virtual organization (VO).

The Open Grid Services Architecture (OGSA) developed by a joint effort of the research community and industry in the framework of the Global Grid Forum (GGF) is intended to create a standard base for building scalable VO and virtualizing resource management. In OGSA every resource is represented as a Web Service that conforms to a set of conventions and supports standard interfaces. OGSA provides a well-defined set of Web Service interfaces for the development of interoperable Grid systems and applications. Recently the WS-Resource Framework (WSRF) has been adopted as an evolution of early OGSA implementations. WSRF defines a family of technical specifications for accessing and managing stateful resources using Web Services. The composition of a Web Service and a stateful resource is termed as WS-Resource. The possibility to define a "state" associated to a service is the most important difference between WSRF-compliant Web Services, and pre-WSRF ones. This is a key feature in designing Grid applications, since WS-Resources provide a way to represent, advertise, and access properties related to both computational resources and applications.

Computer Grids emerged from the research area where the complexity of problems and need to share special unique resources required virtualization of the collaborative environment, but currently the Grid concept can also be used for spatial data management where the large-scale and heterogeneous spatial data sharing is a key problem.

2.2 Spatial Information Grid

Spatial data is the data that can be associated with location on Earth. It also is the dominant form of data in terms of data volume, and has been widely used in many fields of social and economic activities, ranging from mines exploitation to mobile application. Nowadays China has accumulated large-scale, heterogeneous spatial resources, which include established and establishing fundamental spatial database, spatial data processing and application software, spatial facility and instrument, etc. But these spatial resources were distributed over related departments for long, and this has caused lots of obstacle when other users want to share, integrate spatial resource across departments and regions dynamically. This difficult problem can be characterized as grid problem defined as "Flexible, secure, coordinated resource sharing among dynamic collections of individuals, institutions, and resources". Spatial Information Grid (SIG) is proposed as an effective solution of this grid problem based on the latest grid theory and technology, this solutions involved adoption of a service-oriented model and attention to metadata. SIG is defined as an infrastructure and framework, which enables us to share distributed, large-scale, and heterogeneous spatial resources across dynamic virtual organizations cooperatively, organizes and deals with them systematically. It aims to make any kinds of users acquire spatial information on any kinds of spatial resources, especially the metadata about spatial resources, and possesses the capability of service-on-demand. The sharing that we are concerned with is not primarily file exchange but rather direct access to computers, software, data, and other resources. In previous work, we had proposed a spatial data grid architecture and implemented some key services. In order to enable users and agents can discover knowledge across those discrete information isolates effectively and efficiently, we should provide appropriate and efficient mechanism to discover spatial information.

On the basis of these motivations, the evolution of data grids are represented by knowledge grids that offer high level tools and techniques for the distributed mining and extraction of knowledge from data repositories available on the grid. The development of such an infrastructure is the main goal of our research activities that are focused on the design and implementation of an environment for geographically distributed high-performance knowledge discovery applications. The spatial knowledge grid can be used to perform data mining on spatial data grid.

3 Spatial Knowledge Grid Architecture

Spatial data mining is a complex process, which can be deployed by means of multiple approaches. The distributed nature of spatial data and the extension of information sharing makes the Grid a suitable scenario in which data mining applications can be executed .The spatial knowledge grid architecture is shown in Figure 1. The spatial data grid services let users register spatial datasets (registration services), situate the data in spatial, temporal and conceptual contexts (indexing services), execute complex queries (mediation services) and multi-step processing chains (workflow services), and visualize intermediate and final results as tables, maps, and graphs (visualization and mapping services). The spatial data services execute on top of core grid services which implement basic data grid capabilities providing mechanisms for data movement, user access control and security, load balancing and network weather service, and communication with physical grid infrastructure. The core grid services include: authentication, monitoring, scheduling, catalog, data transfer, replication, collection management, databases.

Spatial data mining grid services, which are based on the spatial data grid services, include pre-processing data mining services, data mining services and post-processing services. Fig. 1 shows the three stages: pre-processing stage, data mining stage and post-processing stage. All these spatial data mining services use both basic data and spatial grid services. Spatial data grid services are services oriented to spatial data management in a grid. Besides spatial data grid services, spatial data mining grid services also use generic and standard grid services. Generic services offer common functionalities in a grid environment. Spatial data mining grid services are intended to provide specialized and new data mining services. Whenever a spatial data or generic grid service can be used for a purpose, it will be used. Only if the typical behavior of a service must be modified, a specialized service must be placed at the spatial data mining grid services level. For instance, if we need to transfer files in the pre-processing stage, we can make use of the data mediation service. New services oriented to spatial data mining applications are included in the architecture. These kinds of services are usually linked to spatial data mining techniques and algorithms.

Specialization services are the following:

- ✓ SDFS, Specific Data Filtering Service: due to the huge amount of data involved in the process of spatial data mining, filtering data is an important task, solved by means of this service.
- ✓ SDCS, Specific Data Consistency Service: its main purpose is maintaining the spatial data consistency in the grid.
- ✓ SDDS, Specific Data Discovery Service: this service improves the discovery phase in the grid for mining applications.
- ✓ Apriori services: it implement Apriori algorithm.
- ✓ KMS, Knowledge Management Service: the service manages the result of knowledge discover.
- ✓ KVS, Knowledge Visualization Service: it shows the knowledge.
- ✓ Services slot: additional specific services can be defined for the management of other features offered by the generic or data grid services, without changes in the rest of the framework.

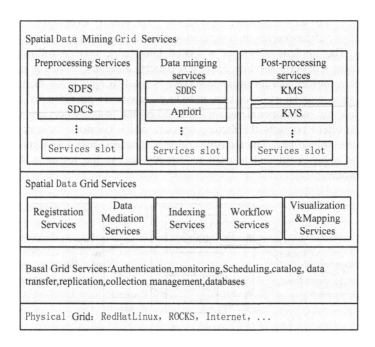

Fig. 1. Architecture of Spatial Knowledge Grid

4 Spatial Knowledge Grid Services

4.1 Implementation of Spatial Knowledge Grid Services

This section describes the design and implementation of the spatial knowledge grid in terms of the OGSA and WSRF models. Fig. 1 proposes a view of the spatial knowledge grid architecture in which each knowledge grid service is exposed as a Web Service that exports one or more operations, by using the WSRF conventions and mechanisms. The operations exported by knowledge grid services are designed to be invoked by user-level applications, whereas operations provided by spatial data services are thought to be invoked both users and grid services. As shown in Fig. 2, users can access the spatial knowledge grid functionalities by using a client interface located on their machine. The client interface can be an integrated visual environment that allows for performing basic tasks (e.g., searching of data and software, data transfers, simple job executions), as well as defining distributed data mining applications described by arbitrarily complex execution plans. The client interface performs its tasks by invoking the appropriate operations provided by the different spatial knowledge services. Those services may be generally executed on a different Grid node; therefore the interactions between the client interface and spatial knowledge services are possibly remote. All spatial knowledge services export three mandatory operations (CreateResource, Subscribe and Destroy) and one or more service-specific operations. The CreateResource operation is used to create a stateful resource, which is then used to maintain the state (e.g., results) of the computations performed by the service-specific operations. The subscribe operation is used to subscribe for notifications about computation results. The destroy operation removes a resource. The implementation of a spatial knowledge service follows the WS-Resource factory pattern (see Fig. 2). In this pattern, a factory service is in charge of creating the resources and an instance service is used to operate on them. Thus the CreateResource mandatory operation introduced above is provided by the factory service, while the other operations are exported by the instance service. To create a resource the client contacts the factory service, which creates a new resource and assigns to it a unique key. The factory service will return an endpoint reference that includes the resource id and is used to directly access the resource through the instance service.

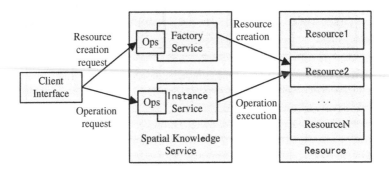

Fig. 2. Implementation of Spatial knowledge Grid Services

4.2 Application Example

We illustrate in Figure 3 the following stages in the life of a data mining computation, which we use to illustrate the working of basic remote service invocation, lifetime management, and notification functions. The environment initially comprises (from left to right) four simple hosting environments: one that runs the user application; one that encapsulates computing and storage resources (and that supports two factory services, one for creating storage reservations and the other for creating mining services); and two that encapsulate database services. The "R"s represent local registry services. An additional VO registry service presumably provides information about the location of all depicted services.

The user application invokes "create mining service" requests on the two factories in the second hosting environment, requesting the creation of a "data mining service" that will perform the data mining operation on its behalf, and an allocation of temporary storage for use by that computation. Each request involves mutual authentication of the user and the relevant factory (using an authentication mechanism described in the factory's service description) followed by authorization of the request. Each request is successful and results in the creation of a Grid service instance with some initial lifetime. The new data mining service instance is also provided with delegated proxy credentials that allow it to perform further remote operations on behalf of the user. The newly created data mining service uses its proxy credentials to start requesting data from the two database services, placing intermediate results in local storage. The data mining service also uses notification mechanisms to provide the user application with periodic updates on its status. Meanwhile, the user application generates periodic "keepalive" requests to the two Grid service instances that it has created.

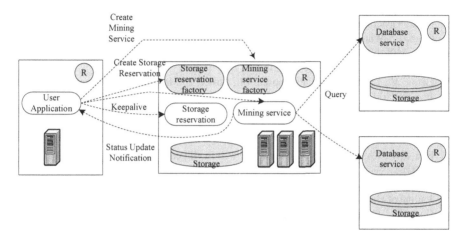

Fig. 3. An application example of spatial data mining

5 Conclusion

Today's data mining process is one of the hot topics in the scientific and business fields. Nevertheless, current data mining approaches are not scalable and do not fit into complex scenarios in which spatial data is heterogeneous and distributed. The Grid infrastructure is growing up very quickly and is going to be more and more complete and complex both in the number of tools and in the variety of supported applications. Along this direction the grid services are shifting from generic computation-oriented services to high level information management and knowledge discovery services. The spatial knowledge grid architecture we discussed here is a significant component for information management and knowledge discovery. It integrates and completes the spatial data grid services by supporting distributed data analysis and knowledge discovery and management services that will enlarge the application scenario and the community of grid computing users.

Currently, we have developed a prototype of spatial data mining, in which some key services have been implemented. In next work, our research will focus on grid services of spatial data mining algorithms.

Acknowledgments. The research was supported by the Research Foundation for Outstanding Young Teachers, China University of Geosciences (Wuhan), which number is CUGQNL0645.

References

1. Foster, I., Kesselman, C., Tuecke, S.: Grid Services for Distributed System Integration. Computer 35(6), 77–78 (2002)
2. Wang, Y., Wu, X.C.: Service Architecture of Urban Data Sharing Based on OGSA. Earth Science 31(5), 635–638 (2006)
3. Antonio, C., Domenico, T., Paolo, T.: Distributed data mining services leveraging WSRF. Future Generation Computer Systems 23, 34–41 (2007)
4. Chen, L., Wang, C.L., Lau, F.C.M.: A grid middleware for distributed Java computing with MPI binding and process migration supports. Journal of Computer Science & Technology 18(4), 505–514 (2003)
5. Getov, V., Laszewski, G., Philippsen, M., Foster, I.: Multi-paradigm Communications in Java for Grid Computing. Communications of the ACM 44(10), 118–125 (2001)
6. Ester, M., Kriegel, H.P., Sander, J.: Spatial data mining: A database approach. In: Scholl, M.O., Voisard, A. (eds.) SSD 1997. LNCS, vol. 1262, pp. 47–66. Springer, Heidelberg (1997)
7. Han, J.W., Chang, K.C.: Data mining for web intelligence. IEEE Computer 35(11), 64–70 (2002)
8. Stork, H.G.: Webs, grids and knowledge spaces: Programmes, projects and prospects. Journal of Universal Computer Science 8(9), 848–867 (2002)
9. Guo, D., Gahegan, M., MacEachren, A.M., Zhou, B.: Multivariate analysis and geovisualization with an integrated geographic knowledge discovery approach. Cartography and Geographic Information Science 32(2), 113–132 (2005)
10. Curcin, V., Ghanem, M., Guo, Y., Köhler, M., Rowe, A., Syed, J., Wendel, P.: Discovery net: towards a grid of knowledge discovery. In: Proc. 8th ACM SIGKDD International Conference on Knowledge Discovery and Data Mining, pp. 658–663. ACM Press, New York (2002)

A Clustering Scheme for Large
High-Dimensional Document Datasets

Jung-Yi Jiang, Jing-Wen Chen, and Shie-Jue Lee

Dept. of Electrical Engineering,
National Sun Yat-Sen University, Taiwan
{jungyi,a140425,leesj}@water.ee.nsysu.edu.tw

Abstract. Scalability and high dimensionality are two common problems associated with document clustering. We present a novel scheme to deal with these problems. Given a set of documents, we partition the set into several parts. We use one part and cluster the constituent documents into groups. By the obtained groups, we reduce the number of features by a certain ratio. Then we add another part, cluster the documents into groups based on the reduced features, and further reduce the number of the remaining features. This process is iterated until all parts are used. Experimental results have shown that our proposed scheme is effective for clustering large high-dimensional document datasets.

Keywords: Document clustering, scalability, high dimensionality, K-means, information gain.

1 Introduction

Recently, document clustering has attracted more and more attention. However, two challenges are usually encountered, scalability and high dimensionality. In this modern world, more and more documents are generated. Research papers, book chapters, recreation articles, etc., are published everyday. Furthermore, a lot of keywords are used for distinguishing one document from another. For example, more than 15000 keywords are used in the real-world document set Reuters21578. As a result, clustering large high-dimensional document datasets is not an unusual task. Most of the existing clustering algorithms have difficulties in dealing with such datasets.

Document collections are usually represented by a *vector space model*, alternatively known as a *bag-of-words* [5]. The words appear in the documents are treated as features and each document is represented as a vector of certain weighted word frequencies in this feature space. Document clustering is a common unsupervised learning technique used to discover group structure in a set of documents. The challenge for document clustering is to minimize the computation cost as well as to maximize the difference among clusters. Several algorithms based on K-means were proposed [1,2,3,6] for text clustering. These approaches are based on distance or similarity measures. A combination of the batch k-means and the increment k-means algorithms [7] was introduced for document clustering . A spherical k-means algorithm for clustering text data was

L. Kang, Y. Liu, and S. Zeng (Eds.): ISICA 2007, LNCS 4683, pp. 511–519, 2007.

proposed in [6]. A hierarchical clustering algorithm based on divisive partitioning was proposed in [4].

Feature reduction is a useful technique to reduce dimensionality and speed up document processing tasks. Many approaches for feature reduction have shown their good performance for document classification tasks [8,9,10,11,12,13]. Since they work on classification, it is required that the class of each document be known. A well-known feature reduction approach is based on Information Gain [8], which is an information-theoretic measure defined by the amount of reduced uncertainty given a piece of information. However, these approaches cannot be used for document clustering.

We propose a novel scheme to deal with the challenges of scalability and high dimensionality associated with document clustering. Given a set of documents, we partition the set into several parts. We use one part and cluster the constituent documents into groups. By the obtained groups, we reduce the number of features by a certain ratio. Then we add another part, cluster the documents into groups based on the reduced features, and further reduce the number of the remaining features. This process is iterated until all parts are used. Experimental results have shown that our proposed scheme is effective for clustering large high-dimensional document datasets.

2   Background and Related Work

As we mentioned earlier, the bag-of-words model [5]is usually adopted to represent documents for document processing. Each document is expressed as a vector of weighted word frequencies. Let d_i be a document and $D = \{d_1, d_2, \ldots, d_n\}$ be a set containing n documents. Let the word set $W = \{w_1, w_2, \ldots, w_f\}$ be the feature set of the documents. Each document d_i, $1 \leq i \leq n$, can be represented as $d_i = < w_{i1}, w_{i2}, \ldots, w_{if} >$ where each w_{ij} denotes the number of occurrences of w_j in document d_i.

2.1   K-Means Clustering

Let $\{P_j\}_{j=1}^k$ be a partition of D where k is a user-specified constant. The goal of the k-means clustering algorithm is to maximize the following objective function:

$$\sum_{j=1}^{k} \sum_{d_i \in P_j} Sim(d_i, P_j) \tag{1}$$

where $Sim(d_i, P_j)$ is the similarity measure between document d_i and P_j. A popular similarity measure is defined as

$$Sim(d_i, P_j) = cos(d_i, m_j) = \frac{d_i \cdot m_j}{\|d_i\|\|m_j\|} \tag{2}$$

where m_j is the centroid of P_j defined as

$$m_j = \frac{1}{|P_j|} \sum_{d \in P_j} d. \tag{3}$$

2.2 Feature Reduction

The feature reduction task is to find a new word set $W' = \{w'_1, w'_2, \ldots, w'_r\}$, $r < f$, such that W and W' work equally well for all the desired properties with D. After feature reduction, each document d_i is converted to a new representation $d'_i = <w'_{i1}, w'_{i2}, \ldots, w'_{ir}>$ and the converted document set is D' $= \{d'_1, d'_2, \ldots, d'_n\}$. If r is very much smaller than f, computation cost can be drastically reduced.

One popular feature reduction approach uses information gain to select W' from W, and W' is a subset of W [8]. This approach only uses the selected features as inputs for classification tasks. It measures the reduced uncertainty by an information-theoretic measure and gives each word a weight. The bigger the weight of a word is, the larger is the reduced uncertainty by the word. Let $\{c_1, c_2, \ldots, c_p\}$ denote the set of classes. The weight of a word w_i is calculated as follows:

$$
\begin{aligned}
G(w_i) = &- \sum_{l=1}^{p} Pr(c_l) log Pr(c_l) \\
&+ Pr(w_i) \sum_{l=1}^{p} Pr(c_l|w_i) log Pr(c_l|w_i) \\
&+ Pr(\overline{w}_i) \sum_{l=1}^{p} Pr(c_l|\overline{w}_i) log Pr(c_l|\overline{w}_i).
\end{aligned}
\tag{4}
$$

The words of top r weights in W are selected as the features in W'. Information gain is applied to compress the complexity of the document set from $O(nf)$ to $O(nr)$. If r is much smaller than f, the computation cost associated with document processing can be drastically reduced.

3 Proposed Method

As mentioned, scalability and high-dimensionality are two challenges for document clustering. Existing clustering algorithms have difficulties in processing all the documents of a document set at one time. Feature reduction algorithms can be used for reducing dimensionality, but they work well only for document classification in which the class label of each document is known. In document clustering, the class labels are not known for documents. We propose a novel approach to deal with these two challenges. A given document set is partitioned into parts. We use one part and cluster the constituent documents into groups. Then we give an appropriate label to each document. Note that all documents have their class labels now. Then we use a feature reduction algorithm to reduce the number of features by a certain ratio. Then we add another part, cluster the documents into groups based on the reduced features, and further reduce the number of the remaining features. This process is iterated until all parts are used. The flowchart of our approach is shown in Fig. 1.

To use feature reduction algorithms efficiently, documents under consideration must be labeled. For this purpose, we cluster parts of documents, and use the assigned cluster labels to perform the feature reduction work. To avoid losing too much information at one time, dimensionality is reduced progressively. We repeat clustering and feature reduction, with more and more documents but less and less features, until all documents are clustered. For clustering, K-means is

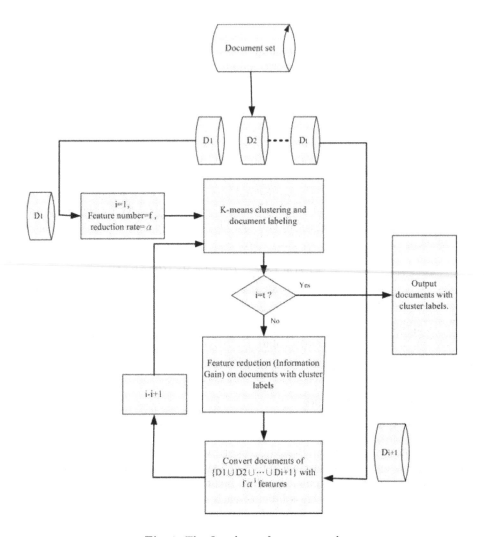

Fig. 1. The flowchart of our approach

adopted, and for feature reduction, information gain is used. The whole process of our approach is detailed as follows:

Input: The document set D, the desired number of cluster K, the feature reduction rate α, the number of partitions t, and the number of original features f.

Step 1: Divide D into partitions $\{D_1, D_2, \ldots, D_t\}$. Let $i = 1$, $g = f$, and $P = \emptyset$.

Step 2: Let $P = P \cup D_i$. Apply K-means on P with g features. K clusters are obtained.

Step 3: Label each document in P with the cluster to which it belongs.
Perform feature reduction with these labeled documents, and the
number of features is reduced by α, i.e., $g = \alpha g$.

Step 4: If $i == t$, go to Step 5; otherwise $i = i + 1$, and go to Step 2.

Output: The document set D' with g features, and the resulting clusters.

Note that $\alpha \in (0, 1]$. When $\alpha = 1$, feature reduction is not applied. After the ith
iteration, the number of features is reduced by $f\alpha^i$.

To evaluate the performance of our proposed method, an entropy based cluster
validity measure is adopted. Let K be the number of clusters obtained by our
clustering approach and L be the number of classes given by the data source.
For each cluster i, we can calculate an entropy e_i of the cluster as follows:

$$e_i = -\sum_{j=1}^{L} p_{ij} log_2 p_{ij} \tag{5}$$

where $p_{ij} = m_{ij}/m_i$ is the probability that a member of cluster i belongs to
class j. Note that m_i is the number of objects in cluster i and m_{ij} is the number
of objects of class j in cluster i. The **entropy weighted sum** E is defined as
the sum of the entropies of each cluster weighted by the size of each cluster, i.e.,

$$E = \sum_{i=1}^{K} \frac{m_i}{M} e_i \tag{6}$$

where M is the total number of data points. If E is smaller, the performance of
a clustering method is better.

4   Experimental Results

To show the effectiveness of our proposed method, experiments on a well-known
data set, Reuters-21578, are performed. The documents of Reuters-21578 are
divided, according to the "ModApte" split, into 9603 training documents and
3299 testing documents. The number of training documents per class varies
from 1 to about 4000, with top 10 classes containing 77.5% of the documents
and 28 classes have fewer than 10 training documents. Our experiments use the
documents of the top 10 classes. In Experiment 1, we use the training documents of Reuters-21578 Top 1. The number of words involved in Experiment 1
is 16283 and the number of documents is 7390. The data set is large and high
dimensional. We will show the performance of our method on this large and
high-dimensional data set in Experiment 1. In Experiment 3, we use the testing
documents of Reuters-21578 Top 10. The number of words involved in Experiment 3 is 16283 and the number of documents is 2787. The data set is small and
high-dimensional. We will show the performance of our method on this small
and high-dimensional data set in Experiment 3. Data sets in Experiment 1 and
3 are formed by converting documents from files to vectors. To provide data sets

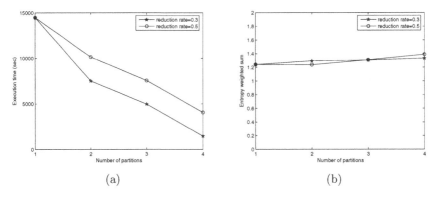

Fig. 2. (a) Execution time for large and high-dimensional data sets. (b) Entropy weighted sum for large and high-dimensional data sets.

with different dimensionality, we use a word clustering approach proposed in [13] to convert the data sets from high dimensionality to lower dimensionality. After converting, two data sets with dimensionality 5000 and 1000, respectively, are obtained from Reuters-21578 Top 10. In Experiment 2, we use the training documents of Reuters-21578 Top 10. The number of words is 5000 and 1000, respectively, and the number of documents is 7390. The data sets are large and lower-dimensional. We will show the performance of our method on large and lower-dimensional data sets in Experiment 2. In Experiment 4, we use the testing documents of Reuters-21578 Top 10. The number of words is 5000 and 1000, respectively, and the number of documents is 2787. The data sets are small and lower-dimensional. We will show the performance of our method on small and lower-dimensional data sets in Experiment 4. In each experiment, two reduction rates, 0.5 and 0.3, are used and the number of partitions is varied among 1, 2, 3, and 4. When the number of partitions is equal to 1, no feature reduction is applied. In other words, K-means clustering with full features is applied when the partition number is 1. For all experiments, K is set to 10.

4.1   Experiment 1: Large and High-Dimensional Data Set

Fig. 2 shows the execution time and entropy weighted sum of our method with different numbers of partitions on a large (7390 documents) and high-dimensional (16283 features) data set. The star-marked line and circle-marked line denote the results with reduction rates 0.3 and 0.5, respectively. As shown in Fig. 2(a), our method reduces the execution time significantly, especially when the partition number is larger and the reduction rate is smaller. When the partition number is 4 and the reduction rate is 0.3, the execution time is reduced from 14474 to 1464 seconds. As shown in Fig. 2(b), our method works well with a little increase in entropy weighted sum.

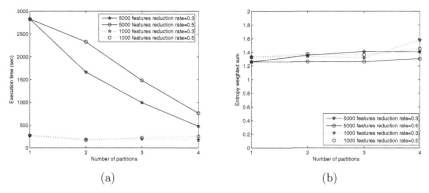

Fig. 3. (a) Execution time for large and lower-dimensional data sets. (b) Entropy weighted sum for large and lower-dimensional data sets.

4.2   Experiment 2: Large and Lower-Dimensional Data Set

Fig. 3 shows the execution time and entropy weighted sum of our method with different numbers of partitions on large (7390 documents) and lower-dimensional (5000 and 1000 features, respectively) data sets. The star-marked solid lines and the circle-marked solid lines denote the results for the data set with 5000 features, while the star-marked dotted lines and the circle-marked dotted lines denote the results for the data set with 1000 features. As shown in Fig. 3(a), the decrease of execution time our method achieves on the data set with 5000 features is more slower than on full features shown in Fig. 2(a). On the data set with 1000 features, our method spends almost the same time as clustering without feature reduction. As shown in Fig. 3(b), our method works well with a little increase in entropy weighted sum on both data sets with 5000 and 1000 features. Compared to Experiment 1, our method is obviously more suitable for high-dimensional data sets.

4.3   Experiment 3: Small and High-Dimensional Data Set

Fig. 4 shows the execution time and entropy weighted sum of our method with different numbers of partitions on large (7390 documents) and high-dimensional (16283 features) data sets. The star-marked line and circle-marked line denote the results with reduction rate 0.3 and 0.5, respectively. As shown in Fig. 4(a), the execution time reduced on small data sets is more slower than on large data sets shown in Fig. 2(a). As shown in Fig. 4(b), our method works well with a little increase in entropy weighted sum. Compared to Experiment 1, our method is more suitable for large data sets with high-dimensionality.

4.4   Experiment 4: Small and Lower-Dimensional Data Set

Fig. 5 shows the execution time and entropy weighted sum of our method with different numbers of partitions on small (2787 documents) and lower-dimensional (5000, 1000 features) data sets. The solid lines with star marks and circle marks

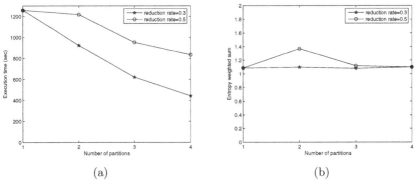

Fig. 4. (a) Execution time on small and high-dimensional data sets. (b) Entropy weighted sum on small and high-dimensional data sets.

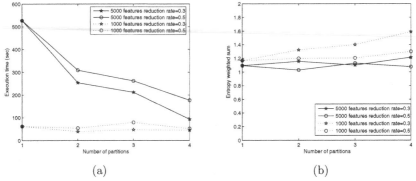

Fig. 5. (a) Execution time on small and lower-dimensional data sets. (b) Entropy weighted sum on small and lower-dimensional data sets.

denote the results for data sets with 5000 features, while the dotted lines denote the results for data sets with 1000 features. As shown in Fig. 5(a), execution time reduced on the data set with 5000 features is more slower than on large data size (compared to Fig. 3(a)) or large high-dimensional data set (compared to Fig. 2(a)). On the data set with 1000 features, our method spends almost the same time as clustering without feature reduction. As shown in Fig. 5(b), our method works well with a little increase in entropy weighted sum both data sets woth 5000 and 1000 features, respectively. Compared to the previous experiments, our method is more suitable for large high-dimensional data sets.

5   Conclusion

Scalability and high-dimensionality are two common problems encountered in document clustering. We have proposed a novel approach for clustering large high-dimensional document datasets. Traditionally, feature reduction is a powerful tool to reduce dimensionality of data. However, they only work well for classification problems. That is, the information about the class label for each document has to be known in advance. To apply feature reduction algorithms to the clustering work, we propose a framework which utilizes the results of clustering on a part of the dataset as class labels, and then perform feature reduction based on the obtained cluster labels. Documents are clustered in a progressive way. To test the effectiveness of our approach, K-means and information gain are adopted for clustering and feature reduction, respectively. Experiments on four different datasets have shown that our approach can work efficiently and get acceptable results on large high-dimensional document datasets.

References

1. Dhillon, I.S., Guan, Y., Fan, J.: Efficient Clustering of Very Large Document Collections. In: Data Mining for Scientific and Engineering Applications, pp. 357–381. Kluwer Academic Publishers, Dordrecht (2001)
2. Dhillon, I.S., Kogan, J., Nicholas, M.: Feature Selection and Document Clustering. In: A Comprehensive Survey of Text Mining, pp. 73–100. Springer, Heidelberg (2003)
3. Kogan, J., Teboulle, M., Nicholas, C.: Data Driven Similarity Measures for K-Means Like Clustering Algorithms. Information Retrieval 8(2), 331–349 (2005)
4. Boley, D.: Principal Direction Divisive Partitioning. Data Mining and Knowledge Discovery 2(4), 325–344 (1998)
5. Salton, G., McGill, M.J.: Introduction to Modern Retrieval. McGraw-Hill Book Company, New York (1983)
6. Dhillon, I.S., Modha, D.S.: Concept Decompositions for Large Sparse Text Data using Clustering. Machine Learning 42(1), 143–175 (2001)
7. Kogan, J.: Means Clustering for Text Data. In: Proceedings of the workshop on Text Mining at the First SIAM International Conference on Data Mining, pp. 54–57 (2001)
8. Yang, Y., Pedersen, J.O.: A Comparative Study on Feature Selection in Text Categorization. In: Proceedings of 14th International Conference on Machine Learning, pp. 412–420 (1997)
9. Baker, L.D., McCallum, A.: Distributional Clustering of Words for Text Classification. In: Proceedings of 21st Annual International ACM SIGIR, pp. 96–103 (1998)
10. Slonim, N., Tishby, N.: The Power of Word Clusters for Text Classification. In: Proceedings of 23rd European Colloquium on Information Retrieval Research (ECIR) (2001)
11. Bekkerman, R., El-Yaniv, R., Tishby, N., Winter, Y.: Distributional Word Clusters vs. Words for Text Categorization. Journal of Machine Learning Research 1, 1–48 (2002)
12. Pereira, F., Tishby, N., Lee, L.: Distributional Clustering of English Words. In: 31st Annual Meeting of ACL, pp. 183–190 (1993)
13. Dhillon, I.S., Mallela, S., Kumar, R.: A Divisive Infromation-Theoretic Feature Clustering Algorithm for Text Classification. Journal of Machine Learning Research 3, 1265–1287 (2003)

Self-tuning PID Control of Hydro-turbine Governor Based on Genetic Neural Networks

Aiwen Guo[1,2] and Jiandong Yang[1]

[1] State Key Laboratory of Water Resources and Hydropower Engineering Science , Wuhan University, Wuhan 430072, China
[2] Department of Automation, College of Power&Mechanical Engineering, Wuhan University, Wuhan 430072, China
aiwenguo@163.com

Abstract. A genetic neural networks (GNN) control strategy for hydro-turbine governor is proposed in this paper. Considering the complex dynamic characteristic and uncertainty of the hydro-turbine governor model and taking the static and dynamic performance of the governing system as the ultimate goal, the novel controller combined the conventional PID control theory with genetic algorithm (GA) and neural networks (NN) is designed. The controller consists of three parts: GA, NN and classical PID controller. The controller is a variable structure type; therefore, its parameters can be adaptively adjusted according to the signal of the control error. The results of simulation show that the presented control strategy has enhanced response speed and robustness and achieves good performance when applied to the hydro-turbine governing system.

Keywords: PID control; genetic algorithm; neural network; hydro-turbine governor.

1 Introduction

In the last decades, conventional PID control has been largely applied to hydro-turbine governors and has achieved valuable results. The main reason is due to their simplicity of operation, inexpensive maintenance, low cost [1], [2]. But many researches were shown that classical PID control was unable to perform optimally over the full range of operating conditions and disturbance, due to the highly complex, non-linear characteristic of hydro-turbine governing system. The plant models are always inaccurate and the governing system is a non-minimum phase system. The unstable zeros can influence the stability and performance of the system, therefore the design of a controller is known to be very difficult. The early power systems were small, often regionally far from the consumer loads and individually isolated. Consequently, classical methods were adopted to adjust governor parameters for acceptable performance by using a single-input-single-output control structure [3]. But conventional governors did not respond satisfactorily over the whole range of plant operation. Moreover, non-linearity exists in the models of the hydro-turbine and the electric generator. A linear controller designed to work optimally around a

L. Kang, Y. Liu, and S. Zeng (Eds.): ISICA 2007, LNCS 4683, pp. 520–528, 2007.

nominal operating point of a non-linear system may be degraded at other operating points. And now, with the development of science and technology, more and more hydro-power units with large installed capacity and high parameters are designed. These problems are more serious than the past, so that the effect of control is dissatisfied and the quality of power is affected. In order to overcome these difficulties, in recent years, many experts take advantage of modern control theory to design controllers for hydro-turbine governor [4], [5]. In this article, a novel PID controller combined with GNN for hydro-turbine governing system is expounded. The GNN PID controller is applied to improve the overall performance of hydro-turbine governing system and adjusts adaptively the parameters as the operating conditions change. Simulation results show that the GNN governor gain a satisfactory characteristic.

2 Implement of Proposed Control Algorithm

In this paper the novel control strategy based on GA and NN is adopted. The GA to optimize the initial weights of the BP neural network (BPNN) and BPNN is used to tune the parameters of the PID on line. It can overcome the faults of reference [8], which didn't give the method that optimizes the initial weights of the neural network.

2.1 Principle of the Control System Figures

The structure of the control system is shown in Fig. 1. The controller is consisted of four parts: 1) Conventional PID controller, it is to control directly the plant and tune the parameters of PID on line; 2) BPNN, it tunes the parameters of PID on line under the situation of the system; 3) GA, it optimize the initial weighs of the BP neural network; 4) RBF neural network (RBFNN), it identifies the model of the plant and get the Jacobian information.

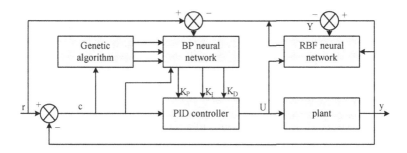

Fig. 1. The structure of intelligent self-tuning PID controller

The algorithm of conventional increment PID is:

$$u(k) = u(k-1) + K_p[e(k) - e(k-1)] + k_i e(k) + k_d[e(k) - 2e(k-1) + e(k-2)] \quad (1)$$

Where k_p, k_i, k_d is the coefficient of proportion, integral and differential.

2.2 Optimization of the Initial Parameters BPNN Based GA

GA is optimization methods inspired by natural genetics and biological evolution. They manipulate strings of data, each of which represents a possible problem solution. These strings can be binary strings, floating-point strings, or integer strings, depending on the way the problem parameters are coded into chromosomes. The strength of each chromosome is measured using fitness values, which depend only on the value of the problem objective function for the possible solution represented by the chromosome. The stronger strings are retained in the population and recombined with other strong strings to produce offspring. Weaker ones are gradually discarded from the population. The processing of strings and the evolution of the population of candidate solutions are performed based on probabilistic rules [9]. References [11], [12] provide a comprehensive description of genetic algorithms.

In this study, we adopt GA to optimize the initial weighs of the BPNN. The flow chart of optimization BPNN weighting initial value with GA is shown in Fig.2.

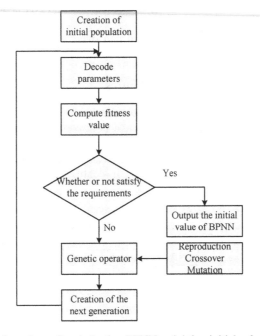

Fig. 2. The flow chart of optimization BPNN weighting initial value with GA

2.3 The BPNN

The BPNN is a three-network which consist of an input layer, a hidden layer, and an output layer (Fig.3). Input layer has M nodes. Hidden layer has Q nodes. Output layer has three nodes and respectively represent the parameters k_p, k_i, k_d of PID.

BPNN tunes the three parameters of PID on line. The detail of the learning algorithm is shown in reference [6].

2.4 The RBFNN

Generally, feedforward neural networks can be used to approximate the continuous nonlinear functions. It is advantage to the outputs of dynamical systems. When used as approximation, the multilayer perceptron can be regarded as a special case of the generalized radial basis function (RBF).

Fig.4 shows the structure of the RBFNN. It has n input nodes, s hidden nodes, and one output node.

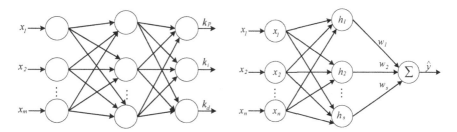

Fig. 3. The structure of BPNN   **Fig. 4.** The structure of RBFNN

Where the radial basis vector is:

$$H = [h_1 \ h_2 \ \cdots \ h_s]^T ,$$

$$h_i = \exp(-\|x_i(k) - a_i\|^2 / 2b_i^2) \quad (i = 1, 2, \ldots, s) \tag{2}$$

Where a_i is the center of node i, b_i is the variance of node i, and $b_i > 0$.

The weight vector of output layer is:

$$W = [w_1, w_2, \cdots, w_s]^T \tag{3}$$

Reference [10], the Jacobian information of the controlled plant is:

$$\frac{\partial y(k)}{\partial u(k)} = \frac{\partial y_s(k)}{\partial u(k)} = \sum_{i=1}^{s} w_i h_i \frac{a_i - x_i}{b_i^2} \tag{4}$$

2.5 The Learn Algorithm of the PID Controller

According to the optimization initial value of BPNN, we should adjust three parameters of PID on line based on the algorithm as follows:

1) BPNN has three layers. The input nodes $M = 5$, and the hidden layer nodes $Q = 8$, and the output layer nodes is3. Learning rate $\eta = 0.3$, and the momentum coefficient $\alpha = 0.2$.

2) RBFNN has three layers. The input nodes $n = 4$, and the hidden layer nodes $s = 4$, and output layer node is 1, and learning rate $\eta = 0.3$, and the momentum coefficient $\alpha = 0.05$.

3) GA optimizes the initial weights of BPNN and gets the initial value of parameters.

4) Sampling and get $y(k)$ and $r(k)$, $e(k) = y(k) - r(k)$.

5) From (13)-(18) of reference [6], the inputs and outputs of every layer of BPNN is computed, and the three parameters of PID, k_p, k_i, k_d is worked out.

6) Get the output $u(k)$ of PID from (1), and the controlled plant and RBFNN get $y(k+1)$.

7) From (4)-(11) of reference [6], we can get the output of the RBFNN, $\hat{y}(k+1)$, then the centers, variances and weights of RBFNN are rectified.

8) From (23)-(26) of reference [6], the weights of BPNN are rectified.

9) Let $k = k + 1$, then return 4).

3 Hydro-turbine Governing System Models

The hydro-turbine governing system includes conduit system, turbine, governor, electro-hydraulic servo system and generator etc. [14]. The structure of the governing system is shown in Fig. 5.

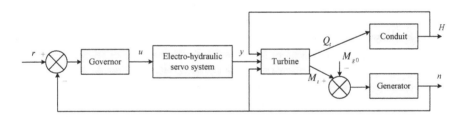

Fig. 5. The structure of hydro-turbine governing system

3.1 Model of Turbine

Here, the Francis turbine is taken as example and the flow and torque equations are described as:

$$\begin{cases} M_t = M_t(H, n, \alpha) \\ Q_t = Q_t(H, n, \alpha) \end{cases} \tag{5}$$

Where Q_t is flow rate, M_t is turbine torque, H is hydro-turbine water head, α is gate opening, n is hydro-turbine speed. When the parameters of turbine vary in the small range at a stable operating point, the above two equations can be linearized as:

$$\begin{cases} m_t = e_x x + e_y y + e_h h \\ q_t = e_{qx} x + e_{qy} y + e_{qh} h \end{cases} \tag{6}$$

Where m_t is the torque deviation of turbine, q_t is the flow deviation of turbine, y is gate opening deviation, x is hydro-turbine speed deviation, and h is the hydro-turbine water head deviation. $e_x = \dfrac{\partial m_t}{\partial x}$ is partial derivative of the torque to speed of turbine.

$e_y = \dfrac{\partial m_t}{\partial y}$ is partial derivative of the torque to gate opening. $e_h = \dfrac{\partial m_t}{\partial h}$ is partial

derivate of the torque to the water head of turbine. $e_{qx} = \dfrac{\partial q}{\partial x}$ is partial derivative of

the flow to speed of turbine. $e_{qy} = \dfrac{\partial q}{\partial y}$ is partial derivative of the flow to gate

opening. $e_{qh} = \dfrac{\partial q}{\partial h}$ is partial derivative of the flow to the water head of turbine. The plant time-vary parameters result in the transmission coefficients of turbine $(e_x, e_y, e_h, e_{qx}, e_{qy}, e_{qh})$ variety with the operating situation [7].

3.2 Model of Conduit System

The conduit system may be modeled as:

$$h = -T_w \frac{dq_t}{dt} \tag{7}$$

Where T_w is water inertia time constant, given as:

$$T_w = \frac{LQ_r}{gFH_r} \tag{8}$$

Where L and F are the length and cross-sectional area of the conduit, H_r and Q_r are the per-unit base values of the water column head and that of the water flow rate, respectively, g is the gravitational acceleration. The water inertia increases with T_w increasing. $\dfrac{dq_t}{dt}$ is derivative of flow to time.

3.3 Model of Generator and Load System

The mathematical model of generator and load is given [15] as:

$$(T_a + T_b)\frac{dx}{dt} + e_g x = m_t - m_{g0} \qquad (9)$$

Where T_a is unit inertia time constant, T_b is load inertia time constant, m_t is turbine torque, m_{g0} is load torque at rate frequency. $e_g = \dfrac{dm_g}{dx}$ is derivative of load torque to speed.

3.4 Model of Electro-hydraulic Servo System

The servo system model may be given as:

$$T_y \frac{dy}{dt} + y = u \qquad (10)$$

Where T_y is servo system time constant, and u is the output of governor.

In practice, many components of a plant model are precisely unknown. Furthermore, most plants are inherently non-linear, and can be approximated by linear models only in the near hood of the operating point. These and other factors lead to uncertainty. A GNN controller whose design is based on the models of governing system should provide stability and performance requirements in the presence of uncertainty.

4 Simulation and Results

Simulation studies have been performed on the governing system using the GNN PID controller.

Several simulations are run after changing a number of system parameters in order to ascertain the sensitivity of the control to these changes. Assuming that the power system was exposed to small changes in load during its normal operation, the model would adequately represent its dynamics. The parameters used in the simulation and their values [13] are: $T_y = 0.02, T_w = 1.27, T_r = 0.15, T_a = 9.06, e_x = -0.761, e_y = 1.190, e_h = 0.835,$ $e_{qx} = -0.163, e_{qy} = 0.930, e_{qh} = 0.359$

The load test is an important design criterion and a step load change may provide an indication about the system stability. A load change of 10% was applied to the governing system and the time-domain responses for classical PID controlled and genetic neural networks controller were shown in Fig.6 (a) and (b), respectively.

From the Fig.6 (a) with classical control, we can find the setting time is 22.9s and the overshoot is 2.64%. But in Fig.6 (b) with GNN control, the setting time is 12.7s and overshoot is 1.78%.

A 2Hz frequency disturbance is applied to the speed reference and the response curves are obtained in Fig.7 (a)-(b). The setting time is 17.8s and overshoot is 0.68% in Fig.7 (a), however, the setting time is 15.3 and overshoot is 0.42% in Fig.7 (b).

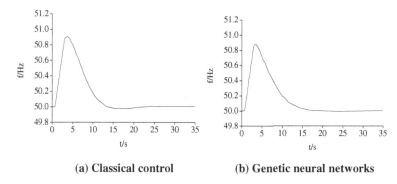

(a) Classical control   (b) Genetic neural networks

Fig. 6. The dynamic curves for 10% load rejection

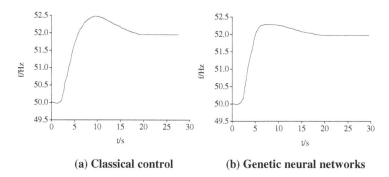

(a) Classical control   (b) Genetic neural networks

Fig. 7. The response for 2Hz frequency disturbance

From Fig.6 and Fig.7, we can observe that the response with the proposed controlled is less oscillatory and has a small frequency swing. The performance of the turbine governing system is improved by the novel controller.

5 Conclusion

Up to date, the PID controller has been applied to operate under much different conditions in hydro-turbine governor system, but tuning the parameters of PID is very difficult.

In this article, an efficient approach to the hydro-turbine governor is presented. Taking the dynamic and static performance as the ultimate goal, the PID controller based GA and NN improves the level of intelligent decision making and can adapt to various working requirements. GA can automatically tune the crossover probability and mutation probability and overcome the premature. When the disturbance is added, the initial parameters of RBFNN have deep affect on control system. A series of simulations for a hydro-turbine governor show that the control strategy not only achieves dynamic performances of the governing system under different run condition, but also has strong robustness. The satisfactory simulation results prove that the control method proposed in the paper is very useful and practical.

References

1. Tao, Y., Yin, Y., et al.: The New PID Control and Its Applications. China Machine Press, Beijing (1998)
2. Astrom, K.J., Hagglund, T.: PID controller: Theory, Design and Tuning. Instrument Society of America, Research Triangle Park, NC (1995)
3. Ye, L., Wei, S., Li, Z.: Field Tests and Operation of a Duplicate Multiprocessor-based governor for Water Turbine and Its Further Development. IEEE Trans. on Energy Conversion 5(2), 225–231 (1990)
4. Liang, H., Ye, L., Meng, A.: Self-adaptive Fuzzy-PID Controller and Its Application to the control system of Hydroturbine Unit. Hydropower Automation and Dam Monitoring 27(6), 26–29 (2003)
5. Wang, Y., Shang, G., Li, Y.: The Adaptive PID Control of the Hydraulic Turbine Regulation System on the Neural Networks. Journal of Hydroelectric Engineering 1, 93–99 (1998)
6. Gao, J., Mou, X., Yang, J.: PID Controller Based on GA and Neural Network. Electric Machines and Control 8(2), 108–111 (2004)
7. Lansberry, J.E., Wozniak, L.: Optimal Hydrogenerator Governor Tuning with a Genetic Algorithm. IEEE Transactions on Energy Conversion 7(4), 623–630 (1992)
8. Li, Z., Xiao, D., He, S.: A Fuzzy Adaptive PID Controller Based on Neural Network. Control and Decision 11(3), 340–345 (1996)
9. Stefopoulos, G.K., Georgilakis, P.S., et al.: A Genetic Algorithm Solution to the Governor-Turbine Dynamic Model Identification in Multi-Machine Power Systems. In: Proceedings of the 44th IEEE Conference on Decision and Control, and the European Control Conference, pp. 1288–1294. IEEE Computer Society Press, Los Alamitos (2005)
10. Liu, J.: Advanced PID Control and MATLAB Simulation, 2nd edn. Publishing House of Electronics Industry, Beijing (2004)
11. Li, S.: Fuzzy Control, Neurocontrol and Intelligent Cybernetics. Harbin Institute of Technology Press, Harbin (1998)
12. Ding, Y.: Intelligent Computing: Theory, Technique and Application. Science Press, Beijing (2004)
13. Zhang, Z., Li, Z., Li, Z.: Fuzzy PID Control for Hydroelectricity Generating Unit. Hydropower Automation and Dam Monitoring 26(1), 34–36 (2002)
14. Shen, Z.: Governing of Hydroturbine, 3rd edn. China Water Publishing Press, Beijing (1998)
15. Hannett, L.N., Fardanesh, B.: Field Test to Validate Hydroturbine Governor Model Structure and Parameters. IEEE Trans. Power Systems 9, 1744–1751 (1994)

Adaptive Rate Selection Scheme Based on Intelligent Learning Algorithm in Wireless LANs

Rongbo Zhu[1] and Maode Ma[2]

[1] College of Computer Science, South-Central University for Nationalities, Minyuan Road 708, Wuhan 430073, China
rongbozhu@gmail.com
[2] School of Electrical and Electronic Engineering, Nanyang Technological University, Singapore 639798, Singapore
emdma@ntu.edu.sg

Abstract. Although IEEE 802.11 wireless local area networks (WLANs) physical layers (PHYs) support multiple transmission rates, it not specifies how and when to switch between the permitted rates. In this paper, a novel method is introduced for rate selection in IEEE 802.11 WLANs by proposing an intelligent learning algorithm, which is simply based on local acknowledgement frames to indicate the correct reception of the packet. Based on the intelligent learning algorithm, a rate selection scheme is also proposed, which aims to improve the system throughput by adapting the transmission rate to the current link condition. Numerical and simulation results show that the proposed intelligent learning algorithm is convergent quickly, and the rate selection scheme is closely approximates the ideal case with the perfect knowledge about the channel, which significantly improves quality of service (QoS) in terms of throughput and delay metrics.

1 Introduction

In recent years, much interest has been devoted to the design of IEEE 802.11 wireless local area networks (WLANs) [1]. IEEE 802.11 WLANs physical layers (PHYs) provide multiple transmission rates by employing different modulation and channel coding schemes. The original 802.11 standard specifies three low speed PHYs operating at 1 and 2 Mbps, and three high speed PHYs were additionally defined as supplements to the original standard: the 802.11b PHY [3] supporting four PHY rates up to 11 Mbps at the 2.4 GHz band, the 802.11a PHY [2] supporting eight PHY rates up to 54 Mbps at the 5 GHz band, and the 802. 11g PHY [4] supporting 12 PHY rates up to 54 Mbps at the 2.4 GHz band. Although IEEE 802.11 WLANs support multiple transmission rates, it only specifies which transmission rates are allowed for which types of medium access control (MAC) frames, but not how and when to switch between the permitted rates. Due to the conservative nature of the link adaptation schemes implemented in most 802.11 devices, the current 802.11 systems are likely to show low bandwidth utilization.

L. Kang, Y. Liu, and S. Zeng (Eds.): ISICA 2007, LNCS 4683, pp. 529–538, 2007.

In practice the network performance is degraded by the interference and time varying property of the wireless channel. In order to address this problem, there are two threads of research to improve the performance of link adaptation: Based on accurate channel estimation and based on local acknowledgement (ACK) information. For the first thread, the transmitter may estimate channel condition. Then the transmitter knows when the channel condition improves enough to accommodate a higher rate, and adapts its transmission rate accordingly. In [5], a link adaptation scheme was presented, which is based on channel estimation without requiring any standard change by utilizing receiver signal strength indicator measurements and the number of frame retransmissions. However, since such scheme operates under the assumption that all the transmission failures are due to channel errors, it does not perform well in multiuser environments, where many transmissions might fail due to collisions. The authors of [6] proposed a hybrid CSI (Channel State Information) rate control scheme, which is based on statistics approach under stable link conditions and based on SNR (Signal-to-Noise Rate) approach under volatile conditions to meet the strict latency requirements of streaming applications. Although the scheme is very simple and easy to implement in devices, it is a heuristic method. In [8], a MPDU-based (MAC Protocol Data Unit) link adaptation scheme was proposed based on the theoretical analysis, which pre-establishes a best PHY mode table by applying the dynamic programming technique. However, such approach requires extra implementation effort and modifications to the current 802.11 standard. The authors presented a joint channel adaptive rate control and randomized scheduling algorithm in [7], which is performed at the MAC layer whereas the rate selection takes place at the Physical/Link (PHY/LINK) layer. However it is just for high-speed downlink packet access networks, not suitable for WLANs. For the second thread, the approaches based on local ACK information. The authors described the ARF (Auto Rate Fallback) link adaptation scheme in [9], which alternates transmission rates based on the ACK counts. If two consecutive ACK frames are not received correctly by the transmitter, the second retry of the current frame and the subsequent transmissions are made at the lower rate 1 Mbps and a timer is started. The timer expires or the number of successfully received ACK frames reaches 10, the transmission rate is raised to 2 Mbps and the timer is canceled. It is obviously that such scheme cannot react quickly to fast wireless channel variations for it attempts increasing the transmission rate at a fixed frequency of every 10 consecutive successful frame transmissions. The authors of [10] proposed an enhancement scheme of the ARF to adaptively use a short probing interval and a long probing interval to deal with the fast-fading and slow-fading wireless channels. However, since the two probing intervals are heuristically set, this scheme works well only with certain patterns of the wireless channel variation. In [11], a VSVR (Variable Size and Variable Rate) scheme was proposed, which uses a goodput regulator to maintain the committed goodput for non-greedy applications and an optimal frame size predictor for maximizing the goodput for greedy applications. Although the scheme maximizes the goodput in a real-time manner, however it will increase rate by receiving one

ACK frame and decrease rate by not receiving one ACK. Such method is a heuristically method and will increase the computation burden.

From the above observations, we can see that for the approaches based on accurate channel estimation require extra implementation effort and modifications to the current 802.11 standard. On the other hand the approaches based on the local ACK information, which are simple but just heuristic methods and lack theoretical analysis. In this paper, we blaze a novel way for rate selection in IEEE 802.11 WLANs. We propose an intelligent learning algorithm based on the algorithm in [7], and a novel adaptive rate selection scheme is proposed which is able to adjust the transmission rate dynamically to the wireless channel variation. The proposed scheme is simply based on local ACK frames to indicate the correct reception of the packet and is compatible with the current IEEE 802.11 protocol. The key idea is to direct the transmitter's rate-selecting attempts in an adaptive learning manner.

The rest of this paper is organized as follows. Section 2 describes the details of the proposed intelligent learning algorithm for adaptive rate selection. Section 3 presents and assesses the simulation results. Finally, section 4 concludes the paper.

2   Proposed Adaptive Rate Selection Scheme

In the following analyse, we just consider the IEEE 802.11a PHY for convenience. Actually, the similar way can be used to analyse other PHY without much difficulty. In IEEE 802.11a PHY, eight different data transmission rates are supported by using M-ary QAM (Quadrature Amplitude Modulation). In order to reduce the error probability, convolutional coding and data interleaving techniques are used for forward error correction (FEC). There are three different code rates used in the standard.

2.1   Bit Error Rate over AWGN Channel

The symbol error probability $P(\gamma)$ with SNR γ for M-ary QAM is (M=4,16,64):

$$P(\gamma) = 1 - \{1 - [2 \cdot (1 - \frac{1}{\sqrt{M}}) \cdot Q(\sqrt{\frac{3}{M-1} \cdot \gamma})]\}^2. \tag{1}$$

With Gary coding, we can get the BER P_e for M-ary QAM:

$$P_e = P(\gamma) \cdot \frac{1}{\log_2 M}. \tag{2}$$

QPSK (Quadrature Phase Shift Keying) and 4-ary QAM are identical, and the BER for BPSK (Binary Phase Shift Keying) modulation is:

$$P_e = Q(\sqrt{2 \cdot \gamma}), \tag{3}$$

where the Q-function is the tail of the Gaussian distribution:

$$Q(x) = \int_x^\infty \frac{1}{\sqrt{2\pi}} e^{-y^2/2} dy. \tag{4}$$

Let P_R be the BER for IEEE 802.11a with transmission rate R, With FEC, under the assumption of binary convolutional coding and hard-decision Viterbi decoding with independent errors at the channel input the union bound for P_R is [12]:

$$P_R < \sum_{k=d_{free}}^\infty c_k P_k(\gamma), \tag{5}$$

where d_{free} is the free distance of the convolutional code, c_k is the total number of error events with k bit errors, and P_k is the probability that an incorrect path at distance k from the correct path being chosen by the Viterbi decoder.

$$P_k(\gamma) = \begin{cases} \frac{1}{2} \binom{k}{i} (P_e)^{k/2}(1-P_e)^{k/2} \\ \quad + \sum_{i=1+k/2}^k \binom{k}{i} (P_e)^i(1-P_e)^{k-i}, k \text{ is even} \\ \sum_{i=(1+k)/2}^k \binom{k}{i} (P_e)^i(1-P_e)^{k-i}, \quad k \text{ is old} \end{cases} \tag{6}$$

2.2   Proposed Intelligent Learning Algorithm

The proposed intelligent learning algorithm maintains a control probability vector to select an action among a set of actions at time. As introduced in [13], a good policy to update the probability vector is a pursuit algorithm that always rewards the action with the current minimum penalty estimate and that stochastic learning control performs well in speed of convergence. In this system, the probability vector is the rate selection probability vector $p(n) = [p_1(n), ..., p_K(n)]$, where n is the index of the sequence of transmission period. The SNR of the channel during nth transmission period is expressed by $\gamma(n)$. The set of rate available are $\{R_i : i = 1, 2, ..., K\}$. At beginning the $p(n)$ are assigned equal values:

$$p(0) = [1/K, ..., 1/K]. \tag{7}$$

So we can get the throughput approximatively:

$$S_i(n) = R_i[1 - P_{e,i}(\gamma(n))] \tag{8}$$

where $i \in [1, K]$.

In order to maximize the throughput, the transmitter is required to find the index of the best transmission rate. Such an approach requires the knowledge of channel state during each transmission period. The intelligent learning algorithm presented in this paper based on the algorithm in [7], randomly selects a transmission rate prior to transmission of a frame. The rate selection probability vector is altered by an iterative updating process, which maximizes the probability of assigning the best transmission rate. Then, the rate adaptation and transmission proceeds with the fixed $p(n)$ until every rate is selected at least M number of times after which $p(n)$ is augmented at each n. Following each transmission, the transmitter receives an ACK signal indicating the successful reception of the data packet. If no ACK or error ACK (NACK) is received, it indicates the packet transmission failure. The current and the past ACK signals are used in augmenting the probability vector $p(n)$ toward the optimum. This is done by maintaining a time-varying estimation of throughput values $S(n)'$ for each R_i. Following each transmission period, an update of $S(n)'$ and $p(n)$ are carried out considering the last M ACK signals of each rate. Then we can get [7]:

$$S(n)' = \frac{R_i}{M} \sum_{j=L_i(n)-M+1}^{L_i(n)} I_i(j), \qquad (9)$$

where $I_i(j)$ is an indicator function:

$$I_i(j) = \begin{cases} 1, if\ ACK\ is\ received \\ 0, else \end{cases} \qquad (10)$$

$L_i(n)$ is the number of transmission periods for which the rate R_i is selected during the time from the start till the nth transmission period.

In order to minimize the bias against estimation of throughput values $S(n)'$, a factor μ is introduced to smoothen the estimated values. Then the estimation of throughput values is estimated as:

$$\overline{S(n)'} = \mu S(n)' + (1 - \mu)S(n - 1), \qquad (11)$$

where $S(n-1)$ is the last actual throughput.

From (9), the intelligent learning algorithm finds the index m' of the estimated best rate $R'_m(n)$ maximizing the throughput $S'_m(n)$ at time n:

$$m' = \arg\max_i \{ R_i \sum_{k=L_i(n)-M+1}^{L_i(n)} I_i(k) \}. \qquad (12)$$

For the probability of making the right decision, let the best rate at time n, $R_m(n)$ be unique. Let $\phi_m(n)$ be the probability that the estimated best rate is the actual best rate. Then we can get:

$$\phi_m(n) = \Pr\{ \sum_{k=L_i(n)-M+1}^{L_i(n)} I_i(k) < \frac{R_m(n)}{R_i} \sum_{k=L_m(n)-M+1}^{L_m(n)} I_m(k) \forall i \neq m(n) \}. \qquad (13)$$

The above probability is readily obtained by using binomial probability distribution.

Since $M \geq \sum_{k=L_i(n)-M+1}^{L_i(n)} I_i(k) \geq 0$ for all $i \in [1, K]$, when $\sum_{k=L_i(n)-M+1}^{L_i(n)} I_i(k) >$ $\sum_{k=L_m(n)-M+1}^{L_m(n)} I_m(k)$, $\frac{R_m(n)}{R_i} \sum_{k=L_m(n)-M+1}^{L_m(n)} I_m(k)$ may exceed M. Let's take into account the fact that $\sum_{k=L_i(n)-M+1}^{L_i(n)} I_i(k) < \frac{R_m(n)}{R_i} \sum_{k=L_m(n)-M+1}^{L_m(n)} I_m(k)$ in such cases.

Let ε_i be the largest nonnegative integer less than $\alpha(R_m(n)/R_i)$, where α is a nonnegative integer. Define the indicator function $O(\cdot)$ which value is 1 when condition within parentheses is satisfied, else is 0. Define the parameter θ_i for $i \in [1, K]$:

$$\theta_i = \varepsilon_i \cdot O(\varepsilon_i \leq M) + M \cdot O(\varepsilon_i > M). \tag{14}$$

Then we have:

$$\phi_m(n) = \sum_{\alpha=1}^{M} [\Pr\{ \sum_{k=L_m(n)-M+1}^{L_m(n)} I_m(k) = \alpha\} \prod_{i=1, i \neq m}^{K} \sum_{\beta=0}^{\theta_i} \Pr\{ \sum_{k=L_i(n)-M+1}^{L_i(n)} I_i(k) = \alpha\}]. \tag{15}$$

Considering that:

$$\theta_i = M, \quad \varepsilon_i \geq M, \tag{16}$$

we can get:

$$\phi_m(n) = \sum_{\alpha=1}^{M} \binom{M}{\alpha} Q_m^\alpha (1-Q_m)^{M-\alpha} \cdot \prod_{i \neq m} \sum_{\beta=0}^{\theta_i} \binom{M}{\beta} Q_i^\beta (1-Q_i)^{M-\beta} \tag{17}$$

where Q_i is the probability of successful transmission of a packet using the rate R_i, which can be expressed as:

$$Q_i(\eta_m) = \int_{\gamma=0}^{\infty} [1 - P_{e,i}(\gamma) f(\gamma | \gamma \in \eta_m)] d\gamma \tag{18}$$

where $i \in [1, K]$, η_m is the range of γ for which the rate $R_m(n)$ is the optimum, $f(\gamma | \gamma \in \eta_m)$ is the probability density function of γ conditioned on $\gamma \in \eta_m$, and it can be calculated as:

$$f(\gamma | \gamma \in \eta_m) = f(\gamma)/(\int_{\gamma \in \eta_m} f(\gamma) d\gamma), \quad \gamma \in \eta_m \tag{19}$$

where $f(\gamma)$ is the unconditional probability density function of SNR γ and can be modelled with Rayleigh or other probability density function . $P_{e,i}(\gamma)$ can be get form (6).

The adaptive rate selection procedure can be summarized as follows:

Step 1. If it is the first transmission period, the station initialises the probability vector as in equation (7). Else the station selects a rate R_i $(i \in [1, K])$ according to probability distribution $p_i(n)$.

Step 2. Update $I_i(j)$ and $L_i(n)$ on receiving an ACK or NACK signal. Then update $S(n)'$ according to (11).

Step 3. If for all $L_i(n) \geq M$ for all i go to next step, else go to step 1.

Step 4. Detect the index m' of the estimated best rate and update according to the following equations:

$$p_i(n+1) = \begin{cases} p_i(n) - \Delta p, & i \neq m' \\ 1 - \sum\limits_{j=1,j\neq m'}^{K} p_i(n+1), & i = m' \end{cases} \tag{20}$$

where Δp is a tunable penalty probability parameter.

The convergence of the proposed algorithm can be deduced following the steps in [7].

3   Numerical and Simulation Results

The well-known simulation tool NS-2 [15] is used to validate the proposed scheme. All the parameters used in the simulation system are as in table 1. Instantiations of the fading channel were generated using the model in [14]. Two different transmission schemes are used for comparison with the proposed link adaptation system. The first scheme is constant rate. The second scheme is the well-known ARF [9] protocol.

Table 1. System parameters

Parameter	Value
Channel rate	6,9,12,18,24,36,48,54Mbps
PHY header	192 bits
ACK, CTS	112 bits + PHY header
RTS	160bits + PHY header
Slot time	9 us
SIFS	16us
DIFS	34us
Propagation Delay	1us
Packet payload	1000 bytes
RetryLimit	5
Δp	0.08

Shown in Fig. 1 is the trend of the probability of detecting the best rate as the selection time M, the number of ACK/NACK signals used in the estimation increases. This curve has been derived using (15) with the transmission rates about 9Mbps, 12Mbps, 18Mbps and 24Mbps at 15 dB respectively. It is clear that when M larger than 4, the probability of detecting the best rate larger than 0.95, which demonstrates that the intelligent learning algorithm performs well in speed of convergence. In the following scenarios, we set M to 4.

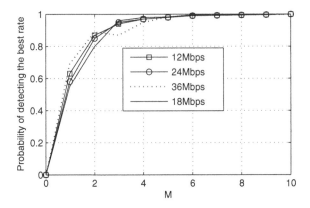

Fig. 1. Probability of detecting the best rate against M

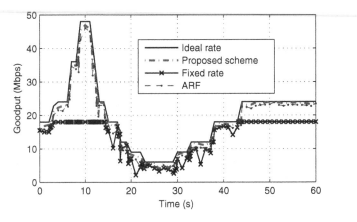

Fig. 2. Goodputs of different schemes against time

In the first scenario, there is only one mobile station and access point (AP) in the WLAN and the station moves with a speed of 0.3 m/s. Fig.2 shows the goodput of the ideal scheme, proposed scheme, fixed rate at 18 Mbps and the ARF scheme.

From Fig.2, it shows that the proposed rate selection scheme can fast adapt to the channel change. Though the channel is very good, the fixed rate can't make use of the good channel state in the period of 3 to 15 second and period of 43 to 60 second for it just fixed the transmission at 18 Mbps. Its goodput frequently goes down when the channel state change from good to bad, especially when the channel is bad. At 22s, its goodput decreases to 2.83 Mbps. For ARF scheme, it cannot adapt to the change of the channel state as fast as the proposed scheme. The average goodput for the proposed scheme is 18.73 Mbps, which is higher

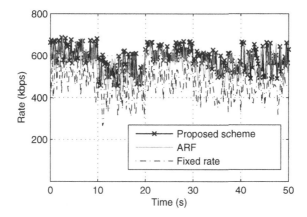

Fig. 3. Transmission rates of different schemes against time

than that of the ARF scheme about 0.68 Mbps, and higher than that of the fixed rate 2.47 Mbps. As a result, the proposed system has the best performance among all the schemes.

In the second scenario, there are 20 stations around an AP, and each station transmits a 600 kbps video traffic to the AP at a speed of 0.25 m/s in time-varying channel. We observe one of the stations state. Fig. 3 shows the transmission rates for different schemes of the selected station.

From Fig. 3, we can see that when the SNR decreases, the rate of the proposed scheme adaptively decreases quickly. Meanwhile the average rate of the proposed scheme is about 598.73 kbps, which is higher than those of fixed rate and ARF schemes about 121.64 kbps and 45.63 kbps respectively. Which demonstrates that the proposed scheme can adaptive alter rate and highly improve the goodput of the system in time. Note that in Fig. 3, the transmission rate achieves the requested rates within first few iterations and remains at the requested values. There is no effort by the algorithm to regulate the transmission rate of the best effort queue. Which also validates that the intelligent learning algorithm is effective.

4   Conclusions

In this paper, we proposed a rate selection scheme based on intelligent learning algorithm. Based on the proposed scheme we have studied the goodput performance in noise mobile environment in detail. Numerical and simulation results proved the validity of intelligent learning algorithm can fast detecting the best rate and help the rate selection scheme quickly adapt to different channel states and improve the whole system goodput. The proposed scheme also is compatible with the current IEEE 802.11 protocol and can be conveniently implemented in the devices.

Acknowledgment

This work was supported in part by Natural Science Foundation of South-Central University for Nationalities under Grant YZZ07006.

References

1. IEEE Std. 802.11-1999: PART 11: Wireless LAN Medium Access Control (MAC) and Physical Layer (PHY) specifications, IEEE Std.802.11, 1999 edn. (1999)
2. IEEE 802.11a WG: Wireless LAN Medium Access Control (MAC) and Physical Layer (PHY) specifications: High-speed Physical Layer in the 5Ghz band. IEEE 802.11a standard (2000)
3. IEEE 802.11b, Part 11: Wireless LAN Medium Access Control (MAC) and Physical Layer (PHY) Specifications: High-speed Physical Layer Extension in the 2.4 GHz Band. Supplement to IEEE 802.11 Standard (September 1999)
4. IEEE 802.11g, Part 11: Wireless LAN Medium Access Control (MAC) and Physical Layer (PHY) Specifications: Further Higher Data Rate Extension in the 2.4 GHz Band. Supplement to IEEE 802.11 Standard (June 2003)
5. del Prado Pavon, J., Choi, S.: Link Adaptation Strategy for IEEE 802.11 WLAN via Received Signal Strength Measurement. In: Proc. IEEE ICC'03, Anchorage, AK (May 2003)
6. Haratcherev, I., Tall, J.R., Langendoen, K., Lagendijk R., Sips, H.: Automatic IEEE 802.11 rate control for streaming applications. Wireless Communications and Mobile Computing 5(4), 421–437 (2005)
7. Haleem, M.A., Chandramouli, R.: Joint Adaptive Rate Control and Randomized Scheduling for Multimedia Wireless Systems. ICC'04 1, 1500–1504 (2004)
8. Qiao, D., Choi, S., Shin, K.G.: Goodput Analysis and Link Adaptation for IEEE 802.11a Wireless LANs. IEEE Trans. on Mobile Computing 1(4), 278–292 (2002)
9. Kamerman, A., Monteban, L.: WaveLAN-II: A High-Performance Wireless LAN for the Unlicensed Band. Bell Labs Technical Journal 118–133 (1997)
10. Chevillat, P., Jelitto, J., Barreto, A.N., Truong, H.: A Dynamic Link Adaptation Algorithm for IEEE 802.11a Wireless LANs. In: Proc. IEEE ICC'03, Anchorage, AK, pp. 1141–1145. IEEE Computer Society Press, Los Alamitos (2003)
11. Song., C.I., Hamid, S.: Improving goodput in IEEE 802.11 wireless LANs by using variable size and variable rate (VSVR) schemes. Wireless Communications and Mobile Computing 5, 329–342 (2005)
12. Ziemer, R.E., Peterson, R.L.: Introduction to Digital Communication, 2nd edn. Prentice Hall, Englewood Cliffs (2001)
13. Oommen, B.J., Lanctôt, J.K.: Discretized pursuit learning automata. IEEE Trans. Syst., Man, Cyebern. 20(4), 931–938 (1990)
14. Jakes, W.C.: Microwave Mobile Communications, Piscataway, NJ. IEEE Computer Society Press, Los Alamitos (1974)
15. Network Simulator, URL: http://www-mash.cs.berkeley.edu/ns

The Research on Generic Project Risk Element Network Transmission Parallel Computing Model*

Cunbin Li and Jianjun Wang

Institute of Business Management
North China Electric Power University, Beijing, China
lcb999@263.net, wangjianjunhd@gmail.com

Abstract. This paper constructs a risk element network(REN) transmission model of generic project. Firstly, input the risk samples probability distribution of the lowest level risk elements in the risk element network; Secondly, use the risk element transmission probability distribution matrixes to express the probability distribution relationship layer by layer, finally, the predict probability distribution values of last risk elements can be acquired from the parallel computing. And the model can back-propagation adjusts the probability distributions matrixes automatically as the BP artificial neural network, whose purpose is to make the predict output exactly. REN can solve the quantitative analysis calculation between the risk elements, and has parallel and high accurate characteristics. As the instance proved, the model is applied in generic project risk element network transmission theory.

Keywords: risk element; network computing; parallel computing; risk element transmission; project risk management.

1 Introduction

Because the existence of uncertainties in nature, society and economy, it is difficulty for people to carry out the projects and related analysis exactly. In order to form a scientific methodology, risk management has been incorporated in the rank of project management which is called project risk management[1,2,3,4]. the plan usually contains some forecasting properties because it is based on historical experiences and data, the implementation of projects can not been proceeded exactly according to the plan owing to the uncertainties factors happening in reality. Currently, how to quantitative analyze the project risks so that decreasing the discrepancy between the forecasting and actual values is a tough problem.

John Raftery[5], Stephen Ward and Chris Chapman[6] analyze risks in the project management. However, the researches of these literatures treat of the project risk attributes, the transmission between the risk factors is not considered. In comprehensive evaluation fields, Felix T.S[7] regarded the risk factors as a branch of evaluation factors in the global suppliers' evaluation, which means the influence of the risk factors in each element can not been expressed for the evaluating score exactly. XU[8] and Li

---

* Project Supported by National Natural Science Foundation of China(NSFC)(70572090).

L. Kang, Y. Liu, and S. Zeng (Eds.): ISICA 2007, LNCS 4683, pp. 539–546, 2007.

Dadong[9] use the interval number theory converting the risk factors' values into interval numbers, Similarly, Bellman[10] and Marchalleck[11] use fuzzy theory converting the risk factors into fuzzy memberships to calculate the overall goal's membership. These literatures consider the risk factors by uncertain theories in each element, but the transmission relationship between each element layers is not considered.

LI[12] presents generic project risk element transmission theory, along with its thought of and classifying the sub risk elements can infer that there should be some quantitative mathematic relations with the overall risk element and its sub risk elements. That means the transferring of risk element's uncertainties can result in the overall goal's variation in the event that the risk of the project goal or its quantitative express was obtained. Although Wang[13] and Ma[14] further the generic project risk element transmission theory, but they have not realize the transferring process in true mean. In order to make up their shortcomings, a risk element network transferring model was constructed referring to the theories of risk element. At first, input the risk samples probability distribution of the lowest level risk elements in risk element network(REN). Second, use the risk element transmission probability distribution matrixes to express the probability distribution relationship layer by layer, finally, the predict probability distribution values of last risk elements can be acquired from the parallel computing. And use the ANN-inspired, REN can back-propagation adjusts the probability distributions matrixes automatically by error between the each layer.

2   Summary of Risk Element Network Transmission Theory

The study of this paper is based on the following assumptions:

1. The influences of risk include both positive and negative sides
2. There should be a state transition process among the risk element where every transition level is called a risk element, and the range of affected results is limited discrete distributed
3. A risk element is a stochastic variable, which is expressed by R, with the sample space of $V_i(i = 1 \cdots n)$
4. The influence from a sample value V of risk element R to the sample space V_i of risk element in the next layer expresses as Figure 1.

The probability of R_i can be obtained by equation(1):

$$p(V_i) = p(V) * p(V_i|V) \tag{1}$$

$$where : \sum_{i=1}^{n} p(V_i|V) = 1$$

Based on the risk element theory above, the risk element network transferring illustration constructed is showed as figure 2,

where, $P(RI_i)(i = 1 \cdots n)$denotes as the probability distribution of risk element in the input level, $P(RO_j)(j = 1 \cdots s)$ is the probability distribution of risk element

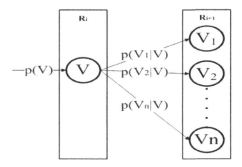

Fig. 1. The basic risk element transmission model

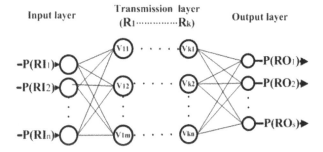

Fig. 2. An illustration of risk element transmission network

in the final output level, and $P(R_i)(i = 1 \cdots k)$ is the probability distribution of risk element in the middle transferring level. Each layer transmission pots is the probability values of the layer, which is expresses by V_{ij}. i expresses the layer, j expresses the number of the sample values, the probability matrix $P(R_i)_{mn}$ represents as the jointing relationship between two levels, which is shown as Equation 2,

$$p_{ij} = p(V_{nj}|V_{mi}) \qquad (2)$$

$$And : \sum_{j=1}^{n} p(V_{nj}|V_{mi}) = 1$$

in which, p_{ij} expressed the risk element conditional probability values m in i layer's influence to the risk element values n in j layer.

And the probability transmission matrix of risk element in each layer can be calculated as equation (3):

$$P(R_{i+1}) = P(RI) \prod_{k=1}^{i} P(R_k) \qquad (3)$$

3   The Learning Model of Risk Element Network Transferring Theory

It is hard to get the conditional matrix $P(R_i)$ in reality, thus the algorithm of BP ANN can be referred to determine the optimal $P(R_i)$ by training the risk element network by sample data training set. Selecting the prior probability of the training set as the probability of network transition matrix, and comparing the probability distribution of the calculating result by network with probability distribution of the overall goal in the training set can conclude that there should be a certain variance between the forecasting result and the training result, which can be calculated as:

$$D = \frac{1}{2}[\sum_{i=1}^{s}(P(RO)_i - P(O)_i)]^2 \tag{4}$$

In which, $P(O)$ denotes as the probability of the overall goal in the training set, and $P(RO)$ denotes as the probability of the calculating result by network. It can be clearly seen that the error is determined by prior layer probability matrix, Use ANN-inspired, a adjustment index ΔP_{kj} is defined to adjust the last layer.

$$\Delta P_{kj} = -\eta \frac{\partial D}{\partial P(RO)_i} V_i = -\eta [P(RO)_i - P(O)_i] V_i \tag{5}$$

where, $k = 1 \cdots n, i = 1 \cdots s$ and V_i expresses the sample space values of the last layer.

Based on adjustment variable, the adjusted probability matrix $P_1(RO) = P(RO) + \Delta P$ can be acquired:

Then, Normalized the $P_1(RO)$ by equation 6 can get $\tilde{P}(RO)$:

$$\tilde{P}(RO)_{ij} = \frac{P_1(RO)_{ij}}{\sum\limits_{j=1}^{s} P_1(RO)_{ij}} \tag{6}$$

Generally,

$$D_t = \frac{1}{2}[\sum_{i=1}^{m}\sum_{j=1}^{n}(P(R_t)_{ij} - P_1(R_t)_{ij})]^2 \tag{7}$$

where: $P_1(R_t)$ expresses the adjusted transmission probability matrix

$$\Delta P_{ij} = -\eta \frac{\partial D_t}{\partial P(R_t)_{ij}} V_{tj} = -\eta [P(R_t)_{ij} - P_1(R_t)_{ij}] V_{tj} \tag{8}$$

Similarly, use ΔP to adjust $P(R_t)$ can get $P_2(R_t)$,
Normalized, $\tilde{P}(R_t)$ is acquired

$$\tilde{P}(R_t)_{ij} = \frac{P_2(R_t)_{ij}}{\sum\limits_{j=1}^{n} P_2(R_t)_{ij}} \tag{9}$$

By equation (9), the feedback learning process can go on layer by layer started from the final layer still the ideal learning precision is obtained. Then a probability transmission matrix is acquired to predict , and it can be used test set to test.

4   Example

The prior probability transferring network of a construction project is showed as Figure 3, where: A,B,C,D denote as the different risk levels respectively, and H (High), M (Middle) and L (Low) denotes as the final forecasting risk level.

Table 1. The risk probability distribution of the final results

	H	M	L
Training Results	0.222	0.568	0.210
Actual Results	0.250	0.550	0.200

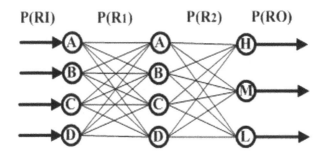

Fig. 3. Risk element transmission picture of a project

The prior probability transition matrix of the training set obtained from the historical data is showed as followings:$P(RI) = (0, 0.6, 0.4, 0)^T$

$$P(RX_1) = \begin{vmatrix} 1 & 0 & 0 & 0 \\ 0.1 & 0.9 & 0 & 0 \\ 0.05 & 0.25 & 0.7 & 0 \\ 0 & 0.1 & 0.3 & 0.6 \end{vmatrix} \quad P(RX_2) = \begin{vmatrix} 1 & 0 & 0 \\ 0.2 & 0.8 & 0 \\ 0.05 & 0.2 & 0.75 \\ 0 & 0.1 & 0.9 \end{vmatrix}$$

Then the risk probability distribution of the final results is showed as table 1 and Figure 4.

Defining $\eta = 0.1$ and precision is 0.0001, it can calculate the probability matrix $P_1(RX_1), P_1(RX_2)$,and training output results when the calculation is stable (table 2 and Figure 5).

$$P_1(RX_1) = \begin{vmatrix} 1 & 0 & 0 & 0 \\ 0.116 & 0.884 & 0 & 0 \\ 0.053 & 0.257 & 0.700 & 0 \\ 0 & 0.101 & 0.304 & 0.595 \end{vmatrix} \quad P_1(RX_2) = \begin{vmatrix} 1 & 0 & 0 \\ 0.217 & 0.783 & 0 \\ 0.079 & 0.193 & 0.728 \\ 0 & 0.008 & 0.992 \end{vmatrix}$$

Verifying it by the testing set $P(RI) = (0.1, 0.2, 0.5, 0.2)^T$ can obtain table 3 and Figure 6. As we seen, the testing results is very close to the actual results.

Table 2. A comparison of the learned training results with actual results

	H	M	L
Learned Training Results	0.25001	0.54906	0.200093
Actual Results	0.250	0.550	0.200

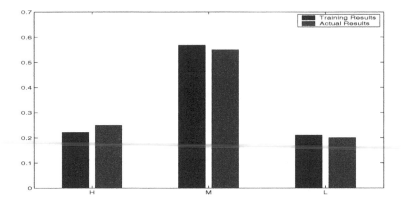

Fig. 4. A comparison between the training probability and actual probability

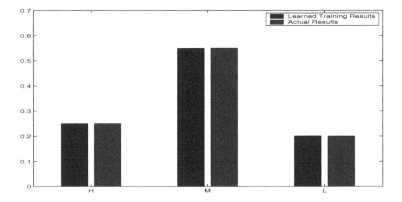

Fig. 5. A comparison between the learned probability and the actual probability

Table 3. A comparison of the verified testing results with the actual results

	H	M	L
Testing results	0.2524	0.3341	0.4135
Actual Results	0.250	0.333	0.417

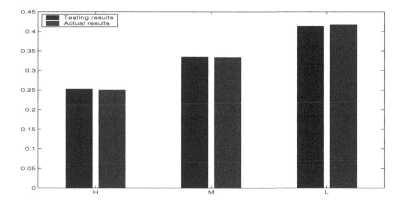

Fig. 6. A comparison between the verified testing probability and the actual probability

5   Conclusion

In this paper, the discrete risk element theory is introduced. Based on the theory, a risk element system is constructed and a self-adaptive risk network learning algorithm is brought out consulting on the BP neural network gradient descending method. In order to achieve the ideal forecasting precision, the deviation value of variance is used to modify the risk conditional probability matrix from the back level to the former level. Because the risk network model with the collateral and high precise properties introduced in this paper makes clear the analytic relationship between two risk elements, it can be not only applied in the research field of general project risk element transferring theories, but also in other risk network transferring calculating problems, it enriches the risk analysis computing.

References

1. Perry, J.G.: Risk management - An approach for project managers. International Journal of Project Management 4(4), 126–211 (1987)
2. Baker, R.W.: Handling uncertainty. International Journal of Project Management. 4(4), 205–210 (1986)
3. Antl, B.: Currentcy risk and the corporate. Economy Publications, London (1980)
4. Hughes, M.W.: Why project fail: The effects of ignoring the obvious. Industrial Engineering 18(4), 14–18 (1986)
5. Raftery, J.: Risk Analysis in Project Management. E & FNSPON (1994)
6. Ward, S., Chapman, C.: Transforming project risk management into project uncertainty management. International Journal of Project Management 21, 97–105 (2003)
7. Chan, F.T.S., Kumar, N.: Global supplier development considering risk factors using fuzzy extended AHP-based approach. Omega 35, 417–431 (2007)
8. Xu, Z.-s., Sun, Z.-d.: Priority method for a kind of multi-attribute decision-making problems. Journal of Management Sciences in CHINA 5(3), 35–39 (2002)
9. Li, D.: Interval Number and its Application:Master Degree Thesis. Southwest Jiaotong University (2004)

10. Bellman, R.E., Zadeh, L.: A.Decision making in a fuzzy environment. Management Science 17, 141–164 (1970)
11. Marchalleck, N., Kande, A.: Fuzzy logic applications in transportation Systems. International Journal on Artificial Intelligence Tools 4(3), 413–432 (1995)
12. Li, C.B.: A new thought of common generalized project's risk analytic theory model. Technology Economics. 7, 56–58 (2004)
13. Wang, J., Ma, T.: The applying of the project risk based on data mining. In: The ninth CSEE Youth Symposium, Article 3–39 (2006)
14. Ma, T.T., Li, C.B.: The applications of risk analysis in the comprehensive evaluation of construction projects. In: Proceedings of CRIOCM 2006 International Research Symposium on Advancement of Construction Management and Real Estate (November 2006)

On the Performance of Metamodel Assisted MOEA/D

Wudong Liu[1], Qingfu Zhang[1], Edward Tsang[1], Cao Liu[2],
and Botond Virginas[3]

[1] Department of Computer Science, University of Essex, UK
[2] Faculty of Computer Science, China University of Geosciences, China
[3] BT Research Laboratories, UK

Abstract. MOEA/D is a novel and successful Multi-Objective Evolutionary Algorithms(MOEA) which utilises the idea of problem decomposition to tackle the complexity from multiple objectives. It shows better performance than most nowadays mainstream MOEA methods in various test problems, especially on the quality of solution's distribution in the Pareto set. This paper aims to bring the strength of metamodel into MOEA/D to help the solving of expensive black-box multi-objective problems. Gaussian Random Field Metamodel(GRFM) is chosen as the approximation method. The performance is analysed and compared on several test problems, which shows a promising perspective on this method.

1 Introduction

Multi-Objective Evolutionary Algorithms(MOEA) is an important stochastic search method for Multi-Objective Optimization(MOO), gaining more and more attention from both research and real application fields. However, when applying to real problem, one main defect of most MOEA methods is that they need great amount of function evaluations before find a good Pareto set, while it is not uncommon to see a black-box and expensive function(or functions) in real-life application solving. Here, we say a function to be expensive and black-box means the function has no analytic form; function value can only be obtained by evaluation; and it needs lots of resource(time, computation power, etc.) for that evaluation. With every evaluation costs a lot, MOEA, which normally need thousands of, cannot be directly applied in those applications.

The hardness of expensive block-box function had been confronted and well handled in many fields. The essential idea behind is to use an approximated cheap function to estimate the real function value, thus reduce the pricy evaluations. The concept was widely adopted in fields such as global optimization[9], experimental design[16] and the large varieties of Response Surface Method[16,3], when facing expensive black-box functions. Many researches[4,8,23,20,18] had also been done in Evolutionary Computation(EC) for single objective problem(seeing [7] for a good survey on this topic) on this matter. However, because of the difference between single and multiple objectives problem solving, those

L. Kang, Y. Liu, and S. Zeng (Eds.): ISICA 2007, LNCS 4683, pp. 547–557, 2007.
© Springer-Verlag Berlin Heidelberg 2007

methods cannot be readily applied in MOEA. Recent researches on MOEA with metamodel assistance also appeared, notably in [11] and [5], which we think represent two kinds of methods to transform metamodel assistance from single objective to multiple objectives.

The first kind is to adapt the metamodel itself. In [5], the underlying statistics model of Gaussian Random Field Metamodel(GRFM) was extended from single objective to multi-objectives. Various screening methods from global optimization were naturally augmented to multi-objective case. Then the individual passing through the screening was evaluated. Another kind of approach decomposes the Multi-Objective Problem(MOP) into Single Objective Problems(SOPs) using weighted-sum[14] or Tchebycheff[14] aggregation function. As those aggregation functions' optimal are also Pareto solutions(and vice verse), this approach can make use of the existed large amount of algorithms and techniques from single objective optimization theory. In [11], a method called ParEGO was introduced. In every iteration of ParEGO, it uniformly randomly picks a aggregation weight and then using an augmented Tchebycheff to form a singular function from the multiple objective functions. Then EGO[10], a global optimization method based on GRFM, was used to approximate the singular function based on the existed evaluation point set and find the next evaluation point. Currently no comparison has been done on the performance of the above two approaches. As we also take the second approach in this paper, we will put our attention on the comparison between our method and ParEGO.

MOEA/D was first proposed and developed by Zhang and Li in [24]. It explicitly made use the concept of decomposition to simplify a MOP into a serial of subproblems in forms of SOP, using weighted-sum, Tchebycheff aggregation or other more advanced decomposing techniques such as Penalty-based boundary intersection(PBI)[2]. It exploited the benefit of that decomposition further by a neighbourhood relationship defined on the subproblems. Subproblems in a neighbourhood can exchange information to accelerate the convergence. It exhibited better performance on most standard test problems. However, no research have been done on the performance of MOEA/D with metamodel assistance on the black-box expensive functions. This paper will show that MOEA/D is so flexible that it still works very well in that situation. In our experiments, we also choose to use GRFM for the purpose of function approximation.

The paper is organised as follow: section 2.1 and section 2.2 give some description of the background on MOEA/D and the metamodel GRFM we used; section 3 details the algorithm, then some experiment results are given in section 4. We conclude the paper in section 5 and give some the future research directions on this matter.

2   Background

2.1   MOEA/D

Most MOEAs do not directly involve in decomposition of the problem; they normally treat a MOP as a whole. In scalar optimization, all solutions can be

---

Algorithm 1. MOEA/D pseudo code

$weights[]$ ▷ The subproblem weights
$neighbours[][]$ ▷ The neighbourhood table
$pop[]$ ▷ The main population
P, N, D ▷ The population size, neighbourhood size and dimension size respectively

procedure MOEAD($objs$) ▷ The main routine
    INITIALIZE
    while Not Terminated **do**
        for $i = 1$ to m **do**
            $ind \leftarrow$ GENETICOP(i) ▷ Generate new individual
            EVALUATE(ind) ▷ Evaluate the new indivisual
            UPDATE(i, ind) ▷ Update neighbourhood
        end for
    end while
end procedure

---

compared based on the single objective function. However, that is not the case in MOP, as domination does not define a complete order among the solutions in objective space. To use the readily made techniques for scalar objective optimization, many MOEAs focus on how to sort the solutions in objective space in a sensible order, and assign fitness according to that order to solutions. They normally aims to generate the whole Pareto front(PF).

MOEA/D took a different approach. On the consideration that in normal case, only a limited number of solutions are needed for decision making, MOEA/D aimed to solve only a subset of all the possible subproblems, which deemed to be more computational efficient and practical. A set of weights were fixed ahead, with each weight corresponding to one subproblem defined by weight-sum, Tchebycheff or any other decomposition approach. Then the neighbourhood of every subproblem was constructed based on the distance between weights. Later, all the new solution generation and updating were performed on the neighbourhood. The rationale behind the neighbourhood is: for closer weights with a less distance, they have a larger chances to share similar good solutions. The using of neighbour greatly sped up the convergence to subproblems' optimal.

Those functions not defined in the pseudo code are implemented straightly forward and omitted from it to save the page. More detail on this method can be found in [24]. This algorithms worked so well that it outperformed the mainstream algorithm NSGA2 on a majority of test problems, as reported in [24].

2.2 Gaussian Random Field Metamodeling

GRFM has been long recognised as a powerful framework to build metamodel for arbitrary unknown functions, especially those are simulated by computer experiment[21]. It's essentially a statistics method based on Bayesian reasoning. It have been widely used[19,12,23,1] in Evolutionary Computation for metamodel assistance. Both the papers we mentioned in the introduction used this method.

Algorithm 2. MOEA/D pseudo code continue

---

procedure INITIALIZE
    $weights[] \leftarrow$ WEIGHTGEN(P, D)
    $neighbours[][] \leftarrow$ NEIGHBOURGEN$(weights)$
    $pop[] \leftarrow$ INITPOP(m)
    for $i = 1$ to P **do**
        EVALUATE$(pop[i])$; UPDATE$(i, pop[i])$
    end for
end procedure
procedure UPDATE$(index, ind)$
    for $i = 1$ to N **do**   ▷ Update the neighbour's solution
        $nindex \leftarrow neighbour[index][i]$; $nweight \leftarrow weights[nindex]$
        $newval \leftarrow$ SCALARPROBLEM$(nweight, ind)$
        $oldval \leftarrow$ SCALARPROBLEM$(nweight, pop[nindex])$
        if $newval \leq oldval$ **then**
            $pop[nindex] \leftarrow ind$
        end if
    end for
end procedure
procedure GENETICOP$(index)$
    $neighbour[] \leftarrow neighbours[index]$
    for $i = 1$ to N **do**
        $j \leftarrow$ RANDOM(neighbour); $k \leftarrow$ RANDOM(neighbour)
        $ind \leftarrow$ CROSSOVER$(pop[j], pop[k])$
        $ind \leftarrow$ MUTATE(ind)
        return ind
    end for
end procedure

In our implementation, a simplified version is used. We are not intended to gives the full derivation of the formula, but just gives the result here. To see more detail on this, please refer to[10,22].

Formally speaking, for an unknown function f, given the training set $T = \{\langle \mathbf{x}_i, f(\mathbf{x}_i) \rangle \mid i = 1..n, \mathbf{x}_i \in D \subseteq R^m\}$ of size n, to predict the function value at a new point $y^* = f(\mathbf{x}^*), \mathbf{x}^* \in D$, the following formulae are used to estimate the value of y^*:

$$\hat{y} = \hat{\mu} + \mathbf{r}'\mathbf{R}^{-1}(\mathbf{y} - \mathbf{1}\hat{\mu}) \tag{1}$$

In the above equation, $\mathbf{y}$ is the vector of the training point's value: $y = \{f(\mathbf{x}_1), ..., f(\mathbf{x}_n)\}$; $\hat{\mu}$ is the expected value of the estimated value, which will be given below shortly, $\mathbf{r}$ and $\mathbf{R}$ are the so called correlation vector and matrix respectively, their elements are computed as:

$$R(i, j) = \exp(-d(\mathbf{x}_i, \mathbf{x}_j)); \ r(i) = \exp(-d(\mathbf{x}_i, \mathbf{x}^*))$$

where $d(\mathbf{x}_i, \mathbf{x}_j)$ is the distance function:

$$d\left(\mathbf{x}_i, \mathbf{x}_j\right) = \sum_{h=1}^{m} \theta_h \left|x_{ih} - x_{jh}\right|^{p_h} ; \theta_h \geq 0 \wedge p_h \in [1, 2] \tag{2}$$

where x_{ih} is the hth element of vector $\mathbf{x}_i$; θ_h and p_h are the hyper parameters that need to be computed first; and that is actually where the learning procedure of Gaussian Process occurs. They are computed by maximise the following likelihood function:

$$\frac{1}{(2\pi)^{\frac{n}{2}} (\hat{\sigma})^n |\mathbf{R}|^{\frac{1}{2}}} \exp\left(-\frac{(\mathbf{y} - \mathbf{1}\hat{\mu})' \mathbf{R}^{-1} (\mathbf{y} - \mathbf{1}\hat{\mu})}{2\hat{\sigma}^2}\right) \tag{3}$$

The $\hat{\mu}$ and $\hat{\sigma}$ are the expectation and standard deviation separately, which are obtained also by maximise the above likelihood when treat θ and ρ as constant:

$$\hat{\mu} = \frac{\mathbf{1}'\mathbf{R}^{-1}\mathbf{y}}{\mathbf{1}'\mathbf{R}^{-1}\mathbf{1}} ; \hat{\sigma} = \frac{(\mathbf{y} - \mathbf{1}\hat{\mu})' \mathbf{R}^{-1} (\mathbf{y} - \mathbf{1}\hat{\mu})}{n}$$

When substituting the above equations into the equation 3, we can get the so-called "concentrated likelihood function", which depends only upon the parameter θ_h and p_h. After maximise this likelihood function, we can get the value of θ_h and p_h, then put them back in the equation 1, we can get an approximation of y^*, namely $\hat{y}$.

One key feature of Gaussian prediction is that, not only a prediction is given, the estimation of the accuracy of this prediction is also given by:

$$s^2 = \sigma^2 \left(1 - \mathbf{r}'\mathbf{R}^{-1}\mathbf{r} + \frac{\left(1 - \mathbf{1}'\mathbf{R}^{-1}\mathbf{r}\right)^2}{\mathbf{1}'\mathbf{R}^{-1}\mathbf{r}}\right) \tag{4}$$

In this way, the prediction of $f\left(\mathbf{x}^*\right)$ is estimated to follow a normal distribution: $y^* \sim N\left(\hat{y}, s^2\right)$. The additional deviation information is of great value as it provides guidance for the search of the solution space, which will be seen in the next section.

3   MOEA/D with GP Model Assistant

If we want to solve a black-box expensive MOP, the target is to limit the number of evaluations as much as possible while can still get a reasonable closeness and distribution of the Pareto front. Here, we present an analysis on the performance of MOEA/D with GP model assistance.

We take a traditional way to do the job, i.e., using model as a prescreening mechanism[5]. Similar to ParEGO, at every iteration, MOEA/D runs on a learnt model, then the best result is selected for evaluation. However, our approach is rather different than ParEGO's. In ParEGO, at every iteration, a GP model is constructed for one selected scalar subproblem. However, as in MOEA/D, we maintain a set of subproblems, it may not be appropriate or practical to construct GP models for every subproblem when the normal size of subproblems ranges from 20 to 100.

Taken the above consideration, we model every objective function instead, and construct the model for the subproblem model based on the objective models. Given a MOP with objective function $\mathbf{f} = \langle f_1, .. f_n \rangle$, we construct the GP model at every generation as $\hat{\mathbf{f}} = \langle \hat{f}_1, .. \hat{f}_n \rangle$, with every $\hat{f}_i$ is estimated to follow normal distribution $\hat{f}_i \sim \mathbf{N}\left(\hat{y}_i, s_i^2\right)$. Then in weight-sum decomposition, the scalar subproblem corresponding to weight $\mathbf{w} = \langle w_1, .., w_n \rangle$ of form $sf = \sum_{i=1}^{n} w_i f_i$ can be estimated as:

$$\hat{sf} \sim \mathbf{N}\left(\sum_{i=1}^{n} w_i \hat{y}_i, \sum_{i=1}^{n} (w_i s_i)^2\right) \tag{5}$$

Also, as suggested in [9], the optimal of the model cannot be directly evaluated because of false convergence or local optimal. There must be some balance between exploit and explore. Thus, as in ParEGO, the concept of Expected Improvement[15] for the normal distribution is also considered here. Formally speaking, in the training set T, if the minimal value for f is f_{min}, then for every other point $\mathbf{x}$, which $f(\mathbf{x})$ is estimated to have a mean $\hat{y}$ and standard deviation s(just as given by the GP model), the *Expected Improvement* over $f_{min}$is defined as[9]:

$$EI(\mathbf{x}) = E\left(\max\left(f(x) - f_{min}, 0\right)\right) = s\left(u\Phi(u) + \phi(u)\right) \tag{6}$$

where $u = \frac{f_{min} - \hat{y}}{s}$, Φ and ϕ are the normal cumulative distribution function and density function respectively. Thus for every maintained scalar function, we need to find a point that maximise the formula 6; The objective is changed from finding its minimal to finding the maximal expected improvement. This should be reflected in MOEA/D when updating the main population. If we maintain n subproblems, at every iteration we could get n points that could act as candidates for real evaluation. To choose one from them, we simply choose the point that has a maximal sum of all the EI over all the subproblems.

As a good solution from an iteration of MOEA/D could possible still be a good solution in the next iteration, a good portion of main population of MOEA/D should be kept between iterations. In our implementation, a k-mean cluster[13] algorithm is used. The number of k is chosen to be $S/5$(the size of main population divided by 5). Then, from each cluster, the best individual that is closet to the cluster's centre is reserved. All the other individuals in the main population are randomised. The cluster technique proved to work well in our settings.

Algorithm 3. MOEA/D with GP model

$evalpop[]$ ▷ The already evaluated points
$gpmodel[]$ ▷ The GP models learnt from the evaluated points

procedure GPMOEAD
    INITGPMODEL
    while Not Terminated **do**
        MOEAD($gpmodel$)
        $ind \leftarrow \max_{ind \in mainpop}(\sum_{sf} EI(ind, sf))$
        EVALUATE(ind)
        $evalpop[] \leftarrow evalpop[] \cup ind$
        UPDATEGP
        CLUSTERPOP($mainpop, S/5$)
    end while
end procedure
procedure UPDATEGP
    for $i = 1$ to d **do**
        $gpmodel[i] \leftarrow$ GP($evalpop, obj[i]$)
    end for
end procedure

After taken the above decisions, all other things are rather straight forward. Thus we could have algorithm 3.

4 Experiment and Comparison

This section deals with experiments on several standard test problems. Even thought this test problems are not black-box expensive, we limit the number of evaluation to 100, 150, or 250, which makes them "expensive" essentially. On every problem, we compared three settings of test: MOEA/D without evaluation limit, MOEA/D with GP assistant, and ParEGO. The test problem chosen are ZDT1(150), KNO1(100), OKA1(100); the number in parenthesis indicates the evaluation limit.

Two metrics are chosen for measurement of the quality of result: D-metric and I-metric. D-metric measure the distance from the Pareto Front(PF) and is calculated as: $D(A, P^*) = \frac{\sum_{v \in P^*} d(v, A)}{|P^*|}$, where A is the approximation to the PF, P^* is the set of uniformly distributed points along the PF, and $d(v, A)$ is the minimal Euclidean distance between v and A. While $|P^*|$ is large enough, D-metric measure both diversity and convergence in a sense. I-metric measure the hypervolume difference between the PF and the approximation set, it's calculated as: $I(A, P^*, R) = H(P^*, R) - H(A, R)$, where H is the hypervolume indicator, or S-metric defined in [25], measures how much volume dominated between the set and R(a reference point). Both D and I are the smaller, the better. All the problems are tested for 10 times and the average and deviation are computed on D and I. The experiment result are listed in table 1.

Table 1. D and I Metric

instance	MOEA/D GP		ParEGO		NSGA2	
	D	I	D	I	D	I
ZDT1	0.01(6.61e-4)	0.02(0.03)	-	-	-	-
KNO1	0.54(0.12)	21.25(5.70)	1.66(0.31)	47.51(8.32)	1.70(0.54)	53.80(16.62)
OKA1	1.43(1.06)	5.24(4.75)	0.65(0.05)	6.10(0.55)	0.95(0.15)	13.64(2.77)

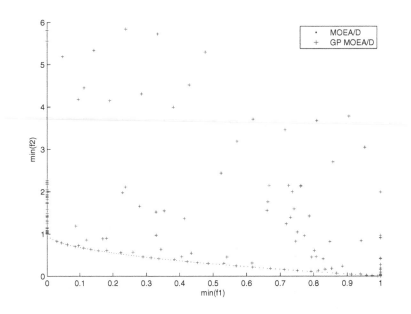

Fig. 1. ZDT1 with 4 parameters, 150 evaluations

Experiment shows MOEA/D GP works extremely well on ZDT1(4 parameters), not only in the metric D and I, but in the figure. On KNO1, MOEA/D GP works also very well, comparing to ParEGO and NSGA2. On OKA1, both the metric cannot show exactly which is better(MOEA/D GP is better in I but worse in D). The reason may lie behind the complexity of OKA1 itself[17], which make it very hard for any algorithm to approximate with mere 100 evaluations.

The experiments demonstrate that on the selected test problems, MOEA/D GP exhibits good and consistent performances that at least no worse than ParEGO and NSGA2 on the black-box and expensive condition. To be noted that, as most settings used in this initial experiments are rather primitive and reserved, MOEA/D GP should gain better result after refinement such as normalisation, advanced decomposition techniques(both was taken by ParEGO), etc.

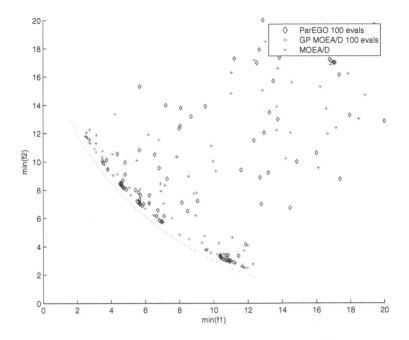

Fig. 2. KNO1 with 100 evaluation

5   Conclusion and Future Study

The paper shows some initial result from the experimenting of a new method to solve black-box expensive multi-objective problems using the newly introduced algorithm MOEA/D.

Although MOEA/D GP's performance was good on the selected test functions, the research is only in primitive and tentative status, which at least can be improved in the following way:

- Advanced decomposition techniques: MOEA/D GP does not work well on concave problem such as VLMOP2 due to the used weight-sum decomposition. The choosing of weight-sum here is just for simplicity because the distribution as solved in equation 5 is hard (but not impossible) to solve for Tchebycheff.
- Alternative exploration techniques: in recent research of global optimization as surveyed in [9], new techniques to replace Expected Improvement have been introduced and exhibited better performance. In [6], augmented EI was proposed. Those advance in single objective optimization can be readily borrowed in the framework of MOEA/D and better performance could be expected in MOEA/D GP.

- Better selection strategy: when choose the next point for evaluation, a simple strategy is applied to choose the best sum of EI over all subproblem. However, as every subproblem has its own improvement measurement, the strategy may not works well especially when no normalisation is done beforehand. Some other smarter strategies can be imaged to speed up the convergence, such as to maximise the best improvement percentage.
- Dynamic weights control: the weights used to decompose a MOP in MOEA/D were fixed, which made it difficult to generate an evenly distributed Pareto set when the weights set size is limited for some known test problem(OKA1). Some dynamic control mechanism of weights may help solve this problem after observing the population's distribution from generation to generation.

The future work will continue on the directions suggested above, and try it on some realistic problems such as BT's multi-objective workforce scheduling problem.

References

1. Buche, D., Schraudolph, N., Koumoutsakos, P.: Accelerating evolutionary algorithms with gaussian process fitness function models. IEEE Trans. Syst., Man, Cybern. C 35(2), 184–194 (2005)
2. Das, I., Dennis, J.E.: Normal-boundary intersection: A new method for generating pareto optimal points in multicriteria optimization problems. SIAM J. Optim. 8(3), 631–657 (1998)
3. Donnelly, T.A.: Response-surface experimental design. Potentials, IEEE 11(1), 19–21 (1992)
4. El-Beltagy, M.A., Keane, A.J.: Evolutionary optimization for computationally expensive problems using gaussian processes. In: Proc. Int. Conf. on Artificial Intelligence ICAI 2001, Las Vegas, pp. 708–714 (2001)
5. Emmerich, M.T.M., Giannakoglou, K.C., Naujoks, B.: Single- and multiobjective evolutionary optimization assisted by gaussian random field metamodels. Evolutionary Computation, IEEE Transactions on 10(4), 421–439 (2006)
6. Huang, D., Allen, T.T., Notz, W.I., Zeng, N.: Global optimization of stochastic black-box systems via sequential kriging meta-models. Journal of Global Optimization 34(3), 441–466 (2006)
7. Jin, Y.: A comprehensive survey of fitness approximation in evolutionary computation. Soft Computing-A Fusion of Foundations, Methodologies and Applications 9(1), 3–12 (2005)
8. Jin, Y., Olhofer, M., Sendhoff, B.: A framework for evolutionary optimization with approximate fitness functions. Evolutionary Computation, IEEE Transactions on 6(5), 481–494 (2002)
9. Jones, D.R.: A taxonomy of global optimization methods based on response surfaces. Journal of Global Optimization 21(4), 345–383 (2001)
10. Jones, D.R., Schonlau, M., Welch, W.J.: Efficient global optimization of expensive black-box functions. Journal of Global Optimization 13(4), 455–492 (1998)
11. Knowles, J.: Parego: a hybrid algorithm with on-line landscape approximation for expensive multiobjective optimization problems. Evolutionary Computation, IEEE Transactions on 10(1), 50–66 (2006)

12. Liang, K.H., Yao, X., Newton, C.: Evolutionary search of approximated n-dimensional landscapes. International Journal of Knowledge-Based Intelligent Engineering Systems 4(3), 172–183 (2000)
13. MacQueen, J.: Some methods for classification and analysis of multivariate observations. In: LeCam, L.M., Neyman, J. (eds.) Proc. of the 5th Berkeley Symp. on Mathematics Statistics and Probability (1967)
14. Miettinen, K.: Nonlinear Multiobjective Optimization. Kluwer Academic Publishers, Norwell, MA (1999)
15. Mockus, J.: Bayesian approach to global optimization. Kluwer, Dordrecht (1989)
16. Montgomery, D.C.: Design and Analysis of Experiments, 2nd edn. John Wiley and Sons, Chichester (1984)
17. Okabe, T., Jin, Y., Olhofer, M., Sendhoff, B.: On test functions for evolutionary multi-objective optimization. In: Yao, X., Burke, E.K., Lozano, J.A., Smith, J., Merelo-Guervós, J.J., Bullinaria, J.A., Rowe, J.E., Tiňo, P., Kabán, A., Schwefel, H.-P. (eds.) Parallel Problem Solving from Nature - PPSN VIII. LNCS, vol. 3242, pp. 792–802. Springer, Heidelberg (2004)
18. Paenke, I., Branke, J., Jin, Y.: Efficient search for robust solutions by means of evolutionary algorithms and fitness approximation. Evolutionary Computation, IEEE Transactions on 10(4), 405–420 (2006)
19. Ratle, A.: Accelerating the convergence of evolutionary algorithms by fitness landscape approximation. In: Eiben, A.E., Bäck, T., Schoenauer, M., Schwefel, H.-P. (eds.) Parallel Problem Solving from Nature - PPSN V. LNCS, vol. 1498, pp. 87–96. Springer, Heidelberg (1998)
20. Regis, R.G., Shoemaker, C.A.: Local function approximation in evolutionary algorithms for the optimization of costly functions. Evolutionary Computation, IEEE Transactions on 8(5), 490–505 (2004)
21. Sacks, J., Welch, W., Mitcheel, T., Wynn, H.: Design and analysis of computer experiments(with discussion). Statist. Sci. 4, 409–435 (1989)
22. Schonlau, M.: Computer Experiments and Global Optimization. PhD thesis, University of Waterloo (1997)
23. Ulmer, H., Streichert, F., Zell, A.: Evolution strategies with controlled model assistance. Evolutionary Computation, 2004. CEC2004. Congress on 2 (2004)
24. Zhang, Q., Li, H.: A multiobjective evolutionary algorithms based on decomposition. Evolutionary Computation, IEEE Transactions on (2007)
25. Zitzler, E.: Evolutionary Algorithms for Multiobjective Optimization: Methods and Applications. PhD thesis, Swiss Federal Institute of Technology (ETH), Zurich, Switzerland (1999)

A High Precision OGC Web Map Service Retrieval Based on Capability Aware Spatial Search Engine

Nengcheng Chen, Jianya Gong, and Zeqiang Chen

State Key Lab of Information Engineering in Surveying, Mapping and Remote Sensing
Wuhan University,129 Luoyu Road, Wuhan, China, 430079
cnc_dhy@hotmail.com, jgong@lmars.whu.edu.cn, czq0119@163.com

Abstract. Recent advances in open geospatial web service, such as Web Map Service as well as corresponding web ready data processing service, have led to the generation of large amounts of OGC enabled links on Internet. How to find the correct spatial aware web service in a heterogeneous distributed environment with some special criteria, such as coincidence of type, version, time, space and scale has become a "bottleneck" of geospatial web-based applications. In order to improve the accessing precision of OGC Web Map Service (WMS) on WWW, a new methodology for retrieving WMS based on extended search engine and service capability match is put forward in this paper. Major components include the WMS search engine, WMS ontology generator, WMS catalogue service and multi protocol WMS client. Here we proposed the architecture and interaction between these components before focusing on the service capability match and design of the WMS ontology. The process of WMS link detection, capability matching, ontology modeling, and automatic registry are reported in this paper. Then the precision and response time of WMS retrieval is evaluated, results show that the mean execution time during per effective hit of proposed method is 0.44 times than that of traditional method, moreover, the precision is about 10 times than that of traditional method. WMS ontology record could be generated by ontology reasoning, registered and served by CSW.

1 Introduction

OGC has been developing OGC Web Service (OWS), a seamless framework for a variety of online geo-information processing and location services based on the web service technology. OWS contains a set of abstract and implementation specifications [1,2], such as Web Map Service (WMS) [3], Web Feature Service(WFS) [4] and Web Coverage Service (WCS) [5], their interface can allow the spatial information share and interoperability on WWW.

Meanwhile, GIS vendors extend their Web GIS software to support OGC Web Service, Such as ArcIMS9.0, MapXtreme4.5.7 and MapGuide. LAITS in US accomplished a set of network processing service for NASA such as the Web Coverage Service, Web coordinate transforming service and the Web Image classification Service [6]. A set of standards in OGC Web Service framework are used in the national spatial database infrastructure of Canada [7].

L. Kang, Y. Liu, and S. Zeng (Eds.): ISICA 2007, LNCS 4683, pp. 558–567, 2007.

Moreover, the WWW holds vast amounts of WMS information. For example, Google has 7360000, Yahoo has 3500000 and Baidu has 172000 links for "WMS" keyword respectively. Recently a few search engines that are specialised with respect to geographic space have appeared. However, users do not always get the effective OGC WMS link information they expect when searching the Web.

This paper focuses on the problems of geospatial information service retrieval in the OGC community, especially for the precision of "WMS" retrieval. Geospatial ontology, semantic web and spatial data mining are described. The architecture and components of system is explained in details. Then WMS service link detection based on capability match, WMS service Ontology modelling and reasoning is addressed. Thirdly, the precision and time cost of crawl and detect is illustrated. Finally, some conclusions are summarized and outlook is put forward.

2 Related Work

Semantic web is to provide a new vision to manage geospatial information and discover geospatial knowledge. It is widely recognized that ontology is critical for the development of the Semantic Web. Ontology has originated from philosophy as a reference to the nature and the organization of reality. In general, ontology is a "specification of a conceptualization" [8]. To provide formal semantic descriptions of NASA data set and scientific concepts, several projects are underway to develop a semantic framework. Described in the OWL language, the ontology within the Semantic Web for Earth and Environmental Terminology (SWEET) [9] contain several thousand terms spanning a broad extent of Earth system science and related concepts (such as NASA GCMD, ESML, ESMF, grid computing, and OGC). The SWEET provides a high-level semantic description of Earth system science. The ontology of geographic information metadata (ISO 19115 and FGDC) being developed in [10] add the semantic meanings to the data description by which data sets are explicitly associated with providers, instruments, sensors and disciplines, and the relationships across these concepts. In [11,12], Geospatial data mining ontology oriented toward the research themes in NASA Earth-Sun system is put forward. However, the mapping between the geospatial information service metadata ontology is still in investigation and definition of OGC compatible geospatial information service ontology is short of.

Local spatial data mining tools are adopted widely, such as GeoMiner, the requirement of spatial data knowledge mining based on internet or intranet is increasing more and more [13]. SPIN[14] project supported by European Commission is a web based spatial data mining system, which combines existing geographic information system with data mining function into a completely coupled, open and extensible system. A Web Based Spatial Data Mining (WBSDM) platform based on XML/J2EE is designed in [15], including Multi-Agent System (MAS), but it is just based on intranet where database structure and contents are relatively fixed known.

In [16], a spatial data search engine – "CIHU" is implemented, which could get the corresponding raster pictures provided by servers using the place name from WWW page. Google have recently introduced a demonstration of locational web search based in the USA. Like the Vicinity search tools it allows the user to specify the name of a place of interest using an address or zip code, which is then matched against relevant documents.

3 Architecture

In this section, we discuss design strategies for WMS search engine. Below is a set of criteria to guide our design. 1) The tool should be able to handle different WMS version, from 1.0.0 to 1.3.0. 2) The costs involved in link detection and service capability match must be kept to a minimum to ensure suitability for an Internet environment. 3) The components involved in prototype must be loosing couple.4) The components must be machine and platform independent to allow for worldwide use on the Internet.

Just as Fig1 shows, five components are included in the architecture of high precision approach to WMS retrieval: distributed GIService, WMS search engine, WMS ontology generator, Catalogue service for WMS and multi protocol WMS client.

Distributed GIService component is the data layer of the architecture, including a great number of geospatial service (such as Web Map Service, Web Feature Service, and Web Coverage Service). A contrived example is created with just four pages to understand the steps involved in the search process. The part I in figure 1 illustrates the links between pages. Page A has three links and one similar "WMS URL" text (page D), the three links contain one "WMS Link" (page F). Page A acts as the input in the crawl, the output is D, E and F. B and B-dup (B-duplicate) have identical content.

WMS search engine is the core business logic of the architecture and responsible for searching and discovering appropriate WMS by means of WMS link detection, service capability and content matching. The links of appropriate WMS are stored in WMS description file.

WMS ontology generator is in charge of generating WMS ontology from service capability information and WMS type information. Ontology files can be stored in files or in database. Regardless of whether they are kept in the database or as files, the entire ontology content has to be registered into a CSW that is kept in persistence before it can be used for querying or manipulation.

Catalogue service for WMS: The URL, description of WMS discovered by engine, and the annotation generated by ontology generator is automatically registered into catalogue service for WMS, which is implemented by OGC catalogue service standard and semantic web technology.

Multi Protocol WMS Client: It is a user interface layer to query and call Geospatial Web Map Service. It is implemented by OGC Web Map Context (WMC) standard and Java portal technology. Similar to the engine layer, the import and export of data are differentiated within the user interface layer. Therefore, a clear separation of the user interface for modifying and accessing the information is part of the client layer. On the import-side interoperability is of great importance between the engine layer and the client layer. On the export-side a high level of interoperability between the system and various web-browsers is required.

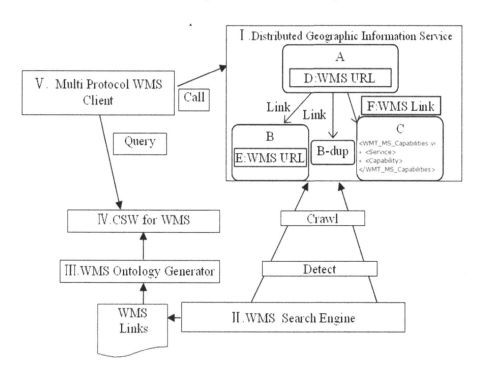

Fig. 1. Architecture of high precision WMS retrieval

4   WMS Discovery Based on Capability Matching

4.1   The Capability of WMS response

The content of WMS response has not only "version" and "updateSequence" attributes, but also "Service" and "Capability" elements. The value of "version" may be 1.0.0, 1.0.7, 1.1.0, 1.1.1, and 1.3.0; "Service" includes the mandatory "Name", "Title", "Abstract" and "OnlineResource" elements, and may contain optional "KeywordList", "ContactInformation", "Fees", "Access Constraints", "LayerLimit", "MaxWidth" and "MaxHeight" elements. "Capability" includes the mandatory "Request" and "Exception" elements, and may contain one or more "ExtendedCapabilities" and "Layer" elements. The above attributes and elements are adopted by the following capability detection. In this paper, the "WMS_Capabilities" tag and the "version" attribute is used to find the "WMS" URL from the "Potential URL Database" in section 4.2; the "Service" and "Capability" elements are adopted to generate the "WMS" ontology instance in section 5.

4.2   Approach

Figure 2 shows the procedures of the Web Map Service crawl and detect. The procedures are listed in the following:

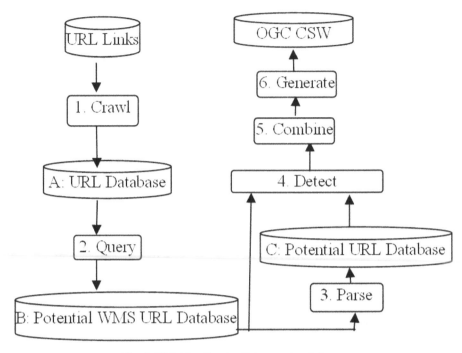

Fig. 2. Web Map Service links crawl and detect

1) Crawl: We use the popular open source search engine (such as Nutch) to track every known page and relevant link, generate the URL database (A) from the specified URL links.

2) Query: Once we have the web content, we can get ready to process queries. The indexer uses the content to generate an inverted index of all terms and all pages. We divide the document set into a set of index segments, each of which is fed to a single searcher process. We use the "WMS" and "Web Map Service" key words to query the indexed web content, the potential "WMS" URL database can be achieved (B).

3) Parse: The potential "WMS" URL database has the links whose content has the "WMS" and "Web Map Service" key words. The content is parsed using html document parser, some "WMS" relevant links can be found and stored in "Potential URL Database" (C).

4) Detect: We send get or post "WMS GetCapabilities " request to the above "WMS" links and get the response. We can get the URL and metadata of "WMS" if the response contains "WMS_Capabilities" element and the corresponding information.

5) Combine: The above "WMS" URLs are compared with each other, a uniform "WMS" URL database is generated.

6) Generate: Each WMS service is registered as a service ontology record in OGC CSW, the record is generated through the capability of WMS.

To verify the WMS links crawl and detect techniques proposed, we have carried out some experiments. In the following section, we will demonstrate how capability matching techniques are used in prototype to improve search results, we also report on the experiments which were carried out to study the time cost for performing WMS retrieval. WMS retrieval techniques are implemented using Java, and they interact with Nutch (an open source text search engine). All experiments were carried out on a Pentium 4 PC with a 2.00 GHz processor and 1.0 GMB of memory, running Microsoft Windows/XP.

4.3 Precision Analysis

This section demonstrates the effectiveness of the WMS retrieval techniques proposed. This is achieved by comparing the search results obtained by the prototype when WMS detection option is either switched on and off. When WMS detection is on, the system performs crawl, query, parse and combine operation proposed in section 4.2, search is carried out basing on both the capability matching and the textual index of web collection. When WMS detection is off, simple query expression is used to perform a textual based search.

The experiment is carried out using a set of crawl and detection in WMS-SITE ("http://wms-sites.com/"), the crawl depth is varied from 5 to 100, and the top 100 results are selected to store and detect. The results produced from running these queries were analysed automatically based on service capability. Table 1 shows the experiment results.

Table 1. Experimental results of WMS-SITE crawl and detection

Depth	TopN	Result Size (M)	Crawl			Detect		
			T(S)	H	E	T(S)	H	E
05	100	6.09	410	93	0	697	87	28
10	100	14.48	1520	353	16	2118	117	54
20	100	34.4	1678	451	22	2332	171	69
40	100	86.4	1724	479	21	2510	152	65
60	100	128	1815	478	22	2797	171	69
80	100	160	1920	478	22	2896	166	67
100	100	203	2088	480	23	3193	174	72

We first studied the hits variation with different depth and the result is shown in Figure 3. As we can see, that hits increases with number of crawl depth -the more depth a crawl has, the more hits is retrieved for crawl or detect, the increase displays a linear tendency. However, when the depth is after 20, the hits are about 65 and tendency to stable.

We then studied the hits precision of the method proposed in this paper. In table 1, the "H" item is the number of hits in craw or detection and the "E" item is the effective number of hits. It is obvious shown from table 1 that the search system performed better when "Detect" option was switched on. Moreover, if the ratio of "E" to "H" can be seen as the representation of retrieval precision, the precision of detection is shown in figure 4 and varied from 0.32 to 0.46, however, the crawl is varied from 0.0 to 0.05.The mean precision of detection is 0.40, while the crawl is just 0.04, the precision of detection is about ten times of crawl. The capability match based detection enabled us to retrieve links which describes "WMS" not only in the Crawl database but also in the corresponding page document content like "http://". Since these links are similar to "WMS" request, the retrieved documents are "WMS" relevant to the query. When "Detect" option was off, it appeared that all retrieved links involve the "WMS" link. Unfortunately, many of these links are not actually "WMS" link.

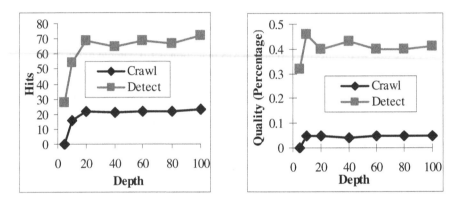

Fig. 3. Hits in WMS-SITE   **Fig. 4.** The percentage of effective hit

4.4 Time Cost Analysis

From figure 5, we can observe that it requires more CPU time to execute retrieval using "detect" than using "Crawl". This is mostly due to the time cost of capability match in "detect". From the total mean response time, the "detect" is about 2363 seconds and the "crawl" is about 1594 seconds, the "detect" is 1.5 times than that of the "crawl". From the mean response time in the per effective hit, the "detect" is about 39 seconds and the "crawl" is about 89 seconds, the "detect" is 0.44 times fast than that of the "crawl".

We then studied the time cost of crawl or detect with different depth, i.e. crawl composed of different numbers of depth, and the result is shown in figure 5. As we can see, that response time increases with number of crawl depth-the more depth a crawl has, the more CPU time is required for detect, the increase displays a linear tendency. This is mostly due to the hits- the more potential "WMS" link has, the more CPU time is required for detect.

The time cost of detect with different hits is studied and the result is shown in figure 6. As we can see, that response time increases with number of hits, the more hits a crawl has, the more CPU time is required for detect, the increase displays a linear tendency.

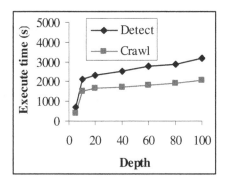

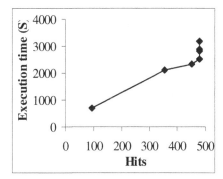

Fig. 5. Execute time of Crawl and Detect   **Fig. 6.** Execute time of Detect with different hit

5 WMS Ontology Modeling Based on OWL-S

Ontology is an explicit specification of a shared conceptualization. Their usefulness for information presentation, information integration and system development has been demonstrated recently. The role of ontology is the capturing of domain knowledge in a generic way and the provision of a commonly agreed understanding of a domain, which may be reused and shared within geospatial service. Ontology serves as metadata schemas, providing a controlled vocabulary of concepts, each with explicitly defined and machine processible semantics. Ontology also provides a structure to mark up the instance data so that it can be used in querying relevant information stored in RDF files.

In this paper, we use Protégé 3.1 to implement WMS Ontology. Protege can import and export the file in format of XML, RDF(S) and XML Schema, after installing a plug-in. It also supports DAML+OIL and OWL. Via plug-ins (PAL and FaCT), Protege can make Consistency check. WMS service ontology enables automatic discovery, invocation and composition of all registered services conforming to the OWL-S specification. The "profile" of OWL-S, which describes who provides the service, what the service does, as well as other properties of services, allows the knowledge base to infer whether or not a particular service is appropriate to a given problem. The "process model" of OWL-S, which states the inputs, outputs, preconditions and effects of a service, allows the knowledge base to figure out whether or not a service meets the requirements as well as the conditions to invoke the service. The WMS ontology is divided into post, get and both WMS.

Since all of the ontology is represented by OWL, the inference engine in the prototype is an OWL reasoner built on Prolog. Ontological information written in OWL-S is converted into RDF triples and loaded into the prototype. The engine has built-in axioms for OWL inference rules. These axioms are applied to facts in the prototype to find all relevant entailments such as the inheritance relation between classes that may be not directly in the subclass relationships.

The component of WMS Service ontology tree can be automatically generated through service capability and ontology reasoning, and the result of first record in table1 is shown in figure 7. There are 15 "Both" ontology instance, 13 "POST" ontology instance and 28 "GET" ontology instance. Each ontology instance has its own service profile, service model and service grounding. Therefore, the catalogue service can automatically harvest the WMS instance and we can use the simple or advanced semantic search method to query WMS from catalogue service.

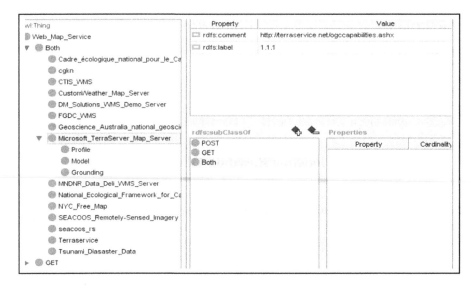

Fig. 7. WMS Service ontology tree is automatically generated by service capability

6   Conclusions

How to find the correct WMS has become a "bottleneck" of geospatial web-based applications. This paper proposes a high precision "WMS" retrieval based on service capability match. The "WMS" aware search engine achieved better properties, which are listed as follows: (1) a flexible deployment method. The search engine adopts the service oriented architecture to package the procedure of crawl, detect and register into service;(2) The better precision of "WMS" retrieval, the "WMS" aware search engine use the "detection" to discover the potential service, the effective hits number of "detection" strategy is 10 times than that of the traditional "crawl"; (3) The better execution of per effective hit, the "detection" strategy is 0.44 times fast than that of the traditional "crawl"; and (4) WMS Service ontology tree can be automatically generated through service capability and ontology reasoning. It is confident that the mature "WMS" aware search engine will play more and more important role in the construction of National Spatial Data Infrastructure (NSDI). The next work is to improve the performance of prototype and continue to study the "WFS" and "WCS" aware search engine.

Acknowledgement

This work is supported by grants from Chinese 973 program (No. 2006CB701305, PI:Jianya Gong) ,NSFC program (No. 40501059, PI: Dr. Nengcheng Chen) and HBSF program (No. 2006ABC010, PI: Qingquan Li).

References

1. Sonnet, J.: OWS 2 Common Architecture: WSDL SOAP UDDI. Open GIS Consortium Inc., Canada, p. 76 (2004)
2. Arliss Whiteside: OpenGIS® Web Services Common Specification. Open GIS Consortium Inc., Canada, p. 106 (2005)
3. OGC 02-069, OpenGIS implementation Specification: Web Map Service, http://www.opengeospatial.org/ docs/01-068r3.pdf
4. OGC 02-058: OpenGIS implementation Specification: Web Feature Service, https://portal.opengeospatial.org/ files/?artifact_id=7176
5. OGC 03-006r3: OpenGIS implementation Specification: Web Coverage Service, https://portal.opengeospatial.org/ files/?artifact_id=3837
6. Zhao, P., Deng, D., Di, L.: Geospatial Web Service Client[J]. In: ASPRS 2005 Annual Conference, Baltimore, Maryland, March 7-11, 2005 (2005)
7. Bernard, L.: Experiences from an implementation Testbed to Set up a National SDI. In: Proceedings of 5th of the Association of Geographical information Laboratories in Europe(AGILE), Lyon, France, April 25-27, 2002 (2002)
8. Gruber, T.: A translation approach to portable ontologies. Knowledge Acquisition 5(2), 199–220 (1993)
9. Raskin, R.: Enabling Semantic Interoperability for Earth Science Data, http://sweet.jpl.nasa.gov/EnablingFinal.doc
10. Islam, L.B., Beran, B., Fellah, S., Piasecki, M.: Ontology for Geographic Information - Metadata (2003), http://loki.cae.drexel.edu/ wbs/ontology/iso-19115.htm
11. Zhao, P., Di, L.: Semantic Web Service Based Geospatial Knowledge Discovery. In: Proceedings of 2006 IEEE International Geoscience And Remote Sensing Symposium, Denver, Colorado, USA, July 31- August 04, 2006. IEEE Computer Society Press, Los Alamitos (2006)
12. Dill, L., Zhao, P., Yang, W., Yue, P.: Ontology-driven Automatic Geospatial-Processing Modeling based on Web-service Chaining. In: Proceedings of the Sixth Annual NASA Earth Science Technology Conference, College Partk, MD, USA, June 27-29, 2006, 7 pages (2006) (CD-ROOM)
13. Li., D., Wang, S., Shi, W., et al.: On Spatial Data Mining and Knowledge Discovery. Geomatics and Information Science of Wuhan University 26(6), 491–499 (2001)
14. May, M., Savinov, A.: SPIN - An enterprise architecture for spatial data mining. In: Palade, V., Howlett, R.J., Jain, L. (eds.) KES 2003. LNCS, vol. 2773, pp. 510–517. Springer, Heidelberg (2003)
15. Fu, M.: The Research of Web-Based Spatial Data Mining, Doctor's Degree Dissertation, Central South University (2004)
16. Bai, Y., Yang, C.: Research on Spatial Information Search Engine. Journal of China University of Mining and Technology 33(1), 90–94 (2004)

Analysis of the Performance of Balance of Digital Multi-value Based on Chebyshev Chaotic Sequence

Yinhui Yu[1], Shuxun Wang[1], and Yan Han[2]

[1] College of Communication Engineering, Jilin University, Changchun, 130012, P.R. China
[2] College Mathematics, Jilin University, Changchun, 130012, P.R. China

Abstract. As the widely investigation of chaotic sequence in the application of communication, binary and quaternary phaseshift keying modes have been extensively adopted in spread spectrum communication. Based on Chebyshev chaotic mapping, binary-phase and quadric-phase chaotic spreading sequences are generated, and the relationship among the performance of balance and fractal parameter, initial value and the period of sequence are analyzed. The non-balanced points which should be avoided in the chaotic sequence when applying in CDMA system are proposed. Aiming at the problem which the conventional degree of balance functions only investigates the balance of the 2-valued sequence, the maximum balanced difference method is proposed. The definition of maximum balanced difference is proposed. We analyze the balance performance of quadric-phase Chebyshev sequence. Experimental results show that the maximum balanced difference function can replace the traditional binary-phase degree of balance function to research the balance of digital 2-valued and multi-valued sequence, which has better universality.

Keywords: chaotic sequence, spread spectrum, maximum balanced differences, performance of balance.

1 Introduction

In present CDMA systems, they almost use linear or non-linear shift register pseudo random sequence as spread spectrum sequence universally. Presently mature QPSK technologies have been widely applied, whereas conventional function of the degree of balance can only investigate the performance of balance of 2-valued sequence. Chaos is a kind of special kinetics system, which is non-periodic, non-convergent but bounded. It has very sensitive dependence on the initial value and external parameter[1-4]. It can produce chaotic phenomenon one dimensional deterministic non-linear difference equation. It can also provide signals which are numerous, unrelated, analogous random, definite, easy to generate and regenerate. These properties just meet the demands of spread spectrum sequence based on DS\CDMA system of spreading technologies. Thus the investigations on chaotic sequences are more and more concerned. One requirement of spread spectrum sequences in CDMA system is that it should have preferable balance of sequence, which has a large effect on the quality of communication. Appropriate sequence performance of balance

L. Kang, Y. Liu, and S. Zeng (Eds.): ISICA 2007, LNCS 4683, pp. 568–574, 2007.

decreases direct-current of sending signals and gains better characteristic of frequency spectrum. It can not only be achieve in the practical project, but also restrain carrier effectively, reduce the sending power and has a better secrecy. On the contrary, sequences that have non-balance performance will cause carrier of the spread spectrum communication system to leak greatly, destroy the secrecy of system, the capability of anti-jamming and anti-intercepting[6,8]. In this paper, chaotic sequences are generated by chaotic mapping. The relationship of fractal parameter that produce sequence, initial value and the length of sequence to the performance of balance is discussed. Effects caused by all kinds of elements to the balance of sequence are analyzed. The definition of maximum balanced difference function used to analyze the balance of multi-valued sequences is given. We utilize the maximum balanced differences function to analyze the balance of 4-phase sequence. It supplies definite theoretic attestation for the dependability of the choice of spreading frequency address codes.

2   The Degree of Balance of Binary-Phase Sequence

Many literatures indicate that chaotic sequences generated by Chebyshev mapping are regarded as preference proposal of the spreading frequency codes in CDMA system [2][5]. So in this paper we adopt Chebyshev chaotic mapping as source mapping sequences to analyze. Chebyshev chaotic mapping with exponent number 2^k is given by:

$$x_{n+1} = \cos\left(2^k \cos^{-1} x_n\right), \quad x_n \in (-1,1), \quad k=1,2,3,4,L \tag{1}$$

The probability density function of Chebyshev chaotic sequence is the even symmetry about initial point[7], so we use a threshold function:

$$c_n(x) = \begin{cases} -1 & (x < 0) \\ 1 & (x \geq 0) \end{cases} \tag{2}$$

to quantify the chaotic sequences, obtain the 2-phase sequence $\left\{\{c_n\} \mid c_n \in \{-1,1\}\right\}$, composed of ± 1. The generated sequence based on this method is with equal probability of finding +1 and -1 are equal. It can ensure that the performance of balance of sequence is ideal, namely:

$$\lim_{N \to \infty} \frac{L}{N} = \lim_{N \to \infty} \frac{1}{N} \sum_{n=0}^{N} c_n(x) = \int_{-1}^{1} c_n(x) \rho(x) dx = 0 \tag{3}$$

Because the performance of balance and DS system have a close relationship with carrier suppression degree, the non-balance of the spread spectrum sequence may cause the system carrier to leak greatly, and easily arouse error code or loss of information [2-3]. Usually we investigate the performance of balance of 2-phase sequence using the function of the degree of balance. As (4) shows, P and Q represent the amount of the sequence ' 1 ' and ' -1 ' respectively, N represents the length of the sequence, E is defined as the degree of balance of the sequence. When E=0, namely the sequence achieve the balanced state; when E=1, it indicates that sequence becomes unit-valued sequence, namely non-balanced sequence; the more near to 0 the value of E, the better the performance of balance of sequence. That is

$$E = |P - Q|/N \tag{4}$$

The relationship between Chebyshev chaotic mapping power series K and performance of balance is that when K>3, the balance of the sequence tends to remain stable and ideal. Fig 1 shows the related curve between power series K and the degree of balance E. Therefore in this paper we adopt the eighth level Chebyshev sequence to analyze. As Fig 1 shows, the fractal parameter in Chebyshev sequence has a slight effect on the performance of balance.

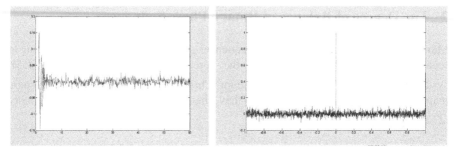

Fig. 1. The relationship between the degree of balance E and power series K

Fig. 2. The relationship between the degree of balance E and initial value X_0

On Chebyshev 2-phase chaotic spread spectrum sequence, the influence caused by the initial value X_0 towards the degree of balance. Most literatures claimed that X_0 influenced the degree of balance slightly, there is no peak point. However, according to the analysis of this paper, the initial value X_0 in Chebyshev 2-phase chaotic spread spectrum sequence influenced the degree of balance assuredly. When initial value X_0 =0, E=1, the graph in Fig 2 is obtained by means of computer simulation, the graph showed that there is a peak point which should be avoided when constituting the sequence.

3   The Maximum Balanced Difference Method Weighing Digitized Multiple-valued Sequence

Multilevel modulation has the advantage of improving frequency band, therefore Quadric Phase Shift Keying (QPSK) has been applied widely now. We need to generate quadric-phase chaotic sequence as the spread spectrum sequence of the system to satisfy the requirement of 4-phase DS/CDMA system. Similar to the generation of 2-phase sequences, we need to obtain reasonable quantization standard through dividing probability in order to get the sequence of the ideal performance of balance. According to the method in [6], from (3), regarding $d_1 = -\sqrt{2}/2, d_2 = 0, d_3 = \sqrt{2}/2$ as the dividing points, we divide interval $(-1,1)$ into 4 subintervals to make the probability that x_n is located all the subintervals to be equal. Evaluating in all the subintervals, we obtain the 4-valued sequence $\{\{c_n\}|c_n \in \{0,1,2,3\}\}$,

then it can generate 4-phase sequence based on plurality:

$$u_n = j^{c_n}, c_n \in \{0,1,2,3\} \tag{5}$$

The performance of balance is still an important performance measure[4], for the 4-phase spreading frequency sequence, however in the related literatures the functions about investigating the performance of balance have never been given. So this paper offers the maximum balanced difference function that can investigate the performance of balance of m sequence to analyze the sequence performance of balance:

$$\delta_{MAX} = |u_{c_n} - 1/m|_{MAX}, \quad m \in \{2,3,4\Lambda\} \tag{6}$$

where, $1/m$ is the proposal of one symbol in m value balanced sequence, u_{c_n} is the proportion of one symbol in candidate investigated sequence, $|u_{c_n} - 1/m|$ is the balanced difference. The maximum balanced difference is the maximum value in all balanced differences of m symbols. When $\delta_{MAX} = 0$, the sequence become the balanced sequence; when δ_{MAX} is the nearer to zero, it shows the better of the balance of sequence; when $\delta_{MAX} = (m-1)/m, u_{c_n} = 1$, the sequence becomes single-valued sequence, namely non-balanced sequence.

As in (6), the maximum balanced difference function of 4-phase sequence can be defined as:

$$\delta_{MAX} = |u_{c_n} - 1/4|_{MAX} \tag{7}$$

We use the maximum balanced difference function to weigh the effect caused by the length of sequence. Theoretically speaking, the longer the period of sequence is, the better the performance of balance is [5-6]. However in CDMA direct-sequence spread spectrum system, over-long spreading frequency codes are not suitable, otherwise the long sequence will add complexity to the device, and bring technical difficulty in the practical application. From the above simulation result, when k=3, and the initial value $x_0 = 0.145$, we can gain the diversification situation of the maximum balanced differences as the length of sequence. The simulation result in

Fig.3 shows that when the period increases, the sequence tends to remain balanced. When the length of sequence is above 1000, the effect on sequence performance of balance by the length will decrease, the performance of balance tends to be stable and ideal. So we can adopt 4-phase chaotic sequence whose length is 1024-2048 as the spreading frequency sequence in CDMA system.

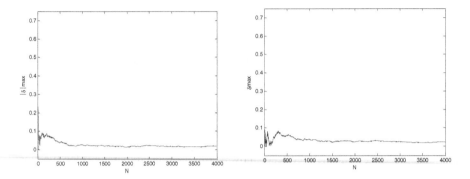

Fig. 3. The relation between 4-phase sequence maximum balanced difference and the length of sequence

Fig. 4. The relation between 2-phase sequence maximum balanced difference and the length of sequence

To investigate the practicability of the maximum balanced differences, from (6), the maximum balanced difference of 2-phase sequence can be defined as:

$$\delta_{MAX} = \left| u_{c_n} - 1/2 \right|_{MAX} \tag{8}$$

To investigate the relationship between the maximum balanced difference of 2-phase sequence and the sequence N in same parameters, according to the simulation result in Fig 4, it has the similar functional orbit compared to the conventional degree of balance function. According to the above mentioned, the maximum balanced difference can obtain the same observational and analyzing result for the performance of balance of 2-phase sequence ; but the traditional degree of balance only investigate the performance of balance of 2-phase sequence. The maximum balanced differences can analyze the m-value sequence ($m > 1$ and m is the integer number). Thus the maximum balanced difference proposed in this paper has the better universality compared with traditional degree-of-balance function when used for analyzing balance performance.

4   Conclusions

In this paper, we examine the concept of the performance of balance and analyze the influence of balance caused by various parameters of Chebyshev chaotic mapping. Experimental when results show that power series $k \geq 3$, initial value $x_0 \neq 0$, sequence period $N > 1000$, we can gain the sequence whose characteristic meets the demand of spreading frequency. Then we propose the maximum balanced difference formula that can investigate the sequence of multi-valued sequence performance of balance. The above analysis indicates that the maximum balanced difference function is an effective method used for analyzing the multi-valued sequence performance of balance. It can replace conventional function of 2-phase degree of balance to investigate the balance of 2-valued and multi-valued sequence, so it appears well universality. At the same time, because the generation of it is simple and it has the well feature of secrecy, so it can be regarded as a candidate scheme of spreading frequency sequence in the third generation of mobile-communication DS/CDMA system.

References

1. Kohda, T., Tsuneda, A.: Pseudnise sequences by chaotic nonlinear maps and their correlation properties [J]. IEICE Trans. Commun. E76-B(8), 855–862 (1993)
2. Mazzini, G., Setti, G., Rovatti, R.: Chaotic complex spreading sequence for asynchronous DS-CDMA-part I: system modeling and results[J]. IEEE Trans. Circuits and system CAS-I-44(10), 937–947 (1997)
3. Makoto, I.: Spread spectrum communication via chaos [J]. International Journal of Bifurcation and Chaos 9(1), 155–213 (1999)
4. Heidari-Bateni, G., McGillem, C.D.: A Chaotic Direct-Spread Communication System [J]. IEEE Trans. On Communication 42(2/3/4), 1524–1527 (1994)
5. Yu, Y.-h., Ma, S.-z., Liu, W.-d.: The performance of balance base on Chebyshev2-phasc Chaotis spreading Sequence [J]. Journal of Jilin University(Information Science Edition) 22(3), 228–231 (2004)
6. Cai, G.-q., Song, G.-w., Zhou, L.-l.: Quadriphase spread-spectrum sequences based onchebyshev map. [J]ACTA Electronica Sinica 27(9), 74–77 (1999)
7. Ling, C., Li, S.-q.: Chaotic Spreading Sequences with Multiple Access Performance Better Than Random Sequences [J]. IEEE Trans on Circuit and System-I: fundamental Theory and Application 47(3), 394–397 (2000)
8. Zhang, Q., Zheng, J.-l., Jian, X.-c.: A-CDMA system simulation with chaotic spreadingsequences. [J]Journal of Tsinghua University(Science and Technology) 42(7), 901–904 (2002)
9. Frey, D.R.: Chaotic digital encoding: an approach to secure communication [J]. IEEE Trans. Circuits and System CAS-I-40(10), 660–666 (1993)
10. Abarbanel, H.D.I., Linsay, P.S.: Secure communication and unstable periodic orbits of strange attractors [J]. IEEE Trans. Circuits and System CAS-I-40(10), 643–645 (1993)
11. Bernhardt, P.: A Chaotic frequency modulation [A]Chaos in communications, Washington, pp. 62–18 (1993)
12. Wang, H., Hu, J.-d.: The Improved Logistic-Map Chaotoc Spread Spectrum Sequences. [J] Journal of CHINA Institute of Communications 18(8), 71–77 (1997)
13. Zhang, Q., Zheng, J.-l., Sun, S.-y.: The Optimization of Quaternary Chaotic Spreading Sequences and System Simulation. [J]Chinese Journal of Radio Science 17(6), 614–619 (2002)

14. Heidari-Bateni, G., McGillem, C.D.: Chaotic sequence of spread spectrum: An alternative to PN-sequences. In: IEEE International Conference on Selected Topics in Wireless communication, Vancouver,Canada, pp. 437–440. IEEE Computer Society Press, Los Alamitos (1992)
15. Karkainen, K.H.: Meaning of Maximum and Mean-Square Cross-Correlation as a Performance Measure for CDMA Code Families and Their Influence on System Capacity [J]. EICE Trans. on Commun. E76-B(8), 848–854 (1993)
16. Yu, Y.-h., Liu, W., Zhu, J.: Balance of binary-phase and quadric-phase oversampled chaotic map sequences. [J]Journal of Jilin University(Engineering and Technology Edition) 36(5), 799–802 (2006)

The Transformation Between Fuzzy Cognitive Maps and a Class of Simplified Dynamical Cognitive Networks

Yuan Miao

School of Computer Science and Mathematics,
Victoria University,
PO Box 14428, Melbourne, VIC 8001, Australia
Yuan.Miao@vu.edu.au

Abstract. Cognitive Map (CM), Fuzzy Cognitive Map (FCM), and Dynamical Cognitive Network (DCN) are three related tools for modeling human being's cognition and facilitate machine inference accordingly. A main desirable feature of FCM is that it is easy to use for modeling and visualizing human knowledge. Different models in the CM family have different capabilities in capturing human knowledge. However, a recent theoretical result shows that a CM can be transformed into a FCM and vice versa. This paper reports that a class of simplified DCN can also be transformed into FCM.

Keywords: Fuzzy Cognitive Map, FCM, Dynamical Cognitive Network, DCN, equivalence.

1 Introduction

Cognitive Map (CM) was proposed by Axelrod [1] for visualizing causal relationships among factors to capture human cognition. CM uses binary concepts to model important factors and binary links to model their causal relationships. CM has been applied in various application domains such as analysis of electrical circuits [12], analysis and extension of graph-theoretic behavior [13], and plant control modeling [5].

As CM does not differentiate different strength of causal relationships, it is not suitable to model complex systems. For an instance, when terrorism threats scare international traders from traveling, business needs urge them not to delay trips too much. CM is not able to differentiate the strength of different relationships and thus not able to provide a logic recommendation in such a case. The situation could be even worse when many factors connect into a complex causal network.

Fuzzy Cognitive Map (FCM) [6,10] is an extension of CM by introducing fuzzy weight to differentiate the strength of different causal relationships. The extension enables FCM to model many real systems and the application has been widen to entertainment [11], games[2], multi-agent systems[7], social systems[3], ecosystems[4] and etc. However, Miao et. al. [9] has pointed out that FCM leaves the strength of causes and the impact of causal relationships un-modeled. Additionally, FCM is not able to model the dynamics of how a causal impact is built up. Therefore, FCM may have contradictory inferences in modeling complex real systems.

L. Kang, Y. Liu, and S. Zeng (Eds.): ISICA 2007, LNCS 4683, pp. 575–582, 2007.
© Springer-Verlag Berlin Heidelberg 2007

Dynamical Cognitive Net (DCN) [9] is a general cognitive map which not only models the strength of causes, the impact of causal relationships, but also models the dynamics of how causal impacts are built up. DCN has avoided the structural non-robustness of CM and FCM. It is more capable in modeling large complex causal systems [7]. However, general DCNs are complex.

Although FCM and CM appear different, a recent report [14] proves that there exist certain theoretical links between the two models. For any FCM, there exists a CM which includes the exact inference pattern of the given FCM. It is also true vice versa because each CM can be viewed as a simplified FCM. This paper shows that such a result also exists between a class of Dynamical Cognitive Networks and FCMs.

The rest of the paper is organized as the follows: Section 2 describes a class of simplified DCN. Section 3 proves that the transformation from such a simplified DCN to FCM is always possible. Section 4 concludes the paper.

2   Simplified Dynamical Cognitive Network

A Simplified Dynamic Cognitive Network- Scalar + Dynamic (DCN-SD) is defined as a tuple,

$$M = <V, A>,$$

where V is the set of vertices representing the concepts of the DCN and A is the set of arcs representing the causal relationships among concepts:

$$V = \{< v_1, f_{v_1}, S(v_1)>, <v_2, f_{v_2}, S(v_2)>,..., <v_n, f_{v_n}, S(v_n)> \} .$$

In DCN-SD, the state space of each concept is a fuzzy set that describe the state of the concept. It is a finite value set:

$$S(v_i) = \{ x^1_i, x^2_i,..., x_i^{\|S\|_i} \} .$$

Each concept has a decision function to decide its state according to the causal inputs:

$$x(k+1) = f_{v_i} = f_{v_i}\left(y_{i1}(k), \quad y_{i2}(k), \quad ..., \quad y_{in}(k)\right),$$

where $y_{i1}, y_{i2}, ..., y_{in}$ are the causal impacts from concepts $v_1, v_2, ..., v_n$.

The causal relationship model of DCN-SD includes a discrete description of the dynamics of how the causal impact is built up:

$$A = \{< a(v_i, v_j), w^1_{ji}, w^2_{ji},..., w^{P_{ji}}_{ji}, y_{ji} > | v_i, v_j \in V \} ,$$

where

$$y_{ji}(k) = \sum_{p=1}^{P_{ji}-1} w^p_{ji} \times [x_i((k-p+1)) - x_i((k-p))]$$
$$+ w^{ji}_{ji} \times x_i((k-P_{ij}+1))) .$$

In general, it is:

$$y_{ji}(k \times \Delta t) = \sum_{p=1}^{m-1} w^p_{ji} \times [x_i((k-p+1) \times \Delta t) - x_i((k-p) \times \Delta t)] + w^m_{ji} \times x_i((k-m+1) \times \Delta t)).$$

Here $m = P_{ji}$. When the time unit Δt of the whole DCN-SD are the same, it can be omitted.

The dynamics of the process of the causal impact is modeled by P_{ji} discrete points in DCN-SD. That is to say, it takes P_{ij} for a causal impact to be fully built. In details,

$$y_{ji}(1) = w^1_{ji} \times x_i(1),$$

$$y_{ji}(2) = w^2_{ji} \times x_i(1) - w^2_{ji} \times x_i(0) + w^1_{ji} \times x_i(2) - w^1_{ji} \times x_i(1),$$
$$= w^2_{ji} \times x_i(1)$$

$$y_{ji}(3) = w^3_{ji} \times x_i(1) - w^3_{ji} \times x_i(0) + w^2_{ji} \times x_i(2) - w^2_{ji} \times x_i(1)$$
$$+ w^1_{ji} \times x_i(3) - w^1_{ji} \times x_i(2)$$
$$= w^3_{ji} \times x_i(1)$$

....

$$y_{ji}(P_{ji}) = w^{Pji}_{ji} \times x_i(1)$$

In case the time units are not unified,

at time $k \times \Delta t = \Delta t$,
$$y_{ji}(1 \times \Delta t) = w^1_{ji} \times x_i(1 \times \Delta t),$$

at time $k \times \Delta t = 2 \times \Delta t$,
$$y_{ji}(2 \times \Delta t) = w^2_{ji} \times x_i(1 \times \Delta t) - w^2_{ji} \times x_i(0 \times \Delta t)$$
$$+ w^1_{ji} \times x_i(2 \times \Delta t) - w^1_{ji} \times x_i(1 \times \Delta t),$$
$$= w^2_{ji} \times x_i(1 \times \Delta t)$$

at time $k \times \Delta t = 3 \times \Delta t$,
$$y_{ji}(3 \times \Delta t) = w^3_{ji} \times x_i(1 \times \Delta t) - w^3_{ji} \times x_i(0 \times \Delta t)$$
$$+ w^2_{ji} \times x_i(2 \times \Delta t) - w^2_{ji} \times x_i(1 \times \Delta t),$$
$$+ w^1_{ji} \times x_i(3 \times \Delta t) - w^1_{ji} \times x_i(2 \times \Delta t),$$
$$= w^3_{ji} \times x_i(1 \times \Delta t)$$

....

at time $k \times \Delta t = P_{ji} \times \Delta t$,
$$y_{ji}(P_{ji} \times \Delta t) = w^{Pji}_{ji} \times x_i(1 \times \Delta t)$$

Note that the above result subjects to the condition that the state of the cause concept v_i does not change. If the state of v_i changes dynamically, the impacts are combined and propagate to v_j according to the definition fomula.

Although DCN-SD is a simplified DCN, it is still a significant extension to FCM. It is able to model not only the strength of causal relationships, but also the strength of the cause and effect. For example, although *terrorism threats* have a *strong negative* causal relationship with the *international trade*, if the strength of the *terrorism threats* is negletable, the causal impact on *international trade* is therefore not much. Such a logic inference could not be modeled by FCM due to the lack of capability to model the scale of the concept. On the other hand, DCN-SD restrict the concept state as a finite value set, which excluded the real interval of general DCNs. It emphsizes on the semantic modeling of concepts, which largely simplified the usage of DCN.

Another significant extension of DCN-SD is the model of the dynamics of causal impacts. To illustrate the extension, consider an example of the war against terrorism. Military actions on a terrorism base *take immediate effect*. Economic sanctions however, *need time before the impact is built up*. Failure to model the dynamics of different types of causal impacts could result in irrational inference results. It is difficult to model the dynamics in CM/FCM as static weights are used for modelling the causal relationships among concepts. DCN-SD gains the capability by using a finite number of points along a finite time frame to model the dynamics. It is able to model a wide range of dynamics, yet not too difficult for domain experts who are familiar with FCM.

DCN-SD can be further simplified by removing the description of dynamics of causal relationships, as DCN-Scalar (DCN-S). It can be viewed as a type of extension to FCM which allows concepts to have multiple values instead of binary/ternary values.

Multi-Value Fuzzy Cognitive Map (MVFCM) or DCN-S is defined as a tuple,

M =<**V, A**>,

where **V** is the set of vertices representing the concepts and **A** is the set of arcs representing the causal relationships among concepts.

$$\mathbf{V} = \{< v_1, f_{v_1}, S(v_1)>, <v_2, f_{v_2}, S(v_2)>, ..., <v_n, f_{v_n}, S(v_n)> \},$$

$$\mathbf{A} = \{< a(v_i, v_j), w(a(v_i, v_j)) > | v_i, v_j \in \mathbf{V} \},$$

where v_i $i=1, 2, ..., n$ are the concepts, n is the number of concepts, $a(v_i, v_j)$ is the arc from v_i to v_j. $w(a(v_i, v_j))$ is the weight of arc $a(v_i, v_j)$; it can also be written as $w(v_i, v_j)$ or w_{ji}.

The decision making function of MVFCM is also defined as f_{v_i},

$$f_{v_i}(u) = f_{v_i}\left(\sum_{j=1}^{n} w_{ij} \times x_j\right).$$

Based on the causal inputs, the decision function decides the following state of the concept. The state spaces of MVFCM concepts are finite value sets:

$$S(v_i) = \{ x^1_i, x^2_i, ..., x^{R_{ji}}_{ji} \}, R_{ji} = \|S(v_i)\|$$

3   Transformation from a DCN-SD to a FCM

It has been proven that CM, FCM and MVFCM can be mutually transformed. This section will show that DCN-SD can be transformed to a MVFCM. Then, accordingly, DCN-SD can be transformed to a FCM, or CM. The vice versa holds as well because MVFCM can be considered as a simple DCN-SD.

Definition 3.1 Inclusive Relationship. of cognitive model instances: *Given two cognitive model instances M and M', M is defined as being included in M' if there exist two constants α and β, for any value x*, of any concept v of M, there exists a value x^+ of a concept v' of M', such that for all $x(k) = x^*$, k=1, 2, ... x'(αk + β) = x^+, k=1, 2, ...*

The inclusive relationship of two cognitive model instances specifies that all the causal knowledge of the included instance exists in the inclusive instance. Any inference pattern of the included instance is actually a sub-sequence of the whole or part of the inference pattern of the inclusive instance.

If M is included in M', it is denoted as $M \overset{e}{\subset} M'$.

Theorem 3.2. *For any DCN-SD, M, there exists a MVFCM M', $M \overset{e}{\subset} M'$.*

Proof

Let

$$M' = \phi, \qquad\qquad \text{// Start with an empty MVFCM model.}$$

For any $v_i, v_j \in M$, if $v_i \notin M'$,

$$M' = M' \cup \{< v_i, f'_{vi}, S(v_i) >\} \quad \text{// If } v_i \text{ does not exist, add the concept}$$
f'_{vi} will be defined later.

If $v_j \notin M'$,

$$M' = M' \cup \{< v_j, f'_{vj}, S(v_j) >\} \quad \text{// If } v_j \text{ does not exist, add the concept}$$
f'_{vj} will be defined later.

For $< a(v_i, v_j), w^1_{ji}, w^2_{ji}, \ldots, w^{P_{ji}}_{ji}, y_{ji}>$, where

$$y_{ji}(k) = \sum_{p=1}^{P_{ji}-1} w^p_{ji} \times [x_i((k-p+1)) - x_i((k-p))] + w^{ji}_{ji} \times x_i((k-P_{ij}+1))),$$

$$M' = M' \cup \{< a(v_i, v_j), w(v_i, v_j) = w^1_{ji} >\} \quad \text{// Add arc from } v_i \text{ to } v_j.$$

If $P_{ij} > 1$,

$$M' = M' \cup \left\{ \left\langle v^{1,2,1}_{ji}, \quad f_{v^{1,2,1}_{ji}}(u) = u, \quad S(v^{1,2,1}_{ji}) = \{x_i \times w^1_{ji} \mid x_i \in S_M(v_i) \right\rangle \right\} \cup$$

$$\cup \left\{ \left\langle a(v_i, v^{1,2,1}_{ji}), w(v_i, v^{1,2,1}_{ji}) = w^1_{ji} = w_{ji} \right\rangle \left\langle a(v^{1,2,1}_{ji}, v_j), w(v^{1,2,1}_{ji}, v_j) = -1 \right\rangle \right\}$$

// Initial the dynamic description concept

$$M' = M' \bigcup_{p=2}^{P_{ji}} \left(\bigcup_{m=1}^{p-1} \left\{ \left\langle v^{p,1,m}_{ji}, \quad f_{v^{p,1,m}_{ji}}(u) = u, \quad S(v^{p,1,m}_{ji}) = \{x_i \times w^p_{ji} \mid x_i \in S_M(v_i)\} \right\rangle \right\} \right),$$

$$M' = M' \bigcup_{p=2}^{P_{ji}} \left(\bigcup_{m=1}^{p} \left\{ \left\langle v^{p,2,m}_{ji}, \quad f_{v^{p,2,m}_{ji}}(u) = u, \quad S(v^{p,2,m}_{ji}) = \{x_i \times w^p_{ji} \mid x_i \in S_M(v_i)\} \right\rangle \right\} \right),$$

// Add two set of concepts to model the dynamics.

$$M' = M' \left(\bigcup_{p=2}^{P_{ji}} \left\{ \left\langle a(v_i, v^{p,1,1}_{ji}), w(v_i, v^{p,1,1}_{ji}) = w^p_{ji} \right\rangle \right\} \right) \cup$$

$$\left(\bigcup_{m=2}^{p-1} \left\{ \left\langle a(v^{p,1,m-1}_{ji}, v^{p,1,m}_{ji}), \quad w(v^{p,1,m-1}_{ji}, v^{p,1,m}_{ji}) = 1 \right\rangle \right\} \right) \cup$$

$$\left\{ \left\langle a(v^{p,1,p-1}_{ji}, v_j), \quad w(v^{p,1,p-1}_{ji}, v_j) = 1 \right\rangle \right\}$$

// Add the links for the first set of concepts

$$M' = M' \left(\bigcup_{p=2}^{P_{ji}-1} \left\{ \left\langle a(v_i, v^{p,2,1}_{ji}), w(v_i, v^{p,2,1}_{ji}) = w^p_{ji} \right\rangle \right\} \right) \cup$$

$$\left(\bigcup_{m=2}^{p}\left\{\left\langle a(v_{ji}^{p,2,m-1},v_{ji}^{p,2,m}),\quad w(v_{ji}^{p,2,m-1},v_{ji}^{p,2,m})=1\right\rangle\right\}\right)\bigcup$$

$$\left\{\left\langle a(v_{ji}^{p,2,p},v_{j}),\quad w(v_{ji}^{p,2,p},v_{j})=-1\right\rangle\right\}$$

// Add the links for the second set of concepts

For any $v_i \in M$, the decision function of v_i in M' is defined as:

$$f_{v_i|_{v_i \in M'}} = f_{v_i|_{v_i \in M}}\left(w_{i1}^1 \times x_1(k) + \sum_{p=2}^{P_{i1}} x_{i1}^{p,1,p-1}(k) - \sum_{p=2}^{P_{i1}-1} x_{i1}^{p,2,p}(k),\ w_{i2}^1 \times x_2(k) + \sum_{p=2}^{P_{i2}} x_{i2}^{p,1,p-1}(k) - \sum_{p=2}^{P_{i2}-1} x_{i2}^{p,2,p}(k),\right.$$

$$\left. ....\ ,\ w_{in}^1 \times x_n(k) + \sum_{p=2}^{P_{in}} x_{in}^{p,1,p-1}(k) - \sum_{p=2}^{P_{in}-1} x_{in}^{p,2,p}(k)\right)$$

Set the initial state

$$x(v_i)\Big|_{v_i \in M'} = x(v_i)\Big|_{v_i \in M},$$

$$x(v)\Big|_{v \in M', v \notin M} = 0$$

It can be verified that

$$x(v_i)\Big|_{v_i \in M'}(k) = x(v_i)\Big|_{v_i \in M}(k)\ ,\ k=1,\,2,\,....\qquad\blacksquare$$

To illustrate the construction of the inclusive MVFCM of a given DCN-SD, Figure 3.1 depicts the equivalent MVFCM of a typical dynamical causal link in DCN-SD, namely

$$< a(v_i,\,v_j)\,,\,w^1_{ji}\,,\,w^2_{ji}\,,....,\,w^{P_{ji}}_{ji}\,,\,y_{ji}>\,.$$

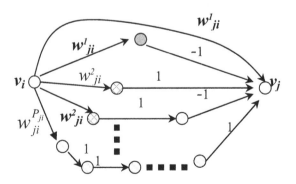

Fig. 3.1. Equivalent MVFCM of DCN-SD

The dynamics of the impact from v_i to v_j is

$$y_{ji}(k) = w_{ji}^1 \times x_i(k) + \sum_{p=2}^{P_{ji}} x_{ji}^{p,1,p-1}(k) - \sum_{p=2}^{P_{ji}-1} x_{ji}^{p,2,p}(k)$$

If the state of v_i is static, it shows that the impact from v_i to v_j is

$$y_{ji}(1) = w_{ji}^1 \times x_i(1),$$

$$y_{ji}(2) = w_{ji}^2 \times x_i(2),$$

....

$$y_{ji}(P_{ji}) = w_{ji}^{P_{ji}} \times x_i(P_{ji}).$$

Therefore, v_j maintains the same state dynamics in the MVFCM as the original concept in DCN-SD does.

4 Conclusion and Discussions

Cognitive Map (CM), Fuzzy Cognitive Map (FCM), and Dynamical Cognitive Network (DCN) are three related tools for modeling human being's cognition and facilitate machine inference accordingly. Different models have different capability in capturing human knowledge. This paper reports that a class of simplified DCN can also be transformed into MVFCM. With a recent report proving the existence of transformation between MVFCM, FCM and CM, all the models are mutually transformable.

It is important to note that the result does not imply that some models, e.g., CM, FCM or DCN-SD need to be abandoned. Neither does the result make extension from CM or FCM unnecessary. On the contrary, it provides domain experts more flexibility to choose the most appropriate models. It is true that if a real system can be modeled by DCN-SD, it can also be modeled by a MVFCM. However, through the constructive proof of the transformation, we can see that it is very impractical for domain experts to come up with the corresponding inclusive MVFCM, not even to say the inclusive CM model. That is to say, if a domain expert models a real system with CM and another domain expert comes up with a DCN-SD, the CM is normally not the CM that is transformed from the DCN-SD. Just like all Computer Software can be represented by an Assembly language. But different levels models (OS, Middle ware, business models) are still of a great significance.

References

1. Axelrod, R.: Structure of Decision: The Cognitive Maps of Political, Elites. Princeton Univ. Press, Princeton, NJ (1976)
2. Chen, M.E., Huang, Y.P.: Dynamic fuzzy reasoning model with fuzzy cognitive map in Chinese chess. In: IEEE International Conference on Neural Networks, vol. 3, pp. 1353–1357. IEEE Computer Society Press, Los Alamitos (1995)

3. Craiger, P., Coovert, M.D.: Modeling dynamic social and psychological processes with fuzzy cognitive maps. IEEE World Congress on Computational Intelligence 3, 1873–1877 (1994)
4. Eccles, J.S., Dickerson, J.A., Shao, J.: Evolving a virtual ecosystem with genetic algorithms. 2000 Congress on Evolutionary Computation 1, 753–760 (2000)
5. Gotoh, K., Murakami, J., Yamaguchi, T., Yamanaka, Y.: Application of Fuzzy Cognitive Maps to Supporting for Plant Control. In: SICE Joint Symposium of 15th System Symposium and 10th Knowledge Engineering Symposium, pp. 99–104 (1989)
6. Kosko, B.: Fuzzy cognitive maps. International Journal of Man-Machine Studies 24, 65–75 (1986)
7. Miao, C.Y., et al.: DCM: Dynamical Cognitive Multi-Agent Infrastructure for Large Decision Support System. International Journal of Fuzzy Systems 5(3), 184–193 (2003)
8. Miao, Y., Liu, Z.Q., et al.: Simplification, Merging, and Division of Fuzzy Cognitive Maps. International Journal of Computational Intelligence Applications 2(2), 185–209 (2002)
9. Miao, Y., Liu, Z.Q., Siew, C.K., Miao, C.Y.: Dynamic Cognitive Net-an Extension of Fuzzy Cognitive Map. IEEE Transaction on Fuzzy Systems 10, 760–770 (2001)
10. Miao, Y., Liu, Z.Q.: On causal inference in fuzzy cognitive maps. IEEE Transactions on Fuzzy Systems 8(1), 107–119 (2000)
11. Parenthoen, M., Tisseau, J., Morineau, T.: Believable decision for virtual actors. In: 2002 IEEE International Conference on Systems, Man and Cybernetics, vol. 3, p. 6. IEEE Computer Society Press, Los Alamitos (2002)
12. Styblinski, M.A., Meyer, B.D.: Fuzzy Cognitive Maps, Signal Flow Graphs, and Qualitative Circuit Analysis. In: Proceedings of the 2nd IEEE International Conference on Neural Networks (ICNN-87), vol. 2, pp. 549–556. IEEE Computer Society Press, Los Alamitos (1988)
13. Zhang, W., Chen, S.S.: A Logical Architecture for Cognitive Maps. In: Proceedings of the 2nd IEEE Conference on Neural Networks (ICNN-88), vol. 1, pp. 231–238. IEEE Computer Society Press, Los Alamitos (1988)
14. Miao, Y., Tao, X.H., Shen, Z.Q., Liu, Z.Q., Miao, C.Y.: The Equivalence of Cognitive Map, Fuzzy Cognitive Map and Multi Value Fuzzy Cognitive Map. In: 2006 IEEE International Conference on Fuzzy Systems, pp. 1872–1878. IEEE Computer Society Press, Los Alamitos (2006)

Cryptanalysis of Two-Round DES Using Genetic Algorithms

Jun Song[1,2], Huanguo Zhang[1], Qingshu Meng[1], and Zhangyi Wang[1]

[1] Computer School, Wuhan University, Wuhan 430079, China
[2] Computer School, China University of Geosciences, Wuhan 430074, China
songjun@cug.edu.cn

Abstract. Cryptanalysis with genetic algorithm has attracted much interest in recent years. This paper presents an approach for the cryptanalysis of two-round DES based on genetic algorithm. However, cryptanalysis of two-round DES using genetic algorithm is usually a difficult task. In this paper, we adopt known plaintext attack and produce a variety of optimum keys based on fitness function. Furthermore, we count every bit of optimal keys one by one, and find some valuable bits, which generate a significant deviation from the other observed bits. Finally, the 56-bit key is successfully gained without searching the whole search space. The experimental result indicates that this is a promising method and can be adopted to handle other complex block ciphers.

Keywords: two-round DES, cryptanalysis, genetic algorithm, fitness function, known plaintext attack.

1 Introduction

John Holland proposed the use of genetic algorithms as an efficient search mechanism in artificially adaptive systems in 1975 [1]. GAs are stochastic adaptive algorithms that start with a population of randomly generated candidates and evolution towards better solutions by applying genetic operators such as crossover, mutation, and selection, modeled on natural genetic inheritance and Darwinian survival-of-the-fitness principle. Generally, evolutionary algorithms are better than conventional optimization algorithms for problems which are discontinuous, non-differential, multi-model and hardly-defined problems [2].

In past years, many researchers in the field of cryptanalysis are interested in developing automated attacks on cipher. The progress remained sluggish until 1993 when numerous papers appeared [3]. R.Spillman etc. showed that simple substitution ciphers [4] and knapsack ciphers could be attacked using genetic algorithms [5]. In 1994, Andrew Clark well researched the use of modern optimization algorithms for Cryptanalysis [6]. Feng-Tse Lin and Cheng-Yan Kao analyzed the genetic algorithm for ciphertext-only attack [7]. In 1998, Clark and Dawson carried out the most extensive investigation on classical cipher cryptanalysis [8]. In their thesis they investigated three search heuristics, simulated annealing, genetic algorithms and tabu search attacks on simple substitution ciphers [9]. In 2002, Julio César Hernández introduced the technique that attacked two rounds TEA based on genetic algorithms

L. Kang, Y. Liu, and S. Zeng (Eds.): ISICA 2007, LNCS 4683, pp. 583–590, 2007.

[10]. In 2005, E.C.Laskari etc. showed how particle swarm optimization method was applied to the cryptanalysis of a simplified version of DES [11]. Other papers [12, 13, 14, 15, 16, 17, 18] introduced more research on genetic algorithms for Cryptanalysis. John A Clark's invited paper [3] showed the comprehensive reviews for cryptanalysis of evolutionary techniques in IEEE CEC2003.

All the past work described that most major optimization techniques have been applied to classical cipher cryptanalysis [3]. It is feasible for classical ciphers to utilize their weak capabilities and low complexities. In general, most classical cryptosystems are based on linearity mappings such as substitution and permutation. It is easily to find a direct or indirect correlation between plaintext and ciphertext in the process of evolution. But for excellent ciphers such as DES, whose designs are of high nonlinearity and low autocorrelation, and can resist corresponding attacks to a certain extent. Thus, cryptanalysis of DES using genetic algorithm is usually a quite difficult task [3].

In this paper, our obvious improvement is using the genetic algorithm to successfully break two-round DES. We adopt known plaintext attack and a new fitness function for cryptanalysis. Firstly, many optimum keys can be produced for different plaintext/ciphertext pairs based on fitness function. We recombined these optimum keys and counted every bit one by one. Then, we can obtain some valuable bits which generates a significant deviation from the other observed bits. Finally, the 56-bit key is successfully gained without searching the whole search space. The experimental result indicates that this is a promising method and can be adopted to handle other complex block ciphers.

The remainder of this paper is organized as follows. In section 2, we give the description of DES, the notion of linear cryptanalysis and differential cryptanalysis. In section 3, we present our method and strategy of cryptanalysis using genetic algorithms. In section 4, the cryptanalysis experimental results of two-round DES using genetic algorithms are reported. In section 5, conclusions are derived and directions for future work are suggested.

2 Preliminaries

In order to describe the problem with cryptanalysis using genetic algorithms, we briefly introduce the structure of DES and the notion of linear cryptanalysis and differential cryptanalysis.

2.1 DES

DES (Data Encryption Standard) is the best known and most widely used Feistel block cipher. A representative characteristic of Feistel-based ciphers is that the encryption function is approximately identical to the decryption function except a reversal of the key schedule. The DES cipher's input is plaintext P and the output ciphertext C, each block with 64 bits. The encryption goes through a 16-round process, where the current 64-bit word is divided into 32-bit parts, the left part L_i and the right part R_i. The round i, $1 \leq i \leq 16$ is defined as follows

$$L_i = R_{i-1},$$

$$R_i = L_i \oplus F(R_{i-1}, K_i)$$

Where K_i is derived from the cipher key K via the key scheduling algorithm, which is the 48 bits subkey used in the ith round, and F is a round function.

The round function F is the core component of DES and consists of three operations, substitution, permutation and XOR operations ($\oplus$). The S-boxes are nonlinear mappings and the only nonlinear part of DES. There are eight S-boxes where each one is a substitution mapping 6 to 4 bits. The more detailed description of DES will be shown in [19, 20].

2.2 Linear Cryptanalysis and Differential Cryptanalysis

The linear cryptanalysis and differential cryptanalysis are among the most powerful cryptanalysis attacks for DES.

In 1993, Matsui introduced the linear cryptanalysis method for DES cipher. As a result, it is possible to break 8-round DES cipher with 2^{21} known-plaintexts and 16-round DES cipher with 2^{47} known-plaintexts [21, 22]. This method can obtain a linear approximate expression by constructing a statistical linear path between input and output of each S-box.

$$P[i_1,i_2,...,i_a] \oplus C[j_1,j_2,...,j_c] = K[k_1,k_2,...,k_c] \tag{1}$$

Where $i_1,i_2,...,i_a$, $j_1,j_2,...,j_b$ and $k_1,k_2,...,k_c$ denote fixed bit locations, and equation (1) holds with probability $p \neq 1/2$ for randomly given plaintext P and the corresponding ciphertext C. The magnitude of $|p-1/2|$ represents the effectiveness of equation (1). Once succeed in reaching an effective linear expression, it is possible to determine one key bit $K[k_1,k_2,...,k_c]$ by the algorithm based on the maximum likelihood method.

In 1990, Eli Biham and Adi Shamir introduced the differential cryptanalysis which can break any reduced variant of DES (with up to 15 rounds) in less than 2^{56} operations [23, 24]. This method of attack can be applied to a variety of DES-like substitution/permutation cryptosystems. Differential cryptanalysis is a chosen-plaintext/chosen-ciphertext attack. Cryptanalysts choose pairs of plaintexts so that there is a specified difference between members of the pair. They then study the difference between the members of the corresponding pair of ciphertexts. Statistics of the plaintext-ciphertext pair differences can yield information about the key used in encryption.

3   Cryptanalysis of Two-Round DES Using Genetic Algorithms

As we discuss above, the genetic algorithm has an obvious advantage in reducing the complexity of the search problem. The optimum key search is core of the cryptanalysis using genetic algorithm. The key length decides the difficulty of key search. As what we do in the binary genetic algorithm, the key representation is a chromosome as well, which is also a binary bit pattern. The chromosome's mating, crossover, and mutation processes become the same as the processes of optimizing key. Thus, cryptanalysis using genetic algorithms is stochastic adaptive processes. In this paper, we adopt known plaintext attack and design a fitness measure function to achieve the goal towards the above attacks.

3.1 Fitness Function

The cryptanalysis using genetic algorithms must find a satisfying fitness function. Furthermore, the function should measure the relation between given plaintext/ciphertext and target plaintext/ciphertext, and help us to find better solutions. It is acceptable that there are possible errors in generated solutions, and what is more the most of solutions are better and are expected to often behave in accordance with the definition. A fitness functions is defined by

$$\theta = (\zeta_0 \times \xi[\gamma^0(C_s)/\gamma^0(C_t)] + \zeta_1 \times \xi[\gamma^1(C_s)/\gamma^1(C_t)]/64 \qquad (2)$$

Where M represents the plaintext, C_s the corresponding original ciphertext, C_t the encryption ciphertext using trial key, and θ the fitness function, $0 \le \xi \le 1$ and $0 \le \zeta_0$, $\zeta_1 \le 64$; The $\gamma^0(C_s)$ denotes the number of 0 in original ciphertext C_s, and $\gamma^0(C_t)$ denotes the number of 0 in the generated ciphertext C_t. And the related $\gamma^1(C_s)$ and $\gamma^1(C_t)$ denote the number of 1 in C_s and C respectively. ξ is the function whose value is less than one. If ξ's value is above zero, ξ is itself. If not, ξ is its reciprocal. The ζ_0 denotes the number of 0 that is in the same bit position for C_s and C_t. Accordingly, ζ_1 denotes the number of 1 that is in the same bit position for C_s and C_t.

3.2 The Initialization Process

The initialization process determines how a population is initialized. In our experiment, the initialization population isn't randomly generated. It is initialized as follows. We need to define a plaintext M, an original ciphertext C, and an initialization seed K:

$$K_i = M_i \oplus C_i \qquad (3)$$

where i is defined as the size of individuals in the seeding population. This customizing design is derived from the approximate expression in linear cryptanalysis (see 2.2 above). It can speed up the evolution process so as to find the better optimum key as quickly as possible.

The initialization process must comprise every different individual, such as the best, the average and the weak. Therefore, the magnitude of population is large enough for the diversity of chromosomes. It is recommended that the size is over 100.

3.3 The Mating Process and the Mutation Process

Given a population of chromosomes, each one with a fitness value, the genetic algorithm will progress by randomly selecting two for mating. The selection is weighted in favor of chromosomes with a high fitness value. In general, the mating rate value is better between 0.70 and 0.80.

After the new generation has been determined, the chromosomes are subject to a low rate mutation function which involves different processes in an attempt to vary the genes. The mutation process is inverting the order of a set of bits between two random points. They can help to prevent the algorithm from being stuck in a local optimum point. In experiments, the mutation rate is below 0.001.

3.4 Breeding Strategy

The above processes are combined together into the whole genetic algorithm. At the beginning of a run of a genetic algorithm a large seeding population of chromosomes is created. Each chromosome will represent a guessed key to the problem. The breeding strategy can produce a new optimum key that inherits the preferably characteristic of the parent pair. It is important to notice the distinction between fitness and objective scores. In our experiments, we omit the IP and IP^{-1} transformation. In order to find the ultimate key, the following steps are repeated until an appropriate objective value is found.

1. A seeding population of chromosomes is generated from multi plaintext/ciphertext pairs.
2. A fitness value for each chromosome in the population is determined.
3. If the fitness value is above first limit point α, we can conclude that the chromosome is an optimum key.
4. If not, the crossover operation and mutation operation are applied to the chromosome.
5. The new population is scanned and used to update the list of better keys across the generations.
6. The process will be repeated if the optimum key can't be found. Whereas this process will stop after a fixed number of generations N.
7. If the fitness value is below first limit point α, the new generated trial key must not be a bad key. Hence, we should judge it through other plaintext/ciphertext pairs. If these new fitness values are below second limit point β, we discard it.
8. Repeat the step from 3 to 7 for γ time runs.
9. For all optimum keys, we calculate the sums 1 or 0 for their every bit position. Then every sum divides by γ.
10. From the step 9, we will find that some sums divided by γ have higher values than δ and other sums divided by γ have lower values than ζ in our experiment. Then, we can deduce those bits higher than δ as 1 and lower than ζ as 0. Those bits are regarded as valuable bits which generates a significant deviation from the other observed bits.

Repeat the step from 1 to 10, and continue to find other bits.

4 Experimental Setup and Results

Table 1 records the parameters and results in our experiments for one-round and two-round DES. The value of α is same as δ, and β is same as ζ respectively in Table 1. The optimum rate describes the percentage of optimum keys in all solutions. The success rate based on the method, that shows the circumstances that the deductions are true in step 10, are reported.

Table 1. The parameters and result in experiments

Round	$\alpha,\ \delta$	$\beta,\ \zeta$	N	γ	Opt.rate (%)	Suc.rate (%)
One	0.75	0.25	10,000	100	0.74	0.98
One	0.78	0.22	10,000	200	0.76	0.99
One	0.75	0.25	20,000	100	0.83	1.00
Two	0.70	0.30	10,000	100	0.71	0.98
Two	0.75	0.25	10,000	200	0.72	0.99
Two	0.70	0.30	20,000	100	0.76	0.99

Table 1 shows the success rate of 100-time and 200-time experiments. This algorithm was programmed in Visual C++ and run on Intel P4-M processor. Now, we illustrate with a simple example for the above algorithm. The original key supposed is used to hex: (64676A727479736F).The fixed number of generations N is 10,000, the runs time γ is 100, α and δ is 0.75, β and ζ is 0.25 in this experiments.

The experiment result for cryptanalysis of one-round and two-round DES is shown in Figure 1 which presents the target sums 1 for every bit position for using the above steps once.

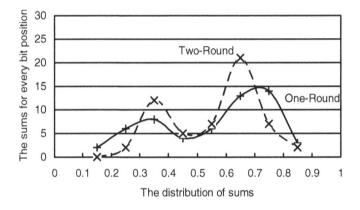

Fig. 1. The result of example in experiments for cryptanalysis of one-round and two-round DES

As what the Figure 1 indicates, the algorithm would find eleven valuable bits for the reduced two-round DES that acts in accord with the Table 1. Subsequently, we could fix those eleven bits in chromosome and repeat the above steps until the whole key is found. In above experiments, the algorithm runs thrice and can find the 56-bit key.

5   Conclusion and Future Work

As we discuss above, the fitness function is the approximate expression for using genetic algorithm to cryptanalyze DES, but it is not certainly the best one in future. We can produce an optimum key for each plaintext/ciphertext pair based on this fitness function. In experiments, the design of actual fitness function should consider the following important factors as well. Firstly, fitness function should possess a better feature of global performance and large numbers of optimum keys can be found. Secondly, it is difficult to produce a high fitness value for each search. In some instances, the difference is relatively obvious between the current chromosome and the actual key. We solve it by using multi plaintext/ciphertext pairs in experiment. Thirdly, we can use the auxilliary fitness function to improve the evaluation measure. Last but not least, this fitness function model doesn't confine to the cipher's inherent structure. Therefore, it can be used for other complex block ciphers.

The above results indicate that this genetic algorithm to cryptanalyze two-round DES cipher is successful and efficient. Contrast with the case of simple ciphers reported earlier, this research result shows that genetic algorithm is also a powerful tool for breaking the modern cipher algorithms. Because the design of the fitness function is unrestrained and flexible, this method can be readily adopted to handle other complex block ciphers. We wish that this observation would be feasible and might discover other better fitness functions for further work in our future.

Acknowledgments

We would like to thank the support by Excellent Young Teachers Program of China University of Geosciences (Grant No.CUGQNL0309), Chinese Natural Science Foundation (Grant No. 66973034, 60373087 and 90104005), and 863 Project (Grant No.2002AA141051).

References

1. Holland, J.H.: Adaptation in Natural and Artificial Systems. University of Michigan Press, Ann Arbor, MI (1975)
2. Gold, D.E.: Genetic algorithms in search, optimization, and machine learning. Addison-Wesley, Reading, MA (1989)
3. Clark, J.A.: Invited Paper. In: Nature-Inspired Cryptography: Past, Present and Future. CEC-2003, Canberra, Australia, December 2003, IEEE, Los Alamitos (2003)
4. Spillman, R.: Cryptanalysis of Knapsack Ciphers Using Genetic Algorithms. Cryptologia XVII(4), 367–377 (1993)
5. Spillman, R., Janssen, M., Nelson, B., Kepner, M.: Use of A Genetic Algorithm in the Cryptanalysis of simple substitution Ciphers. Cryptologia XVII(1), 187–201 (1993)
6. Clark, A.: Modern Optimisation Algorithms for Cryptanalysis, pp. 258–262. IEEE, Los Alamitos (1994)
7. Lin, F.-T., Kao, C.-Y.: A Genetic Algorithm for Ciphertext-Only Attack in Cryptanalysis, pp. 650–654. IEEE, Los Alamitos (1995)
8. Clark, A., Dawson, E.: Optimisation Heuristics for the Automated Cryptanalysis of Classical Ciphers. Journal of Combinatorial Mathematics and Combinatorial Computing, Papers in honour of Anne Penfold Street 28, 63–86 (1998)
9. Clark, A.J.: Optimisation Heuristics for Cryptology, PhD thesis, Queensland. University of Technology (1998)

10. Hernández, J.C., et al.: Genetic Cryptoanalysis of Two Rounds TEA. In: Sloot, P.M.A., Tan, C.J.K., Dongarra, J.J., Hoekstra, A.G. (eds.) Computational Science - ICCS 2002. LNCS, vol. 2331, pp. 1024–1031. Springer, Heidelberg (2002)
11. Laskari, E.C., Meletiouc, G.C., Stamatioud, Y.C., Vrahatis, M.N.: Evolutionary computation based cryptanalysis: A first study, pp. 823–830. Elsevier, Amsterdam (2005)
12. Hernández, J.C., et al.: Easing collision finding in cryptographic primitives with genetic algorithms, pp. 535–539. IEEE, Los Alamitos (2002)
13. Russell, M., Clark, J.A., Stepney, S.: Using Ants to Attack a Classical Cipher. In: Cantú-Paz, E., Foster, J.A., Deb, K., Davis, L., Roy, R., O'Reilly, U.-M., Beyer, H.-G., Kendall, G., Wilson, S.W., Harman, M., Wegener, J., Dasgupta, D., Potter, M.A., Schultz, A., Dowsland, K.A., Jonoska, N., Miller, J., Standish, R.K. (eds.) GECCO 2003. LNCS, vol. 2723, pp. 146–147. Springer, Heidelberg (2003)
14. Morelli, R., Walde, R., Servos, W.: A Study of Heuristic Approaches for Breaking short Cryptograms. International Journal on Artificial Intelligence Tools 13(1), 45–64 (2004)
15. Bafghi, A.G., Sadeghiyan, B.: Finding Suitable Differential Characteristics for Block Ciphers with Ant Colony Technique, pp. 418–423. IEEE, Los Alamitos (2004)
16. Clark, J.A., Jacob, J.L., Stepney, S.: The Design of S-Boxes by Simulated Annealing, pp. 1533–1537. IEEE, Los Alamitos (2004)
17. Albassal, E.A.M.B., Abdel-Moneim, A.W.: Genetic Algorithm Cryptanalysis Of The Basic Substitution Permutation Network, pp. 471–475. IEEE, Los Alamitos (2004)
18. Garici, M.A., Drias, H.: Cryptanalysis of Substitution Ciphers Using Scatter Search. In: Mira, J.M., Álvarez, J.R. (eds.) IWINAC 2005. LNCS, vol. 3562, pp. 31–40. Springer, Heidelberg (2005)
19. Coppersmith, D.: The data encryption standard (DES) and its strength against attacks. IBM Journal of Research and Development 38(3), 243–250 (1994)
20. National Bureau of Standards: Data Encryption Standard, FIPS-Pub.46. National Bureau of Standards, US Department of Commerce, Washington DC (1977)
21. Diffie, W., Hellman, M.: Exhaustive Cryptanalysis of the NBS Data Encryption Standard. IEEE Computer 10(6), 74–84 (1977)
22. Matsui, M.: Linear Cryptanalysis Method for DES Cipher. In: Helleseth, T. (ed.) EUROCRYPT 1993. LNCS, vol. 765, pp. 386–397. Springer, Heidelberg (1994)
23. Matsui, M.: The First Experimental Cryptanalysis of the Data Encryption Standard. In: Desmedt, Y.G. (ed.) CRYPTO 1994. LNCS, vol. 839, pp. 1–11. Springer, Heidelberg (1994)
24. Biham, E., Shamir, A.: Differential Cryptanalysis of the Data Encryption Standard. Springer, Heidelberg (1993)
25. Biham, E., Dunkelman, O., Keller, N.: Enhancing Differential-Linear Cryptanalysis. In: Zheng, Y. (ed.) ASIACRYPT 2002. LNCS, vol. 2501, pp. 254–266. Springer, Heidelberg (2002)

A Novel Artistic Image Generation Technique: Making Relief Effects Through Evolution

Jingsong He[1,2], Boping Shi[1,2], and Mingguo Liu[1,2,*]

[1]Department of Electronic Science and Technology,
[2]Nature Inspired Computation and Applications Laboratory (NICAL),
University of Science and Technology of China
hjss@ustc.edu.cn, {boping, mingguo}@mail.ustc.edu.cn

Abstract. The fascinating effect of applying evolutionary algorithms (EAs) to real-world applications consists in its less requirement of knowledge of the applied domain. Many efforts on this idea lead to a new term *Nature Inspired Computation*, which aims at solving real-world problems with techniques included the nature. Adaptive lossless image compression is one of the most important applications in the field of evolvable hardware(EHW) according to this idea. However, except the adaptive lossless image compression, few extended applications in the field of image processing was reported in the past years. This paper presents a novel evolutionary technique for making relief effects according to the principle of the nature. The proposed method is vary simple and efficient which needs few knowledge about the image precessing, except the common sense that almost all people have. Experimental results show that the proposed method is efficient, and can make quite different effects comparing with conventional method.

Keywords: Evolvable Hardware, Image Processing, Nature Inspired Computation.

1 Introduction

Evolutionary Algorithms (EAs) constitute one of the most important methodologies in the field of computational intelligence[1][2]. The fascinating effect of applying EAs to real-world applications consists in its less requirement of knowledge of the applied domain. That is, except the fitness function that contains more or less human cognitions on the objects, all questions will be solved through evolution. Many efforts on this idea lead to a new term *Nature Inspired Computation*, which aims at solving real-world problems with techniques included in the nature. Evolvable hardware (EHW) is a newly emerged field which reflects this idea of nature inspired computation typically. Initial works related to EHW is to solve some design problems for circuit optimization, and late efforts turn to

---

* This work is partially supported by the National Natural Science Foundation of China through Grant No. 60573170 and Grant No. 60428202.

L. Kang, Y. Liu, and S. Zeng (Eds.): ISICA 2007, LNCS 4683, pp. 591–600, 2007.

automatic circuit design and discovering novel circuit and its self-adaption mechanism for surviving in changing environment[3-6]. Several literatures pioneered routes of the way of and challenges in studying EHW [7-9], several works showed interesting methodologies for extending its applied fields, and perceptibly, many works endeavor to pursue valuable applications[10-16].

Among various applications of EHW in the past years, the adaptive lossless image compression is one of the representative applications of evolvable hardware. T.Higuchi *et.al.*[12] firstly presented an accomplishment of such task of compressing images on a special functional FPGA (F^2PGA) with their special variable length genetic algorithm (VGA), which has been regarded as the merely non-toy problem in the early researches in the field of evolvable hardware[10]. And afterwards, they present a continuous research on the problem of evolving more larger scale images[14]. As a different study, A. Fukunaga *et.al.*[10][15] presented a new prototype system, in which genetic programming (GP) was used as the evolving engine and the Lisp S-expression was used as coding mechanism, to solve the problem of such implementation on conventional FPGA. Both two kinds of systems are belong to extrinsic-EHW mode: VGA[12] for optimizing templates is executed on a host computer[14], as well as the execution of GP engine used in [14]. According to definitions and purposes, the executive mode of EHW can be divided into two types that intrinsic evolvable hardware and extrinsic evolvable hardware[7]. In general, extrinsic evolvable hardware was thought as more suitable for automatic circuit design and optimization(e.g.,[6]), while intrinsic evolvable hardware was thought as more suitable for online adaption and deeply exploring of novel circuit(e.g., [4]). To make the task of adaptive lossless image compression execute on-chip, [17] proposed an intrinsic-EHW mode through converting the predictive function to a kind of binary string, thus the task can be implemented easily by turning on/off a set of switches. No matter what method is, the basic idea of adaptive lossless image compression is to search the optimal predictive function based on the relativity between neighboring pixels.

Image processing is vary important to many vision-related computation and pattern recognition, while image compression is just important one in this field. It is vary important and interesting for us to extend the basic idea included in evolvable hardware to other image-related applications more widely. In this paper, we present a novel evolutionary technique for making relief effects according to the principle of the nature. The proposed method is so simple that it needs few knowledge about the image precessing, except the common sense that almost all people have. Experimental results show that the proposed method can make relief effects rapidly and efficiently. Compare with the conventional method, the proposed method based on evolution can provide more softly and clearly effects, while the conventional method can provide more roughly and blurry effects. The quite difference result between the proposed method and the conventional method shows that the proposed method can be token as a new supplement for artistic image generation.

2   Making Relief Effects Through Evolution

In this section, we employ the idea of *Nature Inspired Computation* which uses mechanisms included in nature, other than conventional techniques which are based on mathematical analysis. In order to make the proposed idea more close to *nature*, we do not intend to talk about conventional artistic image generation techniques, but describe our work as simple as possible.

2.1   Description of Image

For a $m \times n$ size image, we denote the horizontal site as i, and the vertical site as j, and the pixel value of $(i,\ j)$ as $x_{i,j}$. Since those pixels can also be described with roster scan, in 4-neighbor-prediction model, the current pixel value and its 4 neighbors for prediction can be denoted as x, x_1, x_2, x_3, and x_4 (i.e., $x_{i,j}$, $x_{i,j-1}$, $x_{i-1,j-1}$, $x_{i-1,j}$, and $x_{i-1,j+1}$) as in Fig.1 respectively. Conveniently, denote the predictive value of the current pixel as x', the predictive error could be expressed as $\Delta_{i,j}$, thus the error matrix between the original image and the predicted image could be obtained and denoted as Δ.

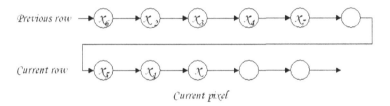

Fig. 1. The current pixel x and the arrangement of its neighboring pixels

2.2   Thinking About How to Make Relief Effects

Here we just tell of what the cognition we have on relief effects in artistic image generation. The motivation is vary simple and intuitive:

- Sensuously, the task of making relief effect is to make borderlands between different visual regions more impressive.
- The degree of relativity between pixels in the same visual region is relative larger than that between pixels in different visual regions.
- The more relativity between pixels, the easier the relationship between pixels can be found, thus the less the prediction errors.
- Usually, values of pixels around borderlands alter acutely, while values of pixels around centers of visual regions alter smoothly. It means, values of pixels around borderlands are relative difficult to be predicted than that around centers of visual regions.

It is apparent, we can employ the adaptive mechanism of lossless image compression directly to make relief effects. That is, we can make relief effects of images through magnifying differences between pixels and the predictive ones. Denote the error matrix Δ as

$$\Delta = \{\Delta_{i,j}|\ i = 1, 2, \cdots, m.\ j = 1, 2, \cdots, n.\}, \tag{1}$$

where

$$\Delta_{i,j} = x_{i,j} - x'_{i,j}, \tag{2}$$

and $x_{i,j}$ denotes the value of pixel at the site (i, j), and $x'_{i,j}$ is predicted value which can be calculated by Eq.(1). Then the magnifier of error matrix can be designed as a remap by normalizing the error matrix to R with bound $[0, 255]$ as follows.

$$R_{i,j} = round\left(\frac{\Delta_{i,j} - \Delta_{min}}{\Delta_{max} - \Delta_{min}}\right) \times 255, \tag{3}$$

where $\Delta_{min} = min(\Delta)$, and $\Delta_{max} = max(\Delta)$. Obviously, borderlands in the original image can be protruded by this way, thus relief effects can be made in R.

2.3   Obtain Predictive Function and Error Matrix Through Evolution

The interpolate function can be used to approach the predictive function. Therefor, the 4-neighbor predictive mode[15][17] can be written as

$$x'_k = \sum_{q=1}^{4} x_{k,q} \exp\left[-\alpha_q d(x_k, x_{k,q})\right], \tag{4}$$

where x'_k denotes the kth predicted value, $x_{k,q}$ denotes the pixel value of the q-th neighbor of x_k, α denotes the parameter of the interpolating function, and $d(x_k, x_{k,q})$ denotes the distance between the current point and its q-th neighbor. Therefore, $\Delta_{i,j} = x_k - x'_k$, and the task of prediction becomes to minimize the following objective function

$$f(x) = \sum_{k=1}^{N} ||x_k - x'_k||^2$$

$$= \sum_{k=1}^{N} \left[x_k - \sum_{i=1}^{4} x_{k,i} e^{-\alpha_i d(x, x_i)}\right]^2 \tag{5}$$

where N is the number of pixels for prediction. Conveniently, assume the effect of interpolating function only affects its neighboring pixels, i.e. $d(x_k, x_{k,q})$ equals to one uniformly. Hence Eq.(5) becomes

$$f(x) = \sum_{k=1}^{N} \left[x_k - \sum_{q=1}^{4} x_{k,q} e^{-\alpha_q}\right]^2. \tag{6}$$

Hence the task of searching the optimal function becomes a parameter optimization problem with four variables α_1, α_2, α_3, α_4.

2.4   Miscellaneous

To make the problem of Eq.(6) be easily implemented by intrinsic EHW, we can code parameters in Eq.(6) with a set of binary string. For example, we can use six-bit to code each parameter in Eq.(6) as

$$101101, \ 011100, \ 101100, \ 010101.$$

Since the length of chromosomes is fixed, and the value of exponential function can be calculated on chip by querying an embedded LOOK-UP Table (LUT), this kind of task can be implemented on chip without any difficult.

On the other hand, Genetic Programming algorithms (GPs) can also be used to do this work with Eq.(3) and the Lisp S-expression. Since GP model is symbol-based approach, its chromosome is relative complex than binary coded expression used in this paper, and it is not vary suitable for on-chip evolution. Here we don't discuss it furthermore.

3   Experimental Results of Making Relief Effects

In this section, experimental results on a set of images are presented to show the performance of the evolutionary method in all aspects, the strongpoint as well as weakness.

Firstly, we use a nature picture to show the performance of the evolutionary method through comparing with the conventional method (here we use the tools of *Photoshop* v6.0). The original picture is shown in Fig.2 (*a*), in which the foreground are vivid birds with white feather, and the background is pine trees interlaced with light and shade. Fig.2 (*b*) and Fig.2 (*c*) show the relief effects made by the proposed method and the conventional method respectively. By comparison, we can see that the evolutionary result looks more softly at borderlands, while the conventional result looks more roughly that makes regions around borderlands attract more attentions.

Secondly, we use a picture of traditional Chinese painting to show the merit of the proposed method. Results of the proposed method and the conventional method are shown in Fig.3 (*a*) and Fig.3 (*b*) respectively. By comparing these two results, we can see that the conventional method makes Chinese characters too roughly to be recognized easily, while the evolutionary result looks much clear. As a whole, the evolutionary result is more softly and clearly, while the conventional result is more roughly and blurry.

Thirdly, we use a picture of city with many buildings to show the demerit of the proposed method. Fig.4 and Fig.5 show the evolutionary result and the conventional result respectively. Compare these two results, we can see that the evolutionary result is still with the merits of softness and definition. However, we can find that the evolutionary method make the distance so weak that it can't be seen almost.

With above experimental results, we can affirm the contribution of the evolutionary method in making relief effects. That is, the proposed evolutionary method has the ability for making relief effects, and its result is quite different to that of the conventional method: the former looks more softly and clearly while the later looks more roughly and blurry. On this point, the proposed evolutionary method is obviously profitable to be a complementarity for the current technique using conventional methods. The merits and demerits shown in experiments are both come form the evolutionary computation: the stronger the predictor made by evolutionary computation, the clear the borderlands showed, and the more the smooth region lost.

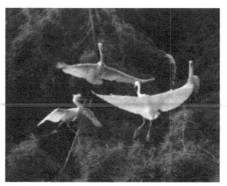

(a) The original picture: birds and trees.

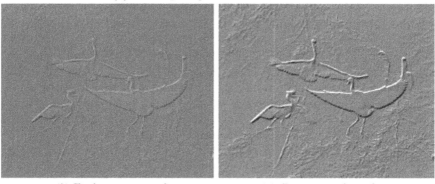

(b) Evolutionary result   (c) Conventional result

Fig. 2. Experimental results of the picture "birds and trees"

In addition, some settings and the computing time costed in experiments are summarized in Table 1. Since the parameter of the number of generation does not affect the final result distinctly, we just set it to 30-generation simply according to our experience. All results are obtained by execution with software. This result of software simulation shows that the proposed method is qualified to be used for real-world applications. As to hardware implementation, such as on DSP Chip (e.g., TMS320C6711) and FPGA (e.g., Altera Cyclone EPEP1C12Q240), the speed will be 30-times and 100-times faster respectively at least. Since here

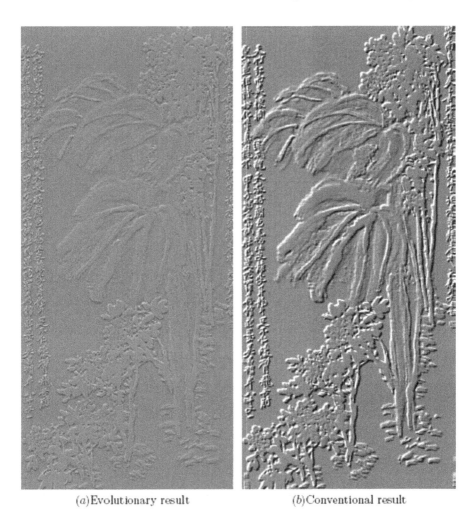

(a)Evolutionary result   (b)Conventional result

Fig. 3. Experimental results of the picture of Chinese painting

Table 1. Recorders of computing time with the proposed evolutionary method, where the number of generations is set to 30, the population size is 5. All results are averaged on 30 runs.

Image name	Image Size	Mean Time	Std. Deviation
Birds and Trees	472×600	0.5406 *sec.*	0.0781 *sec.*
Chinese panting	800×378	0.5660 *sec.*	0.0738 *sec.*
Buildings	534×800	0.6280 *sec.*	0.1053 *sec.*

the main purpose is to show the efficiency of the idea inspired by the nature, something about the details of hardware implementation will not be introduced.

Fig. 4. Evolutionary result of the picture of buildings

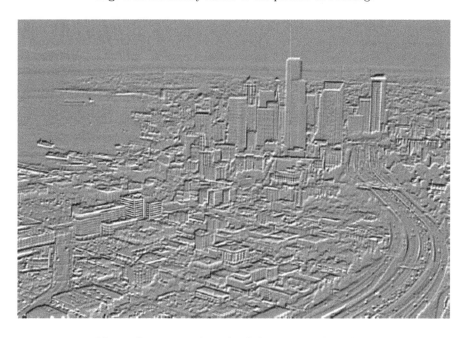

Fig. 5. Conventional result of the picture of buildings

4   Conclusions

Evolvable hardware (EHW) is a newly emerged field which reflects this idea of nature inspired computation typically. Among various applications of EHW in the past years, the adaptive lossless image compression is one of the representative applications of evolvable hardware, not only because it was the merely non-toy problem in the early researches in the field of evolvable hardware, but also because image processing is vary important to many vision-related computation and pattern recognition. It is vary important and interesting for us to extend the basic idea included in evolvable hardware to other applications more widely.

The fascinating effect of applying EAs to real-world applications consists in its less requirement of knowledge of the applied domain. Many efforts on this idea lead to a new term *Nature Inspired Computation*, which aims at solving real-world problems with techniques included the nature. In this paper, we present a novel evolutionary technique for making relief effects according to the principle of the nature. The proposed method is vary simple and efficient which needs few knowledge about the image precessing, except the common sense that almost all people have. Experimental results show that the proposed method can make relief effects rapidly and efficiently.

Compare with the conventional method, the proposed method based on evolution can provide more softly and clearly effects, while the conventional method can provide more roughly and blurry effects. The quite difference result between the proposed method and the conventional method shows that the proposed method can be token as a new supplement for artistic image generation.

References

1. Eiben, A.E., Smith, J.E.: Introduction to Evolutionary Computing. Springer, Heidelberg (2003)
2. Konar, A.: Computational Intelligence: Principles, Techniques and Applications. Springer, Heidelberg (2005)
3. Thompson, A., Layzell, P., Zebulum, R.S.: Explorations in Design Space: Unconventional Electronics Design Through Artificial Evolution. IEEE Trans. On Evolutionary Computation 3(3) (September 1999)
4. Thompson, A., Wasshuber, C.: Evolutionary Design of Single Electron Systems. In: Evolvable Hardware, 2000. IEEE Proceedings. The Second NASA/DoD Workshop, July 13-15, 2000, pp. 109–116. IEEE Computer Society Press, Los Alamitos (2000)
5. Stoica, A., Zebulum, R., Keymeulen, D., et al.: Reconfigurable VLSI Architectures for Evolvable Hardware: from Experimental Field Programable Transistor Arrays to Evolution-oriented Chips. Very Large Scale Integration (VLSI)Systems, IEEE Trans. On 9(1) (February 2001)
6. Hemmi, H., Hikage, T., Shimahara, K.: AdAM:a hardware evolutionary system. In: IEEE International Conference on Evolutionary Computation, pp. 193–196. IEEE Computer Society Press, Los Alamitos (1997)
7. Yao, X., Higuchi, T.: Promises and challenges of evolvable hardward. IEEE Trans. On Systems, Man, and Cybernetics - Part C: Applications and Reviews 29(1) (February 1999)

8. Stoica, A., Zebulum, R., Keymeulen, D.: Progress and Challenges in Building Evolvable Devices. In: Evolvable Hardware, 2001. IEEE Proceedings. The Third NASA/DoD Workshop, July 12-14, 2001, pp. 33–35. IEEE Computer Society Press, Los Alamitos (2001)

9. Montana, D., Popp, R., Iyer, S., Vidaver, G.: EvolvaWare: Genetic Programming for Optimal Design of Hardware-Based Algorithms. In: Genetic Programming 1998. Proceedings of the Third Annual Conference, University of Wisconsin, Madison, Wisconsin, USA, July 22-25, 1998, pp. 869–874. Morgan Kaufmann, San Francisco (1998)

10. Fukunaga, A., Hayworth, K., Stoica, A.: Evolvable Hardware for Spacecraft Autonomy. In: Aerospace Conference, vol. 3, pp. 135–143. IEEE, Los Alamitos (1998)

11. Stoica, A., Zebulum, R., Keymeulen, D., et al.: Reconfigurable VLSI Architectures for Evolvable Hardware: from Experimental Field Programable Transistor Arrays to Evolution-oriented Chips. Very Large Scale Integration (VLSI)Systems, IEEE Trans. on 9(1) (February 2001)

12. Higuchi, T., Murakawa, M., Iwata, M., Kajitani, I., Liu, W., Salami, M.: Evolvable hardware at function level. In: IEEE International Conference on Evolutionary Computation, pp. 187–192. IEEE Computer Society Press, Los Alamitos (1997)

13. Highchi, T., Iwata, M., Keymeulen, D., et al.: Real-World Applications of Analog and Digital Evolvable Hardware. IEEE Trans. on Evolutionary Computation 3(3), 220–235 (1999)

14. Sakanashi, H., Iwata, M., Higuchi, T.: A Lossless Compression Method for Halftone Images using Evolvable Hardware. In: Liu, Y., Tanaka, K., Iwata, M., Higuchi, T., Yasunaga, M. (eds.) ICES 2001. LNCS, vol. 2210, pp. 314–326. Springer, Heidelberg (2001)

15. Fukunaga, A., Stechert, A.: Evolving Nonlinear Predictive Models for lossless Image Compression with Genetic Programming. In: Proc.of the Third Annual Genetic Programming Conference, Winsconsin (1998)

16. He, J., Wang, X., Zhang, M., Wang, J., Fang, Q.: New Research on Scalability of Lossless Image Compression by GP Engine. In: The 2005 NASA/DoD Conference on Evolvable Hardware, The Westin Grand, Washington DC, USA, June 29 - July 1, 2005, pp. 160–164 (2005)

17. He, J., Yao, X., Tang, J.: Towards intrinsic evolvable hardware for predictive lossless image compression. In: Wang, T.-D., Li, X., Chen, S.-H., Wang, X., Abbass, H., Iba, H., Chen, G., Yao, X. (eds.) SEAL 2006. LNCS, vol. 4247, pp. 632–639. Springer, Heidelberg (2006)

Data Genome: An Abstract Model for Data Evolution*

Deyou Tang[1,2], Jianqing Xi[1], Yubin Guo[1], and Shunqi Shen[1]

[1] School of Computer Science & Engineering, South China University of Technology,
Guangzhou, 510641, China
[2] Department of Computer Science & Technology, Hunan University of Technology,
Zhuzhou, 412008, China
deyou_tang@126.com,jianqingxi@163.com,guoyubin@people.com.cn

Abstract. Modern information systems often process data that has been transferred, transformed or integrated from a variety of sources. In many application domains, information concerning the derivation of data items is crucial. Currently, a kind of metadata called data provenance is investigated by many researchers, but collection of provenance information must be maintained explicitly by dataset maintainer or specialized provenance management system. In this paper we investigate the problem of providing support of derivation information for applications in dataset itself. We put forward that every dataset has a unique data genome evolving with the evolution of dataset. Data genome is part of data and records derivation information for data actively. The characteristics of data genome show that the lineage of datasets can be uncovered by analyzing theirs data genomes. We also present computations of data genomes such as clone, transmit, mutate and introject to show how data genome evolves to provide derivation information from dataset itself.

1 Introduction

Data provenance [1][2][3](also referred to as data lineage and data pedigree), which describes how data is derived from its source, including all processes and transformations of data from original measurement to current form [4], has been gaining interest in the past decade. The two major approaches to representing provenance information use either annotations or inversion. In the former, data provenance is represented as annotations and every piece of data in a relation is assumed to have zero or more annotations associated with it [5]. Annotations are propagated along, from the source to the output, as data is being transformed through a query. Alternatively, the inversion method uses the property by which some derivations can be inverted to find the input data supplied to them to derive the output data [3, 14].

Data provenance can be used to trace the lineage of a piece of data in a warehousing environment [6~7] or scientific data management [8, 9]. In recent years, a number of teams have been applying data provenance to data and information

---

* This work is supported by Guangdong High-Tech Program (2006B80407001, 2006B 11301001), and Guangzhou High-Tech Program (2006Z3-D3081).

L. Kang, Y. Liu, and S. Zeng (Eds.): ISICA 2007, LNCS 4683, pp. 601–610, 2007.

generated within computer systems, including applications in software development, organ transplantation management [10], scientific simulation[11], stream filtering, and e-science[12, 13]. However, provenance information must be maintained explicitly by dataset maintainer or specialized provenance management system [18]. Though automated provenance recording support is feasible [15], however, dataset under such scenario is passive. On the other hand, considering data provenance is a kind of metadata [14] makes provenance meaningless if it was isolated from dataset. Therefore, it would be desirable if supporting for provenance can be provided in dataset itself.

In order to support providing provenance by dataset itself, we put forward that 'data' includes not only an abstraction of information (we call it data object), but also the abstraction of derivation information representing evolution of data and its metadata (we call it genetic information [17]). The genetic information isn't invariable, i.e. it evolves with the evolution of data. Either transformation of data object or genetic information will change the semantics of 'data'. As the process of semantics change has similarity with the evolution of life, we call it data evolution (Due to the limit of space, we have to illustrate data evolution in another paper).

Data evolution is different from data provenance. Data provenance is a kind of metadata describing where does dataset come from and why dataset has the current form [19]. Data evolution, however, focuses on representing the process of semantics change for data, and data provenance can be used as the metadata describing the origin of data in data evolution. The most important of all is that 'data' in data evolution is active, which means that 'data' can interact with users' behavior and detect the semantics change actively. In this paper, we present an abstract model— data genome model to manage the genetic information. We use the term dataset to refer to data in any form, such as files, tables, and virtual collections. In our model, every dataset has a unique data genome, which evolves with the evolution of dataset.

The rest of this article is structured as follows: Section 2 gives a new abstraction of data; Section 3 presents the concepts of data genome model (DGM) firstly, then analyzes the characteristics and constrains of DGM. Section 4 presents computations procedures of data genome; Section 5 contains our conclusions and outlines the directions for future research.

2   A New Abstraction of Data

The semantics of data is about associating meaning to data, and understand what data represents, and improve the value of data. Generally speaking, data may be processed many times in its lifecycle, but we think the semantics has been changed if only the content of data has been changed. Unfortunately, it isn't an unalterable truth. For example, when a dataset has been duplicated, can the security agency think the source dataset has the same meaning as before? Definitely, it isn't the same one. The reason is that the copy operation has caused the change of extension semantics of dataset. In fact, for any kind of operation, once an operation is carried out by an entity, we can think the semantics of data has been changed. Some operations cause the change of intrinsic semantics, while the others lead to the change of extension semantics. However, current mechanism of producing data doesn't support representing both intrinsic semantics and extension semantics. To provide such support, we must explore new mechanism of expressing semantics of data.

Since the semantics of data changes after each operation and data in different phase has different semantics, we divide the lifecycle of data into many phases and call the process of semantics change among different phases data evolution. Both intrinsic semantics and extension semantics may change in data evolution. The intrinsic semantics lies on the content of data and many successful data models can be applied to. The extension semantics is about information created in the switch of different phases. Such information has the following features: (1) it closely correlates with data and is meaningless if isolated from data; (2) it changes with the evolution of data and is different in different phase of lifecycle; (3) it is transmissible, i.e. some of the information created in previous phase will be transmitted to the next one in the evolutionary process. As information created in phase switch has the above features, we call such information the genetic information of data.

Though metadata can be used to describe information about data in the given condition, its role limits to be a flat conceptual model and we need a multidimensional data model to manage genetic information. In the next section, we'll introduce a genetic information model named *data genome model (DGM)* using the terms of *genome* in biology as reference [16]. As shown in figure 1, data genome attaches to every dataset, and every dataset has a unique data genome that evolves with the propagation and update of dataset. Data genome is created when dataset is created, deleted when dataset is deleted. Furthermore, Data genomes in different phases are different, but they are correlative in content.

Fig. 1. Data object and data genome

3 Data Genome Model

The genetic information of data includes the origin of data object (where does data object come from), transformations that data object has experienced (why data object has the current form), and external environment that the transformation has taken place (what events or entities have caused the transformation, which entities are related, when is the transformation taken place, etc). Such information can be organized in different granularities. In this section, we introduce the data genome model, which carries genetic information for data hierarchically.

3.1 Basic Concepts

Suppose A is a set of attributes, and V is a set of values, we have the following definitions [17]:

Definition 1. *A data gene fragment f is a 2-tuple: f=(a,v) $a \in A$, $v \in V$. Data gene is a positive closure: $g=F^+$, where F is a set of data gene fragments.*

Data gene fragment is the minimal unit for carrying genetic information. The selection of attributes used in data gene fragment depends on the metadata model of data genome. Genetic information can be cataloged into many kinds, and data gene is used to carry one kind of genetic information in different phase. The catalog depends on the granularity that developer has imposed on the genetic information. Data gene is the basic unit for transformation of genetic information.

Definition 2. *A leading data gene describes the instant information of data as a whole, denoted as g_a.*

g_a is created when the data is created, it holds the global information about data like identifier, title, creator, and so on.

Definition 3. *Data gene sequence is a 2-tuple, $s=<\$sid, D+>$, where $\$sid$ is the identifier of the sequence, D is a set of data genes that describe the same characteristics of data and its evolution.*

The same kind of data genes in difference phases forms a data gene sequence.

Definition 4. *A main data gene sequence s_m is a data gene sequence, which describes the global characteristics of data and its evolution.*

In different phases, g_a may different from the previous phase. If so, the history of data gene fragments that are different with the previous phase must be recorded in a data gene and attached to the main data gene sequence. Therefore, sequence s_m consist of a leading data gene and data genes carrying the history of those data gene fragments in the leading data gene.

Definition 5. *Data genome is a 3-tuple: $dg=<\$gid, S^+, DG'^*>$, where $\$gid$ is the identifier of data genome, S is a set of data gene sequence, DG' is a component data genome set.*

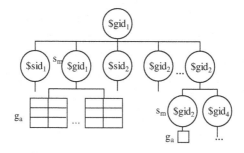

Fig. 2. The structure of Data Genome

In definition 5, $\$gid$ is globally unique to denote different data. The component data genome set DG'^* carries genetic information about the correlated data. For each $dg' \in DG'^*$, dg' is a data genome mutated from the data genome of the correlated data (suppose its data genome is dg'', we also call dg' is a sub data genome of dg''). In addition, there is a sequence in S^+ to record the correlation, whose identifier is equal to the identifier the component data genome. Figure 2 is a description for the structure of data genome.

3.2 Characteristics of Data Genomes

Data genomes in different phases are correlative in content. In fact, it is the correlation of data genome in content implies the lineage of data. In this subsection, we'll show how data genomes express the lineage of datasets. Herein, we introduce some symbols firstly.

$ID(s)$ denotes the identifier of data gene sequence s, $\$(dg)$ denotes the set of data genome identifiers dataset has been used, $T(s)$ denotes the data genes set of data gene sequence s. $I(s_m)$ denotes the leading data gene of the main data gene sequence. $S(dg)$ denotes the set of data gene sequence in dg, $s_m(dg)$ denotes the main data gene sequence of dg, $\psi(dg)$ denotes the component data genome set of dg.

Definition 6. *Data genes* g_1 *and* g_2 *are equivalent iff* $(\forall f \in g_1 : f \in g_2) \wedge (\forall f' \in g_2 : f' \in g_1)$ *, denoted as* $g_1 \equiv g_2$ *. Data gene sequence* s_1 *and* s_2 *are equivalent iff* $(\forall g_1 \in T(s_1), \exists g_2 \in T(s_2): g_1 \equiv g_2)$ *, and vice versa, denoted as* $s_1 \equiv s_2$ *. Data genomes* dg_1 *and* dg_2 *are equivalent iff* $(ID(s_m(dg_1)) = ID(s_m(dg_2)) \wedge (\forall s_1 \in S(dg_1), \exists s_2 \in S(dg_2): s_1 \equiv s_2)$ $\wedge (\forall dg_k \in \psi(dg_1), \exists dg_l \in \psi(dg_2): dg_k \equiv dg_l)$ *, and vice versa, denoted as* $dg_1 \equiv dg_2$ *.*

The fact that two data genomes are equivalent indicates that theirs datasets are duplicates of the same dataset.

Definition 7. *Data genes* g_1 *and* g_2 *are isogenous iff* $(\exists f \in g_1 : f \in g_2)$ *, denoted as* $g_1 \approx g_2$ *. Data gene sequence* s_1 *and* s_2 *are isogenous iff* $(\exists g' \in T(s_1), \exists g'' \in T(s_2): g' \approx g'')$ *(if* s_1 *and* s_2 *are main data gene sequences, it also has* $I(s_i) \approx I(s_j)$ *), denoted as* $s_1 \approx s_2$ *. Data genomes* dg_1 *and* dg_2 *are isogenous iff* $(\exists s' \in S(dg_1), \exists s'' \in S(dg_2): s' \approx s'') \wedge (s_m(dg_1) \approx s_m(dg_2)) \wedge (\$(dg_1) \cap \$(dg_2) \neq \varnothing)$ *, denoted as* $dg_1 \approx dg_2$ *.*

All isogenous data genomes of the same data genome form a data genome family, denoted as $\xi(dg)$, where dg is any data genome in the family. Data genome family describes the versions and distributions of data. Two data genomes are isogenous means that the corresponding dataset evolves from the same source or one datum is the ancestor of the other datum or they belong to the same copy. Two types of typical isogeny are *orthologs* (we also call it *cover*) and papalogs (we also call it *derivation*).

Definition 8. *Data gene* g_1 *covers* g_2 *iff* $(\forall f' \in g_2 : f' \in g_1)$ *, denoted as* $g_1 \succ g_2$ *. Data gene sequence* s_1 *covers* s_2 *iff* $\forall g' \in T(s_2), \exists g'' \in T(s_1): g'' \succ g'$ *, denoted as* $s_1 \succ s_2$ *. Data genome* dg_1 *covers* dg_2 *iff* $(\forall s' \in S(dg_2), \exists s'' \in S(dg_1): s'' \succ s')$ $\wedge (\forall dg' \in \psi(dg_2): dg' \in \psi(dg_1)) \wedge (ID(dg_1) = ID(dg_2))$ *, denoted as* $dg_1 \succ dg_2$ *.*

Orthologs describes the relationship of data genomes with the same $\$gid$. In definition 8, we call g_2 is a sub data gene of g_1, s_2 is a sub data gene sequence of s_1, dg_2 is a sub data genome of dg_1, and dg_1 is a cover data genome of dg_2.

Definition 9. *The minimal cover data genome of dg is the data genome dg′ that satisfies* $(dg' \succ dg) \wedge (\neg \exists dg'': dg' \succ dg'' \wedge dg'' \succ dg)$, *denoted as* $dg' \geq dg$.

All sub data genomes of dg form a sub data genome set, denoted as $\rho(dg)$. All cover data genomes of dg form a cover data genome set, denoted as $\vartheta(dg)$. Data genomes in $\rho(dg)$ or $\vartheta(dg)$ carry genetic information for the same data with different quality level.

Proposition 1. *Data genomes in* $\rho(dg)$ *can be expressed as a tree, called directed propagation tree (DPT for short),* $T \asymp V, E >$, *where* $V = \rho(dg), E = \{< dg_2, dg_k > | (dg_2, dg_k \in \rho(dg)) \wedge (dg_k \geq dg_2)\}$.

DPT expresses the premier disseminator and all receivers of the same copy. This feature can be used to trace the dissemination of data in P2P network, intrusion detection, etc.

Definition 10. *Genome* dg_2 *derives from* dg_1 *iff* $dg_1 \approx dg_2 \wedge ID(s_m(dg_1)) \neq ID(s_m(dg_2))$ $\wedge \$(dg_1) \subset \$(dg_2) \wedge (T(s_m(dg_1)) - \{I(s_m(dg_1))\} \subseteq T(s_m(dg_2)) - \{I(s_m(dg_2))\}) \wedge ((\forall s' \in S(dg_1) - \{s_m(dg_1)\}) \wedge (\forall s'' \in S(dg_2) - \{s_m(dg_2)\}) \wedge ID(s'') = ID(s')) : s'' \succ s')$, *denoted as* $dg_1 \rightarrow dg_2$.

Papalogs describes the relationship of data genomes with different *$gid*. Here, we say dg_2 is the *derived data genome* of dg_1, and dg_1 is the *ancestor data genome* of dg_2. The derived data genome inherits genetic information from the ancestor data genome selectively.

Definition 11. *The direct derived data genome of dg is the data genome that satisfies* $(dg \rightarrow dg') \wedge (\neg \exists dg'': dg \rightarrow dg'' \wedge dg'' \rightarrow dg')$, *denoted as* $dg \mapsto dg'$.

Proposition 2. *For data gnome family* $\xi(dg)$, *let* DG_i $(i = 1..n)$ *be the subset of* $\xi(dg)$, *and* $\forall dg', dg'' \in DG_i : \vartheta(dg') \cap \vartheta(dg'') \neq \phi$, *then* $\xi(dg)$ *is a directed tree called family tree,* $T = < V, E >$, *where* $V = \{DG_i | DG_i \subseteq \xi(dg)\}, E = \{< DG_1, DG_2 > | (DG_1, DG_2 \in V, \forall dg' \in DG_1, dg'' \in DG_2) dg' \mapsto dg''\}$.

In the family tree, the root node is the sub data genome set of the meta data genome and child nodes are those cover data genome sets of its derived data genomes, so do the derived data genomes. The family tree can be treated as a version tree, in which different nodes denote different version of data. Inside each node, it is a *DPT*, which expresses the distributions of data.

Relationships of equivalence, isogeny (including cover and derivation) describe relationships of data genomes whose data have the same provenance. Relationships of data genomes belonging to different data, which have different origin but relevant, are defined as follows.

Definition 12. *For data genomes* dg_1 *and* dg_2, *if* $dg' \in \rho(dg_1) \wedge dg' \in \psi(dg_2) \wedge (\exists s \in S(dg_2) \wedge ID(s) = ID(s_m(dg_1)))$, *then* dg_1 *is a correlated data genome of* dg_2, *denoted as* $dg_1 \triangleright dg_2$.

In definition 12, data genome dg_2 introjects genetic information from data genome dg_1. Suppose A's data genome is dg_A, B's data genome is dg_B. If data A is correlated to data B, $dg_A \rhd dg_B$.

3.3 Constrains of Data Genome

With the evolution of data, the volume of its data genome may become large. We here make some constrains on data genome to keep the structure as concise as possible.

1) Each main data gene sequence has a unique leading data gene.

2) Each data genome has a unique main data gene sequence.

Since each data genome has a unique main data gene sequence identifier, we also use the identifier of the data genome to identify the main data gene sequence.

3) Each dataset has a unique data genome.

4) In data gene sequence, any data gene can't be a sub data gene of the others.

5) In data genome, any data gene sequence can't be a sub data gene sequence of other data gene sequences.

6) For different component data genomes dg' and dg'' in $\psi(dg)$, dg' or dg'' can be a derived data genome of the other, but they can't be sub data genome of the other.

7) For each component data genome dg' in $\psi(dg)$, it has the main data gene sequence only, and the main data gene sequence has the leading data gene only. So do the component data genomes of dg'

8) For each component data genome dg' in $\psi(dg)$, it must have one and only one data gene sequence s in $S(dg)$, which satisfies $ID(s)=ID(s_m(dg'))$, otherwise dg is invalid.

4 Computations of Data Genomes

Dataset has different genetic information in different phases. The variety of genetic information in different phases is performed by data genome itself. In the lifecycle of data, the corresponding data genome changes its structure using the following transformations to accommodate the semantics change of dataset.

1) Clone: $dg' = C(dg) \Rightarrow dg' = \langle \$gid', S(dg'), \psi(dg') \rangle$ where $\$gid' = ID(dg), S(dg') = S(dg),$ $\psi(dg') = \psi(dg)$.

This transformation creates a data genome exactly the same as source data genome, which is the basis of the following transformations.

2)   Add-tail: $dg' = At(dg) \Rightarrow dg' = \langle \$gid', S(dg'), \psi(dg') \rangle$ where $\$gid' = ID(dg), \psi(dg') = \psi(dg),$ $S(dg') = S(dg) - \{s\} + \{s'\}$ where $s \in S(dg) \wedge s \neq s_m(dg) \wedge T(s') = T(s) + \{g\}$.

This transformation doesn't change the identifier of data genome, and it only appends a data gene to the corresponding data gene sequence in data genome. If there doesn't exist a right data gene sequence to manage the data gene, a new data gene sequence is created. This transformation is used to record operations such as data browsing, transferring and replicating. The common features of these operations are that they don't change the content of data object.

3) Mutate: $dg' = M(dg) \Rightarrow dg' = \langle \$gid', S(dg'), \psi(dg') \rangle where \$gid' \neq ID(dg), \psi(dg') = \psi(dg),$

$$\begin{cases} \begin{cases} \begin{cases} S(dg') = S(dg) - \{s_m(dg)\} + \{s_m(dg')\} + \{s\} \ where \\ s \notin S(dg) \wedge s \neq \varnothing \end{cases} or \\ \begin{cases} S(dg') = S(dg) - \{s_m(dg)\} - \{s\} + \{s_m(dg')\} + \{s'\} where \\ s \in S(dg) \wedge T(s') = T(s) + \{g'\} \end{cases} \end{cases} where \\ \begin{cases} T(s_m(dg')) = T(s_m(dg)) - \{I(s_m(dg))\} + \{I(s_m(dg'))\} + \{g\} \wedge \\ I(s_m(dg') \neq I(s_m(dg)) \end{cases} \end{cases}$$

This transformation operates on the main data gene sequence. First, it creates a new data genome dg' as the *clone* transformation and changes its identifier and the identifier of its main data gene sequence. Second, it replaces some data gene fragments if necessary and creates a new data gene g in the main data gene sequence to record those replaced data gene fragments. In addition, a data gene g' is used to record the operation that causes this transformation. If there doesn't exist a right sequence to take up g', a new sequence is created.

Usually, change of metadata manually or update of data content will activate a mutation transformation on the corresponding data genome. This transformation involves in the following transformations.

4) Transmit: $dg' = T(dg) \Rightarrow dg' = \langle \$gid', S(dg'), \psi(dg') \rangle where \ \$gid' \neq ID(dg), \psi(dg') = \psi(dg) ,$

$$\begin{cases} \begin{cases} S(dg') = S(dg) - S - \{s_m(dg)\} + \{s_m(dg')\} + \{s\} \ where \\ s \notin S(dg) \wedge s \neq \varnothing \end{cases} or \\ \begin{cases} S(dg') = S(dg) - S - \{s_m(dg)\} - \{s\} + \{s_m(dg')\} + \{s'\} where \\ s \in S(dg) \wedge T(s') = T(s) + \{g'\} \end{cases} \end{cases} where S \subset S(dg) \wedge S \cap S(dg') = \varnothing$$

This transformation transmits genetic information from ancestor data to derived data selectively. Suppose dg is the data genome of the ancestor data. First, this transformation creates a new data genome dg' exactly the same as dg and adds data genes to dg' to record the content update of data object. Secondly, a data gene g' carrying the operation information triggering the transformation is inserted into one data gene sequence in dg' whose data genes have the same structure with g'. If the data gene sequence doesn't exist, a new data gene sequence is created. Thirdly, it discard some discardable sequence in $S(dg)$. Finally, it performs a mutation on the leading data gene of dg', that is, it changes the identifier of the main data gene sequence, replaces some data gene fragments if necessary and creates a new data gene in the main data gene sequence to record those replaced data gene fragments.

5) Introject: $dg' = I(dg, dg'') \Rightarrow dg' = \langle \$gid', S(dg'), \psi(dg') \rangle where \$gid' \neq ID(dg) \wedge \$gid' \neq ID(dg''),$

$\{\psi(dg') = \psi(dg) \ where \ \exists dg_i \in \psi(dg) | ID(dg_i) = ID(dg'') or \psi(dg') = \psi(dg) + \{dg_j\}$

$where \ dg'' \succ dg_j\}, S(dg') = S(dg) - \{s_m(dg)\} - \{s\} + \{s_m(dg')\} + \{s'\} where s \in S(dg) \wedge$

$ID(s) = ID(dg'') \wedge T(s') = T(s) + \{g'\} or S(dg') = S(dg) - \{s_m(dg)\} + \{s_m(dg')\} + \{s\}$

$where \ ID(s) = ID(dg'') \wedge s \neq \varnothing .$

This transformation manipulates data genomes as dataset correlates other data. First, it creates a new data genome dg' as the *clone* transformation. If there doesn't exist a sub data genome of dg'' in $\psi(dg)$, it filters out a sub data genome of dg'' as a component data genome of dg' and adds a new data gene sequence s to record the correlation of the correlated data genome (suppose s doesn't exist initially). If s does exist and $ID(s)=ID(s_m(dg''))$, it appends a new data gene to s. This transformation doesn't work independently, it often follows by a *transmit* transformation.

6) Prune: $dg' = P(dg, dg'') \Rightarrow dg' = \langle \$gid', S(dg'), \psi(dg') \rangle\, where\, \$gid' \neq ID(dg) \wedge \$gid' \neq ID(dg'')$,

$\psi(dg') = \psi(dg) - \{dg_j\}\, where\, dg'' \succ dg_j, S(dg') = S(dg) - \{s_m(dg)\} - \{s\} + \{s_m(dg')\} +$

$\{s'\} where\, s \in S(dg) \wedge ID(s) = ID(dg'') \wedge T(s') = T(s) + \{g'\}$

This transformation is the inverse of transformation introject. First, it also clones a new data genome dg'. As operator I correlates data genome to other data genome, this operator cancels the correlation. A data gene carrying the cancel information is inserted into the corresponding data gene sequence. Second, if the component data genome can be removed from dg', we remove it. Usually, a *transmit* transformation follows one or multi *prune* transformation.

With the aforementioned transformations, we can simulate the evolution of any kind of data. For instance, when a file is created in an editor, its data genome is created accordingly. Afterwards, if we append some words to the file, the data genome performs a *transmit* transformation to memorize our behavior. If we copy and paste data from other data sources in the process of edit, the data genome performs an *introject* transformation to correlate other data genomes. If we open the file only for browsing, the data genome only needs to perform an *add-tail* transformation after the end of browsing operation.

5 Conclusion and Future Works

In this paper, we give a new abstraction of dataset, which includes not only data object but also data genome. Data genome represents the extension semantics of dataset, evolves with the evolution of data. We present the abstract model of data genome and analyze its characteristics. Six transformations are presented to show how data genome can evolve to reflect the evolution of data and metadata. Due to the limit of space, we don't describe them in details. As this research is in its early stages, more works are required to make data evolution not to be an armchair strategist. Our works can be extended as following.

In the modeling side, we only give coarse-grained constrains on the structure of data genome, future works need to give fine-grained constrains on the semantics of data genome. This includes investigating the schemas of data gene sequences and theirs relationships.

In the performance evaluation side, though we have applied data genome in our peer-to-peer data integration and exchange platform to trace the data flows and source of ill-defined data, the role of data genome is limit to be complementary tools instead of a support technology for data quality, audit trail, etc. We plan to apply data genome to more complicated environment such as DBMS so that we can improve the performance of some algorithms in data genome computation.

References

1. Lanter, D.P.: Design Of A Lineage-Based Meta-Data Base For GIS. Cartography and Geographic Information Systems 18, 255–261 (1991)
2. Greenwood, M., Goble, C., Stevens, R., et al.: Provenance of e-Science Experiments - experience from Bioinformatics. In: Proc. of the UK OST e-Science second All Hands Meeting (2003)
3. Woodruff, A.G., Stonebraker, M.: Supporting fine-grained data lineage in a database visualization environment. In: ICDE '97. pp. 91–102 (1997)
4. Bose, R., Frew, J.: Lineage retrieval for scientific data processing: A survey. ACM Computing Surveys 37(1), 1–28 (2005)
5. Bhagwat, D., Chiticariu, L., Tan, W.C., et al.: An annotation management system for relational databases. VLDB Journal 14(4), 373–396 (2005)
6. Cui, Y.W., Widom, J., Wiener, J.L.: Tracing the lineage of view data in a warehousing environment. ACM Trannsactions on Database Systems 25(2), 179–227 (2000)
7. Cui, Y.W., Widom, J.: Lineage tracing for general data warehouse transformations. VLDB Journal 12(1), 41–58 (2003)
8. Buneman, P., Khanna, S., Tajima, K., et al.: Archiving scientific data. ACM Trannsactions on Database Systems 29(1), 2–42 (2004)
9. Jagadish, H.V., Olken, F.: Database Management for Life Sciences Research. SIGMOD Record 33, 15–20 (2004)
10. álvarez, S., Vázquez-Salceda, J., et al.: Applying Provenance in Distributed Organ Transplant Management. In: Moreau, L., Foster, I. (eds.) IPAW 2006. LNCS, vol. 4145, pp. 28–36. Springer, Heidelberg (2006)
11. Petrinja, E., Stankovski, V., Turk, Ž.: A provenance data management system for improving the product modeling process. Automation in Construction 16, 485–497 (2007)
12. Foster, I., Vockler, J., Wilde, M., et al.: Chimera: a virtual data system for representing, querying, and automating data derivation. In: Proc. 14th International Conference on Scientific and Statistical Database Management, pp. 37–46 (2002)
13. Zhao, Y., Wilde, M., Foster, I.: Applying the Virtual Data Provenance Model. In: Moreau, L., Foster, I. (eds.) IPAW 2006. LNCS, vol. 4145, pp. 148–161. Springer, Heidelberg (2006)
14. Simmhan, Y.L., Plale, B., Gannon, D.: A Survey of Data Provenance in e-science. ACM SIGMOD Record 34(3), 31–36 (2005)
15. Braun, U., Garfinkel, S.L., Holland, D.A., et al.: Issues in Automatic Provenance Collection. In: Moreau, L., Foster, I. (eds.) IPAW 2006. LNCS, vol. 4145, pp. 171–183. Springer, Heidelberg (2006)
16. Tang, D.Y., Xi, J.Q., Guo, Y.B.: Model and Algebra for Genetic Information of Data. In: Ali, M., Dapoigny, R. (eds.) IEA/AIE 2006. LNCS (LNAI), vol. 4031, pp. 1071–1079. Springer, Heidelberg (2006)
17. Tang, D.Y., Guo, Y.B., Xi, J.Q.: The Concept of Data Genome and Its Applications. In: Proceeding of the First International Symposium on Pervasive Computing and Applications(SPCA06), pp. 866–871 (2006)
18. Buneman, P., Cheney, J., VanSummeren, S.: On the expressiveness of implicit provenance in query and update languages. In: Schwentick, T., Suciu, D. (eds.) ICDT 2007. LNCS, vol. 4353, Springer, Heidelberg (2006)
19. Buneman, P., Khanna, S., Tan, W.C.: Why and where: A characterization of data provenance. In: Van den Bussche, J., Vianu, V. (eds.) ICDT 2001. LNCS, vol. 1973, pp. 316–330. Springer, Heidelberg (2000)

Intrinsic Evolution of Frequency Splitter with a New Analog EHW Platform

Wei Zhang[1], Yuanxiang Li[2], and Guoliang He[1]

[1] School of Computer, Wuhan University, Wuhan, 430072, China
[2] State Key Lab. of Software Engineering, Wuhan University, Wuhan, 430072, China
maghero@163.com, yxli@whu.edu.cn, guoliang.he@126.com

Abstract. A new analog intrinsic EHW platform based on AN231E04 FPAA device is described. AN231E04 is in the third generation of the Anadigm FPAA series and is based on switched capacitors technology. Its flexible dynamic reconfiguration ability and function-level design element ensured that more practical and large-scale circuits can be evolved. A frequency splitter is evolved in this research on the EHW platform composed of three parts: Computer, AN231K04-DVLP3 development board and external A/D circuit. GA was employed as the main algorithm of this EHW system and proved to be a good selection when topology of circuit is out of consideration.

1 Introduction

Intrinsic Evolvable Hardware (EHW) was put forward by the combination of Evolution Algorithm (EA) and reconfigurable hardware technology. An evolvable hardware system is composed of two main components: reconfigurable hardware and reconfiguration mechanism [1]. The reconfigurable hardware is the implementation and evaluation platform for EHW [2]. Reconfiguration mechanisms are usually certain branches of EA, such as Genetic Algorithm (GA) and Genetic Programming (GP). Analog circuits are of great importance in electronic system design, since the world is fundamentally analog in nature. Although the amount of digital design activity far outpaces that of analog design, most digital systems require analog modules as interfaces to real world. Approximately 60% of CMOS-based application-specific integrated circuit (ASIC) designs incorporated analog circuits [3]. With challenging analog circuit design problems and fewer analog design engineers, there are economic reasons for automating the analog design process. Analog EHW is an important branch of evolvable hardware research.

From the previous description we can see that the reconfigurable hardware technology is critical for EHW development. Basically, the research in this area goes into two directions: on one hand you can have specialized analog integrated circuits with only limited programming capabilities available. Leading manufacturer in that field is Lattice Semiconductors, makers of the inexpensive but relatively high-performance ispPAC family of programmable arrays, restricted mostly to continuous-time analog filters and lately also to mixed signal power

L. Kang, Y. Liu, and S. Zeng (Eds.): ISICA 2007, LNCS 4683, pp. 611–620, 2007.

supply controllers [4]. On the other hand there are general-purpose Field Programmable Analog Arrays (FPAAs) [5] which are in basic analog versions of probably more known Field Programmable Gate Arrays (FPGAs). There are several approaches to produce FPAA devices, whereof many implementations are designed in switched-capacitor technology. The dpAsp series of Anadigm Inc.® is the most typical production among them. In this research, we use one of the latest productions of dpAsp series to fulfill our work.

The goal of this research is to provide a method to develop intrinsic analog evolution hardware with AN231E04 device. This paper is organized as follows. Section 2 introduces the hardware platform AN231K04-DVLP3 development board. Section 3 briefly describes the implementation of a frequency splitter with AN231K04. The experimental results are given in Section 4. Section 5 draws some conclusions and presents our future works.

2   Hardware Platform

The hardware platform used in this research is the AN231K04-DVLP3 development board from Anadigm Inc.® [6], which is the latest development kit production of this company. The layout of AN231K04 is shown in Fig.1–(a). This board can be divided into three parts: 1) DpAsp chip, which is the most important component of the board, the analog circuit is configured by the program inside this chip and then the analog signal is processed here. 2) Digital interface, which is connected to PC through a RS232 CABLE. This interface is employed to download the configuration data to the chip and to control the course of signal processing. 3) Analog I/O interface. Analog signal is inputted and outputted through this interface.

The device on this development platform is an AN231E04. AN231E04 contains a 2x2 matrix of configurable analog blocks (CAB's) which in turn can hold a number of predefined analogue functions provided as configurable analog modules (CAM's) with the Anadigm Designer 2 software installation. I/O capabilities are provided through two dedicated analogue Type1 I/O cells and two analogue Type1A I/O cells as input or output. All this I/O cells can be programmed to satisfy the requirement of different I/O type. A Look up Table (LUT) is provided to build certain CAM's, see Fig.1–(b).

This device is based on switched capacitors technology. Switched capacitors circuits [7] are realized with the use of some basic building blocks, such as opAmps, capacitors, switches and non-overlapping clocks. The operation of these circuits is based on the principle of the resistor equivalence of a switched capacitor [7]. This principle is illustrated in Fig.2, where ϕ_1 and ϕ_2 are the non-overlapping clocks. In Fig.2–(a), the average current is given by:

$$I_{avg} = \frac{C_1(V_1 - V_2)}{T} \tag{1}$$

where T is the clock period. This is equivalent to the resistor Req, shown in Fig.2–(b). By this technology, integrators, filters, and oscillators can be implemented.

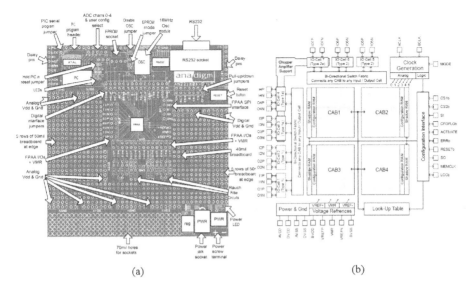

(a)   (b)

Fig. 1. The layout of the AN231K04-DVLP3 development board and architecture of AN231E04 dpAsp chip

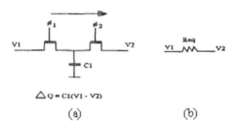

$$\triangle Q = C_1(V_1 - V_2)$$

(a)   (b)

Fig. 2. Switched Capacitor/Resistor equivalence principle

AN231E04 is in the third generation of Anadigm FPAA series. It has a lot of advantages to its older counterparts. The most important one is the flexible dynamic reconfiguration ability. Old devices can be configured any number of times, but an intervening reset is required between each configuration load, so it is hard to develop any EHW application with these devices. AN231E04 is more flexible, allowing for 'on-the-fly' reconfiguration - that means we can reconfigure the device freely without reset. Furthermore, any configuration data can be downloaded in less than 15 ms and this high speed ensured the possibility to develop complex circuit.

3   Implementation of Frequency Splitter

Another advantage of AN231E04 is that AN231E04 uses configurable analog modules (CAMs) rather than basic analog components such as resistors and transistors as its basic design element. Since anadigm's EDA tool Anadigm Designer 2 has offered a lot of various CAM models which are designed by highly experienced analog circuit engineers and tested strictly for hundreds of times, almost all predefined CAMs can offer perfect performance. Therefore, we can place more attention on evolving more practical and large-scale circuits by modifying the connection and sizing of CAMs. In this sense, AN231E04 is more appropriate than FPTA to develop EHW applications.

In this research we will evolve a frequency splitter. Frequency splitter is usually employed to separate signals of certain frequencies from original signal and distribute them through different channels. Frequency splitter is an important basic analog component in electronic communication productions and its performance can exert a pivotal influence on the quality of the whole system. Usually, a good frequency splitter requires precise splitting ability and high fidelity, so we designed the fitness function according to these two factors.

The whole EHW platform is composed of three parts: Computer, AN231K04-DVLP3 development board and external A/D circuit. The computer is employed to generate initial embryo circuit and to fulfill the evolution algorithm. AN231K04-DVLP3 development board actually runs each individual circuit during the course of evolution. The external A/D circuit transforms the analog output of AN231K04-DVLP3 development board into digital signal and feeds it back to the computer in order that the fitness of circuit individuals can be calculated.

3.1 Frequency Splitter

Figure 3 is the schematic circuit diagram of a typical frequency splitter. All GNDs in the diagram are abbreviated. This is a simple typical two channel frequency splitter which is used to separate high and low frequency signals from the original one. All analog signals in AN231E04's switched capacitor network are implemented as differential ones. Therefore, each I/O cell has two signal lines.

Figure 4 is the initial embryo circuit of frequency splitter created by anadigm's EDA tool Anadigm Designer 2. This embryo circuit can be divided into three channels and each channel is composed of two CAMs. The low frequency channel includes an invert gain stage (labeled as *GainInv1*) and an invert bilinear filter stage (labeled as *FilterBilinear1*). The high frequency channel also includes an invert gain stage (labeled as *GainInv2*) and an invert bilinear filter stage (labeled as *FilterBilinear2*). The composed channel includes an invert sum stage (labeled as *SumInv1*) and an invert gain stage (labeled as *GainInv3*).

3.2 EHW Platform

As the previous description, the whole EHW platform is composed of three parts (Fig.5).

Computer. Computer is employed to generate initial embryo circuit and fulfill evolution algorithm. Anadigm Inc.® has offered powerful EDA tool Anadigm

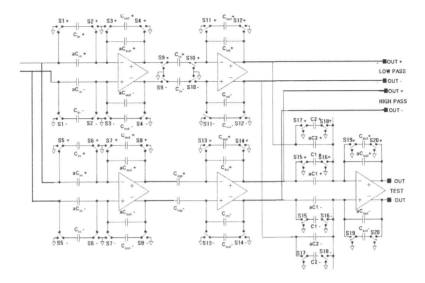

Fig. 3. The schematic circuit diagram of frequency splitter

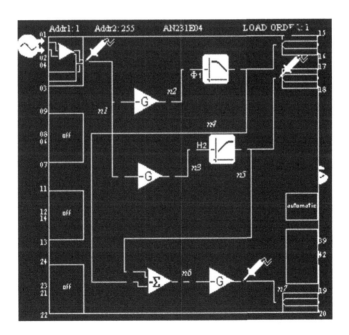

Fig. 4. The initial embryo circuit of frequency splitter

Designer 2 for its dpAsps with the ability to create embryo circuit, to simulate with embedded signal generator and to create c++ code prototype for dynamic configuration.

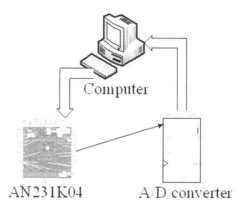

Fig. 5. EHW Platform

AN231K04. AN231K04-DVLP3 development board can actually run each individual circuit during the course of evolution. But in practice, not all individuals were downloaded to development board out of consideration for evolution speed. We employed some tricks of high speed pre-evaluation to find out the individuals which may present awful fitness or even injure the dpAsp chip. The program eliminates these individuals from download array and directly set their fitness to ∞ (worst fitness).

A/D Converter. External A/D circuit transforms analog output of AN231K04-DVLP3 development board into digital signal and feeds it back to the computer in order that the fitness of circuit individuals can be calculated. We employed AD673 8-Bit A/D converter produced by Analog Devices Inc.® to fulfill this work.

3.3   Algorithm

Genetic Algorithm is search algorithms based on the mechanics of natural selection and natural genetics [8], which are gradient-free, parallel and are well adapted to handling complex and irregular solution spaces [9]. GA's operation by generating an initial random population of individuals and progressing to subsequent generations by applying genetic operators, rewarding more suitable individuals with a higher chance of supplying genetic material to the next generation. They have been applied by many to the problem of optimization (for a full review see [10]), and have proven to be a successful tool.

Due to the limitation of anadigm's EDA tool Anadigm Designer 2, the method to dynamic reconfigure the topology of target circuit is to be investigated. So, In this research, we only concern about the evolution of parameters, and GA is an appropriate tool to fulfill pure parameter evolution.

Encoding. The frequency splitter was designed with a standard genetic algorithm operating on its 6 parameters. These parameters are:

- Gain of GainInv1, labeled as $GainInv1Gain$
- Gain of GainInv2, labeled as $GainInv2Gain$
- Gain of FilterBilinear1, labeled as $FilterBilinear1Gain$
- Gain of FilterBilinear2, labeled as $FilterBilinear2Gain$
- Corner frequency of FilterBilinear1, labeled as $FilterBilinear1Freq$
- Corner frequency of FilterBilinear2, labeled as $FilterBilinear2Freq$

We make use of binary encoding method and combine these parameters together to build the chromosome. (Fig.6)

Fig. 6. Chromosome

Operators. The genetic algorithm in this research employed roulette-wheel selection, crossover and mutation operators.

Fitness Evaluation. Candidate solutions are evaluated directly in AN231E04. In the whole process, all possible combinations input signals are applied to each candidate and the response is detected at the output. The fitness function is designed as follow:

$$Fitness = \sum_{i=1}^{n} [(Ar_H - As_H)^2 + (Ar_L - As_L)^2 + (Ar_C - As_C)^2] \qquad (2)$$

where Ar is the amplitude of response signal and As is the amplitude of original signal. 'H' indicates the high frequency channel, 'L' the low one and 'C' the combined one. A smaller fitness represents a better individual.

4   Experimental Results

Figure 7 shows the result of experiment. These curves were generated by our program, which represents original signals and responses from three channels. In these figures, the blue curves represent the signals from low channel, the red ones from high channel and the black ones from combined channel. Bold lines represent the input signals and hair lines their responses. Since the original input is stochastic periodic signal and these three figures were captured in different moments, the outlook of original curves in these figures may look different.

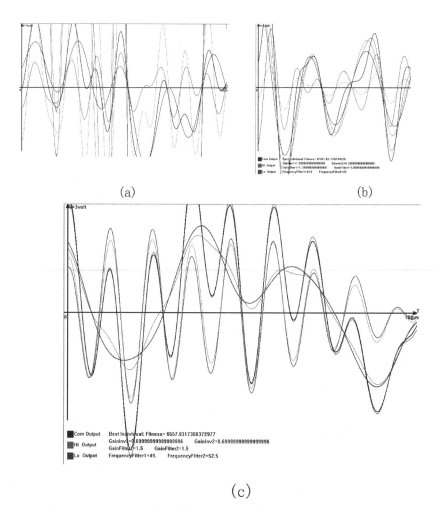

(a)   (b)

(c)

Fig. 7. Experimental Results

Figure 7–(a) was the curves of a discretionary individual in the process of evolution, this figure shows that the difference between original signals and there responses is big and the circuit is far from global optimization. Figure 7–(b) was the curves of the best individual of a discretionary generation. It's obvious that the difference is smaller and the fitness is much better. Figure 7–(c) was the curves of the final result after 1000 generations of evolution. The whole algorithm converges, and the fitness became smallest (6657.93).

5   Conclusion

This research discussed how to develop EHW system with AN231E04 dpAsp device, the third generation of the Anadigm FPAA series. Its flexible dynamic reconfiguration ability and function-level design method ensures that we can evolve more practical and larger-scale circuits.

The EHW platform in this experiment was composed of three parts: Computer, AN231K04-DVLP3 development board and external A/D circuit. The computer was employed to generate initial embryo circuit and execute evolution algorithm. AN231K04-DVLP3 development board actually ran each individual circuit during the course of evolution. The external A/D circuit transformed analog output signal from AN231K04-DVLP3 development board into digital signal and fed it back to the computer in order that the fitness of circuit individuals could be calculated.

GA was employed as the main algorithm of this EHW system, which proved to be a good selection when topology of circuit is out of consideration.

In future work, we will pay attention to the evolution of AN231E04's topology. We will try to find an appropriate encoding skill to realize combination code of connection and sizing. At the same time, we hope we can solve topology dynamic reconfiguration problem with Anadigm's help. We also plan to evolve more complex system such as multi-input controller and devices employed in extreme environments such as high or low temperature analog devices.

Acknowledgment

We would like particularly to thank State Key Laboratory of Software Engineering, Wuhan University, for supporting our research work in this field. This work is also supported by the National Natural Science Foundation of China under Grant No. 60442001 and National High-Tech Research and Development Program of China (863 Program) No. 2002AA1Z1490.

References

1. Stoica, et al.: Evolving Circuits in Seconds: Experiments with a Stand-Alone Board Level Evolvable System. In: 2002 NASA/DoD Conference on Evolvable Hardware. Alenxandria, VA (July 2002)
2. Higuichi, T.: Real-World Applications of Analog and Digital Evolvable Hardware [J]. IEEE Trans. on Evolutionary Computation 3(3), 220–235 (1999)
3. Gielen, G., Sansen, W.: Symbolic Analysis for Automated Design of Analog Integrated Circuits. Kluwer, Boston, MA (1991)
4. IspPAC programmable analog ICs: Lattice semiconductor datasheet, www.latticesemi.com
5. Lee, E.K.F., Hui, W.L.: A Novel Switched-Capacitor Based Field Programmable Analog Array Architecture. Analog Integrated Circuits and Signal Processing, Special Issue on Field-Programmable Analog Arrays 17(1–2) (September 1998)
6. Analog dpASP company: AnadigmApex FPAA Family User Manual (2006), www.anadigm.com
7. Johns, A.D., Martin, K.: Analog Integrated Circuit Design. John Wiley and Sons Inc., Chichester (1997)

8. Goldberg, D.: Genetic Algorithms in Search, Optimization and Machine Learning. Addison-Wesley Publishing Company, Reading, Massachusetts (1989)
9. Chambers, L.D.: The Practical Handbook of Genetic Algorithms, 2nd edn. Applications, vol. 1. Chapman&Hall, Sydney, Australia (2001)
10. Xin, Y.: Evolving artificial neural networks. Proceedings of the IEEE 87(9), 1423–1447 (1999)

Towards the Role of Heuristic Knowledge in EA

Yingzhou Bi[1,2], Lixin Ding[1], and Weiqin Ying[1]

[1] State key laboratory of software engineering, Wuhan University, Wuhan 430072, China
[2] Department of Information Technology, Guangxi Teachers Education University,
Nanning 530001, China
byzhou@163.com

Abstract. Evolutionary Algorithm (EA) is a stochastic search algorithm and widely used in various real world problems. Classic EA uses little problem specific knowledge, so it is called lean knowledge approach. Because of the randomicity of crossover, mutation and selection, its' searching strategy is semi-blind, and the efficiency is usually low. In order to acquire an efficient and effective EA that suits difficult real-world problems, we try to best incorporate heuristic knowledge into an EA to guide the search focusing on the most promising area. By comparing different EAs for solving the traveling sales man problem (TSP) and auto-generating test paper problem, we investigate the role of heuristic knowledge in EA.

Keywords: heuristic knowledge, coding of individual, guided variation operator, traveling salesman problem, auto-generating test paper.

1 Introduction

Evolutionary Algorithm (EA) is widely used as robust searching scheme in various real world applications, including function optimization, optimal scheduling, and many combinatorial optimization problems. It is a stochastic search algorithm, which draws inspiration from the process of natural evolution and adaptation, and solves problem by large computation in a stochastic trial-and-error (also known as generate-and-test) style [1]. Classic EA uses little problem specific knowledge, so it is called lean knowledge approach. Because of the randomicity of crossover, mutation and selection, its' search is semi-blind, and the efficiency is usually low [2]: although spending lots of time, a high quality solution can hardly reach in last stage in EA, and the solution always traps in local optimal. Faced to "exploding" solution space, it is tough to find high quality solution just by increasing the population size, diversity of searching, and the number of iteration.

In order to acquire an efficient and effective EA that suits difficult real-world problems, we try to understand the mechanism of EA and have other methods or data structures incorporated into it. This category of algorithm is very successful in practice and forms a rapid growing research area with great potential [3-6]. According to the no free lunch theorem [7], there is no single evolutionary algorithm (EA) which can solve all possible optimization problems efficiently. For a given optimization problem, we need to focus on designing problem-oriented algorithms in order to

L. Kang, Y. Liu, and S. Zeng (Eds.): ISICA 2007, LNCS 4683, pp. 621–630, 2007.

improve the algorithm's performance. The problem-specific knowledge is important in designing efficient EAs. For the same problem, different EAs can be designed based on different knowledge.

However, for any NP-hard problem, we only have limited heuristic knowledge to use when designing EAs that solve it. In this paper we aim to discover the role of heuristic knowledge in EAs. Through a comparative study of different EAs for solving the traveling sales man problem (TSP) and auto-generating test paper problem, which is a multi-objective optimization problem with practical applications, we investigate the role of heuristic knowledge in EA to find a feasible solution and the optimal solution: for TSP, we utilizes the near neighbor principle to design a guided mutation operator ; for auto-generating test paper problem, we employ matrix coding to represent the individual that exhibits some regularities of the problem and utilize the structural properties of problem to design the guided variation operators.

In this paper, we firstly analyze the destructive and constructive aspects of crossover in section 2, and then design different genetic algorithms for TSP in section 3. In section 4, we compare different coding of individuals for auto-generating test paper problem and finally are the conclusions.

2 Schema-Based Analyses of Crossover Operators

Holland's original formulation of schema theorem shows that schemata of above-average fitness would increase their number of instances within the population from generation to generation. Based on schema theorem, Spear investigates the role of crossover and mutation in EA [6].

The disruption of crossover refers to the offspring of an individual in a schema H_k no longer belong to schema H_k after crossover. It can be depicted as fig.1: P_1 is a member of a third-order schema H_3, and P_2 is an arbitrary string. If the two parents have not matching alleles at position d_2 or they have not matching alleles at poison d_1 and d_3, the schema H_3 should be disrupted.

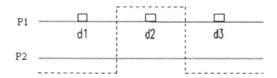

Fig. 1. An example of two-point crossover. P_1 is a member of a third-order schema H_3, and P_2 is an arbitrary string.

The construction of crossover refers to having crossover create an instance of a schema H_k from both parents, where one parent is in schema H_m ,while the other parent is the schema H_n, and $k=m+n$. Figure 2 provides a pictorial example. P_1 is a member of a second-order schema, and P_2 is a member of another second-order schema. A higher schema H_4 is constructed by crossover.

Fig. 2. An example of construction of two-point crossover. P_1 is a member of a two-order schema H_2, and P_2 is a member of another two-order schema H_2.

On one hand crossover operator select low-order schemata and progressively combine them to create higher order schemata; however, on the other hand, crossover operator often breaks the schemata. Based on the systemic analyze of crossover operator, Spear presents some general results: all forms of crossover are more disruptive of high-order schemas and become less disruptive when the population converges. In order to improve the algorithmic efficiency, it is crucial for us to utilize some characteristics and knowledge in the pending problems for restraining the disruption during evolution.

3 Different Algorithms for TSP

TSP can be described as follow: given a set of n nodes and distances for each pair of nodes, find a roundtrip of minimal total length visiting each node exactly once. It can formularize as searching a permutation of integers $\pi=\{V_1, V_2, \ldots V_n\}$, which makes the following express reach the minimum:

$$D_\pi = \sum_{j=1}^{j=n-1} d(V_j, V_{j+1}) + d(V_n, V_1) + d(V_j, V_k) \qquad (1)$$

Where $d(V_j, V_{j+1})$ means the distance between the nodes V_j and V_{j+1}, and D_π is the total length. TSP was proved to be NP-hard and it is very difficult to gain a solution with normal algorithms.

3.1 Canonical Genetic Algorithm for TSP

In canonical genetic algorithm, the permutation of the order of visiting the nodes is usually used as the coding of TSP and $1/D_\pi$ is regarded as the fitness function. Crossover operator is the Order Crossover designed by Davis and Mutation operator is the Inversion Mutation [1].

Algorithm 1
1) Initialize the population randomly or with greedy algorithm.
2) Evaluate the individuals, if stop criterion is satisfied, stop; otherwise, continue the following.
3) Crossover: select two parents and perform crossover.
4) Mutation: mutate the resulting offspring, and insert the new candidates into next generation. Go to step 2.

3.2   A Novel Algorithm with the "Guided Mutation Operator"

Based on the analyze of fitness function (formula (1)), a solution with high fitness should be composed of most short links between nodes, it means the usher and subsequence of a node in a roundtrip is always its near neighbor, and here, we called it "principle of near neighbor". If the principle is disobeyed, the roundtrip will be composed of lots of long links, and the total length must be very long. So we should find the near neighbor of all nodes at the beginning of algorithm and the near neighbor of nodes are selected by the distance between them. Usually, we only consider five nodes as near neighbor. The principle of near neighbor and the near neighbor of every node are the heuristics knowledge for TSP. Here we utilize this knowledge to design the guided mutation operator.

The Guided Mutation Operator as following:

1) Analyze all links in the candidate roundtrip: record each pair of nodes of the long links (if a solution composing n links, we usually consider $n/2 \sim n/3$ longer links) and the distance between them with a structure array. Here we called the structure array *update-array*; every element in update-array A has three parts: two nodes and a distance, where one node is called resource-nodes and another node is called destiny-nodes.

2) Examine all the resource-node in update-array A whether the destiny-node is its near neighbor. For example, for any element in update-array A, suppose its resource-node is V_a ,and its destiny-node is V_b ,if V_b is not near neighbor of V_a ,it means something is wrong in this gene, repairing this gene may increase the fitness of the candidate solution. Inspired by *2-change* algorithms [8], we present *2-repair* operators as the guided mutation operator to repair the malfunction genes in candidate solution.

3) 2-repair mutation operator: for a long link (V_a, V_b),if the destiny-node V_b is not 2-nearest neighbor of resource-node V_a, then select a node V_c from the near neighbor of V_a , obviously we have $d(V_a, V_b) > d(V_a, V_c)$. Suppose node V_d is subsequence of node V_c and the solution is expressed as $\pi=\{V_1, V_2, \ldots V_a, V_b \ldots V_c, V_d \ldots V_n\}$,its total length is

$$D(\pi) = \sum_{i=1}^{r} d_i + d(V_a, V_b) + \sum_{i=r+2}^{t} d_i + d(V_c, V_d) + \sum_{i=t+2}^{n} d_i \ . \tag{2}$$

if $d(V_a, V_b) + d(V_c, V_d) > d(V_a, V_c) + d(V_b, V_d)$,then we reverse all the nodes between V_b and V_c(including V_b, V_c), then get a new solution $\pi'= \{V_1, V_2, \ldots V_a, V_c \ldots V_b, V_d \ldots V_n \}$, its total length is

$$D(\pi') = \sum_{i=1}^{r} d_i + d(V_a, V_c) + \sum_{i=r+2}^{t} d_i' + d(V_b, V_d) + \sum_{i=t+2}^{n} d_i \ . \tag{3}$$

for the symmetry TSP, we always have $d_i = d_i'$, so under the condition $d(V_a, V_b) + d(V_c, V_d) > d(V_a, V_c) + d(V_b, V_d)$,it is obvious that $D(\pi) > D(\pi')$,which means the new solution is better.

According to the analysis in section 2, we present a novel algorithm for TSP: the "blind" mutation operator in algorithm 1 is substituted by the "guided" mutation operator, which utilizing the near neighbor principle for refraining the destruction of crossover while make useful of its construction during evolutionary process.

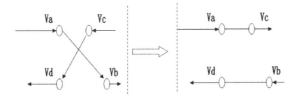

Fig. 3. An example of 2-repair mutation operator for TSP

Algorithm 2
1) Initialize the population with greedy algorithm.
2) Compute the near neighbors of all nodes.
3) Evaluate the individuals, if the solution quality is satisfied then stop, otherwise, continue the following.
4) Crossover: select two parents and perform crossover.
5) Guided mutation: repair the resulting offspring with 2-repair operator and insert the new candidates into next generation.
6) Go to step 3.

3.3 Experimental Results

We examine algorithm 1 and algorithm 2 with 4 TSP instances: two of them are *TSP200* and *TSP500*, where the nodes in *TSP200* and *TSP500* are 200 and 500 respectively; The location of all the nodes are generated randomly in two-dimension space: $x \in [0,1000], y \in [0,1000]$, the distance between nodes is Euclidean distance. Another two instances are *fl417*, *att532*, which come from TSPLIB [9].

In order to compare the performance between algorithm 1 and algorithm 2, we design the parameter as following: in algorithm 1, the size of population is 100;the crossover and the mutation probabilities are 0.5 and 0.3,respectively; the number of generations is 20000; in algorithm2, the size of population is 50; the crossover and the mutation probabilities are 0.8 and 0.85,respectively;the number of generations is 100.The experimental results of algorithm 1 and algorithm 2 over 20 independent runs respectively are listed in Table 1, where n is the number of nodes, D_0 is the known minimal tour length: the results are gained by Concorde package [10] and *time* denotes the average run time, while δ denotes the relative difference [11] which is defined in Equation (4)

$$\delta\% = \left(\sum_{i=1}^{T} (D_i - D_0) \right) / (T \times D_0) \times 100\% \qquad (4)$$

Where D_i denotes the minimal tour length obtained from the i-th run, T is the number of independent runs.

Based on the results in table 1, we can make some general observations: evolutionary algorithm can reach a relatively good solution from a bad solution after a stochastic trial-and-error. Because of the destruction of crossover and mutation, a better solution quality can hardly reach in last stage in EA, and the solution always trap in local optimal. Faced to "exploding" solution space, it is tough to find high quality solution just by increasing the population size, diversity of searching, and the number of iteration. However, with "guided" mutation we can restrain the destruction of crossover by repairing the malfunction genes in candidate solution, and utilize its construction in our novel algorithm.

Table 1. Compare the performance between algorithm 1 and algorithm 2

TSP Instances	D_0	δ		Time(second)	
		Algorithm 1	Algorithm 2	Algorithm 3	Algorithm 4
TSP200	10610	7.9	**0.4**	136	**9**
TSP500	16310	4.6	**1.3**	425	**95**
Fl417	11861	3.3	**1.3**	315	**38**
Att532	86729	6.5	**1.9**	473	**102**

4 Deferent EAs for Auto-generating Test Paper

In order to examine student's situation of mastering to knowledge , the question mainly includes the following attributes [12]: (1)Knowledge range, the level range of knowledge with affiliated content of the examination question ; (2)Question type; (3)Mark; (4)Cognition class, reflecting this question's requirement for content of courses level; (5)Item difficulty, reflecting how challenging assessment questions are for students; (6)Discrimination degree; (7)Time expected to finish the question. So generation of question paper with n questions means extracting n questions from the test bank, whilst n questions and 7 attribute of each question compose a matrix S with n rows, 7 columns.

$$
S = \begin{bmatrix} a_{11} & a_{12} & \cdots & a_{17} \\ a_{21} & a_{22} & \cdots & a_{27} \\ \vdots & \vdots & \cdots & \vdots \\ a_{n1} & a_{n2} & \cdots & a_{n7} \end{bmatrix}
$$

The distribution of the column element in matrix S meets the general requirement for the paper that users appointed separately in advance, which include: (1)Mark proportion that every chapter accounts for; (2)The number of questions belonging to various question type; (3) Mark proportion that the question reflecting the request of different competence levels accounts for; (4) Mark proportion that the questions of different difficulty account for; (5) Mark proportion that questions of different differentiation degrees account for; (6) The total time; (7) The total mark. A qualified question paper should meet the entire requirement that users appointed, or the error is accepted. In practice, we establish a function F_e to express the overall error that synthetically reflects the error between the users' request and these 7 indexes. The overall error can be formulated with the error function F_e:

$$F_e = \sum_{i=1}^{i=7} f_i w_i \tag{5}$$

Where f_i refers to the absolute value of error between i-th index and the user's requirement, w_i refers to the weight of the i-th index.

Automatic generation of question paper is extracting the examination question from the test bank, such that the overall error between the distribution of questions' attributes and the user's requirement is minimized. Therefore, Automatic generation of question paper can be regarded as a parameter optimization problem under restriction condition, where restriction condition is that selected questions should be in the bank. However, the relations among different attributes are hard to describe, we can not know it until we query in the bank, which makes it difficult to solve by adopting classical mathematics method.

4.1 Simple Evolutionary Algorithms for Auto-generating Test Papers

The representation of individuals in simple EA is usually binary string: given a test question bank with m questions, where $X_1, X_2, \ldots, X_m$ expressing these m questions, auto-generating question paper is selected n questions from $X_1, X_2, \ldots, X_m$, such that the whole index Fe(defined in formula (5)) is minimum. Here, we express the candidate solution with a m-bit binary string: $F_1 F_2 F_3 \ldots F_m$. If F_i ($1 \leq i \leq m$) equals to 1, it means this question has been chosen, otherwise, it means this question has not been chosen. Suppose this question paper has been required to have n questions, it means there are exactly n F_i who equals to 1. The simple evolutionary algorithm can be described as follows.

Algorithm 3. simple evolutionary algorithm based on binary coding

   Step 1: Initialize the population randomly.

   Step 2: Evaluate the individuals, if stop criterion is satisfied, stop; otherwise, continue the following.

   Step 3: Crossover. Crossover means choosing a random number in the range [*1, m*], and then splitting both parents at this point and creating two children by exchanging the tails.

   Step 4: Mutation. In mutation operation, we consider each individual separately and allow each bit to flip (i.e., from 1 to 0 or 0 to 1) with a small probability P_m

   Step 5: Correcting invalid solution. After performing these evolutionary operations, the candidate solution may be invalid. For example, after crossover operation, the number of F_i valued 1 in the string $F_1 F_2 F_3 \cdots F_m$ may be not equal to n, which means the candidate solution corresponding to the individual is invalid and it needs correction. Go to step 2.

4.2 Guided Variation Operators for Auto-generating Test Papers

During the process of search, simple crossover operator chooses a crossover point randomly and a good solution can be destroyed. In order to achieve better result, we consider adopting better coding, so we can utilize problem specific knowledge while performing variation operation. Furthermore, auto-generating question paper is generating a question paper with an accepted error, which need not be the optimum usually, so we decide to select the known solution as initialization population to increase efficiency. Based on the analysis above, we decide to adopt matrix coding, which means the variation operator can perform directly on the parameter matrix S, whilst the initialization population is generated by the greedy approach.

Algorithm 4

The process of algorithm 4 is the same as algorithm 3; the only difference between them is coding and variation operator. Here, a solution is a matrix with n rows and 7 columns, where n is refers to the number of questions in a test paper and the different columns in matrix refer to different property distributions of questions.

According to the domain expert, a qualified test bank needs to meet the following requirement: all the knowledge points and relative question type should be covered; furthermore, the type is close related with mark in every question. While generating question paper applying the greedy approach, we can find all the questions in the bank, whose attributes meet the first three columns of the parameter matrix S, but do not meet the last four columns. So what the evolutionary operator should do is to assign the last four attribute. With the heuristics knowledge from domain expert, we design the guided variation operators as follows:

① Guided crossover operator: let *locat:=random(4)*, where random(4) refers to randomly select an integer between 0 to 3. If the crossover probability P_c is satisfied, then split both parents at column *3+locat* and create two children by exchanging the tails, otherwise replicate two parents.

② Guided mutation operator: If the mutation probability P_m is satisfied, we randomly select a row from the candidate solution, i.e. the parameter matrix S, and get the value of the *question type* from the second field of this row. With the value of *question type*, we select another question with the same *question type*, and substitute the selected row of the parameter matrix S with the new question's attribute.

4.3 Experimental Results

The algorithm 3 and algorithm 4 have been performed 10 times respectively. Some results are list as follows, where *Gen* refers to generation, $F_e i (i=1$ to 80) refers to the overall error of the i-th individual, which is defined in Equation (5); P_r refers to replication probability, P_c refers to crossover probability, P_m refers to mutation probability, n refers to the size of population. In practice, we accept the solution if its' overall error is less than 10, however, the condition is usually relax to less than 16 in order to generate a test paper in reasonable time in web-application.

Table 2. Partial result in the 6th run of algorithm 3 $P_r=0.33$, $P_c=0.59$, $P_m=0.08$, $n=80$

Generation	$F_e 10$	$F_e 30$	$F_e 40$	$F_e 50$	$F_e 60$	$F_e 80$
Initial	295	232	141	131	124	128
400th	24	23	25	27	23	25
800th	12	18	19	15	13	17

Table 3. Partial result in the 3rd run of algorithm 3 P_r=0.35, P_c=0.6, P_m=0.05,n=80

Generation	F_e10	F_e 30	F_e 40	F_e 54	F_e 60	F_e 80
Initial	25	32	21	31	24	21
7th	14	21	24	22	16	25
16th	12	14	19	9	13	15

Base on the binary string, classic evolutionary operators utilize little problem specific knowledge, so it has to run a long time, and the quality of the solution is not good enough. The algorithm 3 is quick to reach the near –optimal solution, but the process of finding the last optimal is slow, since the choice of the points for crossover and the genes for mutation is random. When searching for good scheme, a good solution can be destroyed by an inappropriate choice of crossover points.

Our experiments try to solve the auto-generating test paper in three different methods: method 1 is the algorithm 3, while method 2 is the greedy approach and method 3 is the algorithm 4. In practice, the algorithm 4 has been performed 10 times, the size of population being 80, which means that greedy approach runs 800 times, but only 1 solutions' error equals to or less than 16. Based on matrix coding and initialized population using greedy approach, algorithm 4 achieves promising result. In our experiments, it can usually achieve a qualified solution before 20th generation.

5   Conclusions and Future Work

Our experiments show that the incorporation of domain knowledge with EAs can greatly improve the quality of solution in shorter time. It is very important to know how to best incorporate the specific problem knowledge into an EA, for example, utilizing heuristics knowledge to design guided variation operator, and designing appropriate coding that correlate with the problem's structure. Furthermore, working with domain experts is another way to acquire knowledge for designing efficient and effective EA.

Our experiments cover only the problems whose fitness function can be analyzed. In the future we will employ machine learning methods to extract knowledge from search points and their fitness-values. By utilizing the gained knowledge, we may design a wiser EA for more difficult problems.

Acknowledgments. This work is supported in part by the National Natural Science Foundation of China (Grant no. 60204001), and Natural Science Foundation of Guangxi (Grant no. 0679018).

References

1. Eiben, A.E., Smith, J.E.: Introduction to Evolutionary Computing. Springer, Heidelberg (2003)
2. He, J., Yao, X., Li, J.: A Comparative Study of Three Evolutionary Algorithms Incorporating Different Amounts of Domain Knowledge for Node Covering Problem. IEEE Transactions on Systems, Man, and Cybernetics—Part C: Applications and Reviews 35(2), 266–271 (2005)

3. Yao, X., Xu, Y.: Recent Advance in Evolutionary Computation. Journ. of Comput. Sci. & Technol. 21(1), 1–18 (2006)
4. Blum, C., Roli, A.: Metaheuristics in Combinatorial Optimization: Overview and Conceptual Comparison. ACM Computing Surveys 35(3), 268–308 (2003)
5. Jiao, L., Wang, L.: A novel genetic algorithm based on immunity. IEEE Transactions on Systems, Man, and Cybernetics 30(5), 552–561 (2000)
6. Spears, W.M.: The role of Mutation and Recombination in Evolutionary Algorithms. George Mason University, Virginnia (1998)
7. David, H., William, G.: No Free Lunch Theorems for Optimization. IEEE Transactions on Evolutionary Computation 1(1), 67–82 (1997)
8. Helsgaun, K.: An Effective Implementation of the Lin-Kernighan Traveling Salesman Heuristic, http://www.akira.ruc.dk/ keld/
9. http://www.iwr.uni-heidelberg.de/groups/comopt/software/TSPLIB95/
10. Concorde TSP solver for windows, http://www.tsp.gatech.edu/concorde/index.html
11. Gong, M., Jiao, L., Zhang, L.: Solving Tranveling Salesman Problem by Artificial Immunity Responese. In: Wang, T.-D., Li, X., Chen, S.-H., Wang, X., Abbass, H., Iba, H., Chen, G., Yao, X. (eds.) SEAL 2006. LNCS, vol. 4247, pp. 64–71. Springer, Heidelberg (2006)
12. Wang, T., Wang, K., Wang, W.: Web-based Assessment and Test Analyses (WATA) system: development and evaluation. Journal of Computer Assisted Learning 20, 59–71 (2004)

Using Instruction Matrix Based Genetic Programming to Evolve Programs

Gang Li , Kin Hong Lee, and Kwong Sak Leung

Department of Computer Science and Engineering
The Chinese University of Hong Kong
Shatin, N.T. Hong Kong
{gli, khlee, ksleung}@cse.cuhk.edu.hk

Abstract. In Genetic Programming (GP), evolving tree nodes separately would be an ideal approach to reduce the huge solution space of
GP. We use Instruction Matrix based Genetic Programming (IMGP) to
evolve tree nodes separately while taking into account their interdependencies in the form of subtrees. IMGP uses an Instruction Matrix (IM) to
maintain the statistical data of tree nodes and subtrees. IMGP extracts
program trees from IM, and updates IM with the information of the extracted program trees. The experiments have verified that the results of
IMGP are better than those the related GP algorithms in terms of the
qualities of the solutions and the number of program evaluations.

Keywords: Genetic Programming, Instruction Matrix based Genetic
Programming.

1   Introduction

Genetic Programming (GP) [5][1] automatically constructs computer programs
by evolution. In Canonical Genetic Programming (CGP) proposed by Koza [5],
an individual is a LISP-like program tree. The tree is composed of tree nodes
of either functions or terminals. If tree nodes are viewed as nominal variables,
CGP can be treated as a combinatorial optimization problem. Therefore, CGP
has a very large solution space and it is NP-hard. To make things worse, the
number of the tree nodes in CGP is not fixed, so the size of the solution space
may increase exponentially during evolution. It is thus quite common that CGP
has to evaluate a large number of individuals before it finds the optimal program. In addition, evaluating an individual in CGP is usually time-consuming,
because it needs to run the program tree for each training case. Therefore, the
time complexity of CGP is very high. This paper uses Instruction Matrix based
Genetic Programming (IMGP) [6] to search for the optimal programs. To reduce
the size of the solution space, IMGP evolves tree nodes and subtrees independently. Unlike CGP, There is no explicit population to store individual trees in
IMGP. Instead, it uses Instruction Matrix (IM) to maintain the fitness of the
tree nodes and the subtrees. IMGP extracts a new program tree from IM and
updates IM with the fitness of the program tree. Between generations, IMGP

L. Kang, Y. Liu, and S. Zeng (Eds.): ISICA 2007, LNCS 4683, pp. 631–640, 2007.

also modifies IM by reproducing good tree nodes and removing poor tree nodes according to their fitness in IM.

Section 2 reviews the related algorithms. Section 3 describes the algorithm of IMGP in detail. Section 4 presents the empirical results on benchmark GP problems. Section 5 is the conclusion.

2   Related Work

There are some related work which keep the statistical data of evolution and generate individuals from data structures other than populations.

Probabilistic Incremental Program Evolution. (PIPE) [7] maintains a probability tree. A tree node is a vector of the probabilities of the functions and terminals on the tree node. In each generation, PIPE creates a population by constructing trees according to the probability tree, and updates the probability tree with the information of the best individual in the population. However, updating the probability tree only with the best individual may be unable to express the information of the rest of the population. Besides, it ignores the interdependencies between the tree nodes.

Grammar Model-based Program Evolution. (GMPE) [9] evolves programs with Probabilistic Context-free Grammar. It associates each grammar rule with a production probability. It uses the grammars to generate a population of new individuals, and updates the grammars with the good individuals in the population. A grammar generates a single node or a subtree, so it is able to maintain the information of subtrees as well. However, the grammar has no information of the position in the whole tree, so the position of its derivative (node or subtree) in the tree is not fixed.

Program Evolution with Explicit Learning. (PEEL) [8] uses Search Space Description Table (SSDT) to describe the solution space. Ant Colony Optimization [2] is the learning method to update the stochastic components of SSDT. Grammar refinement by splitting certain rules in SSDT is employed to make individuals focus on the promising solution area.

3   Instruction Matrix Based Genetic Programming

IMGP evolves tree nodes separately while taking into account their interdependencies in the form of subtrees. IMGP maintains an Instruction Matrix (IM) to keep the fitness of functions and terminals in tree nodes, and it uses a kind of fixed length expression to represent a program tree. IMGP extracts a program tree from IM by selecting a function or terminal of high fitness for each possible tree node. After the program tree is evaluated, IMGP updates the fitness of corresponding functions or terminals in IM. Between generations, IMGP also modifies IM by reproducing good tree nodes and removing poor tree nodes according to their fitness in IM.

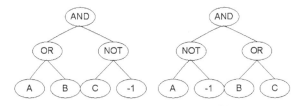

Fig. 1. The trees represented in hs-expression (AND OR NOT A B C -1) & (AND NOT OR A -1 B C)

3.1   Representation

Rather than using *s-expression* as CGP [5], IMGP uses a new program tree representation, i.e. *hs-expression*. It is similar to the array used in Heap Sort, but the "larger-than" relation is changed to the "parent-of" relation. It is a $2^{D+1} - 1$ long array to store a binary tree of depth D at most. Every possible node in the tree has a corresponding element in the array, and vice versa, even if the node does not exist. The tree root is element 0 in the *hs-expression*. For the kth element in the *hs-expression*, its left and right children are the $2k + 1$th and $2k + 2$th elements respectively. If it has no children, the corresponding elements are set to -1 instead. Therefore, the elements in the first half of the array can be either functions, terminals or empty, but the elements in the second half of the array must be either terminals or empty. Fig. 1 shows two examples. Unlike the trees represented by *s-expression*, the trees represented by *hs-expression* of the same length have exactly the same shape if -1 is viewed as a virtual node. Another difference is that the elements at the same locus in *hs-expressions* always correspond to the nodes at the same position in trees.

In IMGP, the information of the population is stored in IM. The cells of IM are data structures consisting of instructions and related information. A row of IM corresponds to an element of *hs-expression* and a tree node. The row contains multiple copies of functions and terminals. Compared to the single copy in the related algorithms, the multiple copies of instructions are good as they have different fitness and subtrees of the corresponding tree nodes. The height of IM is $2^{D+1} - 1$, where D is the tree's maximum depth, and its width is the number of functions and terminals in a row. Fig. 2 shows an example of IM and an *hs-expression* extracted from it. Basically, the element at locus k in the *hs-expression* is extracted from row k in IM. The details are described in Section 3.2.

Besides an instruction of function or terminal, a cell also keeps some auxiliary data. *instruction* is the operation code of the instruction. *best_fitness* and *avg_fitness* are the best and the average fitness of the instruction. *left_branch* and *right_branch* are the left and right branches of the best subtree. *eval_num* is the number of times that the instruction has been evaluated. Please note that in IMGP, the smaller the fitness is, the better it is. These fields keep some information of the fitness landscape of the tree node, and they are used in the evolution of the tree node. Their specific usage is explained in detail in Section 3.2.

A	B	C	D	E	AND	OR	NOT	AND	OR	NOT	→	AND
A	B	C	D	E	***AND***	OR	NOT	AND	OR	NOT	→	AND
A	B	C	D	E	AND	***OR***	NOT	AND	OR	NOT	→	OR
A	B	C	D	E	AND	OR	NOT	AND	OR	***NOT***	→	NOT
A	B	C	D	E	AND	OR	NOT	AND	OR	NOT	→	AND
A	***B***	C	D	E	AND	OR	NOT	AND	OR	NOT	→	B
A	B	***C***	D	E	AND	OR	NOT	AND	OR	NOT	→	C
A	B	C	D	E	AND	OR	NOT	AND	OR	NOT		-1

Fig. 2. Instruction Matrix (IM) and an *hs-expression* extracted from it. IM keeps multiple instructions for each element of the *hs-expression*. An element of the *hs-expression* is extracted from the corresponding row in IM. The cells in bold typeface are the extracted elements.

3.2 Algorithm

Algorithm 1 is the main process of IMGP. It divides a complete tree into separate tree nodes, calculates the fitness of the tree nodes so as to evolve them independently, and combines the optima of the tree nodes into a complete program tree. In each generation, IMGP runs the following steps repeatedly. Firstly IMGP extracts two individuals from IM and calculates their fitness. Then IMGP performs crossover and mutation on them and calculates the fitness of their offspring. After evaluating an individual, IMGP updates the corresponding cells in IM with the fitness of individual. At this point, IMGP deletes all of the individuals because their information has already been stored in IM. A generation finishes after IMGP evaluates P individuals. Then IMGP replaces cells of low fitness with those of high fitness in IM. The best individual is reported as the optimal program after N generations.

Individual Extraction. IMGP extracts the tree nodes from IM probabilistically and combines them into a complete tree. Firstly IMGP constructs an empty *hs-expression* filled with -1, and aligns it vertically with IM. Then it starts to extract the instruction of the tree root from row 0, and puts it at locus 0 in the *hs-expression*. The instruction of a tree node is extracted from the corresponding row using binary tournament selection, and then the extracted instruction is placed at the corresponding locus in the *hs-expression*. Binary tournament selection is comparing the respective average and best fitness of the two randomly selected instructions and selecting one of them probabilistically. If the extracted instruction at locus k is a function, IMGP proceeds to extract its left child from the $2k + 1$th row, and its right child from the $2k + 2$th row. It does so recursively until all the branches are completed. For instance, in Fig. 2, the words in bold italic typeface are the extracted instructions, and the completed *hs-expression* is on the right. The corresponding tree is depicted on the left in Fig. 1. The details of extracting (AND OR NOT A B C -1) from IM is shown in Fig. 3

The best subtree of an instruction is its subtree in the best individual that it has ever been extracted in. After a tree node is extracted, IMGP occasionally checks whether the best subtree of the selected instruction should be extracted as a whole so that the tree nodes in the best subtree are extracted directly without further binary tournament selections. How often it does so depends on the best and the average fitness of the instruction. Eq. 1 is the probability of

---

Algorithm 1. The Main Program of IMGP

Output: the best individual
initialize IM;
for *gen from* 0 *to* N **do**
    while *num* $< P$ **do**
        num $\leftarrow$ 0;
        extract two individuals i and j from IM;
        calculate their fitness respectively;
        update their cells in IM with the fitness;
        if *crossover i with j successfully* **then**
            evaluate the offspring and update its cells;
        else if *mutate i successfully* **then**
            evaluate the offspring and update its cells;
        end
        if *crossover j with i successfully* **then**
            evaluate the offspring and update its cells;
        else if *mutate j successfully* **then**
            evaluate the offspring and update its cells;
        end
        num $\leftarrow$ *num* + *the number of individuals evaluated*;
    end
    shuffle IM;
end

---

1. extract AND from row 0 as the tree root
  2. extract OR from row 1 as the left child of the tree root
    3. extract A from row 3 as the left child of OR
    4. extract B from row 4 as the right child or OR
  5. extract NOT from row 2 as the right child of the tree root
    6. extract C from row 5 as the left child of NOT
7. stop as NOT has no right child

Fig. 3. The steps of extracting (AND OR NOT A B C -1) from the IM in Fig. 2. For each tree node, IMGP selects two instructions in the corresponding row of IM randomly, compares their average and best fitness, and extracts one of them probabilistically. After extracting a tree node, IMGP recursively extracts its left and right child.

extracting the best subtree. The bigger the difference between them is, the more likely its subtree is selected. The reason is that if the best fitness is much better than the average fitness, the tree constructed with the best subtree is likely to be much better than the tree constructed without it. Since tree nodes are highly interdependent with respect to the fitness in GP, best subtrees keep part of the interdependence information between the tree nodes in IM.

$$prob_{best} = 1 - \frac{best\_fitness}{avg\_fitness} \qquad (1)$$

Regarding the binary tournament selection, an instruction is extracted either separately or together with its best subtree. Therefore, when we compare the fitness of two candidate instructions, we should not only compare their average fitness, but consider their best fitness as well. Eq. 2 calculates the expected fitness of an instruction. It considers the probability of selecting its best subtree, and in that case, we should use the best fitness. Then we can use Roulette Wheel Selection [4] to select one of the two instructions based on their expected fitness.

$$E(fitness) = prob_{best} * best\_fitness + (1 - prob_{best}) * avg\_fitness$$
$$= 2 * best\_fitness - \frac{best\_fitness^2}{avg\_fitness} \tag{2}$$

Extracting individuals makes IMGP avoid being trapped in a small solution area. In CGP, when an individual is changed by crossover or mutation, it replaces only a subtree with a new one, so the offspring is still in the neighborhood of the parent. Therefore, the solution space that CGP searches is largely determined by the initial population. However, IMGP does not generate an individual from an existing parent. It extracts a completely new individual from IM, and thus the new individual bears slim similarity with the previous individuals. Therefore, IMGP searches a relatively large solution space, and the extracted individuals together have high diversity. In addition, there are multiple copies of an instruction in a row of IM, and each copy has different fitness and subtrees, so IMGP is relatively robust to local optima.

Instruction Evaluation. To evolve instructions, we need to evaluate their fitness first. In IMGP, the individual is evaluated using the post-order recursive routine. To evaluate a function node, it takes the evaluation of its left and right children as the inputs. To evaluate a terminal node, it evaluates the corresponding program input. Since the individual is discarded right after evaluation and reproduction, it cannot carry along its fitness as in CGP. Instead, the fitness is fed back to its cells in IM so that it can be used in extraction later. The feedback comes in two ways:

1. In Eq.3, the new fitness, $fitness'$, is averaged out with the old fitness, $fitness$. The evaluation number, $eval\_num$, is incremented by one. With this method, we know how good the instruction is on average.

$$fitness = \frac{fitness * eval\_num + fitness'}{eval\_num + 1} \tag{3}$$

2. If the new fitness is better than the best fitness of the instruction, its best fitness is updated, and its left and right branches are changed to those in the current individual accordingly. This actually keeps good subtrees in IM so that they can be extracted in new individuals as described in Section 3.2.

The second point is very important. As pointed out in [10], a new building block is unlikely to survive in the next two generations even if the individual containing it has an average fitness. In IMGP, whenever a good subtree is identified, it is remembered immediately.

We think all the individuals contain useful information of the problem. Therefore, IMGP updates IM not only with good individuals, but all the extracted individuals, no matter how their fitness are. In the related algorithms in Secton 2, they update their models only with good individuals and ignore poor individuals, this would make some of the poor tree nodes spuriously good in the models because they happen to be in the good individuals. In contrary, updating IM with poor individuals decreases the fitness of the actual poor tree nodes.

Genetic Operators. In IMGP, crossover and mutation are similar to those in CGP. However, as IMGP keeps the fitness of the tree nodes in IM, it is able to perform crossover and mutation in a heuristic way. According to the *building block hypothesis* [4], small good building blocks are spread among the individuals and recombined into large good building blocks. Therefore, we think combining subtrees of high fitness is likely to produce individuals of high fitness. When IMGP performs crossover on individuals, it replaces a subtree in one parent only with a better counterpart in the other parent so that the offspring is likely to be better. In mutation, IMGP selects a mutation point in the current individual, and replaces the original subtree with a new subtree extracted from IM.

The crossover is similar to the context preserving crossover [3] because the two subtrees of the parents must be in the same position to reduce the macro-mutation effect of the standard crossover [1]. However, unlike the crossover used in other GP algorithms, the crossover in IMGP is asymmetric. When IMGP tries crossover between individual i and individual j, it picks one of the two branches in both individuals (in the same position), and replaces the subtree of i with that of j if the latter has a better fitness than the former. Otherwise IMGP recursively tries crossover on the picked branches. Please note that crossover would fail when it could not find a better subtree to replace the original one.

Matrix Shuffle. CGP converges by spreading good instructions over the population to reproduce good individuals. When many individuals have good instructions, there may be some individuals consisting of most of the good instructions, and such individuals are likely to be good. IMGP uses matrix shuffle to propagate good instructions in IM, and consequently to increase the probability of selecting them together in the same individual. It shuffles IM row by row. It selects a certain number of pairs of cells in a row, and for each pair it replaces the worse one with the better one in terms of both the best and the average fitness. Therefore, while IM evolves, good instructions emerge to dominate the rows in IM, and the copies of poor instructions decrease.

CGP converges at the cost of diversity. As the population converges, the majority of the individuals have more or less the same instructions, while they have few of the other instructions. In CGP, it is hard to maintain the diversity of the population because measuring the distance between individuals is difficult. However, IMGP evolves on the level of instructions, so it is possible to maintain the diversity of the instructions. In matrix shuffle, when a good instruction $IM[row, i]$ replaces a poor instruction $IM[row, j]$, where i

Table 1. The Experiment Results of IMGP, PIPE and CGP on 6-Bit Parity Problem. It shows the success rate, the number of program evaluations, and the number of tree nodes.

Testing Algorithm	Success Rate	Program Evaluation			Tree Node		
		min	median	max	min	median	max
IMGP	100%	2000	2000	18000	10	29	113
PIPE	70%	9432	52476	482545	22	61	100
CGP	60%	64000	120000	396000	24	90	161

and j are the indices for the two instructions, IMGP needs to check two constraints: $Count[IM[row, i]] < CONVERGENCY$ and $Count[IM[row, j]] > DIVERSITY$.

Basically, matrix shuffle prohibits good instructions from reproducing themselves too many times, and reserves a minimum number of poor instructions. This thus maintains the diversity of the instructions in IM easily and effectively, so it is unlikely for the individuals extracted from IM to have mostly the same instructions.

4   Experiment

As mentioned in Section 2, there are some algorithms related to IMGP. We ran IMGP on the problems tested in other papers and compared the results. However, it is difficult to compare the results precisely, as some of the papers gave the results only in the figures without the exact numerical values. Therefore, we will have to resort to the problems whose results were reported in numbers in other papers. The experiment settings were the same as in the related algorithms.

6-Bit Parity Problem has 6 boolean arguments, and it returns true if the number of true arguments is odd and false otherwise. However, other than using the boolean function set, it uses a real-valued function set $\{+, -, \times, \%, sin, cos, exp, rlog\}$, and its output is mapped to true if it is larger than 0 and false otherwise. We executed IMGP for this problem in 20 independent runs. The result is compared to that of PIPE [7] and CGP in Table 1. As can been seen, IMGP achieved 100% success rate and it required much smaller number of program evaluations.

Max Problem has a single input with value 0.5 and two functions, $+$ and $\times$. The purpose is to find a tree with maximum fitness under the tree size constraint. Obviously the optimal tree is a full tree, whose nodes on the two levels right above the terminals are $+$ to produce the value of 2, and the other nodes on the top are $\times$ to multiply all of the 2s. In this experiment, the maximum tree depth is 7 and so the maximum fitness is 65536. The result compared to GMPE [9] and CGP is reported in Table 2. IMGP achieved a much higher success rate, the number of the program evaluations is bigger than that of GMPE though. However, both IMGP and GMPE are much faster than CGP.

Table 2. The Experiment Results IMGP, GMPE and CGP on Max Problem. It shows the success rate and the number of program evaluations.

Algorithm	Success Rate	Program Evaluations
IMGP	95%	36400
GMPE	60%	13590
CGP	<60%	100000

Table 3. The Experiment Results of IMGP, PEEL and GGP on Function Regression. It shows the fitness of the best individual.

Algorithm	mean	std. dev.	min	median	max
IMGP	5.25	2.79	2.34	5.09	13.56
PEEL	6.79	4.91	0.68	5.21	18.90
GGP	7.87	3.54	0.95	7.56	14.00

Function Regression is to search for the function shown in Eq. 4. The fitness cases were sampled at 101 equidistant points in the interval [0,10]. The fitness is the sum of the differences between the outputs and the correct answers. The fitness of the best individuals of IMGP are compared to those of PEEL [8] and Grammaticaly-based Genetic Programming (GGP) [11] in Table 3. IMGP got a smaller error and a smaller standard deviation. Although the minimum errors of PEEL and GGP are smaller than that of IMGP, their median and maximum errors are much larger than those of IMGP, so IMGP is more stable than PEEL and GGP on this problem.

$$f(x) = x^3 \times e^{-x} \times \cos x \times \sin x \times (\sin^2 x \times \cos x - 1) \qquad (4)$$

5   Conclusion

We have proposed a new Instruction Matrix based Genetic Programming (IMGP). It maintains an Instruction Matrix (IM) to store the fitness of instructions and the links between instructions and their best subtrees. IMGP extracts program trees from IM, updates IM with the fitness of the extracted program trees, performs crossover between the extracted program trees, and shuffles IM to propagate good instructions. The experimental results verified its effectiveness and efficiency on the benchmark problems. It was not only superior to CGP in terms of the qualities of the solutions and the number of program evaluations, but it also outperformed the related GP algorithms.

References

1. Banzhaf, W., Nordin, P., Keller, R.E., Francone, F.D.: Genetic Programming – An Introduction. In: On the Automatic Evolution of Computer Programs and its Applications, Morgan Kaufmann, San Francisco (January 1998)

2. Bonabeau, E., Dorigo, M., Theraulaz, G.: Swarm Intelligence: From Natural to Artificial Systems. Oxford University Press, New York (1999)
3. D'haeseleer, P.: Context preserving crossover in genetic programming. In: Proceedings of the 1994 IEEE World Congress on Computational Intelligence, Orlando, Florida, USA, June 27-29, 1994, vol. 1, pp. 256–261. IEEE Press, Los Alamitos (1994)
4. Goldberg, D.E.: Genetic Algorithms in Search, Optimization & Machine Learning. Addison-Wesley, Reading, MA (1989)
5. Koza, J.R.: Genetic programming: On the programming of computers by natural selection. MIT Press, Cambridge, Mass. (1992)
6. Li, G., Lee, K.-H., Leung, K.-S.: Evolve schema directly using instruction matrix based genetic programming. In: Keijzer, M., Tettamanzi, A.G.B., Collet, P., van Hemert, J.I., Tomassini, M. (eds.) EuroGP 2005. LNCS, vol. 3447, pp. 271–280. Springer, Heidelberg (2005)
7. Salustowicz, R., Schmidhuber, J.: Probabilistic incremental program evolution. Evolutionary Computation 5(2), 123–141 (1997)
8. Shan, Y., McKay, R.I., Abbass, H.A., Essam, D.: Program evolution with explicit learning: a new framework for program automatic synthesis. In: Sarker, R., Reynolds, R., Abbass, H., Tan, K.C., McKay, B., Essam, D., Gedeon, T. (eds.) Proceedings of the 2003 Congress on Evolutionary Computation CEC2003, Canberra, December 8-12, 2003, pp. 1639–1646. IEEE Press, Los Alamitos (2003)
9. Shan, Y., McKay, R.I., Baxter, R., Abbass, H., Essam, D., Nguyen, H.: Grammar model-based program evolution. In: Proceedings of the 2004 IEEE Congress on Evolutionary Computation, Portland, Oregon, June 20-23, 2004, pp. 478–485. IEEE Press, Los Alamitos (2004)
10. Thierens, D., Goldberg, D.E.: Mixing in genetic algorithms. In: Forrest, S. (ed.) Proceedings of the Fifth International Conference on Genetic Algorithms, San Mateo, CA, pp. 38–45. Morgan Kaufman, San Francisco (1993)
11. Whigham, P.A.: Grammatically-based genetic programming. In: Rosca, J. (ed.) Proceedings of the Workshop on Genetic Programming: From Theory to Real-World Applications, San Mateo, July 1995, pp. 33–41. Morgan Kaufmann, San Francisco (1995)

Fuzzy Pattern Rule Induction for Information Extraction

Jing Xiao

Department of Computer Science, Sun Yat-Sen University, Guangzhou, China, 510275
jingsxiao@gmail.com

Abstract. The World Wide Web's vast growth contains a great variety and quantity of on-line information. People need to have the computing systems with the ability to process those documents to simplify the text information. One type of appropriate processing is called Information Extraction (IE) technology. Information extraction can be regarded as one kind of classification problems and one of the main methods to deal with the problem is pattern rule induction. Due to the uncertainty during the induction of pattern rules from natural language texts, in this paper we introduce a Fuzzy pattern Rule Induction System (FRIS) to obtain fuzzy pattern rules for information extraction from semi-structured webpages and free texts.

1 Introduction

The WWW is swiftly becoming a vast information resource that contains a great variety and quantity of on-line information. People encounter a large amount of fast growing information in the form of structured, semi-structured and free texts. This creates a great need for computing systems with the ability to process those documents to simplify the text information. Information extraction is one of such technologies. Generally, an information extraction system takes an unrestricted text as input and "summarizes" the text with respect to a pre-specified topic or domain of interest: it finds useful information about the domain and encodes the information in a structured form, suitable for populating databases [1]. For example, in the seminar announcement domain, the task of IE is to extract slots on "speaker", "starting time", "end time" and "location" from a seminar announcement. Different from information retrieval systems, IE systems do not recover from a collection a subset of documents which are hopefully relevant to a query (or query expansion). IE systems depend largely on relations of relevant items to surrounding context to find the slot information. The goal of information extraction is to extract from the documents facts about pre-defined types of events, entities and relationships among entities. These extracted facts are usually entered into a database, which may be further processed by standard database technologies. Also the facts can be given to a natural language summarization system or a question answering system for providing the essential entities or relationships of the events which are happening in the text documents.

In recent years, machine learning approaches have used as an attractive choice in building adaptive IE systems since manually constructing useful extraction pattern rules is time-consuming and it is tedious to port them to a new domain. There are many IE systems which are based on rule-based relational learning methods that

L. Kang, Y. Liu, and S. Zeng (Eds.): ISICA 2007, LNCS 4683, pp. 641–651, 2007.
© Springer-Verlag Berlin Heidelberg 2007

target at domains with rich relational structure. Such methods generate rules to extract slots either bottom-up [2, 3] or top-down [4]. Some methods combine the bottom-up and top-down approaches [5, 6]. One of the difficulties in rule induction learning systems is that it is difficult to select a good seed instance to start the rule induction process. Some systems simply use randomly selected instance as the starting point [4], and many false starts are often needed in selecting a good seed in order to learn good quality, high coverage pattern rules. Another difficulty we encounter is that the uncertainty problem during we apply the pattern rules for information extraction. Sometimes several pattern rules apply one test instance at the same time with different confidence values. To conquer these two difficulties, in this paper, we introduce a Fuzzy pattern Rule Induction System (FRIS) which emphasizes the use of the feature distribution information on the whole set of training examples in order to select a optimal feature to start the rule induction better. The pattern rules in FRIS take forms of IF-THEN format with certainty values for each rule.

The rest of this paper describes the design, implementation and evaluation of FRIS system. Briefly, the paper is organized as follows. Section 2 discusses related work. Section 3 describes the details of the fuzzy rule generation by examining the distribution of positive training examples. Section 4 presents the experimental results and we conclude the paper with discussion of future research in Section 5.

2 Related Work

Many pattern rule inductive learning systems have been developed for information extraction from free text or structured text [7]. AutoSlog [8] uses a set of predefined linguistic extraction patterns to build a concept dictionary for the terrorism domain. CRYSTAL [9] is a bottom-up covering system that begins with the most specific rule to cover a seed instance, then generalizes the rule by merging similar rules. CRYSTAL takes input that has been processed by a syntactic analyzer and a semantic tagger. It can be used on semi-structured text only if it is supplied with an appropriate syntactic analyzer that allows the text to be treated as if it were grammatical. Both AutoSlog and CRYSTAL rely on prior parsing analysis to identify syntactic elements such as subject, object etc. They identify relevant information at the granularity of syntactic fields.

WHISK [4] induces multi-slot rules from training corpus top-down. It is designed to handle text styles ranging from highly structured text to free text. WHISK performs the rule induction starting from a randomly selected seed instance. $(LP)^2$ [10] is a covering algorithm for adaptive IE system that induces symbolic rules. In $(LP)^2$, the training is performed in two steps: initially a set of tagging rules is learned to identify the boundaries of slots; next additional rules are induced to correct mistakes in the first step of tagging. It proposed to use shallow parsing such as chunking information to improve the precision in identifying slot boundaries when inserting SGML tags into texts.

3 Rule Generation in FRIS

3.1 Pre-processing of Training and Test Documents

FRIS is a supervised covering rule induction algorithm that learns from a training corpus where the users have tagged the sentences containing information of specific slot type such as the name of speaker, venue in a seminar announcement. For each slot type, the tagged instances of that type are regarded as positive examples, while the remaining sentences in the documents are regarded as negative examples. Before learning may commence, both training and testing documents are pre-processed by the same basic NLP modules such as sentence splitter, tokenization, morphological analysis, shallow parsing and named entities extraction. We use the NLProcessor (a shallow parser) from Infogistics company[1] to perform the syntactic analysis to generate information on Part-of-Speech (PoS), noun group and verb group chunking. For example, given a sentence "A bomb was thrown near the house"; after the shallow parsing by NLProcessor, we will get the result "[A_DT bomb_NN] <was_VBD thrown_VBN> near_IN [the_DT house_NN]" in which the "[ ]"s are noun phrases and "( )"s are verb phrases. DT, NN VBD, VBN and IN are the PoS tags for delimiter, noun, verb past, verb past participle and preposition respectively[2]. We use "[ ]" or "( )" as the chunk units in our later experiments. We also employ a rule-based named entity recognition module similar to that used in [11] to derive the semantic classes of some noun groups, such as person, organization, location, and time. The named entity recognition module uses rules which are based on both local sentence-level and global context information from the same document. Sometimes we are unable to identity the semantic type of a noun phrase if we consider only local sentence-level context information. For example, in the sentence of "Herminio Para announced a new system", "Herminio Parra" could be a person or an organization name. Thus we need to employ global information from the whole document to resolve the above ambiguity [12]. The types of global information we used are acronyms, sequence of initial capitals etc.

3.2 The Context Feature Vector

For every tagged training instance, FRIS generates a context feature vector centered around the tagged slot (such as the "starting time" in a seminar announcement) from which to generate the pattern rules. The context feature vector is of the general form:

$$<c_{-k}> ...<c_{-2}> <c_{-1}> <c_0> (tagged\_slot) <c_{+1}> <c_{+2}> ... <c_{+k}> \qquad (1)$$

Here $<c_i>$ *{i=-k to +k; i ≠0}* represents the context units of the tagged slot, and k is the number of context units considered. $<c_0>$ represents the central tagged slot itself. $<c_i>$ can be a token, a noun or a verb phrase or even a syntactic unit such as subject or object and it can be of various feature types, including: words, PoS (if it's a single token), various types of verbs and noun chunks, and semantic classes.

One key characteristic of FRIS is its representation of context feature vector, in which we code all elements (including both the tagged slot and the context elements) at their appropriate lexical, syntactic and semantic representations simultaneously. The context feature vector for a single tagged instance can therefore be represented as follows:

$$<(-k,f_{-k}^1), ..., (-k,f_{-k}^m),..., (-1, f_{-1}^1), ..., (-1, f_{-1}^m), (0,f_0^1),..., (0,f_0^m), (1,f_1^1), ..., (1,f_1^m), ..., (k,f_k^1),$$
$$..., (k,f_k^m)> \qquad (2)$$

where m is the number of linguistic features for each element.

---
[1] http://www.infogistics.com/textanalysis.html
[2] http://www.infogistics.com/tagset.html

As shown in Equation (2), each element is represented as a tuple (g, f_g^i). The first part of the tuple, *i.e.* g, indicates the position of the element within the tagged instance. $g=0$ gives the position of tagged slot, and positive g (or negative g) gives the gth right (or left) hand context element from the tagged slot. If there are m features, and k context elements, then we have a context vector of size $(2k+1) \times m$.

The second part of the tuple, *i.e.* f_g^i, gives the possible appropriate linguistic representation for each element. The overall feature set consists of 12 (*i.e.* $m=12$) lexical, syntactic and semantic features are given in Table 1. The first two representations (Lex. String and PoS) respectively give the original lexical form, and the Part-of-Speech information of the element if it is a single token. The two features are of string type. The next 8 representations (NP_Person, NP_Org., NP_Loc., NP_Date, NP_Time, NP_Perc., NP_Mon., and NP_Num.) cover the general named entities (NE) of type Person, Organization, Location, Date, Time, Percentage, Money and Number ("NP" stands for "Noun Phrase" and "VP" stands for "Verb Phrase"). The last 2 representations (VP_Act. and VP_Pass.) indicate the active and passive voice of VP. The values from the third feature to the twelfth feature are stored as "true" or "false". We also store these representations as string type and for NP and VP, we also store the head noun and root verb for f_g^1, f_g^{11} and f_g^{12}. The set of representations is selected to model all possible patterns used in rule induction. They are selected to capture all essential syntactic and semantic types, and are based partly on related works [8,10] that were demonstrated to be effective. These features are not specific to any domain.

Table 1. Features that FRIS employed

Feature	Description	Feature	Description
f_g^1	Lex. String	f_g^2	PoS
f_g^3	NP_Person	f_g^4	NP_Org.
f_g^5	NP_Loc	f_g^6	NP_Date
f_g^7	NP_Time	f_g^8	NP_Perc.
f_g^9	NP_Mon.	f_g^{10}	NP_Num.
f_g^{11}	VP_Pass.	f_g^{12}	VP_Act.

3.3   Distribution of the Training Examples

In order to find a good seed instance to start the rule induction process, FRIS utilizes a global approach to finding pattern rules by making full use of the feature statistics in the tagged examples. It does not generalize the rule from one single instance like some rule induction IE systems do, such as $(LP)^2$ [10]. Instead, it incorporates the global information in all positive training examples and selects the most prominent feature to construct the rule.

Given the cluster of training instances of a specific slot type, FRIS generates a context feature vector for each instance using Equation (2). By arranging all the instances modeled using Equation (2) in the same table, we obtain the global context feature representation for the whole training corpus as shown in Figure 1. We align these elements according to their corresponding context positions. Note that not all of the feature representations are present for each element. The occurrences of the common element features at a specific position g (g is positive number) are cumulated as e_{gi}. From Figure 1, we can easily obtain the global distribution frequency of any element feature and at any position, and derive the set of instances covered by any feature set f. We consider different features play different importance in various domains, for example, in the free text terrorist attacks corpus, verb features play crucial roles. Thus for each feature, we give it a weight coefficient β_{gi} empirically.

It is important to select a good feature to kick off the rule induction for a covering algorithm. One intuition is to select element feature that has the highest value of $\beta_{gi} \times e_{gi}$ in the active positive training set. By adding this element feature $f_g^{\ i}$ into an active feature set f, we can generate a pattern $r_c(f)$ in terms of the

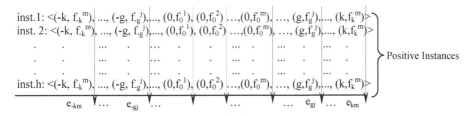

Fig. 1. Overall distribution of instances and representations

feature set f including current $f_g^{\ i}$ so that $r_c(f)$ covers a number of active training instances which have the most prominent feature $f_g^{\ i}$. However, the quality of $r_c(f)$ is determined not only by its coverage in the positive training set but also by the number of instances in the negative set that it covers which would be regarded as errors. Let n_k denote the number of both positive and negative examples covered by the rule $r_c(f)$, and m_k be the number of negative examples or errors covered by that rule. A good measure of the quality of the rule is the *Laplacian* expected error [4] defined as:

$$Laplacian(r_c(\underline{f})) = \frac{m_k + 1}{n_k + 1} \tag{3}$$

The element feature that has the highest $\beta_{gi} \times e_{gi}$ value does not necessarily lead to a rule with the lowest *Laplacian* measure. On the other hand, it is too costly to evaluate the *Laplacian* measures of all possible element features. As a compromise, we evaluate the *Laplacian* measure of the top w element features with high $\beta_{gi} \times e_{gi}$ values in the active positive training set. Our ultimate aim is to select rule that has prominent feature $f_g^{\ i}$ with high $\beta_{gi} \times e_{gi}$ value and whose *Laplacian*$(r_c(f))$ satisfies the pre-defined error tolerance value. It is worth noting that adding more features into f helps to constrain the rule, and ideally lead to improvement in rule precision.

3.4 The Overall Rule Induction Algorithm

The fuzzy pattern rules in FRIS are represented as follows:

IF constraint$_1$ satisfies … AND constraint$_n$ satisfies THEN insert SGML tags (Rule Value : η)

$$\tag{4}$$

where η equals to the value computed by Equation (3). At the left hand side, there are some pattern constraint conditions which could be any feature element we defined in section 3.2. At the right hand side, there is rule action which is to insert opening tag "< >" and closing tag "< />" for each semantic slot. Usually, the SGML tags are inserted beside noun phrase boundaries. For example, a pattern rule of "IF token1 equals to *start* AND token2 equal to *at* and token3 is a named entity with type of *time* THEN the time is seminar's *starting time*" is to insert the tags of "<*stime*>" and "</*stime*>" beside the boundaries of the noun phrase whose named entity type is "time" when the left side of the pattern rule is matched with the instance.

We now present the overall algorithm for FRIS to induce pattern rules as follows:

a) Group tagged instances of the same slot type into one cluster.

b) Generate context feature vectors for all positive instances in every cluster. The resulting k^{th} cluster is C_k, with the positive instance set P_k and negative instance set N_k. Let r_k be the set of rules extracted so far to cover P_k; and set $r_k = null$.

c) For every cluster C_k, perform the followings:

$\quad$ (c$_1$) Loop-1: // to generate new rules

$\qquad\qquad$ Let $f_c = null$ be the current feature set;

$\qquad\qquad$ $r_c(f_c)$ be the current rule; and

$\qquad\qquad$ P_c, N_c be the set of instances covered by $r_c(f_c)$.

$\qquad\qquad\qquad$ Initially, set: $P_c = P_k$, $N_c = N_k$

$\qquad\qquad$ RuleAttempt = 0;

$\quad$ (c$_2$) Loop-2: // to refine current rule $r_c(f_c)$

$\qquad\qquad\qquad$ Find top w element features $\{f_g^i\}$ (based on $\beta_{gi} \times e_{gi}$ values) that covers at least

$\qquad\qquad$ one instance in P_c;

$\qquad\qquad$ Select the f_i^j that minimizes the *Laplacian* measure of the current rule $r_c(f_c \cup f_i^j)$;

$\qquad\qquad$ Add f_i^j to f_c, i.e. $f_c = f_c \cup f_i^j$

$\qquad\qquad$ RuleAttempt++;

$\qquad\qquad$ (c$_3$) IF *Laplacian*($r_c(f_c)$) < σ (error tolerance)

$\qquad\qquad\qquad$ THEN // the quality of resulting rule is good

$\qquad\qquad\qquad$ Add rule r_c to rule set r_k; or $r_k = r_k \cup r_c$;

$\qquad$ Update $P_k = P_k - \{$all instances covered by rule $r_c\}$;

$\qquad$ Go to Loop-1 to generate another rule.

$\qquad\qquad\qquad$ ELSE // more work is needed to constraint rule r_c

$\qquad\qquad\qquad$ Update P_c by removing those instances that are not covered by r_c;

$\qquad$ IF RuleAttempt ≥ λ (max. rule attempt for constraining rules)

$\qquad$ THEN // relaxing error tolerance;

$\qquad\qquad$ Increase σ;

$\qquad\qquad$ Go to Loop-1 to generate new rule with bigger error tolerance;

$\qquad$ ELSE

$\qquad\qquad$ Go to Loop-2 to find new feature f'_i^j to refine rule r_c.

$\qquad$ Repeat until P_k is empty.

The "*RuleAttempt*" is related to the length of the generated rule which the user could pre-specify. For example, if we set the rule length to "4", then "*RuleAttempt*" could be 8.

That is to say, we constrain the rule to the maximum size of 4 contextual units (4 for left side and 4 for right side around the tagged slot respectively). Based on the above algorithm, FRIS will generate rules that incorporate the most prominent features. If using a single feature cannot satisfy the error tolerance for quality, then more features will be added to tighten the constraints until the quality of the resulting rule is good enough. We can also see that FRIS is a local search algorithm. It performs a form of hill climbing and once the rule with current features satisfies the error tolerance it will be output even though adding even more features would result in a lower *Laplacian* value. In case there is noise in the positive training examples, we can apply some "post-pruning" strategies to control the whole quality of the learned rules. For example, after the entire rule set has been generated, some of the rules may have low coverage on the training set. A post-pruning step that discards all rules with *Laplacian* expected error greater than a threshold has the effect of removing the least reliable rules.

During the test phase, we apply the learned FRIS pattern rules to unseen test instances that are also preprocessed by the series of NLP modules as we do for training instances (see Section 3.1). When the left side constraints (see Equation (4)) are matched with the test instances, then the opening tag "< >" and the closing tag "< />" of a slot will be inserted beside a noun phrase boundaries to indicate a detected entity. Sometimes two rules match the same test instance. We select the rule whose η value is higher as the better classification. Moreover, we observe that there may be additional tokens or adverbs within the constraints in the unseen instances. In such cases, the left side constraints of the pattern rules will not be matched and the entity will be missed. To overcome this problem, we perform a flexible matching between the learned pattern rules and the test instances. We allow up to one shift in context of new test instances when matching against the learned pattern rules. For example, the rule of "IF token1 is a noun phrase and token2 is an active verb phrase with verb kill THEN the first noun phrase is *perpetrator*" will match the instance of "FLMN also killed another three persons.", where there is an extra term "also" between the noun phrase and the verb phrase in the "correct" test instance.

We also try the idea of applying edit distance [13] to perform inexact matching of rules. However, the substitution scheme used in edit distance is not appropriate for information extraction task. For example, if we allow one element substitution in the pattern rule, the rule of "IF token1 is a noun phrase and token2 is an active verb phrase with the verb of kill THEN the first noun phrase is *perpetrator*" will also match the instance of "IF token1 is a noun phrase and token2 is an *passive* verb phrase with the verb of kill" since there is only one different element but the noun phrase in the instance is not a "perpetrator" but "victim". Thus we do not use the edit distance to perform partial matching of rules. Instead we found that the simple one shift matching is more effective.

4   Experimental Results

To verify the effectiveness of FRIS, we test FRIS on a number of IE tasks including the semi-structured web page corpora, and the free text corpus such as the MUC-4 corpus [14]. In each experiment, FRIS is trained on a subset of the corpus and the learned rules are tested on the remaining unseen texts, as defined in the respective corpus. The test documents are pre-processed by the same set of NLP tools as described in Section 3.1 and 3.2.

We used 1,500 texts (the standard training documents of MUC-4 plus TST1 and TST2 tasks) for training, in which about 50% of the texts are relevant with their associated answer keys given in the MUC-4 corpus. Our target slots are perpetrator, victim and physical target. During testing, we use 100 texts composing 25 relevant texts and 25 irrelevant texts from the TST3 test set, plus 25 relevant texts and 25 irrelevant texts from the TST4 test set. Table 1 shows the performance of FRIS on MUC-4 corpus based on MUC-4 evaluations. From Table 1 we can see that FRIS can achieve the performance of the state-of-the-art machine learning system called ALICE [15]. Notably, the top performing systems listed in Table 2 are "GE" and "GE-CMU". However, both systems are "manual" systems that involved "10½ person months" manual efforts on MUC-4 evaluations using the GE NLTOOLSET. This is in addition to the "15 person months" manual efforts they spent on MUC-3 evaluations. For a fully automated learning approach, such as FRIS or Alice, the resulting IE system is more portable across domains.

Table 2. Comparisons of FRIS with other systems on the TST3 and TST3 test sets

TST3	Rec.	Pre.	F_1	TST4	Rec.	Pre.	F_1
GE	58	54	56	GE	62	53	57
GE-CMU	48	55	51	GE-CMU	53	53	53
UMASS	45	56	50	SRI	44	51	47
FRIS	45	53	49	Alice	46	46	46
Alice	46	51	48	*FRIS*	45	47	46
SRI	43	54	48	NYU	46	46	46

We report the results of FRIS compared to other reported systems on two publicly available corpora: the CMU seminar announcements and the Austin job listings [16]. For these two tasks, we perform 5 trials validation experiments. In each trial, we randomly partition the data into two halves, using one half for training and the other half for testing. Our results in Table 3 and Table 4 are the average of these 5 trials. The results of the first task in terms of F_1 measure are summarized in Table 3, along with the results of other state-of-the-art IE systems. We extract results for ME_2 and SNoW systems from [17], and those of the other systems from [10]. Considering the average accuracy for all the slots, it can be seen that FRIS outperforms other reported systems on the same task. In this domain, FRIS performs worse on slot of <location>. The main reason is that the named entity recognizer we used is designed to extract the general location types, such as the countries, cities *etc.*, whereas the locations in this corpus are not in general location forms. Most locations refer to meeting room numbers such as "PH 223D". Thus FRIS misses out some of these locations. The second task performs IE on 300 job announcements. The task consists of identifying for each job listing: message id, job title, salary offered, company offering the job, recruiter, state, city and country where the job is offered, programming language, platform, application area, required and desired years of experience, required and desired degree, and posting date. The results of the second task in terms of F_1 measure are presented in Table 4. The results of other reported systems are taken from [10].

Table 3. F_1 measure obtained by FRIS on CMU seminars

	speaker	location	stime	etime	All(ave.)
FRIS	85.7	76.2	99.6	96.0	89.3
ME_2	72.6	82.6	99.6	94.2	87.3
$(LP)^2$	77.6	75.0	99.0	95.5	86.8
SNoW	73.8	75.2	99.6	96.3	86.2
BWI	67.7	76.7	99.6	93.9	84.5
HMM	76.6	78.6	98.5	62.1	79.0
Rapier	53.0	72.7	93.4	96.2	78.8
SRV	56.3	72.3	98.5	77.9	76.3
Whisk	18.3	66.4	92.6	86.0	65.8

Table 4. F_1 measure obtained by FRIS on job listings

slot	FRIS	$(LP)^2$	Rapier	BWI
Id	100	100	97.5	100
Title	45.3	43.9	40.5	50.1
Company	79.1	71.9	69.5	78.2
Salary	80.7	62.8	67.4	
Recruiter	81.2	80.6	68.4	
State	90.1	84.7	90.2	
City	93.5	93.0	90.4	
Country	94.8	81.0	93.2	
Language	88.1	91.0	80.6	
Platform	78.2	80.5	72.5	
Application	76.7	78.4	69.3	
Area	65.1	66.9	42.4	
Req-yeas-e	70.2	68.8	67.1	
Des-years-e	71.1	60.4	87.5	
Req-degree	86.3	84.7	81.5	
Des-degree	73.4	65.1	72.2	
Post date	99.5	99.5	99.5	
AllSlots(averaged)	80.8	77.2	75.9	

Again, it can be shown that FRIS outperforms $(LP)^2$ and Rapier system with respect to the overall average F_1 measure. From the experimental results on these two webpage corpora, we observe that the named entity identification module can improve the accuracy especially in slots of <speaker>, <stime>, <etime>, <salary>, <company> and <recruiter> since it can recognize persons, time, money and organizations. FRIS performs worse on those slots that do not conform to general named entity types, such as <title>, <application>. Some specific domain knowledge can help to improve the precision on these types of slots.

5 Conclusion

The ability to extract desired pieces of information from natural language texts is an important task with a growing number of potential applications. This paper presents a fuzzy pattern rule induction algorithm, FRIS, which combines statistical analysis and rule-based approach. A major difference between FRIS and other pattern rule learning system is that FRIS learns the rule from the whole set of training instances represented at the lexical, syntactic and semantic levels rather than from only one instance. And also the fuzzy rules in FRIS have confidence values which are computed by the Laplacian measurement. Our experimental tests on free texts and semi-structured web pages show that our approach is effective in terms of F_1 measure as compared to other reported systems.

Further works will be carried out in the following directions. First, we plan to conduct further texts and extend our system to other domains, such as the university web pages, rental ads, and the company management changes. Second, the operations among the fuzzy rules are remained to be discussed further. Lastly, we plan to use a text categorization learning system to classify the documents initially before passing them to respective IE system for information extraction.

References

1. Cardie, C.: Empirical Methods in Information Extraction. AI Magazine 18(4), 65–79 (1997)
2. Califf, M.E.: Relational Learning Techniques for Natural Language Information Extraction. Ph.D. Thesis, University of Texas at Austin (1998)
3. Califf, M.E., Mooney, R.J.: Relational Learning of Pattern-Match Rules for Information Extraction. In: Proceedings of the ACL Workshop on Natural Language Learning, pp. 9–15 (1997)
4. Soderland, S.: Learning Information Extraction Rules for Semi-structured and Free Text. Machine Learning 34, 233–272 (1999)
5. Muggleton, S.: Inverse Entailment and Progol. New Generation Computing. Special issue on Inductive Logic Programming, 245–286 (1995)
6. Zelle, J.M., Mooney, R.J., Konvisser, J.B.: Combining Top-down and Bottom-up Methods in Inductive Logic Programming. In: Proceedings of the 11th International Conference on Machine Learning, pp. 343–351 (1994)
7. Muslea, I.: Extracting Patterns for Information Extraction Tasks: A Survey. In: AAAI-Workshop on information extraction (1999)
8. Riloff, E.M.: Automatically Constructing a Dictionary for Information Extraction Tasks. In: Proceedings of the 11th National Conference on Artificial Intelligence, pp. 811–816 (1993)

9. Soderland, S., Fisher, D., Aseltine, J., Lehnert, W.: Crystal: Inducing a Conceptual Dictionary. In: Proceedings of the 14th International Joint Conference on Artificial Intelligence, pp. 1314–1319 (1995)
10. Ciravegna, F.: Adaptive Information Extraction from Text by Rule Induction and Generalization. In: Proceedings of the 17th International Joint Conference on Artificial Intelligence (2001)
11. Chieu, H.L., Ng, H.T.: Named Entity Recognition: A Maximum Entropy Approach Using Global Information. In: Proceedings of 19th International Conference on Computational Linguistics, pp. 190–196 (2002)
12. Chua, T.S., Liu, J.: Learning Pattern Rules for Chinese Named Entity Extraction. In: Proceedings of the 18th National Conference on Artificial Intelligence, pp. 411–418 (2002)
13. Sankoff, D., Kruskal, J.: Time Wraps, String Edits, and Macromolecules, the Theory and Practice of Sequence Comparison. CSLI Publications, Stanford, CA (1999)
14. Proceedings of the Fourth Message Understanding Conference. Morgan Kaufmann Publishers (1992)
15. Chieu, H.L., Ng, H.T., Lee, Y.K.: Closing the Gap: Learning-Based Information Extraction Rivaling Knowledge-Engineering Methods. In: Proceedings of the 41st Annual Meeting of the Association for Computational Linguistics, pp. 216–223 (2003)
16. http://www.isi.edu/info-agents/RISE/repository.html
17. Chieu, H.L., Ng, H.T.: A Maximum Entropy Approach to Information Extraction from Semi-Structured and Free Text. In: Proceedings of the 18th National Conference on Artificial Intelligence, pp. 786–791 (2002)

An Orthogonal and Model Based Multiobjective Genetic Algorithm for LEO Regional Satellite Constellation Optimization

Guangming Dai, Wei Zheng, and Baiqiao Xie

School of Computer Science
China University of Geosciences, Wuhan 430074, P.R. China
gmdai@cug.edu.cn up2uwei@126.com

Abstract. Regional coverage Constellation Optimizing Design is a classical dynamic multi-objective optimizing problem. Against low efficiency of traditional multi-objective evolutionary algorithms and poor utilization of Pareto-optimal solutions distribution regularity etc, in this papera new approach OMEA which bases on the probability-model utilizing Pareto-optimal solutions distribution regularity to obtain a good distribution of Pareto-optimal solutions, we also apply the quantization technique and orthogonal design to generate initial points which spread uniformly in the feasible solution space. Considering coverage rate assessment criterions, we accomplish the design and simulation of Leo Constellation. Compared with NSGA-II, Pareto solutions by OMEA are closer to Pareto-optimal Front. The result of experiments shows a group of Pareto solutions with a uniform distribution can be achieved, which gives strong supports to constellation design determination.

1 Introduction

Multi-objective optimization or multi-criterion programming is one of challenging problems encountered in various astronautic missions, engineering and economic problems. Without the loss of generality, we consider the following multi-objective optimization problem (MOP) in a continuous search space:

$$\min_{X \in \Omega} F(X) = (f_1(X), \cdots, f_m(X))^T \tag{1}$$

with $X = (x_1, \cdots, x_n)^T \in R^n$ are decision vectors, $F(X) = (f_1(X), \cdots, f_m(X))^T$ are corresponding objective vectors, and $\Omega \in R^n$ is the decision space. A solution $X^* \in \Omega$ is called (globally) Pareto optimal if there is no $X \in \Omega$ such that $F(X) \prec F(X^*)$. The set of all Pareto optimal solutions, denoted by Ω^*, is called Pareto optimal set. The set of all Pareto optimal objective vectors, $PF = \{y \in R^m \mid y = F(X), X \in \Omega^*\}$, is called Pareto front. Multi-objective optimization algorithms aim to find an approximation of Pareto front. Meanwhile, there are two goals in multi-objective optimization: (i) to discover solutions as closely to the Pareto front as possible, (ii) to find solutions as diversely as possible in the obtained non-dominated front.

L. Kang, Y. Liu, and S. Zeng (Eds.): ISICA 2007, LNCS 4683, pp. 652–660, 2007.

In last few years, many variants and extensions of classical evolutionary algorithms (EAs) have successfully been employed to tackle MOPs. Such as nondominated sorting genetic algorithm (NSGA-II)[1], strength Pareto evolutionary algorithm (SPEA2) [2], vector evaluated genetic algorithm (VEGA)[3], niched Pareto genetic algorithm (NPGA) [4], Pareto archived evolution strategy (PAES) [5], and so on. Among these, NSGA-II by Deb et al. [1] and SPEA2 by Zitzler et al. [2] are the most popular. Several important techniques, such as the use of a second population (or an archive) [4], [5] have proved to be able to greatly improve the performance of multi-objective evolutionary algorithms (MOEAs).

In contrast to single objective optimization, the distribution of Pareto-optimal solutions often shows a high degree of regularity, but the approximation of the Pareto optimal solutions has not been explicitly addressed in these algorithms, their solutions are often poor in terms of closeness to the Pareto optimal set and uniformity in the decision space.

Recently this regularity has often been exploited implicitly by introducing to a local search by Yaochu Jin et al after global evolutionary optimization [6]. A step further to take advantage of such regularity is the use of a model that captures the regularity of the distribution of Pareto-optimal solutions by Aimin Zhou et al [7], [8]. Their preliminary experimental results are very encouraging.

The orthogonal design method [14] with both orthogonal array(OA) and factor analysis (such as the statistical optimal method) is developed to sample a small, but representative set of combinations for experimentation to obtain good combinations. Recently, some researchers applied the orthogonal design method incorporated with EAs to solve optimization problems. Leung and Wang [14] who incorporated orthogonal design in genetic algorithm for numerical optimization problems found this method was more robust and statistically sound. OMOEA [9] by Sangyou Zeng and -ODEMO by Wenyin Gong[10] adopted the orthogonal design method to solve MOPs. Numerical results demonstrated the efficiency of this way.

Inspired by the ideas from OGA/Q [14] and MEA/HA, MEA/HB [7][8], in this paper, we propose an extension of MEA algorithm based on the K-means method and orthogonal design method for LEO regional covering satellite constellation optimization. Here, we call our improved MEA algorithm as OMEA. And then a linear or quadratic model is used, we adopt a convergence criterion to determine whether the model or the crossover and mutation should be employed for offspring generation ,which is heuristic. The idea is that the algorithm will benefit from using a model-based offspring generation only when the population shows a certain degree of regularity. To evaluate the practicability and efficiency of OMEA, we apply and test it to a satellite constellation optimization problem with three objectives.

The rest of this paper is organized as follows: In Section2, we briefly introduce EMA/HA and EMA/HB which introduced in [7,8].A detailed description of the proposed OEMA is provided in Section 3. Section 4 defines the experimental setup, including the description of satellite constellation optimization problem and the simulation results comparing the performance of the proposed algorithm with NSGA-II, The last section, Section 5, is devoted to conclusions and future studies.

2   Model-Based Evolutionary Algorithm(MEA)

MEA [7] is a new approach for MOPs, the idea comes from Pareto optimal solutions of a multi-objective optimization problem often distributing very regularly in both the decision space and the objective space. In fact, it has been found that Pareto optimal sets can be defined as linear or piecewise functions for most widely-used test problems of multi-objective optimization in the evolutionary computation community [7][8]. In MEA/HA, before the modeling, Local PCA algorithm [10][11] was used for partitioning each odd generations population $P(t)$ into K disjoint clusters $S_1, \cdots, S_K$ (where K is a user-specified algorithmic parameter in the local PCA algorithm). Then for each cluster(sub-population) a probability model is built to model the promising area in the decision space, and then new solutions are sampled from the model .That is closely related to a large class of search algorithms known as estimation of distribution algorithms (EDAs) [9]in the evolutionary computation community. MEA/HA is described as follows:

Algorithm 1. MEA/HA

Step 0 Initialization: Set $t = 0$ and initialize $P(t)$

Step 1 Reproduction:

If $t\%2 == 0$ perform crossover and mutation on $P(t)$ to generate a set of new solutions $P_s(t)$.

Else build a 1-D linear model or 2-D plane surface model, sample N^k new solutions and store them in $P_s(t)$.

Step 2 Selection: Select $p(t+1)$ from $P_s(t) \cup P(t)$.

Step 3 Stopping Condition: If the stopping condition is met, stop; otherwise, set $t = t + 1$ and go to step 1.

MEA/HB is a improved algorithm based on MEA/HA , the difference between these two MEAs is in MEA/HB that Zhou gives a convergence criterion to determine whether the model or the crossover and mutation should be employed for offspring generation, and there is a more sophisticated method to construct the stochastic part of the model in MEA/HB.

3   Our Approach: OMEA

Here, we propose an extension MEA algorithm (OMEA) for MOPs. There are two enhancements of it: (i) using orthogonal design method with quantization technique to generate the initial population, and (ii)adopting K-means clustering algorithm to partitioning the population, and then building probabilistic model based on principal component analysis (PCA) algorithms for each cluster.

3.1   Orthogonal Initial Population

The initial population of evolutionary algorithm (EA) can be considered to be a "experiments". We incorporate experimental design methods into the initialization instead of early random initialization. Before solving an optimization problem, we usually have no information about the location of the global optimal. It is desirable that an algorithm starts to explore those points that are scattered evenly in the feasible solution space. In this manner, the algorithm can evenly scan the feasible solution space only once to locate good points for further exploration in subsequent runs. As the evolution of every population, some points may move closer and closer to global optimal. We apply the quantization technique and the orthogonal design to generate this initial population.

(1) **Design of the orthogonal array:** To design an orthogonal array (OA), in this research, we use $L_R(Q^C)$ to denote the OA with different level Q, where Q is odd and $R = Q^J$ indicates the number of the rows of OA, where J is a positive integer fulfilling

$$C = \frac{Q^J - 1}{Q - 1} \geq N \tag{2}$$

where N is the number of the variables, C denotes the number of the columns. The orthogonal array needs to find a smallest to satisfy,. In this study, we adopt the algorithm described in Ref. [14] to construct an orthogonal array. For convenience, we use $L(R, C)$ to indicate the orthogonal array; and call $a(i, j)$ the j th level of the i th factor. If $C > N$, we delete the last $C - N$ columns to get an OA with N factors.

(2) **Quantization.** For one decision variable x_i with the boundary $[l_i, u_i]$, we quantize the domain into Q levels $\alpha_{i,1}, \cdots, \alpha_{i,Q}$, where the design parameter Q is odd and $\alpha_{i,j}$ is given by

$$\alpha_{i,j} = \begin{cases} l_i & j = 1 \\ l_i + (j - 1)(\frac{u_i - l_i}{Q - 1}) & 2 \leq j \leq Q - 1 \\ u_i & j = Q \end{cases} \tag{3}$$

In other words, the domain $[l_i, u_i]$ is quantized Q fractions, and the difference between any two successive levels is the same.

(3) **Generation of Initial Population.** After constructing a proper OA and quantizing the domain of each decision variable, we can generate the initial population which can scatter uniformly over the feasible solution space. The algorithm of generation of initial population is omitted here, please refer [14] for the detail information. Regularly, the number of the rows of the OA is larger than the population size NP, so we use the ranking mechanism which mention in the NSGA-to select NP solutions from OA.

3.2   Partition and Modeling

Because we haven't any pre-information about the principal curve[8]global distribution regularity, by using clustering algorithm partitioned population into

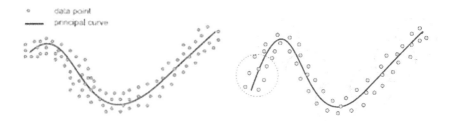

Fig. 1. An example of solution distributed regularity (principal curve) and Cluster analyzing on population

clusters(sub-populations) then we can local search the each cluster for distribution regularity by Principal Components Analysis(PCA)[12,13]. The clustering algorithm we chose is a simple and widely-used k-means, because the Local PCA clustering algorithm mentioned in refer [7,8] it can describe distribution of solutions wonderful but it much time-consuming not suit for the such extreme complex satellite constellation optimization problems.

PCA[13] is a way of identifying patterns in data, and expressing the data in such a way as to highlight their similarities and differences. Since patterns in data can be hard to find in data of high dimension, where the luxury of graphical representation is not available, PCA is a powerful tool for analyzing data. By used PCA, we can project points in a dimensional decision space on to a 2 or 3 dimensional space while preserving broad trends in the data and captures the regularity of the distribution of the data cluster. With the partition of the data, we can build a group of linear models to approximate a principal curve or a principal surface. One model is built in each data cluster. In the k-th cluster C^k, the i-th biggest eigenvalue is $\lambda_i^k, i = 1, \cdots, n$,its corresponding normalized eigenvector is V_i^k,and the mean of cluster C^k is $\bar{X}^k, k = 1, \cdots, K$. So we can calculate the projections on the first and second eigenvectors for each point:

$$s_{1,i}^k = (X_i^k - \bar{X}^k)^T V_1^k, s_{2,i}^k = (X_i^k - \bar{X}^k)^T V_2^k \tag{4}$$

For MOPs with three or more objectives, construct the probabilistic model as follow:

$$H^k(s) = s_1^k V_1^k + s_2^k V_2^k + \bar{X}^k + \xi^k, s_j^k \in [\min_{i=1,\cdots,N^k} \{s_{j,i}^k\}, \max_{i=1,\cdots,N^k} \{s_{j,i}^k\}] \tag{5}$$

where $s_1^k V_1^k + s_2^k V_2^k + \bar{X}^k$ is the deterministic model describing the distribution of the solutions, and ξ^k is a random vector with a normal distribution $N(0, (\delta^k)^2 I)$, it represents a stochastic model which attempts to describe the local dynamics of the individuals. δ^k can be calculated as follow:

$$\delta^k = \sum_{i=1}^{N^k} d_i^k / (N^k \sqrt{n}) \tag{6}$$

Where the d_i^k represents the distance between a point and the reference vector of the k-th cluster.

In the model building stage, we obtain K models, with these models, we can produce the offspring solutions by sampling. For each cluster model $H^k(s)$, we can uniformly create N^k(the number of individual in k-th cluster)new solutions from the model in equation 5.

The main framework of the proposed algorithm is described as follows:

Algorithm 2. The procedure of the proposed OMEA

Step 0 Orthogonal Initialization: Set $t = 0$ and initialize $P(t)$

Step 1 Reproduction:

1.1 Partition $p(t)$ into clusters $C^k, k = 1, \cdots, K$, using k-means.

1.2 For each cluster C^k. If convergence criterion $= \Psi(k) < \rho$ build a 1-D linear model or 2-D plane surface model, sample N^k new solutions and store them in $P_s(t)$; Else perform crossover and mutation on C^k to generate N^k new solutions and store them in $P_s(t)$.

Step 2 Selection: Select $p(t+1)$ from $P_s(t) \cup P(t)$.

Step 3 Stopping Condition: If the stopping condition is met, stop; otherwise, set $t = t + 1$ and go to step 1.

---

4   Our Approach: Simulation Results

4.1   LEO Regional Satellite Constellation Optimization Problems

Optimal geometries for satellite constellations aiming at a continuous global or zonal, regional Earth's coverage have been widely studied over the last twenty years. Partial/ Regional coverage constellation optimizing design is a complex feature mainly due to the dynamic properties of the multi-objective functions. Against with the inefficiency of traditional constellation design method, in the last few years, many scholars introduced evolutionary algorithms (EAs) into constellation optimization[14][15][16]. In our experiment, we apply and test OMEA to a practical example. We assume there are three characteristic ground points $F_1(112.25°, 36.05°)F_2(112.48°, 29.94°)F_1(112.48°, 24.01°)$,the requirements of constellation design is make the constellation at least has 1 satellite can continue uninterrupted 24-hour communications with each of ground station. So the optimization problem is to design a constellation which make the coverage percent of each ground points as max as possible.

Table 1. The decision space of LEO Regional Satellite Constellation Optimization Problems

parameter	E_{min}	h	e	ω	i	Ω	M	N_{plane}	N_{sat}
value	[10,20]	1450	0	0	[20,50]	[0,360]	[0,360]	2	8

We suppose there are N_{plane} circular orbit planes, and each orbit planes have N_{sat} satellites. As we known, an orbit can be described by six classical orbital

Scheme by OMEA	i_{sav}	i	Ω	M_1	M_2	M_3	M_4	M_5	M_6	M_7	M_8	cov%
1	1	35.99	1.7617	34.52	80.15	120.23	167.88	215.15	259.69	304.40	347.50	F_1=100%
	2	36.90	185.37	7.103	55.40	95.51	139.89	191.68	239.12	280.35	321.99	F_2=100% F_3=100%
2	1	43.16	1.729	34.59	80.19	120.91	166.97	214.01	263.31	308.58	357.82	F_1=100%
	2	45.08	184.69	7.333	50.48	98.72	143.81	191.67	236.06	280.19	324.31	F_2=100% F_3=100%
3	1	42.33	3.101	34.60	78.18	124.68	167.85	215.21	260.55	299.71	347.92	F_1=100%
	2	43.77	185.33	7.100	55.63	95.51	144.47	193.37	238.61	279.91	324.41	F_2=100% F_3=100%

Fig. 2. Orbit parameters and coverage property of 3 typical constellation scheme by OMEA. F_i denote the constellation coverage percent of i-th ground point.

scheme by NSGA-II	n_{sav}	i	Ω	M_1	M_2	M_3	M_4	M_5	M_6	M_7	M_8	cov%
1	1	49.94	108.28	26.71	73.12	120.23	163.82	207.42	253.12	298.82	345.23	F_1=99.28%
	2	49.76	287.57	30.23	73.82	118.12	165.93	212.34	256.64	296.01	343.12	F_2=95.54% F_3=82.39%
2	1	37.46	106.17	25.31	70.31	115.31	161.01	206.01	251.71	296.01	340.31	F_1=89.87%
	2	36.58	285.46	31.64	75.93	122.94	166.64	209.53	255.93	301.64	347.34	F_2=97.43% F_3=100%
3	1	41.62	104.76	25.31	70.31	115.31	161.71	205.31	250.31	296.01	340.31	F_1=95.38%
	2	43.20	286.87	40.78	82.26	127.26	172.26	220.07	266.48	312.18	355.07	F_2=99.63% F_3=97.54%

Fig. 3. Orbit parameters and coverage property of 3 typical constellation scheme by NSGA-II

elements $Sat = (a, e, i, \omega, \Omega, M)$, where a is semi-major axis, e is eccentricity, i is inclination, ω is argument of perigee, Ω is longitude of ascending node, M is mean anomaly. The decision space is described in tab1.

E_{min} is minimum observing elevation ,the altitude of all orbits are 1450km, Since all orbits are circular, value of e and ω are equal to 0. So in our experiment have three group of decision variable, there are the inclination of the j-th orbit plane denoted by i_j, the longitude of ascending node of the j-th orbit plane denoted by Ω_j, the mean anomaly of the k-th satellite in the j-th orbit plane denoted by $M_{j,k}$, $j = 1, \cdots, N_{plane}$, $k = 1, \cdots, N_{sat}$. Then the constellation is controlled by 20 parameters, which means the dimension of decision vectors is 20. J2 perturbation is considered in the constellation emulational process. A cycle of emulation is a day (24*3600s), and 10s are set to a emulational step.

Population set to 96. By 100 runs, we get a set of Pareto solutions. Three solutions which have farthest crowded distances are chosen from the set. As fig.2 shows, the covering rate are all 100%, which means the algorithm and constellation we design have a good performance in this experiment. For this test problem in the experiment does not belong to standard test problems, it can't know Pareto-optimal front like other test problems, so we compare our own results to that by NSGA-II 's calculation (fig.3)[15]. The key results of the comparison are:

1. The Pareto solution by the OMEA algorithm is dominate NSGA-II's solution, it's means our results is closer to Pareto-optimal front than known results in NSGA-II in references[15].
2. The Pareto front from results of OMEA shows a uniform distribution, so this algorithm could be applied in constellation optimum design for consulting goals.

In addition, it can get much heuristic information by observed from fig.2 as follow:

- Every two values of i in a solution are very approximate.
- Every two values of Ω_j in a solution have a distinction next to 180deg.
- Satellites in the same orbit have a similar value of phase displacement.

5   Conclusions

In this paper, we proposed an extension MEA algorithm based on k-means for MOPs for satellite constellation design. OMEA implies the orthogonal design method with quantization technique to generate the initial population of points that are scattered uniformly over the feasible solution space, so that the algorithm can evenly scan the feasible solution space once to locate good points for further exploration in subsequent iterations. And it uses k-means and PCA to exploits the regularity in the distribution of Pareto optimal solutions of a MOP and builds a probability model to produce offspring solutions. In experiment, we solve a regional coverage constellation optimizing design mission which has three objective functions.

From the analysis of the results we can conclude that OMEA has a good performance for satellite constellation design, we can obtain a good distribution Pareto solutions which have a highly coverage percent (coverage figure of merit). From Pareto solutions, it can make decisions by special preference or experiences of experts to choose styles of constellations. But satellite constellation optimization is a complex problem, if in the condition of too many characteristic points, calculating time will increase quickly. For reducing optimizing time, distributed computing can be adopted to separate a big problem to smaller problems.

References

1. Deb, K., Pratap, A., Agarwal, S., Meyarivan, T.: A fast and elitist multiobjective genetic algorithm: NSGACII. IEEE Transactions on Evolutionary Computation 6, 182–197 (2002)
2. Zitzler, E., Laumanns, M., Thiele, L.: SPEA2: Improving the strength pareto evolutionary algorithm. Technical Report 103, Computer Engineering and Networks Laboratory (2001)

3. Schaffer, J.D.: Multiple Objective Optimization with Vector Evaluated Genetic Algorithms. Ph. D. thesis, Vanderbilt University (1984, unpublished)
4. Horn, J., Nafpliotis, N.: Multiobjective optimization using the niched pareto genetic algorithm. IlliGAL Report 93005, Illinois Genetic Algorithms Laboratory, University of Illinois, Urbana, Champaign (1993)
5. Knowles, J.D., Corne, D.W.: Approximating the Nondominated Front Using the Pareto Archived Evolution Strategy. Evolutionary Computation 8(2), 149–172 (2000)
6. Jin, Y., Sendhoff, B.: Connectedness, regularity and the success of local search in evolutionary multi-objective optimization. In: Congress on Evolutionary Computation, Canberra, Australia, pp. 1910–1917. IEEE Computer Society Press, Los Alamitos (2003)
7. Zhou, A., Zhang, Q., Jin, Y., Tsang, E., Okabe, T.: A model-based evolutionary algorithm for bi-objective optimization. In: Congress on Evolutionary Computation, Edinburgh, U.K, IEEE Computer Society Press, Los Alamitos (2005)
8. Zhou, A., Jin, Y., Zhang, Q., Sendhoff, B., Tsang, E.: Combining Model-based and Genetics-based Offspring Generation for Multi-objective Optimization Using a Convergence Criterion[C].cec2006 (2006)
9. Zeng, S., Kang, L., Ding, L.: An Orthogonal Multiobjective Evolutionary Algorithm for Multi-objective Optimization Problems with Constraints. Evolutionary Computation 12(1), 77–98 (2004)
10. Cai, Z., Gong, W., Huang, Y.: A Novel Differential Evolution Algorithm based on -domination and Orthogonal Design Method for Multiobjective Optimization. In: Obayashi, S., Deb, K., Poloni, C., Hiroyasu, T., Murata, T. (eds.) EMO 2007. LNCS, vol. 4403, pp. 286–301. Springer-Verlag, Heidelberg (2007)
11. Larranaga, P., Lozano, J.A (eds.): Estimation of DistributionAlgorithms. Kluwer Academic Publishers, Dordrecht (2001)
12. Kambhatla, N., Leen, T.K.: Dimension reduction by local principal component analysis. Neural Computation 9(7), 1493–1516 (1997)
13. Smith, L.I.: A tutorial on Principal Components Analysis [EB/OL] (2000), http:www. cs. otago.ac.nz/cosc453/student tutorials/principal components.pdf
14. Leung, Y.W., Wang, Y.: An Orthogonal Genetic Algorithm with Quantization for Global Numerical Optimization. IEEE Transactions on Evolutionary Computation 5(1), 41–53 (2001)
15. Li, S., Zhujiang, Li, G.: Multi-object optimization of LEO regional communication satellite constellation with GA algorithm. Journal of PLA University of Science and Technology 6(1), 1–6 (2005)
16. Frayss Inhes, E.: Investigating new satellite constellat ion geometries with genetic algorithm [A]. In: AIAA 9623636, AAS/AIAA A strodynamics Specialist Conference [C], San Diego, American Institute of Aeronautics and Astronautics Inc. (1996)
17. Mason, W.J., Victoriav,: Optimal earth orbiting satellite constellation via a pareto genetic algorithm[A]. In: AAS002139, AAS/AIAA A strodynamics Specialist Conference and Exhibit [C], San Diego, American Institute of Aeronautics and Astronautics Inc. (1998)
18. Ferringer, M.P., Spencer, D.B.: Satellite Constellation Design Tradeoffs Using Multiple-Objective Evolutionary Computation. Journal of Spacecraft and Rockets 43(6) (November-December 2006)

Author Index

Lecture Notes in Computer Science

For information about Vols. 1–4566

please contact your bookseller or Springer